W9-CCQ-695

# Remote Sensing of Environment

# Remote Sensing of Environment

Edited by

**Joseph Lintz, Jr.**

*Mackay School of Mines*
*University of Nevada, Reno*

and

**David S. Simonett**

*Department of Geography*
*University of California, Santa Barbara*

1976

**Addison-Wesley Publishing Company**

Advanced Book Program

Reading, Massachusetts

London • Amsterdam • Don Mills, Ontario • Sydney • Tokyo

**Library of Congress Cataloging in Publication Data**

Main entry under title:

Bibliography: p.
Includes index.
1. Remote sensing systems. 2. Geography–Remote sensing.
I. Lintz, Joseph, 1921- II. Simonett, David S.

G70.4.R46 621.36'7 76-47661
ISBN 0-201-04245-2



Printed in the United States of America

*Second printing, 1978*

ABCDEFGHIJK-HA-798

# Contents

# Color Illustrations*



*Color insert follows text page 344

# Contributors

ANTHONY R. BARRINGER — *291-322*
Barringer Research Ltd., Rexdale, Ontario

PETER A. BRENNAN — *412-437*
Aero Services Corporation, Houston, Texas

PETER E. CHAPMAN — *412-437*
Cyprus Mines Corporation, Tucson, Arizona

ROBERT N. COLWELL — *134-154*
Institute of Space Studies, University of California, Berkeley

ROBERT J. CURRAN — *34-84*
Goddard Space Flight Center, Greenbelt, Maryland

JOHN R. EVERETT — *85-127*
Earth Satellite Corporation, Chevy Chase, Maryland

ROBERT S. FRASER — *34-84*
Goddard Space Flight Center, Greenbelt, Maryland

EARL J. HAJIC — *374-411*
Department of Geography, University of California, Santa Barbara, California

FRANK J. JANZA — *194-233*
Department of Electrical Engineering, Sacramento State University, Sacramento, California

DAVID A. LANDGREBE — *349-373*
Laboratory for Agricultural Remote Sensing, Purdue University, West Lafayette, Indiana

JOSEPH LINTZ, Jr. — *323-343, 412-437*
Department of Geology, MacKay School of Mines, Reno, Nevada

DONALD S. LOWE — *155-193*
Environmental Research Institute, Ann Arbor, Michigan

RICHARD K. MOORE *234-290*
Remote Sensing Laboratory, University of Kansas, Lawrence, Kansas

ROBERT J. PEPLIES *483-507*
Department of Geography, Eastern Tennessee University, Johnson City, Tennessee

RONALD J. POGORZELSKI *14-33*
Department of Electrical Engineering, University of Mississippi, University, Mississippi

FLOYD F. SABINS *508-571*
Chevron Oil Field Research Co., La Habra, California

KENNETH A. SHAPIRO *14-33*
TRW Systems Group, Redondo Beach, California

DAVID S. SIMONETT *85-127, 323-343, 374-411, 442-481*
Department of Geography, University of California, Santa Barbara, California

HERBERT SKIBITZKE *572-592*
Water Resources Branch, U. S. Geological Survey, Barrow, Alaska

DON WALSH *593-626*
Institute of Coastal and Marine Studies, University of Southern California, Los Angeles, California

## Units of Measurement in Remote Sensing

In this book quantities are expressed in different ways in different chapters. This lack of consistency results from the editors' giving free rein to the authors of the chapters on units of measurement, rather than adopting a single, rigid system to be followed throughout. While this may be confusing to some readers, it has the advantage of forcing the reader to consider carefully the quantities being measured. This short introduction describes what physical factors are measured in remote sensing and outlines various units of measurement.

In mechanics, the basic quantities which need to be measured are length (L), mass (M), and time (T). There are three fundamental systems in use: mks (meter-kilogram-second), cgs (centimeter-gram-second), and fps or English (foot-pound-second). Other quantities are derived from these units. Force is the product of mass times acceleration. Energy or work is force times length or mass times velocity-squared. Power is the expenditure of energy per time unit. Table 1 summarizes the basic units of measurement in the three systems.

**TABLE 1.**

| | mks | cgs | fps |
|---|---|---|---|
| length (L) | meter (m) | centimeter (cm) | foot (ft) |
| mass (M) | kilogram (kg) | gram (g) | slug |
| time (T) | second (sec or s) | second | second |
| force ($MLT^{-2}$) | newton (N) | dyne | pound (lb) |
| energy, work ($ML^2T^{-2}$) | joule (J) | erg | foot-pound (ft-lb) |
| power ($ML^2T^{-3}$) | watt (W) | erg-$sec^{-1}$ | horsepower |

A system of units known as *Système International d'Units,* abbreviated SI, is rapidly becoming accepted as the form of the metric system to be used. The SI system is essentially the mks system.

In many cases, the basic unit of measurement is either too large or too small for convenience. It can then be multiplied by some convenient factor (almost always a power of ten) to make a derivation unit. Table 2 lists common multiplication factors and the prefixed abbreviations used to indicate them.

TABLE 2.

| *Factor* | *Prefix* | *Abbreviation* |
|---|---|---|
| $10^{12}$ | tera | T |
| $10^{9}$ | giga | G |
| $10^{6}$ | mega | M |
| $10^{3}$ | kilo | k |
| $10^{-2}$ | centi | c |
| $10^{-3}$ | milli | m |
| $10^{-6}$ | micro | $\mu$ |
| $10^{-9}$ | nano | n |
| $10^{-12}$ | pico | p |

For example $10^{-3}$ meter = 1 millimeter (1 mm).

In the SI system, only powers of 10 which are multipliers of 3 are used. That is, the prefix *centi* (as in centimeter) is not used.

In remote sensing, the quantities which we need to measure, and the common units used to express these measurements, are described below.

a) Wavelength of radiation ($\lambda$). Radiation travels through space in electromagnetic waves, and many of the behavioral characteristics of this radiation when interacting with a medium are rel to the wavelength. Common units used are listed in Table 3; preferred units are marked with an asterisk.

TABLE 3.

| *Unit* | *Abbreviation* | *Equivalent* |
|---|---|---|
| centimeter | cm | $10^{-2}$ m |
| millimeter* | mm | $10^{-3}$ m |
| micron or micrometer* | $\mu$m | $10^{-6}$ m |
| nanometer* | nm | $10^{-9}$ m |
| Angstrom unit | Å | $10^{-10}$ m |

b) Frequency of radiation (f). As waves of radiation travel through space, the time between the passage of the crest of one wave and the crest of the next wave is the *period.* The inverse of the period is the *frequency* and describes the number of wave cycles which will pass a given point during one time unit. The frequency is the ratio of the wave celerity ($c = 3 \times 10^{8}$ m/sec$^{-1}$) to the wave length,

$$f = c/\lambda.$$

Since the basic time unit used is the second, the frequency is expressed in cycles per second, called hertz (abbreviation Hz). Since most of the frequencies with which we are concerned are very large (because celerity is large and wavelength is small) we usually use one of the multipliers listed in Table 1, such as megahertz (MHz) or gigahertz (GHz).

c) Energy
Energy in radiation is usually expressed in joules (J) or kilojoules (kJ). In some literature (but nowhere in this book) radiative energy is expressed in calories, a calorie being the amount of energy needed to raise the temperature of 1 gm of water 1° C. 1 calorie = 4.1855 joules.

d) Power (energy per time)
Power is usually expressed in watts (W) equal to 1 joule per second ($Js^{-1}$).

e) Power intensity
Generally we are interested in the amount of radiation incident upon or emanating from a unit area on the earth's surface during some time period. The usual unit for this measurement is watts per square meter ($Wm^{-2}$). The unit $Wcm^{-2}$ also appears in this book, but the *langley* (1 cal $cm^{-2}$) does not.

f) Power intensity per spectral region.
Often the power intensity in some portion of the spectrum is isolated; in this case we might refer to Watts per square meter per micrometer ($Wm^{-2} \mu m^{-1}$).

g) Temperature
Temperature measurements in remote sensing are usually expressed in degrees Kelvin (°K), whereby 0°K is absolute zero, water freezes at 273.16°K and boils at 373.16°K. The Celsius scale is also used where water freezes at 0°C and boils at 100°C. Note that the magnitude of the unit is the same in both scales.

# Preface and Introduction

## I. DEFINITION

Remote sensing is the acquisition of physical data of an object without touch or contact. Fossils of the oldest known trilobite, *Fallotaspis* from the Cambrian Period 575,000,000 years ago, show the presence of compound eyes similar to those of modern grasshoppers or houseflies. Thus remote sensing, in the form of vision or seeing, is nothing new. Niepce is usually credited with the invention of photography (1822), which gives man the opportunity to record vision. The ability to capture transient events with camera and film is an extremely important aspect of remote sensing.

As man has come to understand the physical universe better, there have been intermittent efforts to design tools and instrumentation to serve as further probes and to provide access to still more data. The limitations of the visible region of the total electromagnetic spectrum were early recognized. But it was not until there were important technologic breakthroughs about the time of World War II that man was able to obtain the useful range of data available from the extravisible regions of the electromagnetic spectrum.

In the broadest usage, remote sensing does not limit itself to information derived by means of the electromagnetic spectrum. Although the effects of both gravity and magnetism were known for several centuries, their applications or uses were restricted to the navigator's compass. Again in the nineteenth century the invention of instrumentation permitted the acquisition of knowledge and data concerning physical phenomena of the environment. One of the classic early applications of gravity was the successful determination of the presence of "roots" in the Himalayan Mountains in the 1870's. In the early decades of the twentieth century the increasing sophistication of petroleum exploration techniques led to refinements in instruments for both gravity and magnetics, and applications of the two force fields became commonplace and widely accepted. Petroleum exploration was especially inducive to the development of these applications, because there were wondrous rewards for correctly designed field programs.

Our basic definition of remote sensing lacks restrictions; both the force fields and the electromagnetic spectrum are vehicles for the detection and identification of data concerning physical environmental characteristics. The only limitation on the objectives of remote sensing which we follow here is to restrict our treatment to environmental matter and man's activities on earth. An even broader usage than ours would include applications in astronomy, modern medicine, and metallurgy.

0-201-04245-2

*Remote Sensing of Environment*, Joseph Lintz, Jr. and David S. Simonett (eds.) ISBN 0-201-04245-2

## II. ORGANIZATION AND SCOPE

Given the broad scope of remote sensing, we have not attempted to provide an introduction to the entire range of the topic. Nor do we wish to examine every conceivable application; as pointed out above, the present volume is restricted.

Our presentation is limited to the utilization of the electromagnetic spectrum. We do not include the force fields because there exist a large number of very fine volumes describing the principles and applications of gravimetry and magnetometry. It should be pointed out that the uses of the force fields and the electromagnetic field are not mutually exclusive. There is no reason why they might not be combined to accomplish certain specific experiments or exploration objectives, though some combinations on the same vehicle are incompatible. Commonly, however, both fields may be detected simultaneously without cross interference.

A second limitation concerns the objectives we establish for this volume. Our primary interest focuses on the earth and its natural resources. In this context, our discussions follow paths that are earth oriented. We will continually be "looking" down from some platform, be it a cliff, a "cherrypicker," an aircraft, or a spacecraft. Excluded are applications and instruments oriented toward medicine, internal inspections, and the like. One limitation is especially troublesome, the atmosphere. In this case, we feel obliged to detail for our readers the role that the atmosphere plays in remote sensing, but we distinguish between this functional role and an object role wherein remote sensors are employed to study the atmosphere itself. Even here, an exception could occur, because the pollution of the atmosphere is of concern in that such pollution may very well be a by-product of human activity.

A third limitation is the omission of all military activities. We are well aware that the present state of civilian remote sensing technology has been derived from research funded by the American military establishment in the years after World War II. The military remote sensing programs encompass a variety of wavelengths of the electromagnetic spectrum. Many of their sensors are concealed by military security classification, or at least the performance characteristics are unavailable. The defense establishment's utilization of remote sensing is directed toward the national security of the United States. It is also known that both aircraft and spacecraft are employed on a continuing basis.

At first sight remote sensing appears to be merely an extension of aerial surveying. Once the facade is breached, however, one quickly can perceive the fact that with the new instrumentation and the new techniques available for interpreting the data remote sensing is a unified body of applied science in itself.

Remote sensing, like conventional or force-field geophysics, is based on solid principles of physics. There are no "black boxes" and, as a consequence, no magic. Part I describes briefly the fundamentals of physics from which remote sensing technology draws. Although the laws of physics applicable to electromagnetics are constant and uniform, there are significant changes in emphasis on the various parameters which correlate with the wavelength under consideration. This is a com-

0-201-04245-2

plicating factor which influences every decision from mission planning to data interpretation. Next in importance to the fundamental laws of physics governing remote sensing are the effects of the media through which the signal or electromagnetic energy passes. Problems in this area are often dealt with by mathematics. We have attempted to express the considerations of these problems in concise English as well as in mathematics for those who are more at home with English. Part I concludes with a cogent chapter which establishes a unifying aspect to remote sensing: the concept of the four resolutions—spatial, spectral, radiometric, and temporal.

In Part II will be found the basic papers governing the individual sensors. Although we have concentrated on the more common sensors in the visible, thermal infrared, and microwave regions of the electromagnetic spectrum, all of which are commercially available at present, some sensors which are less well known and less widely available commercially are also included. In these chapters the authors cover instrument design and general characteristics, as well as problems resulting from operating at the selected wavelengths and limited (it is hoped) guesses as to future improvements. These chapters are natural divisions of the overall topic because the techniques required vary as a function of spectral characteristics.

Part III is important to the idea that remote sensing is a unified body of technology in itself. Most practitioners of remote sensing think of it in terms of an integrated system. The systems approach means a total package involving (1) articulation of some problem, either an application or an experiment, which may be amenable to solution by means of remote sensing; (2) definition of a procedure to solve the problem, such as sensor choice, test bed carrier, and other considerations; (3) mission planning, in which *all* details attendant on the acquisition of the data are covered as rigidly as possible, with attention to alternative plans in the event of weather problems or mechanical difficulties; (4) mission execution, during which the data are acquired as planned, together with ancillary data, if required, such as "ground truth" or calibration; (5) data processing, during which the raw data may be subjected to reformatting, rectification, electronic or photographic editing, changes in scale, enhancements, ratioing, and possibly additional intermediate steps as required; and finally (6) data interpretation, which will determine the results of the entire procedure and decide whether or not the problem articulated in the first step has indeed been solved.

At this point it should be noted that there are truly two major aspects to remote sensing. What might be remote sensing technology utilizes remote sensors to solve problems in the disciplinary fields or applications-oriented problems. Activities in this area have generally become commercialized and center about photography of various types and specialities, thermal imaging, and side-looking radar. Remote sensing science, on the other hand, is concerned with the improvement of remote sensing itself to provide a better understanding of the various sensors and how and why they operate as they do. Activities in remote sensing science are almost never commercial, but are in-house experimental or are performed under experimental research contract. A general rule of thumb is that, if the experiment or problem requires acquisition of data over a specific piece of real estate, it is appli-

0-201-04245-2

cations oriented and should be classified as remote sensing technology, whereas if it is independent of specific real estate, it is usually instrument oriented and should be classified as remote sensing science.

The final portion of the volume, Part IV, is oriented toward the environment. Here are the applications and interpretations of the remotely sensed data within traditional earth science disciplines. Five areas of application cover the majority of the environmental aspects, with the possible exception of atmospheric pollution. This last topic, without minimizing its importance, is mentioned in Chapter 3 but does not warrant a separate chapter in Part IV because of the restricted number of sensors useful in its detection and identification. The authors in Part IV have provided information on how remote sensors "see" the physical characteristics of their disciplinary areas and given insights as to the interpretation of these data to provide the understanding that permits a decision as to whether the mission objective has been accomplished.

## III. A PAST PERSPECTIVE ON REMOTE SENSING

Music produced by a single instrument has limitations, despite the fact that the artist may be acclaimed for his talents by the critics and the instrument upon which he plays may be technically perfect. A full orchestra of many instruments will produce a far greater range of harmonies, counterpoints, timbres, and overtones, presenting the listeners with a fuller sense of completeness.

A virtuoso performance on a single instrument is analogous to superb black and white aerial photography. It may well contain great contrast, sharp spatial resolution, and optimum lighting and exposure. But many other characteristics of the ground objective are missing, and although the interpreter can extract much information from the black and white photograph, there are many phenomena he cannot deduce because of the limitations of the evidence.

The recognition that further information might be obtained by utilization of additional wavelengths of the electromagnetic spectrum long preceded the ability to measure and capture electromagnetic energy at wavelengths additional to those of the visible spectrum. Usually the technological breakthrough occurred in some other context, and the idea of using the technique to measure the physical characteristics of the earth's surface was conceived subsequently.

Although Niepce's camera dates from 1822, it was not until 30 years later, when a camera was hauled aloft in a balloon and pointed downward, that the first "aerial" photograph was taken. In an age of miniature and subminiature cameras it is difficult to remember that nineteenth century cameras were awesome apparatuses and that glass plates were coated with emulsion immediately before exposure and developed as quickly as possible thereafter. George Eastman, perhaps more than any other individual, pioneered in the improvements in the photographic process which made photography so widely available to the average person.

0-201-04245-2

By World War II specialized cameras for aerial surveying had improved to the point where they were operationally reliable at the various extremes of temperature to which they were subjected. Color emulsions were still so new, however, that colored photography was a novelty which attracted great attention. The capacity to extend black and white photography from the visible red (0.8 μm) to the solar or reflecting infrared (0.9–1.0 μm) had also been achieved. Newspapers printed occasional photographs of this type in their Sunday rotogravure sections. The armed forces in Europe had a very limited amount of such film, which was useful in camouflage detection work. Essentially, however, in the early 1940's we were still in the single-instrument stage, although we had greatly increased our virtuosity with it.

Extravisible wavelength instrumentation saw its beginnings in the period of World War II. Crude microwave transmitters and receivers became available for the detection of aircraft for approach warnings and also were used for the first time for bombing unseen targets from above the clouds. Once this technology came into existence, the evolution of some sort of visible display was a secondary improvement. But a decade later Royal Air Force SLAR imagery of Scotland, for example, was relatively crude and lacked the spatial resolution to make it attractive to a discipline-oriented interpreter.

By 1960 various American manufacturers were offering radar instrumentation which was beginning to approach current standards. The market was limited, and the military security classification shielded all but a few from these developments and advances. An occasional frame of imagery was declassified and published in the popular press. The APQ-97 system, now commercially available, of which only a single instrument was built, came into existence in the early 1960's. About the same time the APQ-102 system, used in Project Radam in the early 1970's, also evolved; it represented a second-generation side-looking radar, as detailed in Chapter 7.

Although the photovoltaic effect at infrared wavelengths was first discovered in 1904 using lead sulfide or galena, the information lay dormant for at least a decade. Other substances were experimented with during World War I. Again, came a period of quiescence, followed by a revival of interest in World War II, when it was found that the desired reliability and rapidity could be obtained by supercooling the crystals by cryogenic means. Progress in infrared physics and applications moved rapidly. From this start one application lead to the mechanism of the scanning technique. Again these developments were sponsored by the military establishment and took place within the confines of security classification. It was not until 1966 that a manufacturer undertook to fabricate a thermal infrared scanner which would possess design characteristics sufficiently modified to allow it to escape security classification.

An interesting but somewhat restricted remote sensor is the passive microwave system. Passive microwave detectors are based on instrumentation for radio astronomy. The availability of backup instruments for a Martian flyby led to the recogni-

0-201-04245-2

tion that they could be mounted on an aircraft and turned around to look down at the earth. This instrumentation does not yield imaged or maplike data, but rather produces a strip chart or wavy line format which requires extra steps in the interpretation. Hence it has not been used as widely as it might have if a technique for providing images with good resolution were available. Not until the mid-1970's was a good imaging microwave system developed, again by the military.

With the declassification of certain sensors, it became possible to offer remote sensing services on a commercial basis. For infrared scanners/mappers, the cost of the equipment was such that aerial survey companies were able to add thermal IR to their roster of service offerings. However, side-looking radar was so expensive and required such extensive aircraft modifications that the manufacturers themselves entered the contracting business, either alone in one case, or in a joint venture with a large, established aerial surveyor in another case.

In the summer of 1963 Dr. Peter Badgley, a professor at the Colorado School of Mines, was retained by the National Aeronautics and Space Administration on a summer project in Washington, D.C. His assignment was to advise NASA on alternatives to surface exploration of the moon. It was then believed that the lunar surface was so inhospitable to man that surface exploration would never be feasible. What was needed, it was thought, was a system of reconnaissance exploration from a lunar orbiter with spot landings to check out anomalous data. Today one tends to forget that at the time many persons believed that the surface of the moon would possess totally different chemistry and rocks from those of the earth. There were even scientists who believed that the lunar surface was composed of tens of meters of fine dust which would inundate a descending lunar lander and engulf the astronauts.

Dr. Badgley was given access to all government data, and he found that the military intelligence systems were amenable to lunar applications, as well as to many civilian uses pertaining to this planet. At any rate they would have to be tested and calibrated on Earth before they could be used in lunar studies. As more and more scientists became involved in the program in the early 1960's, many saw this body of technology, which became known as remote sensing (the term was coined by a group in the Geography Branch, Office of Naval Research, including Drs. Evelyn Pruitt and Walter Bailey), as a tool for increasing our knowledge of Earth and possibly for solving many of the problems facing mankind. The demand for a review of security classification procedures for remote sensing instrumentation commenced, and liberalization of the rules followed in 1967 and 1968.

A tremendous impetus for remote sensing occurred in 1963 when one of the astronauts smuggled a simple single reflex camera on board a Mercury spacecraft and brought back to Earth the astronaut's tourist-eye view of this planet. These colored photographs were an instant and dramatic success. Enabling the man in the street to view subjectively, in earth-orienting scenes, synoptic views of peninsular India, of the Arabian Peninsula, of Baja California and the southwestern United States, for example, strengthened enthusiasm and political support for NASA's

0-201-04245-2

space programs. The scientific community also saw the value of the synoptic view and immediately began to think in terms of systemic global photographic coverage, rather than occasional dramatic scenes captured by individual astronauts. With the passage of time, lunar programs have been diminished. Lunar orbiters have obtained exquisite photography of most of the lunar surface with the primary objective of selecting landing sites for the astronauts. The U.S. Geological Survey has interpreted these photographs in great detail, collating the data with the rock samples brought back by the astronauts. But today remote sensing activities are very strongly oriented toward Earth and human problems. There is the hopeful expectation that remote sensing will be useful in the solution of the many fundamental environmental problems facing mankind in the immediate future.

The launching of LANDSAT 1 (= ERTS A) in July 1973 and LANDSAT 2 in January 1975 represents the culmination of the objective of obtaining synoptic global data. LANDSAT 1 exceeded its programmed life by 100% and delivered, in the shortest time, the largest increment of knowledge of the globe ever attained. LANDSAT 2 continues this project, and future satellites are in the offing.

## IV. THE IMPORTANCE OF REMOTE SENSING

Sufficient time has elapsed to permit predictions concerning the future of remote sensing. From the beginning persons writing about this new field have faced the difficult task of finding the proper balance between uncritical presentations of remote sensing as a panacea for most environmental problems (of which there have been many), and an equivalent, almost nihilistic, negativism.

Conservatively, it now appears that remote sensing is here to stay and will continue to serve as a useful tool in the management of natural resources of many types. This statement is based on many observations, among them the following.

1. Major petroleum and mining companies are using remote sensing as needed in their exploration operations. In some companies it is still considered to be experimental and is handled out of research laboratories; in others the research laboratory performs only a trouble-shooting, supportive role, and the initiative lies with the exploration offices.
2. The airborne use of Ektachrome IR film has become quite common in forestry. Commercial timber companies are availing themselves of the knowledge provided by this extravisible IR band. The cost is relatively low, and the information obtained cannot be derived by any other means as rapidly, as reliably, and as cheaply.
3. Planners and land-use managers are becoming increasingly sophisticated concerning remote sensing and are demanding that information derived from this source be available to them for the decision-making process. This group includes highway designers, who seek not only optimal routings for roads, but also sources of materials with which to make the roads.

0-201-04245-2

4. The impact of LANDSAT 1 has been global. Every nation has received selected frames of LANDSAT imagery of their terrain at least, and in some instances even complete national coverage on a repetitive basis. The relatively low cost and wide availability of LANDSAT data have aroused the interest of scientists throughout the world.

5. The U.S. government and the United Nations have both promoted the exposure to and the use of remote sensing as an economic tool among the developing nations. Among the areas where remote sensing can support resource management are fisheries, forestry, hydrology, city planning, oceanography, vulcanology, agriculture, range management, and cartography. Some developing nations have adopted remote sensing as an integral tool in their 5-year economic plans. In some cases these adoptions may have been somewhat premature. However, initial interpretation of LANDSAT data has provided them with indications that they are on the right track.

6. An important and attractive feature of remote sensing is that it can be performed at many levels of cost. Multispectral black and white photography can be acquired with cameras and data analysis equipment costing well below $20,000, and this system can utilize existing lightweight aircraft with minor modifications. Additional sensors can be added as money becomes available, as can the desirable interpretation centers, which may become sophisticated with telemetering receiving stations and computer facilities.

7. The U.S. government is planning satellites beyond LANDSAT 2 and has made it known that future satellites will not carry tape recorders but will telemeter data down to the surface in real time. Thus there is a need to build widely dispersed centers to receive these data. The United States expects to share technology for this purpose but does not contemplate funding the erection of these centers except as it might provide some support through AID programs.

8. With a properly designed program, remote sensing is cost effective. For example, for Project Radam, a 4,500,000 km$^2$ survey of Brazil's Amazon Basin, it was less expensive to use side-looking radar than aerial photography. The initial costs of mobilization and preparation were justified by the great size of the total project. The freedom to fly independently of the sun, both as to length of day and as to weather, made it possible to adhere to a well-designed schedule without incurring unpredictable costs and greatly reduced the costs of maintaining flight crews and equipment on standby time.

9. Because of the commonality of operations (aircraft, spacecraft, data processing, and interpretation facilities) remote sensing tends to draw together at the government level various disciplines which are traditionally separated in practice. In economically advancing nations the probability is that centralized remote sensing facilities will evolve at the national or regional level. The output of their data is a powerful tool for resource management, and remote sensing will lead to integrated national policies for resource management.

0-201-04245-2

Each of the nine points above is important in itself. The combination of them is too impressive to permit remote sensing to be ignored. This new field has proved in the marketplace of ideas and the realm of operations that it can significantly contribute to man's knowledge of his environment and can assist in its management.

JOSEPH LINTZ, JR.
D. S. SIMONETT

0-201-04245-2

Part I

# Remote Sensing: Principles and Concepts

0-201-04245-2

## INTRODUCTION

The body of knowledge comprising remote sensing brings together physical principles of properties of objects in our environment as expressed in the energy-matter reactions of the electromagnetic spectrum and weds them to the practices, subject matter, and concepts of the earth, life, and cultural sciences. It is thus in the interaction between these disciplines that this new field is evolving, with remote sensing drawing heavily on both the laws of physics and the general qualitative and quantitative concepts of the "soft" sciences. Part I of this book reflects this pattern.

The opening chapter treats concisely the basic radiation laws and the shifts in emphasis which follow changes in wavelength. The basic laws of radiation, which upon first glance are simple, prove upon closer scrutiny to be quite complex, for the emphasis and significance of individual components varies as a function of wavelength.

In Chapter 2 are discussed in detail the medium through which radiant energy is transmitted (the atmosphere) and its effects on sensing. The energy profile is not simple because there are multiple paths of transmission, and each has its effect on the radiant energy being transmitted. The effects arising from atmospheric variability can radically alter the quantities and character of the energy being measured. This second chapter serves also as a bridge between the principles of physics in Chapter 1 and the explicit concepts of Chapter 3, now being developed and refined in remote sensing. The latter stem from latent concepts or ignored principles in the "soft" sciences.

Remote sensing has achieved increasing recognition as being able to make important contributions to the solution of many natural and resource-use problems. Especially valuable among these contributions are (1) the vast increase in new kinds of data becoming available to managers and decision makers, (2) new techniques of data processing and management which allow the extraction of subtle information heretofore unperceived in the raw data, and (3) concepts of data collection and analysis which are based on the variability of Earth in time and space. Remote sensing has thus evolved into a complete system of data acquisition, processing, and analysis with its own principles and concepts. This unique system is analyzed in Chapter 3 to more fully define and establish the potentials and limits of remote sensing.

0-201-04245-2

*Remote Sensing of Environment*, Joseph Lintz, Jr. and David S. Simonett (eds.) ISBN 0-201-04245-2

Chapter 1

# Introduction to the Physics of Remote Sensing

*Ronald J. Pogorzelski and Kenneth A. Shapiro*

## I. INTRODUCTION

Remote sensing is an interdisciplinary field drawing on the talents of engineers and natural scientists. In such interdisciplinary fields it is important that the central concepts of each discipline be known to both engineers and scientists. This chapter introduces the physics of remote sensing in such a way that persons with a background in natural sciences but only a modest knowledge of mathematics will develop some feeling for the electromagnetic theory which forms the basis for much of remote sensing technology. Engineers and other physical scientists will find that they already know more about this topic than we present here. Although the presentation is less mathematical than descriptive, it is assumed that the reader has a background that includes algebra and elementary trigonometry.

The properties of electromagnetic radiation can be classified into two major classes having (1) wave or (2) particle characteristics. Wave properties are exhibited by radiation in space and material objects, which interacts with matter on a macroscopic scale, whereas particle or quantum properties are exhibited by radiation which interacts with matter on an atomic or a molecular scale. Sections II–VIII cover the wave nature of electromagnetic radiation, while Sections IX and X provide examples of the quantum nature of the radiation.

The units used to quantify the physical quantities which represent electromagnetic radiation will be in the mks (meter-kilogram-second) system, unless otherwise specified. Any of the textbooks listed in the bibliography contains a review of the various sets of units and the methods for conversion between them.

## II. ELECTROMAGNETIC FIELDS

There are two equivalent ways of looking at the interaction of two charged particles. One way is to say that "somehow" the presence of one particle is "felt" by the other some distance away. The second way is to consider that one charged particle sets up a "condition" in the space around it (a field), and it is this "field" which is felt by the second charge. We will treat electromagnetic interaction by means of this second way or "field theory." The field theory provides a convenient and systematic means of keeping track of the interaction of electrically charged objects.

0-201-04245-2

*Remote Sensing of Environment*, Joseph Lintz, Jr. and David S. Simonett (eds.) ISBN 0-201-04245-2

Vectors are the natural mathematical language of electromagnetic theory. Vectors are quantities that have both a magnitude and a direction in space. For example, a car can be moving in a direction 45° east of north with a speed of 80 km/hr. The velocity of the car is a vector and is composed of both its speed and direction. The velocity vector would be represented as an arrow pointing in the northeast direction with a length proportional, on some predetermined scale, to 80 km/hr. All of the equations presented in this chapter will be in ordinary algebraic notation, and vectors will be described only when absolutely necessary.

The source of the electric field is *electric charge. Moving* electric charge or electric *current* is the source of the *magnetic field.* One important relation between current and charge is

$$J = \rho v \text{ amperes per square meter,} \tag{1.1}$$

where $J$ is current density, $\rho$ is charge density in coulombs per cubic meter, and $v$ is velocity of flow in meters per second.

A second mathematical relationship, called the equation of continuity, relates the rate of change of $J$ in space to the rate of change of $\rho$ in time.

The most fundamental set of equations in electromagnetic theory is known as Maxwell's equations. These four equations relate the electric field intensity, $E$, and magnetic field intensity, $H$, to their sources, $\rho$ and $J$, in terms of rates of change of the four quantities in space and time. Two constants are also involved in Maxwell equations. They are $\mu_0$, the permeability of free space, equal to $1.26 \times 10^{-6}$ H/m, and $\epsilon_0$, the permittivity of free space, equal to $8.85 \times 10^{-12}$ F/m.

The solution of Maxwell equations under a given set of boundary conditions yields the electric and magnetic field intensities at all points of interest in a given problem in terms of $J$, $\rho$, $\mu_0$, and $\epsilon_0$. The electric field $E$ exerts forces on charged particles and electrically polarized materials, while the magnetic field $H$ exerts forces on moving charges and currents and on magnetic materials.

In remote sensing, we are interested in those solutions of Maxwell's equations that represent waves: when these waves are emitted by or reflected from objects, we are able to gain information about the objects.

## III. PLANE WAVES

In regions of space which are free of currents and charges, that is, regions where $J$ and $\rho$ are zero, one of the solutions of Maxwell's equations is the so-called plane wave solution, which is represented for $E$ and $H$ as follows:

$$E = E_0 \cos(kx - \omega t) \text{ volts per meter,}$$

$$H = E_0 \left(\frac{k}{\omega \mu_0}\right) \cos(kx - \omega t) \text{ amperes per meter,} \tag{1.2}$$

0-201-04245-2

where

$$k^2 = \left(\frac{2\pi}{\lambda}\right)^2 = \omega^2 \epsilon_0 \mu_0 \text{ per square meter;} \tag{1.3}$$

$x$ is the distance from some plane (e.g., the $y$-$z$ plane) in meters; $\omega = 2\pi f$, where $f$ = frequency of the radiation (i.e., the number of cycles per second); $\lambda$ = wavelength: in a sinuoisdal wave, the wavelength is the distance between two successive "hills" or "troughs"; and $E_0$ is the amplitude, or height, of the wave in volts per meter, at time $t = 0$ and position $x = 0$.

The quantity $(kx - \omega t)$ is by definition the phase of the wave, and Eq. (1.2) is said to describe plane waves because it can be shown that at any given time the phase is constant over a plane, in space, perpendicular to the direction in which the wave is traveling. As time progresses, these planes of constant phase move in the direction of the wave, with a velocity called the phase velocity. This velocity, denoted by $v_p$ or $c$, can be shown to be equal to

$$v_p = c = f\lambda = \frac{1}{\sqrt{\epsilon_0 \mu_0}} \text{ meters per second.} \tag{1.4}$$

By using the values for $\epsilon_0$ and $\mu_0$, $c$ is found to be approximately equal to 300,000,000 m/sec, which is identical with experimental values for the velocity of light!

It must be remembered that $E$ and $H$ are really vectors perpendicular to each other and that they lie in the planes of constant phase. Thus $E$ and $H$ are perpendicular to the direction in which the wave is traveling. Furthermore, the direction in which $E$ or $H$ points in space can change with time. If the direction of $E$ is constant in space, the wave is said to be linearly polarized. If the tip of the $E$ vector traces an ellipse in a plane of constant $x$ as time progresses, the wave is elliptically polarized. When the ellipse degenerates to a circle, the wave is circularly polarized.

The direction of energy flow is in the direction of propagation of the wave, and the intensity of this flow is given by a quantity known as the Poynting vector, $S$, whose magnitude is

$$S = \frac{1}{2} E_0{}^2 \sqrt{\frac{\epsilon_0}{\mu_0}} \text{ watts per square meter.} \tag{1.5}$$

This quantity is the measure of the energy flowing per unit area per unit time in the direction in which the wave is propagating.

When a series of plane waves of slightly different frequencies are traveling together in a given direction, the energy contained in the waves propagates, not with the phase velocity (i.e., the velocity of light in vacuum), but rather with a velocity called the group velocity. The group velocity is generally less than the phase

0-201-04245-2

velocity and is equal to it only when the phase velocity is independent of frequency (as in the atmosphere or in a vacuum).

The wavelengths of plane waves can have any values ranging from the very small ones of gamma rays and X-rays to the very large ones of radio and very-low-frequency (VLF) waves. Figure 1.1 shows the electromagnetic spectrum. The wavelength is given in angstroms (1 Å = $10^{-10}$ m), microns (1 $\mu = 10^{-6}$ m), meters, and kilometers (1 km = $10^3$ m). The frequencies corresponding to various wavelengths are shown, as well as the names of the different portions of the spectrum. Note that, as the wavelength becomes smaller, the frequency becomes larger. This is a consequence of the fact that the product of the frequency and the corresponding wavelength is a constant, namely, the phase velocity, $c$.

In summary, then, Maxwell's equations admit plane wave solutions for the electric and magnetic fields in an electric-charge and current-free vacuum. The electric vector and the magnetic vector are perpendicular to each other and travel in a direction perpendicular to both $E$ and $H$. The source of these waves is a moving electric charge (i.e., current) in another portion of space, and the waves require no medium for their propagation. Energy is propagated in the direction in which the wave is moving, and the phase velocity of the waves is identical to the speed of light. All the different wavelength radiations which comprise the electromagnetic spectrum have these properties.

## IV. SPHERICAL WAVES

Actually, plane waves as described in Section III are almost never encountered in practice. The value of a study of these waves lies in the fact that many important physical parameters are thereby defined. Also, in many cases of interest, plane waves may be employed as an approximation to the real case.

When a source of radiation is localized to a small volume of space such as that occupied by a lens, a mirror, or an antenna, the radiation emitted can be described by the use of a spherical coordinate system with coordinates $r$, $\theta$, $\phi$ and with an origin centered on the source. In this situation the electric field can be expressed as

$$E = \frac{E_0}{r} f(\theta, \phi) \cos(kr - \omega t) \text{ volts per meter,} \tag{1.6}$$

where $f(\theta, \phi)$ represents a generalized mathematical expression depending on the angular position of the observation point relative to the source; this function changes as one moves from point to point in space. The formula for $H$ is obtained by multiplying Eq. (1.6) by $\sqrt{\epsilon_0/\mu_0}$ . These expressions are valid, however, only at distances far from the source, that is, when $r$ is large, where $r$ is the radial distance from the source to the point of interest in space.

The surfaces of constant phase are concentric spheres centered on the source, and hence this type of radiation is called a spherical wave. The energy flow and the

0-201-04245-2

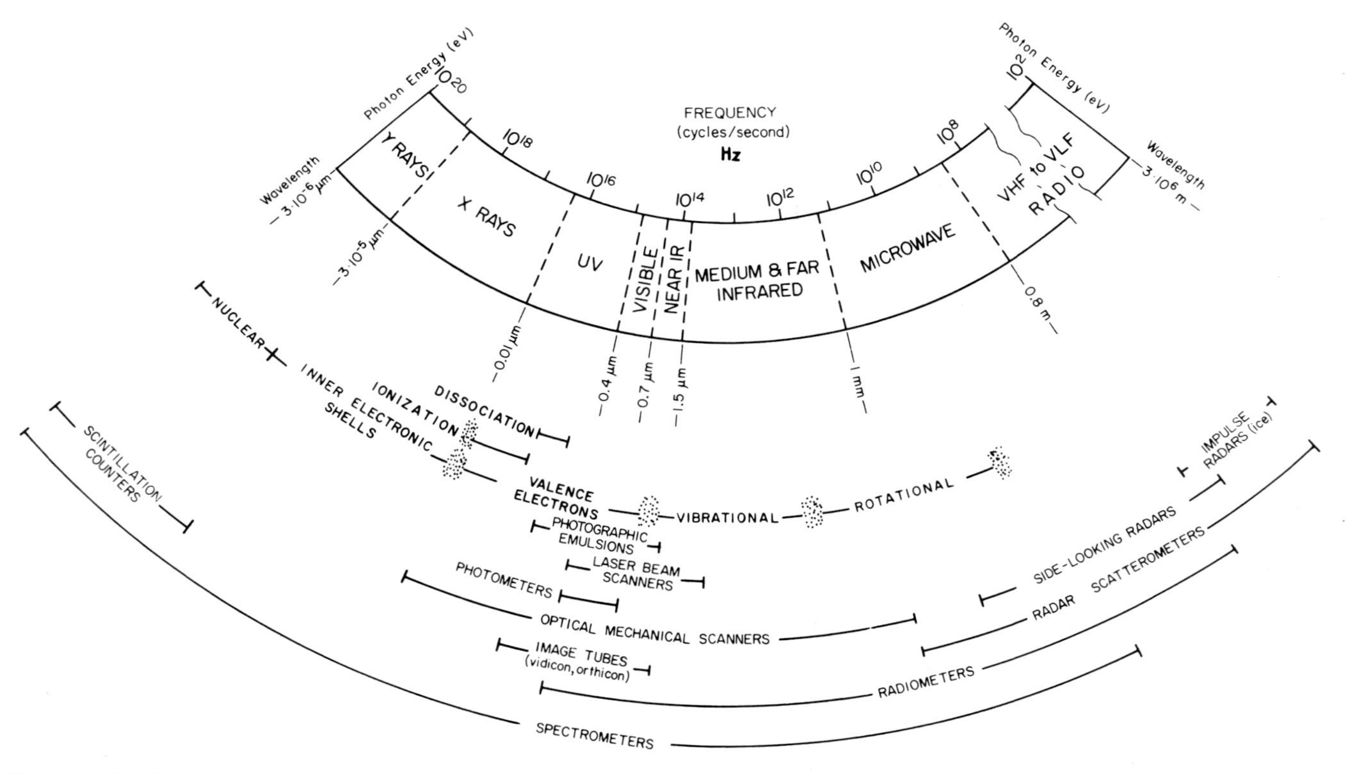


Fig. 1.1. The electromagnetic spectrum.

0-201-04245-2

Poynting vector at great distances from the source depend only on $r$; in particular, the energy flow per unit time per unit area crossing a sphere of radius $r_0$ is proportional to $1/r_0^2$. This yields the famous law that the intensity of spherical waves varies inversely as the square of the distance from the source. This "inverse square" law can be shown to be a consequence of the law of conservation of energy. A simple example of the inverse square law is encountered in moving to and from a fire. At some given distance from the fire one is comfortable. At one-half the distance the radiation is four times greater. At twice the distance the radiation is one-fourth that encountered at the comfortable distance.

The practical consequences of the operation of the inverse square law are seen in remote sensing in the following examples.

*Example I. Thermal Infrared Sensing of Forest Fires*

The accompanying table shows the sharp reduction in energy incident on an infrared scanner-collector from a 1-m$^2$ fire at 800°C, when viewed in an aircraft from progressively higher altitudes. In addition to the inverse square law acting to reduce the energy incident on a detector system, a parallel reduction occurs from the increase in area sensed with a fixed aperture. This decrease also occurs as an inverse square. Thus the product of the two processes acts to reduce the energy incident on the detector as the inverse fourth power, as seen in the table.

| *Height (km)* | *Energy Flow (W/cm$^2$)* | *Energy Ratio* | *Area Ratio* | *Product* |
|---|---|---|---|---|
| 0.5 | $1 \times 10^{-5}$ | 1 | 1 | 1 |
| 1.0 | $2.5 \times 10^{-6}$ | 4 | 4 | 16 |
| 2.0 | $6.25 \times 10^{-7}$ | 16 | 16 | 256 |

*Example 2. Radar Sensing of Terrain Employing a Real-Aperture Antenna*

In this example we note that the energy leaving the radar transmitter diminishes as $1/r^2$ to the target. A portion of the energy incident on the target is backscattered (reradiated) to the antenna. This radiation in turn diminishes as $1/r^2$ to the transmitter. Thus the power leaving the antenna, incident on the target, and reradiated back to the antenna is reduced as $1/r^4 \cdot \sigma$, where $\sigma$ expresses the backscattering coefficient of the target directed toward the antenna.

## V. INTERFERENCE

Two physical manifestations of the wave nature of light are particularly relevant in remote sensing applications. They are the phenomena of interference, which is discussed here, and diffraction, which will be presented in the next section.

Interference occurs when the electromagnetic field at a point is made up of contributions from more than one source. The sources must be coherent; that is,

0-201-04245-2

there must be a deterministic (nonrandom) relation between the phases of the various contributing waves. As a simple example, consider two sources of spherical waves, $S_1$ and $S_2$, and a screen placed a large distance $D$ (many wavelengths) away from the sources, as shown in Figure 1.2. At the screen the surfaces of constant phase will be very nearly plane, and the electric and magnetic fields will be very nearly parallel to the screen. It is almost as if there were two plane waves, illuminating the screen from slightly different directions. At any point $P$ the total electric field is given by

$$E_{\text{total}} \cong E_0 \cos(kx_1 - wt) + E_0 \cos(kx_2 - \omega t) \text{ volts per meter,} \tag{1.7}$$

where $x_1$ and $x_2$ are the distances the two waves have traveled from $S_1$ and $S_2$ to $P$. When $x_2 = x_1 + n\lambda$, where $n = 0, 1, 2, 3, \ldots$, then

$$\cos(kx_2 - \omega t) = \cos(kx_1 - \omega t + 2\pi n) = \cos(kx_1 - \omega t),$$

since $\cos(\theta + 2\pi n) \equiv \cos\theta$ by trigonometric identity. Therefore from Eq. (1.7)

$$E_{\text{total}} \cong 2E_0 \cos(kx_1 - \omega t). \tag{1.8}$$

It can be similarly shown that, when $x_2 = x_1 + \left(n + \frac{1}{2}\right)\lambda$, where $n = 0, 1, 2, \ldots$,

$$E_{\text{total}} \simeq 0. \tag{1.9}$$

Hence, at all points on the screen at which the distances the waves travel from each slit differ by a whole number of wavelengths, there will be a maximum of intensity, proportional to the time average of the square of Eq. (1.8), equal to four times the value of the intensity of each wave arriving there. When the distance transversed differs by odd multiples of a half member wavelengths, the two waves will exactly

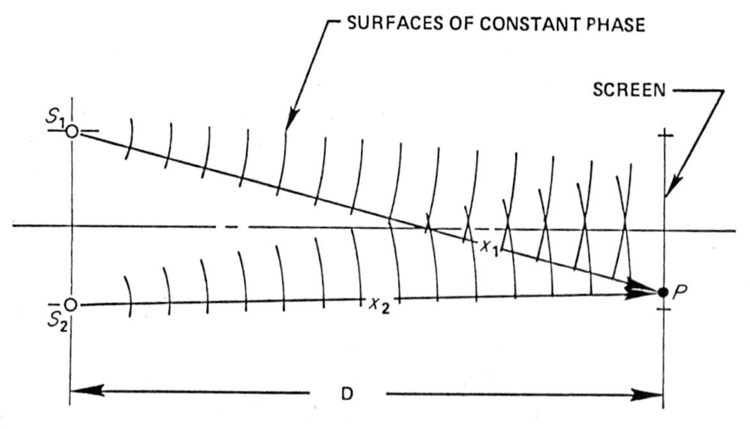


**Fig. 1.2.** Arrangement of coherent sources and screen for viewing interference patterns.

0-201-04245-2

cancel each other. Hence the screen will be covered by alternating light and dark stripes. It should be clearly remembered that interference occurs only when the phases of the waves have a definite (nonrandom) relationship with each other, that is, they are generated by a coherent source.

Interference effects can be of importance in remote sensing whenever devices such as side-looking radar and laser scanners, which have coherent sources of radiation, are employed. The phenomenon known as scintillation (in which small targets are sometimes "seen" and at other times are not "seen") arises in part from interference effects. Thus in observing houses on different radar images one may sometimes miss detecting the houses from these effects.

## VI. DIFFRACTION

The other important manifestation of the wave nature of light is diffraction. Diffraction most commonly occurs when light interacts with an edge of an opaque object. It is the measure of the departure of light from rectilinear propagation.

A common example of this effect occurs when a plane wave is shining through a circular hole or a long thin slit in an opaque screen. One might expect that a cylindrical or a thin, rectangular beam of uniform intensity would result. This is not the case. Rather, the radiation tends to distribute itself over a larger and larger area as it propagates away from the hole. A very large distance $d$ away from the hole or slit, the intensity distribution over the cross section of the beam focused on a screen is not uniform but is approximately proportional to

$$\frac{F^2\,[ka\,(r/d)]}{[ka\,(r/d)]^2}, \tag{1.10}$$

where $k = 2\pi/\lambda$ per meter. Also, in the case of the hole, $a$ = the radius of the hole in meters, $r$ = the distance in meters from the axis of the hole to any point on the screen, and $F$ = an oscillatory mathematical quantity known as the Bessel function of the first order. For the long thin slit, $a$ = *half* of the width of the small side of the slit in meters, $r$ = the distance in meters to any point on the screen from the center line on the screen parallel to the long side of the slit, and $F$ = the trigonometric sine function. In both cases, the expression describing the intensity oscillates between a maximum and zero, and successive maxima have decreasing amplitudes, as shown schematically in Figure 1.3.

In the case of the slit, the points of zero intensity occur when $kar/d = n\pi$, the first or central maximum occurs when $kar/d = 0$, and the successive maxima occur when $kar/d = \left(n + \frac{1}{2}\right)\pi$, $n = 1, 2, 3, \ldots$. Similarly, for the circular aperture, it can be shown that a central intensity maximum occurs with circular light and dark rings surrounding this center bright spot, and more than 80% of the energy is in the central maximum.

0-201-04245-2

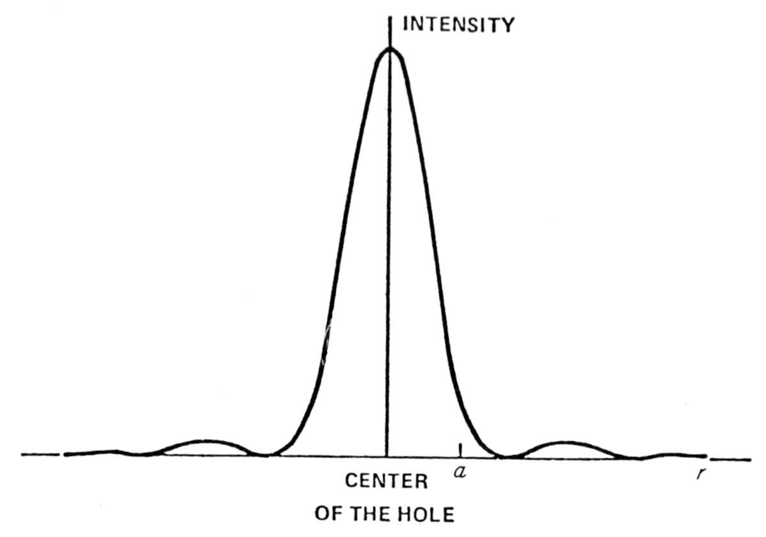


**Fig. 1.3.** Schematic of the intensity of radiation focused on a screen, after passing through a hole of radius *a* in an opaque screen, the intensity pattern is similar to that shown above in the direction perpendicular to the long side.

The reason why diffraction is so important in remote sensing is that this phenomenon ultimately both limits the resolution of sensing systems and defines the transmitting patterns of sources such as radar antennas. It can be shown that for a circular aperture of radius $R$ in a camera two distant point sources of radiation will be resolved only if their angular separation is greater than

$$\theta_0 \simeq 0.61 \frac{\lambda}{R} \text{ radians.} \tag{1.11}$$

For $R = 1$ cm and $\lambda = 5 \times 10^{-5}$ cm (or 500 nm), which is approximately in the center of the visible spectrum, $\theta_0 = 6.3$ sec of arc; that is, two point sources whose angular separation is smaller than 6.3 sec of arc will *not* be resolved.

It should be noted that a perfect lens in the aperture does not affect this result, but only allows the screen, or film, on which the intensity distribution appears to be placed in the focal plane of the lens instead of at a great distance from the aperture.

## VII. PLANE WAVES IN MATERIAL MEDIA

When a plane wave impinges on a material object, some of the radiation is reflected (reradiated without change in wavelength), some is transmitted, and some is absorbed. An analysis of the reflection problem requires an understanding of the manner in which plane waves propagate inside bodies with electric and magnetic

0-201-04245-2

properties. The effects of such media on wave propagation will be presented here, and reflection will be discussed in the next section.

In Section II, Maxwell's equations relating spatial and temporal rates of change of $E$ and $H$ were briefly discussed. These relationships also govern the behavior of $E$ and $H$ in the present case. The major difference lies in the fact that in a homogeneous isotropic medium we must include the effects of conductivity, permittivity, and permeability, denoted by $\sigma$, $\epsilon$, and $\mu$, respectively. A homogeneous material is made up of the same substance throughout, and an isotropic medium has the same properties in all directions. The conductivity term ($\sigma$) is a measure of the ease with which a medium carries an electrical current; $\sigma$ has units of mhos per meter and is related to the current density $J$, introduced in Section II, and $E$ by

$$J = \sigma E \text{ amperes per square meter.} \tag{1.12}$$

The value of $\sigma$ can range from $10^{-17}$ mho/m in media of poor electrical conductivity to $10^8$ mhos/m in the best conductors. No other property of matter has such a wide range of values.

The permittivity $\epsilon$ measures the effects of the electrical polarizability of a medium and can vary from $\epsilon_0 = 8.85 \times 10^{-12}$ F/m in free space to $\epsilon = 1000\epsilon_0$ for ferroelectric materials. The permeability $\mu$ measures the magnetization or magnetic polarizability of a medium and can vary from $\mu_0 = 1.26 \times 10^{-6}$ H/m in free space to $\mu = 5000\mu_0$ in typical ferromagnetic materials.

It can be shown that plane waves are a possible solution to Maxwell's equations in a conducting medium and that they are represented by the formulas

$$E = E_0 e^{-\alpha x} \cos(\beta x - \omega t) \text{ volts per meter,}$$

$$H = E_0 \frac{\sqrt{\alpha^2 + \beta^2}}{\omega\mu} e^{-\alpha x} \cos(\beta x - \omega t + \tan^{-1} \alpha/\beta) \text{ amperes per meter,} \tag{1.13}$$

where

$$\alpha = \omega\sqrt{\frac{\mu\epsilon}{2}\left(\sqrt{1 + \frac{\sigma^2}{\omega^2\epsilon^2}} - 1\right)}, \text{ the attenuation constant}$$

$$\beta = \omega\sqrt{\frac{\mu\epsilon}{2}\left(\sqrt{1 + \frac{\sigma^2}{\omega^2\epsilon^2}} + 1\right)} = \frac{2\pi}{\lambda}$$

and $\lambda$ = wavelength in meters. In Eq. (1.13), $E_0 e^{-ax}$ represents an exponentially decreasing amplitude caused by the work done by the wave in making the current flow in the material. This work ultimately appears in the medium as heat. It should be noted that, when $\sigma = 0$, no attenuation of the wave occurs. There are molecular processes, which we will not discuss here, that can cause attenuation, but conductivity is the only bulk property that can have this effect.

0-201-04245-2

The quantity

$$\delta = \frac{1}{\alpha} \text{ m} \tag{1.14}$$

is called the electromagnetic skin depth because it is the distance at which the electric field penetrating into a conducting medium will be diminished by a factor of $1/e$, where $e$ is the base of natural logarithms. The reduction thus is approximately 37% of the electric field incident on the surface. For geological materials $\delta$ varies from tenths of microns for copper at the frequencies of visible light to about tens of meters for unconglomerated glacial debris and soils such as occur in northeast Canada at frequencies near 200 Hz (cps). The phase velocity of these waves,

$$v_p = \frac{\omega}{\beta} \text{ meters per second,} \tag{1.15}$$

is smaller than $c$. The ratio

$$\frac{c}{v_p} = n, \tag{1.16}$$

where $n$, which is always greater than 1, is called the index of refraction of the medium. The greater the index of refraction, the shallower is the penetration of electromagnetic radiation.

## VIII. REFLECTION OF PLANE WAVES

The reflection of electromagnetic waves from materials is a complicated subject in which the results of the calculations depend on the particular physical situation. The discussion presented here will cover only a single example in order to illustrate the general principles. This example will fall into the category of specular, or mirror-like, reflection. Diffuse reflection, which involves surfaces on which the average dimensions of the roughness are of the order of the wavelength of the radiation, will not be discussed here but is considered in Chapters 6 on microwaves and 7 on radar. Diffuse reflection is a difficult subject and is treated in detail in the book by Beckmann and Spezzichino (1963).

Consider a wave impinging on the plane interface between two media of different $\mu$ and $\epsilon$ ($\sigma = 0$ for both media), as shown in Figure 1.4. Part of the energy in this wave will be reflected, and part will be transmitted. $E$ and $H$ represent the electric and magnetic fields in each situation, and $k$ indicates the direction of propagation of the wave. Subscripts $i$, $r$, and $t$ designate the incident, reflected, and transmitted waves, respectively; subscripts 1 and 2 denote the properties of medium 1 and medium 2. The symbols $\theta_i$, $\theta_r$, and $\theta_t$ represent the angles between the direction of propagation of the incident, reflected, and transmitted waves and the hori-

0-201-04245-2

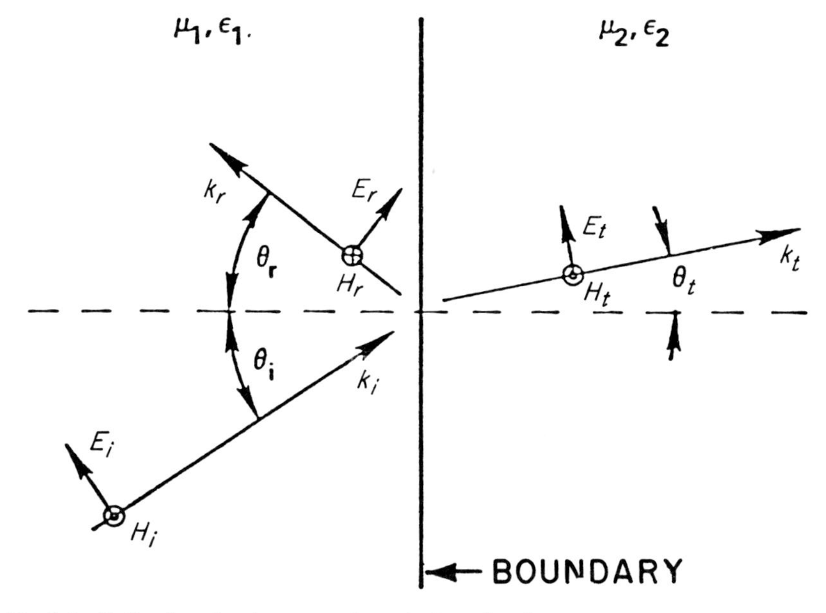


**Fig. 1.4.** Reflection of a plane wave from the boundary between two media.

zontal. The symbol ⊕ means that the positive direction for $H_r$ is taken perpendicular to, and into, the figure; the symbol ◎, that the positive directions for $H_i$ and $H_t$ are taken perpendicular to, and out of, the figure. To obtain the ratio of the electric, or magnetic, field strength which is reflected or transmitted to the respective incident components, conditions on $E_r$, $E_i$, $E_t$, and $H_r$, $H_i$, and $H_t$ at the interface of the two media as derived from Maxwell's equations have to be applied. The mathematics which results is quite involved, and again only the results are presented here. The reflection coefficient, denoted by $R$, is given by

$$R = \frac{E_r}{E_i} = \frac{H_r}{H_i} = -\frac{\sqrt{\mu_1/\epsilon_1}\cos\theta_i - \sqrt{\mu_2/\epsilon_2}\cos\theta_t}{\sqrt{\mu_1/\epsilon_1}\cos\theta_i + \sqrt{\mu_2/\epsilon_2}\cos\theta_t}. \tag{1.17}$$

The transmission coefficient $T$ is given by

$$T = \frac{E_t}{E_i} = \frac{H_t}{H_i} = \frac{2\sqrt{\mu_2/\epsilon_2}\cos\theta_i}{\sqrt{\mu_1/\epsilon_1}\cos\theta_i + \sqrt{\mu_2/\epsilon_2}\cos\theta_t} \tag{1.18}$$

Furthermore, the following relationship exists between the angles $\theta_i$, $\theta_r$, and $\theta_t$:

$$\sin\theta_r = \sin\theta_i, \tag{1.19a}$$

$$\sin\theta_t = \frac{n_1}{n_2}\sin\theta_i, \tag{1.19b}$$

0-201-04245-2

where $n_1$ and $n_2$ represent indices of refraction in the two media. Equation (1.19a) states the equality of the angle of incidence and the angle of reflection. Equation (1.19b) is called Snell's law.

The polarization of radiation is very important in reflection problems. For example, if in Figure 1.4 the polarization is changed from $H_i$ parallel to the interface to $E_i$ parallel to the interface, then in Eqs. (1.17) and (1.18) $\sqrt{\mu_1/\epsilon_1}$ and $\sqrt{\mu_2/\epsilon_2}$ must be replaced by $\sqrt{\epsilon_1/\mu_1}$ and $\sqrt{\epsilon_2/\mu_2}$ . These changes are due to differences in the magnetization and electrical polarization at the interface caused by the different wave polarization. It should be noted that Snell's law holds for any polarization.

By varying $\theta_i$, with $n_2 < n_1$, a situation can occur in which $T = 0$ in Eq. (1.18) and $R = 1$ in Eq. (1.17). This is known as total internal reflection; no energy is transmitted into medium 2. In this case, there is a phase shift upon reflection.

If the effect of nonzero conductivity is considered, Eqs. (1.17) and (1.18) are not valid and are replaced by more complicated formulas. Two new formulas are also needed to express the phase shifts which occur between $E_t$, $E_r$, $H_t$, $H_r$, and $E_i$ and $H_i$. However, again Snell's law remains unchanged. An interesting case occurs when the conductivity of the second medium becomes extremely large. In this circumstance, the reflection coefficient becomes 1, the transmission coefficient becomes 0, and no radiation penetrates the second medium. This agrees with the findings of Section VII, in which the skin depth, given by Eq. (1.14), goes to 0 as the conductivity $\sigma$ becomes extremely large. An even simpler way to think of this relationship is to recall that radiation barely penetrates good conductors, but penetrates poor conductors relatively deeply.

Reflection characteristics are obviously of great importance in remote sensing because in many cases it is precisely by "bouncing" energy off objects at a distance and collecting the resulting radiation that we learn about these distant objects. Thus the study of the moon from Earth has employed both sunlight and radar waves to examine the properties of the lunar surface from afar. In this regard it is instructive to remember that, although the term "remote sensing" was coined by an earth scientist (Dr. Evelyn Pruitt, of the Geography Branch, Office of Naval Research), it was another earth scientist, Dr. Peter Badgley, who provided much of the initial stimulus for remote sensing through the funding of NASA studies of resources for circumlunar satellite and spacecraft data gathering.

## IX. BLACKBODY RADIATION

Blackbody radiation is the first of three topics we will discuss in which the particle, or quantum, nature of light has to be assumed to derive a theory consistent with experimental results. The introduction of the concept of the particle nature of light in its interaction with matter led ultimately to the development of quantum theory, which is the basis of modern physics.

It is a common observation that hot bodies radiate electromagnetic energy. In fact, all bodies at temperatures above absolute zero (−273°C) radiate electromag-

0-201-04245-2

netic radiation over a broad range of wavelengths. Experimentally, it is known that the wavelength near which most of the radiation is emitted depends on the temperature of the object. This fact is represented mathematically in Wein's displacement law:

$$\lambda_{max} = \frac{a}{T}, \tag{1.20}$$

where $\lambda_{max}$ is the wavelength in centimeters at which the radiation is a maximum, $T$ is the absolute temperature of the body in degrees Kelvin (°K: centigrade temperature + 273°), and the constant $a = 0.29$ cm °K. Wein's displacement law thus states that, as the temperature of a body changes, the wavelength at which the maximum radiation occurs also shifts or is displaced. Thus one can measure the temperature of an object remotely by measuring the spectrum of its radiation. The astronomer does this for the stars, the forester may do this in thermal sensing to detect forest fires, and the geophysicist also may do this in infrared sensing of geothermal areas.

It turns out that in equilibrium the fraction of the radiation falling on an object which is absorbed exactly equals the fraction which is emitted. Thus an object that absorbs a large amount of radiation also emits a large amount of radiation. In other words, a good absorber is a good emitter of radiation. Using properties of a body averaged over all wavelengths, we can write

$$\alpha = \epsilon, \tag{1.21}$$

where $\alpha$ is the absorptivity of the body, and $\epsilon$ is its emissivity.

Thus snow, which is a poor absorber of visible radiation, is also a good reflector and poor emitter of radiation in this wavelength range. Both $\alpha$ and $\epsilon$ are dimensionless and have values less than 1. Equation (1.21) is also true at any wavelength and is expressed by replacing $\alpha$ and $\epsilon$ by $\alpha_\lambda$ and $\epsilon_\lambda$, where $\epsilon_\lambda$ is the emissivity at a given wavelength or wavelengths, referred to as spectral emissivity. A word of caution is in order: Do not confuse the emissivity with the permittivity of an object, as discussed earlier. The two are different physical quantities unfortunately represented by the same symbol.

An object which absorbs all the radiation incident on it without any reflection is called a blackbody, and $\epsilon$ (emissivity) for a blackbody is equal to 1. The radiation emitted by a blackbody in equilibrium with its surroundings at any temperature was studied in depth in the nineteenth century. It was not until 1900, however, that Max Planck was able to explain the spectral distribution of blackbody radiation. Planck's radiation law is based on the assumption that radiant energy, $E$, is emitted by molecules and atoms in certain minimum-sized lumps called quanta and is expressed by

$$E = hf, \tag{1.22}$$

0-201-04245-2

where $f$ is the frequency of the radiation, and $h$ is a constant now known as Planck's constant. Planck adjusted this constant to make his theory fit the experimental data. Planck's expression for the energy radiated per unit wavelength per second per unit area of the blackbody, $w_\lambda$ (or spectral radiance), is

$$w_\lambda = \frac{2\pi hc^2}{\lambda^5}\left(\frac{1}{e^{hc/\lambda kT} - 1}\right) \quad \text{watts per square meter per Angstrom,} \tag{1.23}$$

where $h = 6.62 \times 10^{34}$ J-sec (Planck's constant), $c$ is the speed of light in meters per second, $k = 1.38 \times 10^{-23}$ J/deg and is called Boltzmann's constant (caution: Do not confuse this $k$ with the wave number $k = 2\pi/\lambda$, introduced in Section III), $\lambda$ is the wavelength in question in meters, and $T$ is the absolute temperature in degrees Kelvin.

Figure 1.5 shows $w_\lambda$ versus $\lambda$ for several absolute temperatures. The names of the different portions of the spectrum across the top of the figure relate the blackbody curves at different absolute temperatures to these spectrum regions.

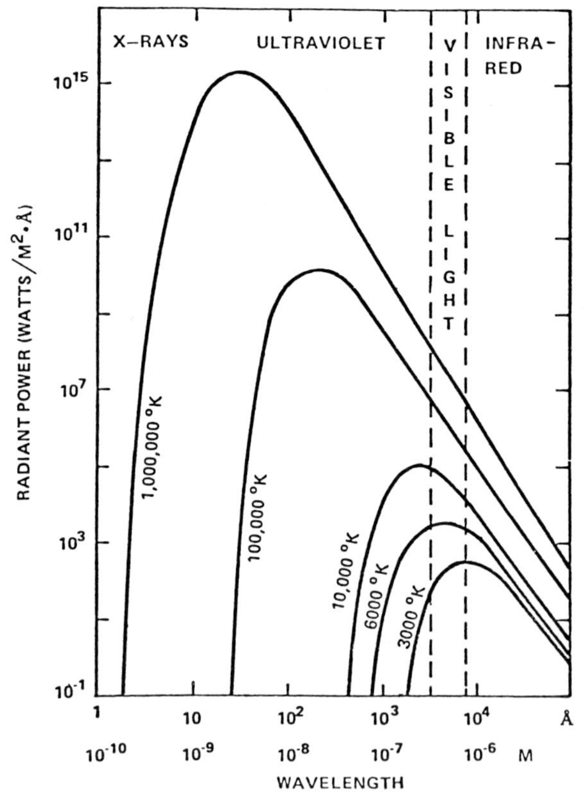


**Fig. 1.5.** Energy distribution from a black body at various temperatures derived from the Planck equation (i.e., Eq. (1.23)).

0-201-04245-2

Planck's law fits all the experimental data and agrees with formulas derived earlier by other investigators in the wavelength domains where these older formulations are valid. For example, when the wavelength $\lambda$ becomes very large—as for the microwave region of the spectrum—it can be shown from Eq. (1.23) that

$$w_\lambda \simeq \frac{2\pi ck}{\lambda^4} T \text{ watts per square meter per Angstrom,} \tag{1.24}$$

which is identical to the so-called Rayleigh-Jeans law derived before Planck's work. The problem with Eq. (1.24) is that it predicts a continual and rapid rise in $w_\lambda$ when the wavelength becomes small. This is impossible physically because it would mean that bodies would radiate *unlimited* amounts of energy as $\lambda$ decreases, a situation known as the "ultraviolet catastrophe." Planck's work showed that Eq. (1.24) is valid only for long wavelengths and thereby avoids this so-called catastrophe.

If we now let the wavelengths become small and substitute this condition into Eq. (1.23), it can be shown that

$$w_\lambda \simeq \frac{2\pi hc^2}{\lambda^5} e^{-hc/\lambda kT} \text{ watts per square meter per Angstrom.} \tag{1.25}$$

This equation is identical to an expression known as Wein's law, which was derived by semiempirical means before Planck's work. Furthermore, it can also be shown that Eq. (1.20) can be derived using Eq. (1.23).

By adding together the power radiated per unit area per unit wavelength, $w_\lambda$, for all wavelengths, one obtains the total power radiated per unit area of a blackbody, $w$, given by

$$w = \frac{2\pi^5 k^4}{15c^2 h^3} T^4 \text{ watts per square meter.} \tag{1.26}$$

Equation (1.26) states that the total power per unit area of a blackbody is proportional to the fourth power of the absolute temperature. This is called the Stefan-Boltzmann law and is another example of Planck's agreement with earlier work. This law states that the total radiated energy, $w$, is proportional to the fourth power of the temperature, that is,

$$w = \sigma T^4 \text{ watts per square meter,} \tag{1.27}$$

where $\sigma$ is the Stefan-Boltzmann constant. This relation applies to all wavelengths of the spectrum shorter than the microwave region. In the microwave area, however, the radiation changes to $w = f(T)$, that is, the power varies directly and linearly as the temperature in degrees Kelvin (caution: Do not confuse $\sigma$, the Stefan-Boltzmann constant, with $\sigma$, the electrical conductivity introduced in Section VII). Comparison of Eqs. (1.26) and (1.25) shows that

0-201-04245-2

$$\sigma = \frac{2\pi^5 k^4}{15c^2 h^3} \text{ watts per square meter per (degree)}^4, \tag{1.28}$$

which numerically agrees with values of $\sigma$ derived earlier from thermodynamic data.

The power radiated per unit wavelength per unit area of nonblackbodies is obtained by multiplying the emissivity of the material at that wavelength, $\epsilon_\lambda$, by Eq. (1.23). Likewise, the total power radiated per unit area of nonblackbodies is obtained by multiplying by $\epsilon$ [cf. Eq. (1.21)] Eq. (1.26), thus:

$$w_{\text{nonblackbody}} = \epsilon w_{\text{blackbody}} = \frac{\epsilon 2\pi^5 k^3}{15c^2 h^3} T^4 \text{ watts per square meter.} \tag{1.29}$$

An understanding of blackbody radiation is important because the radiation from natural bodies affords the opportunity to "observe" materials at wavelengths other than the visible. For example, the spectrum of the sun can be approximated by a 6200°K blackbody ($\lambda_{max}$ = 480 nm), and the sun's radiation reflected from materials and sensed with a camera forms an important part of remote sensing. However, at wavelengths of about 3-5 $\mu$ (1 $\mu = 10^{-4}$ cm) the radiation from the earth's materials at approximately 300°K begins to be greater than the sun's reflected radiation. The power peak for earth emission ($\lambda_{max}$) occurs at 10.6 $\mu$, in the center of the thermal infrared region.

## X. ABSORPTION OF RADIATION

The absorption of electromagnetic radiation by solids, liquids, and gases in the space between an emitter, the object being sensed, and the detection device has a great effect on the characteristics of the received signal. These absorption properties of matter are also utilized in the construction of devices to detect radiation. Two examples of absorption processes will be discussed: the first will illustrate the application of absorption in solids; the second, the change in the character of radiation after it has passed through a gas.

### A. The Photoelectric Effect

Whenever electromagnetic radiation interacts with matter on a molecular or atomic scale, the quantum or particle nature of the radiation is manifest. Perhaps the best-known example of the quantum nature of electromagnetic radiation is the photoelectric effect. When electromagnetic radiation falls on a solid, electrons may be ejected from the surface. These electrons constitute a measurable electric current called a photocurrent. The photocurrent has three properties:

Characteristic 1. The current is proportional to the intensity of the radiation.

0-201-04245-2

Characteristic 2. The kinetic energy of the electrons depends only on the frequency of the radiation and not on its intensity.

Characteristic 3. There is no current below a certain threshold radiation frequency.

Albert Einstein explained this effect, for which he won a Nobel Prize in 1921. His theory begins by stating that electrons require a certain minimum energy to escape from a solid and that this energy is $e\phi$, where $e$ is the electronic charge and $\phi$ is the so-called work function of the material. As in Planck's blackbody radiation theory, radiation is *emitted* in lumps of energy, $hf$, where $h$ is Planck's constant and $f$ is the frequency. In Einstein's description, when a quantum of radiation, or photon, is absorbed by an electron, it gives the electron energy $hf$. The intensity of the radiation (i.e., the energy per second per unit area) is proportional to the number of photons per unit time falling on a unit area of the surface. The photocurrent is proportional to the number of electrons ejected by absorbed photons per unit time per unit area. Thus the photocurrent is proportional to the intensity of the radiation (characteristic 1).

The energy given to a photoelectron is equal to the photon energy, which is in turn proportional to the frequency. Thus the kinetic energy of the *ejected* electrons, KE, is

$$\mathrm{KE} = hf - e\phi \text{ joules} \tag{1.30}$$

(characteristic 2).

When KE = 0 in Eq. (1.30), $f = e\phi/h$, and an electron is emitted with no kinetic energy. If $f$ is less than $e\phi/h$, the electrons will not be given sufficient energy to do the work to escape, and there will be no photocurrent (characteristic 3).

This is the principle upon which the photoelectric detectors used in remote sensing are based. Semiconductor solid-state detectors work on a similar principle, but in that case the electron current caused by the absorption of photons is not external to the material. The photons simply raise the energy of the electron in question from the valence band, where they are immobile, to the conduction band of the semiconductor, where the electrons are free to move under externally applied voltages. Electrons may also be given energy thermally, and this contributes noise to the detection system. Hence many solid-state detectors must be operated at extremely cold (cryogenic) temperatures produced by cooling in liquid nitrogen or helium to achieve maximum sensitivity.

## B. Gaseous Absorption

Molecules in the gaseous state possess three types of *internal* energy states: rotational, vibrational, and electronic. At any time the molecule has a given electronic state, a given vibrational state, and a given rotational state. For any electronic state a variety of vibrational states are possible, and for any vibrational state a variety of

0-201-04245-2

rotational states are possible. However, the total internal energy of the molecules at any time is the sum of the energies of the three states. The nonelectronic states represent the energy of small-amplitude rotational and vibrational motions of the atoms of the molecule about their equilibrium positions.

The energy difference, $E_{ij}$, between any two rotational states, $i$ and $j$, can be characterized by a number, $f_{ij}$, such that

$$E_{ij} = hf_{ij} \text{ electron volts,} \tag{1.31}$$

where $h$ is Planck's constant. The electron volt (eV) is a unit of energy equal to $1.6 \times 10^{-19}$ J. The number $f_{ij}$ will, in general, differ for every pair of rotational states for which state changes are allowable.

The energy difference, $E_{kl}$, between any two vibrational states, $k$ and $l$, can be characterized by a number, $f_{kl}$, such that

$$E_{kl} = hf_{kl} \text{ electron volts.} \tag{1.32}$$

The number $f_{kl}$ will, in general, differ for every pair of vibrational states for which state changes are allowable.

Similarly, the energy difference, $E_{mn}$, between any two electronic states, $m$ and $n$, can be characterized by a number, $f_{mn}$, such that

$$E_{mn} = hf_{mn} \text{ electron volts.} \tag{1.33}$$

The number $f_{mn}$ will, in general, differ for every pair of electronic states for which state changes are allowable.

When electromagnetic radiation of frequency $f$ and intensity $I_0$ passes through a gas of molecules, radiation will be *absorbed* by the gas if the energy $E = hf$ of the photons satisfies any *one* of the following conditions:

$$\begin{aligned}
&E = hf = hf_{ij} = E_{ij} && \text{rotational,}\\
&E = hf = hf_{kl} = E_{kl} && \text{vibrational,}\\
&E = hf = hf_{mn} = E_{mn} && \text{electronic,}\\
&E = hf = h(f_{ij} + f_{kl}) = E_{ij} + E_{kl} && \text{rotational and vibrational,}\\
&E = hf = h(f_{ij} + f_{mn}) = E_{ij} + E_{mn} && \text{rotational and electronic,}\\
&E = hf = h(f_{kl} + f_{mn}) = E_{kl} + E_{mn} && \text{vibrational and electronic,}\\
&E = hf = h(f_{ij} + f_{kl} + f_{mn}) = E_{ij} + E_{kl} + E_{mn} && \text{rotational, vibrational, and electronic.}
\end{aligned} \tag{1.34}$$

The conditions presented in Eqs. (1.34) state that the energy of the photon has to be equal to *one* of the energies corresponding to allowable state changes in the

0-201-04245-2

molecule before it can be absorbed. The intensity $I$ of the radiation leaving the gas will be less than $I_0$ because of the loss of the photons that were absorbed. The energy absorbed by the molecules is converted to heat as these molecules relax back to their equilibrium energy state through molecular collisions after the radiation has passed.

The *electronic* energy levels are separated by energy differences of from 2 to 10 eV. Pure *vibrational* states are separated by energy differences of 0.2-2 eV; pure *rotational* states, by energy differences of approximately $10^{-3}$ to $10^{-5}$ eV. Generally speaking, electronic states are separated by energy differences corresponding to the energy carried by photons in the green, blue, and ultraviolet regions of the spectrum (the visible and ultraviolet portions shown in Figure 1.1); pure vibrational states, by energy differences corresponding to the energy carried by a photon in the yellow and red (visible) and near-infrared portions of the spectrum (Figure 1.1); and pure rotational states, by small energy differences corresponding to the energy carried by photons in the thermal infrared, far-infrared, and microwave portions of the spectrum (Figure 1.1).

When electromagnetic radiation composed of many frequencies, each with its own intensity, passes through a portion of the atmosphere, the various molecular species will absorb photons of different frequencies, and the radiation that reaches the detector will no longer have the characteristics of the object from which it was radiated or reflected. Hence the effects of absorption of the atmosphere at all frequencies of electromagnetic radiation have to be known, so that they can be extracted from the detected radiation. This extraction will allow the real properties of the object being remotely sensed to be ascertained.

0-201-04245-2

Chapter 2

# Effects of the Atmosphere on Remote Sensing

*Robert S. Fraser and Robert J. Curran*

## I. THE ELECTROMAGNETIC SPECTRUM AND THE EARTH'S ATMOSPHERE

If there were no atmosphere above the earth's surface, electromagnetic energy of all wavelengths would interact with the surface and would convey information about the nature of that surface. As it happens, however, the earth's atmosphere is transparent enough for remote sensing in only a small portion of the electromagnetic spectrum. The spectral bands of least attenuation are called windows, but even in the windows atmospheric effects are significant. The gases and larger aerosol particles comprising the atmosphere scatter, absorb, and emit radiant energy. The atmosphere is thus both an attenuator and a source of radiant energy. As a result, information transmitted from the ground to a higher observational platform is attenuated and distorted. The radiant energy scattered and emitted diffusely by the atmosphere adds background noise to the signal. For example, the apparent contrast between an object and its surroundings, or the apparent color of the object, changes with distance from it. Similarly, the apparent temperature of the earth's surface, measured in the infrared or microwave spectrum, varies with height. The diffuse radiant energy leaving the base of the atmosphere is also a source of illumination on the ground. Because the state of the atmosphere is rarely known in full at the time of remote sensing, these complex effects of the atmosphere cannot always be fully accounted for, and the atmosphere is therefore a significant and pervasive complication in remote sensing.

### A. Windows in the Electromagnetic Spectrum

Important windows for remote sensing occur throughout the spectrum; the most significant from space are as shown in Table 2.1. Inspection of this table shows that all wavelengths shorter than 0.30 μm are substantially closed for remote sensing and that the principal windows lie in the visible (and near-visible), infrared, and microwave regions. Atmospheric absorption in the ultraviolet region shorter than 0.32 μm is so intense that the amount of solar energy reaching the earth's surface is insufficient to be used in remote sensing.

The ultraviolet radiation absorbed by the upper atmosphere either dissociates or ionizes gas molecules (Figure 1.1). Photons with energies of 5-9 eV *dissociate* oxygen ($O_2$), ozone ($O_3$), and nitrogen ($N_2$), and those with energies of 12-16 eV ionize $O_2$ and $N_2$. These interactions are so intense in the stratosphere and ionosphere that the ultraviolet spectrum not only is useless for remote sensing, but also

*Remote Sensing of Environment*, Joseph Lintz, Jr. and David S. Simonett (eds.) ISBN 0-201-04245-2


0-201-04245-2

**TABLE 2.1**
**Major Atmospheric Windows Available for Spacecraft Remote Sensing**
**(Clearest Windows Shown in Boldface)**

| | |
|---|---|
| Ultraviolet and visible | **0.30–0.75 μm** |
| | **0.77–0.91** |
| Near-infrared | 1.0–1.12 |
| | 1.19–1.34 |
| | **1.55–1.75** |
| | **2.05–2.4** |
| Mid-infrared | 3.5–4.16 |
| | 4.5–5.0 |
| Thermal infrared | **8.0–9.2** |
| | **10.2–12.4** |
| | 17.0–22.0 |
| Microwave | 2.06–2.22 mm |
| | 3.0–3.75 |
| | **7.5–11.5** |
| | **20.0+** |

shields the surface of the earth from this band of harmful and destructive solar energy.

In the X-ray region, where even more energetic photons ($5 \times 10^1$ to $5 \times 10^5$ eV) arising from transitions in the inner electronic shells of an atom occur, the intensity of interaction with atmospheric gases is such that all X-radiation is absorbed in a few centimeters to tens of meters (depending on the photon energies) of passage through the lower atmosphere. This region is therefore closed both to remote sensing of space and to earth-emitted X-radiation (see Figure 1.1).

The shortest wavelengths (and highest electromagnetic energies) used in remote sensing are associated with gamma radiation of about 1 MeV ($10^6$ eV). Gamma photons are created during transitions between the energy states of the nucleus of an atom. Radioactive materials, such as uranium, in the earth's crust emit gamma photons. These emissions can be detected with devices carried on helicopters or very-low-flying aircraft, but cannot be detected at normal aircraft altitudes or by satellite because the earth's atmosphere strongly absorbs gamma photons. Thus, in the spectral band with wavelengths shorter than 0.3 μm, it is only in the gamma ray region that some restricted remote sensing is feasible.

The first significant atmospheric window begins to open at 0.3 μm and has good transparency in the visible spectrum. Photons are only weakly absorbed in this region, but scattering by both gaseous molecules and particulates of haze and dust is of greatest influence in remote sensing. The visible window continues, but with interruptions, in the near-infrared spectrum, and strong (vibrational) absorption bands, chiefly those of water vapor ($H_2O$), appear more frequently (Figure 2.1). In the thermal infrared band, between 4 and 14 μm, strong absorption bands occur which are caused by vibrational-rotational changes, principally in water vapor and

0-201-04245-2

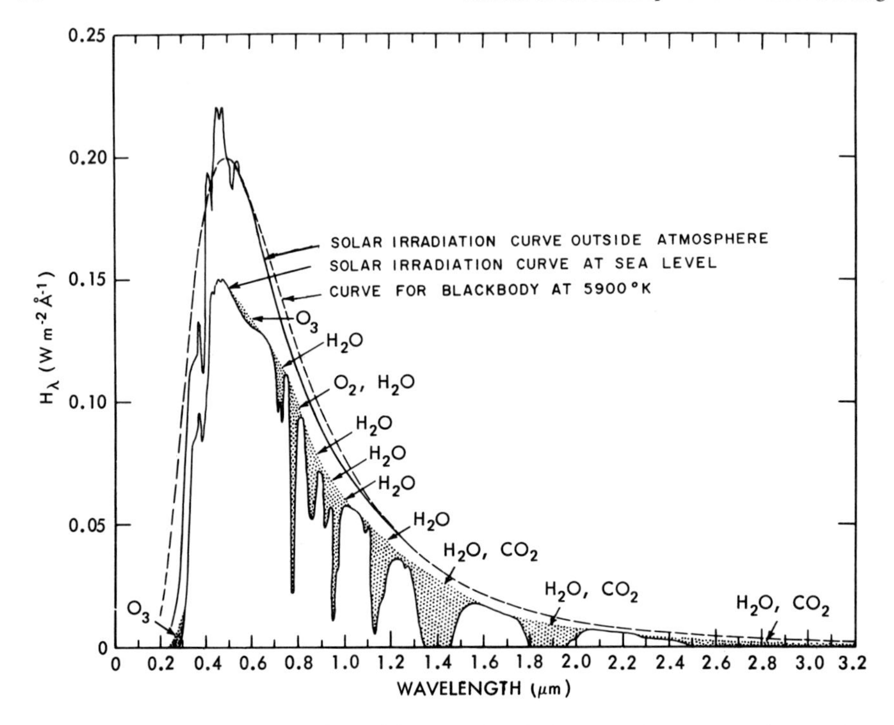


**Fig. 2.1.** Spectral irradiance ($H_\lambda$) of direct sunlight before and after it passes through the earth's atmosphere. The stippled portion gives the atmospheric absorption. The sun is at the zenith. From Valley (1965).

carbon dioxide ($CO_2$). These absorption bands break up the near-infrared and thermal infrared regions into a series of windows of modest transparency, interspersed with regions shuttered by absorption (Table 2.1, and Figures 2.1, 2.19 and 2.20).

Although all of these windows are of importance in remote sensing, those between 1.5 and 1.8, 2.05 and 2.4, 3.5-4.1 and 4.5-5.0, and 8.0-9.2 and 10.2-12.4 μm are of especial importance. Those windows between 1.5 and 1.8 μm and 2.05 and 2.4 μm are of value in detecting changes in crop leaf moisture status, as shown by differential reflection of solar energy. Those between 8.0-9.2 and 10.2-12.4 μm cover the region of normal earth emission. The bands from 3.5 to 5 μm cover the region where forest fires emit their peak energy; this is also a crossover region where *both* solar and earth emissions are important (see Figure 2.20). In the daytime both contribute; at night only the earth emission is important between 3 and 5 μm. Within the 8-14 μm region there is a strong ozone absorption band at 9.6 μm. The long wavelength cutoff at 14 μm is due to $CO_2$ absorption in the region from 14 to 16.8 μm (Figure 2.20).

At wavelengths longer than 22 μm, a whole suite of rotational transitions out to 1 mm and beyond effectively close the long-wavelength infrared and millimeter microwave regions to remote sensing. Most of these transitions are associated with

0-201-04245-2

water vapor. No important atmospheric molecular transitions occur for wavelengths longer than a few centimeters.

In the microwave region the only atmospheric absorption bands lie *between* the windows given in Table 2.1. The rest of the long-wavelength microwave spectrum is very clear. Falling rain interferes significantly with microwave sensing, and this is very strongly wavelength and rainfall-intensity dependent. The effects of clouds are weaker but must be corrected for in programs requiring precise measurements, such as the determination of sea surface temperature. At wavelengths longer than 10 cm only the most intense storms interfere with microwave sensing.

### B. Solar and Terrestrial Radiation

The sun is the strongest source of radiant energy incident on the earth for the 0.3–3 $\mu$m spectral band. Other sources are much weaker in this range. The full moon is the next strongest source, and the radiant energy from it is about $10^{-5}$ of that from the sun (Kondratyev, 1969). Figure 2.1 shows the spectral characteristics of sunlight. The upper continuous curve gives the coarse spectral irradiance above the earth's atmosphere. The maximum irradiance occurs at 4.7 $\mu$m. About 20% of the sun's energy falls in the spectral band $\lambda < 0.47$ $\mu$m, and 44% in the visible band, 0.40–0.76 $\mu$m.

The dashed curve of Figure 2.1 represents the irradiance of a blackbody at $T = 5900^\circ$K; the lowest curve, the irradiance of direct sunlight at sea level. The difference between the two continuous curves represents attenuation by scattering and absorption. The stippled portion shows loss by absorption. Ozone absorption causes a cutoff at 0.29 $\mu$m and also appears weakly in the red part of the spectrum. Except for the strong, narrow oxygen absorption at 0.76 $\mu$m, the infrared absorption is caused principally by $H_2O$ and $CO_2$.

*The earth* and its atmosphere are the principal sources of radiant energy for wavelengths greater than 4 $\mu$m. Figure 2.2 shows an infrared spectrum of radiance measured by the infrared interferometer spectrometer, which was carried by the meteorological satellite NIMBUS 4. The field of view was free of clouds for this spectrum. The smooth curves give the computed radiance that would be emitted by blackbodies at temperatures of 260, 280, and 300°K. The measured radiances which lie between the blackbody curves at temperatures of 280°K and 300°K have penetrated through the atmosphere from the surface with minimal attenuation. The good window between 8 and 14 $\mu$m and narrow windows for wavelengths greater than 17 $\mu$m are apparent by their high temperatures. The temperatures associated with such radiances measured in the 11 $\mu$m window are typically about 5°K colder than the true surface temperature, under clear conditions.

The minimum measured radiances of Figure 2.2 occur at wavelengths of strong atmospheric absorption. Such absorption is caused by strong vibrational and rotational energy changes in trace amounts of principally ozone, carbon dioxide, and water vapor. Ozone not only causes the short-wave cutoff of solar energy reaching the ground, but also absorbs significantly at 9.6 $\mu$m in the middle of a prominent

0-201-04245-2

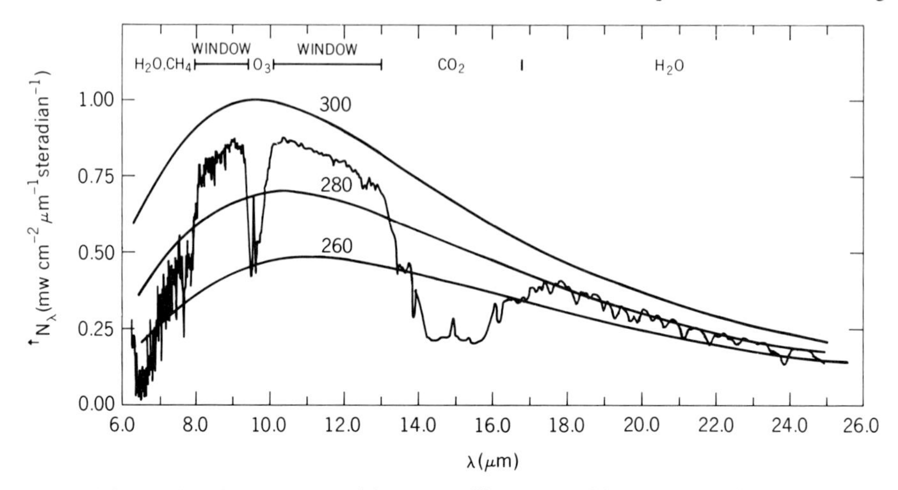


**Fig. 2.2.** Nadir radiance measured from a satellite (curve with structure). The smooth curves give values of blackbody radiances at the indicated temperture in degrees Kelvin. From Kunde et al. (1974).

atmospheric window. Radiation from the warm earth is completely absorbed in the strongest absorption lines. However, the gases that cause the absorption also are strong radiators of energy. Since the atmospheric temperature usually decreases with height, the emitting gases are cooler than the earth's surface and are observed to be colder.

## C. Physical Properties of the Atmosphere

Radiant energy interacts with the atmosphere in the following ways: the energy can be scattered, absorbed, and emitted. To clarify these concepts, we may consider a nearly parallel monochromatic (single-wavelength) beam of radiant energy, say direct sunlight after it passes through an interference filter, which enters a slab of the lower atmosphere at a constant rate. The slab contains atoms, molecules, and large particles. Radiant energy in the beam is attenuated and emerges from the slab at a reduced rate, because the slab absorbs and scatters energy. The scattered photons have the same energy as the incident photons and emerge from the slab in all directions. The absorbed energy either is re-emitted as photons of different energy or increases the internal energy of the slab (i.e., heats it). If the slab absorbs the energy, then by Kirchhoff's law it also emits energy, which increases the radiance of the transmitted light.

All particles in the atmosphere—atomic, molecular, and large—scatter electromagnetic energy. Single atoms are few and of little significance in scattering. Molecular or Raleigh scattering, however, is very important in the near-ultraviolet and visible regions but is negligible at wavelengths beyond 1 μm. This scattering varies as the inverse fourth power of the wavelength and is thus most severe in the ultraviolet

0-201-04245-2

and blue regions, causing the familiar blue color of the sky. Essentially all of the molecular scattering is accounted for by $N_2$ and $O_2$, 78% and 21%, respectively, of a dry atmosphere (Valley, 1965). These percentages remain fixed throughout the lowest 90 km of the atmosphere, since this layer is well mixed. Noctilucent clouds have been observed between 45° and 90° north latitude at a height of 80 km. These clouds attenuate visible light weakly. They may also interact slightly with infrared energy but negligibly with microwave energy.

A few trace gases have significant effects on the transfer of radiation in the infrared. Carbon dioxide, which constitutes 0.03% of the atmosphere and is well mixed with the other permanent gases, has a number of important absorption bands, as noted earlier. The amounts of water vapor are highly variable, in both time and space, in the lower atmospheric layers, varying from a sea level density of $10^{-2}$ g m$^{-3}$ in very cold, dry climates to as much as 30 g m$^{-3}$ in a hot, humid region (Valley, 1965; also see Greaves et al., 1971, for additional climatological data). The average surface concentration is about 10 g m$^{-3}$, giving an average total mass of water vapor in a vertical column of 25 kg m$^{-3}$ or 2.4 g cm$^{-2}$. This amount is 0.3% of the total atmospheric mass. To simplify computations when high accuracy is not required, the water vapor density is assumed to decrease exponentially with height; the scale height is approximately 2.5 km. Methods are being developed to account for the effects of water vapor on remote measurements, one example being the remote measurement of sea temperature (Prabhakara et al., 1975). The other atmospheric constituents are well mixed and substantially stable in amount (except for $CO_2$ near cities).

The density of the dry gaseous atmosphere decreases approximately exponentially with respect to height ($z$), so that the mass of the atmosphere in each successive 5000-m-thick slab is roughly halved: the first 5000 m contains one half the total mass; the next 5000 m, one quarter; the next 5000 m, one eighth, and so on. The total number density of gas molecules per cubic meter ($n$) can be represented approximately as

$$n(z) = n(0) \exp\left(\frac{-z}{H}\right), \qquad (2.1)$$

where $H$ is the scale height, which will be discussed later. The number density is also related to the pressure $[p(z)]$ and temperature $[T(z)]$ by means of the perfect gas law:

$$n = (kT)^{-1}\, p, \qquad (2.2)$$

where Boltzmann's constant $k = 1.381 \cdot 10^{-23}$ J °K$^{-1}$. The number and mass densities at sea level are given in Table 2.2 for the U.S. Standard Atmosphere (Valley, 1965) and also for standard pressure and temperature. Since number densities in gaseous absorption studies are frequently reduced to the values at standard temperature and pressure, $n = 2.69 \cdot 10^{25}$ m$^{-3}$ would then be used in Eq. (2.1).

0-201-04245-2

The scale height ($H$) of the atmospheric gas will be defined for this chapter in terms of the total mass of dry air in a vertical column of the atmosphere. The average mass of one molecule of dry atmosphere equals the average molecular weight ($m = 28.97$) times the atomic mass unit ($m_0 = 1.660 \cdot 10^{-27}$ kg). The mass density is then

$$\rho(z) = n(z)\, m m_0 . \tag{2.3}$$

The total mass $[M(z_1, z_2)]$ in a vertical column of unit cross-sectional area and between a lower ($z_1$) and an upper ($z_2$) level can also be expressed as

$$M(z_1, z_2) = \frac{p(z_1) - p(z_2)}{g(z)} , \tag{2.4}$$

where $g$ is the acceleration of gravity at some level $z$ between $z_1$ and $z_2$. The average total mass from sea level to the top of the atmosphere is

$$\begin{aligned} M(0, \infty) &= (9.81\ m\ s^{-2})^{-1}\ (1.013 \cdot 10^5\ \mathrm{N\ m^{-2}}) \\ &= 1.034 \cdot 10^4\ \mathrm{kg\ m^{-2}} , \end{aligned} \tag{2.5}$$

where the sea level value of $g$ has been used without compromising accuracy. The average mass at sea level varies over the earth's surface from about −6 to +4% of the mean, and the standard deviation of the total mass at an arbitrary place is about 1% (Valley, 1965). The total mass of the dry atmosphere in a vertical column above sea level can be considered constant when computing radiation effects for most remote sensing applications.

The scale height is related to the total mass by substituting Eq. (2.3) in Eq. (2.1), and integrating from sea level to the top of the atmosphere; then

$$M(0, \infty) = \rho(0) H. \tag{2.6}$$

The scale height in Eq. (2.6) is the thickness of a homogeneous layer (at sea level density) containing the entire atmospheric mass. The numerical value of the scale height of the dry atmosphere is obtained by substituting the mass density at standard temperature and pressure (Table 2.2) and Eq. (2.5) in Eq. (2.6) to obtain

$$H = 8.0\ \mathrm{km}. \tag{2.7}$$

If this value of $H$ is used in Eq. (2.1) to calculate the density, the relative deviation of the result from the density of the U.S. Standard Atmosphere for the same height is less than 20%, when the height is less than 20 km. The chief reason for the difference is that an exponential decrease in density (Eq. 2.1)) is strictly valid only for an isothermal atmosphere. The earth's atmosphere deviates from isothermalcy up to

0-201-04245-2

±20%. The temperature of the U.S. Standard Atmosphere, for example, decreases at a rate of 6.5°K/km$^{-1}$ from the ground to a height of 11 km and is constant from 11 to 25 km (Valley, 1965).

The atmospheric pressure also decreases approximately exponentially with respect to height, so that only 1% of the atmospheric mass lies above 32 km. The atmosphere above 32 km can be neglected in the window areas of the spectrum expect for ionospheric and noctilucent cloud effects. The ionosphere, which is above the 80 km level, significantly affects radio propagation for $\lambda < 1$ m, but it does not affect the spectrum for $\lambda < 1$ m.

Scattering of electromagnetic energy is also caused by particles much larger than molecules. These larger particles are composed of either omnipresent haze or dust particles or clouds of liquid and solid water. Haze and dust particles are normally called large particles and have radii between 0.01 and 20 μm. The radii of the cloud particles lie between 1 and 100 μm.

Large-particle scattering (also known as Mie scattering when the particles are spherical) is important in the spectral range of the near-ultraviolet to and including the near-infrared. Dust and haze are naturally concentrated near the earth's surface. Above an altitude of a few kilometers they usually have a small effect. The extinction due to large particles is approximately inversely proportional to the horizontal visibility (Elterman, 1970). Since the ground-level visibility in air without fog varies from about 1 to 100 km, the extinction also varies by 2 orders of magnitude.

**TABLE 2.2**
**Density of the U.S. Standard Atmosphere at Sea Level, Which Is Dry, and Density of the Dry Atmosphere at Standard Temperature and Pressure**

| *Model* | *Temperature (°K)* | *Pressure (N m$^{-2}$)* | *Number Density (m$^{-3}$)* | *Mass Density (kg m$^{-3}$)* |
|---|---|---|---|---|
| U.S. Standard Atmosphere at $z = 0$ | 288 | $1.013 \cdot 10^5$ | $2.55 \cdot 10^{25}$ | 1.225 |
| Standard temperature and pressure | 273 | $1.013 \cdot 10^5$ | $2.69 \cdot 10^{25}$ | 1.298 |

## D. Clouds

The radiative properties of clouds change as a consequence of wavelength. In the visible and infrared regions cloud particles are equal in size to the wavelength of the radiation or larger. The probability of interaction between a particle and an electromagnetic wave is at a maximum under these conditions. Because the wavelength of microwave radiation is much greater than the size of cloud particles, the probability of interaction between microwaves and cloud particles is much smaller in this region than in the visible or infrared. In fact, for many common types of clouds most microwave radiation will pass through the cloud without detectable absorption or

0-201-04245-2

scattering. Therefore the obscuration caused by clouds is considerable in the visible and infrared and much less in the microwave region.

A second radiative effect of clouds in remote observations of the earth is the change in the illumination of the surface caused by clouds. In the visible region this effect is seen either as a shadowing of the direct solar radiation or as a bright reflection, which increases the flux of light incident on the surface (Hulstrom, 1973). At longer wavelengths the emission of the clouds dominates the atmospheric radiative processes and increases the radiant energy incident on the surface. Again, because of the ratios of cloud particle size to radiation wavelength, the effects of surface illumination by clouds are more important at the shorter wavelengths.

The vertical distribution of clouds is involved in the planning of aircraft missions. Here it is important to know at what altitude to make observations in order that cloud obscuration will be minimized. The probability distribution of clouds with altitude varies dramatically with both location and season. In general, cloud formations occur uniformly throughout the troposphere and are usually constrained to altitudes below the tropopause, which varies from 7 to 18 km in height but occurs at about 15 km in midlatitudes. A schematic representation of the probability of cloud obscuration for measurements made at differing altitudes is given in Figure 2.3. This probability function assumes that the surface area observed by the detector is not a function of altitude. The probability value $P_0$ in Figure 2.3 is the limiting value obtained by increasing altitude; this limiting value is soon reached for altitudes above the tropopause. Because of this behavior of the probability function, aircraft measurements made at altitudes immediately above the tropopause and satellite observations suffer nearly identical probability of obscuration by clouds.

Statistics of cloud coverage are used for planning various survey missions. "Coverage" is defined as the fraction of a horizontal surface that is covered by the vertical projections of the clouds. For example, when viewing the earth from satellites, coverages of 0.0, 0.6, and 1.0 mean, respectively, that no clouds are present, that 0.6 of the ground is covered by clouds, and that the ground is completely obscured by clouds. Coverages derived from satellite data are sometimes different from those measured from the ground (Glaser et al., 1968). The satellite coverage depends on the cloud type, on the resolution, and on the spectral band of the sensor. Martin and Liley (1971) have computed statistical relations between cloud coverages observed at the ground and from satellites.

Cloud statistics have to be assembled carefully for mission planning. The statistics should be valid for the region of interest, season of the year, and even time of the day. The mean cloud coverage can change quickly in a short distance, as it does during the summer between the cloudy California coast and 200 km inland, where few clouds occur. The monsoon regions have a dramatic change of clouds between seasons. The daytime cloud coverage for the southeastern United States is about 0.12 higher than the nighttime coverage (Martin and Liley, 1971).

The cloud statistics utilized for mission planning depend on the size of the field of view of the sensor being used, as explained by Greaves et al. (1970). To demon-

0-201-04245-2

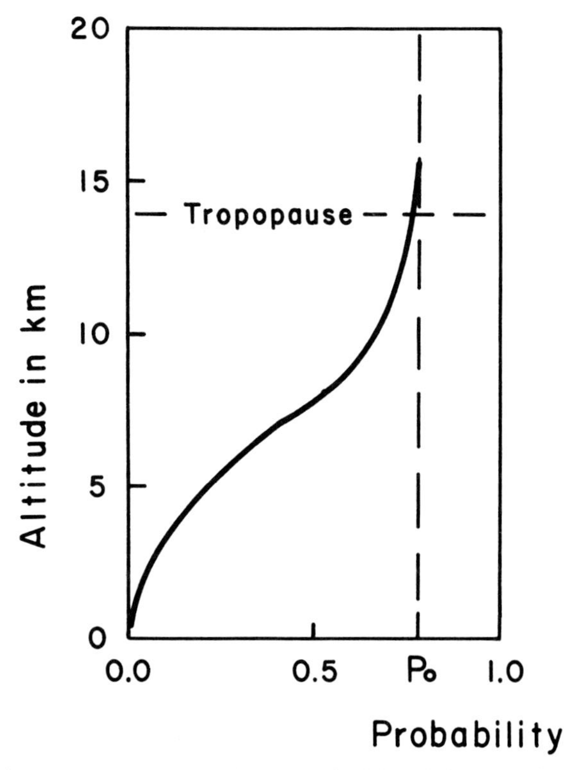


**Fig. 2.3.** Schematic probability, as a function of height, of the ground being obscured by clouds.

strate this they describe the two frequency functions of cloud coverage for a point area and for the entire world. Either a point is covered by at least one cloud, or it is not. As a result, the frequency function is strongly peaked at 0.0 and 1.0 coverage and is zero between these two points. The average cloud coverage for the entire world, on the other hand, is 0.4 (Sherr et al., 1968). The earth is never observed to be either completely clear or overcast. In this case the frequency function is strongly peaked at 0.4. The frequency function of coverage for an area of intermediate size can be completely different from the two values just given as a complex function of location, time of year, time of day, and resolution cell size for the cloud statistics.

The effects of cloud cover on two hypothetical satellite survey missions are discussed by Martin and Liley (1971). The field of view for both missions is a square with sides of 185 km. The area is located in the United States, east of the Rocky Mountains, and north of latitude 38°N. The time is 1000 Local Standard Time in July. The satellite passes over the region of interest at intervals exceeding a few days, which is the approximate time for the cloud statistics to become uncorrelated. The success of the first mission requires at least *one* observation of the

0-201-04245-2

*entire* field of view *without* cloud obscuration. The second mission permits viewing whatever cloud-free areas exist. On successive satellite passes new cloud-free areas may appear, and a mosaic of the field of view can thus be assembled. The basic statistic of cloud coverage is that the probability for the entire field of view to be clear is 0.1. As a consequence, 7 passes are required to have a 50% probability of success for the first mission and 22 passes for 90% probability of success. Probabilities for the second type of mission were computed by Monte Carlo methods (also see Greaves et al., 1970): 90% of the field of view can be seen in 7 passes with a confidence of 94%. Thus almost the entire field of view may be seen by mosaicking with high confidence in 7 passes, whereas the entire field of view will be seen with only 50% probability using the same number of passes.

## II. TRANSFER OF RADIATION THROUGH THE ATMOSPHERE

To obtain a quantitative understanding of the processes by which electromagnetic radiation interacts with the atmosphere several basic quantities must be defined. The first of these is the quantity which describes the amount of radiant energy passing through a unit area of horizontal surface per unit time and per unit wavelength interval. This quantity, called the spectral irradiance, is concerned with a narrow spectral interval centered on the wavelength ($\lambda$). The spectral irradiance, symbolized as $H_\lambda$, describes the flux of radiant energy per wavelength interval passing through a surface whose normal is positioned in a prescribed manner. If the radiation is parallel to the normal of a surface, the irradiance is said to be normally incident. Radiation which is not parallel to the normal is incident at an angle defined by the direction of propagation and the normal to the surface. This angle varies from 0 for normal incidence to $90°$ for grazing incidence.

A second quantity which is useful in describing the radiation field is the angularly dependent radiance. This quantity is easily visualized in terms of the irradiance. Consider a hemisphere as in Figure 2.4 covering a unit area of surface *S*. The irradiance consists of the radiation passing through this hemisphere and incident on the unit surface. We limit consideration to the radiation passing through a very small area $\sigma$ of the hemisphere and incident on the surface. The small fraction of the hemisphere defines a small solid angle which is only a small fraction of the $2\pi$ steradians that is associated with the area of the hemisphere. The irradiance passing through $\sigma$ and incident on the surface is the irradiance for the corresponding solid angle. When this irradiance is defined per unit solid angle and per unit area normal to the beam, the quantity is defined as the radiance. Like the spectral irradiance, the amount of radiance which is confined to a small wavelength interval is called the spectral radiance and is denoted by $N_\lambda$. Generally, spectral radiance and irradiance data will be discussed here; hence the repetitive adjective "spectral" will be omitted.

A final quantity useful in defining the interaction of electromagnetic radiation with the atmosphere is the transmissivity of the atmosphere ($T$) at a given wave-

0-201-04245-2

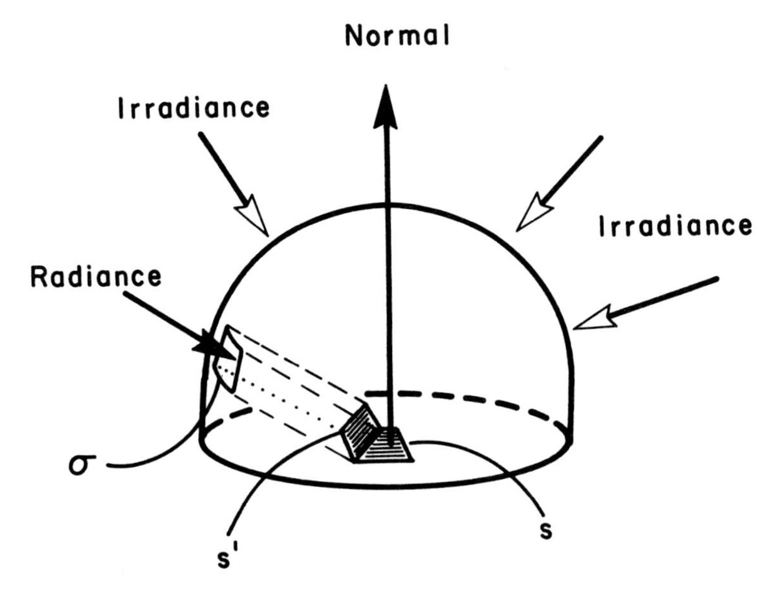


**Fig. 2.4.** Geometric relations between radiances and irradiances of radiant energy incident on a surface *S*.

length (λ). The magnitude of the transmissivity varies in the range from 0 to 1. The transmissivity describes the fraction of the radiance which passes through a portion of the atmosphere without interacting with it.

### A. Upwelling Radiation

The total radiance [$N(z)$] of the upwelling light as a function of the altitude ($z$) from the earth-atmosphere system can be expressed as the sum of two components. The first is the radiant energy that is reflected and emitted by the surface of the earth ($N_0$), of which a fraction ($T$) is transmitted to height $z$. The second component ($J$) is the radiance scattered or emitted by the atmosphere from the surface to height $z$. Sometimes this $J$ component is called the path radiance. Information about the surface is contained principally in the $N_0$ component. The path radiance limits the amount of information that can be extracted from the measured radiation. The sum of these two components gives the total radiance:

$$N(z) = N_0 T + J. \tag{2.8}$$

The radiant energy that interacts with the surface consists, in general, of two parts. The first is the downwelling radiant energy that is reflected from the surface. The second component is the thermal energy that is emitted by the surface. In the visible and near-infrared spectrum just the reflected sunlight is usually important. The thermal radiation is dominant in the mid- and far-infrared. Both reflection and

0-201-04245-2

emission at the surface have to be accounted for in some microwave experiments. Table 2.3 indicates the relative magnitudes of the surface and atmospheric properties.

## B. Upwelling Radiation in the Visible Wavelengths 0.30-0.75 m

Variations in the radiant energy leaving the top of the atmosphere depend on a number of factors, including the solar zenith angle [because the solar irradiance varies as the cosine of the zenith angle ($\theta_0$)], the reflectance of the ground, the particulates in the atmosphere, and the water vapor content, especially in the near-infrared water absorption bands. The effects of clouds will continue to be neglected in this chapter. The dependence of the upwelling energy on the parameters listed above will be demonstrated by a discussion of the radiant emittance of the earth-atmosphere system. Since remote sensing is usually accomplished by measuring radiances, an examination of their characteristics follows. Polarization is beginning to be used in remote sensing, and it will be mentioned briefly. Contrast transmittance and the effect of the atmosphere on measurements of chlorophyll concentration in the ocean and on automatic classification algorithms will be discussed in more detail.

### *1. Radiant Emittance*

The upwelling irradiance of the earth-atmosphere system, which will be called the total spectral radiant emittance ($W$), can be expressed as the sum of two components, similarly to Eq. (2.8). The first component is the light that interacts with the ground or water and eventually emerges from the top of the atmosphere ($W_3$). The second component never interacts with the surface but is scattered entirely by the atmosphere ($W_2$). Information about the earth's surface is contained only in the $W_3$ component, while the $W_2$ component limits the amount of information that can be extracted from the measured radiation. The sum of the two components gives the total radiant emittance: $W = W_2 + W_3$.

In Figure 2.5 the emittances are given as a function of the total optical thickness ($\tau$) of the atmosphere in a vertical column. The optical thickness is related to the transmission by the expression

$$T = e^{-\tau}.$$

The optical thickness of the earth's gaseous atmosphere decreases from 0.3 to 0.03 with increasing wavelength in the visible spectrum. Particulates increase the optical thickness by about 0.2 under normal conditions in the same wavelength range.

The radiant emittances given in Figure 2.5 are calculated for a model atmosphere that is nonabsorbing and scatters light according to Rayleigh's law. Although particulate effects are excluded from this model, Figure 2.5 can be used to illustrate the essential features of the radiation field. The computed radiant emittances in this

0-201-04245-2

0-201-04245-2

**TABLE 2.3**
**Relative Magnitudes of the Terms of Equation (2.8)**

| *Spectrum* | $\lambda$ | *Reflectance* | *Solar Irradiance* | *Thermal Emittance* | *Atmospheric Transmissivity* | *Path Radiance* |
|---|---|---|---|---|---|---|
| Visible | 0.5 $\mu$m | 0.20 | 1942 W $m^{-2}$ $\mu m^{-1}$ | Negligible | 0.8 | Scattering of radiation |
| Infrared | 10.0 $\mu$m | 0.02 | Negligible | $2.5 \times 10^{1}$ W $m^{-2}$ $\mu m^{-1}$ | 0.8–0.9 | Thermal emission |
| Microwave | | | | | | |
| Land | 1.0 cm | 0.1 | Negligible | $7.5 \cdot 10^{-11}$ W $m^{-2}$ $\mu m^{-1}$ | 0.95 | Thermal emission |
| Water | 1.0 cm | 0.5 | Negligible | $3.8 \cdot 10^{-10}$ W $m^{-2}$ $\mu m^{-1}$ | 0.95 | |

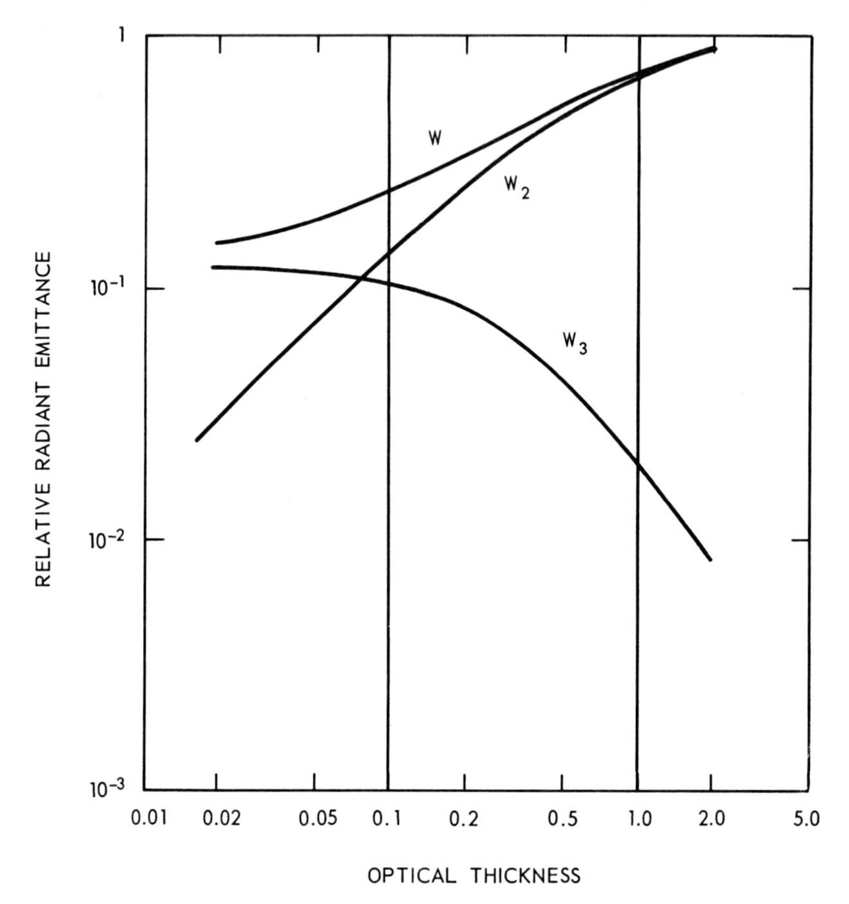


**Fig. 2.5.** Computed radiant emittance at the top of a Rayleigh atmosphere (no absorption or large particles) as a function of its vertical optical-thickness. The reflectance of the ground is $A \doteq 0.1$; the solar zenith angle $\theta_0 = 66.4°$. From Fraser and Walker (1968).

section and radiances in the following section are relative values but can be converted to absolute ones by multiplying them by the monochromatic solar constant $S$ and dividing by $\pi$. In Figure 2.5 the solar zenith is $\theta_0 = 66.4°$, and the ground reflectance is $A = 0.1$. The radiant emittance of just the airlight ($W_2$) vanishes if the atmosphere disappears. At the other extreme of an infinitely thick atmosphere it equals the solar irradiance, $\pi \cos(66.4°) = 1.26$. In the latter case all incident solar radiation is scattered outward by the atmosphere, since no radiant energy reaches the ground. The radiant emittance of the light reflected from the ground ($W_3$) exceeds the radiant emittance of the airlight ($W_2$) only for small optical thicknesses ($\tau_1 < 0.08$, when $A = 0.1$).

If the ground reflectance ($A$) was increased above the value used for Figure 2.5, the relative contribution of the light reflected from the ground would increase. To a

0-201-04245-2

good approximation the total spectral radiant emittance is linearly proportional to the reflectance of the ground: $W = W_2 + \gamma A$, where $\gamma$ is independent of $A$ but depends to some extent on the solar zenith angle and particulates. It is important to realize that in the visible spectrum, for observations of objects on the ground from a large distance, the unwanted light scattered by the atmosphere can *greatly* exceed the light reflected from a surface of low reflectance ($A < 0.1$).

The radiant emittance of an earth-atmosphere system is only weakly dependent on the roughness of the surface as long as its reflectance stays constant (Fraser and Walker, 1968). The roughness has a significant effect, however, on the radiance of the upwelling light, as will be shown soon.

The effect of the solar zenith angle on the radiant emittance at the top of the model Rayleigh atmosphere and smooth ocean is shown in Figure 2.6. The incident solar irradiance is $\pi \cos \theta_0$ and decreases by a factor of 5 as the solar zenith angle increases from 0° to 78.5°. However, at small optical thicknesses the radiant emittance increases as the solar zenith angle increases when $\theta_0 > 37°$. This occurs because the thin atmosphere has only a small effect, and the reflectance of the solar radiation from the smooth surface increases faster than the incident solar irradiance decreases. Surface reflectance is less important when the atmosphere is optically thick. Then the radiant emittance is more nearly proportional to the irradiance of incident sunlight.

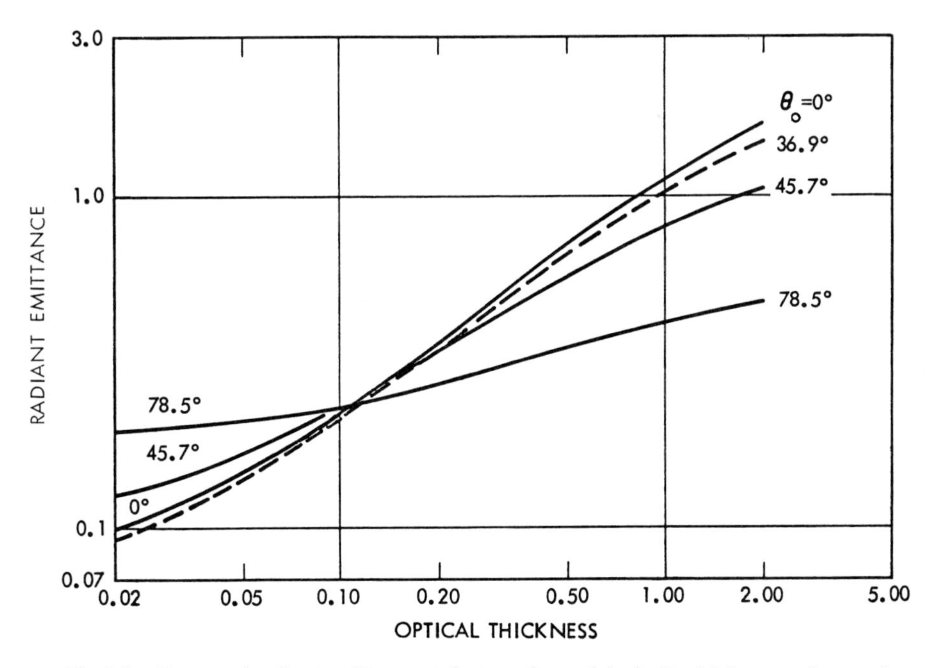


**Fig. 2.6.** Computed radiant emittance at the top of a model of a Rayleigh atmosphere and smooth water surface. The incident irradiance on top of the atmosphere is $\pi \cos \theta_0$.

0-201-04245-2

The effect of particulates on the radiant emittance is not well known, since their optical characteristics have yet to be determined carefully. (For information about large-particle scattering see Bullrich, 1964a, b; Deirmendjian, 1969; Elterman, 1968; Hodkinson, 1963; Kerker, 1969; Plass and Kattawar, 1968, 1969, 1972; and van de Hulst, 1957.) The computed effect of particulates is shown in Figure 2.7 for a model of an earth-atmosphere that contains both the permanent gases and nonabsorbing continental particulates. The optical thickness of the gaseous com-

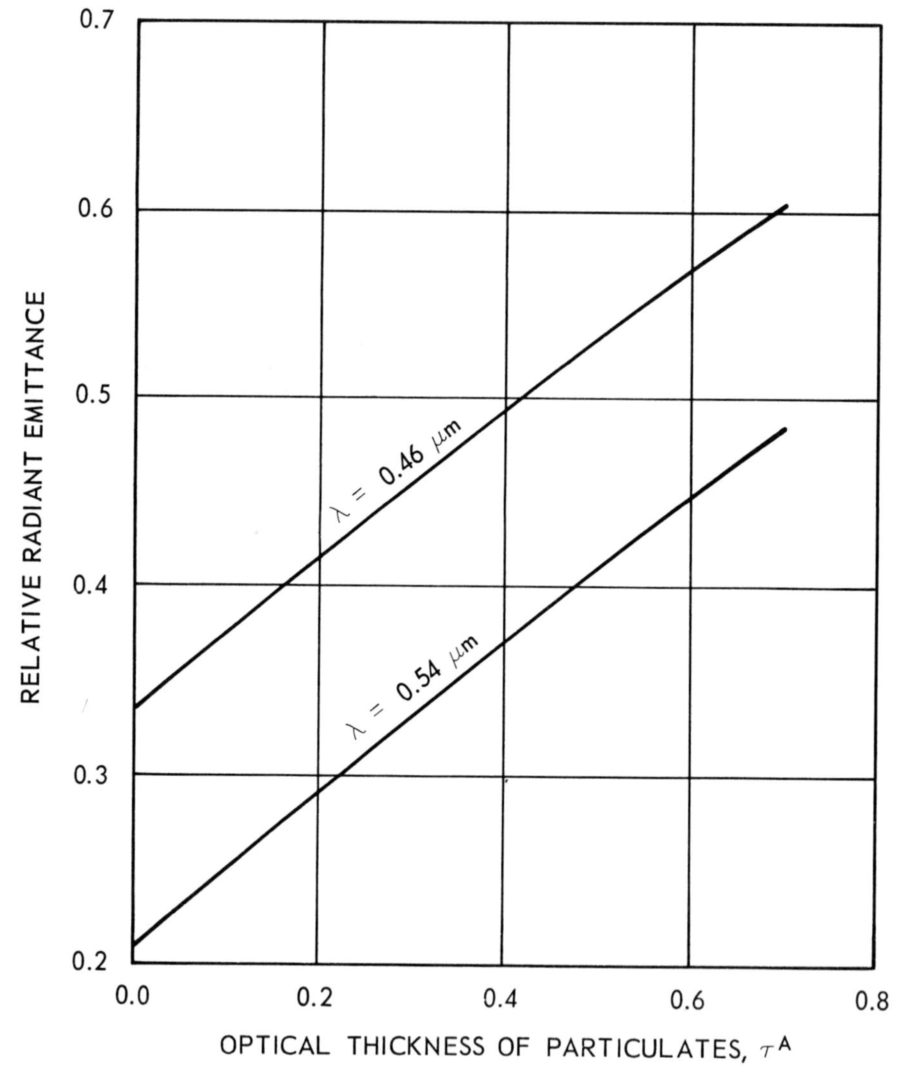


**Fig. 2.7.** Computed radiant emittance at the top of a hazy atmosphere and a surface reflecting light according to Lambert's law ($A$ = 0.03); $\theta_0$ = 40.5°.

0-201-04245-2

ponent is $\tau^G = 0.205$ and 0.101 for $\lambda = 0.46$ and 0.54 μm, respectively. An average value of the larger-particle optical thickness in rural continental regions is $\tau^A = 0.2$ for $\lambda = 0.55$ μm (Shifrin and Shubova, 1964). The solar zenith angle is $\theta_0 = 40.5°$, the incident solar irradiance through a horizontal surface is 2.39, and the ground has the low reflectance of $A = 0.03$. The ground contribution to the radiant emittance at the top of the atmosphere is as small as 0.01 for $\lambda = 0.46$ μm and strong haze and as large as 0.20 for $\lambda = 0.54$ μm and no haze. The radiant emittance increases with increasing particulate optical thickness. However, the particulates are not as efficient as the atmospheric gas in scattering light backwards. For example, the radiant emittance of the upwelling light from an atmosphere whose particulate optical thickness equals that of just the gas ($A = 0.03$) is only 0.25 greater than the radiant emittance from a clear atmosphere without the particulates. As a result, one can expect the characteristics of the upwelling visible light to be dominated by Rayleigh scattering, if the optical thicknesses of the gas and of the particulates are comparable, and the surface reflectances are less than 0.3.

### 2. *Radiance*

The relative radiance at the nadir of an ocean-atmosphere model is given in Figure 2.8. These radiance data are for the same model that was used for the computations of the radiant emittance that are given in Figure 2.5. The total radiance is $N = N_2 + N_3$, where the subscripts have the same meaning as before; $N_3$ is the radiance of light that is reflected from the surface and emerges from the top of the atmosphere. An important point to be learned from Figure 2.8 is that the radiance of the airlight ($N_2$) in the visible spectrum is many times greater than the radiance of the light reflected from the sea.

The roughness of a surface has a strong effect on the radiance of an earth-atmosphere system. The radiances with a subscript 3 in Figure 2.8 refer to light that is reflected from a smooth ocean and eventually escapes from the top of the atmosphere. This reflected light is separated into three components:

$$N_3 = N_{3,1} + N_{3,2} + N_{3,3}. \tag{2.9}$$

Here $N_{3,2}$ is the radiance of the skylight that is reflected from the surface and eventually emerges from the top of the atmosphere. Component $N_{3,3}$ is the radiance of the light that is reflected at least twice from the surface before escaping; it is negligible. Component $N_{3,1}$ represents the radiance of the direct sunlight that is reflected from the ocean surface and eventually emerges in the zenith direction after at least one scattering by the atmosphere. The reflected direct sunlight that is not scattered emerges from the atmosphere at a nadir angle of 66.4°, which is the same value as the solar zenith angle. If the ocean were rough, occasionally the inclined facets of the sea would reflect the direct sunlight toward the zenith. As a result, a larger fraction of the reflected light would escape in the zenith direction. The curve labeled $N^L$ in Figure 2.8 indicates how much the nadir radiance is in-

0-201-04245-2

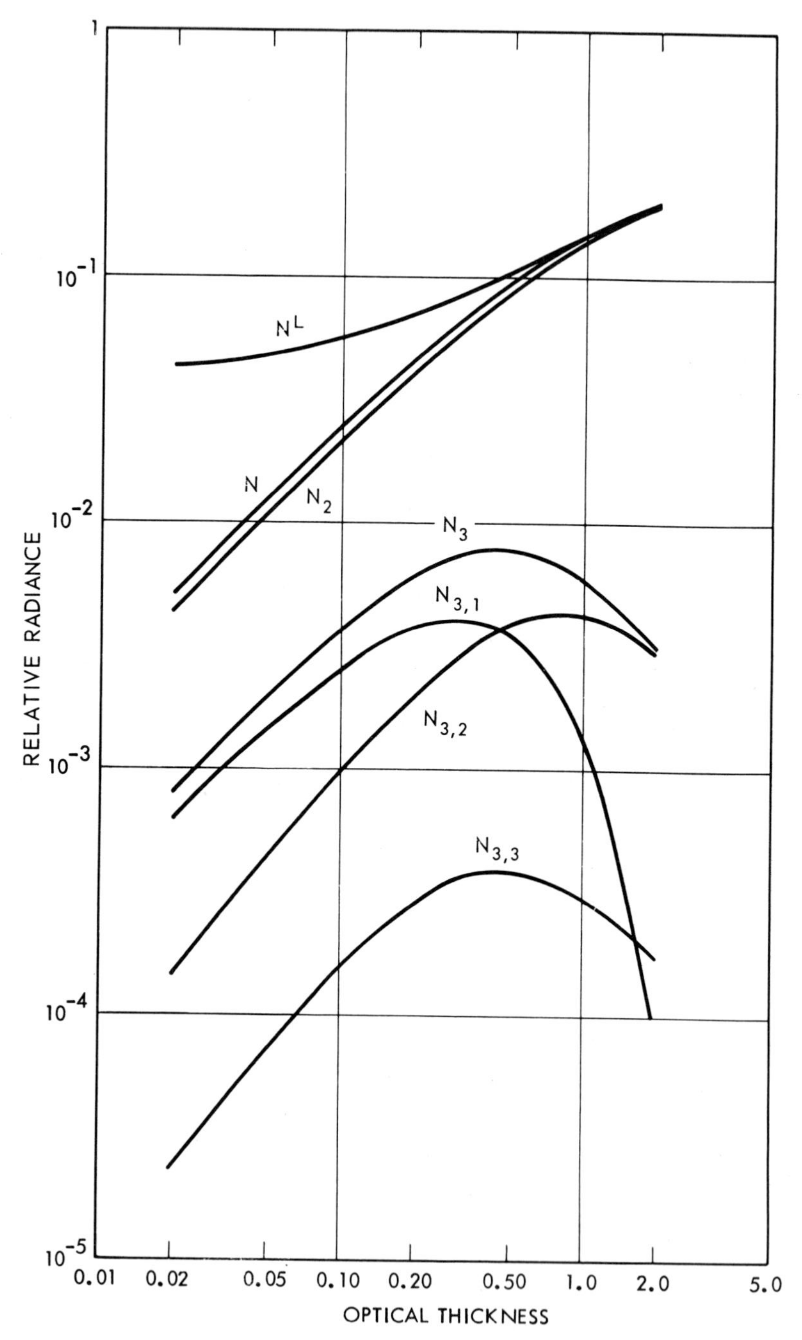


**Fig. 2.8.** Computed relative radiance of the nadir above an atmosphere for a model of a Rayleigh atmosphere and smooth ocean. From Fraser and Walker (1968).

0-201-04245-2

creased by a rough surface that reflects light according to Lambert's law. In this case the total radiance is, to a good approximation, linearly proportional to the surface reflectance: $N^L = N_2 + cA$, where $c$ depends on the solar zenith angle and on the optical thickness.

The computed radiance of a model of an earth-atmosphere containing particulates is given in Figure 2.9. The model is the same one referred to in the discussion of Figure 2.7. The solar constant (the rate at which the solar radiant energy flows across a unit area normal to the direction of propagation and located at the mean distance of the sun from the earth) for just Figure 2.9 is 1, and thus the absolute radiance can be obtained by multiplying by the solar constant. The ground is assumed to reflect light according to Lambert's law with a reflectance of $A = 0.03$.

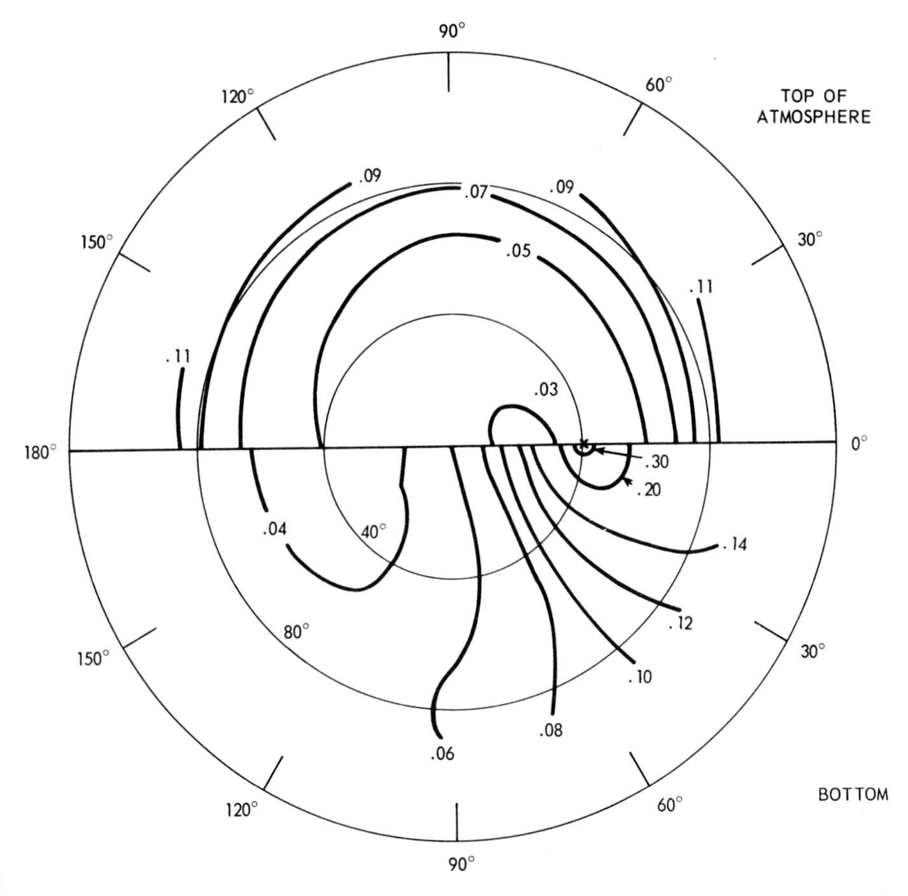


**Fig. 2.9.** Computed relative radiance as a function of spherical coordinates at the base and at the top of a model of a hazy atmosphere with Lambert surface reflection. The incident solar irradiance through a horizontal surface = 0.76; $\lambda = 0.46\ \mu\text{m}$, $\theta_0 = 40.5°$, $A = 0.03$, $\tau = 0.505$. From Curran (1972).

0-201-04245-2

The radiance at the top of the atmosphere is nearly symmetric about the nadir. The radiance of the light incident on the ground, which is given in the lower half of the figure, lacks this symmetry.

The radiances at the base and at the top of the atmosphere are quite different. The radiance at the top of the atmosphere is weakest near the antisolar point, which is indicated on Figure 2.9 by a star. *It is significant that for remote sensing the unwanted airlight is relatively weak near the nadir.* The radiance increases slowly with increasing nadir angle until the nadir angle reaches 60° and then increases more rapidly toward the edge of the earth. To minimize the unwanted contribution of airlight the rim of the earth near the horizon should be avoided. The computations that were just discussed were made for plane-parallel models. These models are satisfactory until the line of sight approaches the horizon (Collins et al., 1972); then a spherical model is a more accurate representation.

The computed nadir relative radiance is given as a function of height in Figure 2.10 for a turbid model atmosphere similar to that used for Figure 2.7. The radiance generally increases with respect to sensing height. The increase is strongest at the shorter wavelengths, because such radiation is scattered more strongly by gases and (usually) atmospheric particulates than at longer wavelengths. If the surface

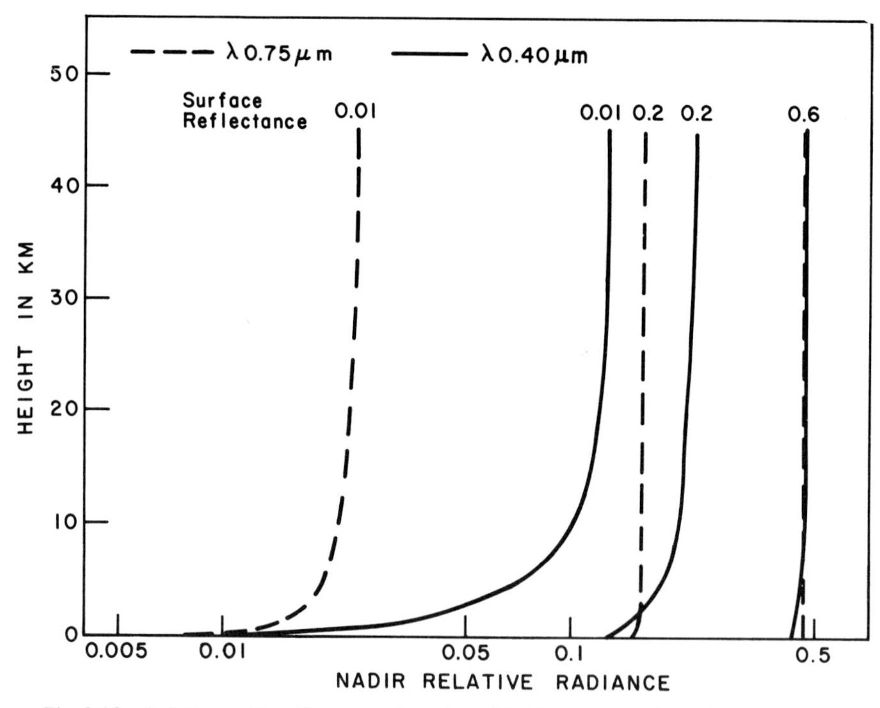


**Fig. 2.10.** Relative nadir radiance as a function of height for $\lambda = 0.40$ and $0.75\ \mu m$. The large particles are nonabsorbing. The incident irradiance on a horizontal surface = 2.41; the solar zenith angle ($\theta_0$) = 40°.

0-201-04245-2

reflection is weak, the radiance increases rapidly with height in the troposphere. As the surface reflectance increases, the radiance becomes more independent of height. The meaning of this for remote sensing is clear: low-reflectivity targets tend to be swamped by path radiance, leading to the familiar reduction in scene contrast with increasing elevation.

Computed nadir radiances are compared with measured values in Figure 2.11. The measurements were made from aircraft over the Atlantic Ocean, 100 km east of Cape Hatteras, in July 1972.* The atmospheric model was the same as that used for Figure 2.10. The surface reflectance in the model is assigned the value that was measured in the nadir direction at a height of 0.3 km. The nadir reflectance decreased from 0.03 at $\lambda = 0.4\ \mu m$ to 0 at $\lambda = 0.75\ \mu m$. The agreement between the computed and measured values is not very good at both the shortest and the longest wavelengths. (The measured values for wavelengths exceeding 0.65 $\mu m$ are not reliable.) The agreement could be improved by adjusting the aerosol parameters in the model. The radiance increases by about 1 order of magnitude from the surface to 11 km. Therefore any variation in color or intensity at the ocean surface not only is attenuated with height, but also is superimposed on a rapidly increasing mean intensity, so that the detectability of these variations decreases rapidly.

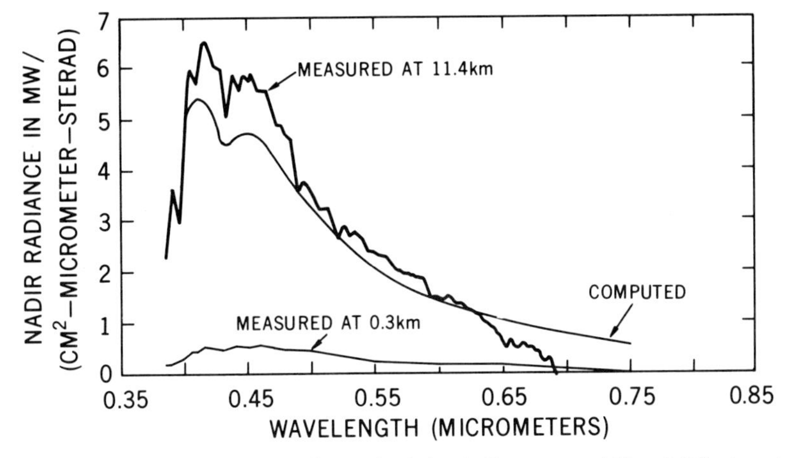


**Fig. 2.11.** Nadir radiance measured over the Atlantic Ocean; $\theta_0 = 44°$ and 51° when the aircraft was at 0.3 and 11.4 km, respectively. Measured values by Hovis (1972).

Very few comparisons of satellite measurements with computed radiances have been made. However, Fowler et al. (1971) found that radiances measured over oceans when the field of view was either completely free of clouds or entirely covered by clouds agreed fairly well with corresponding model computations.

*Additional measured radiances for land and cloud surfaces in the spectral band from 0.4 to 2.4 $\mu m$ are given in the NASA report "Earth Albedo and Emitted Radiation" (1971), SP-8067. Goddard Space Flight Center, Greenbelt, Maryland.

0-201-04245-2

*3. Contrast*

A useful parameter for measuring the quality of an image is the apparent contrast between an object, or target, and its surrounding terrain. The definition of contrast ($C$) that will be used here is that it is the difference between radiances in the directions to the target ($N^t$) and to its surrounding terrain ($N^s$), divided by $N^s$: $C = (N^t - N^s)/N^s$. The airlight radiances ($J$) in the direction to the target and to its nearby surroundings are essentially the same when the target is small. Hence the airlight radiances cancel when taking the difference $N^t - N^s$ at the observer. The apparent contrast then equals the difference at the surface object $(N_0{}^t - N_0{}^s)$ multiplied by the atmospheric transmission ($T$) and divided by the apparent radiance of the surroundings at the observation height:

$$C = \frac{(N_0{}^t - N_0{}^s)T}{N_0{}^s T + J} . \tag{2.10}$$

The ratio of the contrasts at the observer and at the surface objects is called the contrast transmittance ($y$) and can be expressed as

$$y = \left(1 + \frac{J}{N_0{}^s T}\right)^{-1} . \tag{2.11}$$

The value of the contrast transmittance lies between 0 and 1 ($0 \leqslant y \leqslant 1$). If the optical thickness between the observer and the object becomes very large, the transmission approaches 0. Then the contrast transmittance is negligibly small. On the other hand, the contrast transmittance is 1 if the amount of atmosphere between the object and the observer vanishes, that is, the transmission $T = 1$ and the airlight radiance $J = 0$. Unfortunately, the contrast transmittance depends on more than just the optical state of the atmosphere; it also depends on the radiance of the ground surrounding the target. The contrast transmittance increases with increasing radiance of the ground.

The computed effect of airlight on contrast transmission through the entire atmosphere is shown in Figure 2.12 for an optically thin atmosphere ($\tau_1 = 0.02$, which is the approximate lower limit of the optical thickness in the near-IR spectrum), and for $\tau_1 = 1.0$, which is a value encountered for the near-UV. The data of Figure 2.12 apply to an earth-atmosphere model of terrain reflecting light according to Lambert's law with a reflectance of $A = 0.25$ and of a Rayleigh atmosphere. Then the data on the left apply to a wavelength of $\lambda = 0.809\ \mu$m, and those on the right apply to $\lambda = 0.312\ \mu$m. The maximum contrast transmittances are $y = 0.99$ and $y = 0.23$ for the lower and higher optical thicknesses, respectively. Both maxima occur at about the same zenith angle in the principal plane, which contains the observer, nadir, and sun. The contrast transmittance for the large optical thickness approaches 0 at a zenith angle of $\phi = 81°$. In general one can expect that the contrast transmittances will increase with decreasing optical thickness in an arbitrary direction of observation.

0-201-04245-2

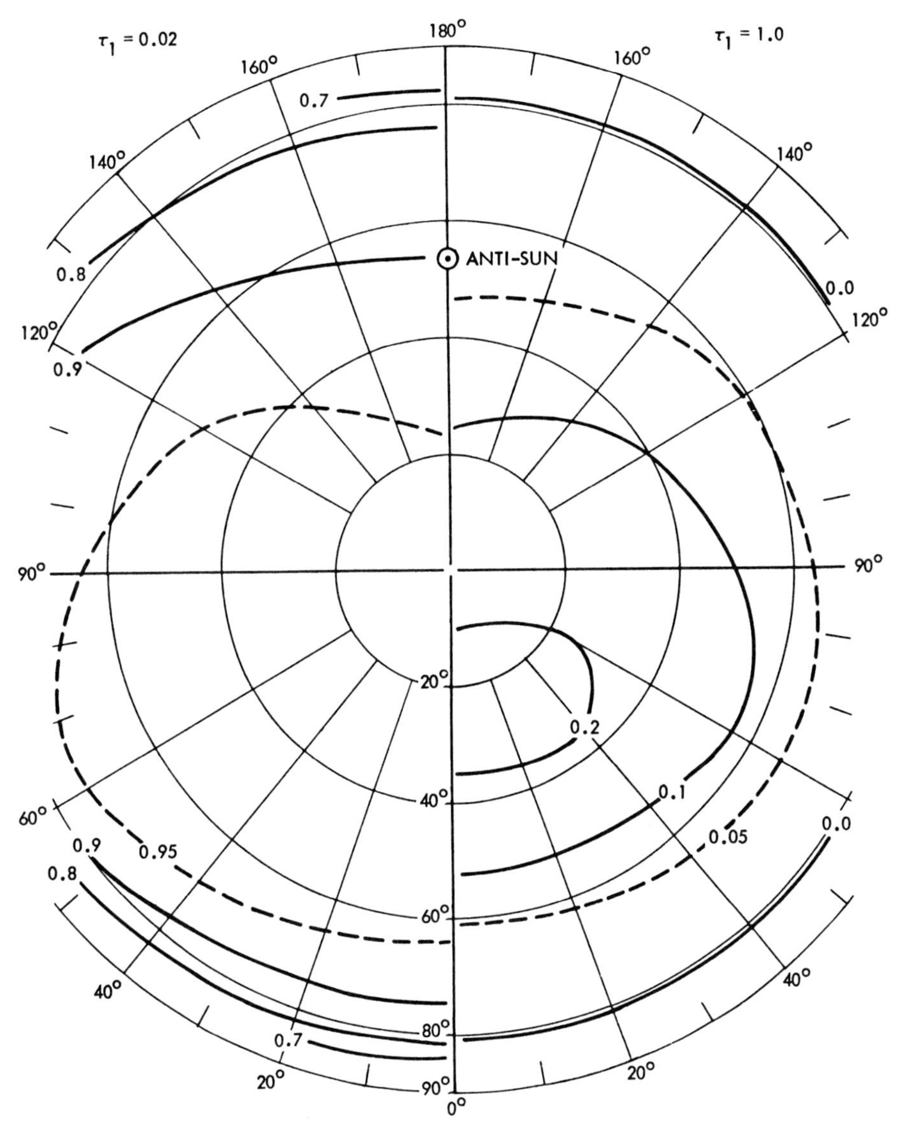


Fig. 2.12. Computed contrast transmittance (Eq. 11) from above a model of a Rayleigh atmosphere and a Lambert surface ($A$ = 0.25). The optical thickness is 0.02 on the left and 1.0 on the right; $\theta_0$ = 53.1°. From Fraser (1964b).

0-201-04245-2

The computed contrast transmittances at the top of different model atmospheres will be given. Lambert's law of reflection is assumed for surface objects, except for the target. The light reflected from the small target can be of arbitrary radiance and polarization. The contrast transmittance in the principal plane is shown in Figure 2.13 for the case in which the terrain surrounding the target has a reflectance $A$ = 0.25 and for $\lambda$ = 0.625 $\mu$m. The dashed lines give the contrast transmittance ($y$). The contrast transmittance is small near the horizons because of both the high radiance of the airlight and the low radiance of the light transmitted from the target. The highest dashed curve gives $y$ for an atmosphere free of particulates and shows the highest attainable $y$ for $\lambda = 0.625\ \mu m$ and $A = 0.25$. The contrast

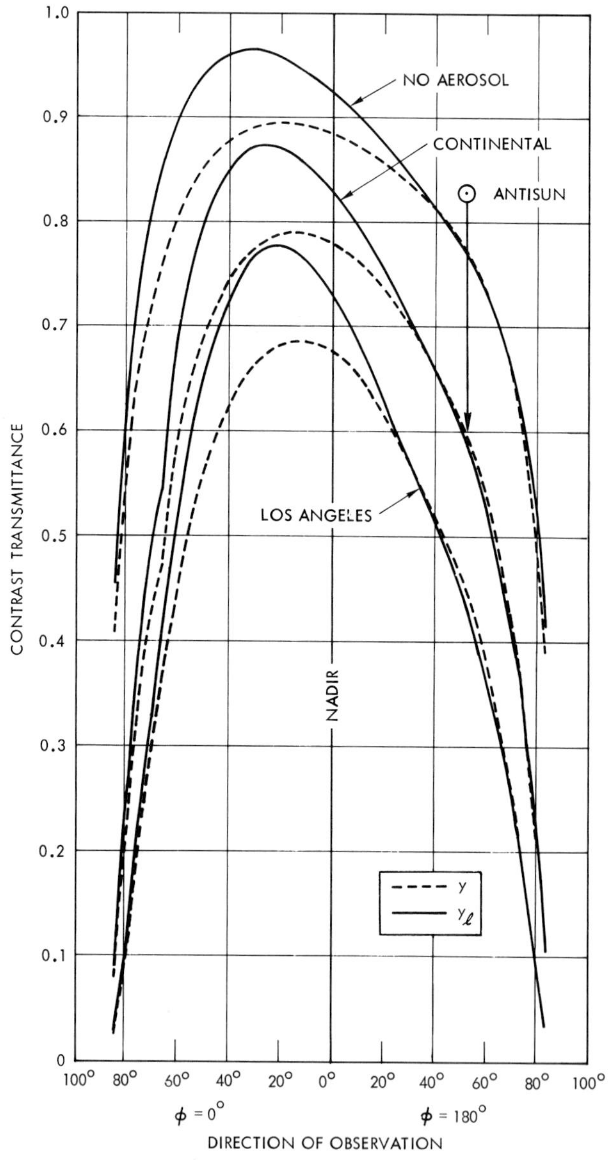


**Fig. 2.13.** Computed contrast transmittance for three model atmospheres. The angular direction is measured from the nadir and in the principal plane. The reflectance $A = 0.25$ for the surface surrounding the target; $\theta_0 = 53.1°$, $\lambda = 0.625\ \mu m$. From Fraser (1964b).

0-201-04245-2

transmittance decreases for the model atmospheres containing continental- and Los Angeles-type particulates.

Some of the unwanted airlight can be removed under certain conditions where it is highly polarized (Coulson, 1966; Fraser, 1964b). The principle is the same as using a polarizer with a camera to enhance the contrasts in a distant outdoor scene. If the scattered airlight within the field of view of a sensor is completely plane polarized, an analyzer in the system such as a piece of Polaroid can prevent this light from reaching the detector. The solid lines ($y_l$) of Figure 2.13 are calculated for the case in which the transmission plane of a perfect analyzer (one that transmits no light through a plane perpendicular to the transmission plane) in the receiver optical system is parallel to the principal plane. Use of an analyzer increases the contrast transmittance, except near the antisolar point, where the polarization of the light is small.

An example of the measured apparent contrast as a function of height is given in Figure 2.14. The intrinsic contrast is not given, but may be 0.99. If so, the values of apparent contrast in Figure 2.14 are only 1% larger than the contrast transmittance ($y$). The apparent contrast decreases *more* with height at the *shortest* wavelength and *least* with the *longest* wavelength. Figure 2.14 reflects the general

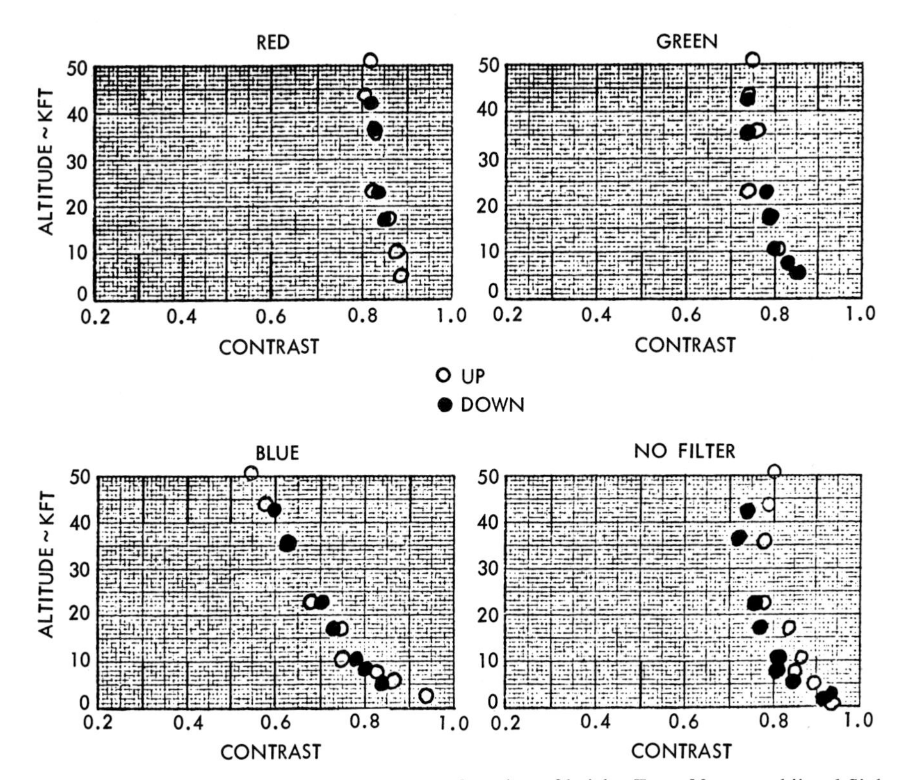


**Fig. 2.14.** Measured apparent contrast as a function of height. From Mazurowskii and Sink (1965).

0-201-04245-2

conclusions from a total of 27 experiments: the apparent contrast shows a small spectral dependence below 3000 m; at an arbitrary height above 3000 m the apparent contrast increases with increasing wavelength. The greatest loss of contrast occurs in the lowest 10,000 m (Mazurowskii and Sink, 1965).

Duntley et al. (1964) have investigated a method of measuring the contrast transmittance from the ground. The ground-based measurements agree well with high-altitude measurements.

### *4. Turbulence*

The effects of atmospheric turbulence on light are apparent when an observer's line of sight to an object passes close to a hot surface such as a highway on a hot day. Then the object appears to move and to be distorted. Another common manifestation of turbulence is the twinkling of stars. Telescopic observations show that a star appears to change position by less than 10 sec of arc, that long-exposure images are blurred into circles whose angular diameters are also less than 10 sec of arc, and that the apparent radiance fluctuates. The computed effect of the turbulent atmosphere on light propagating upward is negligibly small for remote sensing along inclined lines of sight (Farhat and De Cou, 1969; Fried, 1966a, b; Lee and Harp, 1969; Tatarski, 1961; Weiner, 1967). The smallest resolvable size on the ground increases with sensor height up to a height of about 50 km and then remains constant. Above a height of 50 km, turbulence limits the smallest resolvable distance to the order of centimeters, a value much less than that required in current earth resource programs.

### *5. Ocean Color*

The color of the ocean is an important parameter for measuring its productivity, sedimentation, and pollution. It appears that corrections for atmospheric effects may be necessary before high-altitude aircraft or satellite measurements can be used to the best advantage. The process outlined here for making atmospheric corrections of ocean observations is an example of a general approach that may be taken in other remote sensing applications.

As a specific example consider oceanic productivity. In a general way, oceanic productivity is related to the concentration of chlorophyll. Chlorophyll-rich water appears green to the eye. Hence the concentration can be related to optical parameters. Figure 2.15 shows the spectral dependence of the reflectance of light measured just above the ocean surface as a function of the chlorophyll concentration. The explanation for the changes in the curves of Figure 2.15 is that clean water is highly transparent for the visible spectrum, but chlorophyll has a strong absorption band at 0.44 $\mu$m, and both absorb strongly at 0.65 $\mu$m. A band of relatively good transparency lies between 0.44 and 0.65 $\mu$m. Therefore the ratio of the reflectances in the transparent band and in the 0.44 $\mu$m absorption band is relatively sensitive to changes in chlorophyll concentration. The centers of two spectral bands that are

0-201-04245-2

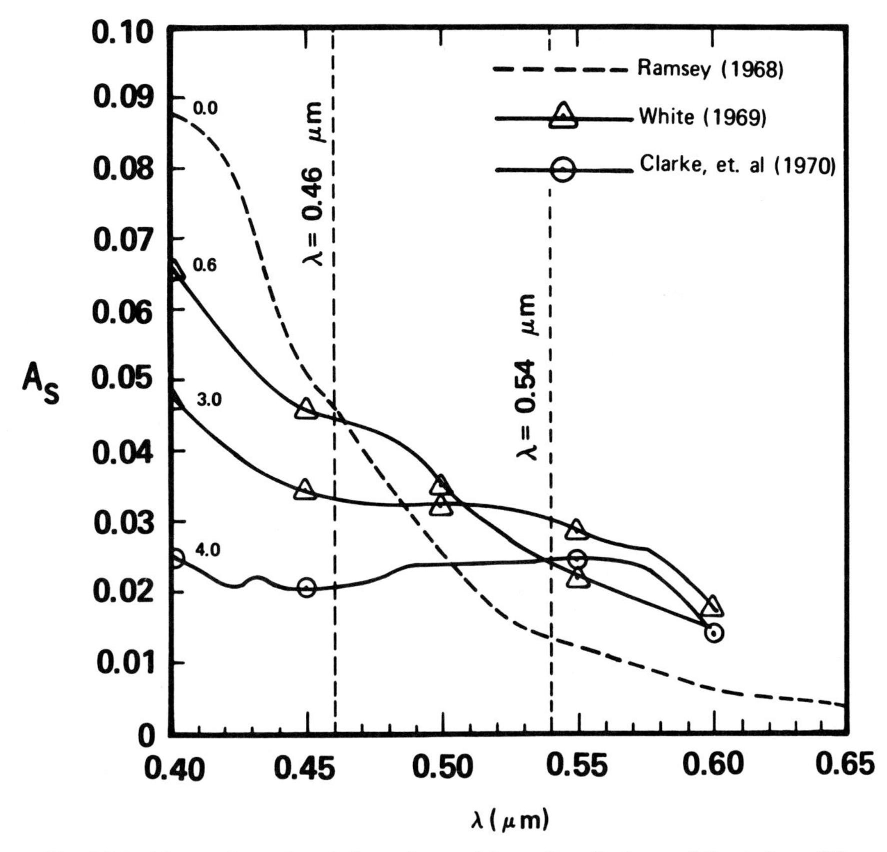


**Fig. 2.15.** Measured wavelength dependence of the nadir reflectance of the surface of the ocean for various chlorophyll concentrations, which appear on the left-hand side of the curve in milligrams per cubic meter of water. From Curran (1972).

used here for this purpose are at 0.46 and 0.54 μm. The ratio of the reflectance in the 0.54 μm band to that in the 0.46 μm band will be called the color ratio. The color ratio increases with the chlorophyll concentration as shown in Figure 2.16. The dispersion of points relating color ratio to chlorophyll concentration is due to optical processes such as scattering from suspended inorganic matter or variation of chlorophyll concentration per organism. As an indication of the accuracy required, biologists would like to remotely determine chlorophyll concentration with an accuracy of 0.1 mg $m^{-3}$ in coastal waters, and with much greater accuracy in midocean, where the average concentration is about 0.1 mg $m^{-3}$.

The effect of the atmosphere on high-altitude observations of the ocean was shown in Figure 2.11. The path radiance contributes about 90% of the total in the visible spectrum. In addition to atmospheric contributions, ocean color can also be modified by the sunlight and skylight reflected from the surface of the ocean. The

0-201-04245-2

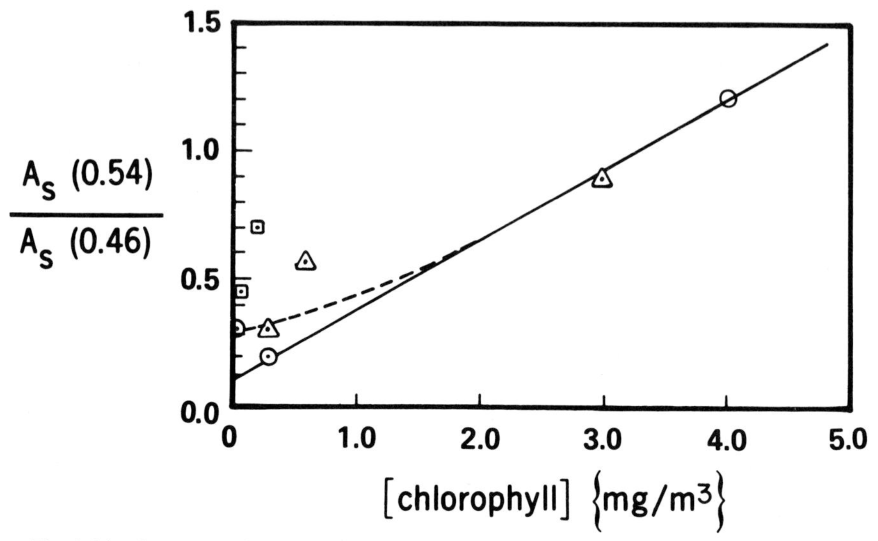


**Fig. 2.16.** Color ratio derived from measured surface nadir reflectance as a function of chlorophyll concentration. From Curran (1972).

reflected sunlight is usually referred to as the sun glint. Figure 2.17 shows the spatially dependent color ratio derived from aircraft measurements of the nadir radiance along a path from Southern California to Catalina Island. The noisy data were measured from an altitude of 0.92 km. The noise appears because the small field of view of the spectrometer resolves the fluctuating glint pattern. The relatively smooth data were measured from a height of 14.9 km. The minimum color ratios

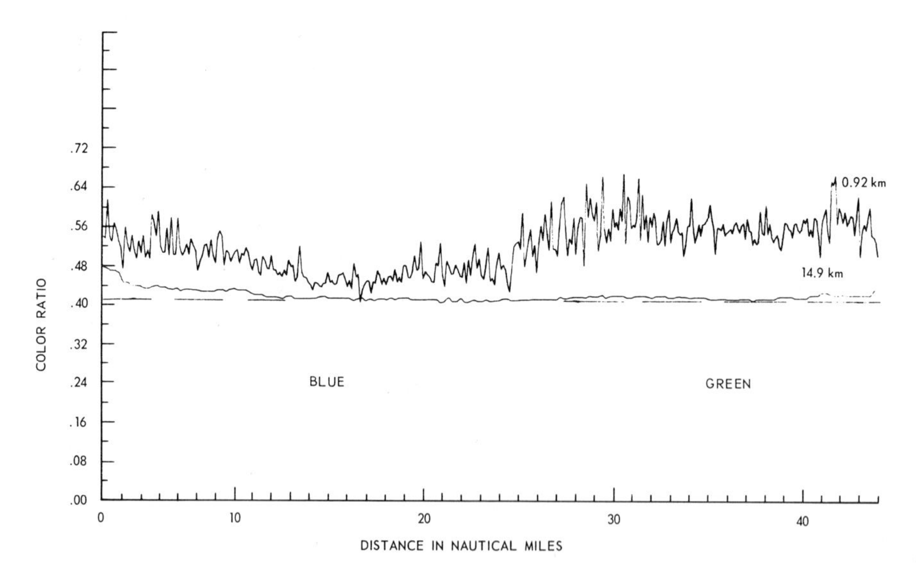


**Fig. 2.17.** Color ratios derived from values of the nadir radiance that are measured at heights of 0.92 and 14.9 km. From Hovis (1971).

0-201-04245-2

appear over sterile waters, which appear to the eye as deep blue. The maxima in the color ratio appear where the chlorophyll concentration seemed to be greatest along the flight path. The chlorophyll concentration at the maxima is 0.25 mg $m^{-3}$, and the water appears green. The atmosphere causes a significant decrease in the color ratio between 0.92 and 14.9 km. If the effect of the atmosphere were constant, it could be corrected for. However, atmospheric particulates cause significant variability.

Curran (1972) and McCluney (1974) have analyzed the effects of atmospheric particulates in the remote observation of chlorophyll concentration. In these studies the relationship between the color ratio at satellite altitudes and the color ratio at the sea surface was computed. Allowing only the particulate optical thickness to vary while keeping all other model parameters constant permitted the variation of color ratio as observed from above the atmosphere to be related to the variation in atmospheric particulate optical thickness. The general results of these studies indicate that differing colors are transferred through the atmosphere with differing attenuations. Furthermore, the atmospheric path radiance contribution also exhibits a dependence on wavelength. The result of these two effects was to distort the color ratio and to diminish the amount of change in it at the top of the atmosphere for a given amount of change in it at the surface.

The computed standard deviation of the chlorophyll concentration is shown as a function of the aerosol optical thickness in Figure 2.18. The average aerosol optical thickness at 0.50 $\mu$m is plotted along the abscissa. One standard deviation in the chlorophyll concentration, as derived from the color ratio in the nadir direction from above the atmosphere, is given by the left ordinate. An average particulate optical thickness for the atmosphere is considered to be 0.2 for continental areas and possibly 0.1 for the central regions of oceans. Since the standard deviation in the particulate optical thickness is not well-known, a global value between 0.1 and 0.2 may be chosen (Shifrin and Shubova, 1964). As indicated by the dashed lines in Figure 2.18, the determination of chlorophyll concentration has an uncertainty of 0.4–0.8 mg $m^{-3}$.

The effects of the atmosphere on satellite observations of chlorophyll concentration can be computed. The accuracy of the corrections is limited, however, by imperfect knowledge of the aerosol characteristics. If only the optical thickness of the aerosol varies, and not its other optical characteristics, the accuracy of the computed correction is limited by the accuracy of the atmospheric optical thickness. The optical thickness can be measured from the ground with an accuracy of 0.01. The third curve of Figure 2.18 relates the chlorophyll concentration standard deviation to the best possible correction for atmospheric effects by the procedure indicated. These computations indicate that the chlorophyll concentration cannot be measured from a satellite with an accuracy of 0.1 mg $m^{-3}$.

0-201-04245-2

### 6. *Computer Classification*

A number of tests have been made to determine the effect of the atmosphere on the accuracy of several algorithms (maximum-likelihood, canonical analysis, etc.) that are used for computer processing of measurements to classify surface features.

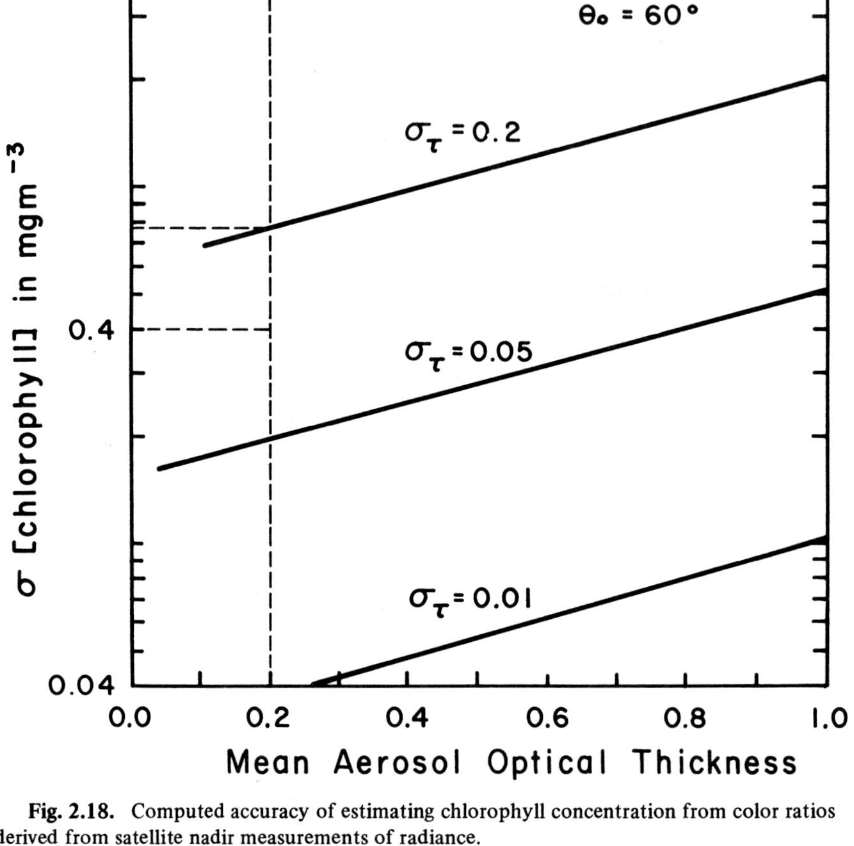


**Fig. 2.18.** Computed accuracy of estimating chlorophyll concentration from color ratios derived from satellite nadir measurements of radiance.

The essence of these algorithms (see Chapter 11) is to compare remotely measured spectral signatures with other remote signatures, which are associated with known features or "ground truth." For example, if a wheat field, bare soil, and water are known to occur in certain areas, their remotely measured spectral signatures are the ground truth data. The known and unknown signatures are measured under essentially the same conditions, such as height of sensor, position of sun, and direction of observation. Since the place and time of the satellite truth sets and the unknown spectral signatures frequently do not coincide, the atmosphere may modify the known and unknown sets differently. Hence one expects that, if the optical states of the atmosphere over the region to be classified and over the region where the truth sets were measured are different, the classification accuracy will be degraded. Numerous studies with aircraft data, principally at Purdue University and the Environmental Research Institute of Michigan (ERIM) over the last decade, have demonstrated the effects of this atmospheric degradation. Algorithms for partial

0-201-04245-2

corrections have been developed. Several preliminary tests reported on here show the magnitude of these effects with spacecraft data.

In two similar experiments spectral signatures measured from the satellite LANDSAT 1 were modified to account for atmospheric effects. In one experiment Potter (1974) computed the modifications; in the other, Fraser et al. (1976) measured them. The unmodified data contained the truth sets. Both experiments yielded similar results: namely, that variations in the optical nature of the atmosphere have a significant effect on the accuracy of classification by means of the maximum-likelihood ratio. (Any algorithm would show the same effect, but not necessarily to the same degree.) When the atmospheric turbidity was increased by two standard deviations, the classification accuracy for fields of soybeans and corn decreased from 99 to 60–70%, depending on the crop (Potter, 1974). On the other hand, it is noteworthy that, when new truth sets were constructed from the modified data, the modified classifications differed by only a few percent from the original, unmodified ones (Potter, 1974).

Additional insight into the effects of the atmosphere on automatic classification algorithms was gained from the experiment by Fraser et al. (1975), employing multispectral scanner (MSS) data from the satellite LANDSAT 1. Radiances in four spectral bands in the visible and near-infrared spectrum were measured along a 185-km-wide strip beneath the satellite. In this experiment such observations were classified according to surface characteristics by the LARSYS classifier (Laboratory for Applications of Remote Sensing, Purdue University) (Swain, 1972). The experiment consisted in fixing the surface scene and the solar zenith angle and changing the atmospheric radiances. The MSS measured the changes over a dark surface, the Atlantic Ocean. The radiance of the atmosphere alone contributed 75% of the total. The radiances became uncorrelated in a distance of 90 km (using ½-mile-square cells, on the western half of the LANDSAT image). The radiances of the picture elements (pixels) of one region were subtracted from the corresponding values (for the same off-nadir angle) of another region separated by a distance of 90 km. These radiance differences represent the change associated with two uncorrelated atmospheric states.

These (representative, uncorrelated) radiance differences were added to a springtime scene of south-central Pennsylvania, containing bare soils, pastures, hardwoods, pines, and water. The satellite measurement of the average reflectivity in the 0.5–0.6 $\mu$m band, for example, was 12.6% for the unmodified data. After modification, it was 11.3%—a decrease of 1.3%. Such a change for a rural area, such as southern Pennsylvania, could be accounted for by a decrease in aerosol mass of two standard deviations. Such a mass change, or larger, would occur only 2.5% of the time in rural regions, but other parameters which are also changeable influence atmospheric radiance too, leading to higher occurrence frequencies than 2.5%.

The classifications of the unmodified data served as the truth set for classifying the modified data. Pixels that had appeared in a given class of the unmodified data could appear in a different class of the modified data. The number of transitions between classes is given in Table 2.4. To understand Table 2.4, consider the first row,

0-201-04245-2

**TABLE 2.4**
**Classification Matrix of Unmodified and Modified Classes, Using the Unmodified Statistics**

| *Class* — *Unmodified* \ *Modified* | *Number of Pixels Classified into Class* *1–6* | *7* | *8* | *9* | *10* | *Number of Pixels in Unmodified Classification* |
|---|---|---|---|---|---|---|
| 1–6 | 750 (75.6) | 218 (22.0) | 0 | 24 (2.4) | 0 | 992 |
| 7 | 0 | 93 (30.8) | 187 (61.9) | 22 (7.3) | 0 | 302 |
| 8 | 0 | | 340 (68.0) | 159 (31.8) | 1 (0.2) | 500 |
| 9 | 0 | | 0 | 959 (100.0) | | 959 |
| 10 | 0 | | 0 | 1 (2.1) | 46 (97.9) | 47 |
| | | | | 0 | | |
| Number of pixels in modified classification | 750 | 311 | 527 | 1165 | 47 | Total = 2800 |

The numbers in parentheses show the transition percentages of a class of unmodified pixels to the various classes of modified classification.

0-201-04245-2

representing bare soils and pastures, which were numbered 1 to 6. The number at the end of the row (992) gives the total number of pixels of unmodified data that were put into classes 1-6. However, only 750 pixels of modified data are given the same classification. Of the unmodified pixels, 218 appear as class 7 of the modified data. Class 7 consists of hardwoods, type 1. Of the remaining pixels in unmodified classes 1-6, none is reclassified into class 8 as hardwood, type 2, 24 are reclassified as pines (class 9), and none is reclassified as water (class 10). The total number of modified pixels that are assigned to classes 1-6 (750) appears at the bottom of the corresponding column. As another example, 159 pixels of the unmodified hardwoods, type 2 (class 8), are reclassified as pines in the modified data. However, all 959 pixels of pines in the unmodified data are still classified as pines in the modified data, but many additional pixels appear as pines in the modified data. The lesson here seems to be that, if the optical characteristics of the atmospheres over an unknown scene and over the region of the truth set are uncorrelated and differ in aerosol mass by two standard deviations, misclassification errors of 20-30% can appear when maximum-likelihood classification algorithms are used.

Another part of this experiment was to construct a truth set from just the modified data, and then classify the modified data with the new truth set. Less than 3% of the pixels appeared in the off-diagonal elements of a classification matrix like that of Table 2.4. These results are basically the same as those obtained by Potter (1974), as noted on p. 65.

Other procedures for atmospheric corrections with LANDSAT 1 MSS data are also described by Malila and Nalepka (1973) and Rogers et al. (1973).

## C. Infrared Radiation

The infrared spectrum to be discussed consists of the spectral band 0.75-22 $\mu$m. For longer wavelengths, between 22 and 50 $\mu$m, the atmosphere is essentially opaque for a typical sea level path longer than a few kilometers. Between 50 and 300 $\mu$m typical atmospheric paths greater than a few meters near ground level are opaque. Thus, because of the strong absorption properties of water vapor, the spectral region between 22 $\mu$m and approximately 0.5 cm is unusable for remote sensing of surface features from platforms at altitudes higher than a few meters.

The spectral band 0.75-22 $\mu$m can be divided into two parts, which are analyzed differently. The first band, 0.75-4 $\mu$m, is dominated by direct and scattered sunlight during the day. The second band, 4-22 $\mu$m, is dominated by thermal emission from the surface *and the atmosphere.* The atmospheric transmittance and spectral radiance are difficult to calculate for the spectral bands where the absorption is strong, because of the myriad of overlapping absorption lines. However, the analysis is simpler for the windows, even though their absorption characteristics are not completely understood. Here the absorption does not have to be computed with great accuracy for many applications because relative differences in radiance for the same scene are more commonly used than absolute radiance values.

0-201-04245-2

Measured transmission data show the spectral positions and relative clarity of the atmospheric windows. Figure 2.19 is an example of measured transmittances in the wavelength interval 0.5–4.2 μm. These data were obtained by Yates and Taylor (1960) over a horizontal path of 5.5 km near sea level. The conditions and path length for these measurements are comparable to those for a vertical path through the entire atmosphere, that is, the scale-height relation enables easily made short-path horizontal measurements to be effectively substituted for vertical measures. Numerous windows may be found between 0.5 and 4.16 μm, as noted in Table 2.1. These windows have found wide application in both aircraft and satellite remote sensing of land, water, and thermal sources such as fires and volcanoes.

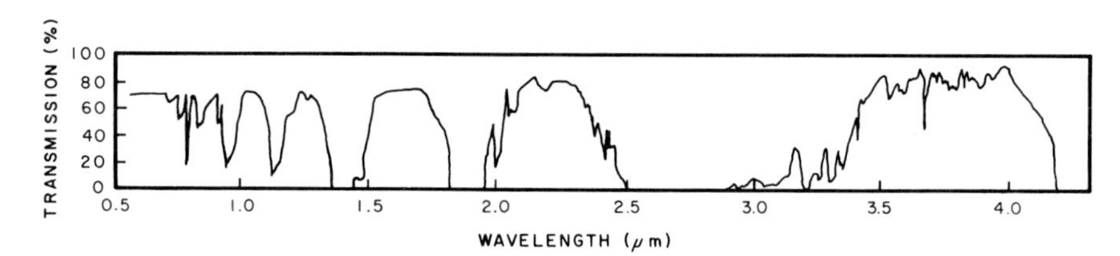


**Fig. 2.19.** Measured transmittance in the spectral band 0.5–4.2 μm over a horizontal path of 5.5 km length. From Yates and Taylor (1960).

A very clear window exists near 3.7 μm, and a somewhat less clear window near 4.7 μm. These two windows would be useful for remote sensing if the radiation sources were stronger. Between 3 and 5 μm, however, the solar irradiance is relatively small and also the thermal emission from terrestrial surfaces is weak. To increase the difficulty of interpreting data in the 3-5 μm interval, both the solar irradiance and the thermal emission are approximately equal in this interval, and during daytime hours confusion between various effects is common.

The wavelength region 5–22 μm contains two windows, one of which finds major use in remote thermal mapping. Figure 2.20 shows measured transmittances through the atmosphere for the 5-25 μm region. These transmittances were derived by Kunde (1973) from measured radiance spectra of the full moon. The clearest window in this figure extends from 8 to 13 μm with a small region at 9.6 μm suffering strong absorption due to atmospheric ozone ($O_3$). Because the major amount of ozone is contained at altitudes above 15 km, this absorption feature is of little consequence in most aircraft observations in the 8–12 μm window. However, it is important in spacecraft remote sensing either of total ozone amount or of restrahlen effects (see Chapter 15). The next two sections discuss in more detail the transfer properties of (1) the solar or near-infrared and (2) the thermal infrared.

### *1. The Near-Infrared*

The wavelength interval 0.75–4.0 μm is dominated by sunlight during the day. In the several windows which exist in this wavelength interval and with the absence of clouds, the total optical thickness is small. The absorption optical thickness is an

0-201-04245-2

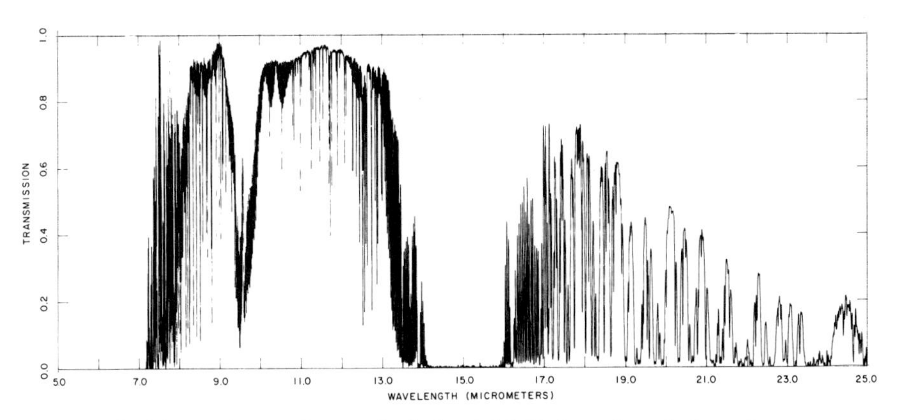


**Fig. 2.20.** Measured transmittances in the band 5.0–25.0 μm through the entire atmosphere. From Kunde (1973).

appreciable fraction of the total optical thickness, thus minimizing the effects of multiple scattering. The atmospheric optical depths for large aerosol particles are estimated in Figure 2.21. The estimates are based on surface transmission measurements in a horizontal direction. Since gaseous absorption occurs even in the windows, its optical depth, which is somewhat uncertain, has to be subtracted from the measured attenuation in order to obtain the optical depth for aerosol particles only. Hence the large-particle optical depths of Figure 2.21 are uncertain. Additional data are given by Barnhardt and Streete (1970), Deirmendjian (1960), and Flanigan and De Long (1971). McClatchey et al. (1970) present nomograms for calculating optical depth.

The wavelength region 3.0–5.0 μm is the transition region from solar radiation dominated to thermal radiation dominated. Figure 2.22 indicates the relative importance of the radiances of the reflected sunlight and of the thermal emissions from the ground for different wavelengths. The radiance of the reflected sunlight depends on the product of the cosine of the solar zenith angle ($\mu_0 = \cos\theta_0$) and the diffuse reflectance $A$ of the surface. For example, when this product is equal to 0.1, with the sun at the zenith, the reflectance is 0.1; with a solar zenith angle of 60°, the reflectance is 0.2. The thermal radiance curves are for three temperatures spanning the average terrestrial temperature range. As may be seen in Figure 2.22, the reflected radiance and the emitted radiance are roughly equal for wavelengths between 3 and 5 μm. Because of the comparable magnitudes of the reflected and emitted radiances, measurements made during daylight conditions of the upwelling radiance for 3.0–5 μm are very difficult to interpret.

In Figure 2.22 the surface was assumed to be diffusely reflecting; thus the incident radiation was reflected with constant radiance into all upward directions. However, many infrared measurements are made over water surfaces which act as nearly specular reflectors. Because a specular reflector returns the incident radiation into a very small solid angle, the average radiance in this solid angle is very high. Thus the effect of infrared observations toward the "sun glint" is to cause the tran-

0-201-04245-2

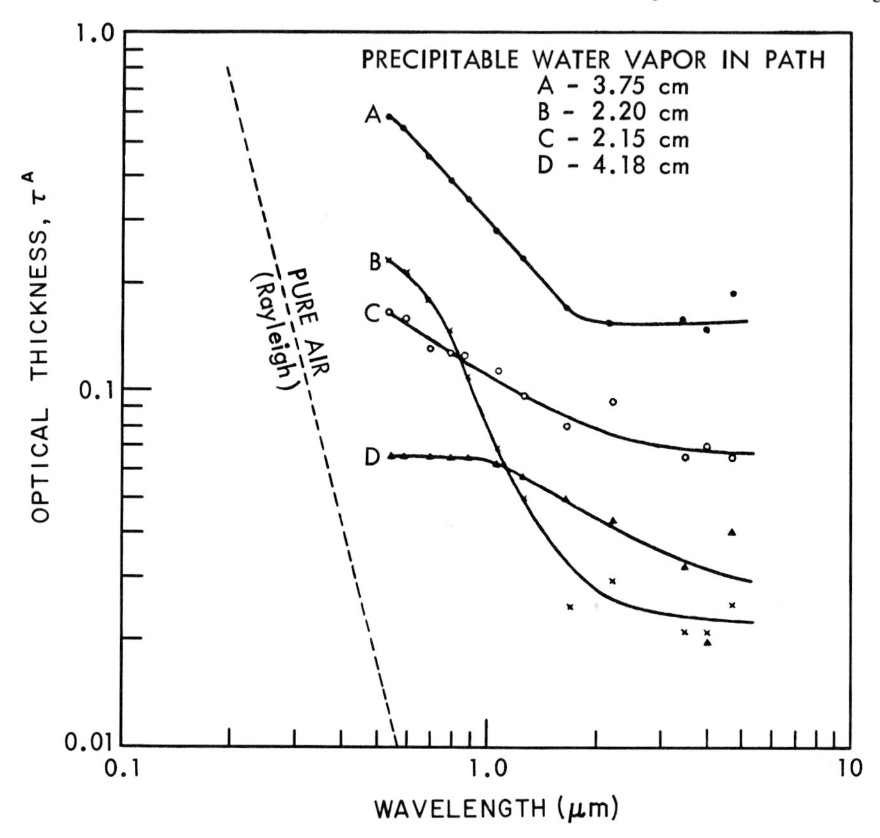


**Fig. 2.21.** Total normal optical thickness of large particles. These data are based on measured values of surface extinction. From Wolfe (1965).

sition from sunlight- to thermal-dominated sources to be shifted to longer wavelengths. *Hence even 11 μm radiances can be substantially affected by specularly reflected solar radiation.*

## 2. *The Thermal Infrared*

In the wavelength interval 4.0–22.0 μm three significant windows occur. As noted in Figures 2.19 and 2.20, these windows are centered near 4.7, 8–13, and 18 μm. The window at 8–13 μm is the most important because the maximum energy radiated by a blackbody in the temperature range 273-300°K occurs there. For this reason the 8-13 μm window finds widespread use in thermal mapping of terrestrial land and water surfaces.

The window regions in the thermal infrared, in the absence of clouds, are obscured primarily by the absorption and emission associated with unresolved water vapor absorption lines and the wings of their nearby lines. However, aerosols also contribute to the absorption in these windows. With the exception of ice and water clouds and very large dust particles, most atmospheric particulates have radii of less

0-201-04245-2

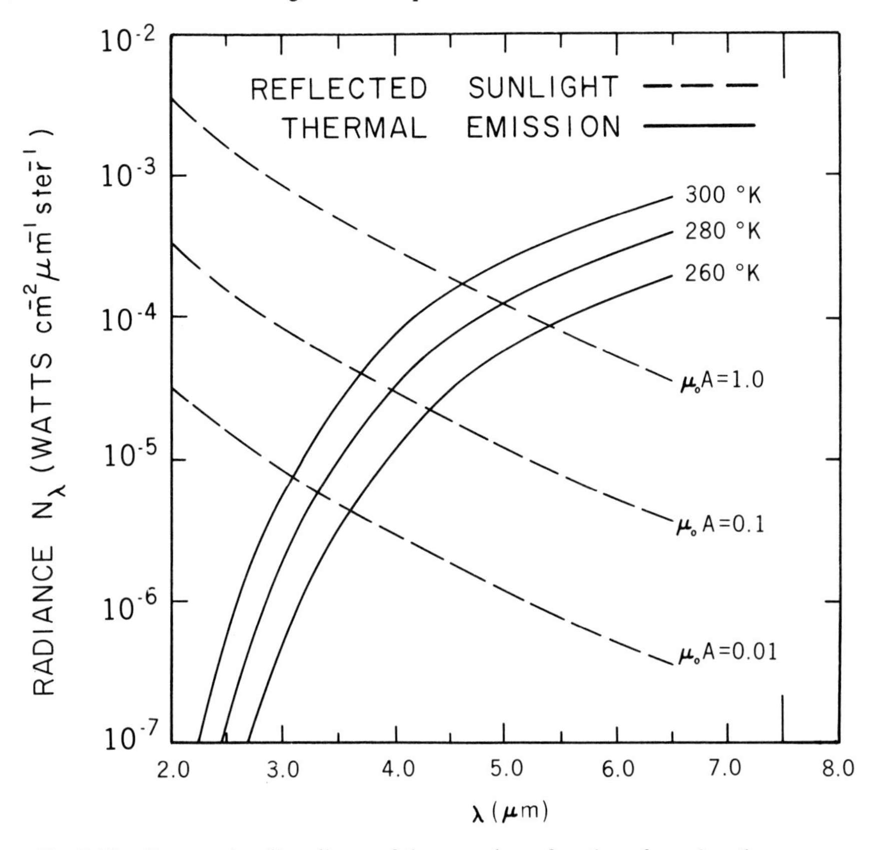


**Fig. 2.22.** Computed nadir radiance of the ground as a function of wavelength.

than a few microns. Solid and liquid materials have strong absorption bands in the infrared. Because aerosols composed of these materials have dimensions *less than* the wavelength of the infrared radiation, the scattering properties of these aerosols can be neglected, as compared to their absorbing properties.

With the assumption that no scattering takes place in the atmosphere, the path radiance of Eq. (2.8) can be simplified as follows:

$$J(z) = [1 - T(z)]B(\theta). \qquad (2.12)$$

The Planck function $[B(\theta)]$ gives the radiance of a blackbody, in this case at a mean atmospheric temperature $\theta$. The factor in brackets represents the emissivity of the atmosphere, which is about 0.2 in the windows (Table 2.1).

In the spectral band 4-22 μm, Planck's law can be approximated by Wien's law:

$$B(\theta) = c_1 \lambda^{-5} \exp\left(\frac{-c_2}{\lambda\theta}\right). \qquad (2.13)$$

0-201-04245-2

The constants in Eq. (2.13) are as follows:

$$c_1 = 1.177 \times 10^{-16} \text{ W m}^{-2} \text{ sr}^{-1} \quad \text{and} \quad c_2 = 0.01432 \text{ m } ^\circ\text{K}^{-1}. \tag{2.14}$$

The units in Eq. (2.13) have been adjusted so that for $\lambda$ in meters and $\theta$ in degrees Kelvin the Planck radiance function $[B(\theta)]$ will have units of watts per square meter per steradian. The error in Eq. (2.13) for atmospheric applications increases with wavelength but is less than 1% at 12 $\mu$m and 8% at 25 $\mu$m. The path radiance, Eq. (2.12), is substituted in Eq. (2.8) to compute the total radiance at an observing platform.

The transmittances for atmospheric windows are not known accurately. Both measured (Bignell, 1970) and computed (Kunde et al., 1974) values show large differences among themselves. Fortunately, accurate transmission data are not required for many current remote sensing calculations, since atmospheric attenuation is compensated for to some extent by emission. For example, if the absorption is overestimated, the emission will also be, and the two errors will tend to compensate.

The mean transmittances for two prominent atmospheric windows, one at 3.5-4.0 $\mu$m and the other at 8-12 $\mu$m, are given in Table 2.5. The transmittances are averaged for a spectral band width of 0.03 $\mu$m in the first window and 0.2 $\mu$m in the second. The transmittances are calculated for two model atmospheres, which are extreme in regard to the amount of water vapor they contain. The total water vapor content in a vertical column is 4.2 g cm$^{-2}$ for the moist tropical model and 0.2 g cm$^{-2}$ for the dry subarctic model. The water vapor transmittance is separated into parts: a continuum, which changes slowly with respect to wavelength, and a selective portion, which accounts for the sharp, narrow water vapor lines. The selective transmittance of water vapor is higher than that of the continuum, except at 8 $\mu$m. The water vapor absorption is negligible in the 3.5-4.0 $\mu$m window and also for the dry subarctic model, but is moderate for the 8-12 $\mu$m window of the tropical model.

Ozone absorption is negligible except near 9.6 $\mu$m. Actually, this absorption band extends from 9.2 to 10.5 $\mu$m, a range for which the mean transmittance is about 0.4. Ozone absorption is weak for altitudes below 15 km. The nitrogen absorption at 4 $\mu$m occurs on the wings of a continuum centered at 4.3 $\mu$m. The transmittances of the remaining gases, chiefly $CO_2$, $CH_4$, and $N_2O$, are listed in the "Mixed Gases" column of Table 2.5.

The aerosol transmittance is given for the condition of a moderate surface visibility of 23 km. The transmittance is separated into scattering and absorption components. Although both cause attenuation, only the fraction absorbed is emitted. Since the large-particle transmittance is poorly understood, it is sometimes neglected (Kunde, 1965; Smith et al., 1970). However, the large-particle attenuation is an appreciable fraction of the total attenuation for the 3.5-4 $\mu$m band. Since

0-201-04245-2

0-201-04245-2

**TABLE 2.5**
**Vertical Transmittances through Atmospheres of Tropical and Subarctic Winter Models**

| Model | Wavelength ($\mu$ m) | *Water Vapor* | | Mixed Gases | Ozone | Nitrogen | *Large Particles* | | Product of All Transmittances |
|---|---|---|---|---|---|---|---|---|---|
| | | *Selective* | *Continuum* | | | | *Scattering* | *Absorptive* | |
| Tropical | 3.5 | 0.99 | 1.00 | 0.97 | 1.00 | 1.00 | 0.94 | 0.98 | 0.88 |
| | 4.0 | 1.00 | 1.00 | 0.97 | 1.00 | 0.90 | 0.95 | 0.98 | 0.81 |
| | 8.0 | 0.52 | 0.62 | 0.69 | 1.00 | 1.00 | 0.98 | 0.99 | 0.22 |
| | 9.0 | 0.88 | 0.67 | 1.00 | 0.99 | 1.00 | 0.98 | 0.99 | 0.57 |
| | 10.0 | 0.92 | 0.58 | 0.99 | 0.69 | 1.00 | 0.98 | 0.99 | 0.35 |
| | 11.0 | 0.91 | 0.55 | 0.97 | 1.00 | 1.00 | 0.98 | 0.99 | 0.47 |
| | 12.0 | 0.90 | 0.42 | 1.00 | 1.00 | 1.00 | 0.98 | 0.99 | 0.37 |
| Subarctic winter | 3.5 | 1.00 | 1.00 | 0.97 | 1.00 | 1.00 | 0.94 | 0.98 | 0.89 |
| | 4.0 | 1.00 | 1.00 | 0.97 | 1.00 | 0.93 | 0.95 | 0.98 | 0.84 |
| | 8.0 | 0.99 | 0.99 | 0.69 | 1.00 | 1.00 | 0.98 | 0.99 | 0.65 |
| | 9.0 | 0.98 | 0.99 | 1.00 | 0.98 | 1.00 | 0.98 | 0.99 | 0.92 |
| | 10.0 | 0.99 | 0.99 | 0.99 | 0.48 | 1.00 | 0.98 | 0.99 | 0.45 |
| | 11.0 | 0.98 | 0.98 | 0.97 | 1.00 | 1.00 | 0.98 | 0.99 | 0.92 |
| | 12.0 | 0.99 | 0.99 | 1.00 | 1.00 | 1.00 | 0.98 | 0.99 | 0.94 |

The horizontal visibility at the ground is 23 km, and the spectral bandwidth is 0.03 $\mu$m for the 3.5 to 4.0 $\mu$m window and 0.2 $\mu$m for the 8–12 $\mu$m window. These transmittances are computed from data given by McClatchey et al. (1970).

the large-particle attenuation is roughly inversely proportional to the surface visibility, a decrease in the visibility from 23 to 2 km would decrease the transmittance to about 0.5 and 0.75 for the 3.5–4 and 8–12 μm bands, respectively.

The effect of atmospheric haze on upwelling radiances in the 11 μm window has been calculated by Jacobowitz (1973). To facilitate understanding the influence of the haze on thermal mapping from aircraft or satellite, the radiances are transformed to equivalent brightness temperatures by using the inverse Planck relation. Figure 2.23 shows the altitude-dependent difference between the equivalent brightness temperature and the true surface temperature. The three curves given in this figure are for three different temperature distributions, considered as mean distributions for the latitude and season indicated. In the calculations shown in Fig. 2.23 it was assumed that the total number of haze particulates was comparable to conditions with a visibility of approximately 10 km. In general the particles cause small differences between measured equivalent brightness temperatures and true surface temperatures. The major effect of the particulates occurs in the first 2 km above the surface because of the fact that an overwhelming majority of the particulates are concentrated in the first 1 or 2 km above the surface.

The intermittent appearance of clouds within all or part of the field of view of a radiometer introduces ambiguity in the interpretation of measured radiances. Smith et al. (1970) relate measured radiances to statistical data to decide whether or not clouds are in the field of view. Anding et al. (1970) suggest another statistical approach. They have made a limited computational study to show that radiometric

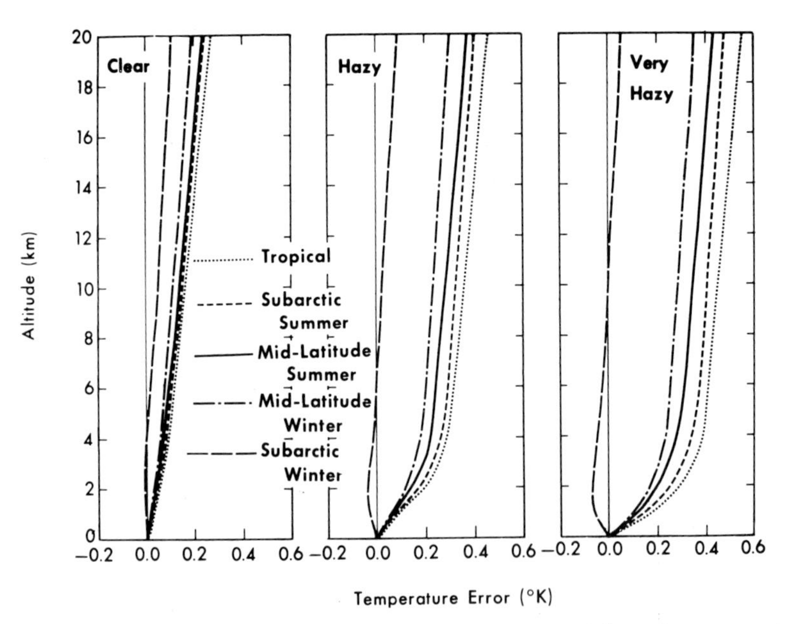


**Fig. 2.23.** Computed effect of large particles on the accuracy of the surface temperature measured in the 11 μm window. From Jacobowitz.

0-201-04245-2

measurements from above the atmosphere in three narrow spectral intervals centered at 4.9, 9.1, and 11.0 $\mu$m can be used to determine, with an error of 0.2°C, the sea surface temperature in the absence of clouds.

*a. Upward Radiance.* The equations for the radiance of the earth and the atmosphere between the earth and the observing platform are obtained by substituting Eq. (2.13) for the ground radiance and Eq. (2.14) for the atmospheric source function in Eq. (2.8). Hovis et al. (1968) have measured the spectral radiance of the nadir in the 9–12 $\mu$m atmospheric window at heights of 0.9, 3.2, and 10.1 km. Although all measurements were completed within an hour and confined to the Mojave Desert, the transmission of the lower atmosphere varied considerably, as shown by the temperatures derived from the measured spectral radiances (11 $\mu$m) at heights of 0.9 and 3.2 km. The difference in temperature between the two heights varied from 0 to 10°C. The temperature derived from the measured radiance decreased about another 10°C at the 10 km height.

Satellite radiometers utilize the 3.5-4.0 and 8–12 $\mu$m windows to measure the temperature of the ground or of cloud tops. In oceanic regions the differences between the apparent temperatures measured in the nadir direction from satellites and the temperatures measured from ships are not large. The computed differences for cloudless skies, if large-particle absorption is neglected, are about 2 and 9°C in the 3.8 and 10 $\mu$m windows, respectively (Kunde, 1965; Smith et al., 1970). Smith et al. (1970) have devised a statistical method of deriving sea temperatures from satellite measurements of radiance in the 3.8 $\mu$m window. Comparison with conventional ship measurements of sea temperature indicates a total error of less than 1°C. Shaw (1970) and Anding et al. (1970) estimate that surface temperatures can be measured above the atmosphere with a greater accuracy of 0.2°C.

The spectral radiances measured from a satellite are given by the continuous lines in Figure 2.24. The measured values are displaced 0.2 radiance unit upward for clarity. The lower, dashed line is computed for a model atmosphere containing the following constituents: 1.58 g cm$^{-2}$ of water vapor, 0.36 atm cm of ozone, 253 cm of carbon dioxide, 1.4 atm cm of methane, and 0.22 atm cm of nitrous oxide. Large-particle effects are neglected. Nevertheless, the theoretical and observed curves show good agreement between 8 and 18 $\mu$m. The higher radiances indicate the most transparent windows of the atmosphere, which are in the bands of 8.2–9.2 and 10.2–12 $\mu$m.

*b. Clouds.* The transmission of infrared energy through the atmospheric windows when clouds are present depends strongly on their thickness and water content, both of which are quite variable. Transmission through ice crystal clouds, such as cirrus, frequently exceeds 50% (Gates and Shaw, 1960). On the other hand, transmission through liquid clouds is frequently much weaker, because their volume extinction coefficient is large, of the order of $\beta$ = 100 km$^{-1}$ (Gates and Shaw, 1960). Gates and Shaw consider liquid clouds of 20-40 m thickness to be opaque. Other estimates of the thickness required to achieve opacity are of the order of 100 m

0-201-04245-2

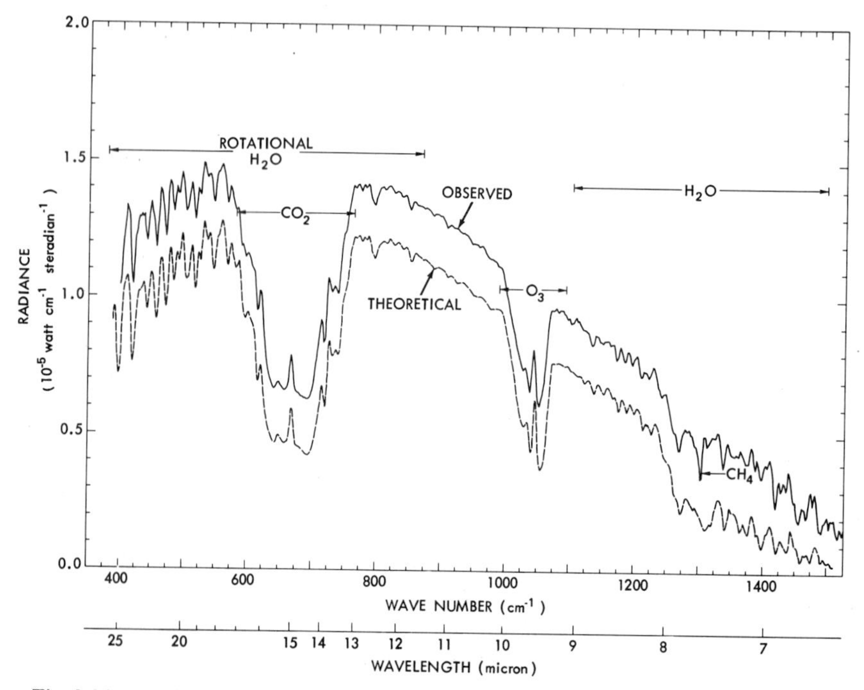


**Fig. 2.24.** A comparison of measured and computed nadir radiances from above the atmosphere for the spectral range 6.7–25 $\mu$m. From Kunde (1974).

(Deirmendjian, 1964; Yamamoto et al., 1966, 1970). Although almost all clouds would seem to be more than 100 m thick, the measured transmittance through liquid clouds frequently is of the order of 50% (Gates and Shaw, 1960).

The infrared emittance of thick clouds approaches 1. We consider just the spectral band 5–40 $\mu$m, for which computed emittances are available. The albedo of single scattering ($\omega_0$) of clouds is less than 0.9 (Deirmendjian, 1964; Yamamoto et al., 1966). If the thickness of a liquid cloud exceeds 100 m, the computed transmittance is less than 0.01, the reflectance is 0.03, and the emittance is 0.97 (Yamamoto et al., 1970). The spectral emittance is least in the band 5–6 $\mu$m, where the emittance equals 0.89 (Yamamoto et al., 1970).

## D. Microwave

The atmosphere, including haze and hydrometeors, is much more transparent to microwave than to visible and infrared transmission (Altshuler et al., 1968; Benoit, 1968; Kerr, 1951). However, the effect of hydrometeors and the atmosphere is sufficient so that they must be taken into account for many remote sensing activities. Since at present wavelengths between 0.8 and 30 cm are actively being used for aircraft and/or satellite sensing, this spectral region will be emphasized in this section. We shall first show how the radiances are used to compute brightness temperatures.

0-201-04245-2

Then the absorption and scattering characteristics of the atmosphere and hydrometeors will be discussed.

### 1. *Theory*

In microwave discussions the term "temperature" is used in place of "radiance." The basis for doing so is given by the Rayleigh-Jeans equation for the Planck function:

$$B_\lambda(\theta) = c_m \theta \lambda^{-4}. \tag{2.15}$$

If $B_\lambda$ is expressed in mks units, but $\lambda$ is given in centimeters, the constant $c_m = 8.28 \times 10^{-7}$.

At a given wavelength the spectral radiance is linearly proportional to temperature ($\theta$). Rather than using $B_\lambda$ and also the spectral radiance ($N$), radio engineers prefer to omit the factor $c_m \lambda^{-4}$ and use just the temperature. This temperature, which will be designated with a subscript $b$ and called the brightness temperature, is related to the spectral radiance as follows:

$$\theta_b = \frac{\lambda^4 N}{c_m}. \tag{2.16}$$

The convention will be adopted in this section that $\theta$ without a subscript and $\theta_s$ are the kinetic temperatures of the atmosphere and surface, respectively.

### 2. *Atmospheric Attenuation and Emission*

Radio engineers express attenuation in terms of a number of decibels ($n$). The transmittance ($T$) through a parallel slab of the atmosphere of thickness $h$ can normally be expressed in either decibels ($n$) or optical thickness ($\tau'$):

$$T(h) = 10^{-n/10} = \exp\,[-\tau'(h)].$$

If the attenuation is measured at an angle $\phi$ to the normal to the slab, then the attenuation in decibels is related to the normal optical thickness by

$$n(h, \phi) = 4.343\tau'(h) \sec \phi. \tag{2.17}$$

The computed attenuation in decibels through the entire clear atmosphere at a zenith angle of 60° is shown in Figure 2.25. The absorption bands are caused by $H_2O$ at 1.35, 0.16, and 0.092 cm and by $O_2$ at 0.50 and 0.25 cm. The entire atmosphere is essentially opaque to vertical propagation at shorter wavelengths (22 μm–1 mm). The atmospheric transparency increases with increasing wavelength for $\lambda > 3$ cm until the longer waves are attained that interact strongly with the ionosphere, which is opaque at normal incidence for $\lambda > 30$ m. Measured atmospheric

0-201-04245-2

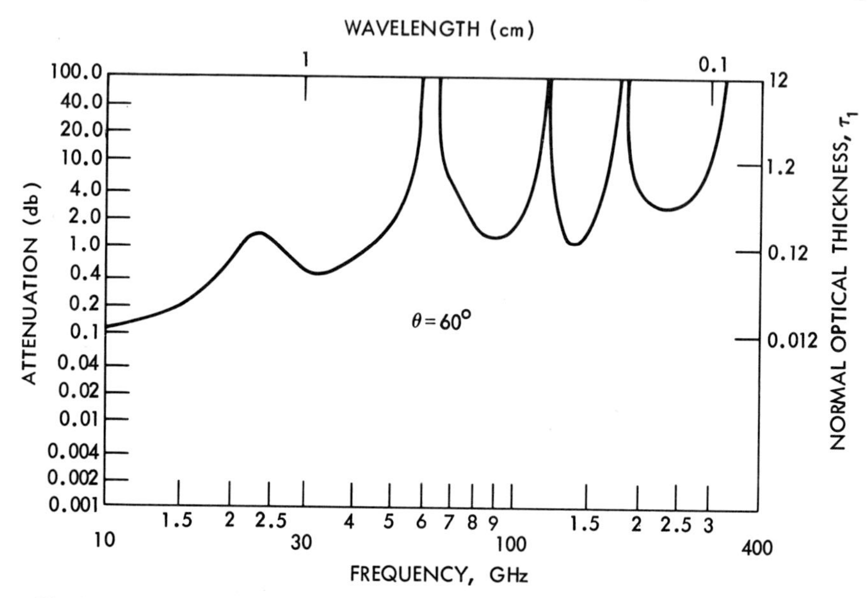


**Fig. 2.25.** Computed total atmospheric attenuation along a slant path inclined 60° from the zenith. From Conway et al. (1963).

transmission data are given by Benoit (1968), Orhaug (1965), Thompson (1971), and Ulaby and Straiton (1969).

Atmospheric hydrometeors such as fog, clouds, rain, hail, sleet, and snow cause attenuation and add microwave noise. Although their effects on remote sensing are usually small at centimeter and longer wavelengths, except for rain, they have to be accounted for if ground temperatures are to be measured with errors of less than a few degrees by high-altitude microwave radiometry. The radiation effects of the hydrometeors can be computed if the attenuation coefficient, which depends on both absorption and scattering, and the absorption coefficient are known at each point between the ground and the receiver. These coefficients depend on the indices of refraction of the particulates, or on their dielectric constants, which depend in turn on temperature and wavelength. Although the size of the particulates and their number cannot be expected to be known accurately, they also determine the strength of these coefficients.

The computed attenuation coefficients of liquid and of solid water clouds are given in Figure 2.26. The new unit in the ordinate refers to the density of water in the liquid or solid phase in grams per cubic meter and indicates that the attenuation coefficients are proportional to the mass of the condensed phase of water. The mass, of course, is the integrated mass of each particulate; as a consequence, the number and size of particulates in a cloud need not be known—just their total mass density is required. The liquid or solid water density in clouds varies from 0.05 to 5 g $m^{-3}$ (Valley, 1965). The attenuation coefficients increase approximately as the

0-201-04245-2

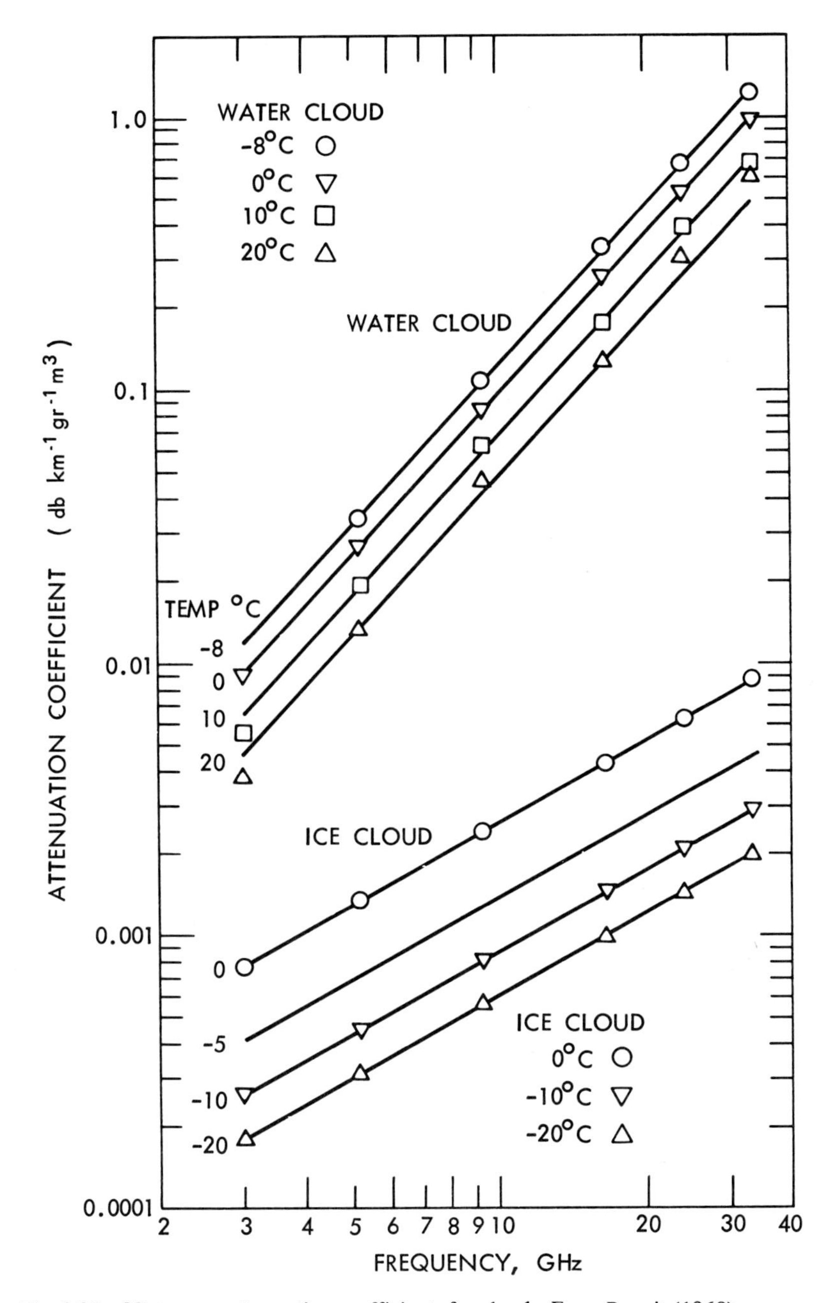


**Fig. 2.26.** Microwave attenuation coefficients for clouds. From Benoit (1968).

0-201-04245-2

square of the frequency for liquid clouds, and linearly for ice clouds. The attenuation of liquid clouds is 1-2 orders of magnitude larger than that of ice clouds.

An example of the use of the data in Figure 2.26 shows that the normal optical thickness of cirrus clouds is usually negligibly small at centimeter wavelengths. The condensed water content of cirrus clouds is quite variable, but an average value is $0.1 \text{ g m}^{-3}$. A thickness that occurs frequently is $h = 200$ m (Hall, 1968). Assume that the cloud temperature is $\theta = -20°$C. The attenuation coefficient from Figure 2.26 is $n = 1.3 \cdot 10^{-3}$ dB km$^{-1}$ g$^{-1}$ m$^3$ for 1.55 cm. Then the optical thickness of the cloud is [Eq. (2.17)] : $\tau'(1.55 \text{ cm}, 200 \text{ m}) = (0.230)\,(1.3 \cdot 10^{-3} \text{ dB km}^{-1} \text{ g}^{-1} \text{ m}^3)\,(0.2 \text{ km})\,(0.1 \text{ g m}^{-3}) = 6 \cdot 10^{-6}$, which is negligibly small.

The computed attenuation coefficients of rain are given in Figure 2.27. The ordinate value has to be multiplied by the precipitation rate with which each set of curves is labeled. For example, if the precipitation rate is substantial at $p =$ 10 mm hr$^{-1}$, the temperature is 0°C, and the frequency is 19.4 GHz ($\lambda = 1.55$ cm), then the attenuation is 0.85 dB for a path 1 km long through the rain. The attenuation coefficient is strongly dependent on frequency and is much less at lower frequencies.

Ionospheric absorption for propagation near to the zenith direction is negligible for $\lambda < 30$ cm. As a consequence, the effect of the ionosphere near the zenith on the brightness temperature is also negligible when the line of sight is near the nadir. The ionospheric kinetic temperatures are quite high, approximately 1500°K, but the absorption of centimeter waves is so weak that the brightness temperature is less than 0.1°K for $f > 1$ GHz (Valley, 1965).

Statistics of the radiance of the sky measured at the ground are given in terms of the brightness temperature in Figure 2.28. These temperatures were measured near Boston during 6 months of a single year for clear skies and also in the presence of clouds or precipitation. At the microwave frequency of 35 GHz and toward the zenith, for example, the effective sky temperature ranged from 10 to 37°K in the 10-90 percentile bracket. Near the horizon the atmosphere is more opaque, and the equivalent temperatures are higher. Wulfsberg (1964) says that cirrus clouds produce a negligible contribution to the sky temperature. Fair-weather cumulus clouds had a brightness temperature of 5-25°K at $f = 35$ GHz and only a few degrees at $f = 15$ GHz. The brightness temperature of rain was high and variable. Toong and Staelin (1970) give additional data on the effect of water vapor and clouds on the brightness temperature.

### *3. Upwelling Brightness Temperature*

The upwelling brightness temperature at some altitude $z$ in the atmosphere consists of the surface brightness temperature, which is attenuated to the altitude $z$, plus the weak contribution of the atmosphere. Since the emissivity of the atmosphere is equal to its optical thickness, the brightness temperature of the atmosphere is of the order of 25°K at centimeter wavelengths (Figure 2.28). The brightness temperature of the surface is the sum of its thermal emission plus the radiant energy falling

0-201-04245-2

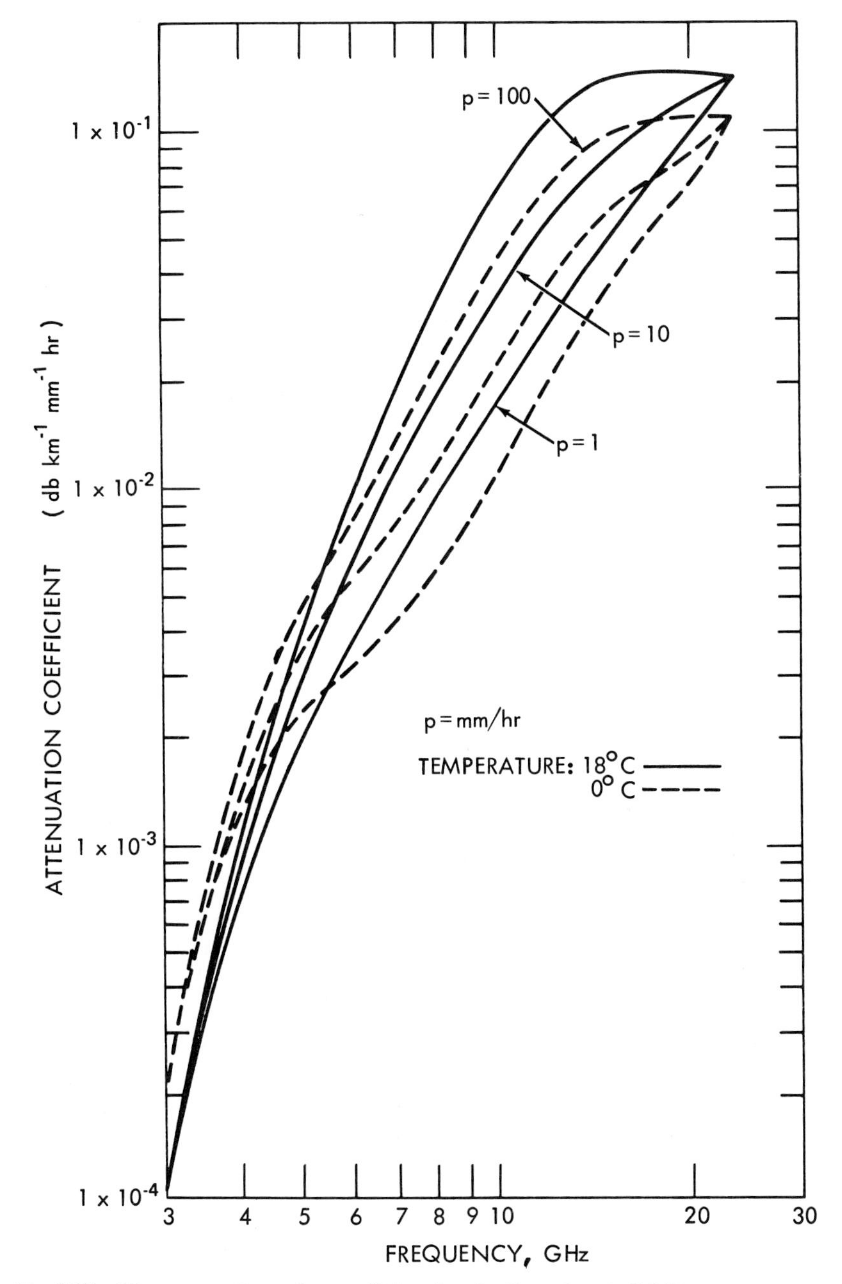


**Fig. 2.27.** Microwave attenuation coefficient in rain. From Benoit (1968).

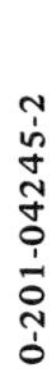
0-201-04245-2

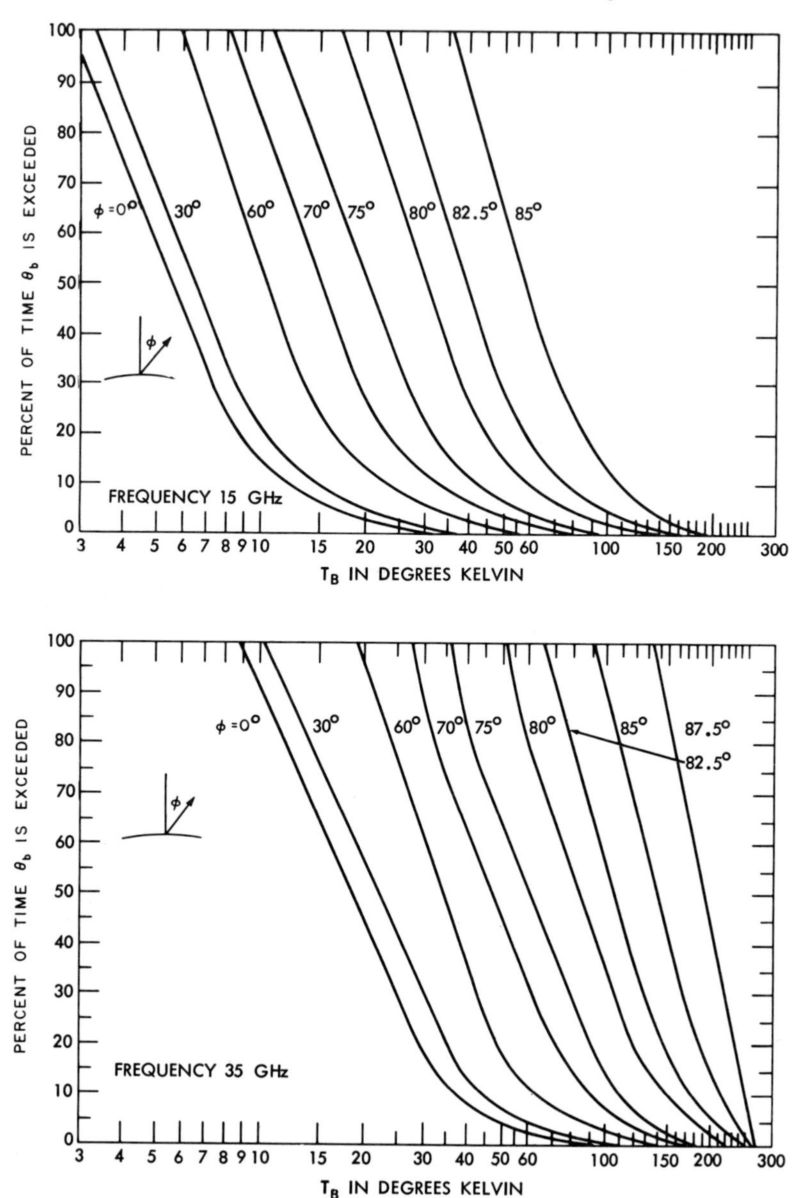


**Fig. 2.28.** Cumulative frequencies of the brightness temperature of the sky, measured near Boston from the ground during the months of February to July 1963. From Wulfsberg (1964).

on the surface and reflected from it. The thermal emission by the surface gives the dominant contribution at an arbitrary altitude, but the other effects must be accounted for to the extent indicated in Figure 2.27.

0-201-04245-2

The apparent brightness temperature of the sea that would be observed from a satellite for representative atmospheric conditions is given in Figure 2.29. The sea temperature is $\theta_s = 285°K$. Curve I gives the brightness temperature of the sea in

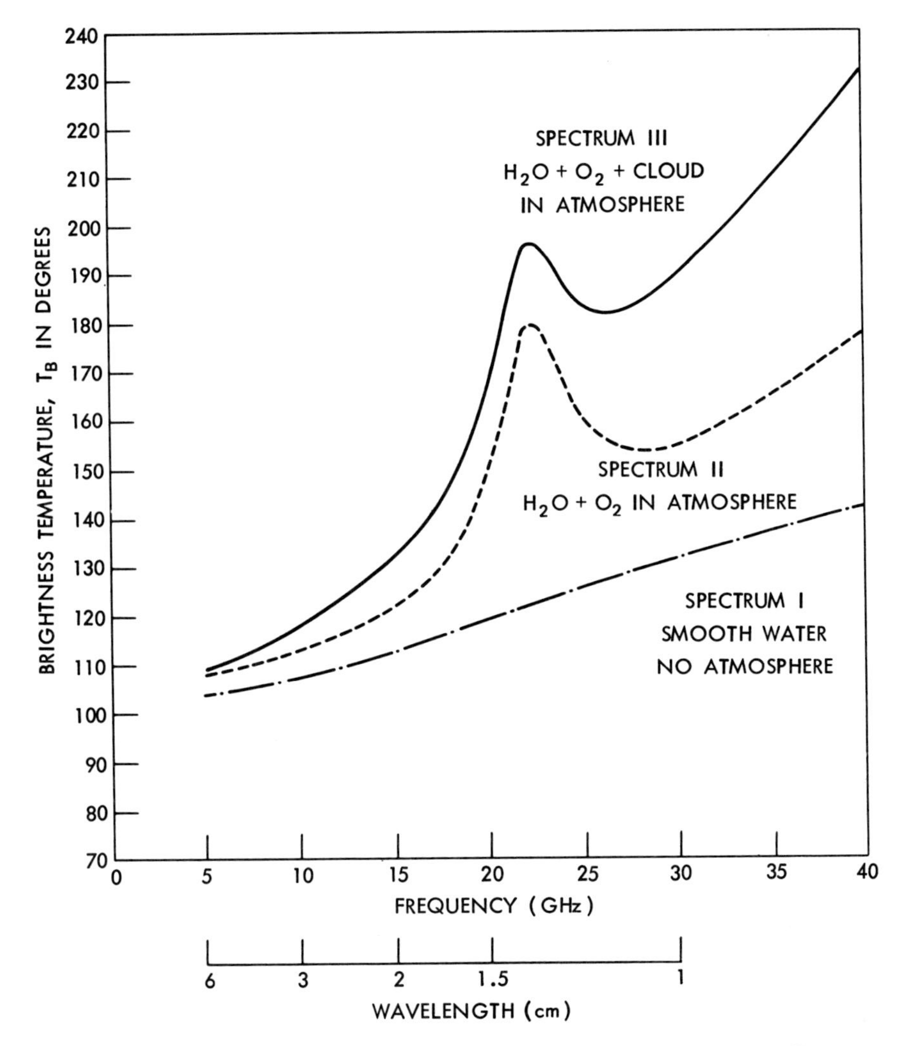


**Fig. 2.29.** Computed brightness temperature as seen from a satellite over water. From Staelin (date unknown).

the absence of an atmosphere. The brightness temperature is low ($T_B = 112°K$ at $\lambda = 2$ cm), because of the low emissivity of the water (0.4). Curve II gives the brightness temperature in the presence of a standard atmosphere and water vapor that has a concentration given by the following formula:

0-201-04245-2

$$\rho_{H_2O}(z) = 10 \exp\left(\frac{-z}{2.2}\right),$$

where $z$ is expressed in kilometers and $\rho$ in grams per cubic meter. Curve III includes the effect of a cloud 1 km thick at a height of $z = 2.5$ km and with a liquid water density of 1 g m$^{-3}$. The effect of the atmosphere on the brightness temperature is considerable at the short wavelengths but increases the apparent sea brightness temperature by only a few degrees at 6 cm. Kreiss (1969) gives additional computed and measured data on the effect of clouds on brightness temperature.

Corrections for water vapor and hydrometeors can be made by taking additional microwave measurements that would not ordinarily be made for observing surface properties. The total mass of water vapor can be measured a little off center of a water vapor absorption band, say at $\lambda = 1.36$ cm. The mass of condensed phase in nonprecipitating liquid clouds may possibly be monitored with observations in the $\lambda = 0.9$ cm window. Rain could be monitored at a longer wavelength of 3 cm, where the clouds are more transparent but the larger raindrops still interact strongly with radiant energy. These methods have not yet been fully developed for making atmospheric corrections.

0-201-04245-2

Chapter 3

# Principles, Concepts, and Philosophical Problems in Remote Sensing

*J. Everett and D. S. Simonett*

## I. INTRODUCTION

Remote sensing is not simply a modest extension of conventional aerial photography. Rather, it represents a revolution in the way we think about and approach resource inventory analysis and management problems. This chapter seeks to emphasize how significant, wide ranging, and comprehensive a revolution in thinking remote sensing constitutes by:

- Comparing a photo interpretation job in 1960 with a remote sensing project circa 1985..
- Summarizing some discussions with university instructors on the educational value of LANDSAT satellites.
- Examining possible contributions to science from satellite remote sensing.
- Giving an example of LANDSAT value to science through its use in geologic applications.
- Outlining key concepts which have emerged in remote sensing, all of which must be used with caution and judgment.
- Discussing some weaknesses in the philosophical underpinnings of remote sensing which must be constantly borne in mind when working in the field.

## II. PHOTO INTERPRETATION (1960) VERSUS REMOTE SENSING (1985)

In 1960 a U.S. geologist, geographer, agronomist, and hydrologist, each with a fresh master's degree, used conventional pan minus blue aerial photographs in their respective disciplines (no other forms of remote sensing were available to them). They obtained the latest 1 : 20,000 scale photographs and uncontrolled photo indexes from repositories run by the U.S. Department of Agriculture or the U.S. Geological Survey. Depending on the area of the U.S., the next-newest photographs could be from 2 to 5 or more years or older. It was not until the late 1960's that the U.S.G.S. began obtaining photographs at a scale of 1 : 60,000, using medium- to high-altitude aircraft. Many urban areas were patchily photographed on an irregular basis, mainly for highway studies. All of these black and white photographs were

0-201-04245-2

*Remote Sensing of Environment*, Joseph Lintz, Jr. and David S. Simonett (eds.) ISBN 0-201-04245-2

acquired between 10:00 A.M. and 2:00 P.M., at times of the year optimal for U.S.D.A. acreage allotment measurements and weather. In 1960 photographs were acquired principally, if not exclusively, for mensuration purposes.

It did not occur to the geologist to wonder whether this customary 4-hour span was the right time of day and provided the right sun angle to emphasize structural features, or whether much smaller scale images in conjunction with the 1 : 20,000 photographs would be helpful—alas, the photos in the mosaic had many different sun angles on them, and these tonal variations from run to run obscured most major features. The geographer may have thought idly about the use of color or color infrared photography or even much lower-altitude and hence larger-scale photography for studying an urban area, but promptly dismissed such possibilities as unreasonably expensive. The agriculturalist, graduated in agronomy, with the barest acquaintance with black and white aerial photography, used the results for simple linear measurements and as plotting bases for farm surveys and soil surveys. The hydrologist may perhaps have wondered whether photographs at other times of the year might be more useful for him—midsummer rather than or as well as early spring, before all deciduous species were leafed out—but the lot of the good hydrologist (or geographer et al.) was simply to accept the prevailing practice and be grateful.

Most 1960 black and white aerial photography was used for its contextual, geometric, and shape information on natural and cultural features for disciplinary interpretations, as shown by any standard photo interpretation text of that vintage, or even in the more advanced *Manual of Photographic Interpretation,* published by the American Society of Photogrammetry (1960). The first color space photographs were not obtained until 1965 by GEMINI astronauts. The first remote sensing Symposium held at the University of Michigan and triggered by the influence of several earth scientists familiar with classified remote sensing devices (Office of Naval Research—Evelyn Pruitt and Walter Bailey; Geography Committee Advisory to the National Academy of Science—Joseph Russell; and U.S. Air Force—Carleton Molineaux) was still a year away. The first formal courses in remote sensing were not given until the mid-1960's, at the University of Kansas, Stanford University, and in the series of National Science Foundation Courses in remote sensing for college teachers at the University of Michigan.

For 1985 we envision a very different situation. Two earth-synchronous resource satellites (36,000 km), two polar orbiting sun-synchronous satellites (800 km): LANDSATS 6 and 7, four meteorological satellites, and a space shuttle are operating along with a DOMSAT, an earth-synchronous domestic relay satellite for U.S. channels. Two major commercial aircraft manufacturing companies are competing for the high-altitude (25,000 m), superhigh-resolution (0.2 m) wide-angle format (100 km) camera market, both domestically and internationally. A third satellite, LANDSAT H, is due for launch in January 1986. Unlike those already in orbit, which have an 11:00 A.M. sun-synchronous orbit over the United States, this will be non-sun-synchronous to take advantage of differential lighting in successive coverage of the same location. Geologists and hydrologists expect to be the principal beneficiaries of this satellite.

0-201-04245-2

The ground receiving station at Sioux Falls, South Dakota, receives data from all these satellites, merges the data into geobased information system format for the same days, and rebroadcasts to the DOMSAT's for retransmission to a series of regional centers (General Electric, 1974).

A data bank at Sioux Falls, along with other data banks, is regularly tapped by geologists in mining and petroleum for a wide array of space and aircraft imagery and for digital tapes of different scales, formats, times of acquisition, and prior weather. The 1960 geologist, now exploration manager for a major mining company, has hired a group of consultant remote sensing geologists to carry out an in-depth analysis of multidate, multiresolution spacecraft-acquired images of radar, camera, and multispectral scanner systems of a major African mineral province as part of a joint venture between the African country and the mining corporation.

The geographer, now chief of the Division of Remote Sensing Estimates in the Bureau of the Census, is preparing a planning budget for contemporaneous high-resolution space shuttle and aircraft flights to take place at the time of the 1990 census for all U.S. cities having populations above 25,000. Local and regional calibrations, using photographs to improve intercensus estimates, proved so successful in 1980 and 1983 that consideration is being given to moving to a systematic 2-year photographic update, census estimation cycle.

The agronomist, now with the branch of foreign crop estimates of the U.S. Department of Agriculture, receives daily reports on wheat and rice production for 18 key foreign countries, based on an integrated meteorological satellite and earth resources satellite program and crop yield and acreage prediction models. He in turn supplies this information to AID, the United Nations, and the World Bank, which are watching closely several potential drought areas in Africa in order to initiate an early response to a potential disaster. He just signed off on continued funding of research on three other crops which still appear to be promising candidates for world crop watch. Two other crops were dropped from further consideration at this time because they proved intractable with LANDSAT 6 and 7 instruments.

In 1984 the hydrologist produced a masterful computer systems analysis of national daily water consumption patterns, employing the full range of meteorological and earth resources satellites, which tipped the scales for his election to the National Academy of Science. At present he is considering an offer to leave his academic position to start a new full-time consultant service with the OPEC countries, who are concerned about their long-term water needs and recognize that only a few decades of oil remain.

Are these examples fanciful? Only in part. The potential is there, and many of the present resource evaluation, management, and monitoring problems are amenable in part to the technology that we see coming. Whether the technology actually arrives as early as 1985 will depend not merely on the importunings of interested scientists and the results of cost/benefit studies, but also on the infighting between federal agencies jockeying for power, the whims of several key figures in Congress, as well as Presidents yet to take office, and the degree to which space remote sensing has been internationalized.

0-201-04245-2

## III. DISCUSSIONS ON THE IMPACT OF LANDSAT SATELLITES ON EDUCATION

*Author 1:* Education from elementary school to the university, but most particularly at the university level, will benefit through use of an earth resource survey satellite akin to LANDSAT 1. Education that takes place in federal and state government agencies, private corporations, and other institutes will also be enriched by the availability of such data. Earth resources satellite (ERS) data in any case will receive increasing use because of the growing number of people who will use these data in graduate programs and later enter government and private industry research and management. New and diverse applications of the ERS systems are likely when such people reach positions of responsibility because they will be sensitized through long training and usage to the efficacy of remote sensing.

Because there were no experiments on LANDSAT to evaluate the impact on education, much of our information was obtained in discussions with investigators and colleagues at major national meetings and LANDSAT symposia. These run the gamut from optimism to deep pessimism. The differences reflect the individual differences between human beings confronted with a new technology. The truth is elusive when dealing with a new technology as complex as that of LANDSAT. Few people involved in management throughout the government have had much daily contact with LANDSAT, and some discount important virtues of these data while others attribute nonexistent qualities to them. Partly for these reasons, both unreasonable hopes and equally unreasonable misgivings about LANDSAT are common in the federal and state governments. This range of attitudes is not present to the same degree within the academic community, which has tended to approach LANDSAT as a rich research tool without having either a professional commitment to its survival or an equivalent vested interest in the status quo. The latter factor may prevent an adequate appraisal of this new tool in many government agencies: it is sufficiently complex that only sustained study and interest will define its advantages and shortcomings.

*Author 2:* Consider the implications of an ERS within the university community, especially in regard to the education of graduate students. During the life of LANDSAT 1 some 300 principal investigators, coprincipal investigators, and other university staff were directly involved in NASA-supported experiments. Working with them have been perhaps 600 graduate students. Indirectly, perhaps double these numbers (600 staff and 1200 graduate students) have been involved in other institutions or in other departments. These studies have had an impact on a wide range of the natural and engineering sciences, including electrical engineering, civil engineering, hydrology, geology, geography, agronomy, and the plant sciences. Thus part of the present generation of young scientists is growing up with LANDSAT and is accepting as commonplace many characteristics of the new system. To appreciate what this acceptance means in terms of mental constructs and of ways of going about problem solving, it is necessary first to review some of the distinctive characteristics of the LANDSAT satellite.

The satellite has a semiangle (half the angle of view) of 5.76°, which gives

0-201-04245-2

images that are essentially orthographic. Thus it is possible at low cost to overlay sequential images both manually and digitally. This capability is possessed by no other remote sensing system, and certainly not by any data source of such volume, uniformity, and worldwide coverage. LANDSAT orthographic, planimetric, and spatial fidelity is such that, once it is corrected to eliminate periodic system errors, it meets national map accuracy standards at a scale of 1 : 500,000 and almost meets them at a scale of 1 : 250,000. Consequently, great diversity of changes can be detected and assessed rapidly by simply overlaying different bands at different dates or the same band at different dates, by computer-ratioing bands of different dates in different fashions, and so on.

In the middle latitudes, although each successive pass differs in solar illumination by a relatively few degrees every 18 days, over the term of a year the solar illumination angle for LANDSAT changes significantly. There is in this changing solar illumination a discriminant which may be used to assist in identifying different lithologies, geologic structures, plant communities, and other natural features in assessing such characteristics of the terrain as roughness, slope angle, and relief. In reasonably cloud-free areas, repeated looks throughout the year enable the changing seasons to fit the series of discriminants to any area through the waxing and waning of vegetation, of snow cover, and of other dynamic features. Even relatively static items in geology are emphasized to different degrees throughout the year through the change in angle and azimuth of illumination, the passage and melting of snow, and the varying rates of progression of the biological greenwave and the senescence of vegetation on rocks of different porosity.

*Author 1:* With the LANDSAT system many of the principles taught in remote sensing and in the biological and natural sciences generally are brought home forcefully: they become part of the mental baggage of students working with such data. Thus hard-won concepts that represent the experience of senior professionals have been transferred to younger people without the same necessity of extensive field and other experience. For example, the concept of "working from the whole to the part," which guides most senior professionals in establishing the context of the problem, or the concept of "a resolution germane to the task" (the distillation of a lifetime of field experience in detailed as well as reconnaissance scale studies) becomes readily apparent to students as they work with imagery and see the different quality and characteristics of data contained in this material in comparison to other sources. Thus very fundamental, important procedures in analysis and in understanding the problems encountered are being transferred in an easily understood fashion much earlier than would be possible without LANDSAT data.

Some of these ideas, as well as several others, are reflected in conversations we have overheard or participated in at national meetings, snippets of which are included here.

## Snippets from National Meetings

- The dynamic aspects of LANDSAT are a real benefit not only in teaching remote sensing courses, but also in teaching advanced ecology and wildlife biology.

0-201-04245-2

- LANDSAT has been immensely valuable in giving students a perspective on the interrelationships between ecosystems in time and space.
- When LANDSAT is used in conjunction with low-altitude aircraft photography, working between multiscale data has made students sensitive to the fact that information is a highly scale-dependent phenomenon.
- LANDSAT has been important in enabling us to observe new areas of modeling biological systems, particularly those in rangeland.
- We have found that it is worth introducing a lot of the LANDSAT material in beginning classes at the undergraduate level. We expect to use it at all levels in graduate work as well.
- LANDSAT has facilitated discussion on our campus between departments as diverse as economics, anthropology, international relations, and political science (particularly the part dealing with public policy).
- LANDSAT has had a tremendous impact on my teaching, and I am convinced it will have a significant role in the future in generating new concepts and theories in science.
- Students confronted with LANDSAT instinctively begin to think in somewhat broader terms. In the past they found it rather difficult to think in an ecosystem context, and fieldwork did not bring this home as forcefully as is now possible when fieldwork is combined with LANDSAT interpretation. The ability to see on a single frame 36,000 $km^2$ with its spatial and temporal fabric has been important in conveying major ecosystems concepts to our students.
- One of the great advantages of LANDSAT is that it forces one to think in a systems context. This is particularly important in the environmental sciences, those in which the interconnectedness between things over large areas is very important. For too long we have segmented the environment for training purposes and have overspecialized our students. At the same time that we are going through the environmental revolution and all the changes in teaching occasioned by it, LANDSAT is available to provide some of the context within which new understanding and discussions can flourish.
- It has been enormously valuable in our studies on this campus to have a diversity of disciplines confronted with and working with this common data base.
- Scientists who work with water resources tend to think in a systems context, and for them LANDSAT is a natural addition to their normal techniques of study and analysis. However, in the other environmental sciences, systems-type thinking in which large problems are examined over large areas has been a relative rarity in the training of students. LANDSAT will help us in this large-area systems-type thinking.
- We are on the verge of many exciting discoveries, as is inevitable whenever a new research tool improves on the resolutions of features which were previously rather obscure. The hardware of the LANDSAT program provides us with such a tool.
- One of the great advantages of LANDSAT is that it has forced our rigid bureaucracies to be less rigid and people to talk across disciplinary and departmental

0-201-04245-2

lines. This will probably turn out to be one of its most useful functions, along with a forced reappraisal of our existing procedures.

*Author 2:* I think that LANDSAT and its successors will revolutionize many of the ways in which we carry out our teaching and research. My students are now much more sensitive to the sources of error in their data, to the ways in which data may be manipulated digitally as well as manually, and to questions of distribution of cost between various means of analysis. They look at multistage sampling as a rather routine procedure. They are sensitive to the many ways of "skinning cats" which this truly diverse technology has provided. They are now conscious of the fact that multitime data may substitute for higher-resolution data, at least in part. They see the value of applying relatively expensive techniques to the analysis of LANDSAT data because the low per unit cost over a very large area justifies such techniques. In short, I think my students will be much less passive in the future in accepting any data which happen to come to them. They see the complex relations between resolution in space, time, quantity, and the multispectral domain. They understand the relationship between these areas in remote sensing and resolution in other data in the spatial, temporal, categoric, and quantitative areas. LANDSAT is not just another form of aerial photography; it is a revolution.

*Author 1:* Major advances in new technology are carried out, not by skeptics, but by those whose creativities are engaged by the new technology and who are prepared to persist in the face of difficulties over an extended period. Although we encountered some strongly negative, indeed caustic portrayals of the shortcomings of LANDSAT in discussions outside the university, it was rare to find within the academic world people who did not subscribe to at least some of the positive viewpoints. This may arise from the natural excitement engendered in a teaching and research-oriented community when confronted with a new tool. However, the reasons may go deeper. As suggested in some of the preceding quotations, LANDSAT must be regarded as an idea generator, a conveyor of mental constructs, and an illuminator of ideas which are fundamental to all of science. The ability to hold many factors constant during several LANDSAT passes in effect collapses the variance of the problem to a manageable form. At the same time, having data available over a year or several years makes it possible to employ a whole arsenal of discrimination techniques. Students therefore become aware of the abundant information available in satellite data, the extraction of which requires extensive massaging and manipulation. Techniques for extracting this information include narrow density slicing, simple and complex edge enhancement, the use of variable sun angles, the employment of rare vegetation enhancement to detect geological features, band ratioing to emphasize vegetation and geological features such as gossans, the search for surrogate data in images, and the inspection of images for lumped parameters for input to a model.

*Author 2:* LANDSAT appears on the scene at a time when important parallel technologies are undergoing major development. Two such technologies of great importance to the growth of earth resource satellites are the development of exten-

0-201-04245-2

sive geobase information systems, along with a diverse array of software for the operation of such systems, and the emergence of hard-wired digital systems for analyzing LANDSAT tapes, or the future emergence of much cheaper digital processing through the use of parallel and microprocessing techniques. The next 5 to 10 years will see startling reductions in the cost of digital processing, so that hybrid and interactive man-computer systems will be commonplace in the United States: rather than relying exclusively on manual interpretation for many problems, combined manual and digital and all-digital interpretation will become normal. Just as these technologies will be important in the future operation of an ERS system, so is the presence of an ERS system of considerable importance to the more cost-effective use of geobased information systems in forestry, water resources management, regional and state planning, and so on.

*Author 1:* To summarize, we see the following benefits to education from LANDSAT remote sensing:

- Improvement in the transfer to students of an understanding of problems and procedures of analysis. This will be especially true in those aspects of the natural and social sciences where comprehension of spatial and temporal variability will be enhanced by the availability of ERS data.
- Contributions in structuring concepts and theories in a teaching environment which relate to ERS spatial, temporal, multispectral, and radiometric resolution.
- Contributions in emphasizing fundamental concepts in science teaching, such as the scale dependence of information, and the importance of establishing the context within which investigations are carried out.
- A significant increase in the diversity of illustrative materials and the flexibility of teaching options in textbook and classroom use.
- Significant increases in the education of professionals in the natural sciences who are familiar with quantitative computer pattern recognition procedures.
- A role as a focus for integrated, multidiscipline teaching and research in environmental sciences, and in stimulating systems-type analysis within and across natural science disciplines.
- The establishment of an easily comprehended context for understanding the complex types of analysis necessary to deal with areally extensive problems such as those arising in ecological assessment.

## IV. THE IMPACT OF REMOTE SENSING SATELLITES ON SCIENTIFIC INVESTIGATION

The diversity of ways in which LANDSAT data have been used foreshadows the type of contributions which future ERS systems may make in the development of new models and new scientific principles and theories. The several examples given below are drawn from the LANDSAT experience.

0-201-04245-2

## A. Use of an Earth Resources Satellite as a Simple Change Detection Device and as a Model for Selecting Areas for More Intensive Study

A number of investigators have employed LANDSAT to detect change within areas which had been previously mapped with high-resolution systems or for which earlier low-resolution images were available for comparison. Thus mapping of major plant communities and rangeland areas, detailed mapping and boundary delineation of land use on the fringe of major urban centers, and forest-type mapping with aerial photography may be used as the first step in a monitoring and updating system. The role of LANDSAT then is to detect where changes have taken place within the classes designated by the detailed photographic analysis. In rangeland areas, plant communities once established and mapped may be monitored at quite coarse resolution to determine the trend and internal variations in pasture response. Similarly, in sensitive areas adjacent to wetlands or to national parks, and on the fringes of urban areas and major water bodies, ERS data can be a first-stage device for monitoring where change takes place. The change may be as simple as clearing forest land for urban development, thereby alerting the planning community to areas where more detailed studies are needed. Finally, as already shown by Aldrich (1975), certain changes in forestland that cannot be observed with the existing national sampling procedures can be monitored with acceptable accuracy with LANDSAT. These models, which may be referred to as change detection models, will become widespread with the advent of an operational ERS system.

## B. Examples of LANDSAT in Generating New Scientific Hypotheses

There are a number of examples of LANDSAT's stimulating new hypotheses, notably in water resource management and geology. In the latter, very large macro- and megalineaments have been detected with LANDSAT, many of them previously unknown. These lineaments are of such dimensions that they will significantly influence geological theories on crustal motion and on methods and characteristics of ore body emplacement. Studies as widely scattered as those of Parizek and Alexander (1973), dealing with regional geological mapping in Pennsylvania, and of Lathram et al. (1973), involving mineral emplacement in relation to major fracture and lineaments systems in Alaska (employing a new hypothesis for ore body emplacement in relation to regional structure), are examples.

In water resource development, the use of a lumped reflectance model by Blanchard (1974) in watersheds in the Chickasha in Oklahoma looks very encouraging. He is developing directly from LANDSAT data which may substitute for laboriously measured data in otherwise ungauged stream systems. The use of such lumped measures as integrated reflectance value over a drainage system as an input to runoff values in the drainage basin is an important possible and continued product of LANDSAT systems. Only when such systems are available over extended periods of time will a diverse group of possible surrogate measures be developed.

0-201-04245-2

## C. Partitioning of Other Data Using an Earth Resources Satellite Remote Sensing System

The high packing density and high-quality spatial information available from an ERS system is such that it may be used to partition aggregated data obtained from other sources. For example, it is possible to use ERS data to partition census statistics. Land-use characteristics are used to disaggregate all lumped data at the census tract level. Partitioning of economic and social statistics may be expected in the future as a typical application of ERS data. Once such partitioning is achieved, the ERS data themselves become calibrated and can be used with change detection and future land-use modeling to provide a variety of social statistics in areas of rapid growth adjacent to the urban fringe. In a similar fashion lumped values from the ERS data may substitute for detailed measurements by extrapolating from other statistics. An example would be the use of forestry statistics on known timber yield values in selected areas where scattered small areas of clearing (as clear cutting) occur outside the framework of reported large-operator planned clearing. The types of clearing would involve small lots, usually individual farm woodlots. Clear cutting could then serve as a simple guide to the amount of timber extracted in an area.

## D. Use of Earth Resources Satellite Data to Monitor the Interaction of Dynamic and Static Components of an Environment

Two examples of this type of modeling come to mind. The first situation is one in which subtle geological features are differentially emphasized at different times of the year through the waxing of the seasons, snow cover, differential wetting, and differential plant response on various lithologies. The same process takes place over a longer period in desert or semidesert areas: very rare widespread rains may be used to fit a vegetation discriminant to the terrain and thus emphasize geological differences. A second example would be the digital overlay of ERS data on geo-based information system containing static high-valued data such as soil type, slope angle, and rock type. In this procedure, an interaction matrix would be set up between the dynamic components of the environment, such as surface water, crops, and land use, and the static ones, which are generally physical features. Through the interaction of these on a cell-by-cell or polygon-by-polygon basis much detailed information on the joint variation of static and dynamic elements of the landscape could be evaluated. The first such models are now beginning to be developed.

## E. Use of an Earth Resources Satellite as a Component of an Integrated System

It is certain that with widespread availability of an ERS operational system, many systems not currently envisaged will be developed in which ERS data are an integral part. Many other data sources may be drawn upon, including meteorological satel-

0-201-04245-2

lites, social and economic statistical data, ground surveys, and supplementary aircraft data. In a number of these instances, ERS data may well turn out to be the linchpin which will raise the system to a high level of cost effectiveness. An example of this may be seen in the studies now underway in Kern County, California (Estes, 1975), in which ERS data are clearly of sufficient quality to drive a regional water balance and salt balance system at almost monthly intervals, far more frequently than existing models can accommodate. Such a system is of potentially great value in management operations. Similarly, the use of surface cover, land-use, and natural plant community change data will provide hydrological model inputs three or four times a year, providing much more detailed information than current hydrological models can employ.

### F. Use of an Earth Resources Satellite System to Calibrate Other Components of a System

As meteorological satellites are further developed and the spatial resolution of these satellites improves gradually, it is likely that many systems will involve meteorological satellite data as an important and integral part of the system. However, future METSAT resolutions (of the order of 0.5 km) will be coarse enough so that a multistage sampling procedure and/or calibration procedure may be necessary. Careful inspection of National Oceanographic and Atmospheric Administration meteorological satellite images and LANDSAT data have already given indications that repeated observations with automatic data collection systems (DCS) and earth resources and meteorological satellites will gradually enable calibration functions to be developed on a region-by-region basis. These functions will relate to surface moisture, snow cover, snow depth, change in land use, and vegetation cover. The role of the ERS would be to provide a continuing check on the meteorological satellite, which would tend to drive the system because of its high frequency of coverage. This would be particularly the case in major snow and water balance studies of large-scale hydrological systems.

### G. Earth Resources Satellite Data May Stimulate Development of New Models with Greater Efficiencies Than Are Now Feasible

One of the major discoveries regarding LANDSAT 1 is that, despite a moderate spatial resolution which makes it unsuitable for some detailed studies, it nevertheless provides data of such spatial detail and specificity that existing models, as noted above, cannot accommodate the data. Among the areas where investigators are now improving or are likely to improve models to accommodate ERS data are the following:

- Larger hydrological models incorporating land-use and land-cover data.
- Models capable of accommodating data from 4 to 12 times a year, depending on cloud cover statistics.
- Models in agriculture related to spatial aspects of crop yield variability.

0-201-04245-2

- Agro-meteorological models in which the ERS performs a confirmatory, weighting, or calibration function.
- Models of land-use change based on more frequent updating and having variable error rates, depending on the planning functions involved.

Many other examples of the influence of an ERS system on model development and system design will emerge through the years. Those mentioned above are representative, *not* exhaustive; many others are now available in a literature already voluminous enough to require extended analysis. The best sources for evaluating these kinds of models are to be found in the reports of the March 1973, December 1973, and June 1975 LANDSAT symposia, and in the Earth Satellite Corporation/ Booz-Allen-Hamilton cost-benefit Study (1974).

### H. Development of a Unique, Internally Consistent Longitudinal Data Set

Even the census of the United States is plagued by internal inconsistencies through time via changes in boundaries, categories, classes, and frequency of observations. There are differences between states and private organizations in all their data. In addition, the census consists of aggregated data; other data sets are characteristically even more highly aggregated.

An ERS system will provide scientific data which are, among other characteristics, highly comparable through time, highly disaggregated, internally compatible, and capable of long-term-change analysis. No other data can provide such a uniform, unique, internally-consistent longitudinal set. The social and natural sciences will find many ways of using such data as new discoveries are made with ERS systems. The probabilities of refinement through hindcasting are open, along with re-evaluation of old surrogates and initial evaluation of new surrogates, and the repeated retesting of quantitative forecasting models.

It is possible that an ERS system could make even larger contributions to science. Many of the constraints which currently exist in the analysis of environmental problems relate directly to the lack of longitudinal data for analyzing and evaluating alternative arrangements and relationships among parameters which operate within ecosystems. Whether the focus is on physical systems such as river basins, or social systems such as metropolitan growth and development, technological innovation most probably will play a significant role in improving basic knowledge of the operation of such systems. Also, although quantitative, tangible benefits are difficult to argue (much like the difficulty in arguing the public benefits of basic research in the short-term), there is likely value in data acquisition systems which permit long-term interrogation concerning questions that may not now even be formulated. For example, within the 10-year frame the use of an ERS datum by the basic scientist exploring the knowledge frontier may show no benefit. On the other hand, a discovery in the twentieth or twenty-fifth year, or even in the following generations, based on the original data collection during the 10-year period, could well produce benefits spread over a large portion of the population. Development

0-201-04245-2

and interrogation of longitudinal data sets having a broad range of response capability would imply the possibility of significant long-term scientific value. The ERS data are raw data of such spatial detail and volume, as well as uniformity, that they are likely to be drawn upon in various time-related scientific studies.

### I. Summary of Satellite Impacts on Science

- Stimulating the development of new hypotheses and theories in the natural and social sciences.
- Providing opportunities for partitioning other aggregated natural and social data (e.g., census data).
- Improving multistage sample designs in research and applications development, the selection of sites for more detailed scientific study, the calibration of other components of an integrated resource management system, and the evaluation of the interaction of static and dynamic components of the environment.
- Providing a basis for extrapolation, interpolation, and refinement of scattered scientific observations, and for the making of unique observations feasible only with an ERS system.
- Stimulating the development of new science models which may be raised with ERS data to higher efficiencies than are now feasible.
- Opening, through storage of a unique, internally consistent longitudinal data set, unparalleled opportunities for developing quantitative hindcasting and forecasting models.

## V. AN EXAMPLE OF LANDSAT VALUE TO SCIENCE: GEOLOGIC APPLICATIONS

LANDSAT images provide a view of the geologic fabric of continents which is compatible with the scales involved in modern theories of global tectonics or plate tectonics. Entire mountain ranges marking paleo plate boundaries or paleo sutures are spanned by a single frame and completely encompassed by a few tens of frames. Entire continents are covered by a few hundred frames. In short, LANDSAT provides geologists with a completely new perspective from which to view the earth. For the first time, geologists have a tool that affords them the opportunity of moving from perception of the whole to understanding of the part.

In the few years since space-acquired imagery has become widely available, it has been applied to a wide spectrum of geologic problems in virtually every environment on earth. Geologists have used data from LANDSAT to explore for resources in South Africa (Viljoen, 1973), Pakistan (Schmidt, 1975), and Arizona (Goetz et al., 1975; Erskine et al., 1973); to map sand seas (McKee and Breed, 1973); to detect coastal erosion (Mairs et al., 1973); to monitor volcanic activity (Stoiber and Rose, 1975); and to better understand regional geology in Iran (Ebtehadj et al., 1973) and central Asia (Molner and Tapponnier, 1975). Despite the diversity of

0-201-04245-2

topics and the geographic scattering of projects, most of the applications involve the furthering of basic geologic knowledge, geologic resource exploration, and assessment of geological hazards. Actually these applications are not isolated functions but are part of a continuum of perceiving and understanding the geologic environment. Mapping is the starting point of almost all geologic investigations, and hypotheses provide the underpinnings of the models used in resource exploration and assessment of hazards.

The literature on the field grows daily. Some of the most important results may never be published, however, because they are closely held by exploration companies.

Most of the work to date has been directed toward evaluating space imagery as a tool for dealing with one or another geologic problem. However, articles are beginning to appear that present space-acquired data as a source of geologic information that can complement, augment, or extend data derived from conventional sources and techniques. This integration marks a welcome advance in the acceptance of this valuable source of information.

## A. Unique Advantages and Characteristics

The synoptic character of space-acquired imagery is probably the most valuable aspect of these new data. For the first time geologists can interrelate the features across an entire fold belt without having to deal with a myriad of photographs, each with a slightly different scale, light angle, exposure density, and print quality. Fault and fracture zones which are discontinuously exposed appear as major lineaments, changes in fold orientation that may signal a major basement fault are obvious, and the large-scale relationships of a sedimentary section onlapping a Precambrian shield that might go unnoticed with standard mapping practices stand out clearly on the small-scale imagery. Other unique characteristics of space-acquired images particularly useful in geology are spatially registered multispectral imagery, repetitive imaging of the same area, and the digital format of several image types. The multispectral and repetitive nature of the space coverage allows spectral characteristics and time to be used as discriminants in geologic analysis. The digital format means that the imagery can be processed routinely in computers which normally would be prohibitively time consuming and expensive.

Taken together, these characteristics allow geologists for the first time to view the same region at different seasons under a variety of lighting and soil moisture conditions and to compare regions with similar rock types and structural styles that may lie half a world apart. These comparisons promote insights that previously were unattainable. Moreover, these space data are acquired without the restrictions imposed by national boundaries or security restrictions and are freely available at minimum cost to geologists the world over. For the first time, geologists everywhere have easy access to the same data—an innovation that undoubtedly will promote better communication among scientists.

0-201-04245-2

## B. Interpretation Techniques

Most geologic applications of LANDSAT data reported through 1975 have relied more or less on standard photo-interpretation techniques. These include viewing imagery unaided or assisted by a variety of magnifying devices and stereoscopes. Stereovision, as with normal aerial photography, is particularly valuable for looking at these small-scale pictures. Most workers have reported unique advantages in using transparencies, greatly enlarged paper prints, or color images. This type of interpretation relies largely on the standard photo-interpretation discriminants such as tone, texture, shape, size, and context. Of course with the large-area, small-scale imagery acquired from orbital altitudes context assumes a much larger role than in normal photo-interpretation.

In addition to standard techniques, some investigators have used a wide variety of optical and digital enhancements of the imagery to improve the recognition of various geologic features (Rowan et al., 1975; Goetz et al., 1975; Vincent, 1973; and many others). Among the most widely used techniques is the simultaneous viewing of several bands of multispectral imagery on a color additive viewer. This is a simple, rapid, and inexpensive technique, and in addition to providing insights into the geology of a region it can lead to the development of photographic techniques for improved image products.

Undoubtedly, the most promising approach to interpretation involves the digital manipulation of the data. Rowan et al. (1975) and Goetz et al. (1975) have developed a series of band-ratioing techniques that can bring out alteration zones as well as a variety of lithologic contacts. There is a suggestion that dealing on a pixel (picture element)-by-pixel basis may make it possible to detect the vegetation anomalies associated with certain types of metal deposits (Lyon and Honey, 1975). Vincent (1973) has developed ratioing techniques that discriminate rock bodies on the basis of iron content. Hord (in press) describes a variety of digital manipulations that can greatly improve the quality of the spatial information contained in the images. Despite this progress, however, it appears that automatic geologic mapping is some distance away and that the great advantage of these techniques is that they substantially improve what an interpreter can extract directly from the data. This is particularly true when the products of digital manipulation are further enhanced by optically or photographically combining them into color-coded images (Rowan, 1975).

## C. Results

Satellite-acquired remote sensing data are extremely useful for geologic mapping and exploration, because they offer a very inexpensive means of delineating many types of geologic features over large areas. This is particularly important in inaccessible areas. The data permit recognition of regional features and relationships unidentifiable by other techniques. The images provide an excellent means of

0-201-04245-2

extrapolating and extending detailed ground information over large areas and of locating areas that hold exploration interest or over which additional data are needed. This ability to perceive regional relationships and identify areas requiring additional examination permits a focusing of attention and resources on successively smaller areas, a concentration which can lead to substantial savings in time and money in geologic mapping and exploration programs (Levin and Everett, 1972; Erskine et al., 1973; Short, 1974; Rowan, 1975).

The results of using these data are as impressive as the applications are varied, and the reports of results are already counted in the thousands of pages. Consistently with the space-acquired imagery being a relatively new tool, the bulk of the results reported are evaluations of the data for mapping and contributions to basic geologic knowledge. However, there are some significant results in the areas of resource exploration and evaluation of geologic hazards.

### *1. Geologic Mapping and Contributions to Basic Geologic Knowledge*

Perhaps the widest application for satellite imagery in the field of geology is as an aid in the preparation of small-scale geologic maps and specialized thematic interpretations such as photolinear maps (Isachsen et al., 1973; Kronberg et al., 1974; Wobber and Martin, 1973; Wier et al., 1973; Drahovzol et al., 1973; and others) and geomorphic maps (Salas et al., 1973; Hallberg, 1973). The mainstay of almost any geologic exploration research program is the geologic map—the medium that permits the geologist to examine the details of spatial relationships between rock types, rock units, and structural features.

Most geologists have found that a surprisingly wide range of geologic discriminations are possible (Houston et al., 1973; Goetz et al., 1975; Collins et al., 1974; Saunders et al., 1973; Viljoen, 1973; Rowan et al., 1975) using LANDSAT and SKYLAB data. The delineation of units of similarly appearing rock types, the mapping of contacts between highly contrasting lithologies, and, in some cases, tentative lithologic identifications have been based exclusively on the imagery. When coupled with the interpretation of aircraft-acquired data and ground work, space imagery is particularly effective and can greatly reduce the time required to map or assess an area, while simultaneously maintaining accuracy. In addition, interpretation of space-acquired imagery has provided the exploration or research manager with a powerful cost-saving tool. He locates and eliminates areas which, on the basis of the imagery, appear to have unfavorable geologic settings. He thus limits expensive techniques, such as detailed geologic mapping and sampling and geophysical surveys, to the areas of greatest promise.

In many regions relatively accurate geologic maps can be prepared with a minimum of ground data. This is especially true in areas of sparse vegetation (Houston et al., 1973; Viljoen, 1973; Abdel-Hady, 1973; Goetz et al., 1973a, b). Most of the units portrayed are "remote sensing formations," but with a modicum of ground information these can be related to conventional map units. In some instances, investigators have subdivided conventional mapping units (Grootenboer et al., 1973; Collins et al., 1973; Goetz et al., 1973a, b).

0-201-04245-2

Almost all geologists that have used space-acquired imagery have noted many more lineaments (inferred by most to be related to fractures) than were previously suspected (Houston et al., 1973; Lathram, 1973; Isachsen et al., 1973; Collins et al., 1974; Mohr, 1973; Goetz et al., 1973a, b; Rowan et al., 1975; Gedney and Van Wormer, 1973; Ebtehadj et al., 1973; Everett and Petzel, 1973; Short, 1973; Drahovzal et al., 1973; Wier et al., 1973; Pickering and Jones, 1973; Abdel-Hady, 1973; Hodgson, 1975). Upon comparison to existing data and field checking, many of these lineaments are found to represent fractures. These discoveries have caused many workers to re-examine basic hypotheses about the origin and history of continents (Hodgson, 1975) and subcontinental features (Collins et al., 1974). This re-evaluation and extension to understanding have occurred even in such reasonably well-studied areas as the Alps (Boriani et al., 1974) and the Grand Canyon region (Goetz et al., 1973a, b). The features revealed in space-acquired imagery have contributed significant new information regarding less well mapped areas such as the Afar Triangle (Mohr, 1973; Kronberg et al., 1974), the Himalayas and China (Molnar and Tapponnier, 1975), and Africa (Mohr, 1973). All of the works cited here have presented new understandings of the interrelationships of lineaments and other major structural features such as fold belts, major strike-slip fault zones, and intrusive centers based on the study of the space acquired data.

In addition to yielding an abundance of structural and lithologic information, space imagery has been used for a variety of other mapping and research purposes. McKee and Breed (1973) developed a descriptive and genetic classification of sand seas on a global basis, using LANDSAT imagery. Morrison and Cooley (1973) studied recent erosion in Arizona that resulted from intensive stock grazing in the area. Williams et al. (1973), Friedman (1973), and Cassinis et al. (1974) studied volcanic features for several purposes, including mapping lava flows of different ages. Gedney and Van Wormer (1973) and Morrison and Hallberg (1973) examined glacial features. Everett and Russell (1975) recommended that examination of LANDSAT or SKYLAB imagery be a first step in investigating any prospective site for highway tunnels.

### 2. *Exploration for Geologic Resources*

When a model of resource occurrence based on theory and previous work is combined with the large-scale relationships visible in orbital imagery, data from satellite platforms become a powerful exploration tool. However, opinions on this subject vary; the reactions among several dozen exploration geologists involved in assessing the value of space-acquired data in geologic exploration with whom we have talked have ranged from outright rejection to substantial enthusiasm. On the positive side, a vice president of exploration for a major minerals company considers the use of space-acquired imagery as the most important new exploration tool since the introduction of airborne magnetometry after World War II.

The unique ability of satellite systems to acquire imagery that portrays large portions of the earth's surface enables geologists to characterize large, widely separated regions (such as shield areas or extensive sedimentary sequences) and to

0-201-04245-2

compare known mineralized areas with as yet unexplored areas as an important first step in the search for mineral deposits (Wilson et al., 1973; Sendwe and Wolyce, 1973). Moreover, once having located an area of general interest (a proper geologic environment), investigators have mapped the tectonic style of the area and delineated the subareas which generally favor economic accumulations of minerals—for example, sedimentary basins, greenstone belts, and fold mountain belts (Saunders et al., 1973; Everett and Petzel, 1973; White, 1974; Nicolais, 1973; Rich, 1973; Viljoen, 1973; Schmidt, 1973; Bechtold et al., 1973; and many others). In several instances, the LANDSAT imagery enabled the geologists to proceed a step further and to identify and map structural features and lithologic associations within the subareas which might serve as keys to economic accumulations of minerals, petroleum, ground water, or geothermal resources (e.g., salt domes, anticlines, intrusions, ultramafic rocks, and fracture intersections). At this level of detail, the geologist is able to correlate known economic deposits with lithologic and structural features perceived in the imagery and to draw additional inferences about factors controlling and localizing economic deposits, thus refining his initial exploration model.

There is some evidence that space imagery can lead to the direct detection of certain geologic resources by virtue of recognizing such features as gossans (Rowan et al., 1975), peculiar tonal-textural features that are strongly correlated with oil and gas production, at least in the Anadarko basin (Collins et al., 1974), depressions under ice caps that mark possible geothermal heat sources (Williams et al., 1973), or terraces along streams known to carry gold (Abdel-Gawad and Silverstein, 1973). However, most often the contribution of space-acquired data is based on inferences drawn from experience or a particular exploration model. In these circumstances, the strength of space data as an exploration tool is greatly increased by the inclusion of ground data and combination with data from other exploration devices such as airborne magnetrometry. Several exploration groups are now doing just such studies. The space imagery is used to locate the proper structural setting, and then aeromagnetic and gravity data serve to refine the details of the model. Targets thus identified are followed up on the ground, using geochemistry, resistivity, conventional mapping techniques, and, finally, drilling.

Schmidt (1973) used LANDSAT imagery to identify potential copper porphyry targets in the Chagi district of Pakistan. Even though he could not detect the strong color anomalies imparted by natrojarosite, field checking of the targets he had chosen revealed that several were mineralized.

In Nevada, Rowan and Wetlaufer (1975), using LANDSAT imagery, have found that mineralization is associated with several major lineations. Once such features are recognized in similar settings, they become potential targets to be investigated by other exploration tools.

Several investigators have used LANDSAT and SKYLAB imagery to tentatively locate structural features of interest to petroleum exploration. In Alaska, near the Umiat oil field, Lathram et al. (1973) spotted a peculiar arrangement of lakes that

0-201-04245-2

geophysical data corroborated as an attractive structural feature for petroleum exploration. Saunders et al. (1973) related lineaments seen in LANDSAT imagery to oil and gas occurrences in west Texas. On the basis of the relationships they saw in the imagery, they have identified several prospective areas.

In the Anadarko basin, Collins and his fellow workers (1974) found that many of the known oil- and gas-producing structural features could be identified on LANDSAT and SKYLAB imagery. In addition, they noted that as yet unexplained tonal-textural features, which they called hazy anomalies, are strongly correlated with oil and gas production. Upon examining the pervasive pattern of lineaments seen in the space imagery and correlating these with geophysical and well data, they also concluded that faulting has played a much larger role in the origin and development of the basin than was previously suspected. From a study of LANDSAT imagery in the Basin and Range provinces of the western United States, Bechtold et al. (1973) and Erskine et al. (1973) noted a complex interrelationship between faulting, intrusion, and mineralization. Interrelationships noted can be used to guide further exploration even in this relatively well explored area. Erskine and his colleagues also noted that there seems to be a regular spacing of lineaments (fractures) that control mineralization and that the highest potential for mineralization is at the intersection of two such regularly spaced sets.

Some of the most promising work being done in the field of geologic exploration involves digital processing of space-acquired imagery in order to improve spatial and spectral information. Rowan et al. (1974) and Rowan and Wetlaufer (1975) have used spectral ratioing of LANDSAT multispectral scanner data to detect hydrothermal alteration and relative iron concentrations. These techniques hold great potential for exploration in arid and semiarid regions. Vincent (1973) has used a similar technique to differentiate rock units on the basis of their ferric iron content. Some aspects of Lyon's (1975) work suggest that it may be possible to detect vegetation anomalies associated with copper porphyry deposits by examining spectral ratios on a pixel-by-pixel basis.

Several investigators have suggested that it may be possible to recognize deposits of industrial minerals such as sand, gravel, and clay on the basis of tone and geomorphology (Pickering and Jones, 1973; Abdel-Gawad and Silverstein, 1973; Morrison and Hallberg, 1973). Shortages of these resources in the vicinity of large cities emphasize the importance of this capability.

As populations grow, the requirement for groundwater increases. Goetz et al. (1973a, b) reports successfully using LANDSAT imagery to guide groundwater exploration. Several other investigations have also noted the value of space-acquired imagery for this purpose (Pickering and Jones, 1973; Gould et al., 1973; Drahovzal et al., 1973). Several organizations, including the Earth Satellite Corporation, are using SKYLAB and LANDSAT imagery to guide regional groundwater exploration programs. The imagery reveals many features, such as fracture zones, alluvial fans, and closed basins, that may be groundwater reservoirs. Once identified, these features are investigated further, using conventional means such as resistivity surveys.

0-201-04245-2

The savings in time and money realized from such applications is already substantial and will increase as these tools are used more widely and become increasingly integrated into exploration programs.

### 4. *Evaluation of Geological Hazards*

The study of space-acquired data has already contributed significantly to understanding and estimating certain types of geologic hazards such as volcanic activity, seismic hazard, collapse, and sand encroachment, and will undoubtedly contribute a great deal more understanding and information in the future.

Abdel-Gawad and Silverstein (1973) and Gedney and Van Wormer (1973) correlated lineaments seen on LANDSAT imagery with recent seismic events. Such correlation, combined with geomorphic evidence, helps to identify seismically active faults and permits one to develop an estimate of seismic risk for certain areas. These types of correlations will also help in guiding the deployment of earthquake-predicting devices to ensure that seismically active features near population centers are adequately monitored. Ward et al. (1973a, b) used ground data collection platforms and the data relay capability of LANDSAT to develop a system of volcano monitoring that is cheap and efficient enough to be employed worldwide. Such a system could greatly reduce the suffering and loss that all too often accompany volcanic eruptions in inhabited areas. Cassinis et al. (1974) used LANDSAT to study and monitor several Italian volcanos, adding substantially to the knowledge of these features. Stoiber and Rose (1975) set out to evaluate the ability of the thermal scanner on SKYLAB to detect heat changes in Central American volcanos that might signal the onset of eruptions. However, cloud cover and lack of nighttime coverage frustrated this promising project. Undoubtedly, as thermal scanners are incorporated on future ERS's, similar experiments will prove to be extremely effective.

McKee and Breed (1973) found that they could map and classify sand seas using LANDSAT imagery. They also gained some insights into the origin and development of these features that may suggest ways to limit their growth and prevent their encroachment onto inhabited areas. By observing the seasonal rise and fall of water levels in Iranian playas during repeated LANDSAT coverage, Krinsley (1973) was able to suggest ways of taking advantage of this transient water resource, as well as to propose a road route across the Great Kavir that would avoid the hazards of seasonally wet salt flats.

Pickering and Jones (1973) were able to detect areas of limestone collapse in Georgia which could pose hazards to building in the area. Wier et al. (1973) used fracture intersections to predict areas of roof-fall hazards in underground coal mines. This information could allow mine managers to plan mines so that accidents resulting from roof falls could be held to a minimum.

Mairs et al. (1973) used LANDSAT imagery to monitor beach erosion in New Jersey, and Morrison and Cooley (1973) employed similar imagery to evaluate surface erosion in Arizona. The understanding of such dynamic processes derived from

0-201-04245-2

these and similar studies can help man to bring his activities into better balance with nature and to anticipate the effect of natural processes on his activities.

This extended analysis and review of geologic applications of LANDSAT (and SKYLAB) imagery and digital tapes has been given to reinforce an understanding of the actual and potential value of space remote sensing. Similar analyses could be made in a dozen or more areas of application. Many of the principles employed in this analysis are now beginning to be formalized as part of the basic constructs of the field of remote sensing. A number of these constructs may now be examined in more detail.

## VI. BASIC CONCEPTS AND PRINCIPLES OF REMOTE SENSING

Remote sensing is so young that it is still difficult to identify basic primitives or principles representing fundamental, primary, or general truths. Most of the so-called principles are, in reality, concepts–thoughts, ideas, or generalizations which have not yet been subjected to close analysis. In the preceding pages we introduced some of these concepts. In this section, four are considered in more detail because of their importance and their possible evolution into principles. In the final section, some of their bases are examined critically.

The four concepts to be examined are as follows:

- Resolution germane to the task.
- Discrimination germane to the task.
- Multiple strategies germane to the task.
- Roles germane to management tasks.

### A. Resolution Germane to the Task

In acquiring data for any kind of analysis, the level of aggregation or resolution strongly influences the quality of the conclusions which may be drawn. This concept, so familiar to the user of census statistics, is equally important in remote sensing. In the former, resolution (fineness of subdivision or level of detail) is involved with respect to space (geographical area), categories (classification, units of subdivision), intensity (quantity values such as numbers or volumes), and time (frequency of observation).

These four dimensions have their parallel in the four interdependent dimensions of resolution in remote sensing: spatial, spectral, radiometric, and temporal. Since these are interdependent, tradeoffs must occur when one attempts to design or use a remote sensing system for a particular application. With fixed resources available to address a problem, a change in one of these four dimensions requires a concomitant alteration in one or more of the others.

The relations between these different resolutions may be expressed in terms of the resources necessary to achieve a given resolution in any one as follows:

$$R = R_s + R_\lambda + R_r + R_t,$$

0-201-04245-2

where $R$ = resources available,

$R_s$ = resources devoted to improving spatial resolution,

$R_\lambda$ = resources devoted to improving spectral resolution,

$R_r$ = resources devoted to improving radiometric resolution,

$R_t$ = resources devoted to frequency of observation.

*Spatial resolution,* as used here, embodies two notions. The first is the technical resolving power of particular instruments. The second is the resolution, as defined above, needed for effective detection and analysis of the features observed with the instruments. The latter concept is clearly the broader one.

In the first usage stated, resolution is the minimum distance between two objects at which the sensor can record them as two distinct entities. Different conventions are employed for photographic, scanning, and radar systems. For the broader usage, classes of natural and cultural features exist on macro-, meso-, and microscales for which progressively finer spatial resolution is required for identification and analysis. Examples of these are given in Figure 3.1, which shows the classes of natural and cultural features for which the resolutions obtainable with LANDSAT, SKYLAB, and aircraft systems are appropriate.

*Spectral resolution* also includes a narrow, instrument-oriented usage and a broader application-oriented usage. It is for the latter that the notion of a spectral resolution "germane to the task" is relevant. In instrumental terms, spectral resolution encompasses both the width of the electromagnetic spectrum that is sensed and the number of channels used. It is dependent on the band pass or width of spectrum admitted to the filter and detector system and the amount (number of photons) of energy present in that particular portion of the spectrum. Instruments designed with narrow bands have the characteristics of high spectral resolution (i.e., accurate color measurements). The tradeoff is, however, between radiometric resolution and spatial resolution. In the data produced by narrow-band instruments it is difficult to separate the signal from the noise in the system (poor radiometric resolution) unless in the imaging process relatively long dwell times (lengths of time particular areas are examined by a sensor) are available. The usefulness of very-narrow-band instruments may therefore be limited. Conversely, broad-band remote sensing instruments produce relatively lower spectral resolutions but have good spatial and radiometric resolutions.

In the broader usage, it is emphasized that there are particular spectral regions which are more useful for certain applications than others. Thus crop growth and development is probably better monitored using very narrow bands in the visible and infrared regions, whereas monitoring sea states is probably best accomplished with multichannel active and passive microwave systems (Figure 3.2).

*Radiometric resolution* can be defined simply as the sensitivity of a sensor to differences in signal strength. The signal may appear in the form of reflected light in the visible portion of the spectrum in photographic systems, or of emitted energy in

0-201-04245-2

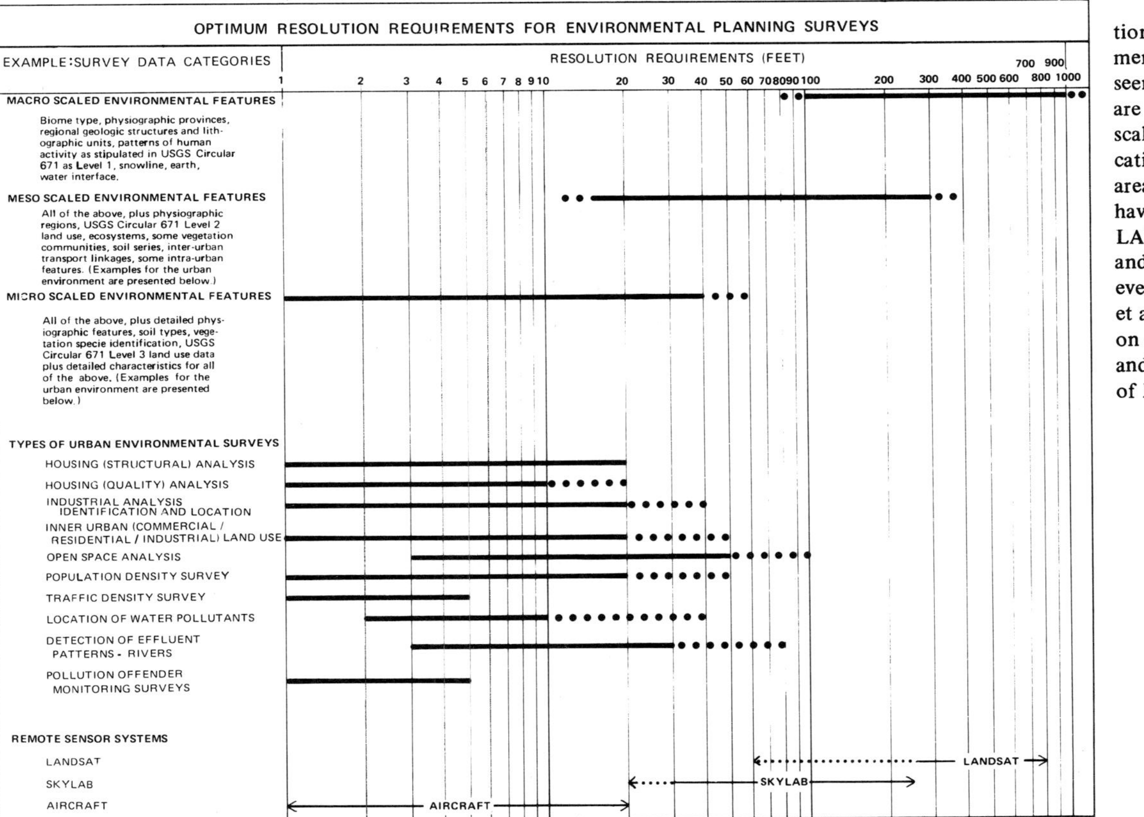


**Fig. 3.1.** Optimum resolution requirements for environmental planning surveys. It is seen that LANDSAT resolutions are suitable mainly for macroscale and some mesoscale applications, principally in nonurban areas. SPACE SHUTTLE will have resolutions (as did SKYLAB) suitable for studying meso- and many microscale features, even in urban areas (see Bale et al., 1975). Based principally on data from Wobber (1970) and Simonett (1969). Courtesy of Earth Satellite Corporation.

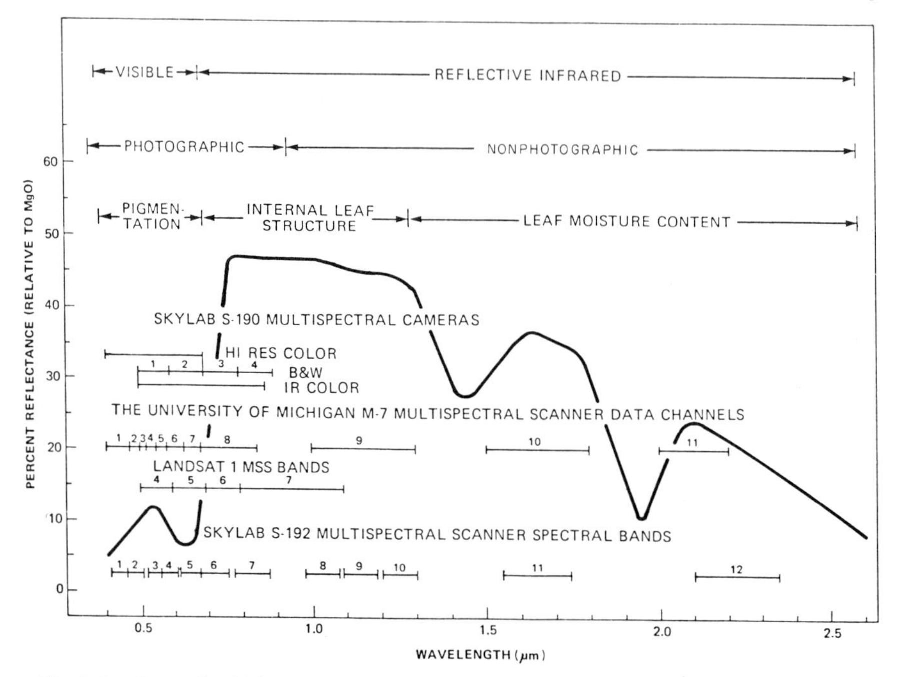


**Fig. 3.2.** Generalized laboratory reflectance curve of green vegetation, superimposed on a diagram showing the spectral coverage of spacecraft, satellite, and aircraft multispectral systems. The diagram covers the spectral range 0.4–2.6 μm in the physical and reflective infrared regions. After Rohde and Simonett (1975).

the thermal infrared and passive microwave portions of the spectrum. In keeping with our tradeoff equation, as radiometric resolution improves, either spatial or spectral resolution or both must suffer.

Since radiometric resolution or sensitivity in effect defines the number of discriminable signal levels, it is deeply involved in the identification of scene objects—the finer the radiometric resolution, the greater is the opportunity to discriminate between objects, commensurate with a given spatial resolution.

Depending on what is to be sensed, instruments are selected with particular types of resolutions. To study the conditions of the roofs of houses, high spatial resolution is necessary. Thus one could use a camera with panchromatic film (poor spectral resolution). However, at any reasonable flying height it is not possible to distinguish cereal crops on the basis of morphology alone (high spatial resolution). In the latter case, one could use a multispectral scanner (good spectral and intensity resolution but relatively poor spatial resolution) to rapidly classify crops over a wide area.

In addition to these three dimensions of resolution, there is a fourth: *temporal resolution.* There are very few aspects of nature that do not change in relation to others and to themselves throughout the course of the year. Very few lack an opti-

0-201-04245-2

mum time of day or year in which to be observed in relation to other aspects of nature. There are also very few man-made or cultural features which do not have an ideal viewing time or viewing duration. Thus, in sensing a substantial number of dynamic events—crop identification and growth, rangeland development, hydrologic processes, drought evaluation processes, earthquake damage, cultural change, marine processes, to name only the most obvious—*time* is often used as a key discriminant.

The use of time for discrimination may permit one to alter the specifications in the other three dimensions of resolution in such a fashion as to permit more economical remote sensing procedures. With a reasonably accurate local crop calendar which details traditional times of planting, growth, and harvest, remote sensing observations can be planned so that a limited number of observations at unique and critical time periods may provide the discriminating information needed to separate one crop type from others. The net effect of using the time dimension and local *a priori* data is that crops may be identified over larger areas with sensors processing spatial and spectral resolutions too poor to identify the same crops on the basis of spectral and morphological characteristics. The introduction and the use of the time dimension in a resource program may have the net effect of sharply reducing the cost of data acquisition and processing.

*Passive* remote sensing systems, operating in the visible or infrared portions of the spectrum, are dependent on the sun for their energy. The power spectrum of sunlight is shown in Figure 2.1. This figure shows that the thermal infrared (heat portion of the spectrum) requires a much larger band width to allow the same amount of energy to reach a detector than the red portion of the visible spectrum. This demonstrates the principle that, for a given spectral and intensity resolution, greater spatial resolution is possible in the visible than in any other portion of the electromagnetic spectrum.

## B. Discrimination Germane to the Task

Three levels of image discrimination are important in resource assessment evaluation: (*a*) detection, (*b*) identification, and (*c*) analysis.

*Detection* means simply that an examination of the data reveals that something is present; for example, in an image of a river the presence of an object which is not water is noted. *Identification* means that enough information is present in the remote sensing data to identify the object or feature perceived; for example, there is a boat in the river. At the *analysis* level of resolution, it is possible to perceive information beyond mere identification of the object; for example, the boat in the river is a dug-out canoe carrying three people.

Detection is adequate for simple dichotomies; for example, something is or is not present. It is used in the first stage of a multistage sampling strategy to define the extent of the universe. Identification is usually sufficient for regional studies (e.g., this is a thorn tree savanna) and for some probabilistic measures based on more detailed work. Image perception at the analysis level is necessary for highly

0-201-04245-2

detailed and operational work (e.g., there are 12 acacia trees per hectare) and forms the third level in a multistage sampling strategy.

In practice about three times the spatial resolution is required to progress from detection to identification. A 1 to 2 order of magnitude (10X to 100X) increase in spatial resolution is necessary to pass from identification to analysis. These are average ratios, and the exact spatial resolution increase will be dependent primarily on the character of the target and the amount of information (type of analysis) desired. As a general working rule there is a scale of acquisition of remote sensing data that is better for each one of the three levels of discrimination and perception; no single scale can serve all three purposes. The scales change with the topic under study.

As a general rule, central planners require information at the first two levels of detection and identification, and implementation activities are conducted at the second and third levels. In almost every remote sensing program there is a tradeoff function between the levels of resolution since the image perception or information content is a resolution-dependent phenomenon. Very high resolution and very low resolution carry their own sets of problems. The higher the resolution, the more difficult it is to generalize on the basis of point and small-area observations. The fact that data of low resolution combine *on average* a great deal of information into each resolution cell or picture element both constitutes a loss of information and implies an averaging response that can produce spurious results. However, small-scale, lower-resolution imagery provides regional perspective and a basis for generalization and allows large areas to be interpreted, analyzed, and understood quickly. Small-scale data provide a framework for identifying needs for specific point or small-area information.

In a more complex situation, it is possible at the analysis level of resolution to perceive information beyond the theoretical limits of resolution of a particular remote sensing system. As an example, items from LANDSAT imagery as small as *8 m in one dimension* have been detected in road systems where the contrast between the object and the background was favorable. In the case of linear features, the human eye and mind help to piece together the concept of "road," which for any single information cell along the system may be ambiguous. The key point is that there tend to be spatial, spectral, and radiometric resolution values appropriate to the particular purposes to be served by the remote sensing survey. To choose the most satisfactory compromise scales, the whole range of uses for which the remote sensing images will be employed must be known at the earliest time possible in planning the survey.

Also contained within these notions of discrimination are the probabilities of correct detection, identification, and analysis. These probabilities will vary from area to area and application to application. For certain applications, a probability of correct identification of 70% is acceptable. For others, accuracies greater than 95% may be needed either because of the nature of the application or because of the accuracy of competing (non-remote-sensing) systems. One of the pervasive difficulties with remote sensing lies in persuading users to accept accuracies which at

0-201-04245-2

first appear to be less than they normally obtain from other systems. Frequently very close analysis is required to show that the competing systems in fact have different and possibly lower accuracies than are initially perceived.

## C. Multiple Strategies Associated with Remote Sensing

Within the past 10 years, a number of new techniques for the design and application of remote sensing systems have evolved. These include the following:

1. Multiple uses of imagery.
2. Multispectral discrimination.
3. Multidate discrimination.
4. Multisensor combinations.
5. Multistage approaches.

When used either separately or combined in an optimum fashion, these techniques can make remote sensing more cost effective as an information gathering technology.

### *1. Multiple Uses of Remote Sensing Imagery*

Multiple uses of remote sensing imagery can improve the cost-effectiveness of remote sensing techniques, particularly now that high-altitude aircraft and spacecraft are capable of producing vast quantities of data for substantial areas of the world. These data are applicable to many disciplines and should be used as frequently as possible as base information source data. They are particularly useful for general reconnaissance purposes and for monitoring events and phenomena over large areas. Low-altitude imagery tends to have fewer multiple-use applications because the specificity of the sensor system increases for more detailed survey work.

An example of multiple use of remote sensing imagery is Brazil's Project Radam (Moura, 1971). In this survey, the government of Brazil acquired side-looking airborne *radar imagery* of 4.5 million km$^2$ of the northern part of the country. The output of the survey included maps of geology, hydrology, vegetation cover, soils, and land-use potential. Interpretations were supported by a limited amount of aerial photography and helicopter-supported fieldwork at selected points within the area. Because of the variety of interpretations for which this imagery was used, the cost of acquisition for any one discipline was very small. By virtue of the multidisciplinary nature of the project, many benefits were derived from the interchange of information among the different scientists working on the project. For instance, it was found that in areas underlain by Precambrian rocks the distribution of vegetation types was a very good indicator of the underlying geology.

### *2. Multispectral Discrimination*

Many applications of remote sensing techniques are based on the concept of multispectral discrimination. Theoretically, every substance reflects or emits a unique

0-201-04245-2

spectrum of electromagnetic energy; with observations in enough narrow spectral bands, most substances can be uniquely identified. Such techniques are effective in the laboratory or in astronomy for identifying mixed gases, but are only moderately successful when working at the earth-atmosphere environmental interface. However, in the real world, almost every sensor records or "sees" a variety of substances present within each resolution element. It is technically difficult to utilize sensing devices with band widths narrow enough to allow accurate and unique sensing of spectra.

However, even with the broad-band approach, it is still possible to divide the spectrum into meaningful bands. The spectral interval is too broad to be descriptive of a particular material but is sufficiently narrow for that material to be predominant in the band in question. In the real world, however, there is too much variability within materials, let alone between materials.

What is required, then, is to *collapse variance.* This can be done in several ways, such as by choosing a particular time of day or year, or by developing constant environmental associations or surrogates (e.g., mapping plant communities which indicate underlying geology, instead of attempting to map geology directly). This is accomplished without relinquishing the search for specific spectral features associated with specific materials. It is encouraging to note that with only four bands several LANDSAT investigators have reported considerable success in identifying spectral signatures for geologic substances (Vincent, 1973; Goetz et al., 1973a, b), and in applying multispectral classification schemes to the identification of crop and land-use types (Draeger et al., 1973; Morain and Williams, 1973).

*3. Multidate Discrimination*

As noted earlier, the time-varying properties of a scene provide a valuable basis for discrimination. In the past, it has rarely proved feasible to use time in this fashion with aircraft, because each new time requires reflying with significant additional costs. Also, the fact that exact repeat coverage of a flight path is not feasible with aircraft compounds the difficulty of analysis. Only the most expensive installations, such as major public works, or the most profitable exploration targets in mining provide sufficient apparent justification for repeated aircraft missions. Studies by Steiner (1970), Brunnschweiler (1967), Colwell (1970), Simonett et al. (1970), and Schwarz and Caspall (1968) showed the possibilities inherent in this technique. However, it was not until the advent of the LANDSAT 1 ERS that time variability could be extensively employed in many situations.

As noted in Chapter 9, the notable advantages of LANDSAT 1 in this regard are its near-orthographic format, its essentially uniform sun angle and azimuth on successive passes (slowly changing throughout the year to add an additional discriminant), and its 18-day repeat cycle in almost precisely the same track. Multidate sensing will become increasingly important as space sensing matures. Indications of possible roles for space sensing may be obtained by examining the coverage cycle needed by federal and other agencies, and capable of being provided by different

0-201-04245-2

spacecraft. Examples of these temporal resolutions are shown in Figure 3.3, taken from a study by General Electric (1974).

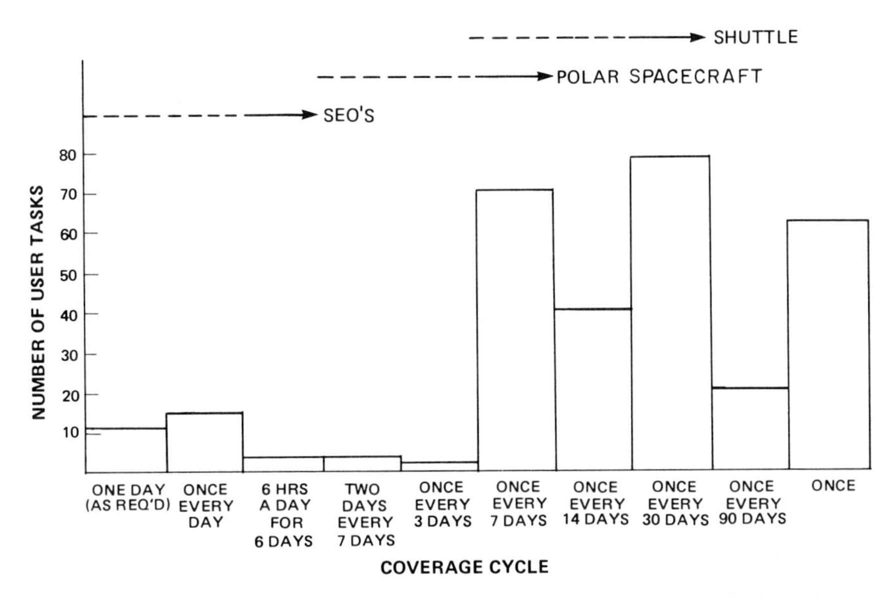


**Fig. 3.3.** Coverage cycle of federal agency tasks, arrayed by number of tasks for each time interval. Shown in relation to coverage potential of SPACE SHUTTLE, POLAR SPACECRAFT, and SEO's in the 1980's. After General Electric (1974).

## *4. Multisensor Combinations*

It may be advantageous to acquire information by using a number of sensors simultaneously. Such an approach permits maximum correlation between different types of remotely sensed data and minimizes system costs attributable to the platform. This provides for mutual support among the sensors. Data from nonimaging sensors can be compared to imagery from the visible portion of the electromagnetic spectrum and to information acquired beyond the visible wavelengths. Such an approach can also reduce the amount of ground support needed, and if the various sensors can be mounted on the same aircraft or spacecraft, a great cost savings in data collection can be achieved. Multiple instruments are now carried on meteorological satellites and will be installed in future earth resources satellites and space shuttles.

## *5. Multistage Approaches*

The combined use of spacecraft, aircraft, and ground-derived information may be needed to provide cost-effective data gathering for resource management programs.

0-201-04245-2

One of the simplest approaches, called "targeting," is simply to locate areas in which change has occurred or which are otherwise of particular significance. Smaller-scale imagery with gross resolution may be used to accomplish these tasks in a cost-effective manner. Once change is detected, planning for more detailed surveys can proceed. The more expensive large-scale-image acquisition missions therefore are concentrated only in the change areas. Bale and Bowden (1972) demonstrated how this detection capability can be utilized to pinpoint small areas (minimum 10–12 acres) of urban change in California desert communities. Subsequent detailed identification and analysis was accomplished by ground survey techniques.

*Multistage Surveys.* These types of surveys use multistage techniques to map the spatial extent and distribution of phenomena that are detectable on small-scale and identifiable on larger-scale images. At the heart of such techniques is a process of extrapolation from one defined "landscape unit" ("photomorphic region" of MacPhail, 1969) to similar or homogeneous units. This type of technique is highly dependent on the availability of capable interpreters and resource scientists to carry out the extrapolation.

The general category of multistage surveys subsumes a large variety of surveys in which multiple image acquisition missions may involve different platforms at different altitudes. The prime characteristic of these surveys, which serves to combine them into a single class, is that base imagery for the entire area is first acquired at the smallest available scale (satellite coverage). Larger-scale imagery is then obtained from predetermined homogeneous landscape units and used to provide calibrated or detailed sample information. Landscape units are selected on the basis of prior information of after an initial period of study of the area. Well-conceived multistage surveys provide a maximum amount of flexibility. Acquisition techniques for the large-scale imagery used for calibration can range from relatively high-cost, controlled aerial photography ($9.00 or more/km$^2$ in the United States) to uncontrolled 35 mm obliques ($0.40/km$^2$) imaged from a high-winged, single-engine light aircraft (e.g., Cessna 170): Such variation makes possible the collection of a maximum amount of information within the constraints of a resource manager's budget (1975 costs).

The latter scheme, using light aircraft and 35 mm transparencies, has been used as an aid in the identification of very localized phenomena depicted on LANDSAT 1 frames. Because these transparencies are relatively inexpensive to obtain, the above-mentioned system constitutes an attractive means for acquiring seasonal information or monitoring dynamic change (i.e., adding a temporal dimension to the management information base).

*Multistage Sampling.* The multistage sampling approach to remote sensing surveys is similar to the multistage survey in that remotely sensed data are obtained at various scales and levels of detail. The key difference is that multistage sampling techniques are designed to provide quantified measurable data, rather than ex-

0-201-04245-2

trapolating by the qualitative transfer of generic recognition (homogeneous landscapes) from area to area and from large- to small-scale imagery. The multistage sampling approach utilizes statistical methods and probability theory to determine where to acquire data. This technique provides very reliable quantitative estimates of resource character, type, quantity, and value. It has been applied commercially to establish timber values, land carrying capacity for cattle, and so forth.

Sampling, in its simplest terms, is the selection of a portion of a universe as representative of the whole. This technique is familiar to social and physical scientists who deal with mass data.

The primary applications of sample surveys in natural resource fields have involved the assessment of the quantity, quality, and distribution of resources such as agricultural crops, range grasses, forest trees, and water resources. Population censuses are now frequently conducted on a sample basis.

Remote sensing can provide the means to obtain reliable estimates concerning the resources of vast areas from relatively small samples. In forest and agricultural surveys, aerial photography has long been used to improve sampling efficiency. Among the sampling techniques used are the following:

1. Stratified sampling (where strata are defined by delineating relatively homogeneous areas on photographs).
2. Double sampling with stratification at the first phase.
3. Double sampling with regression.

These sampling techniques are usually geared to incorporate medium-scale resource photography of a type generally available from U.S. government agencies. Combining information from traditional sources with available LANDSAT data significantly reduces survey costs.

Even a cursory inspection of space imagery strongly suggests that its use in a management-oriented resource inventory requires some form of multistage sample design. With LANDSAT, one can survey vast land areas rapidly, but it is difficult to identify species or even general classes of natural vegetation except in gross physiognomic terms. Furthermore, the location of subjects of relatively small size cannot always be accurately and easily pinpointed on the ground directly from the space imagery, particularly in wildland areas. Although it is not generally feasible to accurately locate small ground plots directly from space imagery, it is possible to build a bridge of successively larger- and larger-scale aerial imagery so that the ground plot may be linked to the space imagery. It is around this linkage process, from the ground plot sample to space, that the sample design is built.

After the space images are selected, they are prepared for subsampling. Subunits may be laid out in the form of a grid, as was done in a forest inventory using Apollo 9 infrared Ektachrome photographs covering 5 million acres in Louisiana, Mississippi, and Arkansas in the United States (Langley et al., 1969). In that survey, the space photographs were partitioned into 4 × 4 mile squares. Regardless of how the space images are partitioned, the subunits should be identifiable from an aircraft during the subsampling phase. More importantly, they should be of such a size that

0-201-04245-2

variation between units is either small or controllable by means of photo interpretation. Variation within units is controlled by appropriate subsampling and photo interpretation. Stratification also should be considered if population characteristics differ among areas that exhibit the same image characteristics from one area to another.

In summary, the recent development of sampling procedures which incorporate available LANDSAT and aerial photography has proved highly cost-effective and promises to make certain phases of resource management and development activity possible on a dynamic basis. As for the use of sampling procedures, no standard sampling procedures are applicable to all resource inventories using space or aerial imagery. The design used in a particular situation must take into account such variables as (*a*) the kinds of parameters being estimated, (*b*) the distribution of the variables used to estimate the parameters, (*c*) the existing information relating to these variables, and (*d*) tradeoffs between needs for statistical survey reliability and additional costs of each level of greater reliability (what error can be tolerated?).

## D. Roles for Remote Sensing as Part of a Data and Management Information System

In the earlier parts of this chapter many of the roles which remote sensing may fill as part of a data and management information system were suggested. These include change detection, calibration of other parts of the system, surrogate relations, substitution for other data after calibration, sample selection guidance, system cost reduction, and development of new science models, to name only the more obvious. These are all what may be referred to as technical roles, that is, the types of role that a remote sensing professional would see from his or her perspective. However, there are other roles which may be seen by an economist, resource manager, or policy maker, and these could be quite different from those of the technical expert. Much of remote sensing technology has been developed by the engineering community with only modest understanding of the complexity and weakness of earth resource models, and even less of the very complex issues involved in management and economic and policy issues, to which remote sensing may or may not make some contribution.

Only recently have these extra-remote-sensing issues been seen to be central to the acceptance or rejection of remote sensing as a suitable component of a resource management system. This issue was confronted repeatedly in the 1974 Earth Satellite Corporation/Booz-Allen-Hamilton cost-benefit study based on LANDSAT 1 for the Department of the Interior. Economists and managers asked such questions as the following: What are the costs compared to competing systems? Will all the present information needs be met by remote sensing? If not, what part will be met, and is it worth the effort? There will be an increasing need for these types of questions to be asked by all parties concerned with the development of remote sensing.

To show the interconnections between remote sensing and all the other components of a management and policy *system,* Figure 3.4 indicates how policy deci-

0-201-04245-2

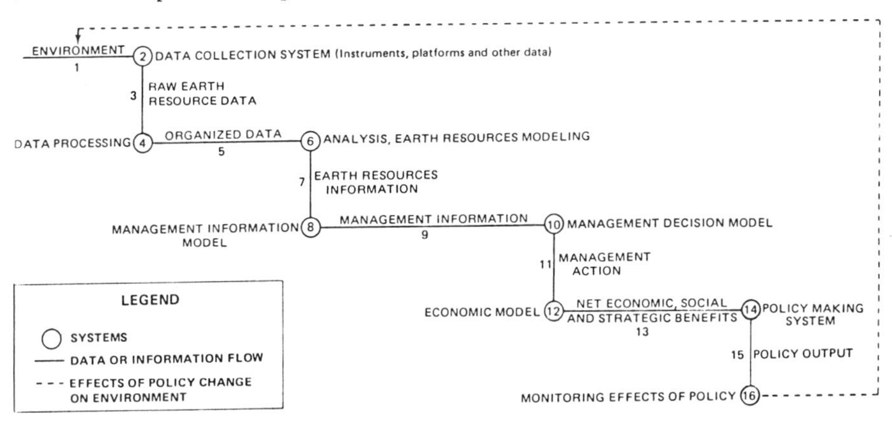


**Fig. 3.4.** Resource management system, showing the links between management and policy decisions and the environment in relation to remote sensing data. Modified from a number of diagrams. Courtesy of Earth Satellite Corporation.

sions and environmental realities are linked in an effective system of which remote sensing data form a part. This diagram is presented to emphasize that remote sensing is *part* of a resource management system, and that for the system to function properly each link in the chain must function properly also. Remote sensing is important principally in steps 2, 3, 4, 5, and 16 of the chart and may be valuable in step 6 as well. In each of these steps, remote sensing may prove to be (depending on the problem) very useful, useful, or of limited use. It may also provide a large, a moderate, or only a small portion of the material collected or processed at a given step. It is helpful to systematically review each of the steps at the beginning of a project where remote sensing is involved. It is particularly important to start from step 16 and work backward through the system, identifying problems and potential uses enroute.

As an exercise in thinking through this diagram, examine Figure 3.5, which shows federal agency information requirements (General Electric, 1974). This figure gives the information grid size, timeliness, and update cycle of federal information needs. These information needs lie in the middle of Figure 3.4 in blocks 8 and 9. With precise information on the actual needs it is possible to track both ways through the diagram to the sensing and policy ends of the system. Given these needs, the ability of a particular remote sensing system (instruments and platform) to satisfy them may be assessed.

### *Data Management Systems*

A country, regional institution, or other entity that proposes to use remote sensing in its development efforts is confronted with the dual issues of (*a*) how to use sensor techniques appropriately, and (*b*) how to incorporate the data acquired into a resource management system. In most instances, an obvious limitation to the establishment of a resource management system is the lack of trained technical

0-201-04245-2

*Federal Information Requirements*

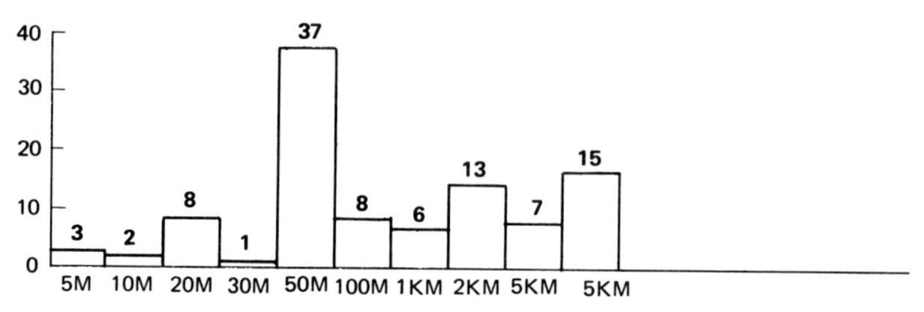


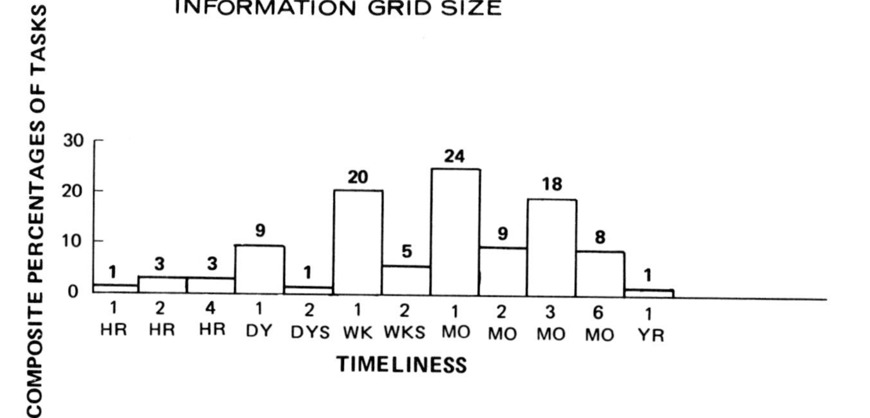


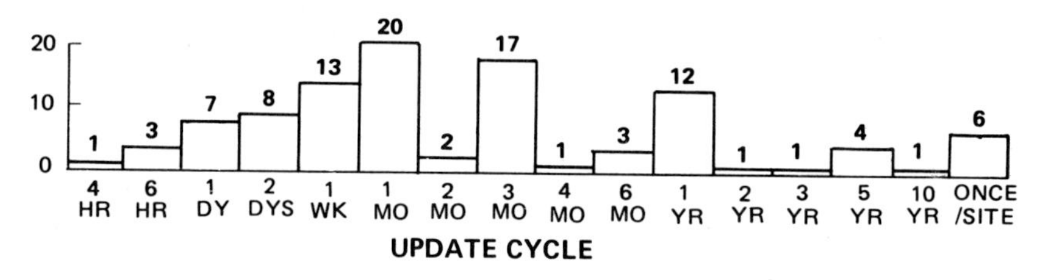


**Fig. 3.5.** U.S. federal agency information requirements with respect to grid size, timeliness, and update cycle. Compiled for U.S. Department of Interior, U.S. Department of Agriculture, and the National Oceanographic and Atmospheric Administration. These information needs lie in the middle of the diagram shown in Figure 3.4. With such detailed information for particular uses it is possible to examine alternative remote sensing systems with respect to their abilities to satisfy the agency requirements and agency decisions. After General Electric (1974).

staffs to handle both the imagery interpretation and the data storage retrieval and analysis. The basic problem, therefore, is one not merely of acquiring and applying the necessary technology, but also of creating the institutional and human capabilities to analyze, store, and retrieve the data collected by the technology and to integrate them into management programs.

Earth resource satellite and meteorological satellite systems are available in conjunction with conventional data sources to establish flexible resource monitoring

0-201-04245-2

and management systems. Before 1973 it would have been impossible for any but the most sophisticated and technologically advanced countries to contemplate the establishment of a dynamic resource data management system. Since then, however, there have been major advances in the development of small, powerful, and relatively inexpensive computers and associated programs designed for geopoint (data stored in a series of polygons identified by their geographic positions) data storage and retrieval systems. These systems are now commercially available, stimulated in some measure by the existence of dynamic data from LANDSAT and by a broadened interest in land-use planning.

Such systems are well suited to the requirements of central planners and resource managers even in the less developed areas of the world, since these smaller computers can handle the smaller-scale data applicable to national resource planning. A base scale appropriate to the use of a system as defined above might be 1:1,000,000 or 1:250,000. Data stored and presented at these scales are sufficiently refined for order-of-magnitude estimates in area resource forecasting and for planning purposes.

## VII. THE PHILOSOPHICAL UNDERPINNINGS OF REMOTE SENSING: SOME PROBLEMS

The concepts in remote sensing outlined in Section V must be used with judgment, for they are only relative, not absolute. Indeed, there are underlying assumptions in their application which, on close scrutiny, may prove questionable.

The major problems with remote sensing relate to the fact that as a discipline it has yet to recognize that there are intractables within the field, just as in any science. Some problems, which may in part be intractable, include the following:

- The degree of consistency of information obtained.
- Artifacting.
- Indeterminacy.
- Improper extrapolation between scales.
- Information as a scale-dependent problem.
- Environmental modulation transfer function.
- Problems of sample design, and system and spatial error budgets.

### A. Degree of Consistency of Information

It is frequently noted that many other sources of data, including censuses, are plagued by inconsistencies, and the *relative* consistency of remote sensing data is emphasized as one of their stronger attributes. Thibault et al. (1972), in a study of land use and environmental quality, suggested that supplementary remote sensing data could improve both the consistency and the aggregation (fineness of subdivision) of existing flood-plain land-use data if used in conjunction with such data (see Table 3.1).

0-201-04245-2

**TABLE 3.1**
**Quality of Existing Flood Plain Land-Use Statistics[a]**
**(After Thibault et al., 1972)**

| | *County Level of Data Collection* | | | | | | | | *Estimated Overall Utility for Environmental Decisions* | *Estimated Gain from Remote Sensing for Spatial and Time Data* | *Total Utility Using Present and Remote Sensing Procedures* |
|---|---|---|---|---|---|---|---|---|---|---|---|
| | *Consistency* | | | | *Aggregation* | | | | | | |
| | *Space* | *Category* | *Intensity* | *Time* | *Space* | *Category* | *Intensity* | *Time* | | | |
| Agriculture | 1 | 7 | 7 | 8 | 2 | 7 | 7 | 6 | 2 | +5 | 7 |
| Urban and rural/urban land uses | 3 | 2 | 2 | 2 | 1 | 4 | 3 | 3 | 1 | +6 | 7 |
| Water bodies | 3 | 6 | 5 | 4 | 3 | 8 | 8 | 3 | 3 | +3 | 6 |
| Forestry | 3 | 8 | 7 | 4 | 4 | 8 | 8 | 5 | 3 | +4 | 7 |

[a] 10, 9 = Very high; 8, 7 = high; 6, 5 = moderate; 4, 3 = low; 2, 1 = very low; 0 = nil.

Table 3.1 assesses the value of existing county-level statistical data for environmental decisions on flood-plain use. Such statistics are collected principally by the U.S. Department of Agriculture and the U.S. Bureau of the Census. Most of the data are published for counties as the lowest level of aggregation and are collected relatively infrequently. Since most flood plains form only a portion of a county, the latter is a very inadequate statistical and decision base. Data consistency as well as aggregation is evaluated in the table for the four dimensions of resolution: spatial, categoric, quantitative, and temporal. Spatial and temporal data are the weakest links and seriously reduce the value of the whole data set. The use of remote sensing, in conjunction with these statistical data, could significantly upgrade the quality of the data base.

A question naturally arises as to the consistency of remote sensing data themselves when used in an operational program. The relations between the system resolution and the level of environmental complexity, on the one hand, and overall consistency, on the other, may be expressed as:

$$C = f[\mathrm{SR}(s, \lambda, r, t), \mathrm{Env.\ Compl.}\ (S, \lambda, Q, T)A],$$

where

| | | |
|---|---|---|
| $C$ | = | level of consistency, |
| SR | = | system resolution with respect to spatial ($s$), spectral ($\lambda$), radiometric ($r$), and temporal ($t$) components, |
| Env. Compl. | = | environmental complexity with respect to the frequency distribution of the spatial ($S$), categoric ($\lambda$), quantitative ($Q$), and temporal ($T$) domains of the environment given for area ($A$). |

Consistency is inversely related to the degree of environmental complexity in space and time and the size of the area being examined, and is directly related to system resolution in the sense that improved resolution will usually give a higher consistency in results. These relations also suggest that a high order of consistency for a very large area of, say, continental dimensions may be obtained only if a small number of very *general* categories are specified. For example, Level I of the Anderson et al. (1972) land-use classification is the level at which high consistency may be obtained with LANDSAT imagery. Level II is distinctly less consistent, and Level III is not at all consistent from locale to locale.

There are no simple answers to this question of consistency. Certainly, persons concerned with major problems distributed over large areas involving a wide range of environmental circumstances would do well to look closely at this question.

## B. Artifacting

Many types of artifacts are produced by aggregation and disaggregation of data in any one of the classes of resolution (space, categories, quantities, and time). As viewed here, an artifact is a structure that is not normally present in the data but is produced by some external agency. The real question is not whether artifacting exists in remote sensing imagery—it does—but rather whether it is of sufficient magnitude and pervasiveness to constitute a severe problem. Particularly, the question arises, Will artifacting be a potential problem when a multistage sample design, which by definition must have a monotonic relationship across scales of observation, is envisaged?

Three general classes of artifact occur:

- Those arising from excessively fine subdivision of categories.
- Those arising from excessive generalization of categories.
- Those arising from other external factors.

0-201-04245-2

*1. Excessively Fine Subdivisions*

This type of problem occurs in level slicing, in which noise may be contoured. Quantization noise, which arises from dividing a continuous distribution into steps, is related. In addition, across-scene gradients external to the scene reflectance properties (e.g., fore- and backlighting) may dominate in level slicing unless care is used. These examples are all drawn from fine slicing in the radiometric domain, but very fine subdivisions in other resolution classes can produce a similar effect. For example, very fine subdivision of categories (after the use of a clustering algorithm) can divide within classes.

*2. Excessive Generalization*

A classic example of excessive generalization from outside the field of remote sensing is the use of monthly average climatic statistics—showing a low pressure centered over northwest India—to generate the hypothesis of the heat-low monsoon. In fact, daily weather charts show moving low-pressure systems which tend to stagnate more often in one location than another; thus averaging shows a semipermanent low which is an artifact of the length of time over which the averaging is carried out.

Equivalent effects in remote sensing include the possibility of major lineations observed by geologists arising from chance alignments detectable only on coarse-resolution imagery, and from the effects of particular sun angle illuminations, leading to the creation of new boundaries, and from lumping of mixtures to create a new nonreal class. (Actually there are two or more separate classes in a mixture.) For example, machine lumping can be comparable to the result produced by an observer who is a notorious lumper, as opposed to one who is an equally consistent splitter.

*3. Artifacts Arising from External Factors*

Many of the artifacts arising from external factors involve the application of a human or machine-based decision rule, appropriate for one environment, but applied willy-nilly in another area for which it is not appropriate. This is particularly a problem with boundary delineation on space images where tone is the principal basis of discrimination, with only limited use of texture, and with virtually no recourse to the other geometric properties and cues used in the interpretation of low-altitude photography. This topic is discussed by Simonett et al. (1969b), in an examination of the boundary delineation problem on space images of the Alice Spring area, Central Australia. They found that the clearer the contrast between adjacent entities, the less ambiguity there is about class boundaries. The more gentle the gradient, the wider is the variance in location and erection of separate classes by both machine and manual clustering. In other words, the process of "classification" by people or people and machine is inherently highly subjective, because few images embody simple structures in simple environments.

0-201-04245-2

### C. Indeterminacy

Closely related to the problem of artifacting is that of indeterminacy. Indeterminacy arises principally because a training sample represents an inadequate sample of the environment or because a clustering algorithm, devised over a very small subset of data, is applied over a larger data set. Within the latter different relations may apply and new classes, not present in the original clustering subset, may be introduced. In essence, both the training and the clustering subsets are geographically bound, while prediction is carried into a larger geographical space, the internal variability of which is not known. This problem may only be partially alleviated by partitioning the universe into homogeneous regions of the natural and cultural environment, as has been demonstrated in a number of studies (MacPhail, 1969; Morain, 1973; Simonett et al., 1969b; Nichols et al., 1974).

Indeterminacy is seen in a very baffling form when training or clustering, performed in well separated areas, produces very dissimilar classes and boundaries in some intermediate location. In which, if either, does the truth lie?

### D. Improper Extrapolation between Scales

As pointed out by McCarthy et al. (1956), relations generated at a given scale with social and economic statistics do not necessarily apply at other scales of generalization. The same principle may also apply with remote sensing data, depending on the problem and the range of scales of generalization. Since the magnitude of the problem depends both on the topic being considered and on the range of generalization, no simple rules for evaluation are available.

An example of this problem may be seen in the classic cell-size problem in cloud statistics. A very small cell size may produce a U-shaped distribution of cloud frequency at a given location. Successive increases in cell size transform the frequency distribution to a broadly Gaussian (or beta) distribution and finally to a narrowly Gaussian distribution. The monsoon example given earlier also constitutes a case of improper extrapolation between the scales of observation and generalization.

### E. Information as a Scale-Dependent Phenomenon

As noted earlier, when a variety of scales of observation are present for an area, different aspects of the environment may be more readily, and properly, observed at one scale than at another. Although this proposition is generally true, it is not equally valid for all objects in a scene, because the sizes and internal characteristics of the individual entities in a scene are themselves variable. Thus a scale appropriate for observing an entity in a scene at a given location may be inappropriate in another scene. Or a scale suitable for a single entity in a scene may be inappropriate for another in the same location. Thus both within- and between-scene variability modulates and influences the quality of information obtainable at a given scale.

0-201-04245-2

Hence the notion that there is a "scale germane to the problem" must be used with caution, because the data in an image are nonhierarchical in a classificatory sense. This conclusion leads to the next, fundamental problem, namely, that the environment modulates the information transferred in a scene.

### F. Environmental Modulation Transfer Function

This relation may be formalized as the *environmental modulation transfer function.* Engineers refer to the effectiveness of an instrument through its modulation transfer function. However, the environment is several orders of magnitude more complex to calibrate than engineering instruments, and the various parameters of the environment are more difficult to quantify than instrumental parameters.

Fundamentally, the environment is not passive to electromagnetic interrogation. Its variability in time, space, and number of categories all notably affect information transfer. This notion may be formalized as below:

$$S = f(I, E),$$

where $S$ = system information (modulation) transfer function,

$I$ = instrument information (modulation) transfer function,

$E$ = environment information (modulation) transfer function.

The instrumental and environmental portions of this equation may be expanded thus:

$$I = f(\lambda_R, R_R, S_R, T_R, N),$$

where $\lambda_R$ = spectral resolution or band width of a single channel,

$R_R$ = radiometric resolution for either reflected or remitted energy or combinations thereof,

$S_R$ = spatial resolution of the system,

$T_R$ = frequency of observations potentially available with the system,

$N$ = number of channels of information.

The contributors to the environmental modulation transfer function may be expressed as follows:

$$E = f(C_R, S_R, T_R, A_R, \ldots, X_R),$$

where $C_R$ = number of distinct land-use or mapping categories present (this is

0-201-04245-2

obviously classification dependent, and the finer the classification cuts, the more complex an environment will be),

$S_R$ = size and spatial arrangement of the homogeneous categories comprising an area,

$T_R$ = a measure of the time-variable nature of the environment,

$A_R$ = a measure of the atmospheric constraints on sensing, principally a measure of the probability of obtaining data in relation to cloud cover,

$X_R$ = other aspects of the environment.

The concept of the environment modulation transfer function simply formalizes the elementary notion that the fitting of a single resolution to a multitude of environments will not produce a uniform class of information for all environments. In short, the environment itself actively modulates the information transfer process. If uniformity of data base is a prerequisite over large areas, the locations of the more intractable environments need to be known in order that supplementary information may be obtained on them, using either aircraft sampling strategies or other data sources.

Studies on environmental spatial complexity (Simonett and Coiner, 1970, 1971) have shown notable differences from area to area in the United States and overseas. Working with aerial photographs and employing clustering techniques on data obtained by laying a series of grids of different sizes over photographs (the smaller nested within the larger), these studies showed that there are at least nine major types of spatial complexity. These range from very simple (in which large rectilinear blocks of a small number of homogeneous categories are found) to exceedingly complex (in which very small, irregular patches of a large number of distinct categories are found). There were numerous situations in which LANDSAT resolution (90 m) would produce a very large number of cells with two, three, or even more entities within a cell. Thus most picture elements (pixels) would be mixed. The problem is compounded in areas of small, complex entities where multidate registration is required as part of an operating procedure. A misregistration of just one pixel is enough to ensure that virtually all entities will be *improperly* mixed in *time.*

A series of sample studies on the nature of within-scene variability and its effect on land-use mapping has also been carried out by Schwarz (1970). These studies, though qualitative, clearly show the magnitude of environmental variability and the difficulty of obtaining a uniform class of information in all locations when a single resolution is fitted to the environment.

Both the Schwarz and the Simonett-Coiner studies deal with the spatial complexity problem. Comparable problems exist with respect to the quantitative, categoric, and temporal domains.

0-201-04245-2

It is clear from the preceding that the root cause of these problems is geographical variability. This is one of the most pervasive problems in remote sensing. The environment and its variability are the least known components of the remote sensing system.

## G. Problems of Sample Design, and System and Spatial Error Budgets

Multistage sampling with aerial photographs is a well-documented, widely applied, and successful technique in forestry, range management, and crop enumeration surveys. A number of studies in forest production (Langley, 1969; Langley et al., 1974; Nichols et al., 1973) have shown that the use of space photographs as the first stage of a multistage sample design can materially reduce the cost of sampling, improve the precision, and reduce the variance of the estimates. The better the resolution employed in the first step of such design, the greater the cost savings of the system.

In working with space-derived data, however, there are intrinsic problems that must be borne in mind in assessing the role of sampling. These relate to the design itself, and to system and spatial error budgets.

In preparing a sample design, a common method is to follow a multistage procedure with samples distributed spatially according to some quantity, such as production volume, monetary value, or acreage. Probability proportional to size (PPS) procedures require that to minimize variance the final allocation be in small sample plots distributed according to some random process. However, in many remote sensing applications, it is more convenient for machine processing to group these plots into a compact area, say a 5 × 5 square mile area, rather than distributing them in small 1 mile square plots as is done in the U.S.D.A. Crop Enumeration Survey. Moreover, PPS sampling requires that each sample be enumerated at each stage of the survey, and that there be no or very few dropouts. This is clearly infeasible in a remote sensing multitemporal survey of crops or any other rapidly varying population where dropouts from patchy cloud cover are inevitable. (This problem does not exist to the same degree with relatively static features such as forests, where cloud-free conditions may be waited for.) Finally, a fundamental uncertainty in the whole process arises from the failure to identify both the acreage and the category in question in areas where it is infeasible to have ground observations as one stage in the design. In normal nonremote sensing surveys, both these items are recorded with high accuracy at the sample plot stage. The latter is the lowest level in the system and the one where the bulk of the variance occurs.

It follows that the *system* error budget in a remote sensing survey (without ground observation) is uncertain and variably distributed between regions. In fact, very pronounced *spatial errors* can accumulate in selected regions. In essence, questions of professional judgment rather than strict sample design and weighting are required at present. Experience will gradually show us better ways of incorporating remote sensing in a sample design. Nevertheless, sampling procedures in situations where ground observations are not available or are only minimally available will remain difficult.

0-201-04245-2

## VIII. CONCLUDING REMARKS

The concerns expressed above, like the earlier concepts, must be viewed with caution. The skeptic will construe evidence of inconsistency in remote sensing data, for example, as an indication of the failure of remote sensing. Neither the uncritical application of a concept nor the equally uncritical rejection is proper. The problems we deal with in the environment are very complex. A mature application of remote sensing requires a thorough systems analysis of all forms of environmental variability. Many systems using remote sensing must be *mixed* systems, concentrating remote sensing in environments where it is appropriate and employing other techniques in areas where it is not.

The world we live in is arrayed in a wide range of time-space-category mixtures. At one extreme are simply structured environments where large blocks of a few easily discriminable, homogeneous entities are arranged in an orderly fashion. At the other extreme are complexly structured environments. In these are found very small entities, many categories, much spatial interfingering, substantial spectral overlap between classes, and little temporal variability between classes. The degrees and locations of environmental complexity are largely unknown except in the most general terms, yet these data are needed to plan large-area analyses with remote sensing. Almost all machine algorithms work best in simply structured environments, and various improvements under development tend to be quite successful in such areas. More complex environments are likely to be essentially intractable for a given remote sensing system.

0-201-04245-2

Part II

# Instrumentation for Remote Sensing

0-201-04245-2

## INTRODUCTION

To the novice one facet of remote sensing which may cause some apprehension concerns the "hows" of the techniques and instrumentation. Just how do sensors operate? What rules or principles of physics serve as their basis of operation? Knowledge of these fundamental aspects is necessary if remote sensing is to be accepted as a valuable tool. The situation is analogous to the conventional geophysics of the petroleum industry, where distinction must be drawn between the "doodlebugger" with his mysterious "black boxes" and bona fide geophysical prospecting, which consists of applications of seismology, gravity, geomagnetics, electrical resistivity, and so forth. The tools of remote sensing are indeed based on classic principles of physics which will be treated in the chapters of this section. What the experienced remote sensing specialist dismisses as "black box" activities is restricted to common circuitry devoted to electrical amplification of weak signals, and similar sorts of circuitry which have been in use by electrical engineers for several decades.

Stated simply, the basic problem in remote sensing instrumentation is to establish devices which can detect and measure energy, usually electromagnetic energy, which occurs at every wavelength, and to convert the resulting signal into a form which can be perceived by human senses. For energy in wavelengths compatible with human senses, that is, the visible portion of the electromagnetic spectrum, we require only a recorder, as little or no wavelength conversion is needed. Beyond the visible sector, however, something more is required, for human senses are limited to a very short region of the total electromagnetic spectrum.

The visible sector of the spectrum, then, becomes a special case. It serves as an excellent starting point for discussions of instrumentation and principles because there is no need to convert data from one wavelength to another, and, at least in color photography, it portrays a world with which we are already familiar. The basic instrumentation, the camera, has been widely and inexpensively available to the majority of the earth's population for the past three or four generations. Cameras for remote sensing systems are very similar to commercially marketed models but possess many sophisticated accessories developed for aerial photography. Some components such as lenses likewise require greater accuracy to prevent distortion.

The current younger generation, in western Europe and North America, is the first to be reared with the full availability of commercial television. Elsewhere in the world, television has lagged only slightly in becoming commercially available. For remote sensing purposes television systems can be substituted for camera sys-

0-201-04245-2

*Remote Sensing of Environment*, Joseph Lintz, Jr. and David S. Simonett (eds.) ISBN 0-201-04245-2

tems with excellent results. There are tradeoffs, of course, to be considered. The camera systems offer slightly better spatial resolution than do the vidicon systems, but for the fullest utilization of data more and more interpreters are turning to computers for assistance in enhancing the recorded data by increasing the contrast between the "background noise" and the desired "signal." With a camera system, it becomes necessary to digitize the data for compatibility with the computer, an extra step already performed by vidicon systems. The tradeoff advantage of the vidicon system, in addition to providing raw data on magnetic tape or via telemetering stations, includes fewer moving parts, with consequent higher reliability. It is also useful for unmanned satellites, as the logistical problems of moving raw film to the camera, exposing it, and removing exposed film to a laboratory are eliminated.

Outside the range of the visible spectrum, there is a need to convert energy forms beyond human limitations to forms that human beings can sense. Two important ways to perform this feat are commercially available: optical-mechanical scanning systems and fully electronic systems with no moving parts.

The optical-mechanical scanning systems, which have proved themselves to be very versatile and highly reliable, form an important group of remote sensors. Their versatility lies in their ability to detect energy at many wavelengths. It becomes a matter of design to build an optical-mechanical scanning system with the capability of detecting any single band of wavelengths (or channel) or to build one with multiband capabilities. Commercially available optical-mechanical scanning systems are commonly offered in one- or two-channel configurations. Four channels are not too difficult to manufacture and undoubtedly will become more common in the future. Experimental multichannel scanning systems with as many as 28 channels have been built and have provided reliable results.

An important aspect of optical-mechanical scanning systems is their versatility. They can be built to operate at a great variety of wavelengths. Shorter wavelengths than the visible sector of the spectrum comprise the ultraviolet region, and scanning devices are the preferred instrument for the detection of energy at these wavelengths. They operate completely across the visible sector, providing full flexibility as to the number of times and ways in which the visible sector may be subdivided. They readily accommodate the collection of data in the solar infrared sector and are the preferred means for collecting data in the thermal infrared sector. Thus a multichannel optical-mechanical scanning system can "see" all the way from 0.37 $\mu$m out to a region of about 14 $\mu$m. Typically, in a multichannel system the incoming energy is divided by prisms which also perform as filters, although additional filtering for narrower bands may be desirable, and the signals of the various wavelengths are converted to electrical energy by means of appropriate detecting devices. The multispectral scanner on LANDSATS 1 and 2 is a very-high-performance optical-mechanical device.

In the microwave region of the electromagnetic spectrum two modes of operation are currently under investigation, and one mode is commercially offered for applications surveys. The microwave region adapts itself more readily to both passive and active techniques of remote sensing than do the shorter wavelengths. To

0-201-04245-2

date, passive microwave systems have not progressed as rapidly, insofar as instrumentation is concerned, as have the active systems. There are greater challenges in building passive microwave instrumentation, which provides its data in a mapped format. The majority of applications-oriented interpreters of remotely sensed data are timid about using graph or strip chart data; they demand the raw data in at least a two-dimensional format. Until relatively recently, most passive microwave data were essentially microwave radiometry, which requires more than average skill and understanding for interpretation. Imaging microwave is only now beginning to become available.

The active microwave system is the widely known side-looking radar (SLR) which is commercially available. Many segments of the earth's surface have now been recorded by this electronic technique, and the largest remote sensing project ever performed, Project Radam, provided data for study of more than 4,000,000 $km^2$ of Brazil's Amazon Basin.

In the microwave sector, the phenomenon of rotational reflectance assumes a more important role than it does in the shorter wavelengths. This yields multiple-polarization data, which can be captured and recorded with additional electronic receivers. It has been found that significant differences in data collected in the plane and cross-polarized modes occur and that these differences are most helpful to the applications-oriented interpreter.

Although the foregoing are the principal instruments for remote sensing developed to date, much interesting additional instrumentation is in an experimental stage. The possibilities with these "exotica" are intriguing. They have performed successfully on a limited basis, perhaps under optimum conditions, but they are not in general use. The creativity and ingenuity of their designers, however, capture the imagination, and activity of this type represents the frontier of remote sensing instrumentation.

0-201-04245-2

Chapter 4

# The Visible Portion of the Spectrum

*Robert N. Colwell*

## I. INTRODUCTION

One of the most common and useful ways of engaging in remote sensing is simply by taking a photograph. By so doing, one can record in permanent form what the human eye is able to see only fleetingly, or perhaps not at all. Once this photographic record has been made available for study, measurements of the images (through a process known as *photogrammetry*) can lead to the construction of accurate planimetric and topographic maps of the area that has been photographed. Such measurements also can permit determinations to be made of distances, directions, areas, heights, volumes, and slopes, with respect to features depicted in the photography. Usually such measurements are much more difficult and expensive to make in the field than on the photograph. Further analysis of the images (through a process known as *photo interpretation*) may permit one to identify an object, determine its condition, and judge its significance. For example, the trained photo interpreter, in his efforts to identify objects, may be able to determine that a certain kind of blob on a photograph is the image of a ponderosa pine tree. From its photographic tone he may be able to deduce that this tree has recently become insect infested (its present condition) and therefore should be removed (a judgment as to its significance) before insect broods that soon will be produced in the pine can infest other trees.

In this chapter, certain recent and highly significant developments that have occurred in the *photographic* aspects of remote sensing will be stressed. It is important to do so because of the great publicity currently being given to certain developments in the *nonphotographic* aspects. There is a strong likelihood that the publicity accorded to developments of the latter type (as applied to sensing in the thermal infrared, microwave, and gamma ray regions, for example) may tend to create the impression that photography is in the process of being replaced by other forms of remote sensing. Such will never be the case, as this chapter will strongly suggest.

One of the most important attributes of a remote sensing record is the spatial resolution (fineness of discernible detail) which it affords. Only rarely can nonphotographic sensors provide spatial resolution approaching that of a photograph. Another important consideration is the ease with which features can be identified from their remote sensing images. Unlike other forms of remote sensing, photography deals with essentially the same wavelengths of energy as are perceived by the

*Remote Sensing of Environment*, Joseph Lintz, Jr. and David S. Simonett (eds.) ISBN 0-201-04245-2


0-201-04245-2

human eye when it views a scene directly. Hence on photographs there usually is a more natural appearance of objects (in terms of the relative colors, tones, or brightnesses with which they are imaged) than on any other form of remote sensing record. Finally, as a collector of radiant energy, the camera lens used in obtaining conventional photographs is much more closely analogous to the human eye than are the "collecting optics" of some of the more exotic remote sensing devices. Consequently, the *geometry* of a conventional photograph usually is more akin to that perceived by the human eye than is the geometry of some other type of remote sensing imagery. This is yet another factor that frequently enables the image analyst to recognize a feature more easily from its photographic image than from its "signature" as obtained in some other part of the spectrum. For these and other reasons photographs will continue to be the basic remote sensing records of the future, and other remote sensing records, in almost every instance, will merely supplement them.

Although terrestrial photographs can yield valuable information, this chapter will deal primarily with photographs taken from aircraft and spacecraft. For more than a century man has been taking aerial photographs. To be sure, the first of these photos were taken from rather uncertain platforms, including kites, balloons, and even homing pigeons to which miniaturized cameras were strapped. But since those early days both the platforms and the cameras mounted on them have been greatly improved. Modern remote sensing platforms range from the hovering helicopter, through a variety of smoothly performing fixed-wing aircraft, to the earth-orbital reconnaissance satellite. Camera mounts have been gyrostabilized against roll, pitch, and yaw motions of the platform and effectively insulated against aircraft vibrations that might otherwise cause image blur. Camera lens aberrations have been greatly reduced to ensure the formation of sharp images at the focal plane. Roll film of very-high-dimensional stability has almost entirely replaced the old emulsion-coated glass plates. But perhaps the most dramatic and significant improvements of all have occurred in the imaging devices themselves.

A primary purpose of this chapter is to describe the most important of these imaging devices, the photographic films used in conjunction with them, and some of the modern equipment available for viewing and interpreting the images thus obtained. Then, by means of specific examples, important uses of such images will be illustrated.

## II. IMAGING DEVICES

Many types of imaging devices for use in photographic remote sensing have been developed since the first aerial photographs were taken near the middle of the nineteenth century. Of these devices, three types presently show the greatest promise for the inventory of man's environment: (1) the conventional aerial camera, (2) the panoramic camera, and (3) the multiband camera. In the succeeding pages these three types of equipment are illustrated and briefly described.

0-201-04245-2

## A. The Conventional Aerial Camera

Most of the basic components of a modern aerial camera are shown diagrammatically in Figure 4.1. These consist of the magazine, drive mechanism, cone, and lens.

The *magazine* is essentially a lighttight box in which the photographic film is held. In most cases it is detachable from the rest of the aerial camera. The film, as received from the manufacturer, is usually in the form of a continuous roll, 240 mm wide and 60 m long, mounted on a supply spool. Approximately 250 exposures, each 227 mm square, can be taken on such a roll. Both the supply spool and the takeup spool fit snugly inside the magazine as illustrated in the diagram.

*The drive mechanism* is a series of cams, gears, and shafts designed to drive the film from the supply spool to the takeup spool. The film, in passing from one spool to the other, is routed, via guide rollers, across the front surface of the locating plate. One of the rollers is so constructed as to meter the amount of film passing from the supply spool to the takeup spool between exposures, thereby assuring a correct and uniform spacing of exposures on the film roll. As a result of this recycling operation, an unexposed portion of film is properly positioned in front of the locating plate between exposures.

While an exposure is being made, suction is created behind the locating plate, either by means of a conventional aircraft venturi tube or a special vacuum cylinder and piston apparatus which is built into the magazine. This suction is transmitted to the film through a network of small perforations and grooves in the locating plate, thereby holding the film in a flat plane against the plate at the instant of exposure. This feature minimizes the distortion of photographic images caused by a warped position of the film at the instant of exposure.

The *cone* is a lighttight element which serves to position the camera lens properly with respect to the exposable portion of the photographic film. The perpendicular distance from the film to the rear nodal point of the lens ($N_2O'$ in Figure 4.1) is known as the focal length. Several commonly used aerial cameras have interchangeable cones of differing focal lengths for the same magazine. Most aerial photography currently taken for the inventory of natural resources employs a focal length of either 152.4, 209.5, or 304.8 mm.

The *lens* consists of several carefully ground and mounted glass elements which sharply focus on the photographic film the light rays reflected to the lens by illuminated objects on the ground. Aerial camera lenses ordinarily are of the fixed-focus type, being focused at infinity and lacking a focusing adjustment. The altitude of the photographic aircraft above the ground usually is such that all objects on the ground will be in sharp focus when the lens is focused at infinity.

On most aerial cameras the shutter is located between the front and rear lens elements and accordingly is termed a "between-the-lens" shutter. In recycling the camera between exposures, the camera-drive mechanism automatically recocks the shutter at the same time that it draws a new portion of unexposed film into position.

0-201-04245-2

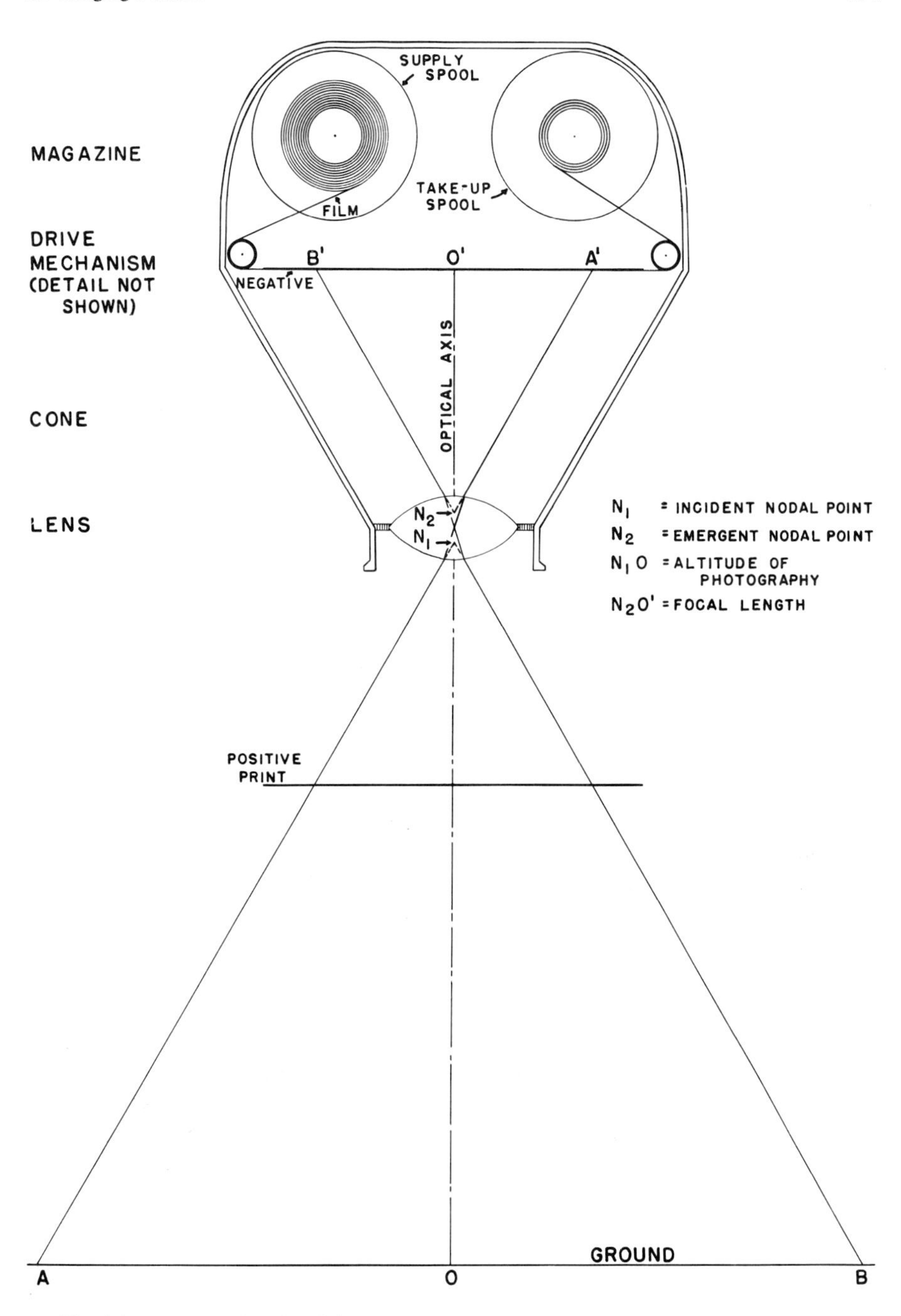


**Fig. 4.1.** A conventional aerial camera.

0-201-04245-2

Aerial camera lenses vary not only in focal length, but also in width of angular coverage. In the United States, many of the most commonly used aerial cameras have a focal length of 209.5 mm, a negative size of 227 mm, and an angular field of approximately 60°.

### B. The Panoramic Camera

The principle on which the panoramic camera operates is indicated diagrammatically in Figure 4.2. With this camera it is possible to photograph a large area in a single exposure at very high resolution (i.e., with a high degree of image sharpness in every part of the photograph). The panoramic camera meets a need but creates some special problems. To get a sharp image when photographing large areas, one needs, paradoxically, a narrow-angle field so as to minimize aberrations of the lens. Such a field is provided in the panoramic camera by a narrow slit in an opaque partition near the focal plane of the camera. The slit is parallel to the camera platform's line of flight. With such a slit, however, one can photograph only a narrow swath of terrain unless the optical train of the camera is equipped to "pan" (move from side to side) as the aircraft advances. The optical train of the panoramic camera is designed to make such movements.

On the other hand, for the panoramic camera to maintain a uniformly clear focus as the optical train moves, the frame of film being exposed must be held in the form of an arc instead of being kept flat as in a conventional camera. With the film in an arc the photographic scale becomes progressively smaller as the distance of objects on the ground increases to the left and right of the flight path. In some applications the scale problems outweigh the advantage of a panoramic field of view, so that it is preferable to use a conventional camera. However, resource managers and others who seek to inventory man's environment with the aid of remote sensing techniques can make very effective use of the panoramic camera to obtain, with only a limited number of flight lines, ultrahigh-resolution imagery of very wide swaths of terrain.

### C. The Multiband Camera

Recent investigations have shown that far more information can be obtained about the various components of man's environment from a study of images taken simultaneously in each of several wavelength bands than from images taken in any one band. Realization of this fact has led to a concept known as "multiband spectral reconnaissance" and to the development of various multiband cameras for the taking of simultaneous photographs in each of several spectral bands. One such camera is illustrated in Figure 4.3. In a camera of this type each film-filter combination employs its own "collecting optics." Such a camera permits exploitation of the concept of multiband spectral reconnaissance throughout the visible and very-near-infrared parts of the electromagnetic spectrum (roughly 400-900 m$\mu$). With it one can take simultaneous exposures of exactly the same area, using several spectral zones. By proper choice of photographic film and filter, the limits of each spectral

0-201-04245-2

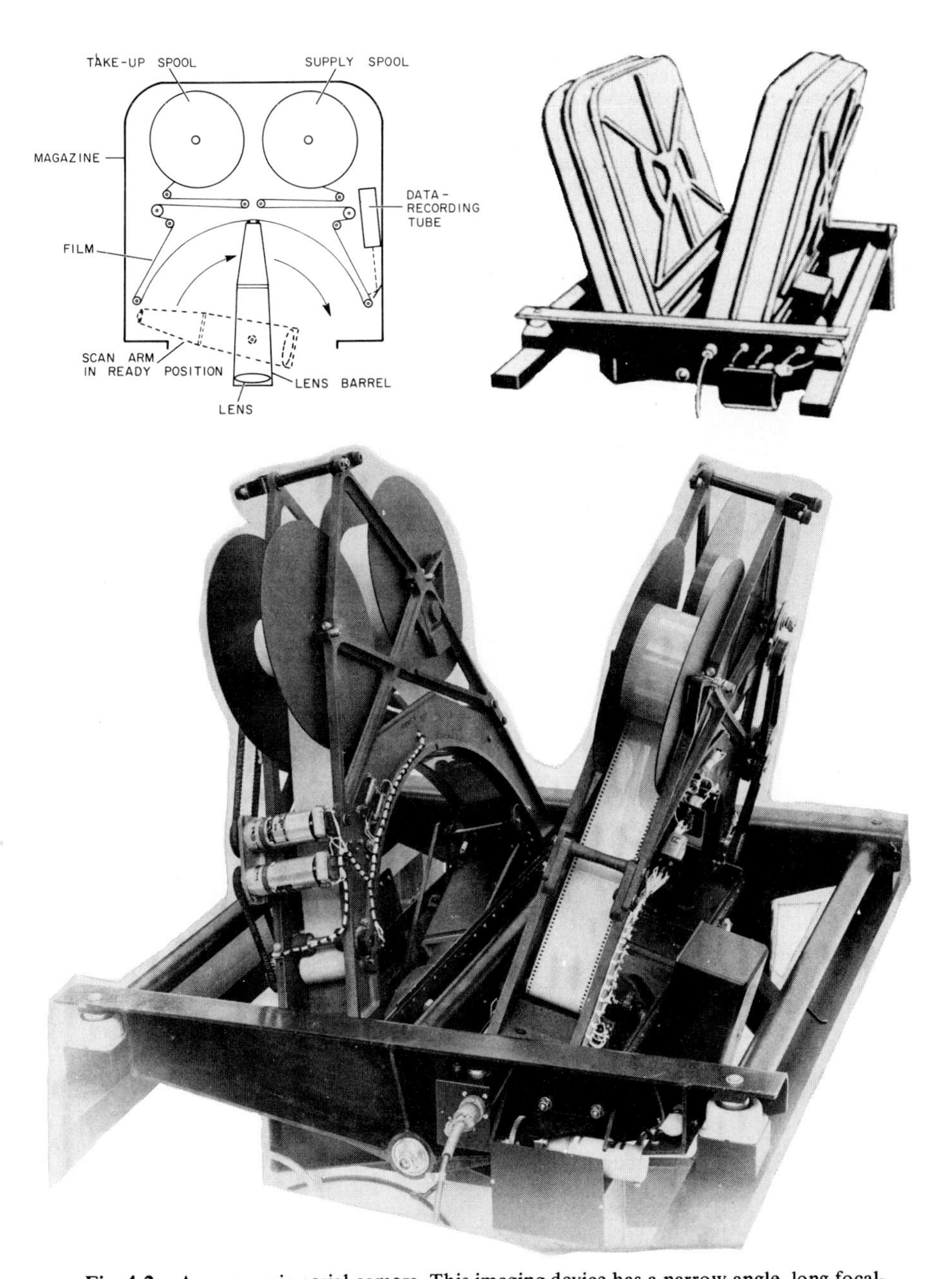


**Fig. 4.2.** A panoramic aerial camera. This imaging device has a narrow-angle, long-focal-length lens system which provides very good spatial resolution. In recording a frame of photography the scan arm pans once from side to side as the aircraft flies along. In this way a very wide swath of terrain is photographed at large scale and with much higher spatial resolution than a conventional aerial camera can provide. When two panoramic cameras are mounted in "convergent mode," as shown, each portion of the swath is photographed from two widely separated points along the flight line, thereby providing stereoscopic coverage on which features can be viewed three dimensionally. The type of panoramic camera shown here is manufactured by Itek Corporation and is known as the "HyAc" (for high acuity). Photographs courtesy of Mark Systems, Inc., Cupertino, California.

0-201-04245-2

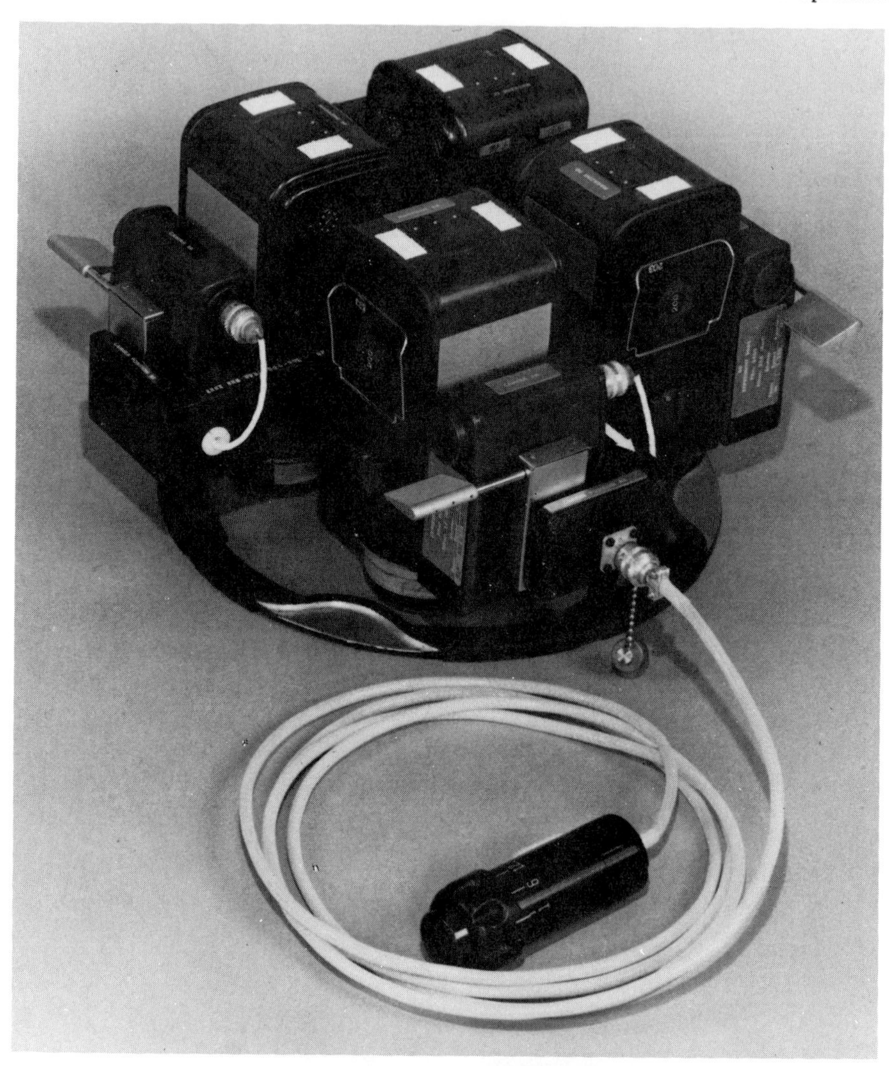

**Fig. 4.3.** The multiband camera assembly used by the Apollo 9 astronauts in conducting the S065 experiment. Photograph courtesy of Allen Grandfield, NASA, MSC, Houston, Texas.

zone can be controlled, as necessary, to obtain the maximum amount of desired information from that zone. Through the use of such equipment, much more information regarding certain objects and conditions within a particular geographic area can be obtained from a study of the tonal values on several of these frames than from a study of the tonal values on any one of them. Examples of the usefulness of multiband imagery obtained with such cameras appear in a later section.

For both conventional cameras and multiband cameras, film transport mechanisms have been significantly improved. Through employment of a principle known

0-201-04245-2

as "forward motion compensation," the film can be made to travel, even during the instant of exposure, at a rate commensurate with the rate of travel of images at the focal plane, caused by the camera's forward motion as the aircraft flies along. By this means sharper images can be obtained, since the image of any particular feature is in effect "frozen" to the same spot on the film during the entire time that the exposure is being made. Such a refinement is necessary to minimize "image blur" when taking aerial photos of the earth's surface from high-speed aircraft at low altitudes.

Even when taking space photography at earth-orbital altitudes (e.g., 160-320 km), failure to compensate for forward motion could limit the sharpness of the image. Consider, for example, an earth-orbital reconnaissance satellite equipped with a camera of long focal length, say 304 mm, and a lens-film system capable of resolving detail as small as 100 lines/mm on the photographic negative. If the satellite were placed in orbit at the conventional altitude of 240 km, photos taken from it of the earth below *theoretically* should "resolve" (show individually) features less than 10 m in diameter. However, if the vehicle were traveling at a conventional earth-orbital speed (e.g., 28,800 km/hr or 8000 m/sec) and the camera were to be set at a conventional shutter speed (e.g., 1/100 sec), a distance of 80 m would be traversed during the "instant" that the camera shutter was clicking. Hence, without forward motion compensation, each small feature photographed would be recorded on the film as a streak which (in terms of the corresponding ground distance) would be roughly 80 m long. This situation obviously would be inconsistent with the objective of achieving 10 m ground resolution. The remedy to such a problem is simply to impart a compensating amount of forward motion to the film during exposure. One device capable of accomplishing this important task consists of a light-sensitive scanner which measures the rate at which images of bright features on the landscape are traveling at the camera's focal plane, where the film is situated. The scanner in turn activates a servodrive unit which can accelerate or decelerate the film's rate of transport to achieve the necessary forward motion compensation.

## III. FILMS FOR RECORDING IMAGES

In the three devices described in the preceding section (i.e., the conventional camera, the panoramic camera, and the multiband camera) images are formed on film that is placed directly in the focal plane of the "collecting optics" or lens system. Some rather delicate choices sometimes must be made in deciding exactly which of many black and white or color films will be used to record images. Hence this section considers the various types of photographic film that are most commonly employed in the cameras just described.

First we will consider black and white panchromatic films, then black and white infrared-sensitive films, and, finally, various color films, some of which are sensitive only in the visible part of the spectrum, whereas others are sensitive in the near-infrared portion as well.

0-201-04245-2

## A. Panchromatic Black and White Aerial Films

Several films of this type that have been developed by Eastman Kodak are compared in Table 4.1. From this table it is apparent that photographic scientists continue to struggle with the inherent tradeoff required between film speed and film definition.

**TABLE 4.1**
**Characteristics of Some Commonly-Used Panchromatic Aerial Films**

| *Film Designation and Type* | *Resolution (Line Pairs per Millimeter)* | *Speed Relative to Film Type 3404* | *Wratten 12 Filter Factor* |
|---|---|---|---|
| High-definition aerial type 3404 | 550 | 1 | 1.5 |
| High-definition aerial type SO-243 | 440 | 1.2 | 1.6 |
| Special fine-grain aerial type SO-190 | 180 | 3.7 | 1.5 |
| Panchromatic X aerial type 3400 | 150 | 9.0 | 1.9 |
| Plus X aerial type 3401 | 100 | 33.0 | 1.7 |
| Super XX aerial type 5425 | 75 | 41.0 | 2.0 |

On the one hand, it is desirable, especially when taking aerial or space photography from unstable, fast-moving platforms, to use films of high speed. With such films one can obtain an adequate photographic exposure in a very short time interval, thereby minimizing blur effects due to various kinds of image motion. But to achieve this high speed a film that has large silver grains in the emulsion is required, so that only a few photons of energy will be needed to form the picture. These few photons, upon striking a given silver halide crystal, even if it is a large crystal, will produce sufficient "latent image nuclei" to raise the crystal's energy level above a critical point. Thereby the crystal will be enabled to accept additional electrons when later placed in contact with a suitable developing agent. Consequently, during the film-developing process, the entire silver halide crystal can be reduced to opaque metallic silver and, because of its large size, can contribute substantially to the production of a negative of proper density in terms of image quality.

On the other hand, it is equally desirable in aerial or space photography (i.e., photography which is necessarily of small scale because of the great distance from camera to ground) to use films of high definition. With such films one can obtain very sharp edge gradients and perceive important photographic detail by viewing the image under suitably high magnification. To achieve this high definition a film

0-201-04245-2

that has very small silver grains in the emulsion is required, so that important details will not be blurred from view as they are in a coarse-grained film. But the smaller the crystal size, the larger is the number of crystals required per unit area of emulsion. It is at this point that one must pay the price in terms of film speed, because, as the crystal size decreases, the number of photons required to activate the crystal does not decrease proportionately. Consequently, the finer the film grain, the larger is the total number of photons required to produce a negative of suitably high density. For this larger number of photons to be captured, the film must be given a longer exposure, that is, the film is "slower." Becuase of this longer exposure, the resulting photograph is likely to suffer from blur effects caused by various kinds of image motion.

The following conclusions are indicated by Table 4.1: (1) consistently with our discussion, there is an inverse correlation between film resolution and film speed, and (2) as shown by the column of filter factors for a Wratten 12 minus-blue filter, the high-definition films are less sensitive to light from the blue end of the spectrum. This fact partly accounts for their slower speeds, although a compensating factor is their extended red sensitivity.

As the term "panchromatic" implies, these films are about equally sensitive to all parts of the visible spectrum, although with some modification, as indicated in the preceding paragraph. The human eye also is "panchromatic" in the same sense Consequently, the relative tones or brightness values of objects are essentially the same on panchromatic photography as those seen directly by the human eye. As previously mentioned, the resulting natural appearance of features may greatly facilitate their photo identification. Herein lies a primary advantage to those who wish to use aerial photography as an aid to remote sensing of the environment.

### B. Infrared Black and White Aerial Film

This film is produced in the United States by only one company, Eastman Kodak. The spectral sensitivity is from 0.36 to 0.90 $\mu$. Consequently, to obtain pure infrared effects with this film, the ultraviolet and visible wavelengths to which it is sensitive should be screened out by use of a Wratten 89B filter or the equivalent.

The resolution of Kodak aerial infrared film (type 8433) is only 55 line pairs/mm. However, when aerial photographs must be taken on hazy days, or even on clear days but from very high altitudes, the superior haze penetration obtainable with this film may enable it to produce sharper images than could be obtained with a nominally high-resolution panchromatic film.

Healthy broad-leaved vegetation has very high infrared reflectance and therefore photographs very light in tone on this film. When such vegetation becomes unhealthy (e.g., because of damage done by diseases, insects, drought, fire, mineral deficiency, or mineral toxicity), it is likely to undergo a loss in its ability to reflect infrared light, even before any other change in its spectral behavior occurs. Consequently, it is of more than passing interest to anyone concerned with the remote sensing of man's environment that infrared photography may provide "previsual

0-201-04245-2

symptoms" of unhealthy conditions that are developing on the broadleaved vegetation within certain parts of a landscape that he seeks to evaluate. Conversely, if the broadleaved vegetation registers only in light tones on infrared photography, there is reasonable assurance that it is not suffering from any of these maladies.

## C. Aerial Color Films

These films are available in three types: (1) color positive films, (2) color negative films, and (3) false color films. As compared with the various black and white films just discussed, each of those color films is potentially more useful for certain purposes. However, the cost of obtaining color photography, when compared to that of black and white photography, is not always justified by the additional information which the former provides.

### *1. Color Positive Films*

Films of this type are sensitized to three primary colors: blue, green, and red, and, when exposed and processed, produce transparencies which appear similar to the original scene when viewed in white light. Two such films currently in common use are Kodak Ektachrome Aero film (E-3 process) and Anscochrome D-200. The former has an ASA film speed of 160, and the latter of 200; both are capable of resolving approximately 100 line pairs/mm. Since the land evaluator is accustomed to seeing and identifying an object, not only by its size, shape, and association, but also by its color, such films provide one more important dimension for use in making aerial photo identifications. For example, Heller et al. (1959) reported that the accuracies with which tree species could be identified were 17% higher on large-scale color transparencies than on panchromatic prints of the same scale. According to Evans (1948), the human eye can separate over 100 times as many color combinations (based on hue, brightness, and saturation) as gray-scale values, although color films cannot discriminate as many colors as the human eye can see.

### *2. Color Negative Films*

In these films the dye coupler components are incorporated in the emulsion layers at the time of manufacture. After the film has been developed and bleached, dye images remain that are not only negative to gradations of the subject, but also complementary to colors of the scene photographed. From such negatives it is possible to make color prints and transparencies as well as black and white prints and transparencies, the latter being almost indistinguishable from those made using panchromatic negatives.

Cooper and Smith (1966) report that color prints made from a negative color film are somewhat lacking in color balance and have considerably poorer resolution than color transparencies. The color balance was found to be most deficient for blue and green objects, including vegetation. Aerial negative films are approximately three-fold slower than aerial positive films, a factor which sometimes further limits the resolution obtainable with them.

0-201-04245-2

One of the most recent developments in color photography is the Kodak Aero-Neg color system. The heart of this system is Kodak Ektachrome MS Aerographic film (Estar Base), type SO-151. Although this film is more commonly known as a reversal color film that can be processed to a positive transparency, it is now possible to process the film to a negative also. The aerial exposure index for this film, when processed to a color negative, is only 8,* and this relatively slow speed may constitute the greatest limitation on its use from aircraft and spacecraft. However, type SO-151 has higher resolution than other color negative films. Probably the most thorough testing of this film to date has been done by various groups of Australian scientists, notably those of the Commonwealth Forestry and Timber Bureau. They find that the film has an unusually wide latitude of acceptable exposure and that the use of proper filtering techniques at the time of printing can eliminate most of the adverse bluish cast and related haze effects that usually plague high-altitude color photography.

### 3. *False Color Films*

In these films objects are purposely imaged with different colors from the ones they exhibit in nature. The purpose of the false color is to accentuate certain features and facilitate the making of certain distinctions, even at the expense of rendering features of lesser importance less interpretable. One such film is Kodak Infrared Aero film (E-3 process); since the film is normally exposed through a Wratten 12 filter, blue light does not contribute to formation of the image. As indicated in Table 4.2, where the spectral characteristics of this film are compared with those of

**TABLE 4.2**
**Spectral Characteristics of a Normal Color Film (Ektachrome Type 8442) and an Infrared-Sensitive Color Film (Infrared Ektachrome Type 8443)**

| *Spectral Region* | *Blue* | *Green* | *Red* | *Infrared* |
|---|---|---|---|---|
| Ektachrome film | | | | |
| Sensitivity bands | Blue | Green | Red | |
| Corresponding colors of dye layers | Yellow | Magenta | Cyan | |
| Resulting colors on photographs | Blue | Green | Red | |
| Infrared Ektachrome film | | | | |
| Sensitivity bands with Wratten 12 filter | | Green | Red | Infrared |
| Corresponding colors of dye layers | | Yellow | Magenta | Cyan |
| Resulting colors on photographs | | Blue | Green | Red |

*Recently a fourfold improvement in this respect reportedly has been achieved.

0-201-04245-2

a normal color film, its three dyes respond to green, red, and infrared wavelengths, respectively, with the net result that green objects (except vegetation which is also highly infrared reflective) appear blue, red objects appear green, and infrared-reflective objects (such as healthy vegetation) appear red. One effect is that the infrared energy of highest intensity produces the brightest reds in the photograph. The predecessor to this film, known as "camouflage detection film," was used as early as World War II, primarily to differentiate green camouflage paint from green (but highly infrared-reflective) foliage. The two looked the same to the naked eye (and also on panchromatic or conventional color films) but conspicuously different on camouflage detection film—hence its name. Tarkington and Sorem (1963) report that the new emulsion, released in 1962, provides better resolution and is three times as fast as the old one. Because of its sensitivity to long wavelengths and the exclusion of short ones through use of a Wratten 12 filter, this film has the ability to penetrate haze exceptionally well.

A false color film containing only two dyes and known as "SN-2 spectrozonal film" is produced by Russian film makers. One layer of the emulsion responds to visible wavelengths of energy; the other, to infrared wavelengths in the 0.7-0.9 $\mu$ region. During film development, color dyes are introduced into both layers to produce images in various colors.

Both of the above-mentioned false color films are of great potential interest for the evaluation of the environment by means of remote sensing. Some features of interest can be distinguished on photographs taken in the visible part of the spectrum, but not on those taken in the infrared part; exactly the reverse is true for other features. These false color films combine in a single composite color image the possibility of distinguishing both types of features.

## IV. DEVICES FOR ANALYZING PHOTOGRAPHIC IMAGES

Broadly speaking, the equipment which a remote sensing scientist needs to extract information from imagery falls into three categories: viewing, measuring, and plotting (see Figures 4.4 and 4.5). The full gamut of equipment in each of these categories is covered in the *Manual of Remote Sensing* (American Society of Photogrammetry, 1975) and in earlier publications of the American Society of Photogrammetry. Consequently, no further details relative to such equipment need be given here. However, with the recently developed capability for acquiring imagery simultaneously in several parts of the spectrum there has arisen a need to develop new kinds of equipment, as described below, that will facilitate the extraction of information from such multiband imagery.

In considering multiband data analysis systems, it is helpful to enlarge slightly on the previously mentioned concept known as "multiband spectral reconnaissance": reflection and emission of electromagnetic energy by a particular kind of natural terrain feature are selective with regard to wavelength and specific for the kind of feature in question. Exploitation of this principle, through multiband

0-201-04245-2

reconnaissance and multiband data analysis, theoretically should lead to positive identification of every feature with which we are concerned as we seek to inventory various components of our environment.

As the number of spectral bands used in multiband remote sensing is increased, the tone signature for each natural resource feature becomes more complete and more reliable. With this increase of spectral bands, however, the task of data analysis can become astronomic unless the image analyst is provided with some kind of image-correlating equipment. Several methods of accomplishing the required image correlation will be discussed presently. First it must be emphasized, however, that to use any of these methods successfully, it is necessary at the outset to "calibrate" the multiband tone signature (response) for each type of object or condition to be identified. This is best done if a suitable "ground truth" test site has been included in the multiband reconnaissance flight. Within such a test site (if it has been properly selected) each type of object and condition that is to be identified operationally is exhibited in each of several accurately known localities. It is from a preliminary study of the multiband images for these test sites that the identifying tone signatures are derived. The task of image correlation is greatly facilitated if all of the multiband frames of imagery covering a given area of terrain have the same geometry. This is a standard attribute of multiband cameras and ordinarily poses no serious problem. Not only is uniform image geometry useful in the calibration phase, but also it is essential in the subsequent operational phase if any of the following methods is to be used.

## A. Optical Color Combiner

Through the use of this method the multiband imagery of a given portion of the terrain is reconstituted as a single color composite. Each type of feature is then identified merely through visual perception of the color exhibited by it on the composite color imagery. When this method is employed, it is common practice to project simultaneously onto a viewing screen all of the black and white images of a given portion of the terrain that have been obtained with the multiband reconnaissance system. The color rendition is achieved by the use of colored filters. Each black and white frame is projected, usually in lantern slide form, through a filter of suitable hue. For any feature, the intensity of that particular hue, as seen on the color composite, is governed by the gray-scale value (tone) exhibited by the feature on the corresponding black and white lantern slide. Laboratory equipment suitable for producing color composites from as many as four spectral bands is illustrated in Figure 4.4D.

The advantages of this optical system are (*a*) it is easy to construct and use (the equipment is inexpensive and readily obtainable, and the enhancements are readily performed, (*b*) the data can be directly interpreted from the screen, and (*c*) there is a large selection of color filters, and thus a large number of color combinations with which to experiment.

0-201-04245-2

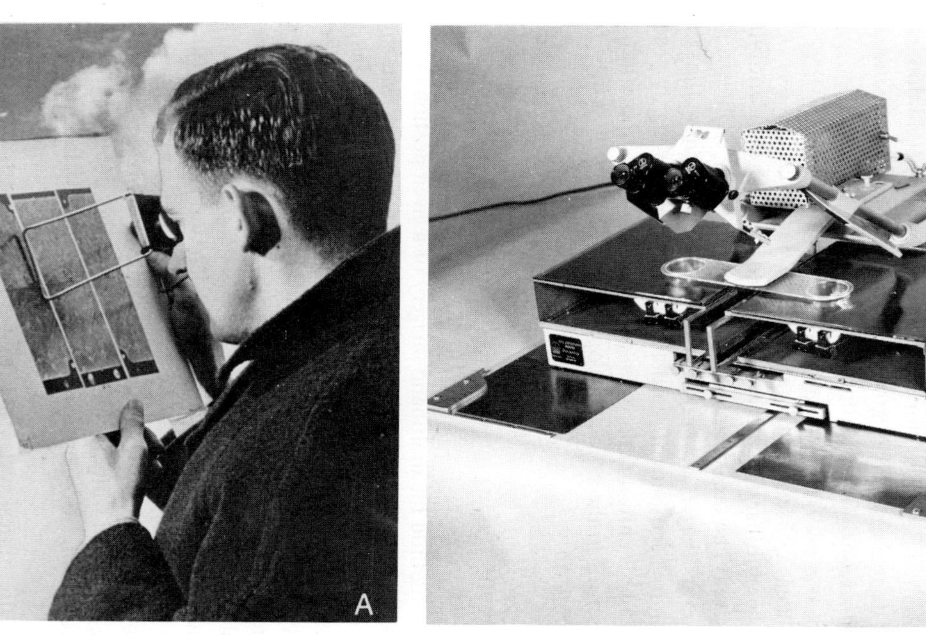


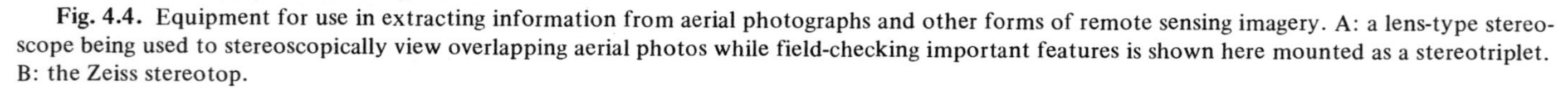

**Fig. 4.4.** Equipment for use in extracting information from aerial photographs and other forms of remote sensing imagery. A: a lens-type stereoscope being used to stereoscopically view overlapping aerial photos while field-checking important features is shown here mounted as a stereotriplet. B: the Zeiss stereotop.

0-201-04245-2

0-201-04245-2

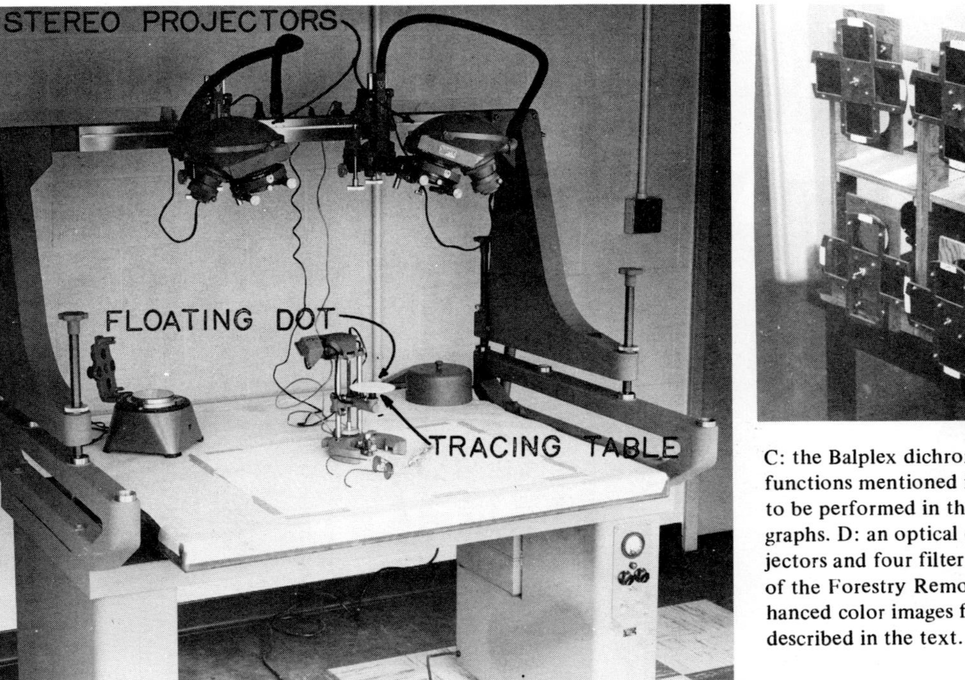

C: the Balplex dichromatic projection plotter permits all three functions mentioned in the text (viewing, measuring, and plotting) to be performed in the process of making a map from aerial photographs. D: an optical combiner consisting of four lantern slide projectors and four filter wheels. This device was assembled by personnel of the Forestry Remote Sensing Laboratory for use in producing enhanced color images from multiband black and white photos, as described in the text.

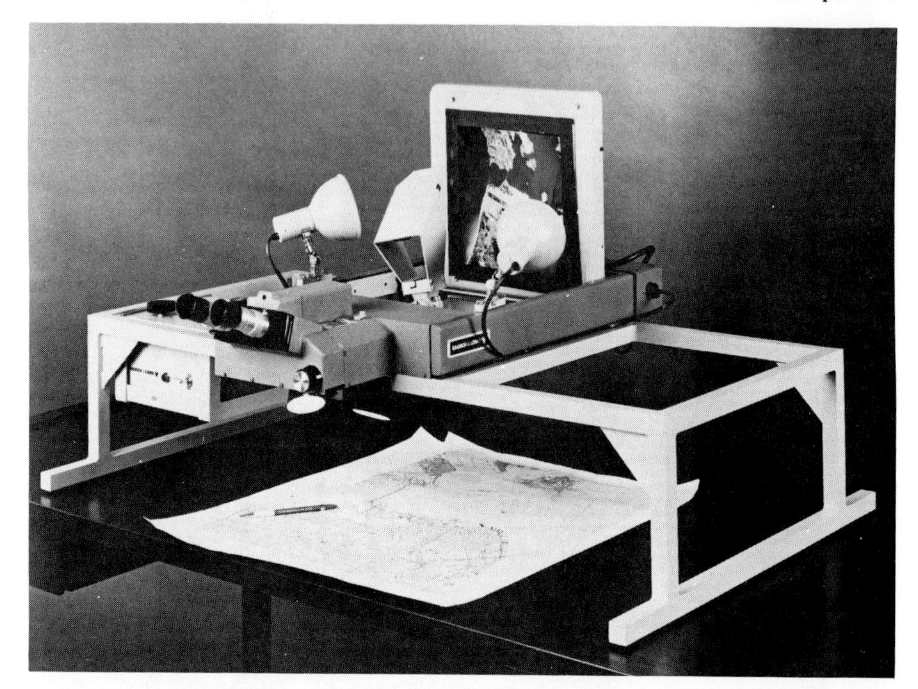

**Fig. 4.5.** Bausch and Lomb Zoom Transfer Scope, used for changing scale and transferring information from photographs to maps.

Limitations are mainly related to the deficiencies encountered in attempting to use stronger light sources for more flexibility of color combinations (since some of the color filters are exceedingly dense for some illuminants). There is also the problem of adequate registration of images for color enhancement; slight differences in geometry between images can cause density combinations that are confusing to interpret. In addition, a problem can result from the use either of uncalibrated data of variable quality or of images from two separate photographic missions. The densities to be enhanced must have some quantitative relationship to the features they represent; hence changes in time of photography, sun angle, film type, or even processing procedure can cause shifts which alter this relationship.

## B. Electronic Color Combiner

Commercially available color combiners operating electronically are offered by several manufacturers. A demand for this type of instrumentation was created with the announcement that LANDSAT 1 would carry multispectral instrumentation. Representative models are illustrated in Figures 4.6, 4.7, and 4.8. The advantages of this instrumentation includes brightness control for each channel of data. Isodensity enhancement can be performed. Output is best viewed on a screen, and documentation or recording is performed by photographing the screen with color film.

0-201-04245-2

**Fig. 4.6.** IDECS console and viewer arrangement. A complete technical and operational description of this system appears in a report by G. W. Dalke and J. E. Estes, prepared for USAETL under Contract No. DAAK 02-67-C-0435, 1968.

**Fig. 4.7.** Philco-Ford two-channel color enhancement console viewer. Not clearly visible in photo is the viewing aperture (approximately 5 × 7 in.), located in the dark area to the left of the console controls. These controls operate the mixing and digitizing of the two channels of imagery.

Electronic color combiners are quite flexible and generally quicker to set up than optical combiners. With the advent of LANDSAT they are widely available and can process large quantities of data more rapidly than other techniques. However, their chief role involves the color additive technique, and this technique does not appear to have achieved its original promise. More advanced equipment for analysis has the capacity to perform the color additive technique, plus others which have subsequently evolved.

### C. Optical-Electronic Combiner

An example of a system which incorporates both optical and electronic capabilities for enhancement purposes is the Philco-Ford console viewer shown in Figure 4.7. This viewer makes possible the combination of two multiband images, which are illuminated by a single source of light for projection and scanning. Beam-splitting optics provide this feature, which eliminates any concern for variability in the relative intensities of two or more illumination sources. This feature also reduces the number of flying spot scanners that the system requires. The scanner employed in the Philco-Ford viewer can resolve a total of approximately 1000 lines, whereas the currently used IDECS system employs scanners which can resolve only about 500 lines (the same as commercial television displays).

A special color wheel employed in this system results in the additive color mixing of red and green hues. Sixteen levels of density can be referenced within each channel and "digitized" for subsequent quantification. The "level-select" feature is nominally linear and can be expanded and contracted as with the IDECS system; each density level can also be modified should the image analyst prefer to weigh its importance in the display.

Among the most advanced of the photoanalyzers is General Electric's "Image 100" system (Figures 4.9 and 4.10). Designed to accept magnetic tapes, the system

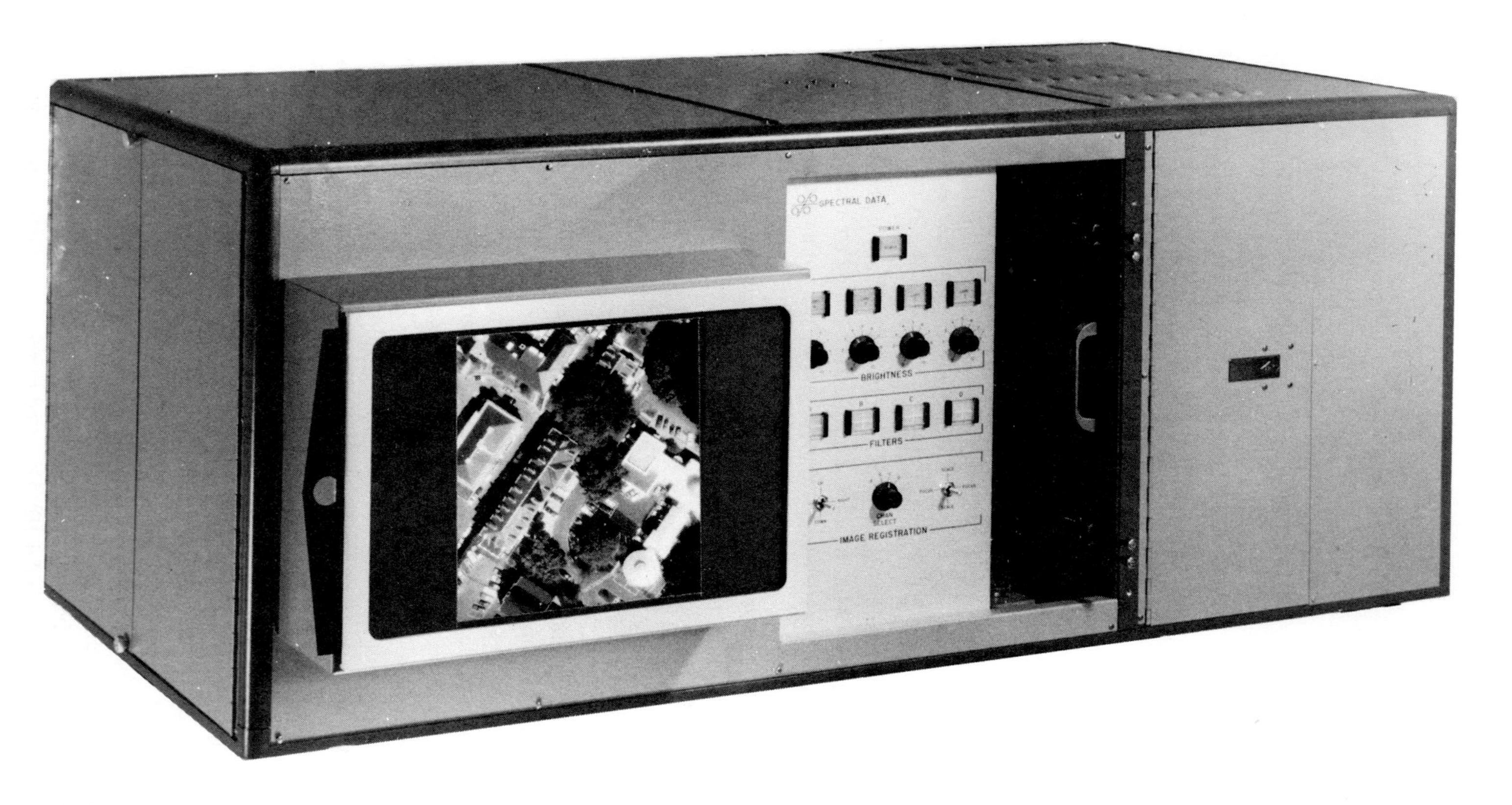


**Fig. 4.8.** Spectral data model 75 multispectral additive color viewer.

0-201-04245-2

**Fig. 4.9.** Image 100 photograph analyzer. This system accepts either magnetic tape or films and features a minicomputer for the implementation of feature extraction algorithms.

also contains a digitizer for use with films. A minicomputer gathers spectral data from known local areas and then compares data from other areas, deciding whether the latter meet criteria established in the local area. Classification decisions appear graphically on the cathode ray tube. Outputs include the colored CRT display, line printer, color film recorder, and digital tape. System schematics are shown in Figure 4.10.

Image 100 represents the current state of the art in photographic analyzers. This instrument is priced out of the reach of most universities but is available to qualified investigators at several points in North America. Among the locations of Image 100 systems are the Canadian Remote Sensing Unit in Ottawa; the U.S. Geological Survey facility at Sioux Falls, South Dakota; Jet Propulsion Laboratory in Pasadena, California; and Johnson Manned Space Center, Houston, Texas. A unit in Beltsville, Maryland, is available on a commercial basis at an hourly rate.

Broadly comparable instruments are offered by $I^2S$ Corporation of Sunnyvale, California and Bendix Aerospace Corporation of Ann Arbor, Michigan.

## D. Other Systems

Not all systems in existence today that utilize color as an enhancement feature for multiple images can be mentioned here. Many scientists are experimenting with devices and techniques that closely resemble the ones described herein. General Electric, RCA, and Hycon, for instance, all have very sophisticated color enhancement systems under development for a number of varied applications. A few "hybrid" optical systems can be found in various research laboratories. Some em-

0-201-04245-2

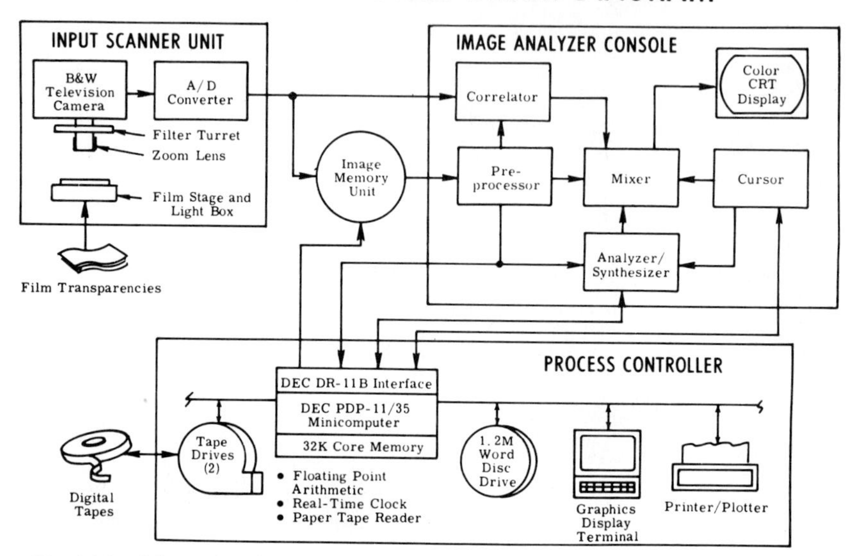


**Fig. 4.10.** Schematics of the Image 100 system.

ploy the same image format (3¼ X 4 in.), while others employ 35 mm transparencies. At the present time, all of these devices or systems can be regarded basically as research tools with which the image analyst can experiment. The need is great for electronic systems which can automate the enhancement procedure and which eventually will provide an on-line computing capability. Such equipment probably will be essential in the near future if the interpreter is to keep pace with the large volumes of imagery requiring analysis.

Although the additive color image enhancement techniques described in this section show great promise for earth resource surveys, they are still in the developmental stage. Additional research is needed in three areas: (1) determining which wavelengths of the electromagnetic spectrum are most useful for identifying each earth resource feature of interest; (2) determining which color combinations, of the many that can be used in producing image enhancements, are most easily and accurately discerned by the image analyst; and (3) determining in representative instances the net gain, if any, resulting from the use of such image enhancement techniques. Once this additional research has been performed, there is a strong possibility that nearly all conventional uses of aerial photographs, including their application for the inventory of various components of man's environment, can be facilitated through the acquisition of multiband photography to proper specifications and the employment of modern image enhancement techniques.

0-201-04245-2

Chapter 5

# Nonphotographic Optical Sensors

*Donald S. Lowe*

The optical region of the electromagnetic spectrum is the region where optical techniques of refraction and reflection can be used to focus and redirect radiation. The region extends from X-rays to microwaves and encompasses the ultraviolet, visible, and infrared. Although its limits cannot be precisely set, the practical limit at the shorter wavelengths is around 2000 Å, where conventional refracting materials and reflecting surfaces have low efficiencies. The long-wavelength bound is set by diffraction limits and detection techniques and is generally considered to be around 1000 $\mu$m. Within this broad spectral region, ranging from 0.2 to 1000 $\mu$m, optical sensing from air- or spaceborne platforms is limited to atmospheric windows between 0.32 and 13.5 $\mu$m. These windows, which are spectral intervals where the atmosphere is relatively transparent, are described in Chapter 2, which also discusses the meteorological factors that affect their transmittance.

## I. SENSOR MEASUREMENTS

With few exceptions, remote sensors, optical or otherwise, do not measure or observe directly the information being sought by the user. Rather, they observe or measure the radiation from elements in a scene, and from these data the user deduces the information he is seeking. For example, an infrared scanner cannot directly observe moisture stress in vegetation or locate mineral deposits. It only senses variations in radiant emission arising from differences in surface temperature or emissivity. The surface temperature may be related, however, to moisture stress or mineral deposits, and in that case a potential for detecting moisture stress or mineral deposits exists. The sensor, then, observes an effect which is exhibited as a change in the radiation characteristics of the material. The user generally is directly interested not in the observed effect (say a high-temperature leaf) but in the cause which produces it. Thus in a remote sensing program one attempts to find a reliably predictable relationship between a cause and a radiation observable effect, the cause being the information the user is seeking and the effect being the phenomenon observed by the sensor.

### A. Radiation Observables

The characteristics of electromagnetic radiation which can be observed are as follows.

0-201-04245-2

*Remote Sensing of Environment*, Joseph Lintz, Jr. and David S. Simonett (eds.) ISBN 0-201-04245-2

*Spatial distribution.* The variation in radiation coming from elements in a scene is the source of contrast between an object, or elements of an object, and its background. This permits recognition of an object through information as to its shape, texture, and environmental context.

*Spectral distribution.* The radiant intensity may vary with wavelength. In the visible region, this produces color.

*Polarization.* The radiation emitted or reflected by an object and its background may have polarization differences.

*Temporal variations.* The radiation from an object may vary with time. These may be changing variations, such as leaves fluttering in the wind, or slowly varying effects, such as seasonal changes.

*Goniometric variations.* All of the above-mentioned parameters may be angular dependent and thus influenced by the geometry of viewing.

The objective in most remote sensing research programs is to find a relationship between the radiation attributes of an object and the information that the user needs. This research may require laboratory or field measurements, or both.

Of the radiation attributes listed above, the most information is to be found in the spatial and spectral characteristics. This chapter will direct its discussion to optical-mechanical scanners for producing images, nonimaging spectrometers, and multispectral scanners, which combine the optical-mechanical scanner with a multichannel spectrometer.

## B. Energy Available for Sensing

Optical sensors can be active or passive. In an active system, the sensor provides the illumination of the object under observation. An example of an active system is a laser ranging device. Most remote sensors are passive, relying on solar illumination or self-emission. Figure 5.1 gives the maximum spectral radiant energy available from a scene during daylight hours. This curve is a composite of the solar irradiance outside the earth's atmosphere (approximated as having the spectral distribution of a 5800°K blackbody) and the emittance of a 300°K blackbody. Below 3.5 $\mu$m the energy is predominantly reflected sunlight, while at wavelengths beyond 4 $\mu$m the energy is predominantly thermal emission. For daylight operation, the minimum spectral radiance lies between 3 and 4 $\mu$m. Figure 5.1 is only approximate, as the radiant energy from a surface is a function of its reflectivity, emissivity, temperature, and solar irradiance. The magnitude of solar irradiance varies with atmospheric conditions, atmospheric slant path, and atmospheric absorption. Figure 5.2 shows the measured spectral distribution of solar irradiance outside the earth's atmosphere, as well as at the earth's surface.

The curve in Figure 5.1, which is peaked at 10 $\mu$m, shows the spectral radiant emittance of a 300°K blackbody. Emission of this magnitude can be observed from most terrestrial objects during night or day, as all objects emit radiation and the magnitude of this emitted radiation is dependent solely on the object's surface temperature and emissivity. The nature of thermal emission will be discussed briefly,

0-201-04245-2

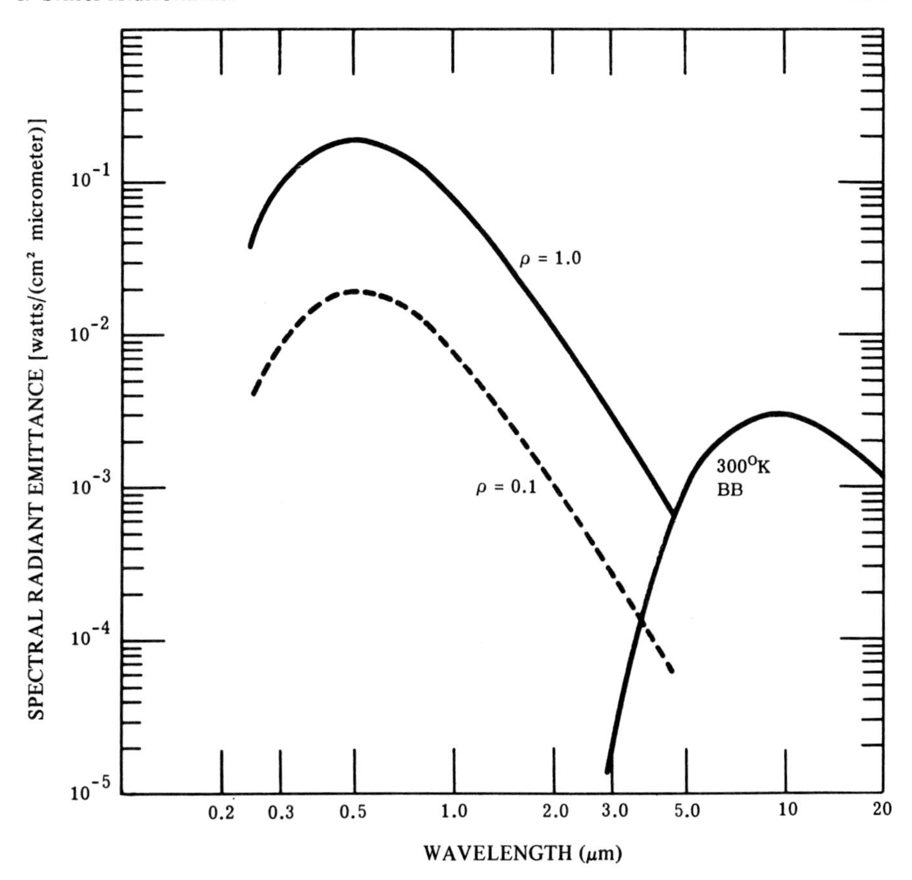


**Fig. 5.1.** Energy available for remote sensing.

along with factors which produce variations in radiance among objects and their backgrounds.

Thermal emission from an object can best be understood and described by the concept of a blackbody, an opaque object with zero reflectance. Although such a body exists only theoretically, it can be approximated by a small opening in a hollow cavity (Wolfe, 1965). The spectral radiant emittance, $M_\lambda$,* of a blackbody is given by Planck's law:

$$M_\lambda = \frac{c_1}{\lambda^5(e^{c_2/\lambda T} - 1)}, \tag{5.1}$$

where $M_\lambda$ = the spectral radiant emittance [W/(cm² · μm)] at wavelength λ,

$c_1 = 2\pi c^2 h = 3.74 \times 10^4$ W/[cm² · (μm)⁴]

0-201-04245-2

*For a description of spectroradiometric symbols and nomenclature see *Jour. Opt. Soc. Amer.* 57, 854 (1967).

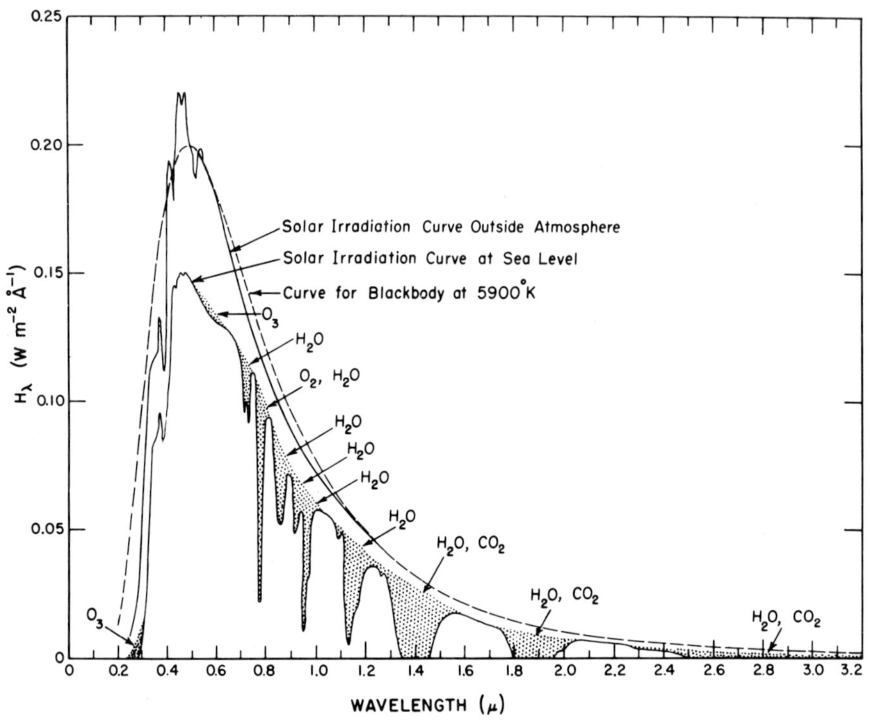


**Fig. 5.2.** Spectral distribution of solar irradiance. Shaded areas indicate atmospheric absorption. From Valley (1965).

$c_2 = hc/k = 1.439 \times 10^4 \ \mu\text{m} \ ^\circ\text{K},$

$T$ = surface temperature (°K).

The total radiant emittance can be obtained by integrating this equation over all wavelengths. This yields the well-known Stephan-Boltzmann law:

$$M = \int_0^\infty M_\lambda \, d\lambda = \sigma T^4, \tag{5.2}$$

where $\sigma = 5.67 \times 10^{-12}$ W/(cm$^2 \cdot {}^\circ$K$^4$)

Another useful expression for describing blackbody emission is Wein's displacement law, a simple expression which gives the wavelength where the peak of the spectral emittance curve lies for any given temperature:

$$\lambda_m T = 2897.9 \ \mu\text{m} \ ^\circ\text{K}. \tag{5.3}$$

Tables (Nagel and Pivovonsky, 1961) and slide rules (Wolfe, 1965) are available from which the spectral radiances of blackbodies can be readily and accurately

0-201-04245-2

determined. For most applications, the General Electric radiation slide rule (obtainable from General Electric Company, Utica, New York) provides the most expedient means for computing or estimating the radiation characteristics of a blackbody.

The self-emission from a nonblackbody of a given temperature can be expressed as a fraction, $\epsilon$, of the emission from a blackbody of the same temperature:

$$M_\lambda \text{ (nonblackbody)} = \epsilon(\lambda) M_\lambda \text{ (blackbody)}, \tag{5.4}$$

where $\epsilon(\lambda)$ is the spectral emissivity of the nonblackbody. As will be shown later, the spectral emissivity of an opaque object is related to its spectral reflectance.

Radiation incident on an object must be reflected, absorbed, or transmitted. Conservation of energy requires that

$$\rho + \alpha + \tau = 1, \tag{5.5}$$

where $\rho$ = the fraction reflected,

$\alpha$ = the fraction absorbed,

$\tau$ = the fraction transmitted.

For an opaque object, $\tau = 0$ and $\rho + \alpha = 1$.

From Kirchhoff's law, the ability of an object to emit radiation, $\epsilon(\lambda)$, is equal to its ability to absorb radiation, $\alpha(\lambda)$:

$$\epsilon(\lambda) = \alpha(\lambda) = 1 - \rho(\lambda). \tag{5.6}$$

The significance of Kirchhoff's law is obvious. In a thermal equilibrium situation (such as an object in a closed room), the fraction absorbed must be equal to the fraction emitted. Otherwise the object would heat up or cool down and become warmer or cooler than its environment.

An airborne sensor observes radiation emitted and reflected by the terrain as modified by the intervening atmosphere. The apparent spectral radiance, $L_\lambda$, of a terrain feature has three contributing factors: a self-emission component, a reflected component, and an intervening atmospheric emission and scattering component:

$$L_\lambda = \frac{\tau(\lambda)\epsilon(\lambda)}{\pi} M_\lambda + \tau(\lambda)\,[1 - \epsilon(\lambda)]\, \frac{E_{\lambda(\text{sky})}}{\pi} + \frac{[1 - \tau(\lambda)]\, M_{\lambda(\text{air})}}{\pi}, \tag{5.7}$$

where $\tau(\lambda)$ = the spectral transmittance of the intervening atmosphere,

$\epsilon(\lambda)$ = the spectral emittance of the terrain,

0-201-04245-2

$E_{\lambda(\text{sky})}$ = the downwelling spectral irradiance of the objects under observation,

$M_{\lambda(\text{air})}$ = the spectral emittance of a blackbody at the temperature of the air path (assumed to be uniform).

In the thermal infrared region, then, the apparent radiance of an object as seen from a distance is made up of three components: an emitted term, a reflected term, and emission from the intervening atmosphere. At shorter wavelengths where self-emission is negligible and reflected solar energy predominates, the apparent radiance of an object is made up of two components: reflected downwelling irradiance and radiation scattered by the intervening atmosphere.

## II. DESIGN CONSIDERATIONS

Most nonphotographic optical sensors can be classified as electro-optical systems since their detectors are transducers which convert electromagnetic radiation into detectable electrical signals. These sensors may be designed to generate images of a scene or merely measure the radiation from an object at which they are pointed. Regardless of their functional requirements, all optical sensors have an optical head, an electronics package, and an output unit. In a simple system, these may be packaged as a single entity. Some of the components and functions to be found in these subsystems are the following:

1. Optical Head

   A *collector* which gathers the incident radiation and condenses it into the detector.

   A *field stop* for defining the field of view.

   A *reticle or chopper* for encoding the radiation.

   A *reference source* for calibration.

   An *optical filter* which, together with the detector, defines the wavelength region of response.

   A *detector* to convert the incident radiation into an electrical signal.

   A *preamplifier* to increase the signal level before further handling or processing.

2. Electronics Package

   A *signal processor* to format the data as required.

   An *amplifier* to increase the signal level to the desired output level.

   A *power distribution unit* to distribute electrical power as needed.

   A *control panel* for initiating and controlling measurements.

3. Output Unit

   An *indicator* or display for monitoring the output signal.

   A *recorder* for data storage.

0-201-04525-2

### A. Radiant Power Transfer in Optical Sensors

The performance of a sensor can be defined in terms of parameters such as field of view, resolution, and signal-to-noise ratio ($S/N$). These performance requirements, in turn, set the design specifications of the sensor. Tradeoff equations will be developed to determine the relationship among performance parameters and design parameters and constraints.

In remote sensing applications, one wishes to measure the radiance or spectral radiance of a source. For these measurements to be meaningful, the object under observation must fill the field of view of the measuring instrument. Otherwise, the instrument is measuring the emission from a mixture of objects. When a source fills the field of view of an instrument, it is said to be an extended source. The radiant power transfer from an extended source to the detecting element of the sensor will now be considered.

Figure 5.3 shows the geometric relationship between a radiating source and a generalized optical sensor. The spectral radiant power, $P_\lambda$, which enters the optical system and ultimately falls in the detector is given by

$$P_\lambda = \frac{\tau(\lambda)\tau_0(\lambda)A_s A_c L_\lambda}{R^2}, \tag{5.8}$$

where $A_s$ = the area of the extended source which is within the sensor field of view,

$A_c$ = the effective area of the collector optics,

$L_\lambda$ = the spectral radiance of the source [W/(cm$^2$ · sr · $\mu$m)]

$R$ = the distance to the source,

$\tau(\lambda)$ = the transmission of the intervening atmosphere,

$\tau_0(\lambda)$ = the transmission of the optics.

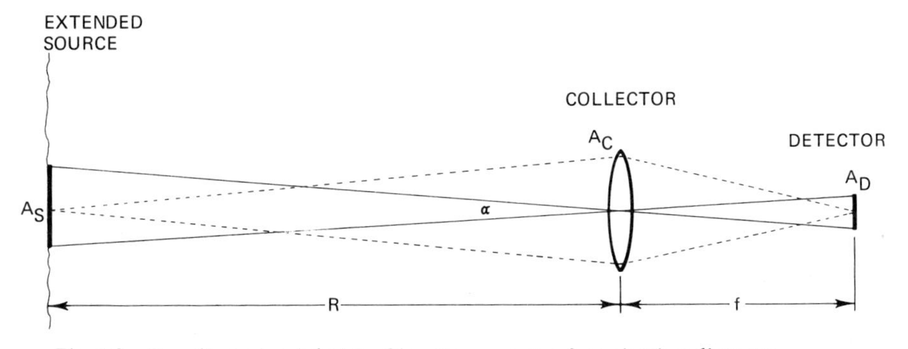


Fig. 5.3. Two-dimensional sketch of imaging geometry for a simple radiometer.

0-201-04245-2

The power incident on the detector in a given wavelength band can be obtained by integrating Eq. (5.8) over the wavelength limits of the band.

This power, $P$, incident on the detector produces an $S/N$ given by:

$$S/N = \int_{\lambda 1}^{\lambda 2} \frac{P_\lambda \, d\lambda}{NEP_\lambda}, \tag{5.9}$$

where $NEP_\lambda$ is the spectral noise equivalent power of the detector.

For many quantum detectors, the $NEP_\lambda$ is given by:

$$NEP_\lambda = \frac{\sqrt{A_D \, \Delta f}}{D_\lambda^*}, \tag{5.10}$$

where $A_D$ = the area of the detector,

$\Delta f$ = the electronic band width (which is inversely proportional to the observation time),

$D_\lambda^*$ = the spectral detectivity.

Figure 5.4 plots the spectral detectivities of a number of infrared detectors. Combining Eqs. (5.8)-(5.10) yields:

$$S/N = \int_{\lambda 1}^{\lambda 2} \frac{P_\lambda D_\lambda^* d\lambda}{\sqrt{A_D \Delta f}} = \int_{\lambda 1}^{\lambda 2} \frac{\tau(\lambda)\tau_0(\lambda) A_s A_c L_\lambda D_\lambda^* \, d\lambda}{R^2 \sqrt{A_D \, \Delta f}}. \tag{5.11}$$

From Figure 5.3, it can be seen that $A_s/R^2$ is the solid angle of the field of view and, for a square detector, is given by:

$$\alpha^2 = \frac{A_s}{R^2} = \frac{A_D}{f^2}, \tag{5.12}$$

where $f$ is the focal length of the collector.

Substitution of Eq. (5.12) into Eq. (5.11) yields:

$$S/N \int_{\lambda 1}^{\lambda 2} = \frac{\tau(\lambda)\tau_0(\lambda)\alpha A_c L_\lambda D_\lambda^* d\lambda}{f\sqrt{\Delta f}}. \tag{5.13}$$

The magnitude of the $S/N$ is a measure of how well a signal can be observed and in an imaging sensor determines the maximum contrast obtainable in an image. The concept of $S/N$ is shown in Figure 5.5. The output of a detector is continuous with time, and noise is that variation in the output level which cannot be predicted.

0-201-04245-2

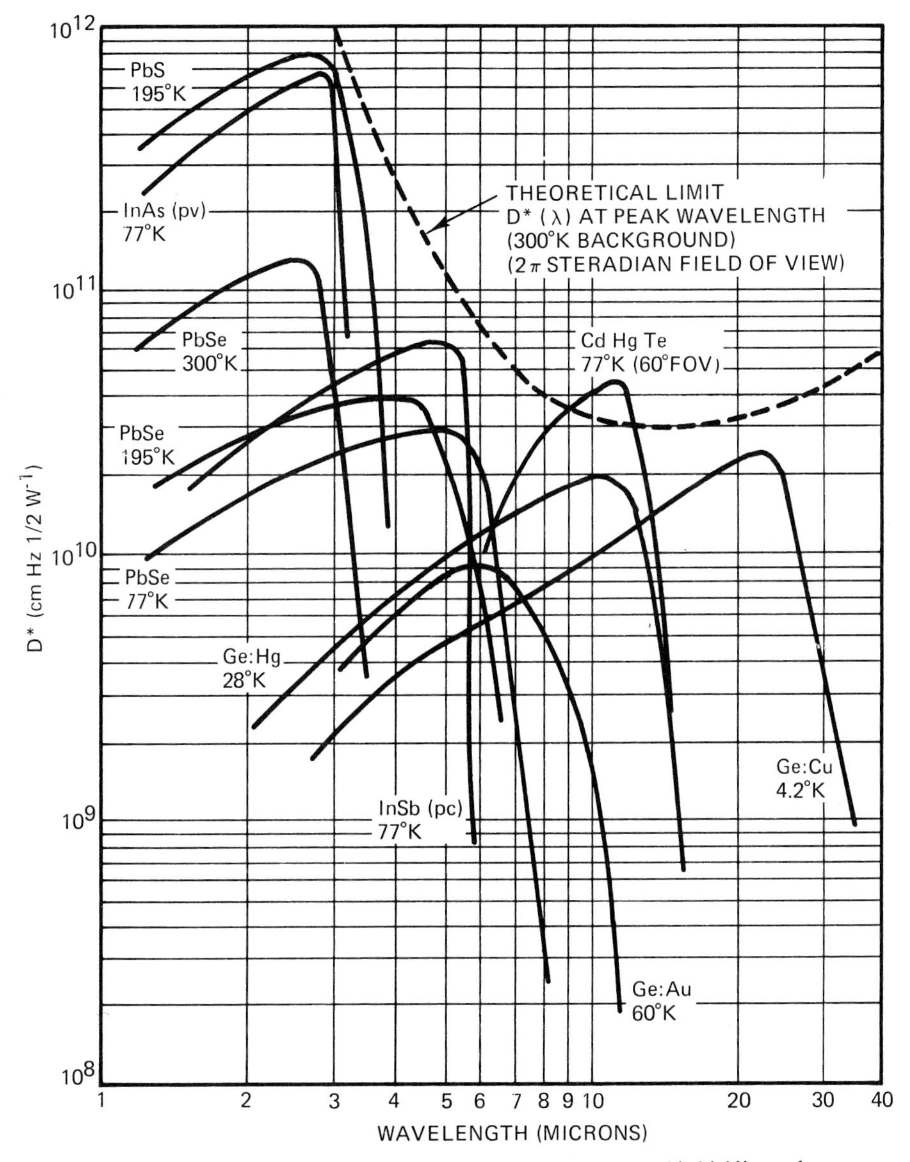


**Fig. 5.4.** Spectral detectivity of infrared quantum detectors. No cold shielding unless indicated.

Hence noise limits the accuracy to which a signal can be measured and determines the minimum detectable signal.

The $S/N$ as given by Eq. (5.13) is valid for all electro-optical sensors, be they radiometers, spectroradiometers, or imaging scanners. Many variations of this equation are to be found in the literature, because many of the design parameters are

0-201-04245-2

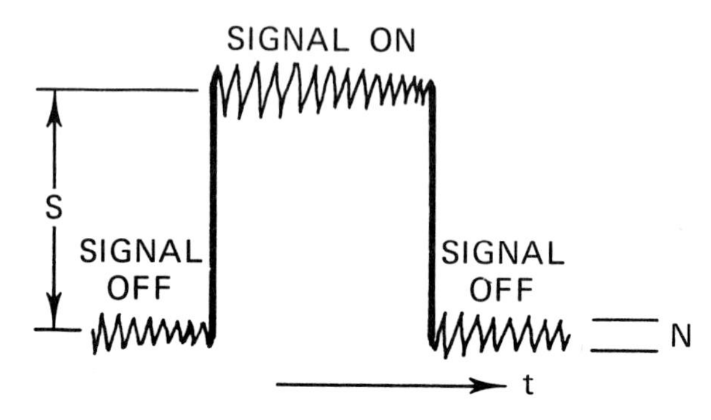


**Fig. 5.5.** Output signal as a function of time.

interrelated and can be expressed in other ways. For example, the collector area can be expressed as $\pi D^2/4$, and the aperture ratio, $F$, of the optical system is $f/D$, where $D$ is the diameter of the collecting aperture. When these substitutions are made, Eq. (5.13) can be written in a more familiar form:

$$S/N = \int_{\lambda 1}^{\lambda 2} \frac{\tau(\lambda)\tau_0(\lambda)\alpha\pi D D_\lambda^* L_\lambda d_\lambda}{4F\sqrt{\Delta f}}. \tag{5.14}$$

Equations (5.13) and (5.14) were derived on the assumption that the detector or its image served as the field stop, that is, $\alpha = \sqrt{A_D}/f$. It can be shown that these equations are valid for any optical system, no matter how complex, provided the optics following the field stop are nonvignetting (Wolfe, 1965). For Eqs. (5.13) and (5.14) to be valid for such systems, the aperture ratio, $F$, must be defined as the angle of convergence of the radiation falling on the detectors.

## B. Radiation Noise-Limited Detectors

In many modern detectors, the limiting noise arises from fluctuations in the radiation falling on them or fluctuations in the generation of conducting electrons. Both of these phenomena occur even when the radiation incident on the detector is considered to be constant. In the first case, the variation occurs because of the quantum nature of radiation and the statistical variation in the flow of quanta arriving at the detector. Such noise is evident in cooled detectors such as the Ge:Hg. In the second case, the probability of freeing an electron by the interaction of radiation with matter is statistical in nature. Thus, in photoemissive detectors where the current flow is proportional to the number of photoemitted electrons, a noise would be generated even if the incident radiation were constant. This noise appears as a variation in current.

Both of these noises are produced whenever radiation is incident on the detector, and the noise is proportional to the square root of the magnitude of the inci-

0-201-04245-2

dent radiation. When using such a detector, one finds that the signal-to-noise relationship expressed in Eq. (5.14) is no longer valid. The $D^*_\lambda$ of a radiation-limited thermal detector is proportional to the square root of its angular field of view as defined by its cold shield:

$$D^{**}_\lambda \text{ (cold shield)} = \frac{D^*_\lambda \text{ (unshielded)}}{\sin \theta/2}, \tag{5.15}$$

where $\theta/2$ is the half-angle of the acceptance beam. When the cold shield is designed to accept only radiation from the last optical element, $\sin \theta/2$ is equal to the reciprocal of twice the effective $f$/no. of the optical system, that is,

$$\sin \frac{\theta}{2} = \frac{D}{2f} = \frac{1}{2F}. \tag{5.16}$$

An example of improved performance through cold shielding can be seen in Figure 5.4, where the only detector with a cold shield (Hg:Cd:Te) exceeds the theoretical limit of unshielded detectors.

When using a cold-shielded detector, $D^*$ in Eq. (5.14) should be replaced by $D^{**}$ as defined by Eq. (5.15). In this case:

$$S/N = \int_{\lambda 1}^{\lambda 2} \frac{\tau(\lambda)\tau_0(\lambda)\alpha\pi D D^*_\lambda L_\lambda d_\lambda}{2\sqrt{\Delta f}}, \tag{5.17}$$

and $D^*_\lambda$ is the detectivity of the detector before cold shielding.

In this situation, the resulting $S/N$ is not directly a function of the aperture ratio of the optical system. This does not mean, however, that the aperture ratio does not figure in the design considerations; physical constraints or limitations in detector size may dictate the aperture ratio of a system. For example, suppose that one wants a small instantaneous field of view (resolution element in a scanning system) and there is a limit as to the smallest available detector size. The higher the resolution, the longer the focal length must be, since $\alpha^2 = A_D/f^2$. If the collector diameter is fixed, the aperture ratio ($f$/no.) must increase with resolution.

## III. INFRARED SCANNERS

The techniques employed in infrared scanners are no different from those used in early television systems employing optical-mechanical scanning techniques (Morton and Zworykin, 1940). Compared with photographic cameras of comparable resolution, the optical-mechanical scanner is a complex instrument. Some of the advantages of scanners are the following:

1. One can obtain imagery outside the photographic region.

0-201-04245-2

2. The output signal is in electrical form and can be readily transmitted, recorded, analyzed, or processed as needed or desired.
3. The detectors generally have a wider dynamic range than photographic film, and the detection process is reversible, that is, the detectors are not consumed in the detection process.
4. The scanner is amenable to calibration, and the resulting data can yield quantitative radiometric data.
5. Collection of data simultaneously in many wavelength channels is possible.

By far, the most successful thermal imaging devices are optical-mechanical scanners which use point detectors. Scanners can be classified as image plane or object plane. In the former, the entire field of view is imaged with wide-angle optics and the image is scanned by, or moved relative to, the detector. In object plane scanners, the optics are used on axis and the object field is scanned by the optical telescope. This difference is illustrated in Figure 5.6, which compares two elementary image and object plane scan techniques.

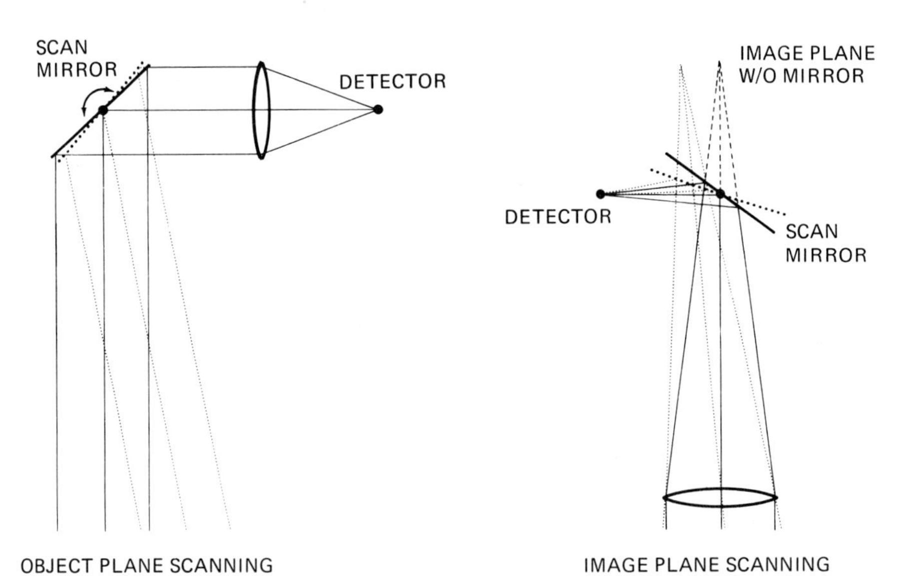


**Fig. 5.6.** Examples of image and object plane scanning.

## A. Design Considerations

Many of the optical-mechanical scanning techniques were used in early television systems. Several examples of image plane scanning devices are shown in Figure 5.7. A major disadvantage of image plane scanning is the need for wide-angle optics. Other disadvantages are governed by the exact design. For example, the Nipkow disk suffers from the relatively large aperture ratio, $F$, of the system and the large

0-201-04245-2

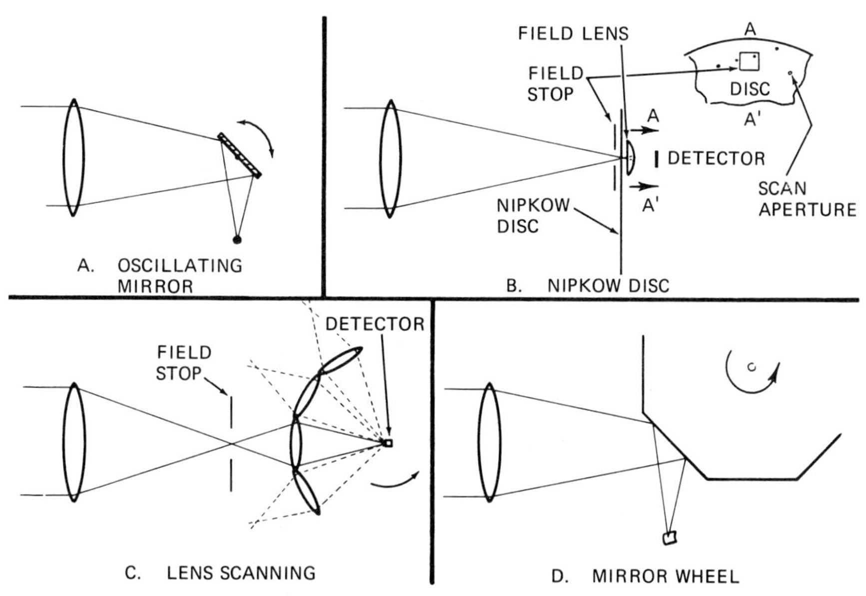


**Fig. 5.7.** Some image plane scanning techniques.

size of the disk. Consider a simple system scanning 100 lines with an *f*/2.54-in. aperture collector, and assume a field view of 5°. The field stop would be 0.87 in. and the circumference of the disk would be 87 in., which corresponds to a radius of 14 in. With an *f*/1 field lens, the holes in a 100-line disk would reduce the effective *f*/no. of the field lens to *f*/100. Such an optical system is extremely inefficient.

By far the most functional infrared imaging device is the airborne reconnaissance scanner. This object plane scanner scans lines normal to the motion of the aircraft at such a rate as to generate continuous scan lines, as shown in Figure 5.8.

The field of view of an infrared telescope (radiometer) is reflected from a rotating mirror that scans the ground in lines perpendicular to the flight path of the aircraft. The scan rate is adjusted so that succeeding scan lines are adjacent or overlapping as the aircraft moves forward. Thus, the rotating mirror provides the scanning motion normal to the flight path, while the aircraft motion advances the scan pattern. The output signal from the detector is amplified and used to intensity-modulate a light source. The result is a graphic image of the scene radiance. Examples of infrared strip maps are given in Figure 5.9. As in photography, objects are recognized in the imagery on the basis of shape, texture, or context within an environment. Since an object must be contrasted with its background in order to be detected, much less recognized, it is important to select the wavelength interval so that the scanner has adequate $S/N$ and the contrast between the object being sought and its background is maximized.

Equation (5.14) contains four terms which are wavelength dependent. In many applications, the integration of the product of these terms is unnecessarily complex, particularly in narrow spectral intervals where the spectral variation may not be

0-201-04245-2

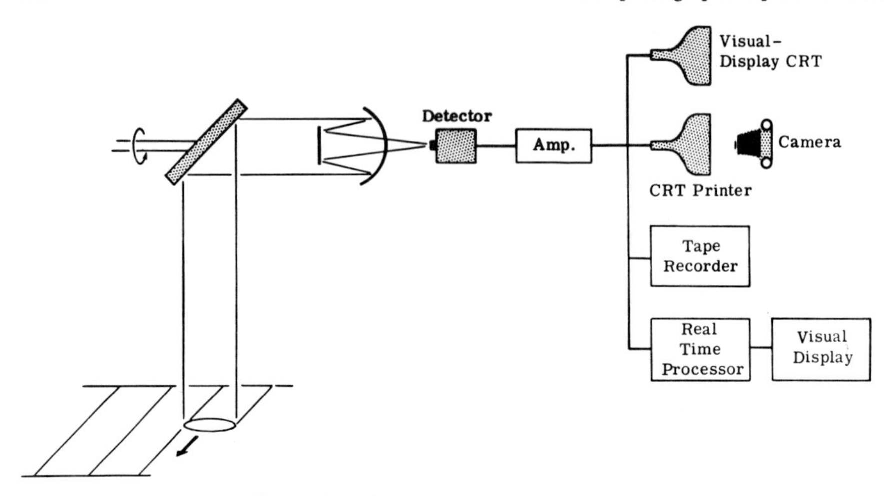


**Fig. 5.8.** Schematic of optical-mechanical scanner.

significant. In the rest of this chapter, average values of the spectrally varying parameters will be used for the sake of simplicity. When using average values, however, caution should be exercised to limit oneself to either very narrow spectral intervals or regions where these parameters are slowly varying. When there is doubt or when precision is required, the integral form should be used.

Equation (5.14) gives the $S/N$ at the detector output of a sensor observing an extended source of a given radiance. The $S/N$ of the sensor would be the detector $S/N$ divided by the noise factor, $n_f$, of the electronics. More important than the $S/N$ when viewing a single-resolution element is the $S/N$ of the signal difference between two adjacent resolution elements, as this is a measure of the ability of a sensor to distinguish between two sources. Using average values for the spectrally varying parameters, dividing by the electronics noise figure, and substituting $\Delta L$, the radiance difference between two sources, for $L$ in Eq. (5.14) yields the $S/N$ for an imaging system:

$$S/N = \frac{\tau\tau_0 \pi \alpha D D^* \,\Delta L}{4 n_f F \sqrt{\Delta f}} . \tag{5.18}$$

The electronic band pass of an imaging scanner is directly proportional to the data rate. For a scanner which scans straight lines without under- or overlap, as illustrated in Figure 5.8, the optimized band width for detecting a point target is given (Braithwaite, 1966) by:

$$\Delta f = \frac{1}{2t} = \frac{(V/h)\theta}{2\alpha^2 p\gamma} , \tag{5.19}$$

where $t$ = dwell time per resolution element,

$V/h$ = sensor platform velocity-to-height ratio,

0-201-04245-2

0-201-04245-2

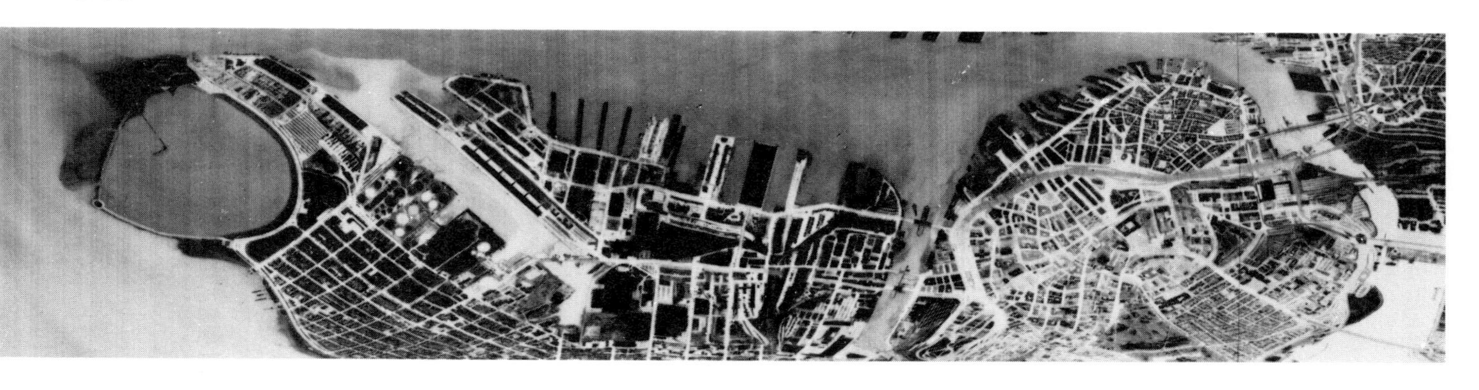

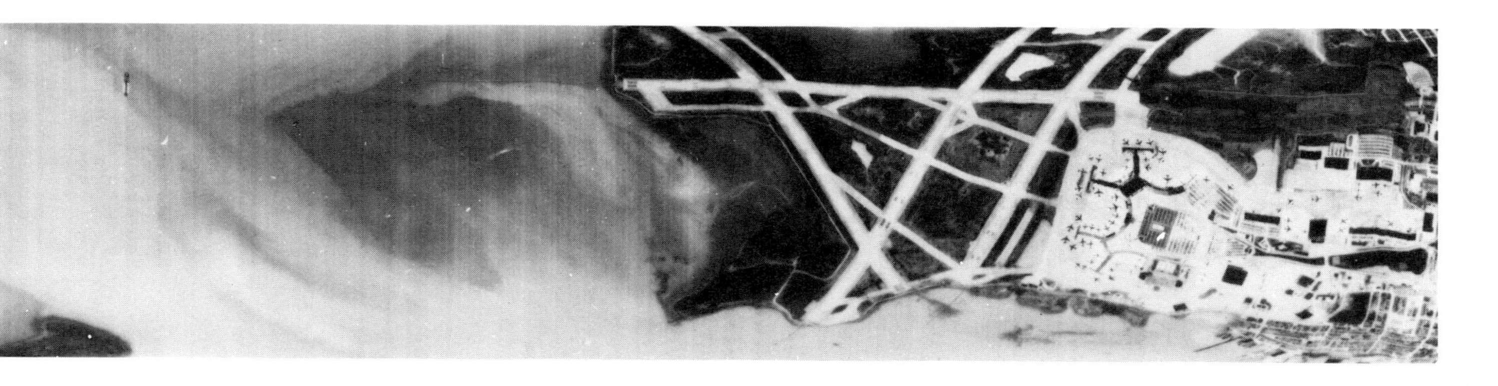

Fig. 5.9. Infrared strip maps of Boston Harbor.

Alt: 3000 Ft
Time: 3 a.m., 11 June 1969
Detector: InSb
Scanner: 3-in. Aperture, Bendix

$\theta$ = angle of coverage lateral to the flight path,

$\alpha$ = angular resolution,

$p$ = number of detecting elements (number of lines scanned per sweep),

$\gamma$ = scan duty cycle or efficiency.

Substitution of Eq. (5.19) in Eq. (5.18) yields:

$$S/N = \frac{\tau\tau_0 \pi\alpha^2 DD^* \Delta L \sqrt{2p\gamma}}{4n_f F \sqrt{(V/h)\theta}} . \tag{5.20}$$

Equation (5.20) gives the relationship between the sensor output $S/N$ and the design parameters. A more useful expression is the noise equivalent temperature difference, $NE\,\Delta T$, which is defined as the temperature difference between two adjacent resolution elements that produces a peak signal-to-RMS noise ratio of 1 at the sensor output. It is commonly assumed that the two sources are radiating as blackbodies at approximately 300°K.

Referring to Eq. (5.20), it is apparent that $NE\,\Delta T$ is the reciprocal of the $S/N$ when observing the radiance difference between two blackbody sources having a temperature difference of 1°K; for example, if a temperature difference of 1°K produces an $S/N = 6$, the $NE\,\Delta T = 1/6$°K. Thus:

$$NE\,\Delta T = \frac{4n_f F \sqrt{(V/h)\theta}}{\tau\tau_0 \pi\alpha^2 DD^* (\Delta L/\Delta T) \sqrt{2p\gamma}} . \tag{5.21}$$

Figure 10 gives values of $\Delta L/\Delta T$ for a number of spectral bands commonly used in thermal imaging.

As an illustration of the current capability of scanners, assume a simple, single-mirror scan system, which is shown schematically in Figure 5.8 and has the following design parameters:

$n_f = 2, \quad F = 2, \quad V/h = 0.2;$

$\theta = 120° = 2.1 \text{ rad}, \quad \gamma = 120°/360° = \frac{1}{3};$

$\tau = 0.8, \quad \tau_0 = 0.6, \quad \alpha = 3 \times 10^{-3} \text{ rad};$

$D = 8 \text{ cm}, \quad D^* (8\text{–}14\ \mu\text{m}) = 10^{10} \text{ cm/W};$

$\Delta L\ (8\text{–}14\ \mu)/\Delta T = 7 \times 10^{-5} \text{ W/(cm}^2 \cdot \text{sr} \cdot °\text{K});$

$p = 1.$

Substitution in Eq. (5.21) yields $NE\,\Delta T = 0.17$°K.

0-201-04245-2

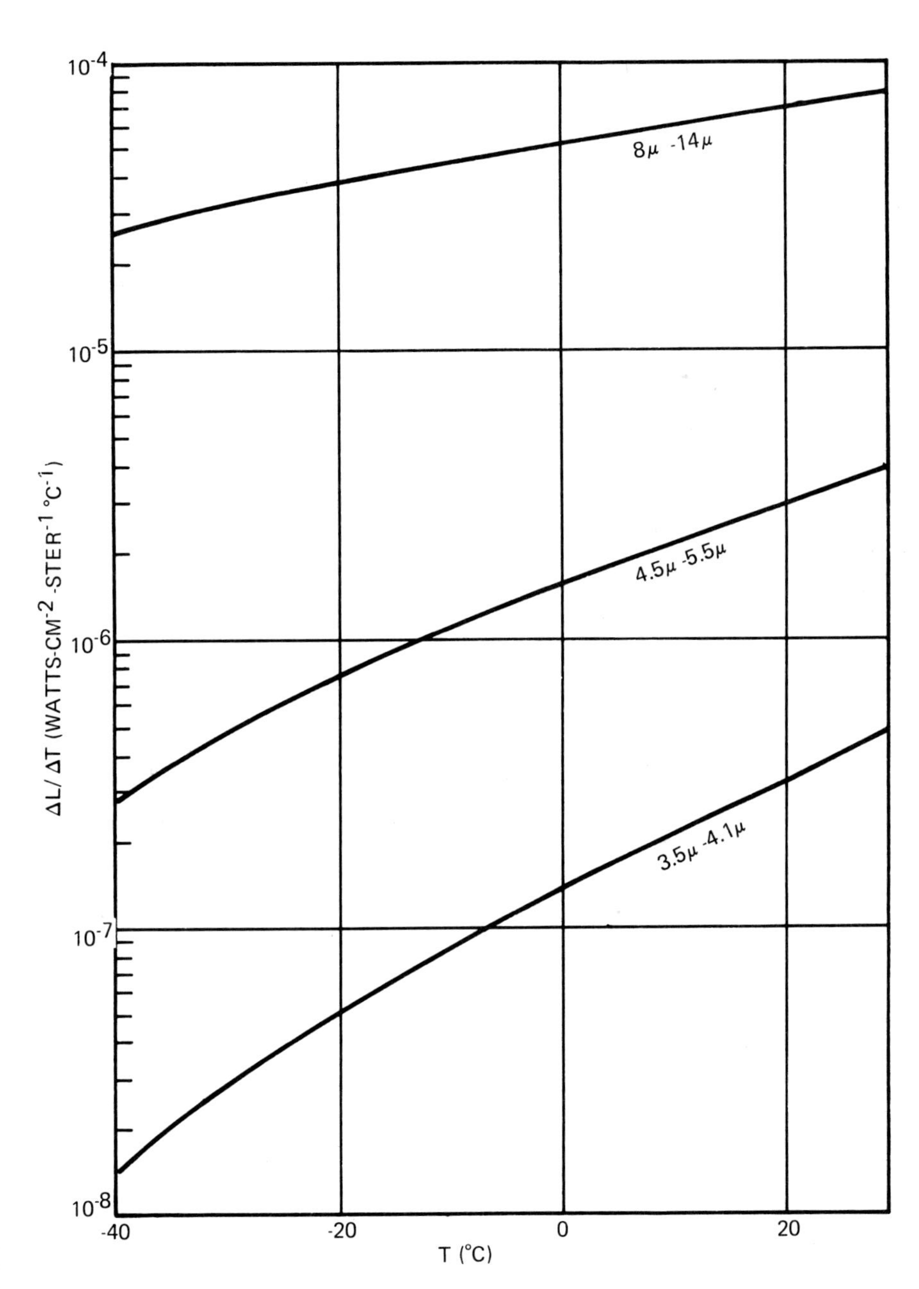


**Fig. 5.10.** Change in blackbody radiance for a 1°K temperature difference, plotted as a function of ambient temperature for commonly used spectral bands.

0-201-04245-2

## B. Image Interpretation

The procedures for interpreting infrared imagery are the same as in photography. One observes shape, notes tone, evaluates texture, and associates an object with its environment. The details of infrared interpretation differ from those of photographic interpretation in the significance of tone, format, and resolution.

### *1. Origin of Tone*

As discussed in Section IB of this chapter, the tone of an object is a function of its surface temperature, emissivity, and downwelling radiation. The surface temperature undergoes diurnal variations and is by far the most significant factor for producing tonal variations in a scene. The driving forces which determine the surface temperature are shown in Figure 5.11 (Geiger, 1957). During daytime, the largest factor is direct sunlight, which heats the earth. At night, the predominant factor is radiation cooling. These prime factors are affected by meteorological conditions, thermal properties of the object, and internal heat generation (such as man-made heat sources or naturally occurring oxidation).

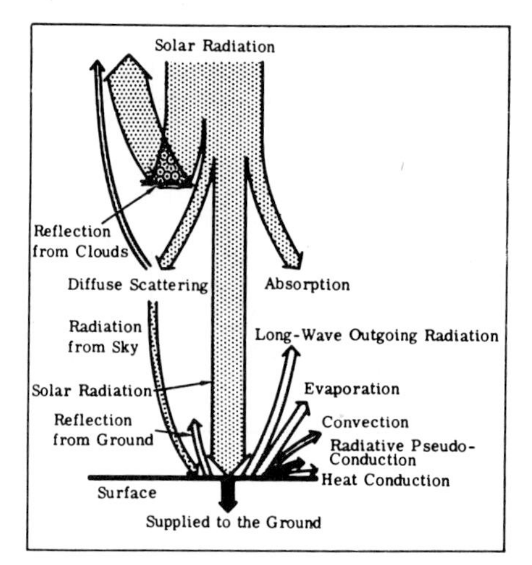


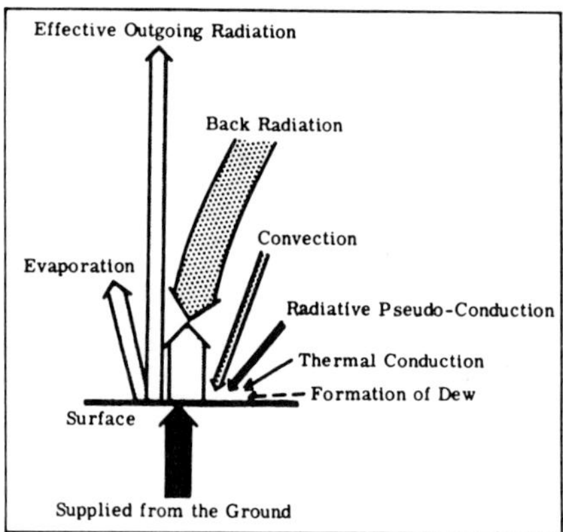


**Fig. 5.11.** Energy budget at the surface of the earth for day (left) and night (right). From Geiger (1957).

### *2. Distortions*

Line scan devices differ from conventional photography in several features of presentation. These differences are all associated with the method of data generation; that is, line scanners collect image data serially. Unlike a framing camera, which has a nadir point, an infrared strip map has a nadir line. Thus, parallax distortions in strip maps occur only normal to the flight line. The top of a flagpole off the center of the strip map will lean perpendicularly to the ground track, as illustrated in Figure 5.12.

0-201-04245-2

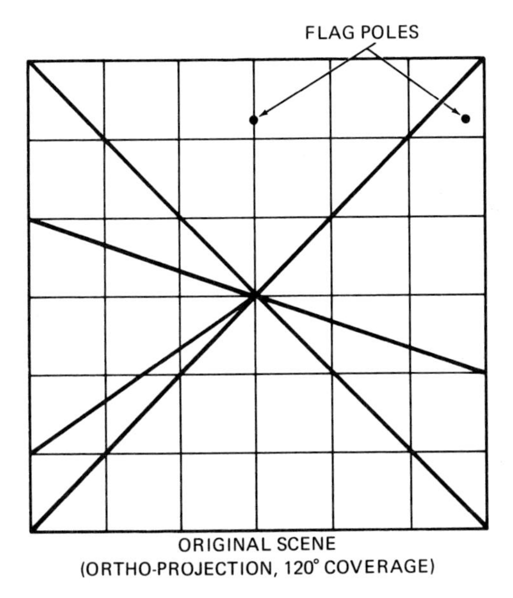


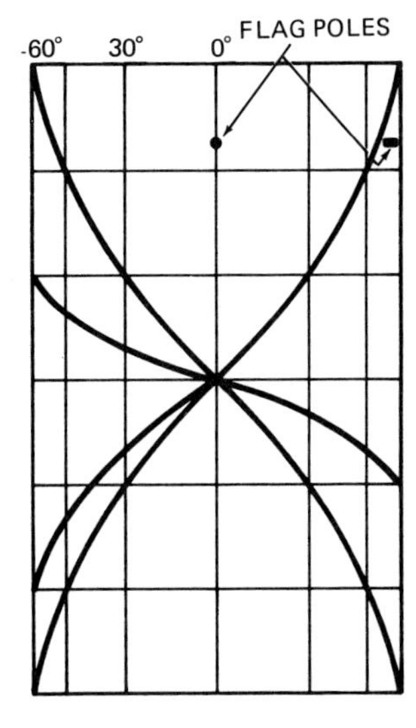


**Fig. 5.12.** Scanner parallax and angular distortion.

In addition to parallax distortion, two other types of distortions in format are present; one arises from the method of display and the other from variation in the aircraft motion. High-speed scanning sensors use rotating mirrors to perform the scanning function. These mirrors rotate at a constant angular velocity.

Since both the distance to the earth and the projection of the scan velocity on the earth vary inversely as the cosine of the angle from the nadir, the velocity of the scanning aperture over the earth's surface is nonlinear. From Figure 5.13 it can be

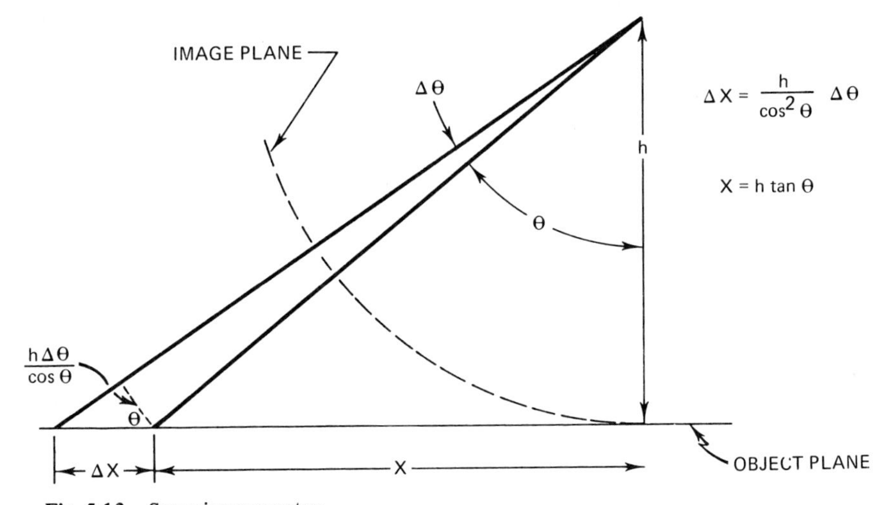


**Fig. 5.13.** Scanning geometry.

0-201-04245-2

seen that the beam velocity, $\dot{X}$, as it sweeps over the earth is given by:

$$\dot{X} = \frac{h\dot{\theta}}{\cos^2\theta} = h\dot{\theta}\sec^2\theta. \tag{5.22}$$

While the beam velocity scanning the earth is nonlinear, the reproduced image is generally created with a constant-velocity beam writing speed. This difference in scanning and beam writing speeds compresses the scale at the edge of the strip map in the scan direction, as illustrated in Figure 5.12. The compression increases exponentially with scan angle, and Figure 5.12 illustrates the distortion for a lateral coverage of 120° (±60° about the nadir). It can be seen that distortions are barely detectable for scan angles less than ±30°. Although this compression is objectionable, it can be readily corrected. It is generally accepted and tolerated for non-mapping purposes because such a format is easiest to generate, and the data are packed most efficiently on the film. The latter advantage arises from the fact that both the angular resolution and angular scan rates are constant, and a constant beam writing speed packs the data on the display at a uniform rate.

Another way of visualizing the format distortion and "efficient" data packing is to consider the image transfer plane to be curved as shown in Figure 5.13. The strip-map scale is constant in angle in a direction perpendicular to the ground track and constant with distance in the direction of the ground track. The difference in scales can be corrected so that the scale is constant with distance. Such a format would require an increase in width of film by $\tan\theta/\theta$. For a 120° scanner (60° half-angle), the recording film width would be 65% greater than that required for a constant angular scale.

Uncorrected motion of the aircraft introduces displacement errors between scene points since resolution elements are scanned sequentially, whereas in a camera all points in a frame are observed simultaneously. $V/h$ distortions elongate or shorten objects in the direction of the flight path, roll tends to shear the image, and drift distorts the angles of objects. All of these parameters can be monitored, however, and appropriate correction made in the display or recording system so that no distortion results.

The distortions mentioned above are concerned with the spatial fidelity of scanner imagery. There can also be distortions in the video signal; these appear as distortions in tones on the strip map. In this case, thermal strip maps cannot be considered as quantitative or semiquantitative thermal maps; that is, there is not a one-to-one relationship between the tone in the image and the radiance of the scene. The most common video distortion arises from the fact that electronic circuits in many scanners are ac-coupled for ease of signal handling and recording. As a result, the average signal level at the output of the electronics is zero, and the instantaneous signal is known only as to its value with respect to this average. In such a system, the tone of an object in an image is dependent on other objects that lie within the scan lines which determine the average signal level.

0-201-04245-2

Consider, for example, a scanner flying parallel to a land/water interface. Assume that the land is warmer than the water, and that two passes are flown—one directly over the shoreline, and the other offshore. The resulting video signals from the detector as a function of time are hypothesized in Figure 5.14, along with the average signal level. In an ac-coupled system, all signals are measured with respect to the average. As shown in Figure 5.14, the signal difference, *B,* between land and water is independent of the flight line. However, the magnitude of the voltage representing land, *A,* with respect to the average signal level is smaller when the scanner is flown over the shoreline than over the water. Thus, when the scanner is flown offshore, both the land and the water appear warmer in the imagery than when the scanner is flown directly over the shoreline. The difference in signal between land and water, however, is the same. By similar arguments, it can be reasoned that the tone of any object in ac-coupled imagery is affected by the signals from other objects within its environment.

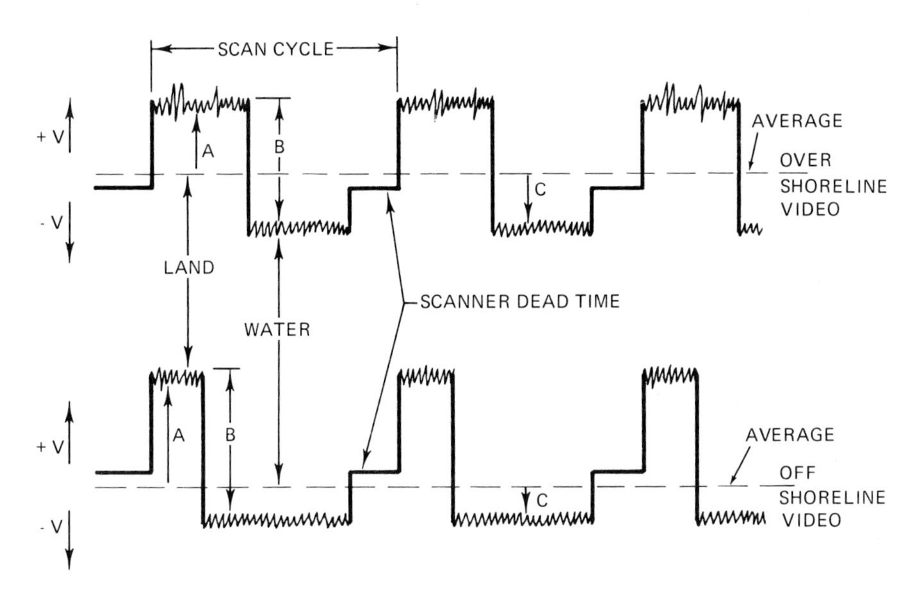


**Fig. 5.14.** Typical signal relationships as ratio of dwell time on land to water changes.

This problem of nonquantitative tones in ac-coupled imaging systems can be readily solved by a technique known as dc restoration, provided the low-frequency response is sufficient to avoid droop, that is, hold a steady-state signal over the period of one scan line. This is possible because the relative magnitudes of the signals are preserved in an ac-coupled system. Direct-current restoration requires that the video signal contain a repetitive signal of constant level from a source whose position in the video is known. One such signal may be the inside of the scanner housing, provided it can be made constant throughout the data collection mission.

0-201-04245-2

If the tone of the source known to be constant is forced to be constant in the imagery, the tones of all other objects are quantitative, provided the amplification gain is constant and the sensor is calibrated. The circuitry for dc restoration is well known and is used in television systems. It requires sampling electronically the output level of the known and constant signal and adding or subtracting a dc voltage to make it constant.

While dc restoration, or clamping, makes the tone of an object independent of its environment, the tone is quantitative if, and only if, the system response (volts out/watts in) is constant and known. If the response is constant, implying fixed detector response and amplifier gains, the response can be computed from the observation of two or more sources of known radiance [W/(cm$^2$ · sr)]. If the response is linear, two will suffice. Because the detector response is temperature dependent, frequent calibration is desirable. This leads to a scanner which carries its own reference and calibration sources.

### 3. *Optimum Band of Operation*

Within the limits of atmospheric windows, two spectral regions can be used for thermal mapping: 4.5-5.5 and 8-14 $\mu$m. The short-wavelength band has some advantages for mapping small, hot objects, but general terrain mapping is best accomplished in the 8-14 $\mu$m band. Since the spectral distribution of emitted energy shifts to the shorter wavelengths with increasing temperature, operating at shorter wavelengths emphasizes small, hot objects, such as combustion engines, cooking fires, and forest fires, or geothermal activity.

For example, consider a scanner with a ground resolution of 2 m$^2$ looking for a small blackbody source (10 cm$^2$) having a temperature of 500°C. Assume that one has the choice of operating in either the 8-14 or the 4.5-5.5 $\mu$m band. It is possible to calculate which band gives the greater increase in signal as compared with terrain having a temperature variation of 1°C. From Figure 5.10, it can be seen that $\Delta L/\Delta T$ (8-14 $\mu$m) is 6.5 × 10$^{-5}$ W/(cm$^2$ · sr · °K) for a 20° C ambient temperature background, whereas $\Delta L/\Delta T$ (4.5-5.5 $\mu$m) is 3 × 10$^{-6}$ W/(cm$^2$ · sr · °K). Two resolution elements having a temperature difference of 1°K have a radiant intensity difference ($\Delta L$ × area) of 2.6 W/sr in the 8-14 $\mu$m band and 4.5-5.5 $\mu$m = 12 × 10$^{-2}$ W/sr in the 4.5-5.5 $\mu$m band. Calculation shows that the small blackbody source has a radiant intensity of $J$(8-14 $\mu$m) = 11 W/sr and $J$(4.5-5.5 $\mu$m) = 9.6 W/sr. Thus, when operating in the 8-14 $\mu$m band, the signal from the small 500°C source is only 4 times larger than a 1°C temperature differential between two sources filling the field of view (2.6 vs. 11 W/sr). In the 4.5-5.5 $\mu$m band, however, the signal from the small, hot source is about 80 times larger than terrain variations of 1°C. To a first approximation, then, the small source is equivalent to a 4°C temperature difference in terrain features which fill the field of view when operating in the 8-14 $\mu$m band. Such a temperature difference readily occurs among natural terrain features, and hence the small source may be indistinguishable from 2-m-square terrain features such as rocks, water puddles, and vegetation. Thus,

0-201-04245-2

detection of the small, hot source may be difficult, if not impossible, in the 8–14 μm region. On the other hand, surface temperature variations of normal terrain could never be large enough to match the signal difference (equivalent to approximately 80°C) observed by the sensor operating in the 4.5–5.5 μm region.

## IV. NONIMAGING SPECTRORADIOMETERS

As discussed in Section I, much of the research in remote sensing programs is designed to establish or verify correlations between radiation observables and a phenomenon which the user is interested in detecting or measuring.

Although operational remote sensing may require sensors which search large areas from fast-moving platforms such as aircraft and spacecraft, much of the research leading toward the development of these sensors can be accomplished with nonimaging instruments in the laboratory or field. In this section, nonimaging spectroradiometers for measuring the spectral characteristics of materials will be discussed.

### A. Spectroradiometer Design

As its name implies, a spectroradiometer is a meter for measuring the radiation of an object as a function of wavelength. A radiometer can be considered to be a degenerate form of a spectroradiometer which is confined to operate in a single spectral interval. A radiometer may be classified as narrow or wide band, depending on the spectral band pass of operation. It is incapable of distinguishing the spectral distribution of the radiation it observes. Spectroradiometers are used to measure the spectral characteristics of radiation, be it radiation which is reflected, scattered, transmitted, or emitted by an object. There are basically three types of spectroradiometers, classified by the technique used in sorting the spectrum:

1. *Dispersing spectrometers,* which use gratings and prisms. In these instruments, the wavelength of operation is a function of the angular positions of the source, grating or prism, and detector.
2. *Filter wheel spectrometers,* of which the circularly variable filter is perhaps the best known. The wavelength of operation is dependent on the spectral transmittance of the filter in the radiation path.
3. *Interferometer spectrometers.* The radiant intensity as a function of wavelength is obtained by a Fourier transformation of the signal produced by the detector in a two-beam interferometer as the path between the two beams is varied.

To determine which of the above types of spectrometers is best suited for a particular measurement program, one must consider a number of design and operational parameters, such as spectral resolution, field of view, complexity, scan rate, environment, data reduction requirements, and sensitivity. The fact that there are hundreds of commercially available spectrometers attests that there is no single

0-201-04245-2

"best" spectrometer. One can, however, generalize on the features of the various types of spectrometers and offer comments which may influence the selection for a given job.

As discussed in Section II of this chapter, the $S/N$ of a sensor is a measure of its ability to measure small changes. In a spectroradiometer, the $S/N$ is a measure of how well the instrument can measure small changes in radiation with wavelength. Equations (5.14) and (5.17) give the $S/N$ of a spectroradiometer for measuring thermal emission with detectors which are detector-noise and radiation-noise limited, respectively. Since the spectrally dependent terms do not vary significantly within the limits of the spectral resolution or band width, $\Delta\lambda$, the integral form can be replaced by a product of average values. In this event, these equations can be written as

$$S/N = \frac{\tau(\lambda)\tau_0(\lambda)\alpha\pi D D_\lambda^* L_\lambda \,\Delta\lambda}{4F\sqrt{\Delta f}} \tag{5.23}$$

for detector noise systems, and

$$S/N = \frac{\tau(\lambda)\tau_0(\lambda)\alpha\pi D D_\lambda^* L_\lambda \,\Delta\lambda}{2\sqrt{\Delta f}} \tag{5.24}$$

for radiation-noise-limited detectors, where $D_\lambda^*$ is the detector detectivity before cold shielding to accept radiation passing through the optical system. From these equations it can be seen that the $S/N$, when observing a source of spectral radiance, $L_\lambda$, is dependent on design parameters such as optical efficiency, field of view, collector diameter, spectral resolution, and electronic band width (data rate). It is beyond the scope of this chapter to describe in detail all the intricacies of spectroradiometer design. Instead, Table 5.1 summarizes some of the more important advantages and limitations, based on the method of spectrum sorting.

## B. Laboratory Versus Field Measurements

Spectral measurements can be made in the laboratory or field. Laboratory instruments are usually designed to measure hemispherical reflectance (or emittance) or transmittance. Laboratory measurements are generally made on small samples such as parts of a leaf, stem, or bark of a plant. Such measurements are excellent for understanding the mechanism involved in spectral changes.

Laboratory measurements have the following advantages:

1. The instruments have controlled sources of illumination.
2. The environment is stable and controlled.
3. Instruments can be bulky and complex.
4. Weight and power considerations for the instrumentation are of secondary importance.

0-201-04245-2

**TABLE 5.1**
**Comparison of Spectroradiometers**

| *Type* | *Advantages* | *Disadvantages* |
|---|---|---|
| Circularly variable filter wheel | 1. Simple with minimum components<br>2. Rugged<br>3. Precise optical alignment not required | 1. A CVF is needed for each wavelength octave<br>2. Relatively poor resolution; maximum resolving power $= \lambda/\Delta\lambda - 100$ |
| Prism | 1. Wide spectral coverage (many octaves) can be obtained | 1. Nonlinear dispersion; i.e., wavelength is not linear with distance in image plane<br>2. Performance is limited by available prism materials; some of the more useful materials have poor physical properties<br>3. Extremely high resolution is difficult |
| Grating | 1. Resolution and wavelength region of operation are limited only by grating ruling (spacing)<br>2. Resolving power is independent of $\lambda$<br>3. Dispersion is linear with $\lambda$<br>4. High dispersion is possible | 1. Need presorter to separate multiple octaves; therefore difficult to operate over wide spectral interval<br>2. For high optical efficiency grating should be blazed; optical efficiency is highly wavelength dependent |
| Interferometer | 1. High throughput<br>2. All wavelengths observed simultaneously; this has $S/N$ advantage when using detectors that are not radiation-noise limited<br>3. Can operate in high- or low-resolution modes | 1. Complex computation required to reduce data<br>2. Spectral data obtained by extremely small movement of mirror; therefore, subject to microphonics and temperature changes<br>3. Data recording requires large dynamic range |

5. Speed of observation is not usually important.
6. Logistical support is at a minimum.

On the other hand, there are also disadvantages:

1. The environment of the laboratory is not natural, and movement of the sample into the laboratory may disturb its spectral properties.
2. The geometry of illumination and observation is not natural. For example, laboratory instruments frequently employ hemispherical illumination or observe

0-201-04245-2

hemispherical reflectance, a situation which does not exist in "real world" observations.

3. In remote sensing, a resolution element contains a conglomerate of components (a view of a plant contains leaves, stems, ground, and bark). These would make difficult samples for laboratory measurements.

Field measurements are ideal for observing an object *in situ.* To do this without disturbing its natural environment, the spectral properties of the object are generally observed by reflected sunlight or self-emission, although one can conceive of special measurements using artificial illumination such as lasers.

Some of the advantages of field observations are as follows:

1. The sample is undisturbed.
2. Measurements are made with natural illumination and backgrounds.
3. Some measurements can be made only in this manner; for example, effective spectral emissivity or surface temperature can be observed only in the natural environment.

On the other hand, field measurements present certain difficulties:

1. The equipment must be portable.
2. The environment may be hostile and prevent measurements; for example, meteorological conditions may be adverse.
3. Variations in environmental parameters may affect measurements, and one must compensate for them; for example, to obtain spectral reflectance from spectral radiance observations requires a knowledge of spectral irradiance.
4. Cryogenically cooled detectors are desirable, if not required.
5. Logistical support is extensive.

Field measurements are not restricted to ground or tower observations but can be made from moving platforms such as aircraft or spacecraft. Airborne measurements are ideal for large-area coverage and for observing relatively inaccessible areas (tops of large trees or icebergs).

Some of the advantages are the following:

1. *In situ* observation is feasible. The geometry of viewing can be identical to that of an operational system.
2. The platform is highly mobile and can range far and wide.
3. Data can readily be collected on a wide variety of backgrounds.

On the other hand, there are disadvantages:

1. Instruments mounted in aircraft are usually more complex;; for example, they may require stabilization or pointing.
2. Measurements are generally less precise than laboratory or ground measurements.
3. Data not in image form are difficult to reduce or associate with specific ground elements unless simultaneous bore-sighted photography is available.

0-201-04245-2

4. Aircraft facilities are expensive and complex, often costing orders of magnitude more than the sensors themselves.
5. Sophisticated data reduction is usually required because of the large number of data collected.
6. Equipment reliability is more stringent. A flight mission is expensive, and failure of the sensor is costly.

The advantages and disadvantages of measurements from unmanned space vehicles closely parallel those of aircraft measurements but differ in several respects:

1. The instrument environment can be extremely quiet.
2. Reliability is at a premium; the price of failure is high as no maintenance is available.
3. Measurement routine (calibration, pointing, etc.) must be automated.
4. Consumable materials, if required, limit the lifetime of the experiment.
5. Data return is limited by the telemetry or data capsule.
6. Weight, volume, and power are at a premium.

## C. Calibration

The prime purpose in making measurements in a research program is to determine whether the spectral attributes can be used to recognize an object or some property associated with it; for example, does moisture stress in vegetation exhibit an identifiable spectral change? To analyze the data, calibration of some form or degree is required to assure that the variation being observed is associated with the object or phenomenon being investigated and not with the spectroradiometer itself. Any variations in signal not associated with the object and not accounted for must be considered as noise.

Some of the parameters which can affect the results and precision of spectral observations are the following:

1. Source illumination or radiance.
2. Instrument field of view.
3. Wavelength calibration.
4. Spectral resolution.
5. Atmospheric and instrument transmission.
6. Detector response (volts out/watts in).
7. Electronic and recording transfer functions and their stability.

Techniques have been developed which can ignore variations in most of these parameters or compensate for them. For example, transmittance and reflectance measurements in the laboratory are relatively simple. By using a double-beam technique, one observes alternately the reflectance or transmittance of a reference (standard) and a sample. The illuminating source is controlled so that its level is constant over the time period between observing the reference and observing the

0-201-04245-2

sample beams. Radiation from the source is modulated so that extraneous radiation or illumination is ignored. The level of radiant energy falling on the detector (and thus the output signal) can be limited by programming the slit width, thereby easing signal handling. Wavelength calibration is maintained by controlling the temperature of the environment. The net result of using these techniques is that the output can be readily and automatically plotted as spectral reflectance or transmittance.

Making reflectance measurements in the field is more complex, however, when natural solar radiation or thermal emission is used. The entire radiation entering the instrument is measured, and this can contain extraneous radiation from outside or within the instrument. One can use a dual-beam technique or sequentially scan the spectral radiance of a sample and a reference. In doing this, however, the sampling rate must be sufficiently high and the reference and sample objects sufficiently close to avoid variations in the illumination level during the measurement of sample and reference. Field measurements are further complicated by the fact that atmospheric absorption imposed on the solar radiation produces an illuminating source whose intensity varies drastically over short wavelength intervals.

Measurement of spectral radiance or brightness quantitatively is generally more complex than quantitative measurements of reflectance or transmittance. Quantitative transmittance measurements require sequential observations with the sample in and the sample out of the beam during a period when the source providing the incident illumination is constant. Reflectance measurements can be made in a similar manner. For quantitative spectral radiance measurements, however, the unknown source must be referenced to a source which not only is constant in radiance but also has a known spectral radiance. The requirement for a source of known spectral radiance eliminates the need for a precise knowledge of the spectral resolution or band width of the spectroradiometer. When viewing a source, the output signal, $V$, is given by:

$$V(\lambda) \sim L_\lambda R_\lambda \, \Delta\lambda, \tag{5.25}$$

where $L_\lambda$ = the source or scene apparent spectral radiance,

$R_\lambda$ = the system spectral responsitivity ($\mathrm{V \cdot W \cdot cm^2}$),

$\Delta\lambda$ = the spectral band pass.

If a known source and sample source are observed alternately with a spectroradiometer with identical slit widths, $\Delta\lambda$, for a given wavelength, and the response has not changed during measurements, then:

$$\frac{V_k}{V_s} = \frac{L_{\lambda k}}{L_{\lambda s}} \quad \text{or} \quad L_{\lambda s} = L_{\lambda k} \frac{V_s}{V_k}, \tag{5.26}$$

where the subscripts $s$ and $k$ stand for sample and known sources, respectively. This

0-201-04245-2

rationing at many wavelengths, although simple, can be tedious and time consuming for a large number of spectra. The process can and should be computerized if large numbers of spectra are made. Once one decides to reduce the spectra with a computer, the data can be automatically reformatted for ready comparison with other data. Furthermore, the spectra can be analyzed statistically for significant variations. The use of spectroradiometers in remote sensing programs is well illustrated in the *Proceedings of the Symposia on Remote Sensing of Environment* (University of Michigan) and the book *Remote Sensing: With Special Reference to Agriculture and Forestry* (National Research Council, 1970).

## V. MULTISPECTRAL SCANNER

As discussed in Section IA of this chapter, it is possible that the spectral characteristics of a material may be correlated to information which a user is seeking. This section will discuss the rationale and design considerations for a multispectral sensor which can perform a wide-area search and classification based on spectral information. Such a sensor combines the technology of line scanners, multichannel spectroradiometers, high-speed signal handling, and automated information/decision processing.

### A. Rationale for Multispectral Sensing

Most remote sensors for operational systems are image forming. The data collection rate of these sensors if quite high compared with what man can handle. Thus, in general, imaging devices collect data at a rate too fast for man to interpret in real time. In many remote sensing applications with extensive coverage, identification through shape from high-resolution imagery requires an excessive number of data. Consider satellite photography of the United States with 0.6 m ground resolution. Roughly 1140 kg of film would be required to produce this coverage, and it would take a 10-MHz telemetry link 121 days to transmit the information to earth (Sattinger, 1966). If identification can be made on the basis of spectral information, it is sometimes possible to reduce the sensor resolution requirements. In agricultural sensing for survey purposes, one may wish to know only what a farmer has in a field. Although resolution of the crop structure itself is not possible with large-area coverage, it may be possible to distinguish the crop on the basis of its spectral signature. If such a detection is possible, a ground resolution of 60 m may be adequate, in which case the resulting imagery of the United States could be recorded on 0.15 kg of film and the telemetry time reduced to about 17 min.

Although single-band scanners (such as an infrared scanner) produce data at a rate too fast for man to assimilate in real time, they throw away spectral information. Let us assume that a sensor can distinguish between 10 brightness levels in each band of operation. A single-band scanner can classify objects, on the basis of tone, into only 10 categories. Consider, however, a scanner operating in 20 wave-

0-201-04245-2

length bands. Such a scanner can potentially distinguish between $10^{20}$ different states; that is, an object can have a tone of 1-10 in 20 different bands. This large number of distinguishable states does not actually exist or cannot be observed in nature for many reasons; there is variance in the illuminating conditions and within a class of objects, and the spectral reflectance at one wavelength is not independent of the reflectance in a neighboring band. Nonetheless, a large number of different and distinguishable states can and do exist, and one should be capable of building a sensor which uses spectral information.

A schematic of a multispectral scanner is shown in Figure 5.15. Whereas most scanners use a detector or a field stop imaged on the detector to define the field of view, the entrance slit of a multichannel spectrometer is employed in this sensor. One possible configuration of the spectrometer is shown in Figure 5.16. In such a system, each detector of the spectrometer observes the same resolution element of the scene but in a different wavelength region. The output signal from each detector element is a video signal corresponding to the scene brightness in the particular wavelength region of operation. This video signal can be used to generate an image of the scene in the wavelength region as defined by the position of the detector in the spectrometer. The output signals from multiple detectors can be combined to determine the spectral distribution of the radiation from each scene point. This spectral information then can be used to enhance or suppress the detection of objects or materials in a scene. If the detection can be repeated with confidence (high reliability with acceptable misclassification), one has a recognition or automatic classification sensor.

It should be noted that the multispectral scanner avoids many of the registration problems inherent in a multisensor system, that is, one which uses multiple sensors—photography, infrared, and radar. The spectral information from the vari-

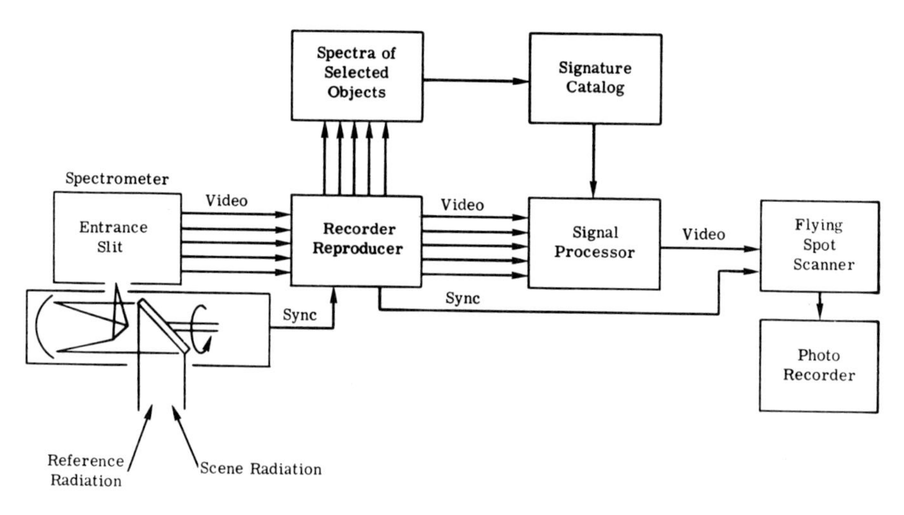


**Fig. 5.15.** Schematic of multispectral scanner and data processor.

0-201-04245-2

$$\alpha = \frac{d_1}{F_1 D_1}$$

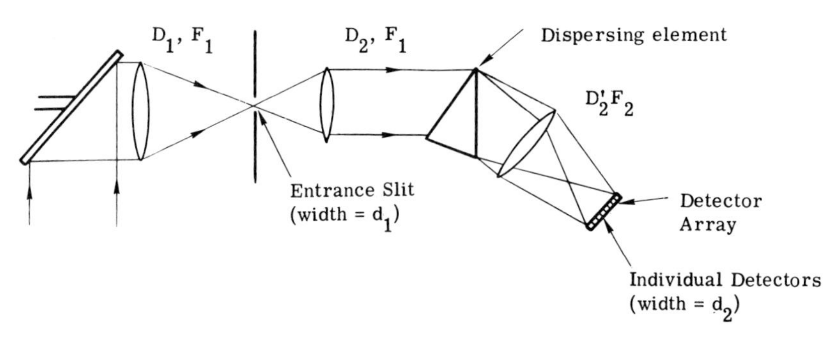


**Fig. 5.16.** Schematic of multispectral dispersive scanner. From Braithwaite (1966).

ous channels of the multispectral scanner is collected with complete registration in both the time and the space domain.

## B. Design Considerations

In a multispectral scanner, one is interested in measuring variations in spectral radiance, spectral reflectance, and/or spectral emissivity. In the spectral region where detector performance is described in terms of detectivity, the $S/N$ is given by Eq. (5.20), where $\Delta L$ is replaced by $L_\lambda \, \Delta\lambda$, the radiance within the wavelength interval, $\Delta\lambda$, of operation:

$$S/N = \frac{\tau\tau_0 \pi\alpha^2 D D_\lambda^* L_\lambda \, \Delta\lambda \sqrt{2p\gamma}}{4 n_f F \sqrt{(V/h)\theta}} . \tag{5.27}$$

For wavelength regions where detector performance is best described by the detector photocathode response, $R_c$, or quantum efficiency, $Q_e$, the $S/N$ can be shown (Braithwaite, 1966) to be given by

$$S/N = \frac{e^2 D}{2 n_f \Delta\rho} \left( \frac{R_c \rho\gamma\tau\tau_0 H_\lambda \, \Delta\lambda}{\rho e (V/h)\theta} \right)^{1/2} , \tag{5.28}$$

where $H_\lambda$ = the downwelling spectral irradiance,

$\rho$ = the average reflectance around which changes, $\Delta\rho$, are sought,

$e$ = the electron charge.

The detector response and quantum efficiency are related through

0-201-04245-2

$$Q_e = R_c \frac{hc}{e\lambda}, \tag{5.29}$$

where $c$ is the velocity of light.

When operating in a multispectral mode, a line scanner has additional design constraints since, as shown in Figure 5.16, the entrance slit of the spectrometer serves as the field stop of the scanner and thus sets the angular resolution. It has been shown (Braithwaite, 1966) that the spectral resolution is determined by:

$$\Delta\lambda = \frac{D_1 \alpha}{D_2 (d\theta / d\lambda)}, \tag{5.30}$$

where $D_1$ and $D_2$ = the diameters of the collecting optics of the scanner and spectrometer, respectively,

$d\theta / d\lambda$ = the angular dispersion of the prism or grating.

## C. Data Processing

Chapters 10 and 11 discuss data-processing techniques and results. As shown there, the complexity of data-processing techniques varies considerably among applications. Some investigators use decision rules formulated from statistical analysis of data from training samples of known ground truth. Others may prefer to subject the data to factor analysis and relate the factors to features in the scene. In this section, the processing techniques will not be considered. Rather, various output forms of the data will be presented to illustrate conceptually the type of information which the multispectral scanner collects.

Figure 5.17 displays A-scope traces from two channels of data collected with the University of Michigan 12-channel scanner. The vertical deflection corresponds to the scene brightness. By sampling all 12 tracks of the tape-recorded video signals, one can extract spectra of scene points. The 12 signals in Figure 5.18 represent a raw spectrum of a calibration panel which was overflown. Figure 5.19 plots the reduced spectral reflectance of the panel, along with the reflectance curves supplied by the manufacturer and measured in the field with a spectroradiometer.

The data from the scanner can be digitized and analyzed with a general-purpose digital computer. Figure 5.20 shows the mean value of the radiance of a large cornfield as a function of viewing and illuminating geometry for 1 of 12 channels of data (Malila, 1968). Figure 5.21 shows the angular variation of all 12 channels for a wheat field. The curves in Figures 5.20 and 5.21 represent the mean values of many scan lines.

While the power in the multispectral scanner lies in automatic processing of the spectral information as discussed in Chapters 10 and 11, there are some cases in which one may wish to view the output of the scanner as multiband imagery. Figure 5.22 shows 18 channels of multispectral imagery as generated by the University of Michigan.

0-201-04245-2

TYPICAL SPECTROMETER VIDEO DATA (U)
California Rice Fields 5-26-66 1400

Calibration
Lights

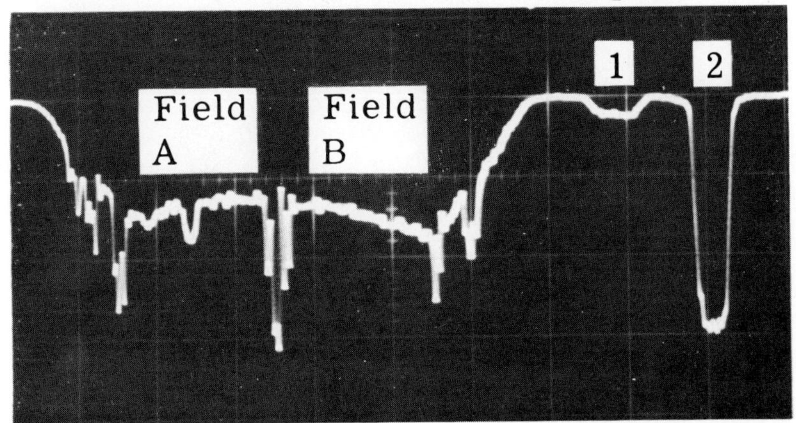


Time ⟶

0.40-0.44 $\mu$

Calibration
Lights

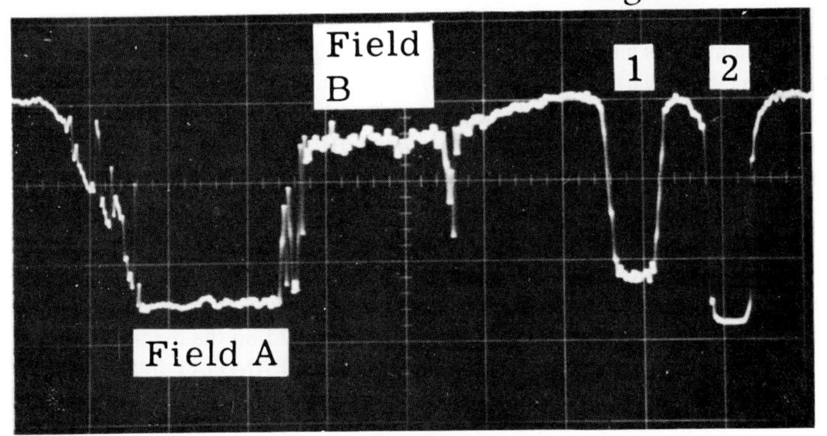


Time ⟶

0.80-1.0 $\mu$

0-201-04245-2

Fig. 5.17. A-scope traces of a common scan line from two spectral bands.

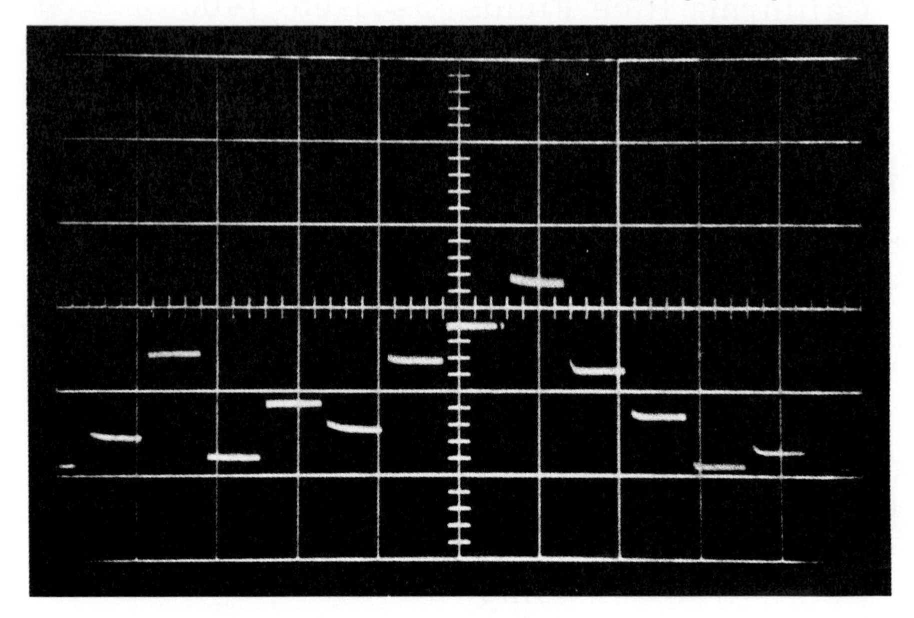

**Fig. 5.18.** Sampled raw spectrum of green painted panel.

## D. Current State of the Art

The concept of combining a scanner and multichannel spectrometer to form a multispectral scanner was an outgrowth of remote sensing programs at the University of Michigan (Braithwaite and Lowe, 1966). Researchers there built a 12-channel unit in 1965, using a prism spectrometer. Shortly thereafter Bendix Aerospace Systems Division constructed a 9-channel grating unit and later built a 24-channel multispectral scanner and data-processing system for NASA (Beilock et al., 1969). The instrument is used by personnel from the Manned Spacecraft Center in support of earth resource sensing experiments. A 4-band multispectral scanner (Norwood, 1969) was orbited under the ERTS-A (Earth Resources Technology Satellite) or LANDSAT Program in mid-1972, and a 10-band scanner was part of the Earth Resources Experiment Package orbited in mid-1972 on the manned SKY-LAB. A six channel scanner is now under consideration for LANDSAT follow-on (LANDSAT D).

0-201-04245-2

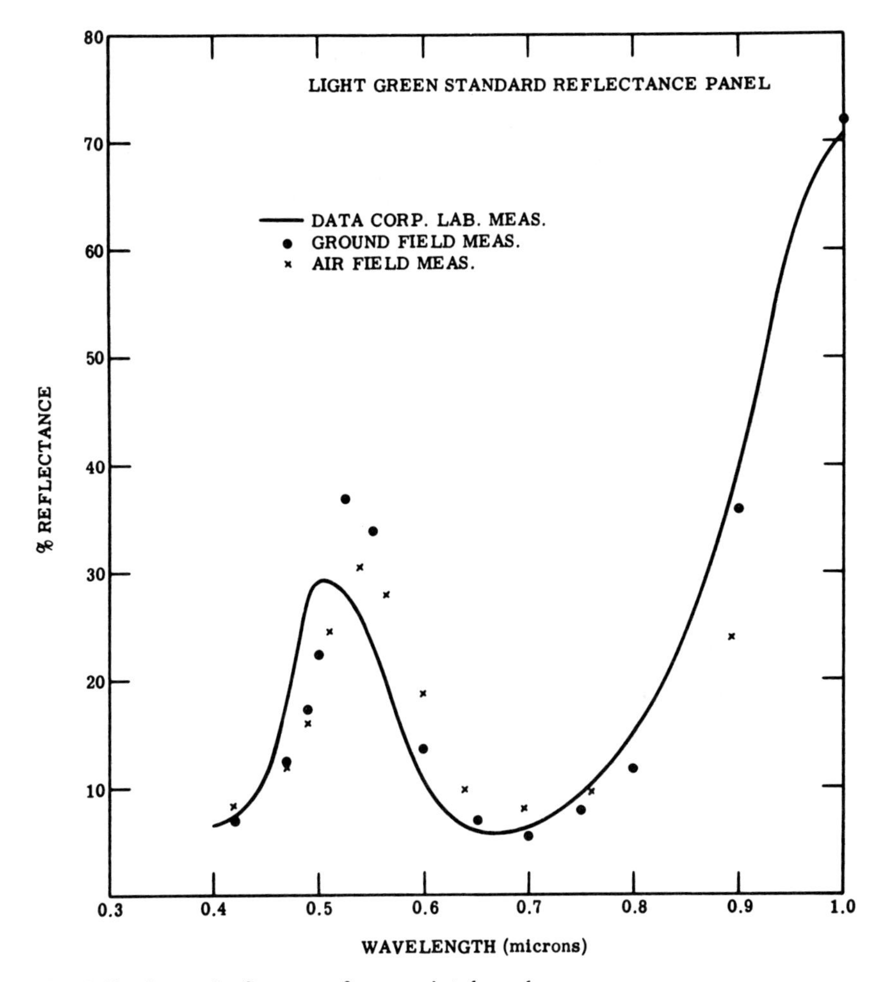


Fig. 5.19. Spectral reflectance of green painted panel.

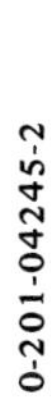
0-201-04245-2

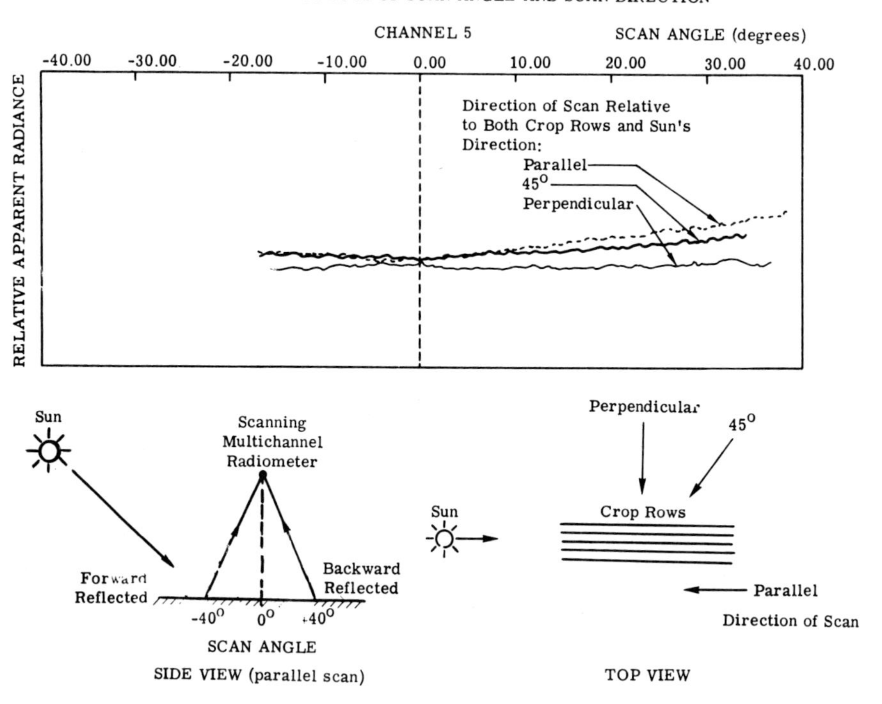


**Fig. 5.20.** Goniometric effects of illumination angle. From Malila (1968).

0-201-04245-2

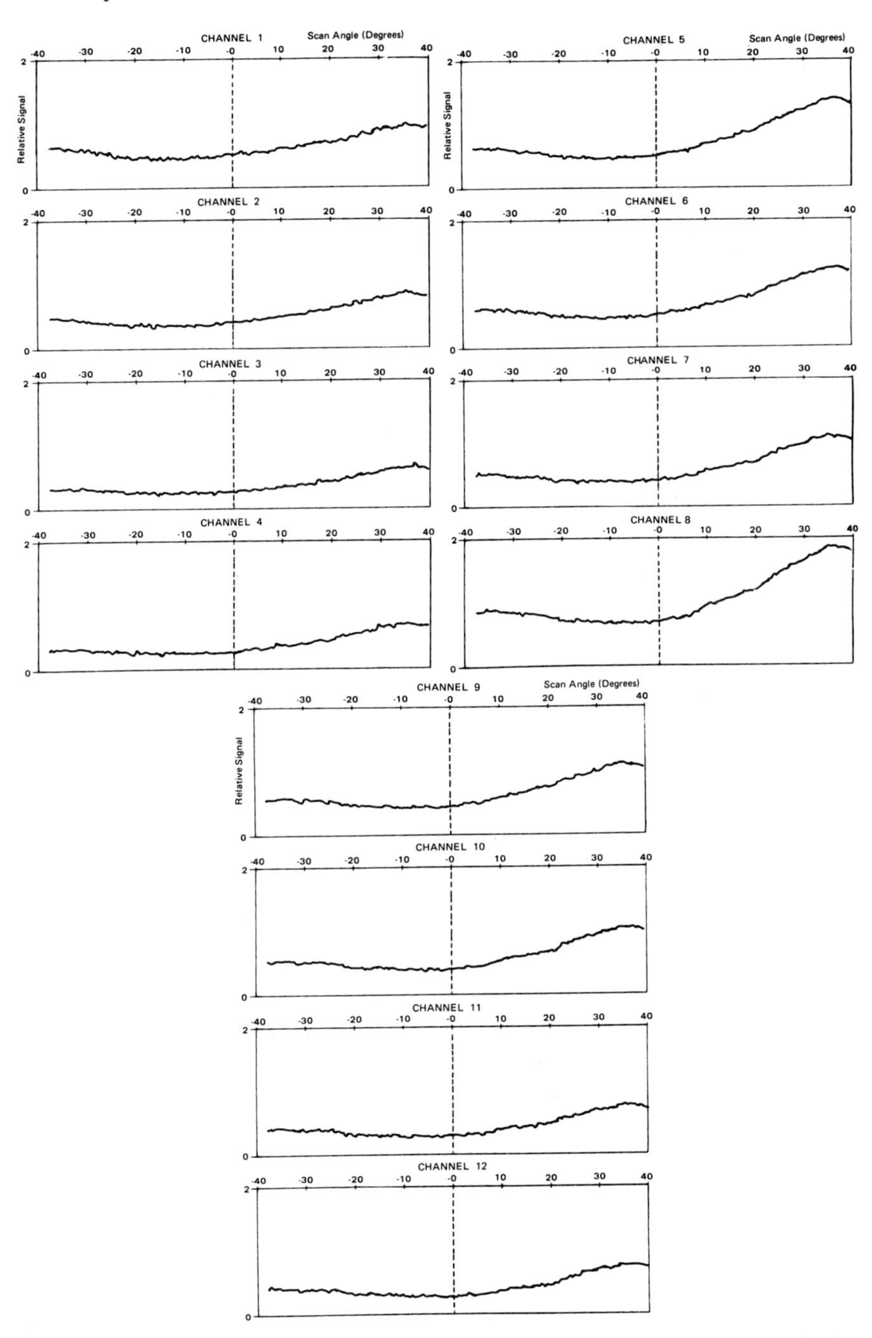


**Fig. 5.21.** Relative goniometric brightness curves for wheat fields, data acquired 6/30/66, 0900 hr. From Malila (1968).

0-201-04245-2

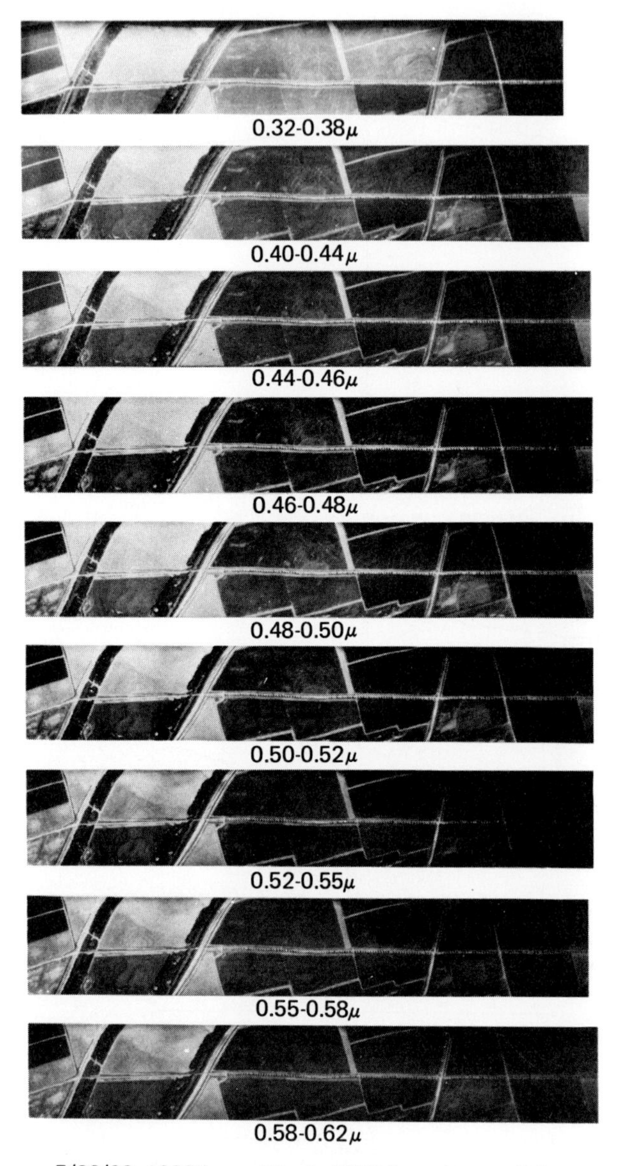


5/26/66; 1600 hrs.; altitude 2000 feet; sky condition, clear and bright, 10% cloud cover at 30,000 feet; surface temperature 27°C

**Fig. 5.22a.** Multispectral imagery of Davis, California, agricultural area, 5/26/66.

0-201-04245-2

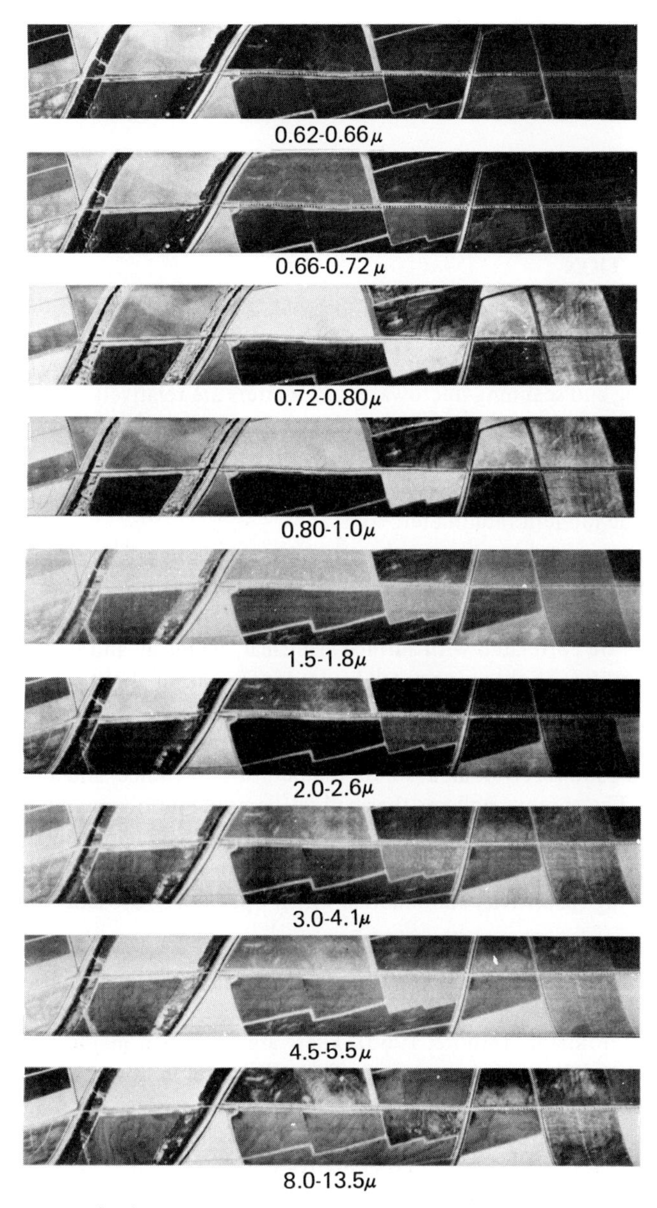


5/26/66; 1600 hrs.; altitude 2000 feet; sky condition clear and bright, 10% cloud cover at 30,000 feet; surface temperature 27°C

**Fig. 5.22b.** Multispectral imagery of Davis, California, agricultural area, 5/26/66.

Chapter 6

# Passive Microwave Systems

*Frank J. Janza*

## I. INTRODUCTION

### A. Background

Passive profiling and scanning microwave radiometers are relatively new instruments in the field of remote sensing. They are usually adaptations of highly developed radioastronomy radiometers. Nearly every significant development or device to increase sensitivity and detectability and to improve stability has found its way into the design of the modern radiometer.

A radiometer is used to measure the effective or radiometric temperatures of land and sea surfaces and the intervening atmosphere. It has the unique advantage of being a passive all-weather remote sensor. The brightness temperatures recorded by radiometers are correlated with surface electrical, chemical, and textural properties. The possibilities of obtaining much-wanted information about subsurface properties with the radiometer appear promising at longer wavelengths.

Basically the radiometer consists of a scanning or nonscanning antenna, a very sensitive broad-band receiver, an absolute temperature reference, and a magnetic tape recorder. The antenna collects thermally generated microwave radiation and concentrates it on the sensitive receiver, where it is detected, amplified, and recorded either as a voltage-time record or as an image on photographic film. Figure 6.1 shows two high-sensitivity radiometers which operate at 13.7 and 31 GHz.

#### *1. Comparison between Radiometers and Scatterometers*

Microwave radiometers may be compared with the microwave roughness-measuring radar, or "scatterometer" (Moore, 1966). There are important major differences, however, between these two instruments. The radiometer is a passive broad-band receiver which collects radiation emitted or reflected from a surface, while the scatterometer generates and transmits its own single-frequency power to the surface, where it is backscattered to the relatively narrow-band receiver. The radiometer is sensitive to surface brightness temperature differences, whereas the scatterometer responds mainly to surface roughness differences. Both sensors are responsive to changes in surface electrical properties, roughness, and composition. They are usually designed for dual-polarization, all-weather, and day-night operation.

Radiometers and scatterometers are capable of looking through heavy clouds and moderate rains below 15 GHz with no appreciable signal error being intro-

*Remote Sensing of Environment*, Joseph Lintz, Jr. and David S. Simonett (eds.) ISBN 0-201-04245-2




0-201-04245-2

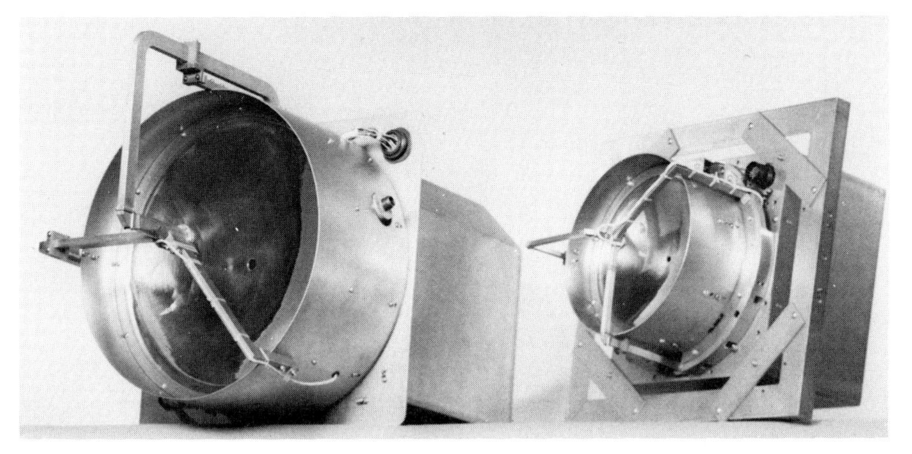

**Fig. 6.1.** 13.7-GHz and 31-GHz dual-polarization radiometers. Courtesy of Teledyne Ryan Company.

duced, and around 3 GHz the atmosphere appears quite transparent. Partly because of this nearly all-weather operation, there has been extensive development of radar systems in the past 35 years. More recently, attention has turned strongly to the development of passive microwave systems.

Two modes of operation may be used in passive microwave remote sensing: a profiling mode and a scanning or imaging mode. For the former the antenna is positioned to collect signals along the flight path at a discrete angle of incidence in the range of 0 to 70°. This mode (employing either vertical or horizontal polarization) is used primarily for profiling or obtaining statistical averages over a flight run. For the imaging mode there are two methods of scanning, transverse and conical. Recently a transverse-scan radiometer developed in the United States has been used for mapping extensive areas.

## *2. Spectrum*

What is the best portion of the microwave spectrum to select for remote sensing? This question is best answered when one knows what information is desired from the terrain or the atmosphere. The proper answer requires also an understanding of the radiation physics of both areas and the performance capabilities of radiometers.

As an example, consider some airborne radiometer experiments in the study of river basin hydrology. Hydrology missions require the collecting of brightness temperature information for land and water surfaces under adverse weather conditions.

The all-weather regions of the microwave spectrum are depicted in Figure 6.2. Regions of high attenuation are caused by water vapor and oxygen. Water has a strong resonance at 22.235 GHz, and oxygen shows a series of prominent resonances in the near-60-GHz region. Below 15 GHz, however, the spectrum is almost transparent, and at 3 GHz even moderate rain causes little attenuation. Therefore the spectrum below 15 GHz can be selected for aircraft experiments. The particular

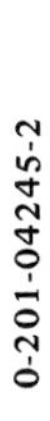

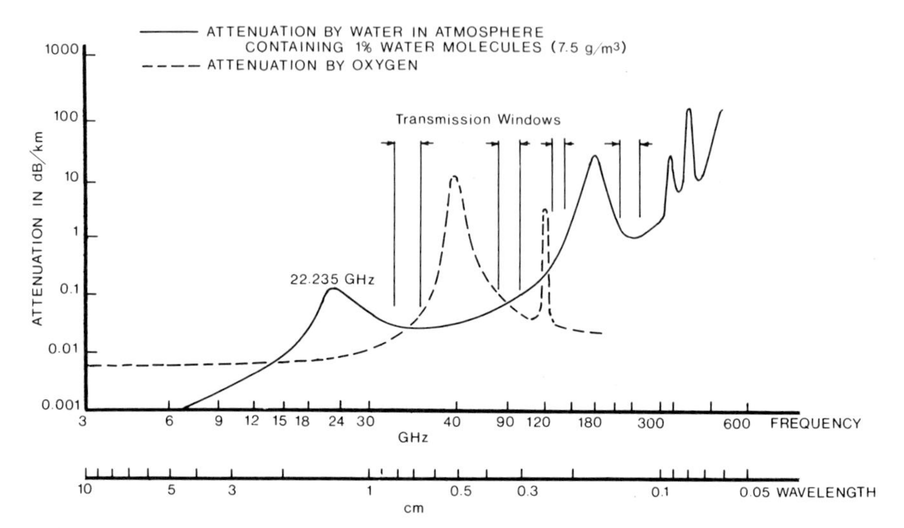


**Fig. 6.2.** Atmospheric attenuation of microwaves due to water vapor and oxygen.

area chosen involves tradeoffs between spatial resolution, band width, available spectrum, and the capabilities of the aircraft.

The following regions of the spectrum are of special interest for microwave sensing: (1) *0.4–1.6 GHz,* a region of good soil penetrability, yielding information related to subsurface materials, moisture, voids, and interfaces; (2) *0.4–15 GHz,* the all-weather region suitable for using multispectral systems to establish surface material properties; (3) *15–22 GHz,* the region for oceanography, showing useful polarization and emissivity effects for measuring surface temperature, roughness, and salinity; (4) *22 GHz,* a region of meteorological information for determination of atmospheric water vapor, using the 22.235-GHz water vapor resonance line; (5) *60 GHz,* another meteorological area for determination of atmospheric temperature profiles; and (6) *windows: 35, 94, 135, and 225 GHz,* regions of high spatial resolution with small physical antenna apertures—however, atmospheric attenuation is high.

### *3. Signals and Signal Compromise*

Ordinarily one wishes to have discrete information concerning a particular area, but may have to settle for statistical information. As seen earlier, obtaining fine spatial detail in a passive microwave image cannot be coupled with fine temperature resolution (requiring long signal integration times), with operation at space altitudes, or with fine spectral resolution.

The collected signals of the profiling or scanning radiometer are recorded as amplitude-time histories which are related to an absolute temperature reference. The accuracy with which the brightness temperature of the ground may be measured is governed by the design and the motion of the radiometer. At times when the terrain is statistically stationary over a large expanse, average brightness tem-

0-201-04245-2

perature values are very useful. The ocean falls into this category, since it is a homogeneous body with the wave structure generally uniform over large areas.

The present state of development of radiometers is such that temperature sensitivities of ±0.05°K rms and antenna beam widths of 0.5° are now possible. Because of size, weight, and power limitations, however, compromises become necessary for each application. The brightness temperature of a discrete target can be degraded by inadequacies in resolution, predetection band width, dwell time, system noise figure, and calibration.

Because of degradation in brightness temperature signals imposed by radiometer limitations, the signal variations are best collected, processed, and analyzed on a relative basis by taking account of the portions of signals common to the entire system. Where absolute temperature values are mandatory, an involved inversion procedure can be applied to remove errors caused by the weighting functions of the radiometer. For a scanning radiometer, the image is compromised by inadequate dwell time over a ground area element.

## B. Applications—Actual and Proposed

Possible areas for which microwave radiometers may be useful include applications in hydrology, meteorology, geology, oceanography, and glaciology. More specifically, radiometers may be used to (1) detect soil moisture and flooding, delineate snow and ice cover, and measure snow depth; (2) identify soil parameters, geological structures, voids, and metal and mineral deposits; (3) find targets in the presence of dense ground cover; (4) conduct zero-visibility aircraft landings with the radiometer system; (5) make brightness temperature maps of the terrain; (6) profile the temperature of the atmosphere in the presence of clouds; (7) measure integrated water vapor in the atmosphere; (8) measure the total water content of clouds; (9) measure altitude passively; and (10) detect pollution in estuaries, in rivers, and on coastal shores.

## C. Future Needs

Microwave radiometry is sufficiently new in its application or potential application to the areas mentioned above that much basic study is still needed. Some of the types of data and experiments required are as follows:

1. Carefully controlled laboratory experiments to show the changes in calibrated-output radiometer temperature-related signals under static conditions when surface (or subsurface) composition, temperature, or structure is changed. Correlations with dielectric constant due to varying amounts of water are of major interest. Equally important are the signal changes due to varying the field of view of the antenna, angle of incidence, wavelength, polarization, predetection band width, and integration time.
2. Field experiments in which *in situ* measurements can be made under prevailing ambient and atmospheric conditions. The same materials can be remotely tested

0-201-04245-2

farther away from the surface (altitudes to possibly 100 m), but still under static conditions.

3. Dynamic measurements from helicopter or aircraft (170–10,000 m) over extensive areas of the same soils tested in the laboratory and the field. The compromises in the correlations due to changes in altitude, atmosphere, and attitude need to be recorded.
4. Dynamic measurements from orbiting satellites to show the extent of correlations between recorded brightness temperatures and the surveyed land or ocean surface, when the influences of the entire depth of the atmosphere are of major importance.

A more comprehensive list of experiments would include studies of the advantages of correlating the outputs of multiple sensors. The possibilities for increasing the confidence with which signature features of the ground are specified, through multisensor correlations using photographic, infrared, microwave, and ground truth inputs, have been investigated for hydrology (Janza, 1969).

## II. THEORY

### A. Microwave Radiation

Future progress in using microwave radiometers will depend largely on how well output signals can be correlated with the information desired from the earth and its atmosphere. The main purpose of this section is to outline the parameters available for correlations and to show whether their sensitivities are adequate for the purpose. We may start by considering bodies in thermal equilibrium.

A body in thermal equilibrium, regardless of material and equilibrium temperature ($T_0$), radiates as much power as it receives. Its absorptivity ($\alpha$) cannot exceed unity; thus the power reradiated is maximum for the perfect absorber, which is defined as a blackbody. Emissivity, $\epsilon$, is the ratio between the radiating capability of a material and that of a blackbody, and for an ideal radiator the ratio is $\alpha = \epsilon = 1$. When the absorptivity is less than unity, the emissivity is correspondingly the same, or $\alpha = \epsilon$, for all conditions where $0 \leqslant \alpha \leqslant 1$ and $0 \leqslant \epsilon \leqslant 1$. The absorptivity is found to be a function of wavelength, $\lambda$, and it can be shown that $\alpha_\lambda = \epsilon_\lambda$ for each spectral component in the electromagnetic spectrum.

For a radiometer, radiation is collected over a band within the microwave spectrum; hence the absorptivity can be expressed as the average over the band, $\alpha_{\Delta\lambda}$, and correspondingly for emissivity, $\epsilon_{\Delta\lambda}$. For the radiometer band widths considered, the average values of these two parameters and their values at band center can be taken to be the same. For large wavelength separations, the wavelength sensitivity becomes a signature for correlations.

For an airborne radiometer survey, the propagation of radiation follows many paths before the radiation reaches the receiver, and path effects must be considered in making correlations. The total radiation incident on a surface, $P_i$, must be

0-201-04245-2

accounted for by absorption, emission, reflection, and transmission:

$$P_i = P_\alpha + P_\rho + P_\tau, \tag{6.1}$$

where $P_\alpha$, $P_\rho$, and $P_\tau$ are the absorbed, reflected, and transmitted powers, respectively. By normalization with the incident power, $P_i$, the following relationship results:

$$1 = \alpha + \rho + \tau, \tag{6.2}$$

where $\rho$ and $\tau$ are the reflectivity and transmissivity, but since $\alpha = \epsilon$ Eq. (6.2) can also be written as

$$1 = \epsilon + \rho + \tau. \tag{6.3}$$

The relationship $\epsilon = 1 - \rho$ is found to apply to most earth surveys when the surface materials are opaque, and thus $\tau = 0$. This relation is useful for simple mathematical model analyses.

The parameters of $\epsilon$ and $\rho$ are found to be dependent on the wavelength, $\lambda$; angle of incidence, $\theta$; vertical polarization, $V$; horizontal polarization, $H$; and actual temperature, $T$ (°K). Their sensitivities are quite pronounced for certain angles of incidence and polarizations. These dependencies are exploited in establishing correlations between radiometer signals and the surface.

## B. Temperature Aspects

### 1. *Brightness and Antenna Temperature*

Microwave radiometers are discussed in terms of the brightness temperature, $T_B$, of either the terrain, the atmosphere, or the sky. Since the temperatures of these areas have direct physical significance, radiometers are calibrated directly in terms of the antenna temperature, $T_A$ (°K), which is the temperature that a blackbody at the antenna port must reach to radiate the same power as is collected. Occasionally the actual rather than the brightness temperature is important; this is determined by an inverse convolution in which the weighting functions of the antenna, radome, and other identifiable sources of stray radiation are removed.

It is important to consider individual radiation sources and their properties for radiometer model developments in establishing correlations. The properties of such radiation and its connections with the radiometer are presented by a review of the basic theory. Starting with an enclosed cavity whose walls are maintained at a uniform temperature, $T$, consider a small object within the cavity. For this object, consider the energy, $\Delta E$, in the frequency interval $f$ to $f + \Delta f$, which falls during the time interval $\Delta\tau$ on a small area $\Delta A$, normal to the direction concerned, and coming toward $\Delta A$ from within a small cone of solid angle $\Delta\Omega$ surrounding the direction.

0-201-04245-2

The brightness given in the specified direction is commonly expressed as

$$B(f, T) = \frac{\Delta E}{\Delta A\, \Delta\Omega\, \Delta f\, \Delta\tau}, \tag{6.4}$$

where $A$, $\Omega$, $f$, and $\tau$ are area, solid angle, frequency, and time, respectively; or the unit of brightness, $B(f, T)$, in the rationalized mks system is watts per square meter per steradian per hertz (hertz = cycle per second).

The brightness is generally expressed as the brightness temperature, which relates to the temperature of a blackbody at a given frequency and direction having the same brightness as the source.

Planck's radiation law defines the general relationship for the blackbody radiation in the enclosure between the brightness, $B(f, T)$, and the brightness temperature, $T_B$:

$$B(f, T) = \frac{2hf^3}{c^2} \left[\exp\left(\frac{hf}{kT_B}\right) - 1\right]^{-1}, \tag{6.5}$$

where $h$ is Planck's constant ($6.63 \times 10^{-34}$ J-sec), $k$ is Boltzmann's constant ($1.38 \times 10^{-23}$ J °K$^{-1}$), $c$ is the velocity of light in vacuum ($3 \times 10^8$ m sec$^{-1}$), and $f$ is the frequency.

In the microwave region, $hf \ll kT$, and Planck's radiation law for an ideal blackbody is given by the Rayleigh-Jeans approximation:

$$B(f, T) = \frac{2kT_B}{\lambda^2}. \tag{6.6}$$

Thus the brightness temperature of a source is known once the brightness is determined.

The value of $T_B$ depends on the radiometer design and is influenced by the antenna collecting aperture, $\Delta A$; the beam solid angle, $\Delta\Omega$; the predetection band width, $\Delta f$; and the integration time, $\Delta\tau$; thus

$$T_B = \frac{\lambda^2 B}{2k}, \tag{6.7}$$

where Eq. (6.4) for $B$ shows these dependencies. The complexity of determining $T_B$ by this procedure has led to the use of relative methods employing a calibration source at a precisely known temperature.

The representation of the power collected by the antenna as a temperature is clarified by introducing some of the parameters that describe the antenna, and showing the connection between brightness temperature, $T_B$, and antenna temperature, $T_A$. Consider the interaction of the receiving antenna with the random radia-

0-201-04245-2

tion from the surveyed source. It is assumed that the incident radiation is randomly polarized and is in the interval $f$ to $f + df$. The radiation is given in terms of its brightness $B$, which is a function of the emission region coordinates, or $B(\theta, \phi)$; the brightness temperature is also so related, $T_B = T_B(\theta, \phi)$. The incident power density at the antenna in the direction $(\theta, \phi)$, within the cone of solid angle $d\Omega$, is given by $B(\theta, \phi)\, d\Omega\, df$. The power at the antenna output port per unit band width can be expressed as

$$dW = \tfrac{1}{2} A(\theta, \phi) B(\theta, \phi)\, d\Omega \tag{6.8}$$

where $A(\theta, \phi)$ is the effective antenna aperture in square meters. The factor ½ comes from the fact that ordinarily the radiation is not polarized and is equally distributed between orthogonal polarizations; the antenna, being usually linearly polarized, accepts only one-half the incident radiation.

Thus the total power within view of the antenna, or within its acceptance solid angle, is

$$W = \frac{1}{2} \iint_{f\ \Omega} A(\theta, \phi) B(\theta, \phi)\, d\Omega df \tag{6.9}$$

The antenna can be conveniently thought of as a transmission line terminated in a matched resistive impedance, or as the receiver input, where the noise power of the antenna is equal to that of the matched resistor, $T_R = T_A = W/k$, with $T_R$ the real temperature of the resistor, and $T_A$ the antenna temperature. Thus

$$T_A = \frac{1}{\lambda^2} \iint_{f\ \Omega} A(\theta, \phi) T_B(\theta, \phi)\, d\Omega df \tag{6.10}$$

The collecting aperture of the antenna can be expressed as $A(\theta, \phi) = A_m F(\theta, \phi)$, where $A_m$ is the maximum value of $A(\theta, \phi)$ and $F(\theta, \phi)$ is the normalized antenna power pattern, $0 \leqslant F(\theta, \phi) \leqslant 1$; thus

$$T_A = \frac{A_m}{\lambda^2} \iint_{f\ \Omega} F(\theta, \phi) T_B(\theta, \phi)\, d\Omega df \tag{6.11}$$

For the condition of thermodynamic equilibrium, $T_R = T_B(\theta, \phi)$, and the matched termination of $T_R$ is at the same temperature as the isothermal enclosure. Thus $T_B(\theta, \phi)$ is independent of $(\theta, \phi)$ and equal power flow takes place from antenna to load, and vice versa; and

$$\Omega_A = \frac{\lambda^2}{A_m} = \int_{\Omega} F(\theta, \phi)\, d\Omega \tag{6.12}$$

0-201-04245-0

For remote sensor applications in which the brightness temperature is a function of direction, finally,

$$T_A = \frac{1}{\Omega_A} \int_f \int_\Omega F(\theta, \phi) T_B(\theta, \phi) \, d\Omega df \tag{6.13}$$

The antenna temperature, $T_A$, is a weighted average of the source brightness temperature incident at the antenna, the weighting function being the antenna power pattern function $F(\theta, \phi)$. The side lobes are shown to play an important part in obtaining "clean" data. The fraction of energy accepted from outside the main beam has been defined as the "stray factor"; it ranges from 0.03 to 0.4 for most narrow-beam antennas.

*2. Temperature Sensitivity*

The ability to collect radiation from a specific ground cell, or a succession of ground cells, depends on the radiometer antenna spatial resolving power, receiver sensitivity, and receiver calibration. The sensitivity refers to the minimum rms temperature interval $\Delta T$ which can be measured by the system. When a radiometer is able to resolve more discrete sources than it can detect, it is sensitivity limited. Conversely, when it can detect more discrete sources than it can resolve, it is resolution limited.

The signal observed at the output of the radiometer fluctuates about a value which is the average for the total noise power in the receiver. The total noise is represented by an equivalent noise temperature $T = T_R + T_A$, or $T = (F - 1)T_0 + T_A$, where $T_R$ is the receiver noise temperature, $F$ the receiver noise figure, and $T_0$ an input reference temperature of 290°K.

Without signal smoothing (integration or RC filtering), the output signal shows a wide variation in amplitudes. By averaging $N$ independent samples of this information, however, the averaged value can be made to converge to the true long-time average to almost any precision, provided one is willing to pay the price in predetection band width, integration time, or receiver noise figure.

The sensitivity $\Delta T$ of a radiometer receiver can be estimated by considering the information contained in the predetection band width $B$, expressed in hertz. Many remote sensing problems with radiometers would be solved if the band width could be increased; however, this causes two changes: the spectral aspects of the radiation are decreased, with $\epsilon_\lambda$ becoming $\epsilon_{\Delta\lambda}$; and the possibilities of radar interference increase. A total of $N = 2B\tau$ independent samples is available for integration time $\tau$. From statistical theory, the rms fluctuation about $N$ independent readings of a random signal decreases as $1/\sqrt{N}$. Thus the accuracy with which the input temperature can be measured is given by

$$\Delta T = \frac{CT}{\sqrt{B\tau}}, \tag{6.14}$$

where the constant $C$ relates to specific modulations.

0-201-04245-2

### 3. *Resolution*

Resolution refers to the ability of the radiometer to distinguish between two identical point sources adjacent to each other. As related to microwave radiometer imagery, this would constitute the problem of not smearing the temperature rendition of one ground cell with that of its neighbor. The angular separation $\gamma$ between two objects that can just be resolved is given by

$$\gamma = \sin^{-1}\left(\frac{0.61\lambda}{r}\right), \tag{6.15}$$

where $r$ is the radius of a circular aperture (Kruse et al., 1962). However, a more common resolution criterion is used; two identical point sources can be resolved if their angular separation is greater than the half-power or 3-dB beam width of the antenna. The 3-dB beam width is nearly always taken as the resolving angle for the antenna (Ko, 1964).

High resolution at microwave wavelengths calls for large physical apertures if beam widths comparable to those found for infrared or optical systems are mandatory (0.5° or less). Some indication of the physical size of antennas can be obtained by again considering a circular aperture which is excited by a constant amplitude and phase source; thus

$$\beta = \frac{1.02\lambda}{D}, \tag{6.16}$$

where $\beta$ is the half-power beam width in radians, $\lambda$ the wavelength, and $D$ the aperture diameter in wavelengths (Silver, 1949). The first side lobe is down 17.5 dB from the main beam. For high spatial resolution, it is necessary to either operate at extremely low altitudes or further decrease the beam width by increasing the antenna size. At $\lambda = 1$ cm, or 30 GHz, a 0.5° beam width requires an antenna diameter of approximately 1.1 m. An average infrared scanner has beam widths of from 0.25 to 0.05°, while a 5 mm lens at optical frequencies has the phenomenal angular resolution of 0.001°.

### 4. *Attenuation*

As the thermal radiation propagates through earth materials and the atmosphere, its brightness changes because of continuous and selective absorption and emission. For aircraft or spacecraft surveys of the earth's surface, the radiometer is many miles from the earth and the radiation emanating from the surface must pass through the intervening atmosphere, in which part of the radiation is absorbed and part is reradiated, while an added amount comes from self-radiation of the atmosphere.

Consider the change in power, in terms of its ground-target brightness temperature, $T_0$, as radiation propagates through an atmosphere with a power attenuation

0-201-04245-2

coefficient $\alpha(s)$ ($m^{-1}$) and a total transmission path depth $d$ to the antenna. The power density incident at a layer of atmosphere of thickness $s$ to $s + ds$ is $P(s) = B(f, T)\, df\, d\Omega$ (W $m^{-2}$), and the power is shown to be a function of position, $s$. The power loss per unit area through the layer $ds$ is $dP = -\alpha(s)\, ds\, P(s)$. The power at some distance from the source is given by

$$P(s) = P_0 \exp\left[-\int_0^s \alpha(s)\, ds\right] \tag{6.17}$$

where $\tau(s) \equiv \int_0^s \alpha(s)\, ds$ is defined as the "optical depth," and is introduced to characterize the radiation attenuation in the layer of atmosphere at distance $s$ from the source.

The brightness temperature at distance $d$ is $T_B(d) = T_0 \exp[-\tau(d)]$. Furthermore, consider a layer, $ds$, at its actual temperature, $T(s)$, emitting thermal radiation governed by its emissivity, $\epsilon$; thus $dP = B(f, T)\epsilon(s)\, df\, d\Omega\, ds$. Each differential layer has a temperature, $T(s)$; an emissivity, $\epsilon(s)$; and an attenuation factor such that the brightness temperature of the emitted radiation at $s = d$ from all along the path is

$$T_\epsilon = \int_0^d T(s)\epsilon(s) \exp\left[-\int_s^d \alpha(s)\, ds\right] ds. \tag{6.18}$$

Thus the total brightness temperature at the receiver when $T(s)$, $\epsilon(s)$, and $\alpha(s)$ are constant, and scattering and variations in the index of refraction in the medium can be neglected, or $\epsilon = \alpha$, is

$$T_B = T_0 \exp[-\tau(d)] + T\{1 - \exp[-\tau(d)]\}. \tag{6.19}$$

The importance of Eq. (6.19) is very apparent when correlations are attempted between the radiometer output and the ground surveyed.

## C. Simulation

With the digital computer, it is feasible to evaluate the success of a proposed complex aircraft radiometer mission by simulation. This is accomplished by using analytical and empirical descriptions of the hardware and the physical environment (terrain, aircraft, computers, and support equipment) to develop mathematical models of entire systems. These mathematical models are formulated into computer programs. By varying the values of the critical parameters of the simulated system, its response is studied and limiting factors are determined. The cost in time and material is far less than that of even a modest experimental program and supplies many of the basic results.

0-201-04245-2

*1. Signal Paths*

The signal paths encountered in gathering brightness temperature data are depicted in Figure 6.3. It is important to note these paths, since the brightness temperature contributions may be erroneously assumed to come only from the ground surface.

A radiometer installed in an aircraft in inclement weather receives radiation via the following paths: (1) direct path from underlying clouds or overhead sky; (2) direct path from ground surface through water vapor and clouds (clouds can contain ice, snow, and rain), (3) direct path from subsurface through water vapor and clouds; (4) indirect path from clouds, sky, or sun as reflected from the surface or subsurface strata; and (5) direct path from miscellaneous sources (foam, spray, dust, etc.).

For low-grazing angles, the brightness temperature observed is high, about 270–320°K, because of the long transmission path through the denser concentration of water vapor at the earth's surface.

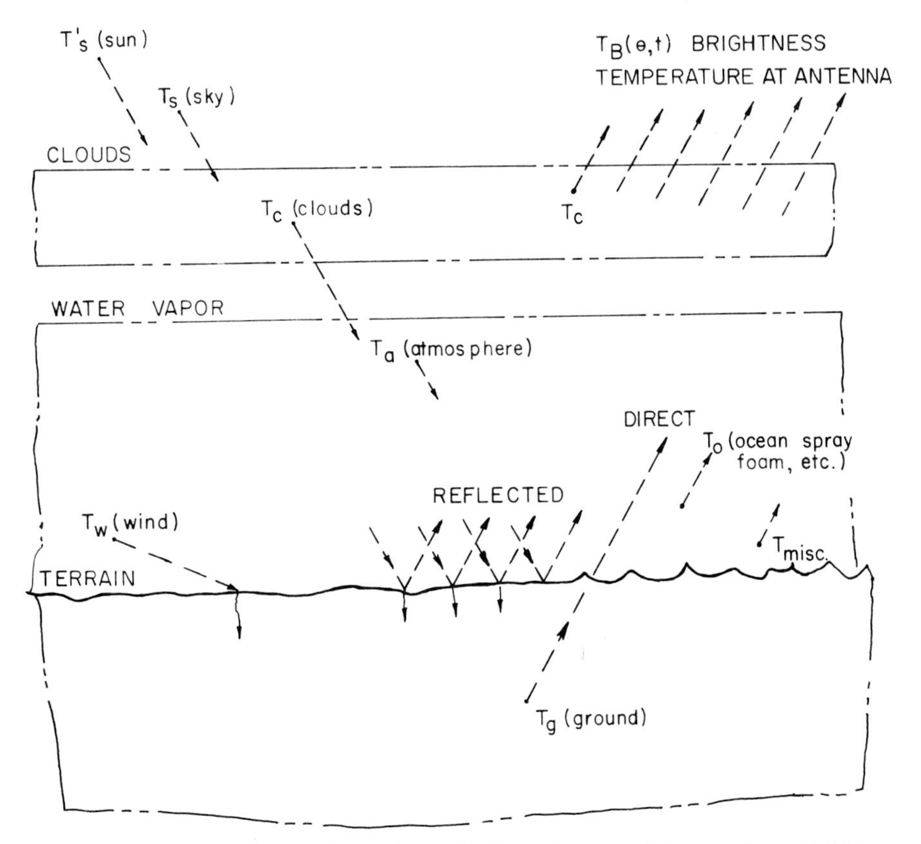


**Fig. 6.3.** Electromagnetic radiation inputs, both steady-state and time-varying, to brightness temperature model.

0-201-04245-2

*2. Models*

Separate mathematical models, analytical or empirical, of surface and subsurface terrain, the intervening atmosphere between aircraft or satellite and ground, the sky, and the radiometer system (radome, antenna, receiver, etc.) provide clearer pictures of all parameters and of the interaction mechanisms involved in obtaining radiometer signature data. Some primary models to consider would be (1) microwave radiometer (wavelength, angle of incidence, polarization, resolution, scan, and nonscan); (2) terrain surface and subsurface (smooth, rough, and layered); (3) atmosphere (temperature, water vapor, rain, snow, ice, oxygen, and other constituents); and (4) flight navigation and control (roll, pitch, yaw, altitude, and velocity).

The brightness temperature incident at the antenna represents a lumping together of all the inputs from the various sources after each one has been modified to some extent by the transmission media. Hence a number of separate models need to be considered for $T_B$, which can be written as the sum

$$T_B = C_1 T_g + C_2 T_w + C_3 T_c + C_4 T_s + C_5 T_m, \tag{6.20}$$

where $C_{1,2,...}$ are the emissivities, reflectivities, and attenuations of each path, and $T_g$, $T_w$, $T_c$, $T_s$, and $T_m$ are the ground, water vapor, cloud, sky, and miscellaneous brightness or actual temperatures, respectively.

Models of land and water surfaces for the simulation of radiation received by the microwave radiometer are assembled from basic expressions of the radiation reflected from a smooth surface and that generated by the surface (Peake et al., 1960). The model is complicated by the fact that most surfaces are rough in terms of the exploring wavelength. However, a model including roughness is realistic and can provide important information about the surface; indeed, roughness is used in models for studying sea states.

A simple one-dimensional model with $T_B(\theta)$ as a function of the angle of incidence is assembled. A summation is made of the radiation reflected from the sky and that emanating from the ground:

$$T_B(\theta) = \rho_g T_s + \epsilon_g T_g, \tag{6.21}$$

where $\rho_g$ is surface reflectivity; $\epsilon_g$, emissivity of the ground; $T_s$, brightness temperature of the sky; and $T_g$, actual ground temperature.

An assembly of the model given by Eq. (6.21), using a sky model, $T_s(\theta)$, and a ground model, $T_g(\theta)$, with a smooth surface, clearly demonstrates the form of the plots and the unique signature aspects of $T_B(\theta)$. For example, starting with a homogeneous lossless material having a relative dielectric constant of 3.2 and a loss tangent assumed to be 0, plots are made of the absolute values of the Fresnel reflection coefficients for vertical polarization, $R_v(\theta)$, and horizontal polarization, $R_h(\theta)$, as given in Figure 6.4*a*. Most dry soils have relative dielectric constants of about this

0-201-04245-2

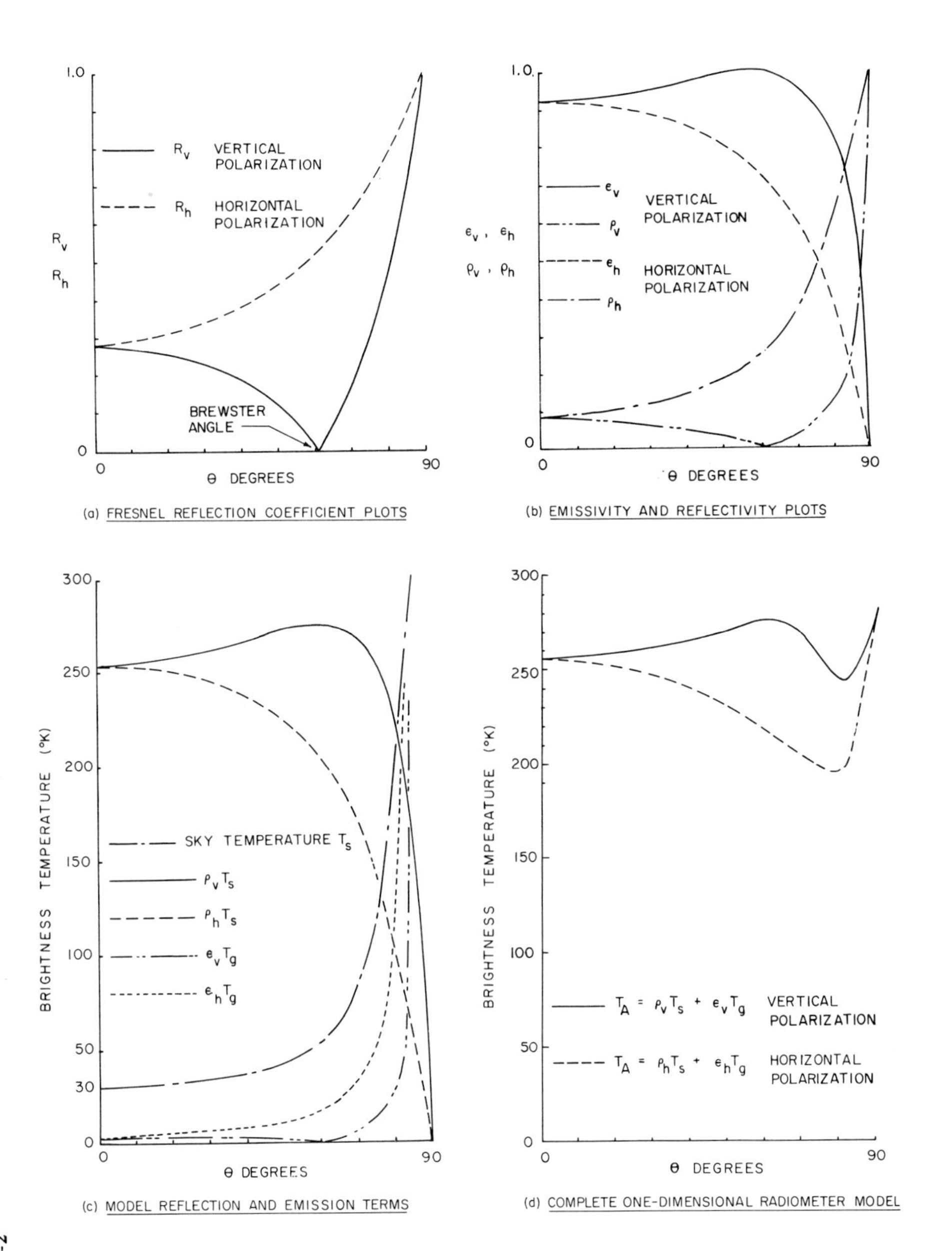


**Fig. 6.4.** Brightness temperature model assembled from plots of emissivity and reflectivity of the ground surface and models of the sky and ground temperatures.

0-201-04245-2

value. For perfect reflectors, such as metals, the reflection coefficients are both unity and independent of $\theta$, and they provide excellent signature sources.

Figure 6.4*b* shows the reflectivities (squared values of the reflection coefficients) $\rho_v(\theta)$ and $\rho_h(\theta)$ for vertical and horizontal polarization, respectively. The more useful emissivity plots are derived from the reflectivity plots $\epsilon = 1 - \rho$, or $\epsilon_v(\theta)$ and $\epsilon_h(\theta)$ for vertical and horizontal polarization, respectively, and are superimposed on the reflectivity plots.

The brightness temperature of the sky, as a function of the angle of incidence, $T_s(\theta)$, (with a zenith temperature of 30°K), is shown in Figure 6.4*c*, together with the first and second product terms of Eq.(6.21), $\rho_g T_s$ and $\epsilon_g T_g$, for both polarizations.

The final set of plots in Figure 6.4*d* shows the sum of the product terms in Eq. (6.21) for an actual constant ground temperature, $T_g$, of 275°K. For the analysis of radiometer data, the shape and slopes of these plots are very significant, since their forms are clearly evident in the experimental data. The Brewster angle effect (where power is incident at the surface but none is reflected; see Figure 6.4*a*) is quite pronounced for vertical polarization and is found to shift with dielectric constant and surface roughness. A major difference between the two polarizations occurs in the range $40 \leqslant \theta \leqslant 70°$. Both plots rise rapidly in the grazing-incidence range.

Ideally, one would like to fly only in clear weather, with air temperature and water vapor known and constant as functions of altitude over the flight path. These ideal conditions are approached in some flights over deserts; however, for orbiting satellites taking synoptic data, such conditions are most unlikely, and onboard methods are desirable to remove the atmospheric radiative biases before any information is telemetered to the ground.

Although the emphasis in this chapter is on terrain radiation, this radiation must propagate through the intervening atmosphere. Therefore it is necessary to include in the total model appropriate auxiliary models of the atmosphere. It is important to compute $T_B$ for the expected range of variations given by these models. Variations in the temperature and water vapor profiles of the atmosphere can only cause uncertainties that must be coped with in the total model.

The emission of microwave radiation as a function of water vapor density at its resonance frequency of 22.235 GHz can have a dominant influence on the design of a simulation program, depending on whether surface information is sought or information on the integrated water vapor in the atmosphere. The results of water vapor and oxygen resonances will significantly affect the choice of optimum frequencies for system operation. Atmospheric data are available as standard models for various latitudes.

The need for modeling the flight dynamics of the vehicle in which the radiometer is installed is also apparent. Since the radiometer is generally not installed on an inertial platform, positional errors due to roll, pitch, and yaw can appreciably compromise the data. Attitude and navigational errors contribute to inaccuracies when the radiometer output is correlated with actual ground positions.

0-201-04245-2

## III. RADIOMETER SYSTEMS

### A. Survey of Instruments

A passive microwave receiver employs a large collecting aperture to measure thermal noise in a spectrum about 10-500 MHz wide. The total noise is made up of two parts: the signal collected by the antenna, which is infinitesimal, and that generated by the receiver, which is many times larger, although both have the same statistical properties. The signal noise is about $10^{-11}$ to $10^{-20}$ W, thus making clear one design goal, namely, to achieve maximum sensitivity within the constraints on power, weight, size, and budget.

Numerous radiometer types, covering the range from below 0.3 GHz, well below the microwave region, to 300 GHz, bordering on the long-wavelength infrared region, have been investigated; theoretical as well as hardware studies have been conducted for a wide range of applications. Of all the types studied, the signal-modulated or Dicke radiometer (Dicke, 1946) is still the one most commonly used, even though it has a low intrinsic performance because of its relative simplicity and insensitivity to receiver gain variations.

#### *1. Basic Receivers*

The receiver consists basically of a high-gain antenna, Dicke switch, broad-band rf amplifier (tunnel diode or parametric amplifiers are required for high sensitivity), square-law detector, band-pass filter, video amplifier, synchronous detector (phase or multiplier detector), and integrator (RC filter).

Dicke was the first to apply modulation for eliminating fluctuations induced by receiver instabilities. The input to the receiver is switched between the antenna port and the reference thermal waveguide load. The thermal loads can be either heated or cryogenically cooled sources, or the zenith sky. A zero output is established by adjusting the thermal load power to $kT_RB$ so as to equal the average antenna temperature power $kT_AB$, after which the signal power changes provide the amplitude modulation. The switching rate is purposely set considerably higher than receiver instabilities, and detection of the modulation is accomplished by a synchronous, or phase-sensitive, detector. The signal output from the integrator is found to be proportional to $\Delta T$ within the uncertainties of the system and its calibrations.

The sensitivity of the Dicke radiometer is given by

$$\Delta T_{\mathrm{rms}} = \frac{2(T_A + T_R)}{\sqrt{B\tau}} . \tag{6.22}$$

For achieving maximum sensitivity, contributions to the output due to reference temperature and gain variations must be small in comparison to changes in the signal, $\Delta T$. This condition is met by minimizing temperature variations of the

0-201-04245-2

reference thermal load, by proper modulator design, and by maintaining the reference temperature, $T_R$, very close to the average antenna temperature, $T_A$.

Since brightness temperature varies with polarization, valuable signature information can be derived from the scene by utilizing the polarization feature; therefore many of the profilers are designed with separate antenna feeds for each polarization, vertical and horizontal. A waveguide switch allows sampling each polarization alternately or the two polarizations simultaneously with a dual-channel receiver.

### *2. Calibration*

Three important requirements must be imposed on the output signal. It must be (1) calibrated in degrees Kelvin (°K), (2) referenced to absolute zero, and (3) linear over the range of cperation (usually 50-350°K).

The degree of uncertainty associated with thermal sources, both hot and cold, depends on the precision with which the reference temperatures can be controlled and the accuracy with which the heat sources (as viewed by the radiometer input) can be measured and calibrated.

Small bridge-controlled heated waveguide ends are advantageous for simplicity, precision, and control and are feasible for space applications. Cryogenic loads are best for precision and for establishing low-temperature calibration points, either 4°K (helium) or 77°K (nitrogen) (Stelzried, 1968; Trembath et al., 1968).

### *3. Developed Instrumentation*

Various radiometers have been developed in the past few years for laboratory, field, and aircraft use in the spectrum 1.4-94 GHz. The possibilities of extending this range are excellent, since laboratory type radiometers have been operated successfully at 140 and 225 GHz (Cohn et al., 1963).

Table 6.1 is based on a number of radiometers of varied design, most of which were intended primarily for aircraft profiling applications. Comparisons can be made of their sensitivities, antenna types and beam widths, polarizations, integration times, and calibration methods. A radio astronomy radiometer with a moderate-sized antenna (18 m dish) is included to indicate generally the ultimate in design, with a $\Delta T$ of ±0.033°K. Also, a recent imaging radiometer is included to show the progress being made in the state of the art.

## B. Imagers

The advantages and great appeal of imagery for studying the earth have led to the development of scanning radiometers for aircraft and meteorological satellites. The scanning radiometer is much the same as the profiler except for a high-gain antenna, which is scanned either mechanically, electronically, or in a multiple-feed arrangement; an ultrasensitive receiver; a synchronized display system; and a strip-film recorder. Systems are designed for both vertical and horizontal polarization, operat-

0-201-04245-2

0-201-04245-2

**TABLE 6.1**
**Comparison of Radiometers, 1.4–225 GHz**

| *Frequency (GHz):* | 1.4 | 8–12 | 10 | 13.4 | 19.4 | 22.235 | 32.4 | 60 | 94 | 138 | 225 |
|---|---|---|---|---|---|---|---|---|---|---|---|
| *Special comment:* | Atmospheric absorption negligible | Atmospheric absorption slight | Atmospheric absorption slight | Profiler | Imager ±50° scan | Absorption line, $H_2O$ | Window | Temperature[a] profiling, $O_2$ complex | Window absorption high | Window | Window slightly opaque |
| *Antenna type:* | Two dimensional phased array | 60-ft dish | Lens-horn | Lens-horn | Phased array | Lens-horn | Lens-horn | Dual-feed | Long horn | Convex lens and horn | Convex lens and horn |
| *Beam width 3 dB (degrees):* | 16 | 0.06 | 5 | 5 | 3 nominal | 5 | 5 | – | 0.5–1.5 | 0.3 | 0.3 |
| *Polarization:* | Linear | – | Linear | Dual | Dual | Linear | Dual | – | Dual | Either land-based | Either land-based |
| *Calibration:* | Noise source | Nitrogen | Noise source | – | Nitrogen Helium | Noise source | Noise source | Cryogenic | – | VL[b] | VL |
| *Amplifiers:* | Mixer, IF | Maser, rf | Mixer, IF | Mixer, IF | – | Mixer, IF | Mixer, IF | Multisection mixers | Mixer, IF | TWT[c] | TWT |
| *Noise figure (dB):* | 6 | 0.06 | 5.5 | – | 6 | 8.5 | 10 | – | – | 20 | 26 |
| *ΔT (K) rms:* | ±0.26 | ±0.033 | ±0.14 | ±1 | – | ±0.8 | ±0.5 | – | ±1 | ±0.3 | ±1.0 |
| *Integration time (sec):* | 1 | 2 | 1 | 0.5–15 | 1 | 1 | 1 | – | 0.5–15 | 10 | 10 |
| *Application:* | ERP[d] | RA[e] | ERP | ERP | ERP | ERP | ERP | ERP, satellite surveys | EM[f] | RA | RA |

[a]No firm data available; radiometers are designed and operational.
[b]VL–blackbody thermal enclosure with variable heat source.
[c]TWT–traveling-wave tubes.
[d]ERP–Earth Resources Program (NASA).
[e]RA–Radio astronomy.
[f]EM–earth materials: soil, snow, and ice.

ing either simultaneously or in a time-share mode. Color image enhancement is applied during data processing to yield an image that brings out the temperature subtleties of the scene, thus facilitating analysis.

*1. Image Generation*

Brightness temperature images can be made of the earth or the atmosphere by any of three basic methods.

1. The first and most easily implemented uses the constant forward velocity of aircraft to provide horizontal scan separation of the image, which is recorded scan by scan, or placed onto moving film by a flying-spot scanner.
2. The second employs multiple beams collecting radiation simultaneously in either a transverse or a conical plane.
3. The third uses a mosaic of beams to collect a brightness temperature picture of the terrain for a short integration time, $\Delta\tau$. The multiplicity of beams requires a correspondingly large number of receivers, which add measurably to the complexity in design.

Mechanical steering of the antenna beam has such advantages as multispectral capabilities, reduced feed losses, and electronic simplicity. The inertia associated with mechanical scanning systems, however, is a major disadvantage for space. In contrast, electronically scanned systems are without inertia and scan more rapidly, making their application to space almost unique. An integral part of the antenna scanning system is the beam-steering computer, which can now be designed with integrated circuits to be small and light.

The antenna must have high beam efficiency, which is defined as the ratio of the power in the main lobe to the power in the side lobes. Radiation received by the side lobes degrades the image and increases the uncertainties in the determination of absolute temperature measurements. The antenna assembly (which usually includes a radome) must be small, compact, light, and isolated from thermal gradients, and, with its steering computer, require little power.

To meet these requirements, electronically scanned phased arrays have been developed and have proved to be highly successful in providing rapid scanning without inertia for both polarizations.

*2. Scanning Requirements*

Restrictions on imaging radiometers are brought to light by considering the geometry of the flight path and the scanning beam.

Consider an airborne scanner operating at from 300 to 11,000 m, and for velocities to 400 knots. The scanning rate of the antenna must be adjustable for both changing altitude and changing velocity. To add to the complexity, attitude instability produces image distortion, and such effects must be minimized.

The geometry of the aircraft and the beam intercepts on the ground are shown in Figure 6.5 for a conical-scan system. The beam can be swept continuously or

0-201-04245-2

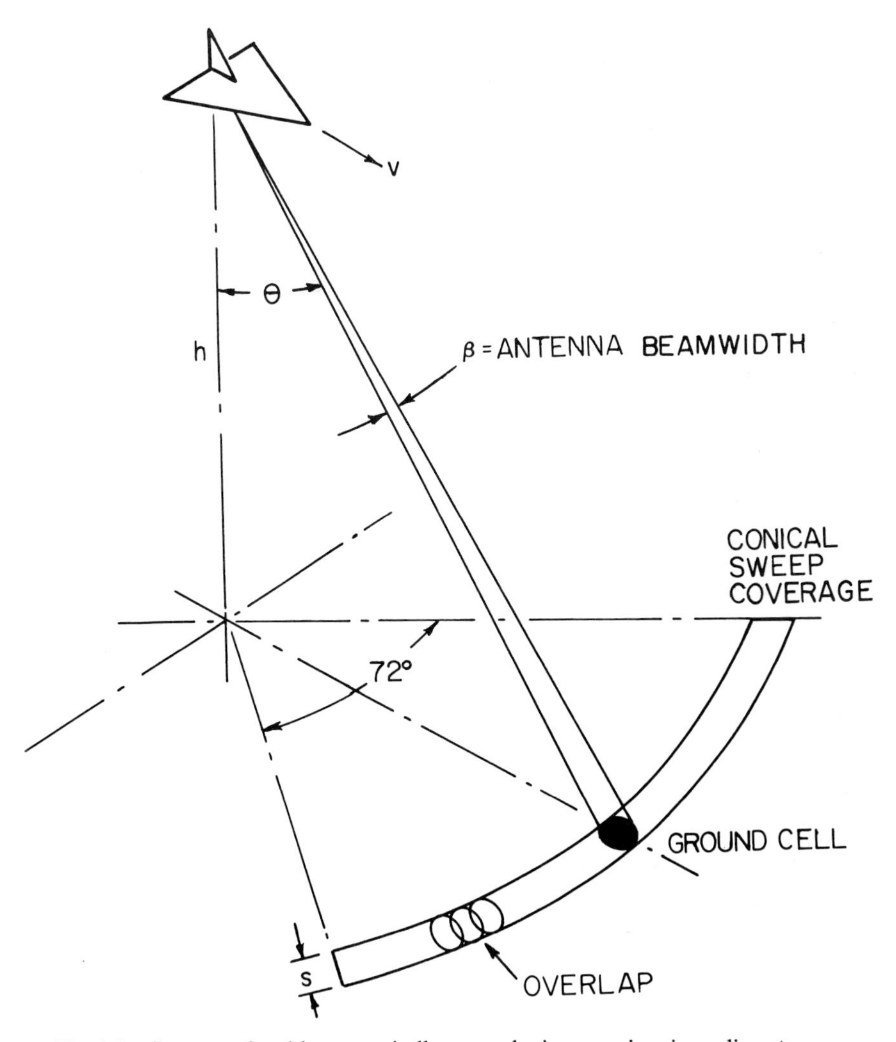


**Fig. 6.5.** Geometry for airborne conically scanned microwave imaging radiometer.

stepped almost instantaneously from one ground cell to the next, and allowed to dwell on the spot for time $\Delta\tau$, to prevent smearing.

The dwell time on the ground cell is determined by noting the time required to fly distance $s$, as functions of the angle of incidence $\theta$, altitude $h$, velocity $v$, and antenna beam width $\beta$, where $s = h\beta/\cos^2\theta$, the time to fly distance $s$ is $t = s/v$; and the time available for dwell, $\Delta\tau$, is given by

$$\Delta\tau = \frac{h\beta}{nv\cos^2\theta}, \tag{6.23}$$

0-201-04245-2

where $n$ is the number of cells in the conical scan. The dwell time must be adjusted to allow for cell overlap in the forward and transverse directions. It can be shown that

$$\Delta T = \frac{C_1 F T_0}{\beta} \sqrt{\frac{v/h}{B}} . \tag{6.24}$$

A number of limitations for the scanning radiometer are apparent. High spot resolution calls for a narrow beam width, which in turn requires a shorter dwell time and a faster beam-stepping rate. Maintaining $\Delta T$ constant for a short dwell time requires an improvement in system noise figure $F$, an increase in the predetection band width, or a smaller $v/h$.

### *3. Scanner System Highlights*

To produce good brightness temperature maps, a complete system is required which can program the scan, give inflight calibration, provide "quick-look" displays (i.e., oscilloscope facsimile or photograph), and record output signals in computer-compatible form for later processing. The design objective of such a system would be to produce high-detail brightness temperature maps to within ±2°K uncertainty over a range of target temperatures of 50–350°K. A radiometer system of this type is shown by an artist's exploded view in Figure 6.6.

Some design goals of such a system would be to develop or include the following:

1. A dual-polarization scanning array controlled by a beam-steering computer with $v/h$ inputs, servocontrolled from aircraft altitude and velocity sensors, with each spot viewed with both polarizations.
2. A special receiver using two separate channels, one for each polarization, gain-stabilized to a high degree by referencing a precision hot load in the rf section to an absolute dc voltage (Hach, 1968).
3. Continuous inflight calibration by switching between two precision hot loads to an expected absolute accuracy of ±0.4°K.
4. Display of images on a CRT monitor, facsimile, or film for "quick look."
5. A ground data-processing system to provide displays of vertical and horizontal polarizations, their differences, and color. Image enhancement would be provided by synthetic resolution, edge-effect compensation, and smoothing, and absolute values by a computer program using inversion techniques.

### *4. Transverse Scan Imager*

A 19.4-GHz transverse-scan system has been developed by NASA Goddard Space Flight Center, employing an electronically scanned narrow conical-beam antenna which scans ±50° in the transverse direction of flight. The antenna has a beam

0-201-04245-2

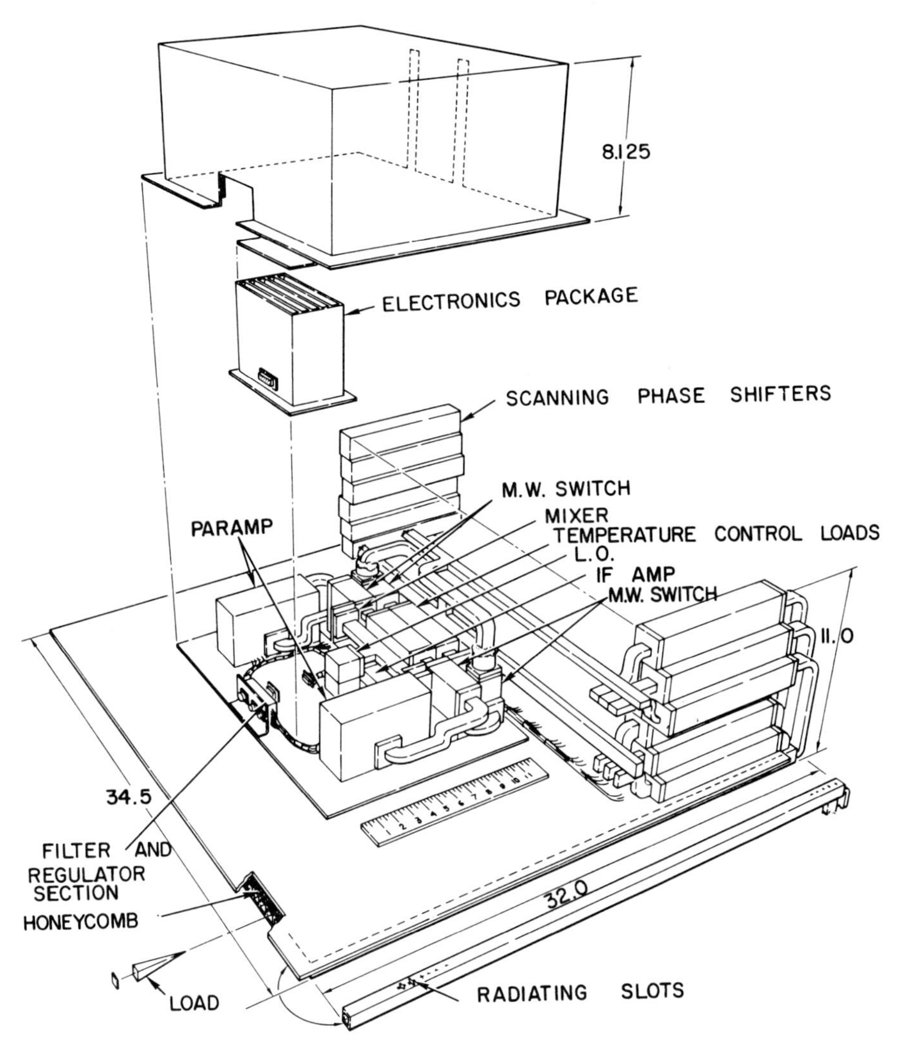


**Fig. 6.6.** Exploded view of microwave imaging system with an electronically swept antenna beam for conical scanning and dual polarization. Courtesy of Teledyne Ryan Company.

width of 2.9°, with a side-lobe stray factor of 0.08 and an insertion loss of 0.6 dB, has a scan rate of 1 or 2 sec, and is 0.45 by 0.45 by 0.08 m in size. Signal calibration uses a hot and a cold reference at the end of each transverse scan. The resolution ($\Delta T$) is ±1°K.

The electronically scanned antenna has a beam width comparable to that of the infrared scanner used on weather satellites (the medium-resolution infrared radiometer, MRIR), and the beam is stepped sequentially from ground cell to ground cell. For example, at an altitude of 10,000 m, the beam sweeps out a

0-201-04245-2

ground strip about 25 km wide, each cell of which is roughly 450 by 450 m (Catoe et al., 1967).

The 19.4-GHz imaging radiometer was installed on a Convair CV-990 jet aircraft, and imagery has been taken of farmland, desert, river areas, tundra, ice, tropical rain forests, cities, dense clouds (snow, ice, and rain), and water (Salton Sea and the Pacific and Atlantic oceans). The resulting brightness temperature maps clearly demonstrate that the imaging radiometer is a significant new all-weather, day-night remote sensor.

The all-weather feature is shown in imagery taken on May 11, 1967, of the St. Louis, Missouri, river junction at 11,370 m through a thick, uniform cloud deck (Figure 6.7). The radiometer image clearly shows the confluence of the Illinois and Mississippi and the Missouri and Mississippi rivers through the clouds. In the image, the brightness temperatures of the rivers and lakes appear very low (black) in contrast to warmer land (lighter shades). The contiguously recorded imagery is enhanced by color quantizing in 32 to 52 shades, greatly improving the brightness temperature contrasts.

Such color-enhanced imagery provides an excellent means of making an immediate qualitative assessment of ground features. Metal roofs, concrete roads, canals, rivers, and lakes appear cold (black to blue); dirt and asphalt roads, plowed fields, and open fields, moderately warm (green to yellow); and grass, heavy vegetation, and trees, hot (red to white).

Another illustration, taken near Calexico, California (near the United States-Mexico border), shows the contrast between populated areas and farmland (Figure 6.8).*

### 5. *Space Radiometers*

The feasibility of microwave systems for space applications has been amply demonstrated by the Apollo moon missions. These microwave systems required "hardening" for the rigors of space (i.e., high thermal gradients, extremes in temperature, and the vacuum of free space). Passive microwave systems, profilers, and scanners have also proved feasible for orbiting spacecraft applications. One such device, the electrically scanned microwave radiometer (ESRM), is presently operating on NIMBUS 5.

The attitude control of spacecraft proposed for remote sensing surveys is within 1° for roll, pitch, and yaw. The attitude rate is found to be slower than the rate of change of the ground radiation surveyed.

Microwave imagery is feasible for satellites even up to 500 nm with a transverse scan of ±45° or with a conical scan.

Satellite passive profilers and imagers are entirely feasible at long wavelengths, 1 meter or more, for studying soil subsurfaces. The larger antenna apertures required for narrow beam widths at these wavelengths in space have been designed as unfurlable, unfoldable, roll-out panels.

*See color insert.

0-201-04245-2

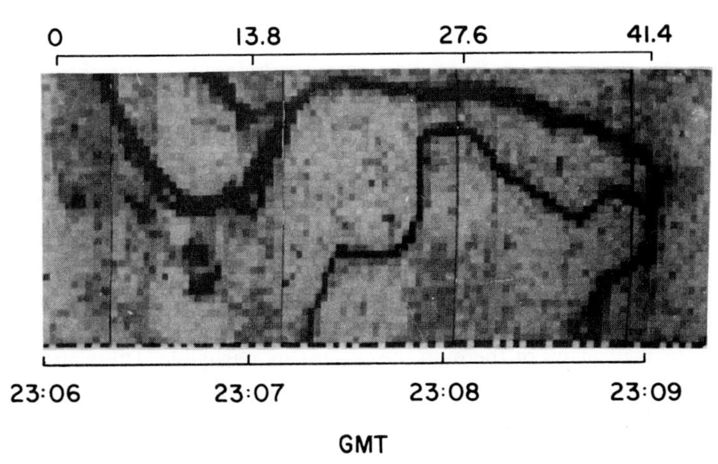


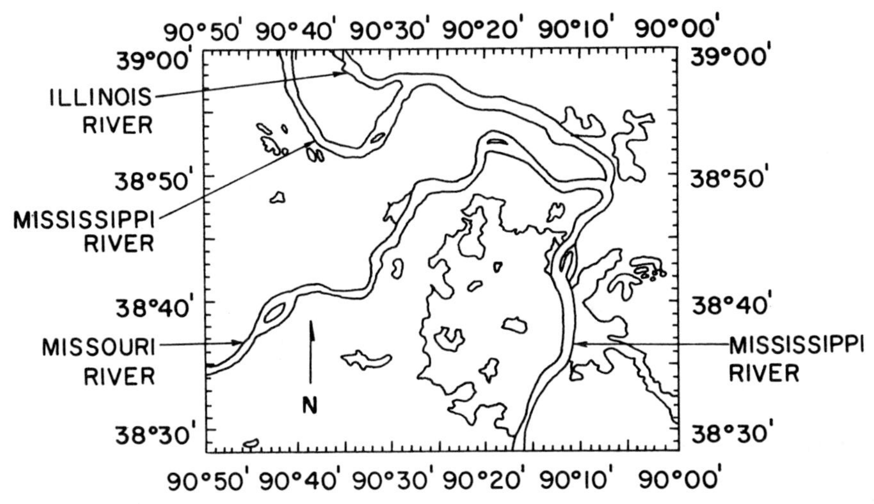


**Fig. 6.7.** Radiometer image of the river junction region by St. Louis, taken with a 19.4-GHz electronically scanned radiometer at 11,370 m. Courtesy of Aerojet-General Corporation.

## IV. INFORMATION

### A. What Information Is Available?

0-201-04245-2

For the passive microwave region it is not an easy task to assemble the vast amounts of collected data into some usable order, and then to compact the result into a simple and understandable form. The undertaking is complicated by the poor spatial resolution of the systems, and also by the lack of simple methods of abstracting data from radiometer signals.

The information available from radiometers includes (1) relative brightness temperature plots (statistically averaged), (2) real-time plots showing spatial detail, (3) brightness temperature maps, and (4) absolute temperatures.

The measured brightness temperatures are functions of the electrical, compositional, and textural properties of the surface and of the particular radiometer used for the measurements: $T_B$ ($\lambda, \theta, p, \epsilon, \rho, \beta$, etc.), where $\lambda$ is the center wavelength of the predetection band width; $\theta$, the angle of incidence; $p$, the polarization ($V$, $H$, and $V$ minus $H$); $\epsilon$, the emissivity; $\rho$, the reflectivity; $\beta$, the antenna beam width; etc. These parameters are not equally sensitive, and under certain operating conditions the effects of many of them on the system are negligible. There are published plots of $T_B(\lambda, \theta, p)$; however, the coverage is very limited.

### *1. Analog Records*

The output of either the profiler or scanner radiometer is a temperature-calibrated voltage-time record, recorded directly or converted to digital form. The recorded continuous signal is closely tied (both temporally and spatially) to the ground cells through accompanying aerial photography (in the absence of clouds) and ground truth. Calibrations are on the same record. The critical functions of timing, attitude, altitude, and velociiy are accurately controlled by the aircraft navigational and guidance system. This information is recorded for correction of the processed data, which appear as (1) brightness temperature values as functions of angle of incidence for polarization, wavelength, and resolution; (2) real-time brightness temperature profiles on chart paper; and (3) imagery or maps in the form of film (black and white or color enhanced), facsimile, or computerized maps.

### *2. Parametric Information*

The parameters involved are separated into those associated with the environment and those associated with the radiometer; the former are uncontrollable, and the latter are controllable to some degree of precision based on the radiometer design. There are three predominant terrain parameters that influence signatures: emissivity, roughness, and polarization.

Emissivity varies with the temperature and composition of an area, including the soil moisture (which lowers the emissivity). Emissivity also varies with the radiometer-controllable parameters of angle of incidence, polarization, and wavelength. The pronounced sensitivity of $\epsilon$ to the angle of incidence and polarization is clearly indicated by Figures 6.4*a* and 6.4*b*.

0-201-04245-2

A gross-trend analysis has been made from a review of theoretical and experimental data on changes in the emissivities of water and ground as functions of temperature, contained soil moisture, and frequency. The trends are plotted in Figure 6.9 for vertical polarization, near vertical incidence.

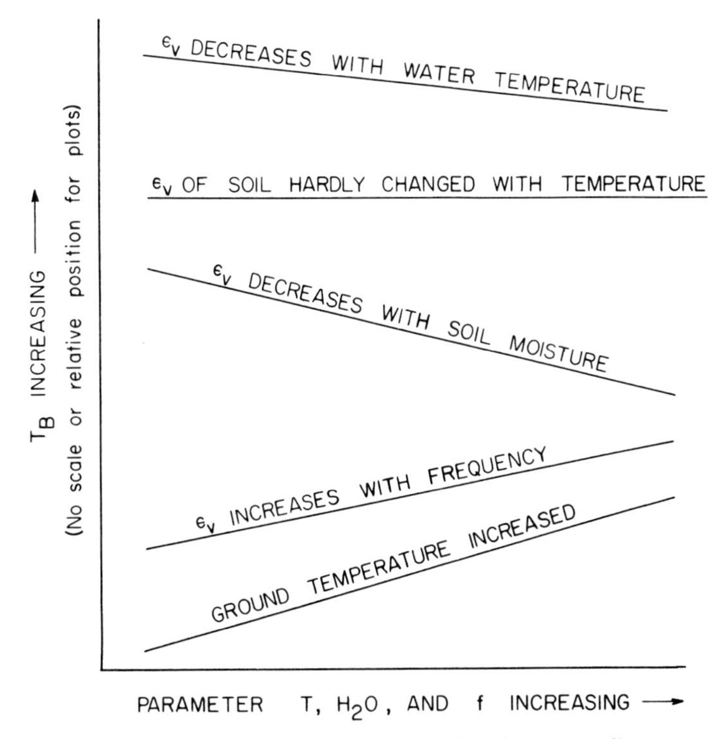


**Fig. 6.9.** Emissivity and brightness temperature trends for microwave radiometer near vertical incidence.

The changes in brightness temperature for changes in soil moisture are indicated by field experiments taken with dual-polarized radiometers at 13.5 and 37 GHz (Kennedy and Edgerton, 1967a). Figure 6.10, for the 13.5-GHz radiometer with vertical polarization, shows the large change in antenna temperature (about 115°K at $\theta = 10°$) between water-inundated mud and dry soil.

## B. Methods of Establishing Information

In establishing correlations between surface materials and their brightness temperatures, the need for a wide range of experimental data, as a function of terrain

0-201-04245-2

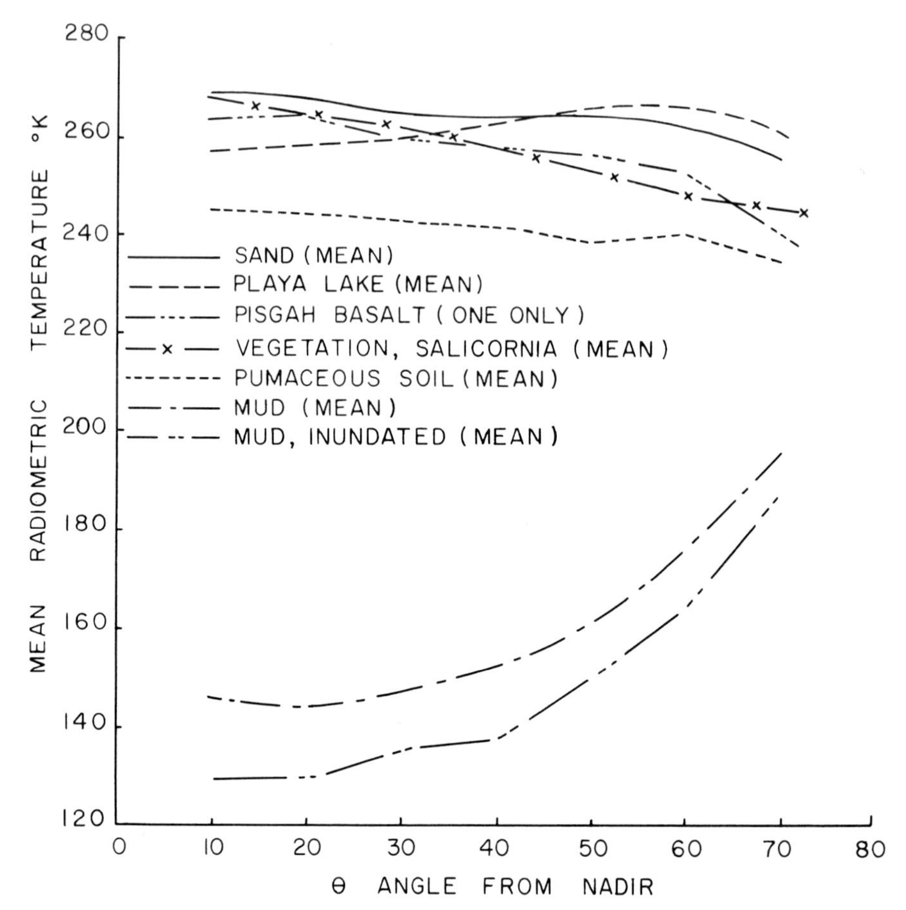


**Fig. 6.10.** Comparison of mean radiometric temperatures, 13.5-GHz vertical polarization.

characteristics, and radiometer variables is soon apparent. For completeness, such information is required for a broad spectrum from, say, 0.3 to 250 GHz.

## *1. Active and Passive*

Laboratory experiments have been conducted using bistatic measurements with active systems from approximately 0.4 to 35 GHz (Lundien, 1966; Peake, 1968; Krishen, 1968) to study the reflectivity, penetrability, roughness characteristics, and dielectric constants (of great significance in affecting brightness temperature) of various terrains, and these experiments have direct relevance to passive microwave studies.

0-201-04245-2

*2. Reflectometer*

The reflectometer (Pozdniak, 1960; Lovell and Thiel, 1968) is an active laboratory bistatic device designed expressly for determining the dielectric constants of materials, particularly of soils (a 31-GHz system is shown in Figure 6.11). If the soil sample is prepared to meet Fresnel's conditions of a smooth surface of infinite extent, the complex dielectric constant can be measured.

Typical measurements of the reflection coefficients are shown for standard Ottawa sand for both polarizations in Figure 6.12. The Brewster angle goes to zero for an angle of incidence of about 56° (vertical polarization). The emissivity is unity at this angle, and no power is reflected to the receiving antenna (Chapman, 1970).

The true meaning of the reflection coefficient is lost when measurements are made of rough surfaces. This is shown by the same measurements made on Ottawa sand the surface of which was made rough by the Rayleigh criterion (Figure 6.13). One must now speak of an effective or "pseudo" reflection coefficient. In nature, one rarely deals with truly smooth surfaces, and therefore simple discriminations related to dielectric variations are seldom feasible.

*3. Computer Simulation*

The sensitivity of the brightness temperature of the ocean to variation of a numbei of surface and radiometer parameters has been simulated by mathematical models (Stogryn, 1967), and a similar computer simulation model was used by Dedman et al. (1968) to determine whether thermal upwellings in the ocean could be detected at various frequencies, wind speeds, and water conductivities (Figure 6.14).

The brightness temperature of the ocean was shown by these studies to be very sensitive to changes in frequency, possibly implying an increase in emissivity with rising frequency. As the conductivity increases, the surface becomes more reflective, emissivity decreases, and the surface appears colder. A change in conductivity of 1 mho $m^{-1}$ results in a 1° change in temperature. The decrease in surface temperature of the ocean due to thermal upwelling, and the extent of the areas involved, are known. From this information, required radiometer sensitivities ($\Delta T$) can be computed.

Two important brightness temperature variations were discovered by the computer program for vertical and horizontal polarization for changes in wind velocity. The winds change the structure, including foam, and the height of the waves. The increase in wave height due to an increase in wind velocity results in a marked change in brightness temperature, but mainly for horizontal polarization in the range of incidence angles from 40 to 60°. For vertical polarization, in the vicinity

0-201-04245-2

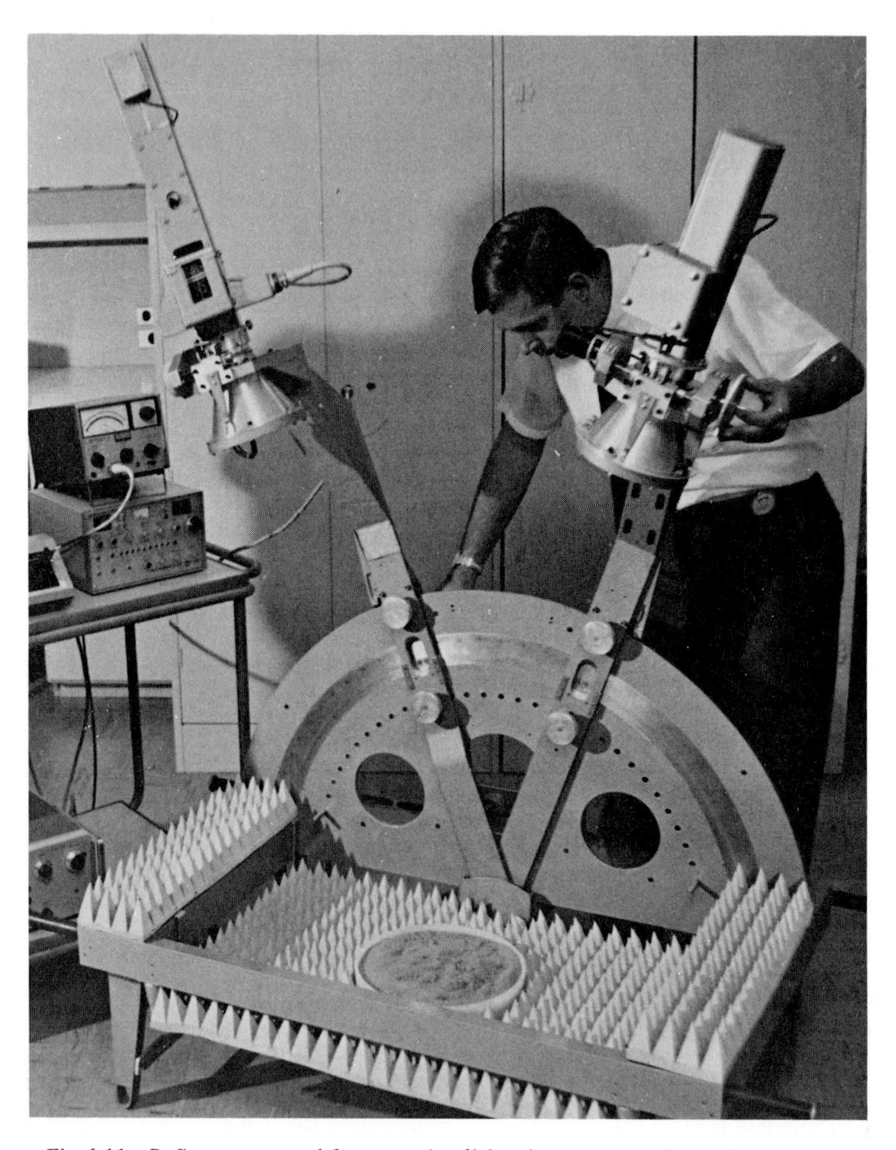

**Fig. 6.11.** Reflectometer used for measuring dielectric constants and emissivity of earth materials. Courtesy of Teledyne Ryan Company.

0-201-04245-2

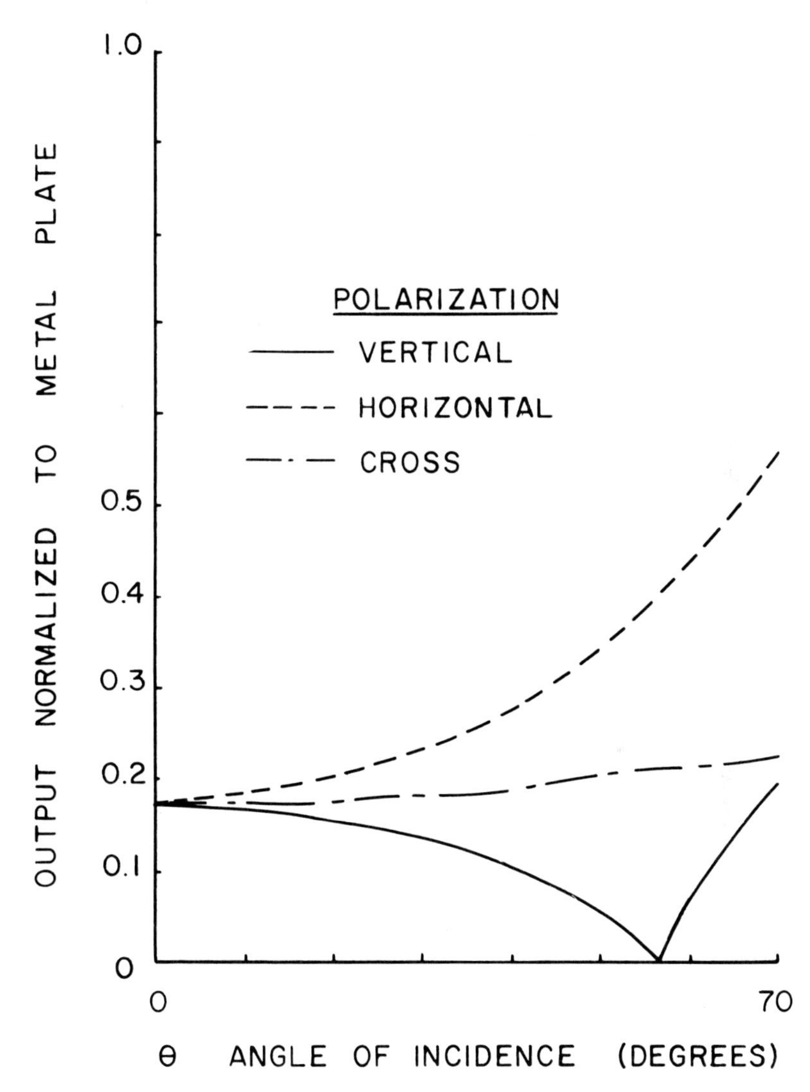


**Fig. 6.12.** Effective reflection coefficient for dry, dense, uniform Ottawa sand with smooth surface.

0-201-04245-2

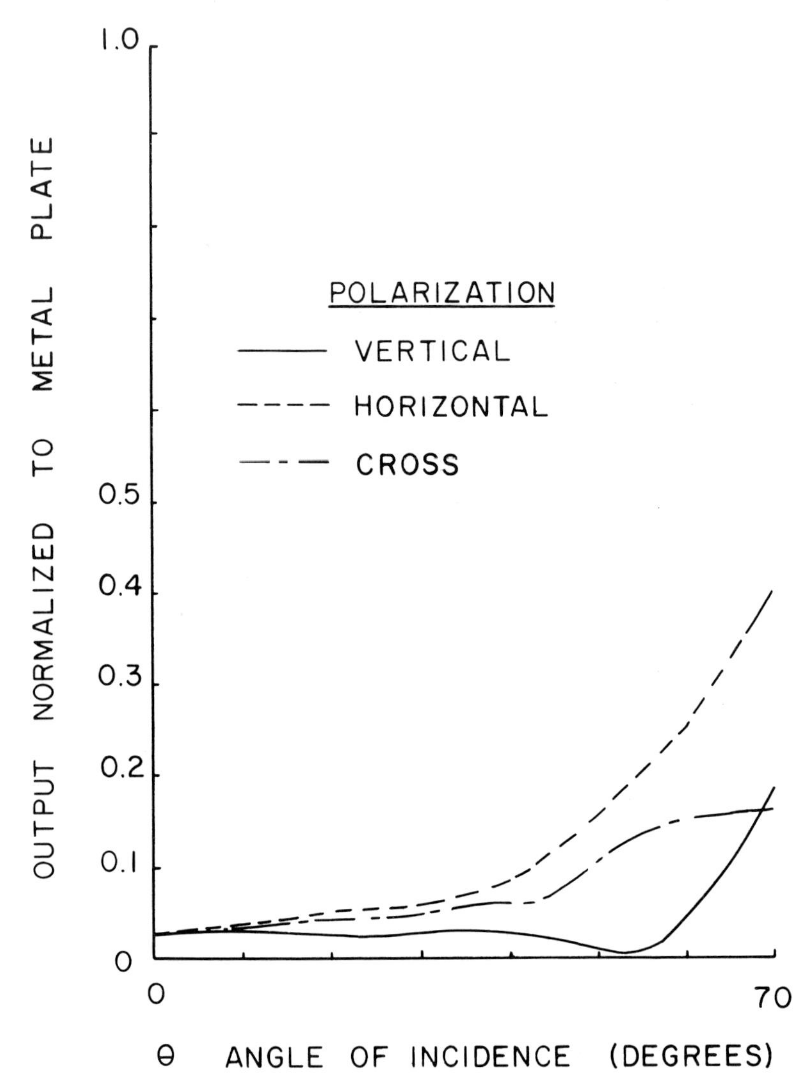


**Fig. 6.13.** Effective reflection coefficient for dry, dense, uniform Ottawa sand with rough surface.

0-201-04245-2

of an angle of incidence of 50°, near 19.4 GHz, and an altitude of about 1 km, the surface brightness temperature of the ocean is predicted to be invariant to wind speeds of up to 14 m $sec^{-1}$.

## V. APPLICATIONS AND RESULTS

Microwave radiometer experiments are being conducted to gather information in the fields of hydrology, geology, oceanography, meteorology, and agriculture. Detections of oil slicks, voids, pollution, thermal upwellings, and fires or hot spots have aroused much government and general public interest. Many of these experiments are conducted to demonstrate feasibility or to prove the hardware. Although results are much less firm than in other areas of remote sensing, the potential with passive microwave systems of being able to monitor large-scale oceanographic and hydrological systems (for example) with satellites is so great that these early results are reviewed in some depth.

### A. Geology

#### *1. Subsurface Void Detection*

In the earlier discussion, it was shown that variations in soil, water, and mineral content produce marked differences in brightness temperatures and are sensitive functions of wavelength, angle of incidence, and polarization. Unfortunately, a great many of the natural resources valuable to mankind lie below an overburden of soil. A multispectral radiometer system (0.3–30 GHz) appears to be a likely candidate for ferreting out radiometric signatures relating to the materials beneath this overburden, particularly when even longer penetrating wavelengths can be utilized. The ability to probe through an overburden has considerable geological significance for delineating structural patterns, mapping gross changes in materials due to changes in their average dielectric constants, identifying thermal gradients and hot or cold anomalies, and showing caves or voids.

The finding of underground voids is of major interest to many people, particularly to road builders and civil engineers, who are concerned with ground stability. Theoretical studies and numerous measurements have been made of microwave soil penetration, and the analyzed data indicated that microwave radiometers should be able to detect subsurface voids associated with karst development beneath quite thick soil cover (Kennedy, 1968).

Since it is very difficult to model the effects of surface roughness, moisture content, vegetation, and a nonhomogeneous sky, microwave radiometer surveys

0-201-04245-2

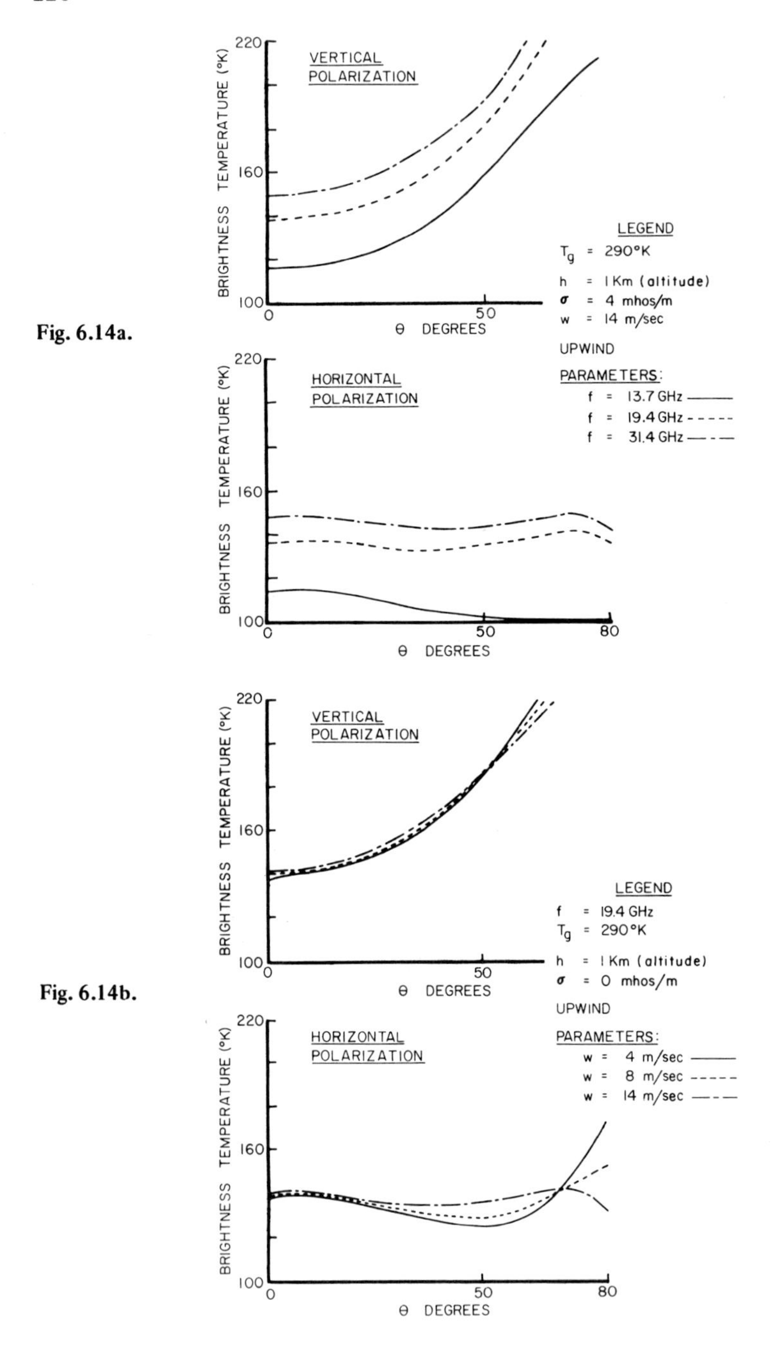


**Fig. 6.14.** Apparent brightness temperature plots obtained from computer program for radiometer model when the parameters of frequency, wind speed, and conductivity were varied.

0-201-04245-2

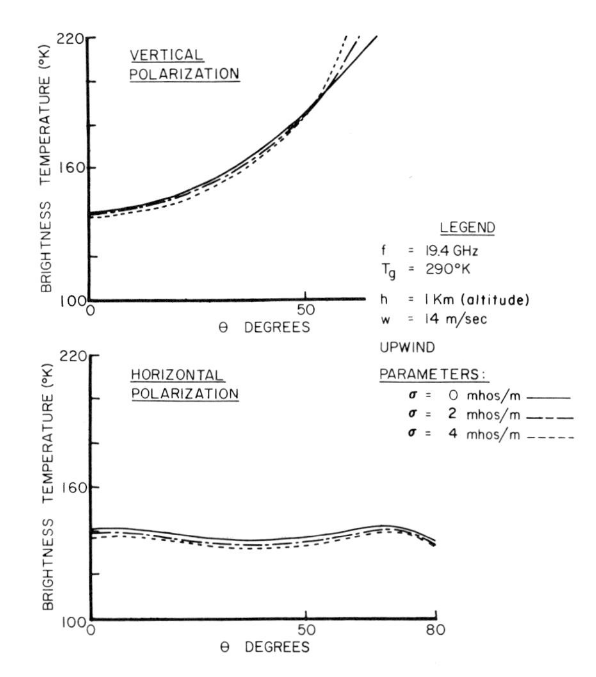


**Fig. 6.14c.**

**Fig. 6.14.** Apparent brightness temperature plots obtained from computer program for radiometer model when the parameters of frequency, wind speed, and conductivity were varied.

were used in lieu of computer methods. A mobile laboratory with 13.4-, 37-, and 94-GHz radiometers, was employed to obtain *in situ* data. "Cold" anomalies were associated with void space development beneath tens of feet of soil cover. These surveys do not establish that radiometers can uniquely identify subsurface voids. Rather, they suggest that, when supported by other geological data, the mapping of subsurface voids may be feasible with multiple-imaging radiometers.

## B. Hydrology

The application of the microwave radiometer to the field of hydrology has received considerable attention, particularly for making hydrological surveys from orbiting satellites. If the radiometer could be used to provide information remotely in any of the following areas, such data would be invaluable to several current urgent needs in the field of hydrology:

1. Forecast of precipitation, temperature, and stream flow on short- and long-term bases.
2. Data on evaporation from water and land, transpiration, and water consumption.
3. Accurate data on soil moisture over large areas.

4. Accurate data on snow cover, snow water content, and rate of snow melt.
5. Information on rainfall distribution and drainage basins.
6. Better information on currents, tidal effects in harbors and estuaries, saltwater intrusions, effects of stream flow, and water pollution.

Unfortunately, there have been very few hydrology experiments with microwave radiometers, and only two with the NASA remote sensor aircraft (Janza, 1969).

### *1. NASA Hydrological Missions*

The two NASA aircraft missions used the following instruments: (1) four JPL microwave radiometers at 9.4, 15.8, 22.2, and 34 GHz; (2) a microwave scatterometer at 13.3 GHz; (3) a Reconofax IV infrared mapper; and (4) an RC-8 aerial mapping camera.

The sites were the South Cascade Glacier in Washington (to study floods and snow and ice melt) and Coachella Valley, California (to study ground and surface water). Only the latter study is reported here.

The main goal of the flights over the Coachella Valley was to obtain remote sensor data for rural and urban land use, that is, information on soil moisture and heat balance. Substantial data were obtained with the three microwave radiometers, photography, the infrared scanner, and the radar scatterometer. Ground temperature and soil moisture data were also obtained to correlate with the sensor data.

It was found that gross radiometric temperature details could be derived by applying some basic reasoning:

1. Radiometric temperatures considerably less than actual ground temperatures indicate three possibilities: that soil emissivity has been lowered by added moisture or soil change, that surface water exists, and that considerable water vapor exists above ground.
2. Radiometer temperatures are equal to ground temperatures or lower; soil temperatures should be very close to radiometric temperatures, since emissivities are greater than 0.95.
3. A combination of the infrared and microwave radiometer sensors shows a strong possibility of being able to identify moist soil regions through their opposite radiometric temperature indications. Wetness contrasts are emphasized by the infrared sensor.
4. The radiometer and scatterometer appear to have value for hydrology, since their observed analog signals are essentially inverted; the radiometer is sensitive to emissivity, whereas the scatterometer is sensitive to reflectivity, $\epsilon = 1 - \rho$.

### *2. Snow Surveys for Hydrology*

Airborne radiometer survey data for snow cover, snow melt, glacier, and ice conditions have been very sparse to date.

0-201-04245-2

Concerted field and laboratory measurements have been made to determine the response of microwave radiometers to various snow conditions (Kennedy and Edgerton, 1967b). Measurements were made at three frequencies (13.5, 37, and 94 GHz) for both polarizations, and for angles of incidence from the nadir to the local horizon. The field program was carried on within Crater Lake National Park with data obtained from new-fallen powder snow and old metamorphosed snow. Similar laboratory measurements were made with artificial snow manufactured in an environmental chamber in which snow temperature and moisture content were controlled.

Further microwave radiometer snow measurements were made near Dillon Reservoir, Colorado, of snow and ice systems (Edgerton et al., 1968). At Dillon the snow was considerably different—mostly old, highly dense, coarse textured, and rippled by wind.

These studies indicate that different microwave frequencies do not respond in the same manner to changes in snow moisture content. This is due to the frequency-dispersive properties of water in the microwave spectrum. It is speculated that measurements from a remote sensing platform, using specific combinations of wavelength, polarization, and angle of incidence, should yield information concerning the melt conditions of snow, and possibly its density.

### C. Oceanography

The microwave profiling and scanning radiometer appears to have all-weather potential in five specific fields for oceanography, but not all necessarily from space platform operation:

1. Monitoring of sea surface temperatures, including thermal upwellings.
2. Monitoring of sea state, ocean currents, floating ice, and ice cover.
3. Monitoring of pollution patterns, river runoff, and shore temperature changes.
4. Measurement of atmospheric effects (cloud cover, rain, etc.) through brightness temperature changes to provide a calibration support function for other remote sensors, such as the scatterometer.
5. Surveillance functions over the oceans for fishing vessel count, military purposes, etc.

Probably the most readily apparent and potentially significant application of the microwave radiometer to oceanography would be the measurement of sea surface temperatures. To add to this, multispectral radiometer systems are available for all-weather operation in transparent or moderately opaque regions of the microwave spectrum (22.2 GHz), or for operation in spectral regions that would yield atmospheric information for oceanographers.

The ocean has interesting properties which have been utilized in ocean mathematical model studies: (1) the surface emissivity is low (reflectivity around 0.6); (2) the medium is essentially homogeneous and isotropic; (3) the emissivity in-

0-201-04245-2

creases slowly with frequency; (4) the emissivity increases with actual ocean temperature for frequencies below approximately 20 GHz (it can decrease or increase above about 20 GHz, depending on the actual ocean temperature); and (5) the brightness temperature is a function of surface roughness, and in turn is relatable to the wind fields. On the basis of these properties, computer-simulated models of the ocean have been developed to predict sea temperatures and sea states.

### *1. Radiometer Ocean Model*

Until recently, the microwave radiometer was solely identified as an instrument for surface brightness measurement, but recent important model analyses (Stogryn, 1967) indicate that the radiometer also has good potential for measuring ocean surface roughness. In the analysis, however, contributions to the brightness temperature from small-scale roughness were excluded.

With the radiometer receiving horizontally polarized radiation, sea state can be determined at one angle of incidence at a time, as depicted by Figure 6.14*b* for a 19.4-GHz radiometer. To establish a plot of sea state, $T_B(\theta)$, the antenna must be stepped to fixed angular positions (20°, 30°, etc.), and the signal must be integrated about 1 sec. Upwind and downwind data require separate flights in each direction, or separate radiometers pointing fore and aft. The temperature resolution possible is also shown in Figure 14*b:* at $\theta = 50°$, the change in $T_B(\theta)$ is about 10°K for a change in wind speed from 4 to 14 m sec$^{-1}$, or 1°K per 1-m change in wind speed. Present radiometers have resolutions of $\Delta T = \pm 0.5°$K rms, or better.

The analysis described above neglected surface features such as foam, breakers, and whitecaps, and instead assumed that the ocean was composed of large facets as compared to the wavelength. Edgerton et al. (1968) have shown convincingly that when foam is present relations are sharply altered.

### *2. Water Temperature and Salinity*

The brightness temperature of water depends on its actual temperature, salinity, and frequency (Saxton and Lane, 1952), as well as surface roughness. The emissivity and absorption of sea water as functions of wavelength, salinity, water temperature, and dissolved gases were recently computed (Geyer, 1968).

The brightness temperature is most sensitive to salinity changes at 3 GHz and is much less affected by such changes above 8 GHz. At 35 GHz the brightness temperature exhibits a nonlinear dependence because of the kinetic water temperature. A frequency of 4 GHz appears optimum for linear response as a function of actual water temperature for salinities of 35 parts per thousand by volume, whereas for wider salinity variations 6 GHz appears optimum. In general, the brightness temperature increases with increasing salinity and with increasing water physical or real temperature.

On the basis of these emissivity sensitivity data, what are the possibilities of measuring ocean temperature with the microwave radiometer? The theory indicates that the microwave sensors would be in a linear and large gradient region ($\Delta T_B/\Delta T_w$, where $T_B$ is the brightness temperature and $T_w$ is the actual water tempera-

0-201-04245-2

ture at nominal salinity) for a frequency of 6 GHz. At 6 GHz, with angles of incidence of about 30°, and in regions where the atmospheric paths are not long, the emissivity of ocean water is around 0.4, and the gradient, $\Delta T_B/\Delta T_w$, is estimated at 0.4°K per degree Kelvin, a relatively small change. This small change could be lost if measurements were made for rough sea or during inclement weather conditions. On the other hand, the invariant temperature point as predicted by model studies is independent of sea state. These two approaches to the measurement of physical ocean temperatures offer opposed predictions (Edgerton and Trexler, 1969).

## D. Meteorology

Since the next few years will see many satellite surveys conducted for remote sensing of geological features of the earth, through an intervening atmosphere, the effects of the atmosphere on earth-generated radiation become highly important. Hence the need, aside from weather forecasting, for atmospheric data on a global basis is apparent.

Microwave radiometers are the only sensors now showing capabilities of measuring atmospheric temperature profiles or atmospheric water vapor in the presence of clouds and are, in fact, in use for this purpose on NIMBUS 5. These observations from satellites offer the possibility of making measurements all the way to the surface, even in some densely clouded areas.

The applications of microwave radiometers to the field of meteorology have advanced to the point where measurements are made of the atmospheric temperature and atmospheric distribution of water and ozone. The results of this complex technology are contained in a number of significant reports (Barrett and Chung, 1962; Staelin, 1966, 1969).

Measurements of the atmospheric temperature and composition profiles can be achieved by utilizing the microwave resonances of the $O_2$ complex near 60 GHz and the $H_2O$ resonance at 22.235 GHz, respectively. The microwave absorption of the oxygen complex near 60 GHz (Van Vleck, 1947; Mizushima and Hill, 1954) provides the structure to measure atmospheric profiles from topside, the satellite, or the ground.

The feasibility of temperature-profiling the atmosphere has been recently established (Meeks and Lilley, 1963; Lenoir, 1965). A mathematical model was developed for the brightness temperature $T$, in the form of an integral, the integrand of which is a weighted product of the atmospheric temperature profile as a function of altitude, $T(h)$, and of $W(h, f)$, which is a weighting function of altitude and frequency $f$:

$$T_B(f) = \int_0^\infty T(h)W(h, f)\,dh + T_0 \exp\,[-\tau(f)], \tag{6.25}$$

where $T_0 \exp\,[-\tau(f)]$ is the background temperature contribution.

0-201-04245-2

The significance of the weighting function is that it shows the sensitivity of the brightness temperature, as a function of frequency, to the actual temperature as a function of altitude. Weighting functions for space applications have been computed for microwave radiometers (Lenoir, 1965).

Statistical inversion techniques have been applied to the radiometric data (Waters and Staelin, 1968). It is indicated that, if temperature profiles can be measured to an accuracy of 3°K on a global scale, radiometric sensors designed to operate on the $O_2$ complex will provide an excellent means of collecting synoptic meteorological information.

In the past, little attention had been paid to the possibility of using the microwave resonance of $H_2O$, 22.235 GHz, as a tool for measuring the composition of the atmosphere. Recently, ground-based microwave radiometers were proposed for the study of $H_2O$ distributions (Barrett and Chung, 1962). Spaceborne radiometers are now under test on NIMBUS 5 for this purpose. Radiometric measurements of composition are more readily made in the semitransparent regions of the microwave spectrum, where the spectrum is more sensitive to the distribution of the constituents than to the temperature distribution. Weighting functions can again be applied for determining the density distribution profiles of constituents having resonances (Staelin, 1966).

## E. Special Applications

### *1. Oil Pollution*

The passive microwave radiometer has been applied to the detection and mapping of oil pollution. One such program, reviewed by Edgerton and Trexler (1969), consisted of a laboratory radiometer measurement phase at 13.4 and 37 GHz of five petroleum pollutants: gasoline, Bunker fuel oil, and 20, 30, and 40 API gravity crude oils. Measurements were performed as functions of the angle of incidence, oil film thickness, and age of pollutants. In addition, the dielectric properties of the pollutants were measured at 37 GHz, using a reflectometer. The real part of the dielectric constant for the petroleum pollutants ranged from 1.85 to 2.41, as compared to approximately 21 for seawater at 23°C. A slight increase in the real part and a large increase in the imaginary part of the dielectric constant were observed as the pollutants aged.

The low dielectric constants observed for petroleum products are related to high emissivities (see Figure 6.4) and warm brightness temperatures. Horizontally polarized 31-GHz brightness temperatures were 7-70°K warmer than the temperatures of calm, unpolluted sea water. Gasoline had lower brightness temperatures, since it is more transparent at this microwave frequency. Film thicknesses of less than 0.1 mm had marginal detectability at 37 GHz.

Airborne measurements at altitudes of 60–240 m off the southern coast of California of controlled spills of small quantities of pollutants were performed at 37 GHz at an angle of incidence of 45°. The measurements indicated that thinner

0-201-04245-2

oil slicks are more detectable than laboratory data suggest. It was concluded that (1) the emissivity of floating petroleum products is appreciably higher than that of a calm sea, (2) crude oils have increasing emissivities (low dielectric constants) with increasing API gravities, (3) time of day and age of oil have little effect on radiometric response, (4) detection improves with increased frequency, and (5) a reliable microwave pollution-detection system can be designed for the types of oil examined.

0-201-04245-2

Chapter 7

# Active Microwave Systems

*Richard K. Moore*

## I. INTRODUCTION

Active systems provide their own illumination, whereas passive systems depend on external sources of illumination or thermal radiation. Since control of illumination offers the system designer many more options than are possible with passive systems, the discussion of active systems must be either much longer than that of passive systems or less exhaustive; here the latter alternative was chosen. Details of individual systems are not listed, and emphasis is placed on the general principles involved, as well as on systems used to sense the surface of the earth.

Resolution is a greater problem for microwave than for optical and infrared systems, because the wavelengths are longer by 3–5 orders of magnitude than those for the latter systems. Active microwave systems are able to achieve much better resolution than their passive counterparts by means of a number of techniques using the known properties of the illumination.

Active microwave systems are relatively independent of weather conditions and time of day. Although heavy precipitation can cause difficulties, the trouble is less severe than for passive systems with the same wavelength; in turn, the difficulties for passive microwave systems are far less severe than for infrared and optical systems.

The term "radar" is used synonymously with the term "active microwave systems" for most purposes. Radar is an acronym for *r*adio *d*etection *a*nd *r*anging, so that a literal use of the term would exclude devices in the optical wavelength region, as well as microwave devices that do not perform a ranging function. Sometimes the term "optical radar" is used, and "radar" may be employed for active microwave systems that do not, in fact, perform a range measurement. Many *scatterometers* fall in this class. Because the word "radar" is short, it is normally used in this chapter in place of "active microwave systems" even when ranging is not involved.

The role of radar in remote sensing operations for geosciences has only recently been recognized. Clearly, radar is particularly important for time-dependent surveys, for a radar survey can be flown almost any time the aircraft can fly, whereas surveys made with other systems are dependent on the weather. A radar survey in Darien Province, Panama (Viksne et al., 1969) was achieved in 6 hr of flying, whereas aerial photography of this perpetually cloudy region was far from complete after 25 years of attempts.

0-201-04245-2

*Remote Sensing of Environment*, Joseph Lintz, Jr. and David S. Simonett (eds.) ISBN 0-201-04245-2

Radar systems carried at normal aircraft altitudes are capable of imaging wide swaths of ground. Systems have been built, for instance, that can cover a width of 50 miles. Surveys with radar can be conducted, therefore, with minimum flying time and on a schedule relatively independent of weather.

Consider, for example, a radar survey of the state of Kansas, which might be used for agricultural census purposes. The state is approximately 320 by 640 km and is rectangular. If a radar with an 80 km swath is used, and 20% of each pass is reserved for overlap with adjacent passes, the state of Kansas can be mapped with only five 640 km-long passes. Thus an aircraft that flies at 800 kph can map the state in about 5 hr, including time for turn-around at the end of each run.

The value of radar has been demonstrated for mapping geologic structure (Wing and Dellwig, 1970) and, to some extent, for mapping lithologies (MacDonald et al., 1967) in lightly vegetated regions. Thus radar also has a place in surveys that are not tied to time-dependent phenomena.

Radar images are not as familiar to photo interpreters as are photographs. Furthermore, in many cases the resolution of the radar image is poorer than that of the photograph. Thus radar often performs a complementary rather than a primary role. It can be used when the weather is wrong for the photographic sensor, or it can serve to provide information in addition to that which can be gleaned from the photographic image. As research in the application of radar continues and its use becomes more widespread, it will grow in importance as a primary independent sensor both because more will be known of what it can do and because more people will be familiar with methods for using radar.

Radar has demonstrated its usefulness for monitoring the surface of the ocean to learn the wind and wave conditions (Moore and Pierson, 1970). Such monitoring from aircraft will never be sufficiently synoptic to provide significant inputs to worldwide forecasting systems, but radars carried on satellites should indeed be able to provide significant inputs to weather and wave forecasting over the world's oceans.

Radars discussed here include imaging radars for ground mapping, scatterometers, and combined microwave radiometer-scatterometer systems. Most users will find the radar imager of much greater value than the other systems. Applications of scatterometers are specialized because these systems cannot present the spatial information available in an image. Applications of combined active and passive systems are also specialized because of the necessarily poor resolution of the passive system.

## II. BASIC PRINCIPLES OF RADAR

### A. The Radar Equation

The fundamental equation showing the amount of signal received by a radar system from a particular *target* is called the *radar equation.* Many of the system tradeoffs

0-201-04245-2

possible with radar can be described in terms of variations of this equation; hence it is the place to start the study of radar.

The radar equation may be written in the form

$$W_r = \left(\frac{W_t G_t}{4\pi R^2}\right) (\sigma) \left(\frac{1}{4\pi R^2}\right) A_r. \tag{7.1}$$

Here $W_r$ is the received power; $W_t$, the transmitted power; $G_t$, the gain of the transmitting antenna; $R$, the slant range to the target (distance from radar to target); $A_r$, the effective aperture of the receiving antenna; and $\sigma$, the effective backscattering cross section.

In Eq. (7.1), the first quantity in parentheses gives the power per unit area at the target. The second contains the scattering cross section in square meters. It includes the effective receiving area of the target, the absorption in the target, and the scattering pattern for the target. (For instance, $\sigma$ is small with no absorption if most of the power is scattered in some direction away from the radar, but is large under the same condition if most of the power is scattered back toward the radar.) Combining the first two parenthetical quantities gives the strength of the target as a source for energy coming toward the receiver. This energy spreads out spherically so that the power per unit area at the receiver is given by the product of the three quantities in parentheses. The power actually received is the power per unit area multiplied by the effective area of the receiving antenna ($A_r$). These quantities may be combined into a more convenient form:

$$W_r = \frac{W_t G^2 \lambda^2}{(4\pi)^3 R^4} \sigma. \tag{7.2}$$

Here the receiving area has been expressed in terms of $G$ and the wavelength $\lambda$.

All of the factors on the right-hand side of Eq. (7.2) except $\sigma$ are under the control of the radar designer; $\sigma$ describes the target itself. The actual total target cross section is commonly used for isolated objects (e.g., aircraft or ships) and is also applicable in imaging the earth for such objects as houses or transmission line poles. Usually, however, in imaging the earth we are concerned about area-extensive targets. For this purpose a differential scattering coefficient is normally defined.

For an area-extensive target containing $N$ separate scattering elements. Eq. (7.2) may be written as

$$W_r = \sum_{i=1}^{N} \frac{W_{ti} G_i^2 \lambda^2}{(4\pi)^3 R_i^4} \left(\frac{\sigma_i}{\Delta A_i}\right) \Delta A_i. \tag{7.3}$$

Normally in electric circuits superposition can be applied to voltage and current, but not to power. If, however, the relative phases of the returns from different elements within the area-extensive target are random, the power from each may be added to that from the others, as shown in Eq. (7.3). In this equation the subscript

0-201-04245-2

$i$ identifies quantities related to target element $i$. The second quantity in parentheses represents a pure numeric which is the ratio of the effective cross section for backscatter in target element $i$ to its actual area. If $N$ is large enough, we may replace the sum of Eq. (7.3) by an integral:

$$W_r = \int_{\text{Illuminated area}} \frac{W_t G^2 \lambda^2}{(4\pi)^3 R^4} \sigma^0 \, dA. \tag{7.4}$$

Here the average value of the effective cross section of the target, divided by its actual area, is $\sigma^0$. This quantity is called the *differential scattering cross section* or *cross section per unit area.* Clearly, it is given by

$$\sigma^0 = \text{mean value of } \frac{\sigma_i}{\Delta A_i}. \tag{7.5}$$

The differential scattering coefficient is commonly used for describing the properties of an area target, such as the ground.

Some authors use a value of differential scattering cross section per unit projected area rather than per unit ground area. This is given by

$$\gamma = \text{mean value of } \frac{\sigma_i}{\Delta A_i \cos\theta}. \tag{7.6}$$

When this is used, the radar equation [Eq. (7.4)] becomes

$$W_r = \int_{\text{Illuminated area}} \frac{W_t G^2 \lambda^2}{(4\pi)^3 R^4} \gamma \cos\theta \, dA, \tag{7.7}$$

where $\theta$ is the angle of incidence (angle with the normal).

Integrals (7.4) and (7.7) are evaluated over the area illuminated at any particular instant. The transmitter power, gain, and range all remain within the integral, for the general case, to permit weighting the strength of the illumination in different parts of the area. In special cases, one or more of these quantities may be considered constant and brought outside the integral.

## B. Factors Governing $\sigma^0$

The differential scattering coefficient is governed by factors under the control of the radar designer and by factors that are properties of the target itself, as indicated in Table 7.1. Thus the wavelength and polarization (direction of the electric field vector) of the source may be selected by the designer.

0-201-04245-2

**TABLE 7.1**
**Factors Governing the Differential Scattering Coefficient**

| |
|---|
| Properties of source: |
| Wavelength |
| Polarization |
| Incidence angle |
| Azimuthal look angle |
| Resolution (size of area averaged) |
| Properties of target: |
| Roughness |
| Slope |
| Inhomogeneity (vertical and horizontal) |
| Permittivity and conductivity |
| Resonant-sized objects |

The resolution is determined by the system design, and the angle of incidence and azimuthal look angle by the flight pattern.

The properties of the target that determine the response are also listed in Table 7.1. As a general rule, scatter from a surface that is rough in terms of wavelength is relatively independent of the incidence angle; a stronger return is observed when the radar looks nearly horizontal than is obtained for smoother surfaces. The slope of the target determines the local angle of incidence, which modifies any general effects of the incidence angle. Clearly a target that is horizontally inhomogeneous, either in roughness or in electrical properties, will give a different return from one that is homogeneous. Less clear is the problem of vertical inhomogeneity. Radar signals penetrate into the surface of the earth to varying degrees, but seldom more than a few wavelengths. Nevertheless, this distance may be great enough so that there are significant differences in the electrical properties. Each of a set of layers within this depth will affect the signal observed. If the region penetrated by the waves contains objects with properties different from those of the general surrounding medium, the signals they scatter will add to those scattered from the surface. Vegetation growing above the surface will itself scatter, but often will also permit the signals to penetrate to a scattering surface.

Numerous theories for radar scatter from surfaces have been advanced in the past 25 years. Scattering is often described as coming from "facets" or "bright spots." In this description the rough surface of the ground is approximated by an assembly of various flat scattering facets with different sizes and slopes. Figure 7.1 shows the effect of facet size on reradiation. On the left is the reradiation pattern from a small facet (in terms of wavelength) illuminated at normal incidence. The reradiated signal spreads out more or less uniformly over a fairly wide range of angles. Larger facets concentrate the scattered energy more at normal incidence, as shown in the next two illustrations. For an infinitely large facet, all the energy is reflected back at the source, as shown on the right. This is the situation for geometric optics, which is based upon zero-wavelength approximation. When the

0-201-04245-2

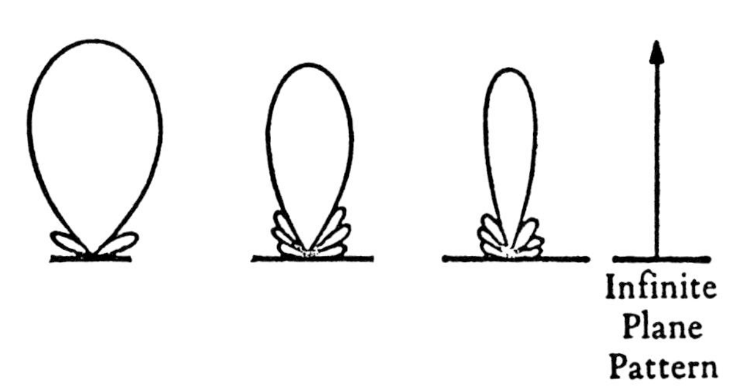


**Fig. 7.1.** Facet patterns: uniform normally incident illumination. Note that small facets are almost nondirective. Large facets give strong signals in a single direction only. By permission, from Committee on Remote Sensing for Agricultural Purposes, Agricultural Board, National Research Council, *Remote Sensing: With Special Reference to Agriculture and Forestry,* National Academy of Sciences, Washington, D.C., 1970.

illumination is from some other direction, the general shape of the patterns remains the same, but the peak of the reradiation pattern is in the direction of reflection from an infinite plane.

This illustration shows that a rough surface containing many small facets with a wide variety of slopes scatters nearly uniformly. A relatively smooth surface containing only large, nearly horizontal facets scatters most of its energy in the direction in which a wave would be reflected from a smooth surface. Thus, for a radar looking well away from normal incidence to the ground, backscatter will be strongest when the facets are small or when the surface has very steep slopes so that larger facets may be nearly perpendicular to the incoming wave. Natural earth surfaces seldom have this property, but vegetation with vertical stems and trunks may satisfy the requirement and give a strong, near-grazing echo. Similarly, many of the works of man contain vertical surfaces (walls, fence posts, telephone poles) which scatter strongly, at near-grazing level.

The facet theory has often been used to describe radar return qualitatively, but quantitative description by this means presents difficulties because it is hard to describe most real surfaces in terms of facets. Various random surface models have been used to describe radar return from the earth and the planets. Most of these are highly idealized compared with the real earth, although there is some hope that such models can actually work for the ocean and even for nonvegetated land surfaces. Many theories use the Kirchhoff approximation (Beckmann and Spizzichino, 1963); this depends on the surface being smooth enough so that the currents flowing in it are the same as would flow in an infinite plane having the same slope as the local tangent. Curves of variation of differential scattering coefficient with angle obtained using this theory can be made to look much like experimental curves by assuming arbitrary statistical properties for the surface. Usually, though, it is not easy to relate these arbitrary properties to the actual properties of the surface.

A perturbation model originally due to Rice (1951) has been revived to describe scattering from the ocean and gives moderately good results in spite of the obvious

invalidity of its fundamental assumption (Wright, 1966). The approximation involved assumes that all heights ($z$) above the mean surface are given by

$$e^{-j2kz\cos\theta} = 1 - j2kz\cos\theta, \qquad z << \frac{\lambda}{4\pi\cos\theta}. \tag{7.8}$$

Here $\theta$ is the angle of incidence (angle with the normal). Since the wavelengths involved are usually only a few centimeters, this says that the heights must be only a few tenths of centimeters except for angles very near grazing incidence. Since the method gives better results than might be expected under these conditions, even for rather high seas, it is believed that the major contributor to radar return from the ocean at centimeter wavelengths must be capillary waves and waves only slightly larger.

The Kirchhoff method and the perturbation method (Fung and Chan, 1969), as well as the Kirchhoff method and the facet method (Wright, 1968), have been combined to attempt an improvement in the description of the surface for theoretical purposes.

A few theoretical and experimental attempts have been made to take into account vertical inhomogeneities (the effect of layers and particles within the surface) (Boles, 1969; Krishen, 1968). Although these do not model actual surfaces very well, they indicate that the phenomenon is an important one for these surfaces.

Peake (1959) developed a model for vegetation with vertical stalks. This appears to be the only theoretical model other than those for surface and volume scatter. Scatter from large isolated objects has been extensively studied.

A "corner reflector effect" occurs when vertical walls and horizontal ground surfaces intersect, so that buildings give strong returns even when the angle of incidence is not near grazing. On the other hand, the azimuthal look direction must be somewhere near normal to a wall for the strong signal associated with the corner reflector effect to be observed. This was noted early with airborne imaging radars and was called the "cardinal point effect" (Levine, 1960), from observations that radar signals pointed along the streets of a city grid (frequently aligned N-S or E-W in the United States) received many building echoes, whereas those pointed at other angles showed fewer such echoes.

Natural surfaces, other than the ocean, and most man-made surfaces appear to be so complex in their geometry that a true theoretical description of the radar return seems almost impossible. Nevertheless, the theoretical work gives us many clues as to the phenomena to be expected in our empirical observations of radar returns from natural surfaces.

Material properties affect the radar return through the permittivity and conductivity. The permittivity, also known as the dielectric constant, is sometimes given as a complex number in which the imaginary term represents the effect of conductivity. The so-called complex dielectric constant is defined as

$$\epsilon_r = \epsilon' - j\epsilon''.$$

0-201-04245-2

Some typical values for the microwave dielectric properties of materials are given in Table 7.2. If $\epsilon''/\epsilon'$ is less than 0.16, the distance at which the penetrating wave is reduced to 37% of its original value exceeds a wavelength in the material. Clearly this is true for all the materials listed, with the possible exception of distilled water. Since radar waves must travel both ways, the factor should actually be less than 0.08 if the signal coming back out of the material is to be reduced to 37% of its ingoing value (assuming 100% reflection). Thus, for the dry soil listed in the table, and certainly for snow, the radar waves will penetrate several wavelengths into the surface, although the distance may not be practically significant.

**TABLE 7.2**
**Microwave Dielectric Properties of Sample Materials**
**(Most at $3 \times 10^9$ Hz)**

| *Material* | $\epsilon'$ | $\epsilon''/\epsilon'$ |
|---|---|---|
| Sandy, dry soil | 2.55 | 0.006 |
| Loamy, dry soil | 2.44 | 0.001 |
| Freshly fallen snow | 1.20 | 0.0003 |
| Distilled water | 77 | 0.157 |
| Mahogany | 1.9 | 0.025 |

All of the permittivities of the materials of Table 7.2 are quite small except the value for water. Other materials may have permittivities up to 6 or 8, but unless water is present the permittivities of most materials are relatively small and differ little from one material to the next. The effect of this difference on the scattering is even smaller, since scatter is given by expressions involving the reflection coefficient, which is determined by $\sqrt{\epsilon_r}$. On the other hand, the presence of water molecules in either soil or vegetation can make a tremendous change in the permittivity. Figure 7.2 shows the variation in permittivity for a corn leaf as a function of the percent moisture. Obviously, the effect of moisture is much greater than the variations from one dry material to another. Figure 7.3 shows the same effect for soil. Thus radar is very sensitive to moisture content, whether it be in soil or in vegetation. Differences in the radar return observed from one material to another are much more likely to be caused by differences in moisture content than by differences in material.

The last target property listed in Table 7.1 is the presence of resonant-sized objects. The radar cross section of such an object can be much greater than that of another object physically much larger in size. For example, a radar image of an airport once showed several bright objects where neither buildings, aircraft, nor vehicles were known to be. Investigation showed that the strong signals came from

0-201-04245-2

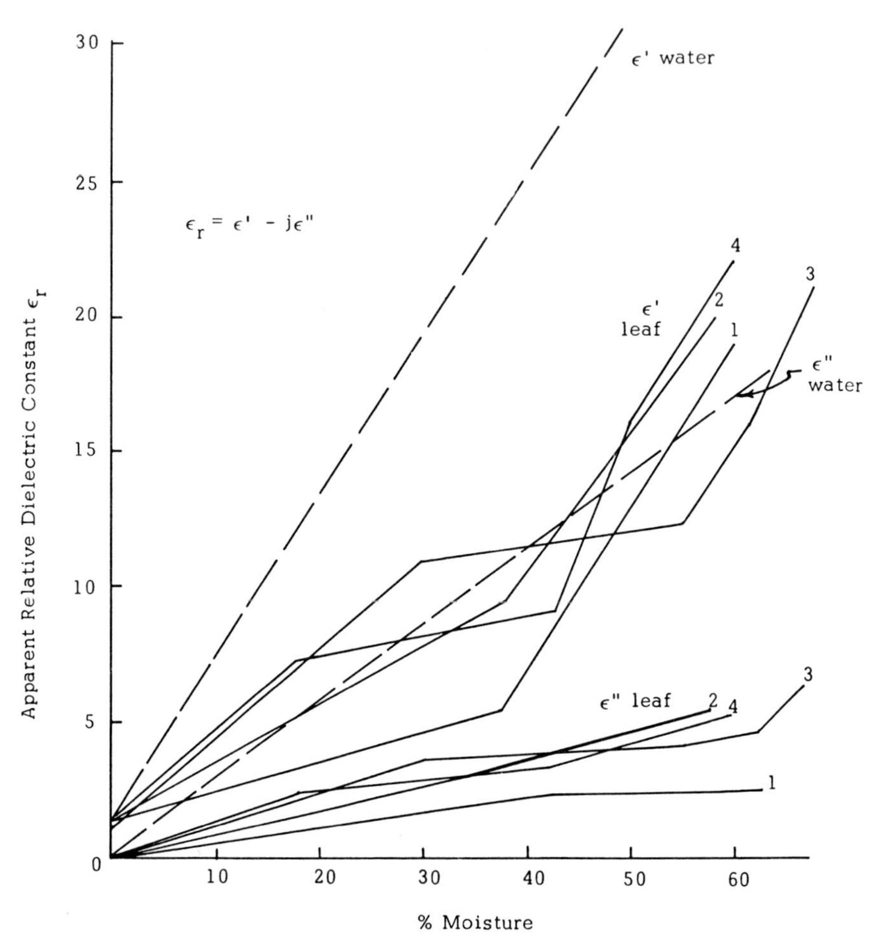


**Fig. 7.2.** The complex dielectric constant of corn leaf plotted as a function of moisture content. Curves 2, 3, and 4 represent samples taken from leaves of the same plant. From Carlson (1967).

resonant-sized storm-sewer gratings! Similar phenomena have been observed with resonant-sized fence posts.

In summary, once the source properties are established, the primary factors determining the radar return are the surface geometry (expressed in roughness and in resonant-sized objects) and the moisture content. The conducting or wet objects may be vegetation, surface layers of the ground, or even materials somewhat beneath the surface.

## C. Basic Elements of the Radar System

Figure 7.4 illustrates the basic elements of a radar system. The signal originates with the transmitter, leaves the transmitting antenna, bounces off the ground, and is re-

0-201-04245-2

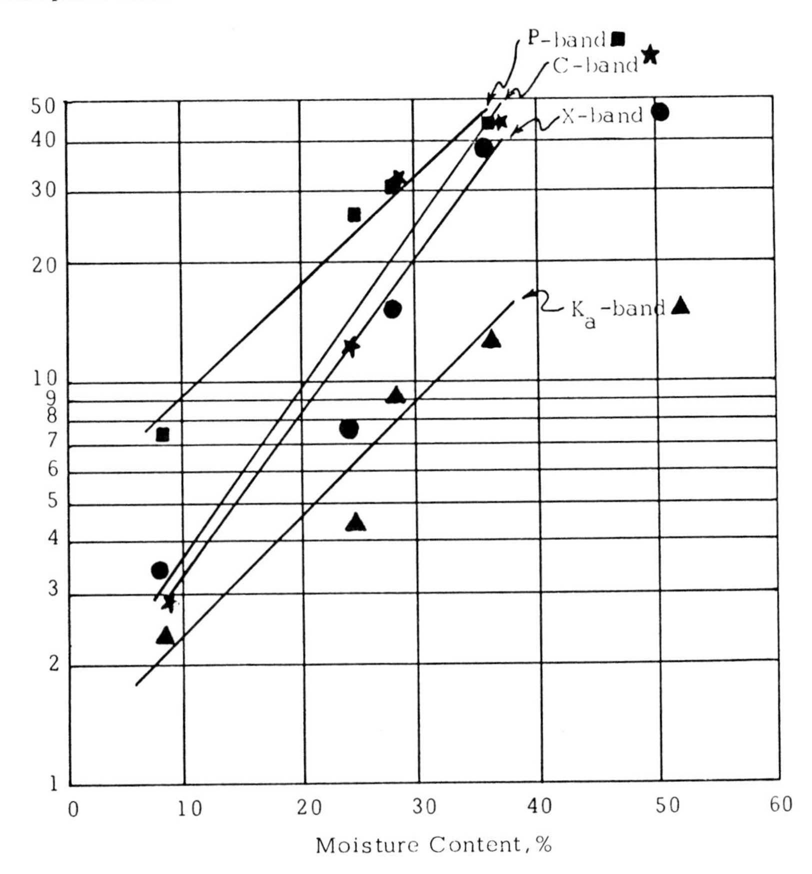


**Fig. 7.3.** Apparent relative dielectric constant versus moisture content. (Richfield Silt Loam.) After Lundien (1966).

ceived at the receiving antenna. In the receiver it is increased in strength to the point where it can be used to produce a display or a recording. A synchronizing system is required to relate the time and the character of the transmitter signal to those of the received signal. For most radars the same antenna is used for transmitting and receiving, and a switch is provided to transfer the antenna from one to the other. If the transmitter operates continuously, however, separate antennas are almost always required because the received signal is many orders of magnitude smaller than the transmitted signal, and isolation of each is very difficult if the two simultaneously use the same antenna.

## D. Radar Resolution

0-201-04245-2

Measurement of the four quantities shown in Table 7.3 is possible with a radar. All sensors measure some kind of signal strength, such as light intensity, temperature, or radio signal strength. Both active and passive sensors also measure the direction

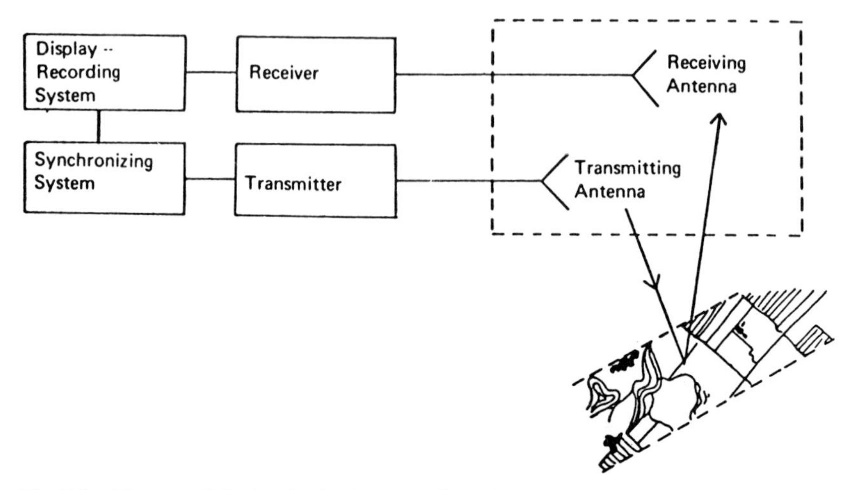


**Fig. 7.4.** Diagram of the five basic elements of a radar system. Note that the antenna system may use separate transmitting and receiving antennas, or a common antenna. The common-antenna system contains a switch to connect the antenna to the transmitter and the receiver alternately. By permission, from Committee on Remote Sensing for Agricultural Purposes, Agricultural Board, National Research Council, *Remote Sensing: With Special Reference to Agriculture and Forestry,* National Academy of Sciences, Washington, D.C., 1970.

from which the signal comes, or they separate the signals coming from different directions.

**TABLE 7.3**
**Quantities Measurable with a Radar System**

|  |
| --- |
| Quantities common to active and passive sensors: |
| Signal strength |
| Direction of target |
| Quantities unique to active sensors: |
| Distance (by comparing with time of transmission) |
| Speed (by comparing with frequency transmitted) |

The active sensor (radar) can perform two other measurements because of knowledge of the properties of the illumination: distance and speed. Distance is measured by comparing the time at which a signal is received with the time at which the same signal was transmitted. Since the speed of electromagnetic waves through space is known, the time delay measurement establishes the distance. Speed along a line joining radar and target is determined by comparing the frequency of the signal received with the frequency of the signal transmitted. Any moving object changes the frequency of the scattered signal. By observing the amount of this change, radar can determine the relative speed.

0-201-04245-2

Resolution techniques for airborne radars sensing the ground are shown in Figure 7.5. The size of the patch of ground illuminated and observed by a radar transmitting a continuous signal without making measurements of either range or speed is shown in Figure 7.5*a.* In this case the only way to establish what part of the ground one is observing is by measuring the angle at which the signal returns. Passive sensors, whatever their wavelength, use only this technique.

In Figure 7.5*b* the angular measurement sets the size of the observed patch in one dimension, but a range measurement is used to reduce the dimension of the patch in the direction radially away from the aircraft. By observing only the signals that come between two limits of range (distance), the cross-hatched region is outlined. It is formed by the intersection of the antenna pattern on the ground (also shown in Figure 7.5*a*) with a ring formed by concentric circles of constant slant range.

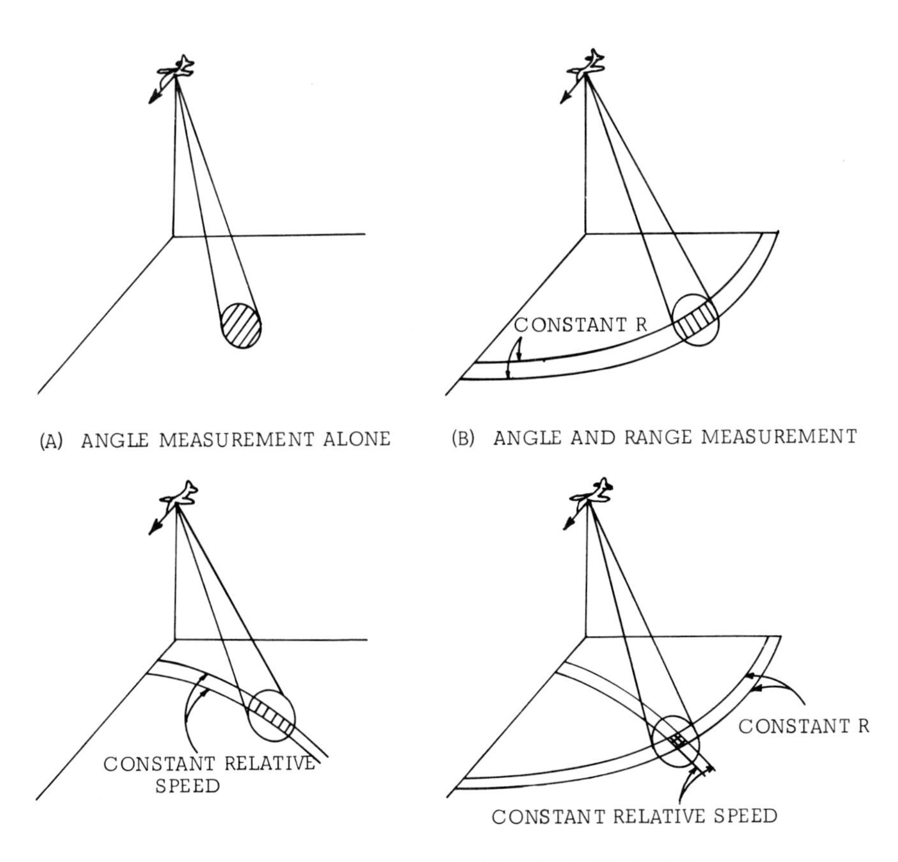


**Fig. 7.5.** Resolution techniques for airborne radars sensing the ground.

If a speed measurement is used instead of a range measurement, the situation is as shown in Figure 7.5*c*. For horizontally moving aircraft, lines of constant relative speed are hyperbolas with their foci along the line of flight. Filtering permits the reception of only signals whose frequency shifts lie between the limits outlined by the two lines of constant relative speed shown on the figure; the region within the antenna beam which can be observed is, therefore, as shown by the hatching. Again, it is smaller than the antenna alone would permit, but the direction of the strip across the antenna pattern is different from that for the range measurement.

If both speed and range measurement are used, as shown in Figure 7.5*d*, the size of the patch observed may be much smaller than that observable using an antenna alone. Thus the observed patch is the intersection of the two strips corresponding to the spacing between two constant range lines and the spacing between two constant relative speed lines.

The independence of antenna beam width indicated by the measurement of Figure 7.5*d* makes possible ground resolutions for an air- or spaceborne radar far superior to those possible with passive microwave equipment. In theory the resolutions obtained can be comparable in most cases to practical resolutions with sensors in the optical and infrared regions. It is not possible, however, to resolve distances of the order of a wavelength or smaller, so that the comparability depends somewhat on the microwavelength used.

## E. Range Measurement

In principle, range measurement always involves comparing the received signal with a delayed replica of the transmitted signal. By finding out how much delay is necessary to line up the replica of the transmitted signal with a given received signal, the radar operator may learn how far away the scattering object that caused the signal is. This is illustrated in Figure 7.6.

Two examples of range measurement systems are shown in Figure 7.7. The simplest and most common radar range measurement system uses short transmitted pulses and presents the returns as a function of time. Thus the signal present at any particular time after the transmission of the pulse represents a return from the corresponding distance. Figure 7.7*a* shows a typical return for an extended target where the antenna beam width permits signals to come from many different distances. The amplitude of the transmitted pulse has been reduced in the figure to a level close to that of the received signal.

Figure 7.7*b* shows the principle of a frequency-modulated range measurement. Here a continuous signal is varied in frequency as shown at the left of the diagram. The signal returning at any particular time is at a frequency transmitted earlier. Thus the slanted lines to the right of Figure 7.7*b* illustrate returns from five different targets. The comparator of Figure 7.6 is simply a device that measures the frequency difference $f_d$ between the outgoing and incoming waves. If this frequency difference is presented as a spectrum, it appears as in Figure 7.7*c*, where the analogy between the spectrum for the FM signal and the amplitude for the pulse signal is clear.

0-201-04245-2

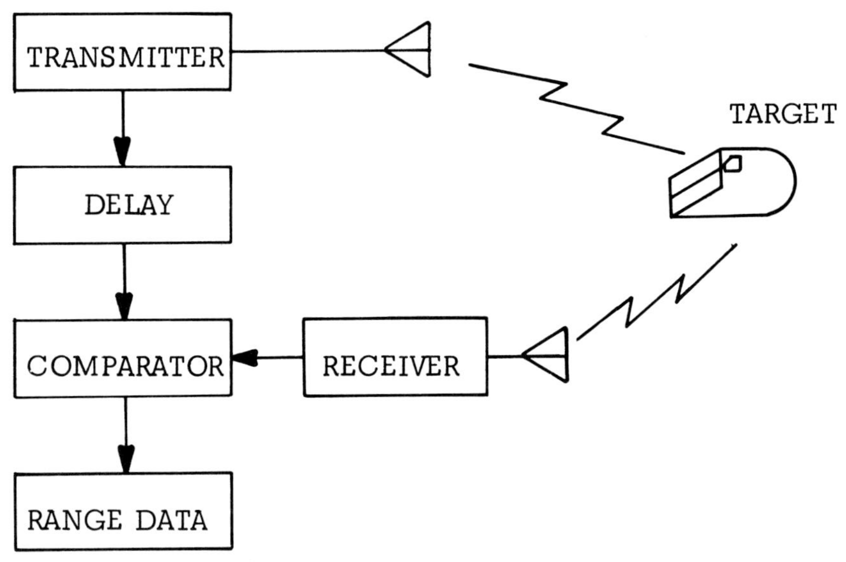


**Fig. 7.6.** Block diagram of basic radar system.

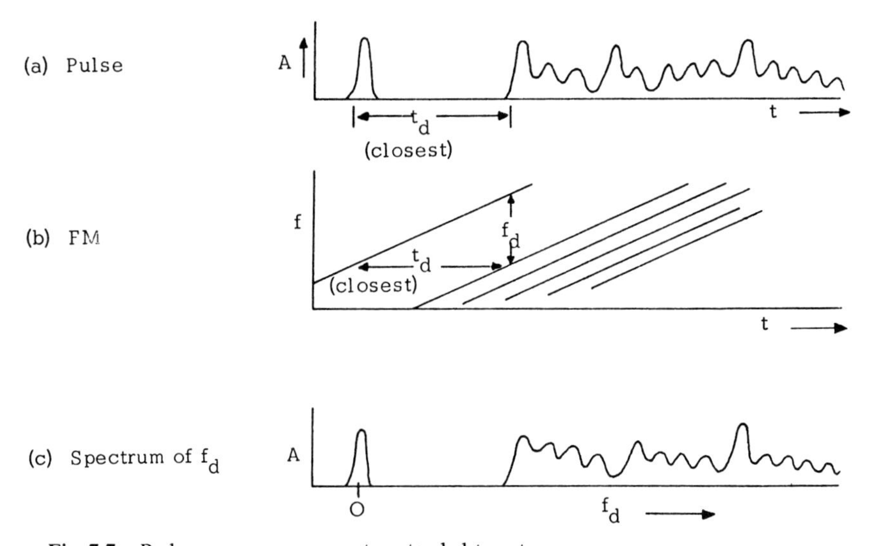


**Fig. 7.7.** Radar range measurement–extended target.

The velocity of propagation of an electromagnetic wave in air is approximately $3 \times 10^8$ m/sec and is usually designated by the letter $c$. Thus the relationship between the time delay associated with a target at range $R$ and the range itself is given by

0-201-04245-2

$$cT = 2R. \tag{7.9}$$

## F. Speed Measurement

The measurement of the relative speed between the target and the vehicle carrying the radar is based on the Doppler principle. The Doppler frequency shift occurs because the phase of a sinusoidal wave is a function of distance, so that changing distance changes phase.

To illustrate the basis for the Doppler frequency shift, we consider the voltage received from a target at range $R$ when the transmitted signal is at a frequency $f_c$ or angular frequency $\omega_c$. This voltage is given by

$$v_r = \cos\phi = \cos\left(\omega_c t - \frac{4\pi R}{\lambda}\right). \tag{7.10}$$

The instantaneous phase angle $\phi$ is made up of two terms: a continuously advancing one associated with the sinusoidal transmitted signal, and a delay term representing the phase shift in transmission from radar to target and back. The instantaneous angular frequency is the time rate of change of phase and is given by

$$\omega_r = \frac{d\phi}{dt} = \omega_c - \frac{4\pi}{\lambda}\left(\frac{dR}{dt}\right) = \omega_c + \omega_d. \tag{7.11}$$

The carrier frequency is modified by a term which is proportional to the rate of change of distance from radar to target. This last term is the Doppler angular frequency, which along with the Doppler frequency is given by

$$\omega_d = -\frac{4\pi}{\lambda}\left(\frac{dR}{dt}\right), \quad f_d = -\frac{2}{\lambda}\left(\frac{dR}{dt}\right). \tag{7.12}$$

The minus sign occurs because a Doppler frequency increase is associated with travel *toward* the target (decreasing range), whereas a Doppler frequency decrease corresponds with an increasing range as the radar goes away from the target. The magnitude of the Doppler frequency is just twice the relative velocity, expressed in wavelengths per second.

Measurement of Doppler frequency is achieved by comparing $f_c$ with $f_r$, the received frequency. If a sample of the transmitter signal (frequency $f_c$) is combined in a mixer with the received signal, the output of the mixer contains the sum and difference frequencies:

$$f_c + f_r, \quad f_c - f_r.$$

A filter is used to eliminate the sum frequency, and the difference frequency is then simply the Doppler frequency:

$$f_d = f_c - f_r.$$

0-201-04245-2

The Doppler frequency is usually a very small fraction of the carrier frequency. In fact, the ratio of the two is just twice the ratio of the relative speed to the speed of light. Thus an aircraft relative speed of 150 m/sec causes $f_d = 10^{-6} f_c$. For many airborne radar purposes the relative speed may in fact be only a few meters per second because of the angle of observation, so the precision of the measurement must be within something like 1 part in $10^8$. Thus the electronics equipment must be extremely stable to perform useful speed measurements.

### G. Angle Measurement

Angular discrimination is essentially an interference phenomenon in both the optical and the microwave regions. The diffraction limitations of both lenses and antennas are determined in part by the amount of phase interference that can be achieved. The beam width of an antenna or the angular discrimination ability of a lens is determined by its size in wavelengths. The larger the lens or antenna, the better is the angular discrimination, provided that it is made with sufficient precision.

Figure 7.8 illustrates the interference principle as applied in the radio-frequency region. Figure 7.8*a* shows a simple two-element, widely spaced antenna array commonly known as an interferometer. If this antenna array is used to receive a signal from a great distance, so that the lines joining points *a* and *b* with the source are essentially parallel, the geometry illustrated in the figure is correct. The phase interference is associated with the difference in path length, $L \sin \theta$. The voltages received at the two antennas *a* and *b* are added to determine the interference pattern. Thus, for a cosine transmitted waveform, we have

$$v_r = v_a + v_b = v_0 \left[\cos\left(\omega t - \frac{2\pi R_a}{\lambda}\right) + \cos\left(\omega t - \frac{2\pi R_b}{\lambda}\right)\right] \tag{7.13}$$

Here the distance from the source to antenna *a* is $R_a$, and that from the source to antenna *b* is $R_b$. Applying the trigonometric identity for the sum of two cosines, we find

$$v_r = 2v_0 \cos \frac{1}{2}\left[2\omega t - \frac{2\pi(R_a + R_b)}{\lambda}\right] \cos \frac{1}{2}\left[\frac{2\pi(R_a - R_b)}{\lambda}\right],$$

which can be rewritten for convenience as

$$v_r = 2v_0 \cos(\omega t - \phi_{ab}) \cos \frac{\pi}{\lambda}(R_a - R_b).$$

Here we see that the first cosine is the time-varying factor, and the second one determines its amplitude. The effect of distance in the first factor is only to shift the phase, but the amplitude is determined by the difference in distance to the two receiving elements, as indicated by the second factor. When the value from Figure 7.8*a* is substituted, this expression for the received voltage becomes

0-201-04245-2

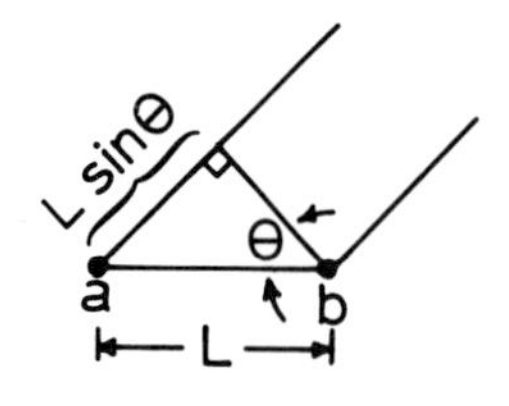


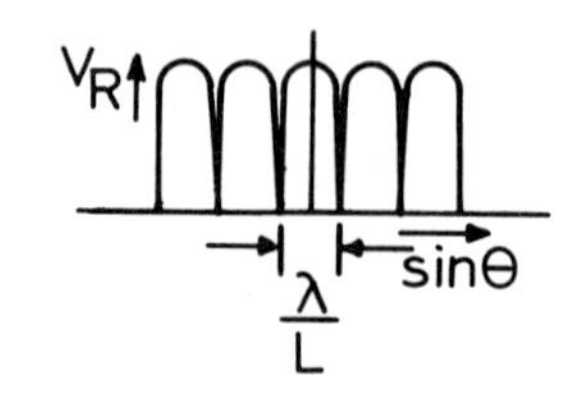


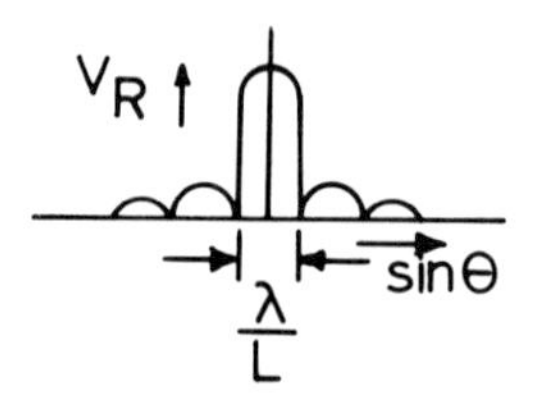


(A) INTERFEROMETER GEOMETRY (B) INTERFEROMETER PATTERN (C) ANTENNA PATTERN

**Fig. 7.8.** Interferometer and antenna of length $L$.

$$v_r = 2v_0 \cos\left(\frac{\pi L \sin\theta}{\lambda}\right) \cos(\omega t - \phi_{ab}). \tag{7.14}$$

The amplitude is a maximum ($2v_0$) whenever the argument of the first cosine factor is an integer multiple of $\pi$, and is zero when it is an odd multiple of $\pi/2$. Thus

$$v_r = 0 \quad \text{when} \quad \frac{\pi L \sin\theta}{\lambda} = \pm\frac{\pi}{2} + n\pi,$$

where $n$ is an integer. This condition therefore occurs when

$$\sin\theta = \pm\frac{\lambda}{2L} + \frac{n\lambda}{L}. \tag{7.15}$$

When $L$ is large enough, the sine of the angle is approximately equal to the angle. Therefore the width of the "lobe" from one zero to the next is $\lambda/L$, as shown in Figure 7.8*b*.

The interferometer of Figure 7.8*a* gives a pattern like that of Figure 7.8*b*. However, when enough elements are present between *a* and *b*, all lobes are suppressed except the central one, resulting in the antenna pattern of Figure 7.8*c*. With most antennas the central lobe is somewhat wider, but the subsidiary lobes are depressed further than shown.

A commonly used antenna that works exactly like its optical counterpart is a parabolic reflector. The signal is transmitted toward or received from the reflector at the focal point, and the rays leaving the reflector are essentially parallel, that is, a parabolic reflector is focused at infinity. Of course, reflector antennas can be built in which the shape of the reflecting surface is adjusted to focus at some closer point, but these are seldom used in radar.

Another common antenna is the linear array. A linear array may be thought of as analogous to a lens in optics, although the analogy is not a perfect one. Figure 7.9 illustrates the comparison. For a lens focused at infinity, the parallel rays coming from the object converge at the image point because of the additional delay

0-201-04245-2

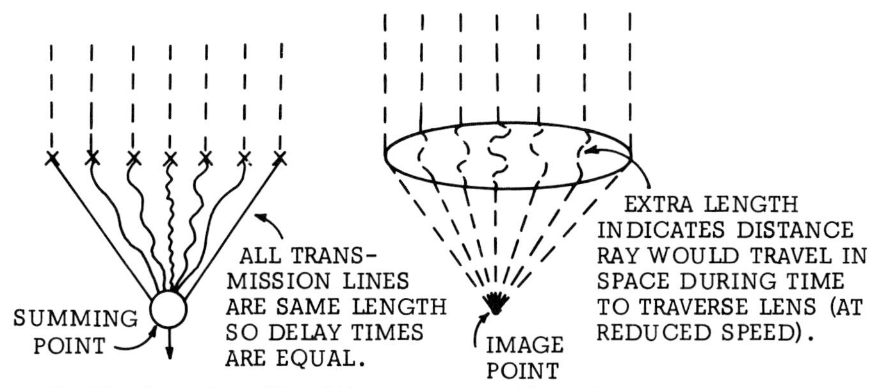


**Fig. 7.9.** Comparison of broadside antenna array with lens focused at infinity.

inserted in the shorter paths by the lower velocity of the light in the glass of the lens. Thus the distance from infinity to the image point is different through the center of the lens than it is through the edge of the lens, but the time of travel is the same for waves coming through the center, the midpoint, and the edge of the lens. The same technique can be used in an antenna array, as shown in the figure. One way to do this is to connect the various elements of the array to a summing point by transmission lines of equal length. This means, of course, that the transmission line that follows a shorter direct path will be curled up somewhat in order that it may be as long as the straight transmission line from the end of the antenna. In practice most antennas achieve the same effect without the transmission line for the central element being increased in length. For phase purposes any transmission line differing in length from another by exactly one wavelength is the same as one differing by two, three, four, or any other integer number of wavelengths. Thus the paths are actually adjusted so that they differ from each other by integer numbers of wavelengths rather than making all paths the same length.

If, instead of having our antenna pick up at right angles to its length, we wanted it to pick up from some direction to the side, we could insert a progressive phase delay from one end of the antenna to the other. The optical equivalent of this is inserting a prism in front of the lens.

Most real-aperture radar antennas are focused at infinity, as illustrated in Figure 7.9, but synthetic-aperture antennas (see discussion below) often are focused at a nearer point, as shown in Figure 7.10. For the lens seen in the figure, the paths from object *O* to image point *I* through *a, b, c, d,* and *e* all experience the same delay, the extra delay for the shorter path through *c* being due to an extra distance traveled in the low-velocity path within the lens. For the antenna the same effect can be achieved by adjusting the lengths of the transmission lines from the various elements to the summing point. Thus the delays along paths *OaS, ObS, OcS, OdS,* and *OeS* are all the same. Hence the antenna is focused at point *O.*

The aperture length for a fine-resolution radar often must be much greater than is feasible to carry on an airplane or spacecraft. For example, a radar may operate

0-201-04245-2

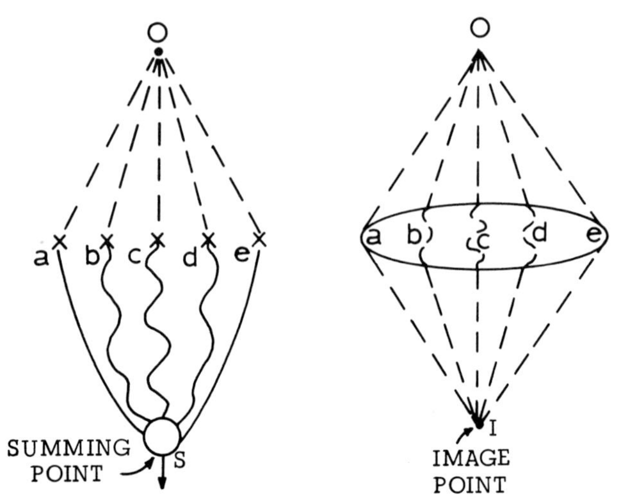


**Fig. 7.10.** Comparison of focused antenna array with lens focused at point *O*.

with a wavelength of 3 cm and an antenna 300 cm in length. At a range of 10 km the beam width on the ground for this radar is 100 m. If we wish to reduce this to 10 m, we must build an antenna 30 m long. This might be feasible on an aircraft, but it certainly would not be convenient. The synthetic aperture was developed to overcome this problem by making it possible to achieve long *effective* antenna lengths with a small *actual* antenna.

The synthetic aperture can be built either in the form of an array focused at infinity, as in Figure 7.9 (called an unfocused synthetic aperture), or in the form of an array focused at some closer point, as in Figure 7.10. The element positions of the array (*a–e* in the figures) are not simultaneously occupied by physical antennas. Rather, the radar travels first to *a,* then *b,* then *c,* then *d,* then *e,* etc. The signal received at each point is compared in phase with the signal transmitted at that point, and the result stored in a memory. Either in the aircraft or on the ground, the signals stored in the memory are combined by some sort of computer to achieve the equivalent of the transmission line lengths shown in Figures 7.9 and 7.10. A common way to store this information is on *signal film.*

Another way to look at the synthetic aperture is in terms of Doppler frequency measurement, as indicated in Figure 7.5*d.* If we consider a radar looking to the side of the flight path of an aircraft, the real-aperture antenna (all elements physically present at the same time) is as shown in Figure 7.11*a.* In the case of the synthetic aperture, the relative velocity measurement is used to improve the resolution of the real aperture, as shown in Figure 7.11*b.* By using a narrow-band-width filter, only the signals having Doppler frequencies very close to zero are permitted to pass to the recorder. Thus much of the territory illuminated by the real antenna is eliminated, and the area actually observed is greatly reduced, as shown. Figure 7.11*b* is equivalent to the antenna focused at infinity, shown in Figure 7.9.

0-201-04245-2

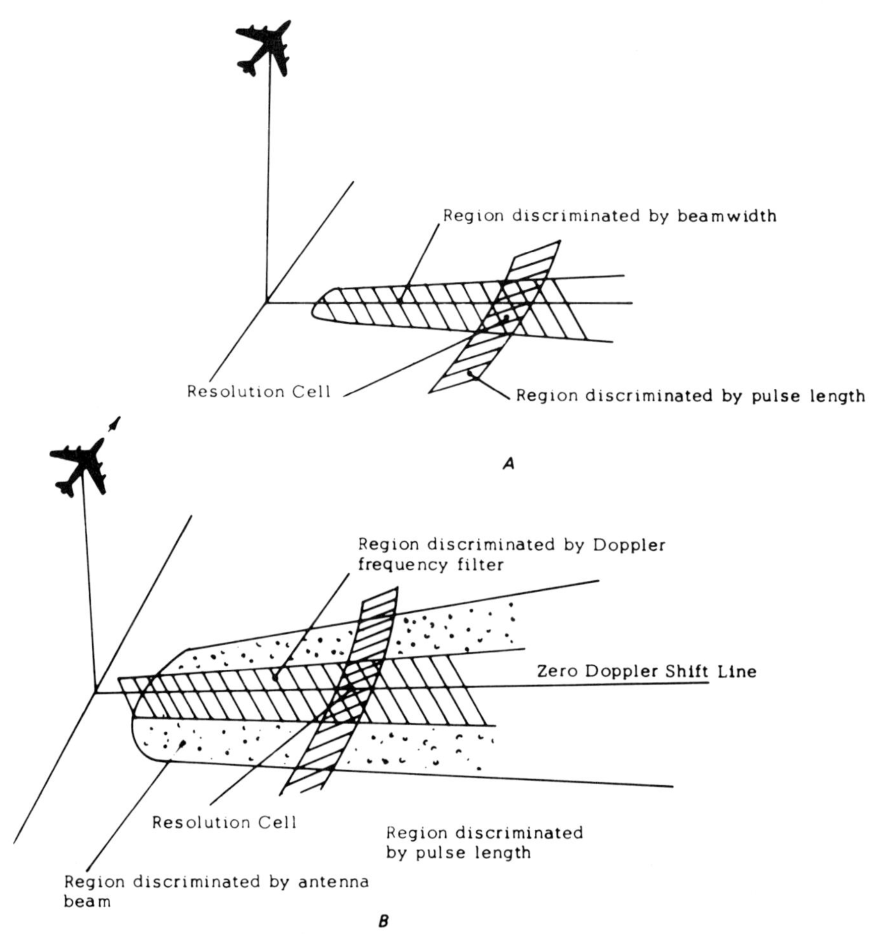


**Fig. 7.11.** Comparison of azimuth and range discrimination in a real aperture radar (A), and in an unfocussed synthetic aperture radar (B).

To focus at a closer point like the antenna of Figure 7.10, a method must be devised to use all the information available from the time the target first enters the beam of the real antenna until it leaves this beam. Not only will this give a better resolution, but also it does not waste power as does the system diagramed in Figure 7.11*b*. Figure 7.12 is a simplified illustration of the way in which the focused synthetic-aperture system operates, as viewed from a Doppler frequency point of view. Here we assume that each of the sketches represents what happens in the general vicinity of one of the points (*a* through *e*) of Figure 7.10. Actually, this means that each of the points shown in Figure 7.10 represents tens or hundreds of elements for this situation; otherwise there would be insufficient samples to achieve the Doppler filtering of Figure 7.12. Figure 7.12*a* shows the situation when the target *O* has just been illuminated by the real aperture. At this time the set of Doppler

0-201-04245-2

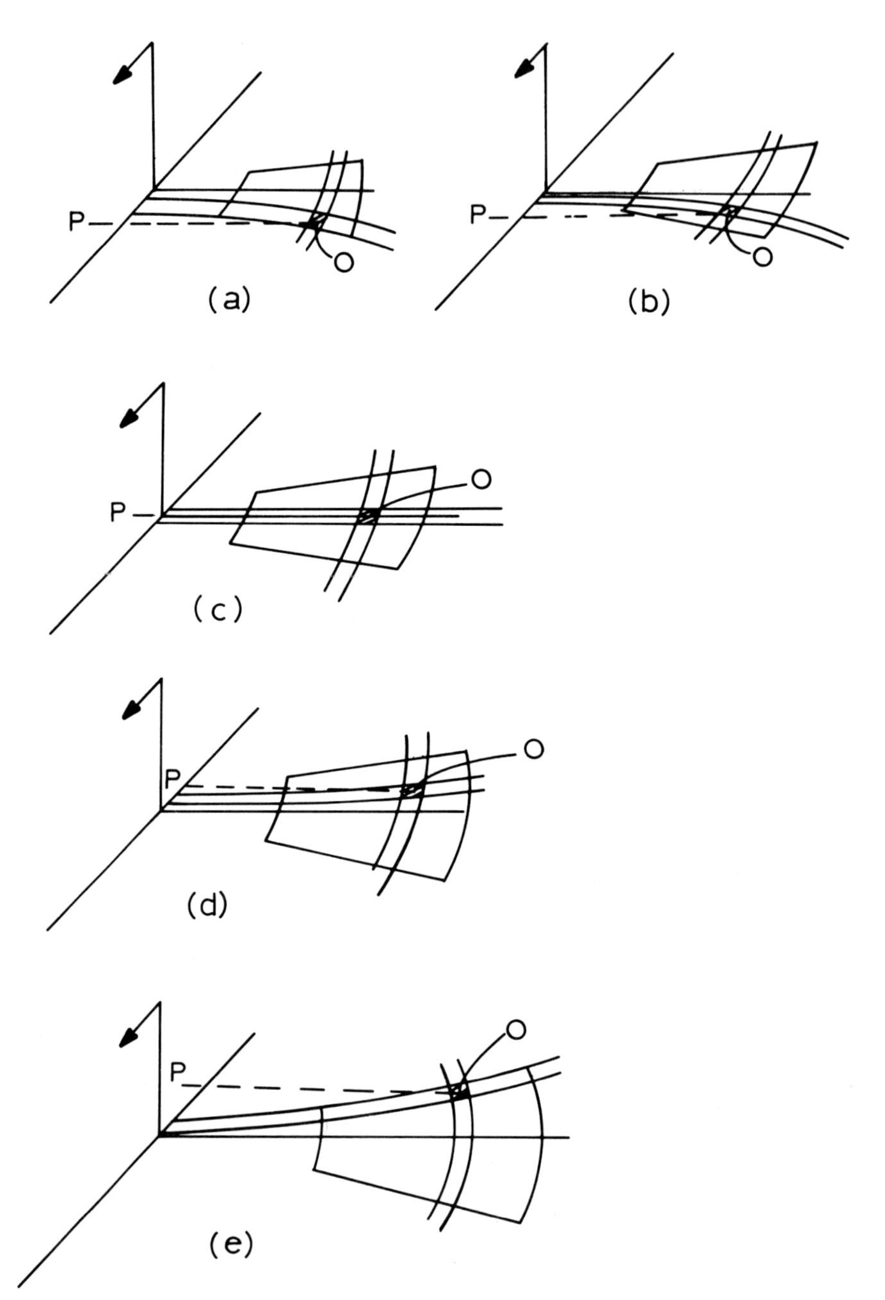


**Fig. 7.12.** Operation of focused synthetic-aperture system from a Doppler point of view (*a-e* have same meanings as in Figure 7.10).

0-201-04245-2

frequencies (relative velocities) corresponding to the forward edge of the real aperture is filtered and used for target *O*. Range lines are also shown. The point *P* on the ground track of the aircraft is the foot of a perpendicular from point *O*. The aircraft has not yet reached this point. In Figure 7.12*b*, corresponding to point *b* of Figure 7.10, the Doppler frequencies used are those midway through the forward part of the beam, and the aircraft is closer to point *P*. In Figure 7.12*c* the aircraft is directly over point *P* and abreast of point *O*. The Doppler filter here is the same as that of Figure 7.11. In Figures 7.12*d* and 7.12*e* the mirror-image situation prevails as the aircraft moves on past *P*. The results of each of these five filtering operations are stored and combined in the processing system to produce the equivalent focused synthetic aperture. In a real system the filter moves continuously rather than incrementally as indicated here. Furthermore, additional filters are at work for each of the elements within the real aperture. Obviously, the mechanization is quite complicated. An illustration of an optical technique for this is given in Section IV on imaging radars.

## III. RADAR SCATTEROMETER

### A. Introduction

A scatterometer measures the scattering of radar signals from the ground. Thus any radar which makes a measurement of the signal strength is a scatterometer. The term is usually applied, however, to special systems designed primarily for measuring signal strength rather than for producing images; therefore the scatterometer is treated here separately from imaging radars.

Scatterometers are used primarily to gather engineering design data to use with imaging and special-purpose radars. Consequently, most scatterometers measure the signal strength as a function of angle, polarization, frequency, or some other variable, rather than simply making a measurement at a single frequency, angle, and polarization. The most common arrangement is a device intended to produce curves of differential scattering coefficient versus angle of incidence. Since more information is contained in such a curve than in the single number associated with the same polarization at a fixed angle of incidence, such scatterometers may turn out to be useful in measuring properties of the earth not discernible with imaging radars.

The radar scatterometer can be used for measuring the roughness of the ocean, and from this the wind speed at the ocean surface. For a scatterometer to be effective in this application, the wavelength must either be very short (of the order of 2 cm) or very long (of the order of 10 m).

### B. Along-Track Fan-Beam Scatterometer

0-201-04245-2

The fan-beam scatterometer system was developed so as to obtain a curve of scattering coefficients versus angle with a single aircraft pass (Moore, 1966). Figure 7.13*a* shows the use of a pulse-modulated fan-beam system. The antenna illuminates a narrow patch along the ground from directly beneath the aircraft to some point far ahead. The part of the patch observed at any instant is determined by the position of a short pulse at that instant, as indicated. The pulse duration indicated is $\tau$, and a short enough $\tau$ will make a quite small illuminated area in the along-track direction.

Figure 7.13*b* shows a system that uses velocity measurement to discriminate among the different parts of the fan beam. With this technique the Doppler frequency shift is positive ahead of the aircraft and negative behind it; hence a velocity-measuring system that can distinguish between positive and negative Dopplers permits measurements both fore and aft. Sometimes this is useful because of differences in the scattering properties of terrain when viewed from the two directions. The velocity of the aircraft is $v$, and the band width of the Doppler filter used is $\Delta f_d$, whose geometric equivalent is as shown in the diagram. By making $\Delta f_d$ small enough, a fine resolution in the along-track direction can be achieved.

The fan-beam system permits gathering information about many different parts of the ground at the same time, although the information for each corresponds to a different angle of incidence. Figure 7.14 illustrates the way in which this occurs. Doppler filters or pulse samplers obtain information at three angles of incidence:

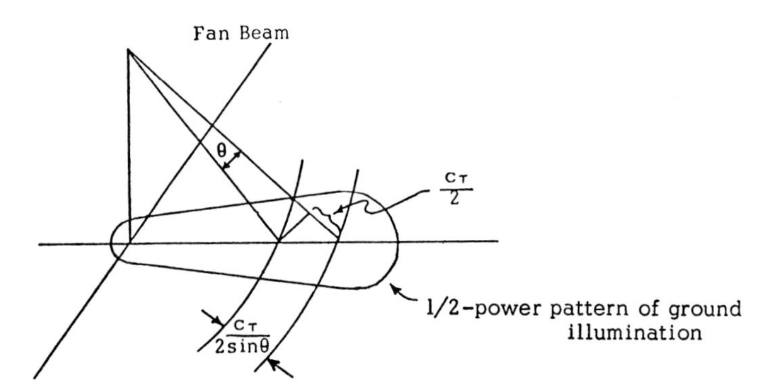


**Fig 7.13*a*.** Operation of a fan-beam pulse scatterometer.

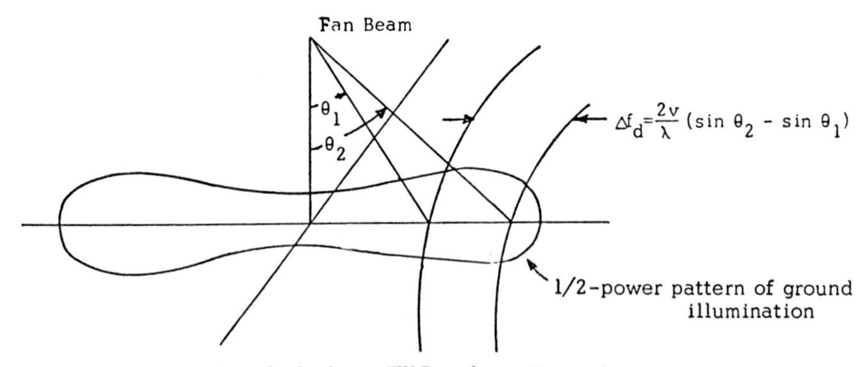


**Fig. 7.13*b*.** Operation of a fan-beam CW Doppler scatterometer.

0-201-04245-2

vertical, $\theta_1$, and $\theta_2$, as indicated. When the aircraft is in position 1, these three outputs correspond to points $A$, $B$, and $C$ on the ground. When the aircraft has moved on to position 2, the vertical now is for $B$, and the three angles shown for position 1 give information for $B$, $C$, and $D$. Also at this time information at an angle $-\theta_1$ is available at $A$. Similarly, when the aircraft moves on to position 3, the original three angles correspond to points $C$, $D$, and $E$ on the ground, and the information for angle $-\theta_1$ comes from $B$. This information may be collated after flying to determine a curve for any particular point, such as $A$, $B$, or $C$. For instance, the curve for the scattering coefficient at point $C$ is made up of observation at $\theta_2$ for position 1, at $\theta_1$ for position 2, and at vertical incidence for position 3. Use of this technique permits determining a set of curves of scattering coefficient versus angle along a continuous track with only a single pass.

The radar equation for the scatterometer can be written to show the measured value of the differential scattering coefficient. For a simplified case in which the antenna beam has a gain $G_0$ and a width in the cross-track direction $\phi_0$, if all quantities inside the integral in Eq. (7.4) are constant across the small region of integration, the value of scattering coefficient becomes for the pulse system

$$\sigma^0 = \left(\frac{W_r}{W_t}\right) \frac{2(4\pi)^3 R^3 \sin\theta}{\lambda^2 G_0 \phi_0 c\tau} . \tag{7.16}$$

The significant measured quantity is the ratio of received to transmitted power. Other quantities are as defined previously, $\tau$ being the pulse duration.

The comparable equation for the Doppler system is

$$\sigma^0 = \left(\frac{W_r}{W_t}\right) \frac{2vR^2}{\lambda^4 G_0{}^2 \phi_0 \, \Delta f_d} , \tag{7.17}$$

where all quantities have been previously defined.

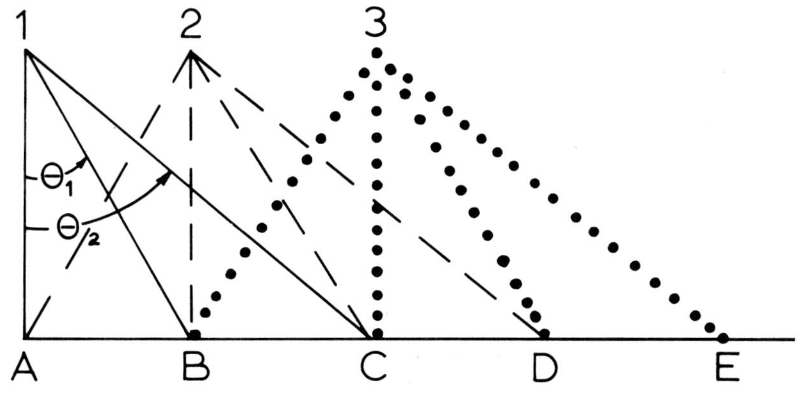


Fig. 7.14. Angular information obtained with fan-beam scatterometer.

0-201-04245-2

Equation (7.16) was derived for angles well away from the vertical and has to be modified within a few degrees of the vertical. Equation (7.17), however, applies for all angles of incidence. The pulse duration in Eq. (7.16) must be set at the time of transmission, but the filter band width of Eq. (7.17) can be established in the data processing and changed at will.

### C. Pulse System

The pulse scatterometer is a simple pulse radar with built-in calibration equipment. Without calibration equipment a radar cannot truly be classified as a scatterometer, although, of course, the calibration system may be external to the actual radar. Figure 7.15 shows a block diagram of a pulse scatterometer. The timing source determines when the transmitter sends out a pulse and also when the received pulse is to be observed on an oscilloscope or some sort of sampling device. The pulse generator modulates the transmitter (often a magnetron at centimeter wavelengths or a triode at longer wavelengths). The transmitted signal goes through the transmit-receive switch to the antenna.

Most pulse systems use the same antenna for transmitting and receiving. The received signal passes through the T-R switch to the receiver, where it is amplified

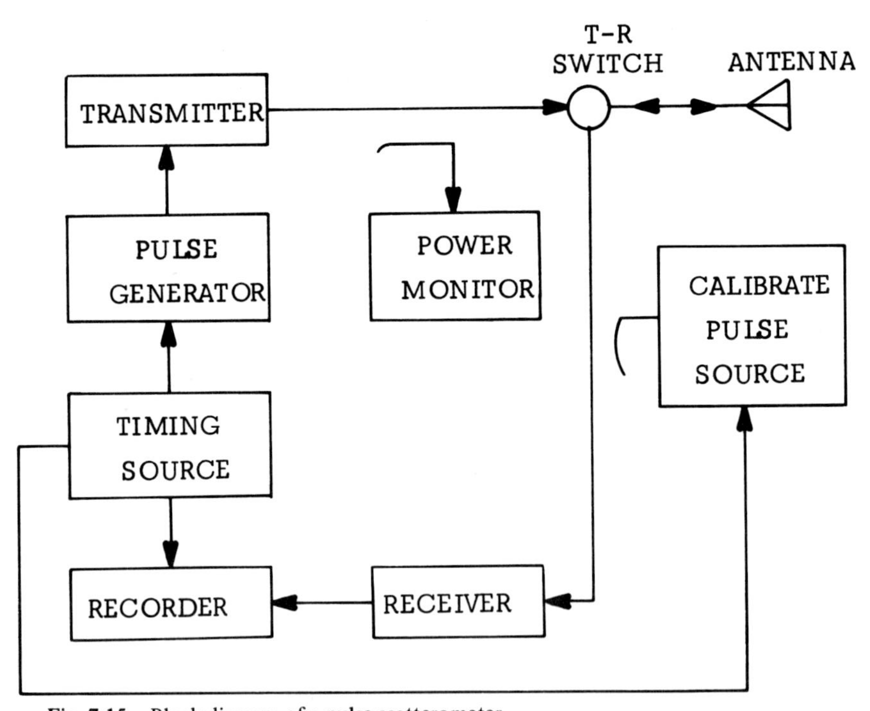


**Fig. 7.15.** Block diagram of a pulse scatterometer.

0-201-04245-2

and detected. The output may be in the form of a trace on an oscilloscope but is more likely to be sampled at various times, with the samples recorded on magnetic tape.

To determine the ratio of received to transmitted power, the transmitted power must be monitored as shown and some sort of signal from a calibrated pulse source must be passed through the receiver. This calibration signal must appear at a time when signals from the ground are absent. At some stage during the calibration of the system, the pulse source must have its amplitude varied over a wide range so that a complete input-output characteristic for the receiving system can be established. Usually this is not necessary on an individual pulse basis, but may be done at the beginning and end of a run or perhaps at the beginning and end of a day.

Figure 7.16 shows a typical oscilloscope presentation of a pulse returned to a scatterometer. The first signal coming back is the one from directly beneath (the altitude signal). One might expect that the signals coming back after the altitude signal would decrease smoothly to zero. In practice, however, the signals vary up and down as indicated on the sketch. (This variation is known as "fading.")

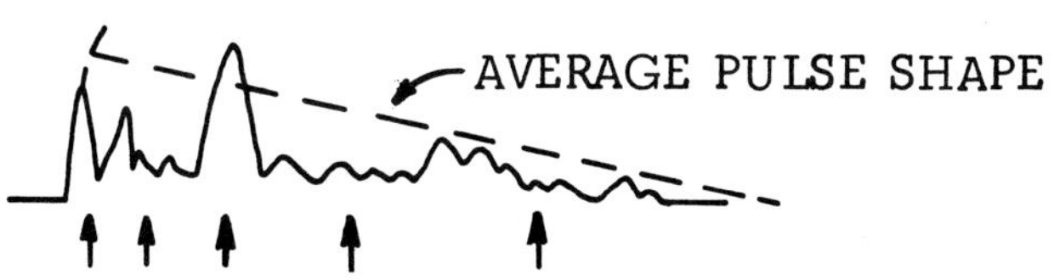


**Fig. 7.16.** Typical oscilloscope presentation of a pulse returned to a scatterometer. Arrows show locations of sampling gates.

A difficulty with the pulse scatterometer is the time compression of the returns from the various angles near the vertical. If we consider the radar at a height $h$, the slant range, angle, and delay time $T$ are related by

$$R = \frac{h}{\cos\theta}, \qquad T = \frac{2h}{c\cos\theta}. \tag{7.18}$$

Normally the total delay time is not important because the recording system is not activated until just before the altitude signal returns. If we label the altitude-signal delay time $T_0$, the time of interest is the difference between the time for a particular range and the time for the altitude signal, as shown in

$$T - T_0 = \frac{2h}{c}\left(\frac{1}{\cos\theta} - 1\right) = \frac{2h}{c}\left(\frac{1 - \cos\theta}{\cos\theta}\right). \tag{7.19}$$

Near the vertical, this equation gives a time difference approximately proportional to the square of the angle of incidence. In other words, the angle of incidence

0-201-04245-2

corresponding to a particular time is proportional to the square root of the time delay from the time for the vertical return. The example shown in Table 7.4 for a radar at a height of 3 km above the ground shows the sort of crowding of pulses that occurs. The entire range from the vertical out to 10° is compressed into the first 0.3 μsec, whereas the next 10° takes a full microsecond. As a result, the resolution of angles near the vertical requires exceedingly short pulses. For example, to confine the first pulse duration to 2.5° the pulse must be only 0.02 μsec (20 nsec) long. This is often hard to achieve. Consequently, the pulse system is not very effective in the immediate vicinity of the vertical. On the other hand, it is relatively easy to operate away from the vertical, and both the pulse radar techniques and the sampling methods for recording the data are well known and easy to implement.

**TABLE 7.4**
**Example of Relation between Time Delay and Incident Angle (3-km Height)**

| *Incident Angle (deg)* | *Time after First Return (μsec)* |
|---|---|
| 0 | 0 |
| 2.5 | 0.019 |
| 5 | 0.077 |
| 10 | 0.305 |
| 20 | 1.30 |
| 40 | 6.1 |
| 60 | 20 |

The resolution of a pulse system at angles away from the vertical is given by

$$\text{Resolution} = R\phi_0 \left( \frac{c\tau}{2 \sin \theta} \right) = r_a r_r . \qquad (7.20)$$

Here $R$ is the slant range as usual, $\phi_0$ is the beam width in the cross-track direction, and the other quantities are as defined in Figure 7.13. The shape of the resolution cell changes from relatively narrow across track and long in the direction of flight near vertical incidence to the opposite at angles away from vertical incidence. At vertical incidence itself, the expression in Eq. (7.20) for the along-track resolution is not appropriate, and, indeed, the cross-track resolution may be determined by pulse duration rather than antenna beam. The radius of the circle corresponding to the first pulse duration of illumination at the vertical is given by

$$\text{Radius} = \sqrt{c\tau h} . \qquad (7.21)$$

0-201-04245-2

## D. Velocity-Measurement Scatterometer System

A scatterometer system that uses velocity measurement (Doppler frequency measurement) is in many ways even simpler than the pulse system. Such a device requires only a low-power continuous-wave transmitter, and in its simpler form the receiver requires only a mixer and an audio amplifier. A block diagram of this version is shown in Figure 7.17.

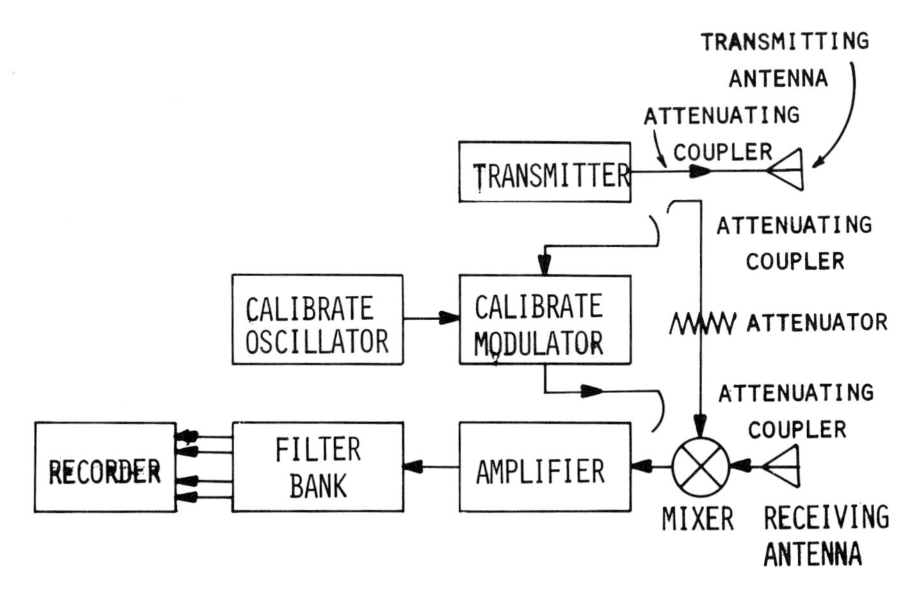


**Fig. 7.17.** Block diagram of CW Doppler scatterometer using homodyne system.

The transmitter need only be moderately stable, but must not emit excessive noise within the Doppler band width. For a klystron transmitter, this means that the power supply must be extremely well regulated. An attenuated sample of the transmitter signal serves as a "local oscillator" supply and is mixed with the received signal. A signal return from a fixed target would, therefore, beat down to zero frequency and appear merely as a dc level dependent on the relative phase of the transmitted and received signals. A signal received from a moving target (or a fixed target when the radar is moving) appears shifted from zero in the output by the amount of the Doppler frequency. The amplifier, therefore, need only have a band width suitable for passing the Doppler frequencies, normally only a few kilohertz or even less; and the signals coming from the different angles are separated by a filter bank, with each being separately recorded. Since the signals are at audible frequencies, headphones provide a useful performance monitor.

0-201-04245-2

The filter bank need not be carried with the radar, for the output of the amplifier may be recorded on magnetic tape. The band width of such a tape recorder need be only a few kilohertz. Filtering can be done later on the ground, using either analog or digital techniques.

This system cannot provide the signal from directly beneath the radar, for the Doppler frequency there is zero. The signal at zero frequency will be confused in the mixer with the sample of the transmitted signal itself; the two cannot be separated.

A more elaborate system is required if the scatterometer is to measure the vertical incidence signal. Figure 7.17 shows a homodyne, which is a superheterodyne using a zero intermediate frequency. Figure 7.18 shows a system using a nonzero intermediate frequency, which can serve to measure signals at vertical incidence.

The resolution of the velocity or Doppler system is given by

$$R\phi_0 \times \frac{R\lambda}{2v}\,\Delta f_d \tag{7.22}$$

Cross track    Along track

As with the pulse system, the cross-track resolution is determined by the beam angle $\phi_0$; the along-track resolution, by the filter band width $\Delta f_d$ associated with a particular Doppler channel.

### E. Accuracy

The measurement accuracy of a radar scatterometer depends on the number of independent samples averaged. One would think that use of a very short pulse or a very narrow Doppler band width would permit an arbitrarily small along-track reso-

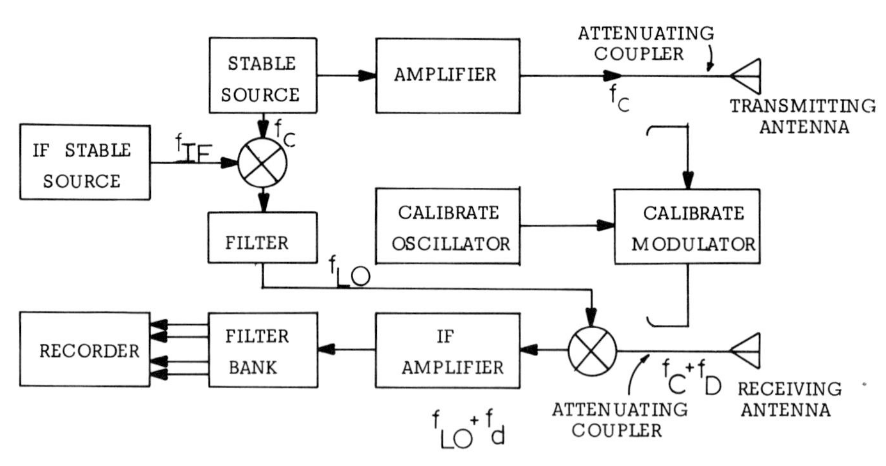


Fig. 7.18. Block diagram of CW Doppler scatterometer using intermediate frequency.

0-201-04245-2

lution. Such a resolution, however, results in a large variance of the returned signal, so that the advantage of the improved resolution is illusory.

A reasonably large target cell, containing many independent scatterers of more or less the same size, produces a signal with noiselike characteristics. Its amplitude is governed by the Rayleigh distribution, and its power is given by an exponential distribution:

$$p(W_r) = \frac{1}{\overline{W}_r} e^{-W_r/\overline{W}_r}, \tag{7.23}$$

where $W_r$ is the mean returned power. For this distribution, the spread between the value exceeded 5% of the time and that exceeded 95% of the time is 18 dB. Expressed this way, it may not seem so great. Expressed as a ratio, however, this means that the 5% power is 63 times as great as the 95% power! With such a tremendous ratio of possible returns from the same target area, the need for averaging is clear.

Calculations show that, for a rectangular spectrum such as might be obtained from a rectangular filter pass band for the Doppler system, four independent samples are obtained after square-law detection for every hertz of predetection band width, or two for every hertz of postdetection band width. This number is the same for both pulse and Doppler systems. With this concept, the number of independent samples can be shown to be (Moore and Waite, 1969)

$$N_i = \frac{8(\Delta\rho)^2 \cos^3 \theta_m}{\lambda h}, \tag{7.24}$$

where $\Delta\rho$ is the along-track resolution-cell width, and $\theta_m$ is the mean value of the angle of incidence to that cell.

The probability density function for the average of $N_i$ exponentially distributed variables can be shown to be that of a $\chi^2$ distribution with $2N_i$ degrees of freedom. For averages of more than 10 independent samples, the standard deviation of the mean is approximately

$$\sigma_m \approx \frac{2\mu}{\sqrt{N_i}}, \tag{7.25}$$

where $\mu$ is the mean value for the actual statistical process. Figure 7.19 illustrates the variation of the 5-95% range of the values taken on by the mean for different numbers of independent samples. It will be noted that a 3-dB range is achieved only after more than 25 samples have been averaged—and this means that the 90% range is still over a ratio of 2 to 1. Thus, reducing the resolution cell length $\Delta\rho$ below the value required for an adequate number of independent samples means only that several resolution cells must be averaged together. It does not mean that significant additional information is resolved!

0-201-04245-2

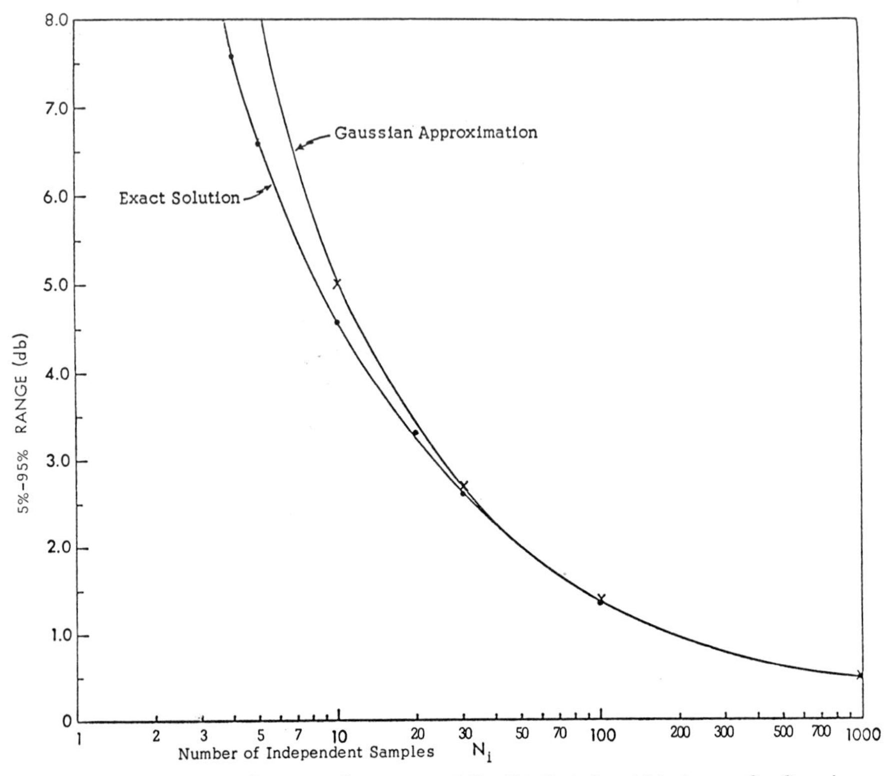


**Fig. 7.19.** Accuracy of averages for exponentially distributed variable (power for Gaussian or Rayleigh voltage fading).

A sort of "uncertainty principle" applies to the relation between precision and resolution for the scatterometer. If only a single resolution cell is averaged for any particular measurement, the uncertainty principle takes the form:

$$\Delta\rho \left(\frac{\sigma_m}{\mu}\right) = \sqrt{\frac{\lambda h}{8 \cos^3 \theta_m}}. \tag{7.26}$$

The quantity in parentheses on the left is the relative precision, as measured by the ratio of the standard deviation of the mean to the value of the mean. Since the quantity on the right is fixed for a particular radar, height, and angle of incidence, the tradeoff indicated by the equation states that greater precision can be achieved only with poorer resolution or conversely. Should $M$ resolution cells be averaged together, the uncertainty principle takes the form:

$$\sqrt{M}\, \Delta\rho \left(\frac{\sigma_m}{\mu}\right) = \sqrt{\frac{\lambda h}{8 \cos^3 \theta_m}}. \tag{7.27}$$

0-201-04245-2

Since the precision improves only as the square root of the number of cells averaged, whereas increasing the size of the cell causes a linear increase in precision with distance averaged, $\Delta\rho$ should be increased wherever possible rather than averaging the results from several cells.

## F. Presentation Problems

Presentation of the multiangle output of a radar scatterometer to a user is a difficult problem. The difficulty is that a collection of curves of $\sigma^0$ versus $\theta$ is difficult to relate to terrain. On the other hand, a collection of outputs at different angles is also difficult to relate. Figure 7.20 shows the traces that could result from a collection of output filters for the Doppler system (or sampling gates for the pulse system). One might be inclined to think that a single such trace would be useful for comparison with terrain photographs, but the output for a single trace could just as easily have been obtained from a calibrated radar imager, and the imager would have given two-dimensional spatial context as well as amplitude. The advantage of the scatterometer is in the multiple-angle information, but to apply it one must use all of the traces of Figure 7.20 together.

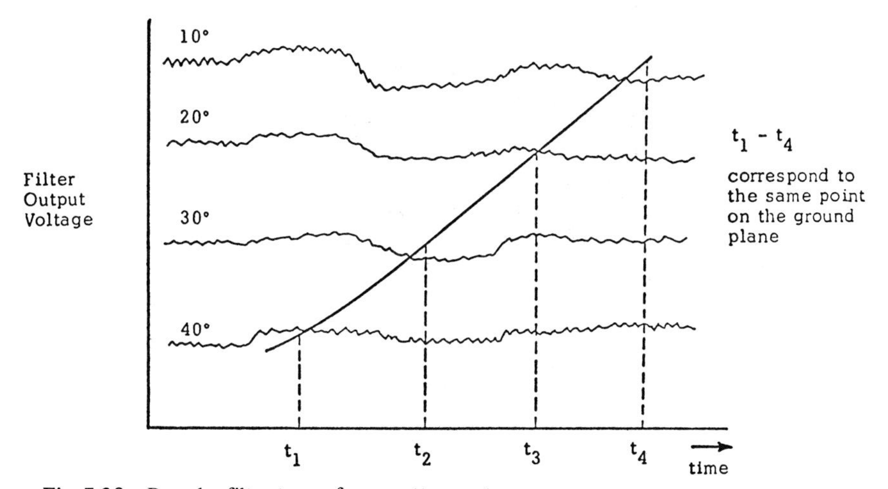


**Fig. 7.20.** Doppler filter traces from scatterometer.

In Figure 7.21 the curves of scattering coefficient versus angle for various areas are compared to the mean scattering coefficient curve for the entire run. Although differences can be readily seen by examining the photograph and noting the different shapes of the small scattering coefficient curves, this format is difficult to use.

A technique was attempted in which three values from the curves of Figure 7.21 were used to set the intensity of light in three colors. The result was a unique color for each set of values. This color display can be used as an overlay over a

0-201-04245-2

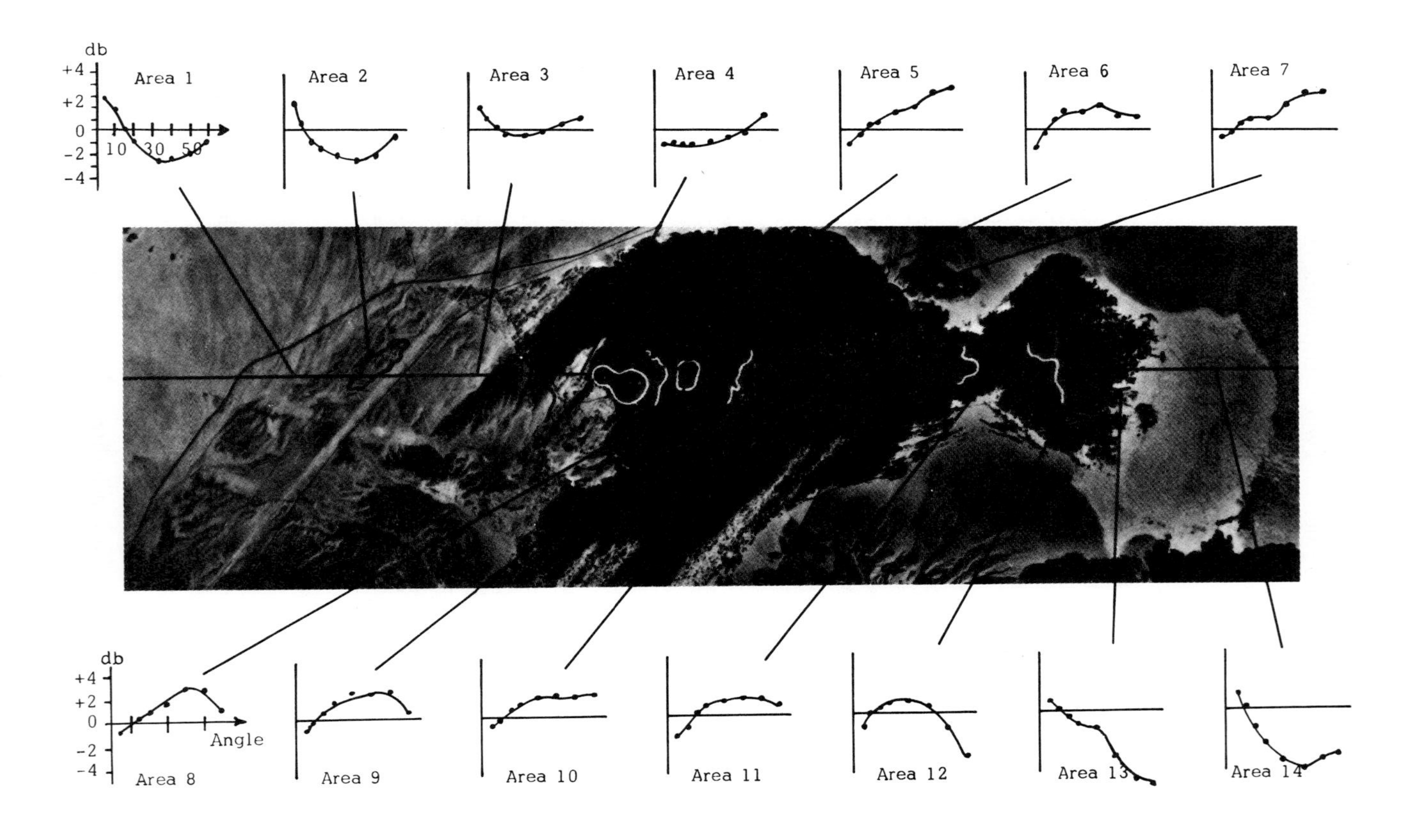


**Fig. 7.21.** Deviation from average curves for a scatterometer, compared with different ground regions in area of Pisgah Crater, California. Each numbered area has significantly different surface properties.

0-201-04245-2

photograph or radar image; it seems to provide the best method for relating scatterometer data over land to terrain. It is not illustrated here because of the difficulties of color reproduction.

A major use of the scatterometer is in determining statistics for the design of radar imagers. Fortunately, presentation problems are not so important for this application. The various classes of terrain can be segregated, and statistical descriptions of the scattering coefficient obtained for each class of terrain.

Nevertheless, the possibility of using multiangle data is intriguing, and presentation methods need further analysis.

## G. Special Systems

A radar scatterometer operating at only a single angle of incidence for each portion of the ocean probably will give sufficient information, when combined with meteorological data, to permit measurement and forecasting of winds and waves at sea (Moore and Pierson, 1970). A system proposed for this purpose uses an electronically scanned antenna carried on a spacecraft. The antenna is tilted somewhat ahead of the spacecraft, so that the minimum angle of incidence is 25 or 30°, and scans are made to either side, as shown in Figure 7.22.

To improve the results when precipitation between spacecraft and ocean significantly attenuates the radar signal, the same system can carry a microwave radiometer. Since the radiometric signal is much more sensitive to the attenuation in the atmosphere than is the radar signal, it can serve to determine the amount of

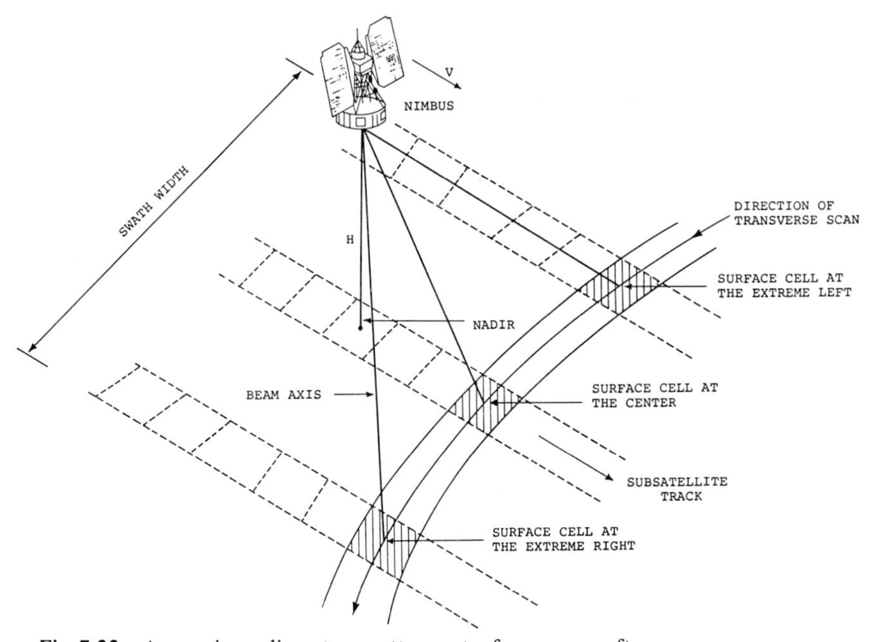


**Fig. 7.22.** A scanning radiometer–scatterometer for a spacecraft.

0-201-04245-2

attenuation present. This attenuation can then be applied to correct the radar signal from the ocean.

Similar combined radiometer-scatterometer systems can be used in aircraft over land to produce images, but they must operate at extremely high frequencies to achieve usable land resolutions with antennas of feasible size.

## IV. IMAGING RADARS

### A. Introduction

Imaging radar systems normally produce pictures of the illuminated terrain by scanning their beams in one direction and using range measurement in the other direction. The earliest imaging radars used rotating antennas and achieved range resolution by transmitting short pulses. Figure 7.23 illustrates two types of presentation used with such systems. When the antenna scans only a small sector, the signals may be presented as intensity modulation on a display of range versus angle, as indicated in Figure 7.23*a;* this is called a B-scan presentation. Because the same angular spread represents different distances at different ranges, such a presentation causes a distorted picture of the ground. This can be improved by the use of the plan position indicator (PPI) in Figure 7.23*b*. Here the sweep scans radially outward from the center of the tube, thus achieving the proper relationship between angle and range. Ordinary PPI presentations, however, suffer somewhat because the time measurement gives a slant range rather than a ground range. The original application of this system was to ship radars, where this made no difference, but with airborne radars corrections may be required.

Modifications of the PPI presentation are commonly used in the weather radars carried by all commercial aircraft. Most ground-based radars also use PPI's. Fine-resolution imaging radars, however, normally are operated in the *side-looking* mode. Such a radar is called a *side-looking airborne radar* (SLAR). With the SLAR, motion

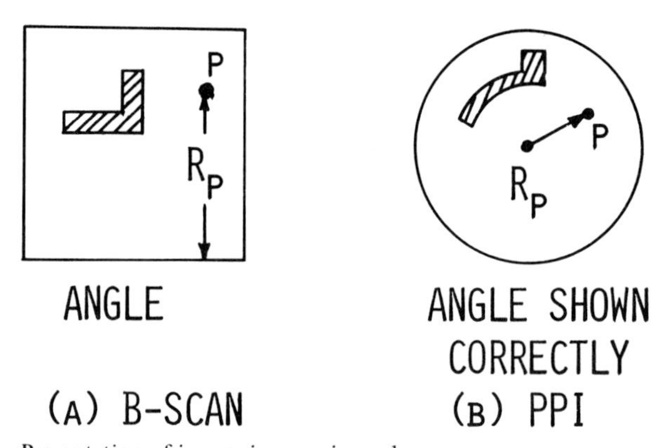


**Fig. 7.23.** Presentation of images in scanning radars.

0-201-04245-2

of the aircraft replaces rotation of the antenna as a scanning technique. Since the antenna can remain fixed in the aircraft, it may be larger than one that must be scanned, and a larger antenna can produce a narrower beam with consequent improvement in angular resolution. Furthermore, the synthetic-aperture technique described earlier cannot be applied to a system that looks directly ahead of the aircraft, but can be used with a SLAR.

Figure 7.24 illustrates the principle of the SLAR. The aircraft carries an antenna whose long direction is in line with the fuselage and (neglecting drift) the line of flight. With continuous illumination this beam would simultaneously illuminate an entire strip at right angles to the flight path, but the radar transmits short pulses, so that different parts of the strip are illuminated at different times; and the returns, therefore, appear at different times as indicated in the sketch. These returns are used to intensity-modulate a single line display on a cathode-ray tube, and a film is moved past this display in synchronism with the motion of the aircraft. When strong targets appear, such as the bridge and the woods indicated, the signal return is strong and the CRT is bright, thus exposing the film. When returns are small, the signal is weak, the CRT is dim, and the film is barely exposed.

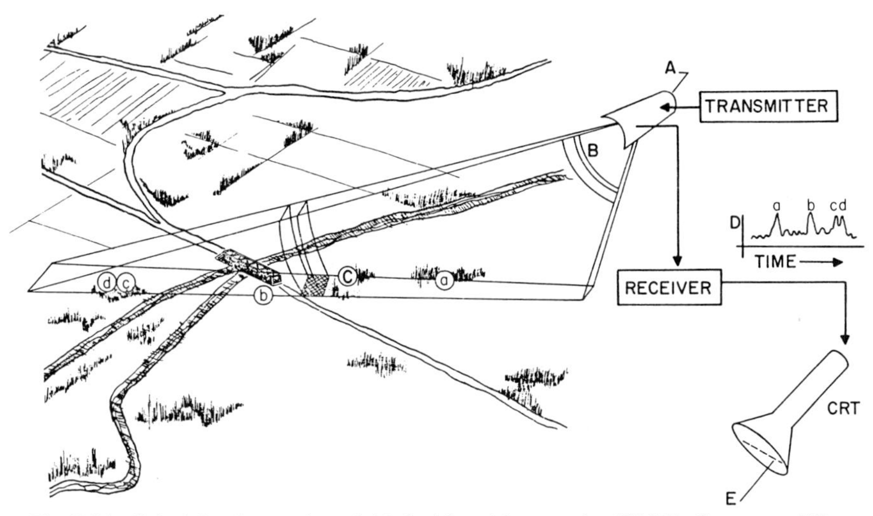


**Fig. 7.24.** Principle of operation of side-looking airborne radar (SLAR). Courtesy of Westinghouse Aerospace.

Figure 7.25 illustrates the film drive mechanism for such a SLAR. Since the film drive must be synchronized with the aircraft motion over the ground, it is normally controlled by the Doppler navigator system on the aircraft. Such a careful control of film speed leads to complexity in the design of the recorder, particularly since it must work in an aircraft vibration environment.

Orthogonal look directions for radar may be desirable to reduce the effect of shadowing and to improve knowledge about the terrain imaged. An ordinary SLAR

0-201-04245-2

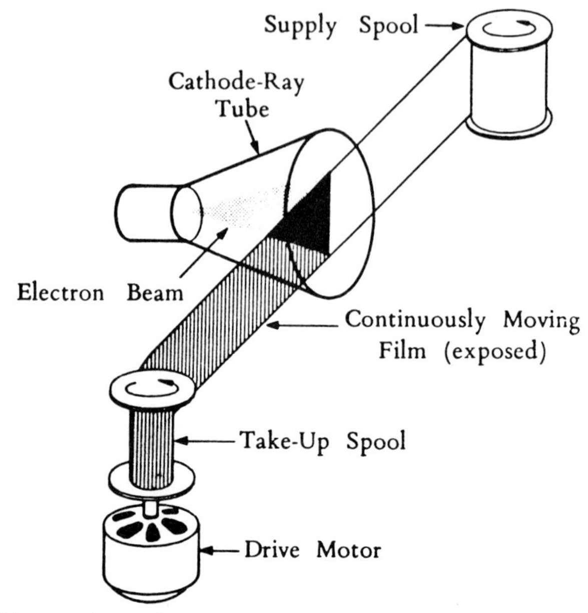


**Fig. 7.25.** Diagram of recording technique for SLAR. Electron beam scans vertically with position proportional to distance from flight track. By permission, from Committee on Remote Sensing for Agricultural Purposes, Agricultural Board, National Research Council, *Remote Sensing: With Special Reference to Agriculture and Forestry,* National Academy of Sciences, Washington, D.C., 1970.

can be used to achieve orthogonal look directions by flying two paths at right angles to each other. However, a modification of the SLAR would permit it to operate in a *squint mode,* in which the beam is ahead of the normal side look or behind it.

Figure 7.26 shows a system (yet to be constructed) that switches alternately between beam 1 behind the sideways direction and beam 2 ahead of it. With such a system, separate recorders are used for beam 1 and beam 2, and the radar is

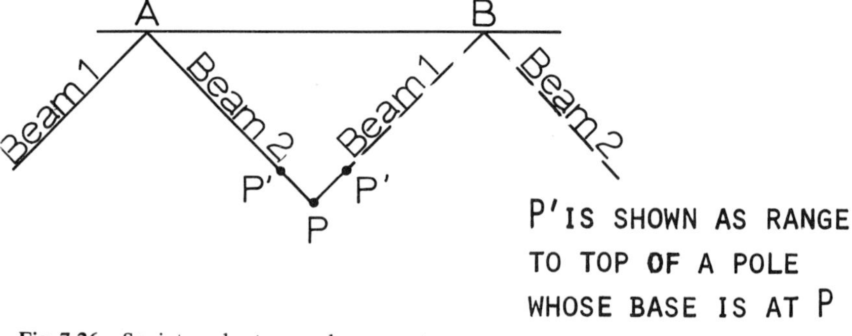


**Fig. 7.26.** Squint-mode stereo radar geometry.

0-201-04245-2

switched alternately between them in synchronism with the antenna switching. The film is moved past the single line recorder at an appropriate slant angle, so that the recorder paints a picture of the ground in the proper geometry. By comparing the pictures made of point *P* with beam 2 for the aircraft at point *A*, and beam 1 for the aircraft at point *B*, the interpreter can learn a great deal more about the ground than if he had only an ordinary SLAR.

This same technique can be used to achieve stereo height measurements. If the point at *P* is above the mean ground level, it will appear closer to point *B* along the line of beam 1 and to point *A* along the line of beam 2. Thus, since the same point appears displaced in different directions on the two images, its height can be determined from its displacement.

### B. Real-Aperture System

Figure 7.24 illustrates the general arrangement for a real-aperture SLAR. Figure 7.27 shows the geometry important in determining the resolution for such a system. Clearly the illuminated area has the dimension:

$$r_a r_r = R\phi_0 \left( \frac{c\tau}{2 \sin \theta} \right),$$

an arrangement identical with that for the pulse scatterometer. As the slant range $R$ increases, the azimuth resolution degrades while the range resolution improves because of the increasing value of $\theta$. Thus, close to the aircraft, such a radar may have a narrow azimuth resolution but a wide range resolution, whereas far from the aircraft the situation is reversed.

The radar equation for an area-extensive target was given by Eq. (7.4). With the dimensions indicated above, this becomes

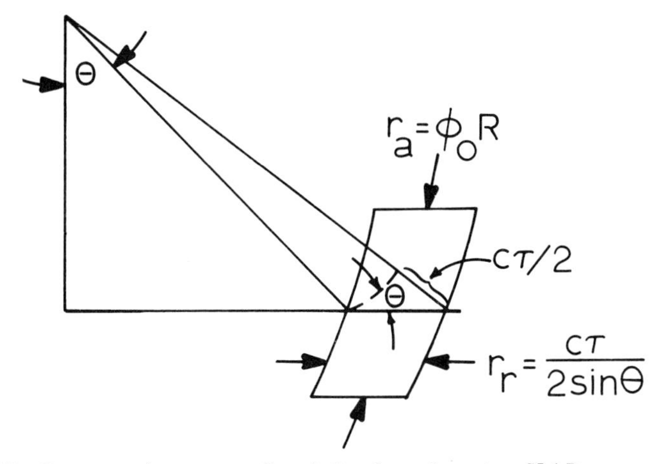


**Fig. 7.27.** Exaggerated geometry of resolution for real-aperture SLAR.

0-201-04245-2

$$W_r = \frac{W_t G^2 \lambda^2 \sigma^0}{(4\pi)^3 R^4} \left( \frac{\phi_0 R c\tau}{2 \sin\theta} \right). \tag{7.28}$$

The radar records the scattering coefficient either qualitatively on an image or, if the system is calibrated, quantitatively. Thus Eq. (7.28) can be rewritten in the form

$$\sigma^0 = \left( \frac{W_r}{W_t} \right) \frac{2(4\pi)^3 h^3 \tan\theta}{G^2 \phi_0 \lambda^2 c\tau \cos^2\theta}. \tag{7.29}$$

To obtain a given value of received power, a larger scattering coefficient is required as the angle of incidence increases toward grazing. This is due to the combination of the effects of increasing range and improving range resolution. Thus, to achieve a constant sensitivity relating received power to scattering coefficient, the antenna gain variation with $\theta$ is ordinarily adjusted to compensate for the angular variation in this equation. For systems optimized for point targets, the gain varies as $\csc^2\theta$. To correct for area-extensive targets as indicated in Eq. (7.29), it would have to vary as $\csc\theta\,(\cot\theta)^{1/2}$. Sometimes this compensation is obtained by varying the gain of the receiver with range, rather than by shaping the antenna pattern. Such a system is called a sensitivity-time-constant (STC) circuit.

Achieving fine azimuth resolution is the major problem with real-aperture radar because

$$\phi_0 \approx \frac{\lambda}{L},$$

where $L$ is the antenna length, and

$$\text{Azimuth resolution} = r_a = \phi_0 R \approx \frac{\lambda R}{L}. \tag{7.30}$$

A practical upper limit for the design and construction of air-borne antennas is 5m, 500 wavelengths at 1 cm. Hence the azimuth resolution is always greater than $R/500$ m. Thus, if the range is 10 km, the azimuth resolution is 20 m or worse.

As an example of *range resolution,* we consider a system that looks only at angles with the vertical larger than 30°. In this case

$$\sin\theta \geqslant 0.5,$$

so that the range resolution $r_r$ is given by

$$\frac{c\tau}{2} < r_r \leqslant c\tau. \tag{7.31}$$

0-201-04245-2

For a pulse 0.1 μsec long, this means that the range resolution varies between 15 and 30 m, from grazing (90°) to 30° respectively.

Pulses significantly shorter than 0.1 μsec can be achieved, but they require a very high peak transmitter power to achieve adequate signal-to-noise ratio. Pulse compression permits the use of longer pulses with lower peak power to achieve the same range resolution. Thus pulse compression is often needed for fine-resolution SLAR's as well as for moderate-resolution high-power radars of any kind. Figure 7.28 illustrates the principle of a pulse compression system that uses a pulse whose frequency is swept linearly (chirped) during transmission. The frequency of the transmitted pulse is shown in Figure 7.28*a* as it would be received from a point target. This pulse is sent through a frequency-sensitive delay line in the receiver. The delay line could theoretically have zero delay at $f_2$ and a delay equal to the pulse duration at $f_1$. Thus the last return to come in from the target is not delayed, but the earliest return is delayed, so that it leaves the delay line at the same time as the last return. Hence all outputs from the delay line appear at the same time, as shown in Figure 7.28*c*.

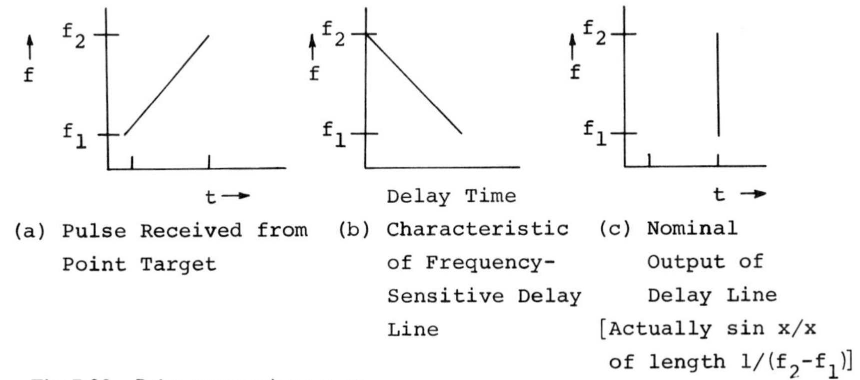


**Fig. 7.28.** Pulse compression system.

The figure idealizes the situation. In fact, using the band width $B$ given by

$$B = f_2 - f_1,$$

the output pulse turns out to be equivalent to a short pulse of duration

$$\text{Equivalent short-pulse duration} = \frac{1}{B}. \tag{7.32}$$

For the linear sweep illustrated, the short pulse has the shape of a sin $x/x$ curve. Various modifications of the characteristics of the system can produce outputs of different shapes.

0-201-04245-2

As an example of such a system, we consider the use of an actual pulse length of 1 $\mu$sec with a band width $B$ equal to 50 MHz. The equivalent pulse length is then $^1/_{50}$ $\mu$sec, which corresponds to a slant range resolution of 3 m instead of the 150 m that goes with the 1-$\mu$sec pulse.

The block diagram for the real-aperture imaging radar is the same as that shown in Figure 7.15 for a pulse scatterometer. In fact, the same basic radar could serve as an imager or as a scatterometer. If the imager is calibrated by using the power monitor and calibrate-pulse source in Figure 7.15, it can indeed be considered a scatterometer. Thus the primary difference between the system discussed in Section III.C and that considered here is the orientation of the antenna and the method of recording. With the imaging radar the received signal is used to intensity-modulate a line on a cathode ray tube, and a recording is achieved by pulling a film past this line.

Good radar images require careful matching of the characteristics of the IF amplifier, the video amplifier, the CRT, and the film density to the expected range of variation of the incoming signal. Figure 7.29 illustrates the elements of this problem in terms of the transfer characteristics at different points in the system.

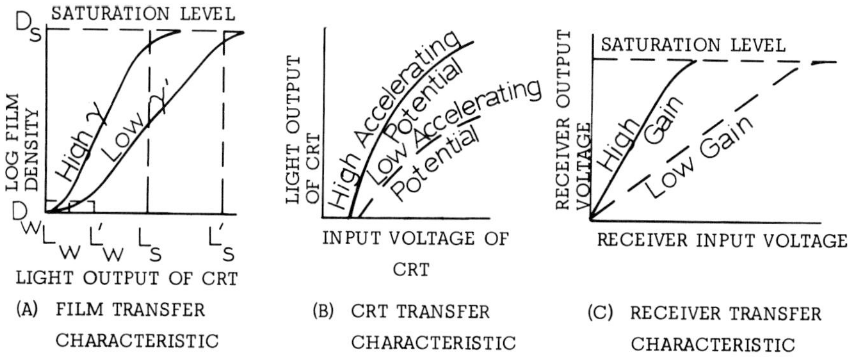


(A) FILM TRANSFER CHARACTERISTIC (B) CRT TRANSFER CHARACTERISTIC (C) RECEIVER TRANSFER CHARACTERISTIC

**Fig. 7.29.** Elements in determining transfer characteristic for recording.

The density of the recorded image on film varies exponentially with the light intensity over some limited range. Thus a plot of the logarithm of film density versus light output, as in Figure 7.29*a*, contains a linear region whose slope $\gamma$ is the exponent relating the film density to light intensity. For weak light the film density increases more slowly, and for bright light a saturation phenomenon takes place. Thus the usable "dynamic range" of film lies between $D_s$ and $D_w$ on Figure 7.29*a*. Depending on $\gamma$, this corresponds to different amounts of light output from the CRT. If radar returns from weak targets are not to be lost, the light intensity of the CRT should be $L_w$ for zero input signal, which can be achieved by adjusting the intensity control on the CRT (just as brightness is adjusted on an ordinary television screen). The equivalent photographic operation is "prefogging."

Figure 29*b* shows typical transfer characteristics of CRT's. These are nonlinear, and the intensity for a given input voltage depends on the acceleration potential

0-201-04245-2

used in the CRT. With proper adjustment of the acceleration potential and of the input voltage level in the actual signal, the CRT transfer characteristic can be matched to that of the film.

The receiver transfer characteristic is shown in Figure 7.29*c*. Most receivers produce an output voltage proportional to the input (or to the square of the input) for weak signals, but strong signals cause them to saturate as shown. The gain setting determines the input signal for which saturation occurs. An appropriate match between the gain and the expected range of signal intensities entering the receiver is necessary if saturation is to be prevented. If the gain is too low, however, saturation may be prevented for very strong signals, but there is little difference in output for signals from targets between which one wishes to differentiate. Therefore some kind of compromise may be necessary. Proper operation of the imaging radar demands that these characteristics be taken into account by the operator or that the system be designed to handle them automatically.

The dynamic range of radar or other electronic systems is, in fact, the ratio of the signal that causes saturation to the noise level. The statement is frequently made that the "dynamic range" is inadequate for one system and adequate for another, when in fact both may be the same. This happens because the numbers of apparent shades of gray discernible may differ, either because of improper matching of the transfer characteristics illustrated in Figure 7.29 to the range of the incoming signal, or because of differences caused by varying amounts of averaging of fading signals, as discussed below.

The dynamic range of the system with linear amplification can be made to fit within the usable range of the film. If this is done, small variations in radar input signal may not be detectable on the film, because most of the film characteristic has been reserved for recording the unusual large signals. Because radar signals vary over such a wide range, a more common arrangement is to use an IF amplifier whose output is proportional to the logarithm of the input. Widely varying signals are compressed into a smaller variation at the output, and any two signals having the same *ratio* at the input have the same absolute *difference* in brightness at the output. This makes more pleasing images than those produced by linear IF amplifiers, but it means that some significant input differences are obscured because they are so slight at the output. Since these differences are recorded on the original film, subsequent photographic processing of copies with different $\gamma$'s may restore the useability of the data. The film can also be scanned electronically, and the significant range expanded and rerecorded. Because of this problem, logarithmic IF amplifiers are almost universally used in real-aperture SLAR's. Unfortunately they cannot be used in the synthetic-aperture systems described below.

### *1. Gray-Scale Variations*

0-201-04245-2

Signals received by the imaging radar have random amplitude distributions similar to those for the scatterometer, and for the same reason: the amplitude of the gray-scale variations for a radar imager using one independent sample from each resolution cell is also given by the Rayleigh distribution. The Rayleigh distribution in amplitude corresponds with the exponential distribution in power given in Eq. (7.23). The number of independent samples required to reduce the variance of the return has been analyzed for the scatterometer. The same results apply here, but the side-looking radar has a different fading band width because of the different geometry.

The Doppler band width for a side-looking radar whose azimuth resolution is $r_a$ is given by

$$\text{Doppler band width} = \frac{2r_a v}{R\lambda}, \tag{7.33}$$

which was derived by applying the geometry of the problem to Eq. (7.12). The period of the fading band width before detection is therefore the reciprocal of this value. This Doppler frequency would be observed with a continuous wave (CW) system transmitting to the resolution cell in question. With a pulse system, the amplitude of each pulse received from the resolution cell is a sample of the fading signal that would be observed by the CW system. If the pulse repetition rate is high, many samples are observed per cycle of even the highest frequency within the Doppler spectrum. Only two samples per cycle are required to specify a signal. Additional samples may not contribute significantly to the averaging since they are not independent. Even if each resolution cell is hit by a hundred pulses while the radar passes, this does not mean that a hundred independent samples have been observed: only one may have been observed, and the consequent brightness is governed by the Rayleigh distribution rather than by the average of a hundred Rayleigh-distributed variables. Thus it is most important to calculate the proper number of independent samples to determine what variance is to be expected for the return.

The time for the radar to pass the azimuth resolution cell is given by

$$\text{Time to pass a cell} = \frac{r_a}{v}. \tag{7.34}$$

It can be shown (Moore, 1966) that the use of square-law detection broadens the band width so that the resulting spacing between independent samples is given approximately by

$$N_i = \frac{4\,(\text{time to pass a cell})}{\text{period of predetection band width}} = \frac{8r_a^{\,2}}{R\lambda}, \tag{7.35}$$

which combines Eqs. (7.33) and (7.34). For an idealized antenna pattern $r_a = R\phi_0$, so we may write Eq. (7.35) as

$$N_i = \frac{8\phi_0^{\,2}R}{\lambda}. \tag{7.36}$$

0-201-04245-2

The number of independent samples for a real-aperture system is, therefore, greater at longer ranges, so the gray scale should be better. This happens only because the azimuth resolution is poor at these larger ranges. In other words, a tradeoff occurs between reduction of gray-scale variability (which requires many samples) and resolution cell size. This is not unlike the uncertainty principle applied to the radar scatterometer.

As an example of the number of independent samples involved in producing a real-aperture radar image, we consider the following: $r_a = 20$ m, $R = 10$ km, $\lambda =$ 1 cm. Applying Eq. (7.35) gives us $N_i = 32$, from which the fading range is, according to Figure 7.19, 2.5 dB. Unfortunately, this derivation and example apply only to the best case, in which the scattering coefficient is uniform across the entire resolution cell, and only the random variability associated with the Rayleigh distribution is involved in determining independent samples. If our target, having an azimuth resolution of 20 m, consisted of a weak background with a few points of strong return, all within 5 m of each other, we would have to use 5 m rather than 20 m for $r_a$ in Eq. (7.35). The result would be only two independent samples averaged! Hence the range from the 5% to the 95% probability for this signal is 11.2 dB or a ratio of power of 13.3.

The difficulties with fading in the imaging radar (or in any other radar for that matter) could be overcome by building radars which average in frequency. A radar which averages returns at different frequencies to reduce the variance of signal intensity is called *panchromatic* (Moore et al., 1969). Commercially available SLAR's operate at single frequencies broadened only by the spectra associated with their pulses. In optics, this is analogous to illumination by lasers, which are also single-frequency devices. In both cases the resulting images are "speckled" because of the random fading. Full-scale examples of this effect are hard to find to illustrate the point, but Figure 7.30 shows a contrived example: a panchromatic aerial oblique photograph and another photograph made by illuminating the negative of the original with laser light. Here the speckled nature is due to variations within the negative rather than to variations in the actual target. The effect is the same. Clearly the panchromatic image is much superior for most purposes because it is not subject to speckling due to fading.

The frequency spacing between independent samples can be shown to be (Waite, 1970; Thomann, 1970)

$$\text{Frequency spacing of independent samples} \geqslant \frac{150}{r_r} \text{ MHz}, \tag{7.37}$$

where $r_r$ is the range resolution in meters. Targets 5 m apart correspond to independent samples every 30 MHz, whereas targets 20 m apart correspond to independent samples every 7.5 MHz. The previous example involved azimuth resolution, but here we refer to range resolution. This tells us that a band width of 1.5 GHz is required if we wish to reduce the 5–95% statistical range for a 1-m distance resolution to 4.6 dB, corresponding to 10 independent samples. On the other hand, if

0-201-04245-2

**Fig. 7.30.** Contrived example to illustrate effect of monochromatic and panchromatic illumination. A photographic transparency was illuminated with (*A*) coherent (6328 Å, laser) light (representing monochromatic radar); (*B*) incoherent (3200°K, incandescent) light (representing panchromatic radar).

0-201-04245-2

our resolution is 20 m, we can achieve the same result in improving the gray scale by going to a band width of 75 MHz, provided that the full 20 m is occupied by significant targets.

Range resolution appears in the equation for frequency spacing for independent samples because the range resolution achieved depends on the band width. A range resolution of 15 m requires a 0.1-$\mu$sec pulse and a receiver band width of 10 MHz. To improve the range resolution to 1.5m, the pulse must be shortened or pulse compression used; in either case a band width of 100 MHz is required. Hence another way to achieve the independent samples associated with the panchromatic radar is to use the band width to obtain finer range resolution than desired, followed by averaging into a single coarser resolution cell. Implementation depends on the system considered; and the averaging may take place in the detector, in the CRT display, on the film, or after the film is returned, depending on the system chosen.

Calibration of the radar imager is desirable, although few such systems have been calibrated in the past. As indicated by Figure 7.15, a sample of the transmitted pulse must be observed and a calibration pulse must go through the receiver. Since the receiver is normally turned off during the time the transmitter operates, a separate device is required for sampling the transmitter signal. The receiver itself cannot be calibrated only with a single amplitude pulse, which would give a measure of the gain of the system. Rather, the calibration must permit reproduction of the transfer characteristics of Figure 7.29. These transfer characteristics may be different at different ranges if the receiver gain is adjusted to compensate for the variation in signal strength with range.

Often this adjustment is made using a shaped antenna pattern, but calibration then requires accurate knowledge of the antenna pattern. This is difficult to obtain for antennas mounted on aircraft.

### 2. *Effect of Aircraft Instability*

The effect of aircraft instability is much the same on the production of real-aperture SLAR images as on the production of images with slit cameras and electromechanical scanners. Fortunately, minor amounts of roll do not significantly affect the radar image, for the transverse position of a target in the image is determined by the actual time delay, not by the attitude of the aircraft. Thus the only effect of roll is to modify the antenna gain in any particular direction, and this effect is usually small. On the other hand, both pitch and yaw of the aircraft cause serious problems comparable to those encountered with line scanners. Short-period rapid variations in yaw produced by turbulence cause the aircraft radar to vary in side-look direction more quickly than motion compensation equipment can adjust. Thus the radar does *not* look consistently at the same angle to the side; some areas may be "seen" twice, and others not at all. Very-high-drift angles caused by a major cross wind may exceed the capacity of the radar compensation system, and skewed or slanted imagery then results.

0-201-04245-2

## C. Synthetic-Aperture Systems

The use of a synthetic aperture in SLAR permits much finer azimuth resolution than is possible with a real aperture. Thus a synthetic aperture is necessary to achieve resolutions suitable for many earth science applications if the radar is to be at a long distance from the target. The concept may be considered in either of two rather different ways. The name "synthetic aperture" implies that the radar sequentially occupies the different elements of an array, storing the phase information, and the stored data are combined later in a computer. Another approach describes the synthetic aperture as "Doppler beam sharpening"* and regards it as a system that combines velocity measurement and range measurement and does not use its real aperture for resolution at all (Cutrona, 1961).

The radar equation for the synthetic-aperture system may be obtained by approximating Eq. (7.4). Thus it is given by

$$W_r = \frac{W_t G^2 \lambda^2 \sigma^0 r_a r_r}{(4\pi)^3 R^4} = kTBFS_i ,$$

where the received power has been shown both in terms of the radar equation and in terms of the signal-to-noise ratio and noise level in the receiver. The central term in this equation is like Eq. (7.4) with everything assumed constant across the illuminated area. The area itself is given by $r_a r_r$, the product of azimuth and range resolutions. The receiver noise is $kTBF$, where $k$ = Boltzmann's constant, $T$ = temperature (°K), $B$ = band width (Hz), $F$ = receiver noise figure. The quantity $S_1$ is the signal-to-noise ratio required for a single pulse return. Because of the way in which the synthetic aperture builds up, using many pulses, this is not the same as the final output signal-to-noise ratio. The output signal-to-noise ratio after $n$ pulses is $S_n = nS_1$. The number of pulses used in making the aperture is equal to the number of pulses per second divided by the velocity and multiplied by the length of the aperture. Performing appropriate substitutions, we can obtain a value for the final signal-to-noise ratio in terms of the radar system and ground parameters:

$$S_n = \frac{W_t p G^2 \lambda^3 r_r{}^2 \sigma^0 \sin\theta \cos^3\theta}{(4\pi)^3 v h^3 kTF} . \tag{7.38}$$

Here the only new parameter is $p$, the pulse repetition rate in pulses per second.

This equation can take many forms, depending on which parameters are held constant and which are considered appropriate to vary. For example, the antenna gain used here is proportional to the length of the antenna and inversely proportional to the swath width. Hence it is possible to write the equation in terms of these parameters. Furthermore it is often more important to know about the aver-

*This term is used by C. A. Wiley, who is the apparent inventor of the technique.

0-201-04245-2

age power rather than the peak transmitted power, $W_t$. A transformation to average power is easy in this form of the equation, since the average power is simply the product of peak power, pulse duration, and pulse repetition rate. On the other hand, the pulse duration determines $r_r$, so even this complicates the equation.

The pulse repetition rate that must be used in producing a synthetic aperture may limit the allowable swath width. On aircraft this is ordinarily not a significant problem, but for a spacecraft the limitation may be severe. The PRF must be at least high enough to obtain two samples for each cycle of the highest Doppler frequency.

The azimuth resolution that can be obtained with a synthetic-aperture system depends on the type of processing used. An *unfocused* synthetic aperture (really focused at infinity) cannot achieve the same resolution as a *fully focused* system. On the other hand, the fully focused system must be separately focused at the different ranges if it is to be used for wide swaths. The theoretically feasible resolutions are given by

$$\begin{aligned} &\text{Real aperture:} && r_a = \frac{\lambda R}{L}, \\ &\text{Unfocused synthetic aperture:} && r_a = \frac{1}{2}\sqrt{\lambda R}, \\ &\text{Focused synthetic aperture:} && r_a = \frac{L}{2}. \end{aligned} \tag{7.39}$$

Here $L$ is the length of the actual antenna.

The resolution for the real-aperture system is included here for comparison purposes. The unfocused synthetic-aperture resolution is only approximate; other values as high as twice the one listed here have been quoted, depending on what assumptions are made.

The unfocused synthetic aperture obtains an azimuth resolution that works out to be essentially the same as the length of the synthetic aperture used. Clearly from the formula it is related to the Fresnel zone; the resolution increases with range, but not as rapidly as that for the real aperture. Surprisingly, the size of the physical antenna (and hence its gain) has nothing to do with the azimuth resolution finally achieved.

The fully focused synthetic aperture theoretically achieves a resolution half the length of the real antenna, and independent of both range and wavelength. Ordinarily azimuth resolution can be improved by using a longer antenna; with the fully focused synthetic aperture, however, the opposite is true: resolution is improved by shortening the antenna. In practice this may not be the case for really short antennas because of stability problems discussed below. Nonetheless, in theory the resolution for a 2-m-long antenna would be 1 m, whether the antenna was used on an aircraft for a range of 2 km or on a spacecraft for a range of hundreds of kilometers! This happens because the length of the processed aperture increases

0-201-04245-2

directly with range so that the beam width of the synthetic aperture decreases with range, and the resolution cell remains the same regardless of distance from antenna to target.

Tradeoffs between resolution, wavelength, and antenna size are illustrated in Figures 7.31 and 7.32 for real-aperture, unfocused, and fully focused synthetic-aperture systems based on the theoretical resolutions of Eq. (7.39). Since the complexity of a radar system increases with the degree of synthetic-aperture focusing required, the user who does not need it is well advised to consider the use of a real-aperture or a partially focused synthetic-aperture system.

The synthetic-aperture resolution of Eq. (7.39) can be achieved only with an ideal stable radar and air-borne platform. The stability of the radar system itself is easier to achieve than that of the platform, since aircraft are subject to both random motions due to turbulence and steady motions such as drift. Hence the actual synthetic-aperture resolution is not likely to be as good as half the antenna length.

The magnitude of the stability problem may be illustrated by considering an example. To do so, we must determine the length of the synthetic aperture, $L_s$. For a synthetic aperture Eq. (7.30) becomes

$$r_a = \frac{\lambda R}{2L_s} \,. \tag{7.40}$$

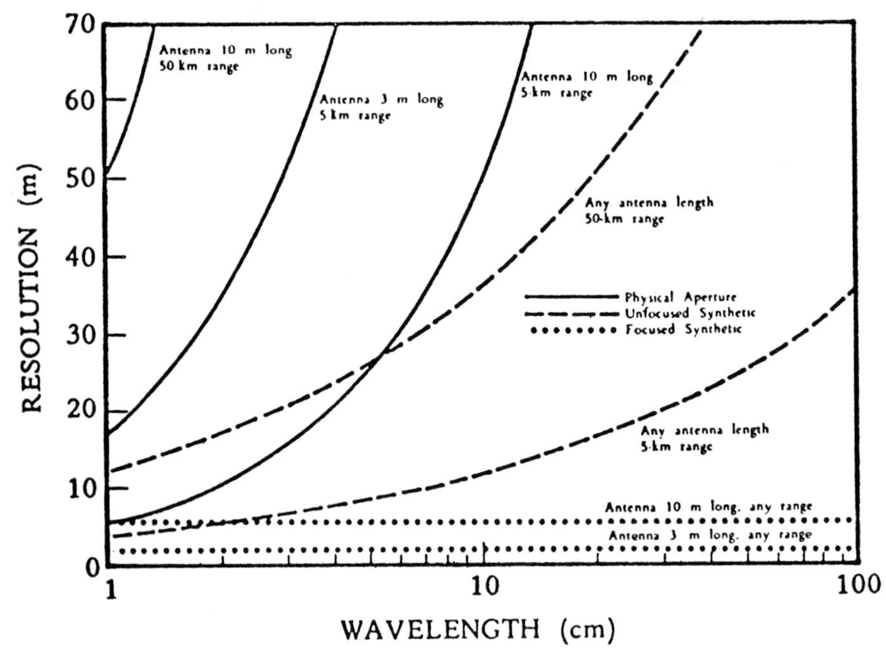


**Fig. 7.31.** Design tradeoff information for SLAR's. By permission, from Committee on Remote Sensing for Agricultural Purposes, Agricultural Board, National Research Council, *Remote Sensing: With Special Reference to Agriculture and Forestry*, National Academy of Sciences, Washington, D.C., 1970.

0-201-04245-2

It will be noted that the azimuth resolution for the synthetic aperture is half that for the real aperture because the synthetic-aperture beam width depends on the round-trip phase difference, while the latter depends on the one-way phase difference (Cutrona, 1961).

An example of the length of synthetic aperture may be based on the following parameters:

| | |
|---|---|
| Range: | 30 km |
| Wavelength: | 3 cm |
| Azimuth resolution: | 10 m |

In this case the synthetic-aperture length is 45 m, and the real-aperture length would be 90 m. The stability requirement is that the phase difference due to random motion along the 45-m path differ by no more than 45°. This means a deviation from a straight path of no more than 3.75 mm! Some compensation for this is possible if information is available from accelerometers or other sensors associated with the motion of the aircraft. Nevertheless, this is clearly a very stringent requirement. If full focusing had been called for and the antenna length had been 4 m, the

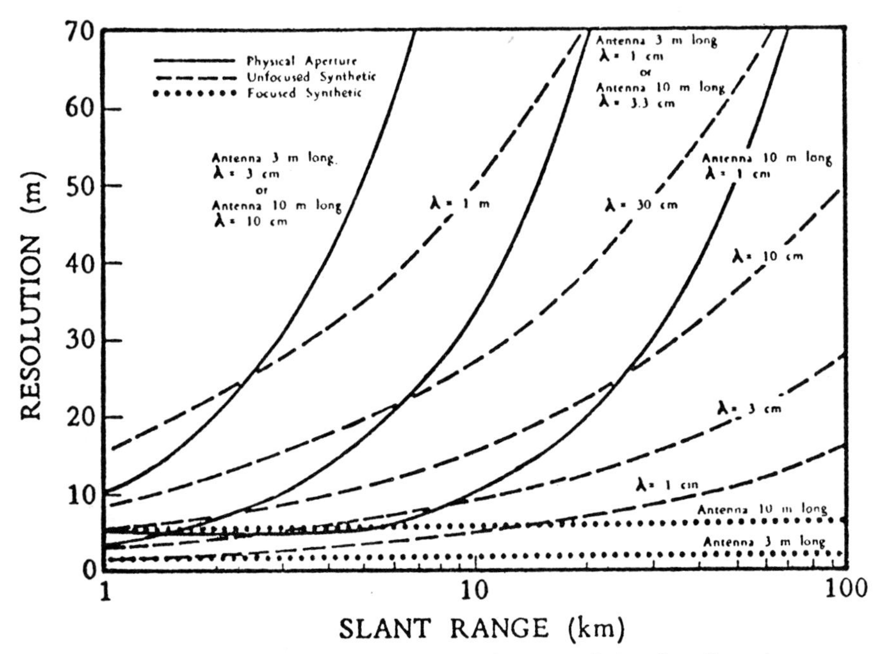


**Fig. 7.32.** Design tradeoff information for SLAR's. By permission, from Committee on Remote Sensing for Agricultural Purposes, Agricultural Board, National Research Council, *Remote sensing: With Special Reference to Agriculture and Forestry,* National Academy of Sciences, Washington, D.C., 1970.

0-201-04245-2

synthetic-aperture length would have been 112.5 m, and the 3.75 mm would have had to be maintained over this distance of travel.

The stability of the radar itself is better treated in terms of the Doppler frequencies. The Doppler band width was given by Eq. (7.33). For this example, if we assume a speed of 180 m/sec, the band width is given by

$$B_d = \frac{V}{L_s} = \frac{180}{45} = 4 \text{ Hz}. \tag{7.41}$$

This would appear to indicate that a stability of the order of 4 parts in $10^{10}$ would be required, because that is the ratio of Doppler band width to transmitted carrier frequency. Actually the situation is not this severe, for what really matters is the drift between transmission and reception of a single pulse, and the time involved for this is much less. For our 30 km, this time is only 200 μsec. If we assume a frequency drift permitting an 18° phase shift in this time, the error comes out to be 1 part in $4 \times 10^7$. Temperature-stabilized crystal oscillators can readily achieve this stability. No such stability requirement exists for the real-aperture system, and thus the synthetic-aperture radar is more complicated.

### *1. Synthetic-Aperture System Implementation (Harger, 1970)*

The elements of a synthetic-aperture system are shown in Figure 7.33. Of course, each block in the system illustrated is in fact a complex subsystem itself. For the synthetic-aperture system to store phase information, a "coherent detector" is required. The coherent detector is in fact not a detector at all, but rather a mixer in which the received signal is mixed either with a sample of the transmitted signal or with a frequency very close to that of the transmitter. The result is translation of the received carrier frequency down to a quite low frequency or to zero frequency. The detailed block diagram for a system with zero output carrier frequency is the same as that of the homodyne Doppler scatterometer (Figure 7.17). The more com-

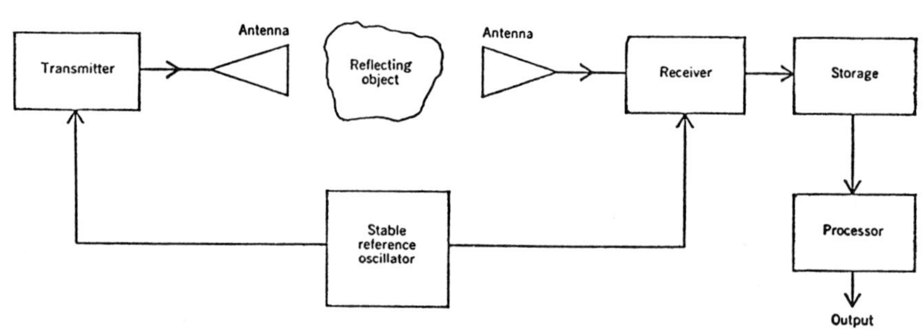


**Fig. 7.33.** These pieces collectively constitute a synthetic-aperture radar. By permission, from William M. Brown and Leonard J. Porcello, An Introduction to Synthetic-Aperture Radar, *IEEE Spectrum* 6(9) (September 1969).

0-201-04245-2

mon arrangement involves mixing to a low-frequency carrier, in which case the system is more like that of Figure 7.18. In each case complications occur because of the use of a pulse transmitter for the synthetic-aperture system rather than the CW transmitter used with the scatterometer. More details concerning this sort of *pulse-Doppler* radar are given in texts on radar systems (Skolnik, 1962, 1970; Barton, 1964).

The type of storage used depends on whether the synthetic-aperture processing is to be performed immediately or on return of the data to a ground station. With immediate processing the information may be stored in a recirculating delay line, which may also serve as part of the processor. Storage can also be achieved by separately gating returns from each range element into a capacitor with a long decay time constant. A technique for temporary storage that appears more suitable for a fully focused processor involves the use of one or more storage tubes.

Storage for later processing can be achieved by recording the signal on magnetic tape if the tape recorder has sufficient band width. This is reasonable with the present state of the art as long as the range resolution is not too fine. Recording the received signal on film with the same type of recorder employed for the real-aperture system permits use of a wider band width than does magnetic tape. At present this is the most commonly used method for focused synthetic-aperture systems.

The film recording of the output of the coherent detector is like a hologram in that the phase information is contained in interference fringes on the film (Brown and Porcello, 1969). This point is illustrated by Figure 7.34, which shows how a hologram is constructed. With the hologram a sample of the illuminating beam is mirrored onto the photographic plate, where it interferes with the scattered light from the object viewed to produce the well-known holographic interference fringes. In the radar the same effect is achieved by the homodyne receiver, in which the interference is between the sample of the transmitter signal and the scattered signal received on the antenna.

Just as an image is reproduced from the hologram by illuminating it with monochromatic light, so an image can be produced from the radar film by similar illumination. Figure 7.35 shows a sample arrangement for this purpose. The conical lens in front of the film takes care of the need to focus the synthetic aperture separately for each of the different ranges across the film (in the $y$ direction). The spherical lens performs the holographic transformation, and the cylindrical lens is necessary because no hologram is present in the range direction. The synthetic-aperture image of a single line transverse to the aircraft is viewed through the slit in the opaque screen at $P_2$.

The film recorded at the output of the synthetic-aperture receiver, which is in effect a microwave hologram, is called *signal film*. Like an optical hologram, it is an interference pattern which is difficult to interpret when viewed by itself. Thus it must be processed in an arrangement like that of Figure 7.35, or by some electronic scheme, to produce an *image film* for use by the interpreter.

0-201-04245-2

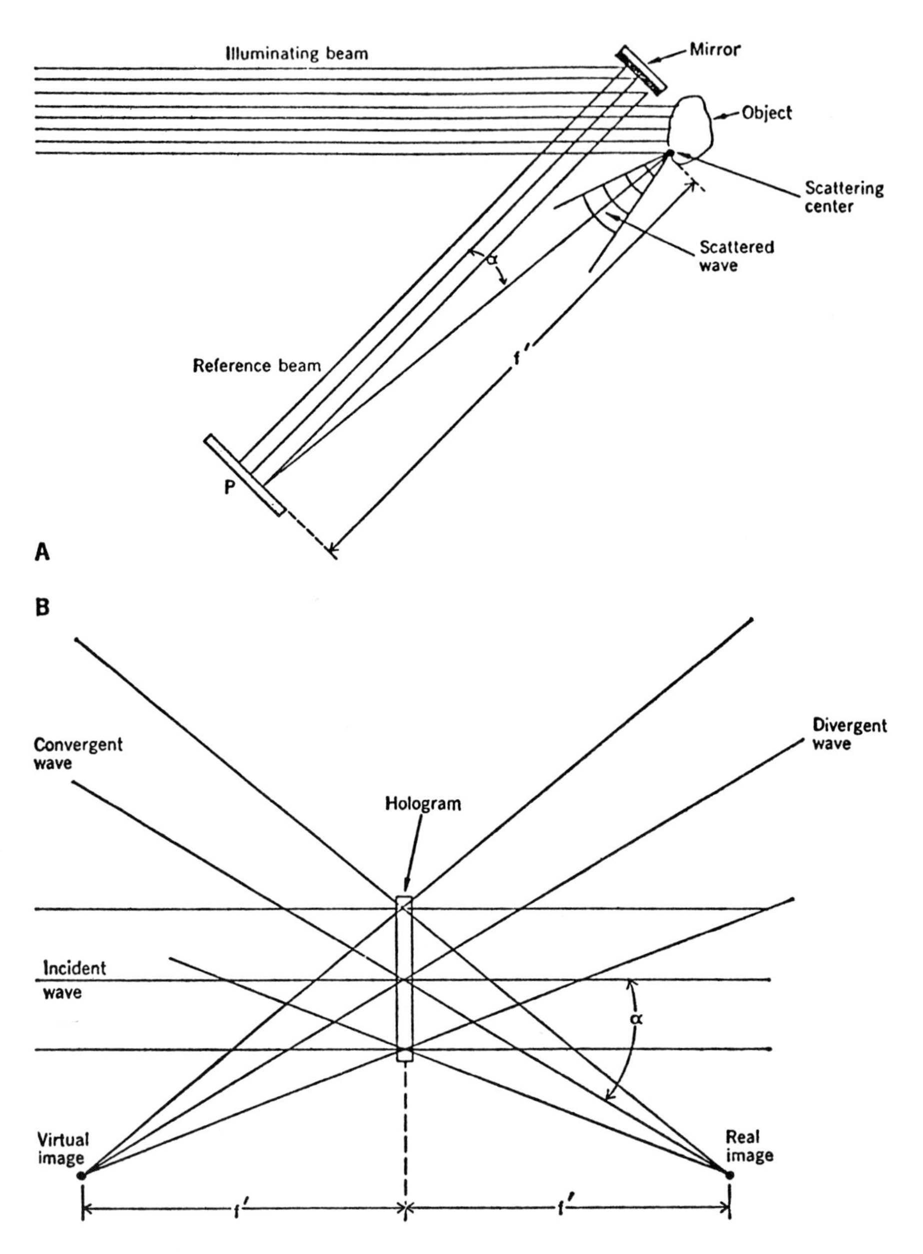


**Fig. 7.34.** Top diagram (*a*) depicts conventional optical setup for producing a hologram. Hologram (*b*) can, in turn, reproduce both a virtual and a real three-dimensional image. By permission, from William M. Brown and Leonard J. Porcello, An Introduction to Synthetic-Aperture Radar, *IEEE Spectrum* 6(9) (September 1969).

0-201-04245-2

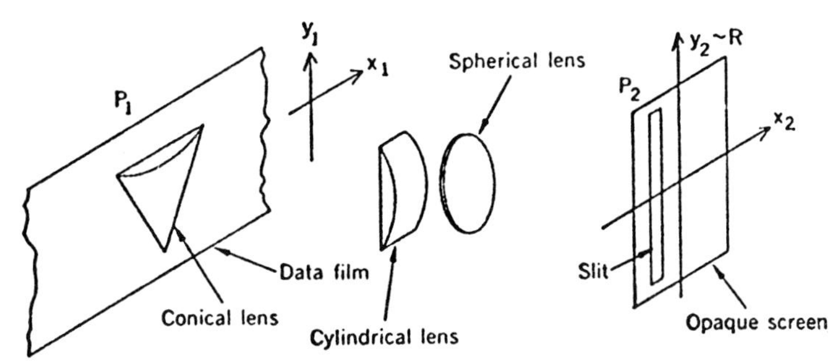


**Fig. 7.35.** With this optical arrangement, a view through the vertical slit at $P_2$ displays the output of the synthetic aperture for all ranges with the same along-track dimension. By permission, from William M. Brown and Leonard J. Porcello, An Introduction to Synthetic-Aperture Radar, *IEEE Spectrum* 6(9) (September 1969).

When pulse compression techniques like those illustrated in Figure 7.28 are used along with the synthetic aperture, the output of the receiver may be recorded on film without use of the frequency-sensitive delay line of Figure 7.28. The signal film in this case has holographic properties in two dimensions. Here a technique similar to that of Figure 7.35, but with a different arrangement of lenses, can be used to process both the azimuth and range interference patterns to produce the image.

### 2. *Gray-Scale Problems*

The problems of gray-scale discussed in connection with real-aperture systems are magnified when a synthetic aperture is used. The synthetic aperture produces only one independent sample per resolution cell. The very process by which the synthetic aperture is developed guarantees that averaging does not take place. Thus, if the Rayleigh distribution applies to the particular target, the image produced by the synthetic aperture necessarily has a gray scale governed by the Rayleigh distribution. With this wide variance of the signal brightness, transitions from one target type to another may be hard to recognize unless a homogeneous region contains many resolution cells; in this case the eye can perform a spatial averaging to come up with a mean gray level for each region.

The methods described to reduce the variance of the image for the real-aperture system can also be applied to the synthetic-aperture system. In essence these methods amount to processing for better resolution than will finally be used, followed by averaging some of these fine-resolution cells to produce a coarser one with a more representative gray value.

Another problem more important with the synthetic-aperture system than with the real-aperture system is detector saturation on strong signals. With the real aperture, a logarithmic IF gain characteristic can be used. The synthetic aperture, however, depends on the use of a linear amplifier to achieve adequate weighting when combining the signals from the various parts of the aperture. Although the dynamic range of the recorded signal is not as great as that of the processed image, a gain control setting that is adequate for open country will probably cause saturation when a city is being imaged. This problem can be handled better by recording on magnetic tape than by recording on film, because the tape has a wider usable range than the film, but frequency response limitations preclude the use of tape for systems with extremely fine resolution.

The difference in signal strength between strong and weak targets is so great that film exposures for *images* adequate to show differences in vegetated regions are bound to result in saturation on strong targets like buildings. Settings adequate for a city are likely to give poor results in the country. Thus multiple optical processing may be necessary to produce results suitable for both purposes, even though saturation may not be severe on the *signal* film.

### *3. Synthetic-Aperture Summary*

Synthetic-aperture radar can produce extremely fine resolution that is nearly independent of range. Thus, for long-range systems requiring fine resolution, the synthetic aperture is the only technique that can be used in the microwave region of the spectrum to obtain the kind of resolution desired. Vehicle stability problems require elaborate compensation schemes and stabilized antennas to achieve fine resolution. The synthetic-aperture system is much more complex to build and to operate than a real-aperture system; consequently, the latter is ordinarily preferred where it can do the job. On the other hand, the synthetic-aperture system is feasible and must be used when high resolution and long range are to be combined.

## D. Multispectral/Multipolarization Radars

Multispectral radar systems are likely to be just as useful as multispectral systems in other parts of the electromagnetic spectrum. Just how far apart the different bands should be remains to be established. Although a few preliminary studies have been performed on systems at different wavelengths, most of the analyses to date have been based on either 1- or 3-cm systems. In one case in which a set of wavelengths ranging from 1 up to 75 cm was available, large differences were found between the results with the various wavelengths (Dellwig, 1969).

A long-wavelength system can be expected to penetrate through a few centimeters to a few meters of soil, and to penetrate through most vegetation except very dense forest. Intermediate wavelengths are most sensitive to soil moisture and to heavy vegetation; the very short wavelengths, to finer vegetation structures and the very top layer of soil. Future systems may be expected to use three or more wavelengths, lying between about 1 m and 3 mm.

In principle such systems are easy to build, but in practice they are very costly. Producing a multiple antenna and duplicating most of the microwave components of the system are not simple tasks. Some of the electronic components can be shared, but others must be separate.

0-201-04245-2

The value of multiple polarization on radar systems operating in the 1-cm region has been well documented (Moore and Simonett, 1967). Clearly, future systems for remote sensing will, wherever possible, use multiple polarization. Many significant results have been obtained using a single transmitted polarization with reception of both the same and orthogonal polarization. This is relatively easy to achieve, since it requires only an additional antenna, receiver, and recorder without a separate transmitter or high-power microwave switch.

## E. Comparison of Imaging Systems

Imaging radar systems are likely to be far more useful to the earth scientist than nonimaging systems like scatterometers. Hence a summary of the features of the different imaging systems is in order.

The real-aperture system is easy to build but is limited in resolution. It must have an antenna whose dimensions are many hundreds of wavelengths if fine resolution is to be achieved at a reasonable distance. This means that the real-aperture system can achieve good resolution only at relatively short wavelengths like 1 cm and can never be expected to give good results at wavelengths as long as 10 cm. The gray scale is intrinsically better with the real-aperture than with the synthetic-aperture system because, under most conditions, each recorded resolution cell signal is the average of several independent samples. Furthermore, it is relatively easy to make a panchromatic real-aperture system.

The synthetic-aperture system is more complicated to build than the real-aperture system. Not only is the former system itself more complicated, but also special pains must be taken to compensate for the motion of the aircraft carrying the radar. On the other hand, a synthetic aperture is required if good resolution is to be achieved with an adequate swath width. Making a synthetic-aperture system panchromatic does not appear to be as easy as making a panchromatic real-aperture system. Nevertheless, if the synthetic-aperture system can be made with better resolution in either range and/or azimuth than is actually required, the fine cells may be averaged together to achieve a better gray scale at a coarser resolution.

Resolution requirements have been the subject of much discussion. A significant point often overlooked is that several cells are required for each category if the interpreter is to be sure of his identification of that category. If only one cell exists for a given category, its brightness may indicate the presence of a truly different category but may also be a random occurrence.

Resolutions of the order of 100 m appear adequate for some geologic and geomorphic structural studies and for determining such major vegetation patterns as the differences between forest and range land and the boundaries of large fields. In urban areas this kind of resolution may permit the location of regional boundaries for different land uses, but certainly not the identification of land use in areas where mixing occurs. This resolution may be adequate also for outlining major transportation networks and drainage systems.

To show agricultural patterns in most parts of the United States, as well as in other areas of the world where similar-sized fields are found, 30-m resolution

0-201-04245-2

should be adequate. This resolution will even provide moderate detail on urban land use and will suffice to detect most transportation networks and drainage patterns. It may also be adequate for information about changes in coastline environment and for shipping inventory. On the other hand, resolutions as good as 10 m or better will probably be required for detecting fields in areas in which many small fields exist. Furthermore, this sort of resolution may be necessary for picking up some of the finer road networks and certainly is required for detailed studies of urban areas.

In regard to long- versus short-wavelength radar imagery, the use of wavelengths of 3 cm or shorter seems to be justified for most purposes. On the other hand, long wavelengths may be quite superior for forestry and some geologic studies, particularly when used with multipolarization. Insufficient data have been collected at the longer wavelengths to permit a definitive statement. Clearly the best system is one that combines both long and short wavelengths with intermediate lengths as well.

The small extra cost associated with receiving an orthogonal polarization seems well justified in terms of the additional information available. Whether two polarizations need be transmitted is somewhat less certain, but the practice is probably also justifiable. If one had to choose between an additional transmitted polarization and an additional frequency, however, the additional frequency would almost certainly win out.

## V. THE FUTURE

Active microwave sensing systems of the future will almost certainly involve polypanchromatic (multispectral with each component panchromatic) systems operating at wavelengths between about 1 m and 3 mm. A typical system may well use four wavelength regions.

The imaging radar of the future will probably be a squint-mode device permitting orthogonal looks at a target with only one pass. Such a system also permits stereo viewing and stereoscopic height measurement.

Similar systems can be expected in orbiting vehicles. The swath width for the orbiting vehicles will probably be limited initially to 40 or 50 km, although that for aircraft may be greater. Later, wider-swath orbiting systems will become available.

Combined active and passive systems scanning beneath the aircraft will undoubtedly make their appearance before long. These systems will provide information in a geometry very similar to that for the electromechanical scanner or the slit camera. They will be usable for imaging purposes only at relatively low altitudes because of antenna size limitations. On the other hand, they will be employed at spacecraft altitudes for monitoring the oceans and clouds.

0-201-04245-2

Chapter 8

# Airborne Geophysical and Miscellaneous Systems

*A. R. Barringer*

Airborne geophysical systems first appeared soon after World War II. They represent one of the first nonmilitary applications of remote sensing techniques from aircraft (apart from aerial photography) and have had a profound impact on geological mapping and mineral exploration during the past 20 years. Present-day exploration usage is confined mainly to various combinations of aerial photography, aeromagnetic mapping, electromagnetometer surveying, and gamma-ray mapping; new sensors are emerging from the laboratory, however, and it is likely that a considerably broader range of techniques will come into use during the next few years. Furthermore, airborne survey methods will be supplemented by orbital surveys from satellites, and it seems probable that there will be close integration of airborne surveys with satellite remote sensing and imaging techniques. This integration will result from the fact that satellites and aircraft are often best suited for different types of measurements, which for many applications are highly complementary to each other.

The broad variety of sensors having potential application in airborne exploration systems includes not only conventional geophysical techniques such as magnetometers and electromagnetometers, but also many imaging devices (e.g., optical line scanning equipment and side-looking radar). Since these devices are dealt with elsewhere, this chapter will be confined to the consideration of nonimaging systems for studying the earth's surface and subsurface as well as the trace composition of the overlying atmosphere. This equipment is applicable not only to mineral exploration but also to many other practical problems, such as studies of engineering geology, location of geothermal areas, permafrost mapping, and surveys of the distribution of air pollution. A chart of the nonimaging techniques described in this chapter is shown in Table 8.1, together with current and potential applications. Salient features of these systems will be described under appropriate headings.

## I. AIRBORNE MAGNETIC SYSTEMS

At present the majority of aeromagnetic surveys are carried out with systems which measure the magnitude of the local terrestrial magnetic field with an accuracy of $\pm 1\gamma$. In the past the principal apparatus used was the fluxgate airborne magnetometer, perfected during World War II for submarine detection and first tested for geological applications by H. E. Hawkes of the U.S. Geological Survey in 1943 (Jensen, 1945). Gulf Research and Development made the equipment available for

0-201-04245-2

*Remote Sensing of Environment*, Joseph Lintz, Jr. and David S. Simonett (eds.) ISBN 0-201-04245-2

**TABLE 8.1**
**Applications of Nonimaging Sensors for Geophysical Exploration, Geological Mapping, and Air Pollution Studies**

| *Application* | *Magnetometer* | *Electromagnetometer* | *VLF H Field* | *VLF E Field* | *Natural EM Fields* | *Gamma-Ray Spectrometer* | *Radiometers–IR and Microwave* | *Air Samplers* | *Correlation Spectrometers* |
|---|---|---|---|---|---|---|---|---|---|
| Porphyry copper exploration | X | | X | X | X | X | | X | X |
| Lead-zinc exploration | | | X | X | | | | X | |
| Precious metal exploration | | | X | | | X | | X | |
| Massive sulfide exploration | X | X | X | X | | | | X | |
| Diamond exploration | X | | X | X | | X | | | |
| Manganese and hematite exploration | | | | | X | | | X | |
| Uranium exploration | | | X | | | X | | X | |
| Geothermal exploration | X | | X | X | X | X | X | X | |
| Bulk mineral exploration | | | | X | | X | | | |
| Placer exploration | | | | X | | | | | |
| Permafrost mapping | | | | X | | | X | | |
| Water resource investigations | | | | X | X | | | | |
| Structural geological mapping | X | | X | X | X | | | | |
| Sedimentary mapping | X | | | X | | X | | | |
| Dam site and engineering geological studies | X | | X | X | X | X | | | |
| Subsurface cavity detection | | | | X | | | X | | |
| Water pollution mapping | | | | X | | | X | | |
| General geological mapping | X | | X | X | X | X | | X | |
| Air pollution mapping | | | | | | | | | X |

commercial aeromagnetic surveys, and large-scale use commenced shortly after the war (Wyckoff, 1948). In 1954 Packard and Varian announced the first prototype nuclear precession magnetometer, and attempts were made to introduce this instrument for airborne surveys (Packard and Varian, 1954). For a long time, however, fluxgate equipment dominated the field because of the preference of oil exploration geophysicists for the smooth profile curve as compared to the stepped output of the proton magnetometer. Early acceptance of the proton precession magnetometer was also hindered by instrumental problems associated with relay switching in the "polarize head." The attitude toward airborne use of the proton precession magnetometer has now changed owing to the greatly improved performance and reduction in size associated with solid-state circuitry, the development of electronic

0-201-04245-2

conversion methods which provide a direct reading of the absolute field in gammas, and the adoption by the oil exploration industry of digital processing techniques. The drift-free and inherently digital output of this instrument has made it very attractive, especially when used in conjunction with computerized data handling (Hood, 1969).

The current trend to digital processing is perhaps the most significant development in aeromagnetic mapping. Introduction of in-flight digital recording allows contoured maps of a total field to be prepared at great speed, as well as other types of map of considerable value. For example, two-dimensional digital filtering can be used to suppress near-surface effects if these are troublesome or, conversely, to accentuate them to assist in surface mapping. Two-dimensional filtering can also be employed to enhance linear features such as faulting, and digital Fourier analysis can be applied to determine the depths to basement rocks (Naidu, 1969; Bhattacharyya, 1965, 1966a, 1969b).

In regard to high-resolution aeromagnetic mapping, the oil industry is now using this approach for the determination of subtle structural features in sedimentary basins where rock susceptibilities are low (Royer, 1967). Such surveys have always been carried out with optically pumped magnetometers, which have noise levels in the vicinity of $^1/_{100}\ \gamma$; effective precision, however, is of the order of $\pm ^1/_{10}\ \gamma$ because of the practical problems of navigation, maintenance of altitude, and so on. High-resolution magnetic sensors have also been used as components in vertical magnetic gradiometers, and in one system, operated by the Aero Service Corporation, two "birds" are flown (positioned one above the other), each containing an optically pumped magnetometer (Slack et al., 1967). Special aerodynamic designs have been employed to achieve stable flight, and many thousands of miles of successful survey have been flown using this technique.

An alternative device for the measurement of vertical magnetic gradients uses high-sensitivity directional cryogenic sensors of the Josephson junction type, mounted a relatively short vertical distance apart (Otala, 1969). Whether these systems become practical will depend largely on finding solutions to the problems of aircraft compensation.

The ultimate value of vertical gradient measurements for both oil and mining exploration remains controversial. One school of thought contends that precision total field recording, used in conjunction with ground control stations and the digital computer, can serve to produce the same type of maps with fewer problems than are associated with actual measurement of the vertical gradient. The counterargument is that the mathematical models are not sufficiently precise to substitute for the actual measurement of gradient fields. It appears that both approaches will continue to be used for some time.

Airborne magnetic surveys have been carried out with vertical fluxgate sensors mounted on gyroscopes (Lundberg, 1947), but the accuracy of the vertical field component measurement is limited to the order of $\pm 25\gamma$ because of the sensitivity of the fluxgate to misalignment when it is not operated in line with the earth's field. Although the use of vertical field measurements has some interpretive advan-

0-201-04245-2

tages in equatorial regions where the magnetic dip is flat, the scope of this magnetometer is severely limited by the high noise level.

Aeromagnetic surveys remain to this day one of the prime airborne geophysical methods. In the field of mineral exploration, aeromagnetics were originally thought of purely in terms of prospecting for magnetic ores, but today their geological significance is fully appreciated. In Precambrian Shield areas, for example, they help to delineate greenstone belts, the pattern of distribution of intrusive bodies, major faults, and many other geological features which often cannot be adequately elucidated with conventional mapping methods (Morley, 1969). The value of this type of geological information to exploration geologists is immense, and experience in Canada has shown that the publication of regional aeromagnetic maps has proved repeatedly to be the precursor of a massive exploration effort by the mining industry in the documented area.

Similarly, in the oil industry aeromagnetic maps have found widespread use as a very inexpensive means of determining the depth of sedimentary columns overlying basement rocks and of delineating structures related to the basement. Aeromagnetic surveys cost so little in comparison with seismic exploration that the oil industry has seen fit to make large-scale use of the airborne magnetometer even though it is a far less definitive structural tool than seismic methods.

The full value of aeromagnetic maps can be achieved only when large areas are covered and localized geology is brought into perspective with the broad regional patterns which appear on large-scale aeromagnetic compilations. The picture of the crustal structures in the Canadian Precambrian Shield that is unfolding as the aeromagnetic surveys are published has been dramatic (Kornick and McLaren, 1966), and it is clear that similar large-scale surveys should be carried out in all countries interested in developing their mineral and oil resources. The impact of magnetic surveying has also been very marked in offshore areas. Publication by Raff and Mason (1961) of the remarkable striped pattern of the ocean floor magnetics off the Pacific Coast of America has been followed by worldwide offshore magnetic studies and the integration of magnetic, bathymetric, and seismic data into a theory of ocean floor spreading and continental drift that has had profound impact on our understanding of the dynamics of the earth's crust. The continuing value of the airborne magnetometer as a remote sensing tool of the greatest importance to the earth scientist cannot be overstressed.

## II. ELECTROMAGNETIC SURVEY SYSTEMS (INDUCTIVE FIELD)

Low-frequency inductive field airborne EM systems have achieved the most spectacular successes to date in the application of remote sensing methods to mineral exploration (Pemberton, 1962; Ward, 1967, 1969). Their use has directly resulted in ore discoveries having a total value of billions of dollars. These methods are unusual among remote sensing techniques in that they have received almost no attention from the military; in fact, the initial reduction to practice of an operational

0-201-04245-2

airborne system was carried out by the International Nickel Company of Canada in 1952 (Cartier et al., 1952). After the early lead by Inco, subsequent years saw the introduction of a whole range of airborne EM systems, all of which depended on creating a low-frequency inductive field in the vicinity of the aircraft, inducing eddy currents in conductive bodies, and detecting the secondary fields reradiated by these conductive bodies.

The principal components of an airborne EM system are a transmitting loop carrying an alternating or pulsed current for generating a powerful primary inductive electromagnetic field, and one or more receiving coils for detecting secondary electromagnetic fields. The primary field induces eddy currents in adjacent conductive bodies, and these eddy currents in turn generate secondary fields. The principal engineering problem in detecting the induced currents in subsurface conductive bodies is the isolation of the primary fields from secondary fields. One approach has been to mount the transmitting and receiving coils in a rigid geometrical relationship so that in the absence of subsurface conductors the primary field from the transmitting coil is detected by the receiving coil as a signal of fixed amplitude. When a conductive body is presented to the transmitting/receiving coil assembly, induced eddy currents generate secondary fields which modify both the amplitude and the phase of the detected signal, an effect which can be sensed at high sensitivity with suitable electronic equipment. Various airborne configurations have been employed, including transmitting and receiving coils mounted on opposing wing tips of aircraft (Figure 8.1) on nose and tail booms, and at the opposing ends of long rigid booms towed by helicopters (Figure 8.2).

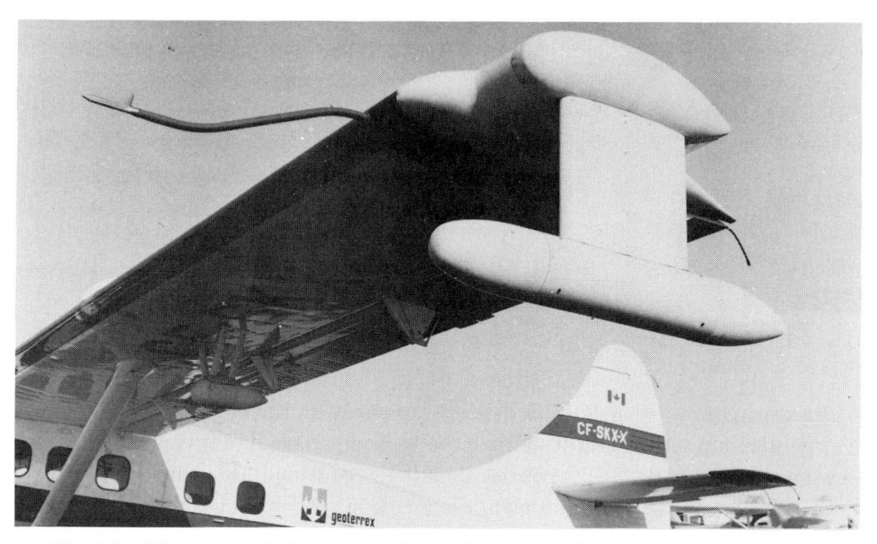


**Fig. 8.1.** Wingtip installation of receiving coil on geophysical prospecting aircraft. (Courtesy Geoterrex Limited, Ottawa).

0-201-04245-2

**Fig. 8.2.** Electromagnetic detection system in 30 foot boom towed 100 feet beneath Helicopter (BRL, Toronto)

The sensitivity of such systems depends primarily on the degree of rigidity of the structures supporting and separating the transmitting and receiving coils. Thus, in a wing-tip system, there is inevitable flexure of the wings during air turbulence. This causes a change in the coil separation, which in turn produces variations in the detected amplitude of the primary field. Under typical conditions the variations may be of the order of 30 ppm of the primary field. It then becomes impossible to detect a secondary field having an amplitude of less than 30 ppm. Problems due to

0-201-04245-2

wing flexure are further aggravated by vibratory movements of the metal skin of the aircraft, which carries large eddy currents induced by the primary field. As a result wing-tip installations seldom achieve noise levels of less than 15–20 ppm in aircraft having a wing span of 70 feet, even under optimum conditions. This tends to limit the maximum achievable depth of penetration at a flying height of 200 ft to somewhere between 150 and 250 ft beneath the surface. To increase penetration below these depths calls for improvements in the technology of compensating for coil movements.

Helicopter EM systems employing a towed rigid boom can achieve somewhat lower noise levels and deeper penetration than aircraft with wing-tip coils providing that the boom is engineered with extreme care. Operational costs are significantly higher, however, than for fixed wing aircraft. A recently developed helicopter system, DIGHEM, systematically operates on survey with noise levels of less than 3 ppm; this corresponds to a coil separation fluctuation in a 10-m boom of less than 10 $\mu$. In addition, this system is the only one employing three orthogonal receiving coils, and it is claimed that the three-component output is valuable in achieving recognition of overburden conductivity and in detailing the attitude of conductive bodies (Fraser, 1970). An example of DIGHEM records is shown in Figure 8.3.

An alternative approach to rigidly fixed transmitting and receiving coils employs a towed "bird," which carries the receiving coil separated a considerable distance from the aircraft-mounted transmitter. Continual variation then exists in the relative positions of the transmitting and receiving coils, and means must be provided for eliminating the effects of such changes. One technique measures only the components of the secondary field which are out of phase with the primary field; this calls for the use of at least two frequencies if estimates are to be made of the conductivity of detected bodies. This approach has been used in a number of successful airborne applications (Puranen and Kahma, 1953, 1956; Paterson, 1961). A variation is provided by the International Nickel Company's system, which records both in-phase and out-of-phase components and uses additional receiving and transmitting coils at different orientations to monitor "bird" movements and to derive compensation for such movements (Figure 8.4) (Cartier et al., 1952; Dowsett, 1969).

In a still different approach the resultant transient fields are measured when the primary field consists of a series of high-powered pulses. Transient secondary fields take a finite time to die out after the termination of the primary pulse, and by sampling the output of a wide-band receiving coil at various delays following the primary pulse it is possible to separate secondary field responses from the primary field. This method, known as the INPUT* system (INduced PUlse Transient), has seen widespread use (Figure 8.5) (Barringer, 1962; Boniwell, 1967; Nelson and Morris, 1969; Becker, 1969). An unusual feature of the INPUT system is that its pulse transmission contains a large number of frequency components, and therefore

*Registered trademark.

0-201-04245-2

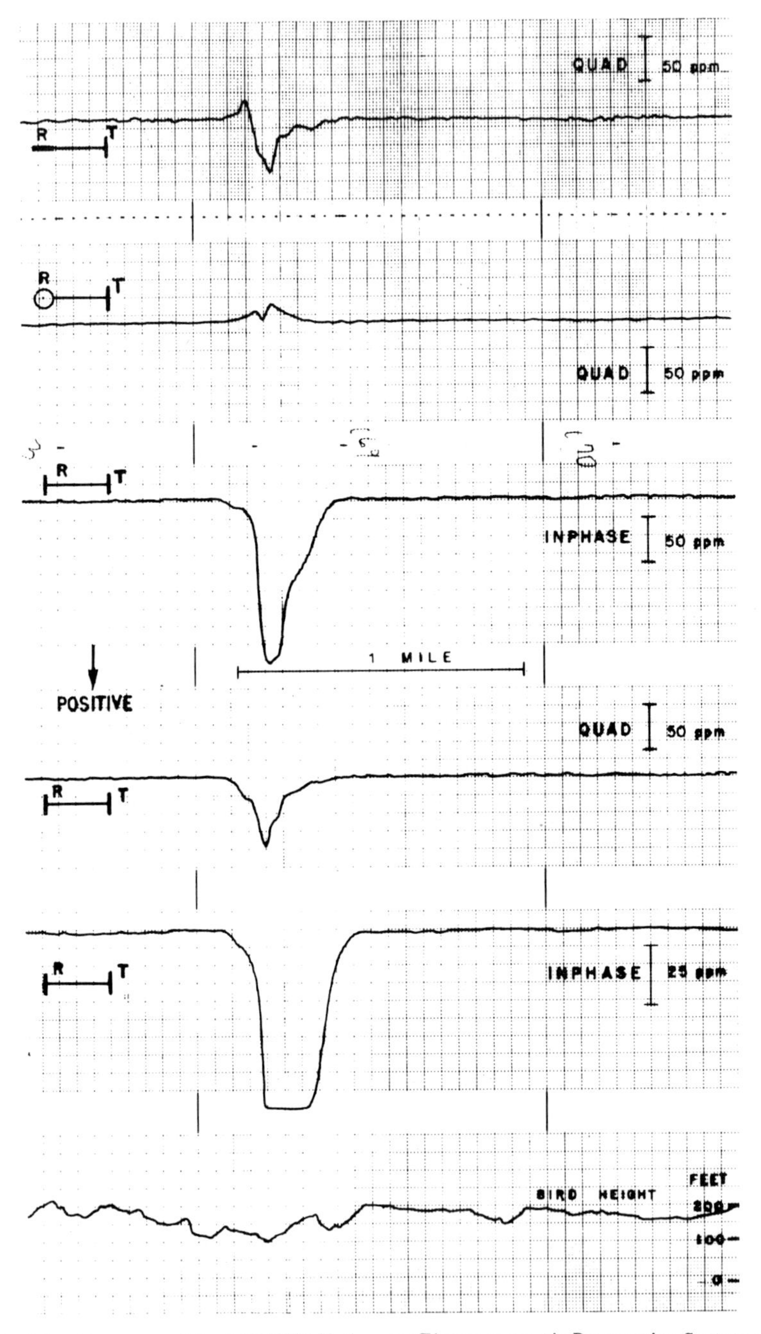


**Fig. 8.3.** Response of the DIGHEM Helicopter Electromagnetic Prospecting System over the Whistle deposit, Ontario, Canada. Symbols indicate orientation of transmitter (T) and receiver (R) coils relating each channel (Courtesy of DIGHEM Limited, Toronto).

0-201-04245-2

**Fig. 8.4.** International Nickel Company's DHC-6 Otter fitted with dual audio-frequency transmitting coils for geophysical prospecting (Courtesy International Nickel Company Limited and DeHavilland Aircraft Limited).

the secondary field response carries an exceptional amount of information as compared with single- and dual-frequency systems. Complex conductors can be recognized by the shape of the decay transient, and the system has potential application for computer analysis and automated interpretation schemes.

An extremely important factor in all airborne EM systems is the effect of conductive overburden. Under many geological and climatic conditions, relatively high surface conductivities are developed in the upper soil layers which not only inhibit penetration of electromagnetic signals into the earth, but also produce strong secondary fields that constitute a form of noise, limiting the ability of the system to detect the presence of subsurface conductors. Development of high surface conductivity can be related to the accumulation of surface salts in semiarid regions of the earth, such as the salt pan areas of western Australia, or it can be associated with highly conductive clays in certain glaciated regions. Tropical weathering can also produce relatively high surface conductivities under some conditions. Such enhanced conductivity is invariably patchy and does not produce a steady secondary field during aircraft traversing, but rather results in a fluctuating response. The more rapidly the fluctuations occur, the more difficult it becomes to identify discrete conductive bodies of the type that may be associated with mineral deposits.

One approach to reducing the effects of conductive overburden is to employ coil configurations which have minimal sensitivity to horizontal conductive sheets and maximal sensitivity to vertical or steeply dipping conductive sheets. Conductive ore deposits such as the so-called massive sulfides frequently exhibit steeply dipping attitudes within the underlying strata, as compared with the flat-lying characteristics of conductive overburden, and geometrical discrimination can be quite effective. Typical of such an approach are helicopter electromagnetic systems in which coaxial transmitting and receiving coils with horizontal axes are mounted in a long

0-201-04245-2

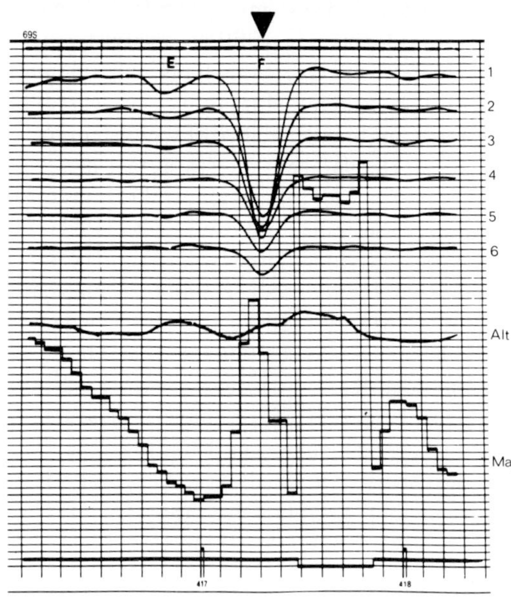


**Fig. 8.5.** INPUT (Induced Pulse Transient) airborne electromagnetic prospecting aircraft showing transmitting loop around aircraft, magnetometer boom and towed EM receiving coil. The record shows the geophysical anomaly responsible for the discovery in 1969 of a major ore-body at Sturgeon Lake, Ontario. Channels 1 to 6 show the amplitudes of transient response at increasing delays following termination of the transmitted pulse. The magnetometer channel indicates exact coincidence of a magnetic anomaly associated with pyrrhotite in the ore and is displaced on the record due to differences in time lag between the magnetometer and EM systems. (Record courtesy of Questor International Surveys, Toronto, Canada) (Photograph BRL, Toronto)

0-201-04245-2

boom towed beneath the helicopter. Similarly, the use of coplanar horizontal-axis transmitting and receiving coils mounted on the wing tips of an aircraft in line with the direction of flight provides minimum coupling with horizontal sheets and maximum sensitivity for steeply dipping bodies.

Unfortunately, geometrical discrimination against overburden response is far from totally satisfactory because of the fact that overburden conductivity is not generally homogeneous. Changes of thickness and in clay content or salinity can continue to cause a strongly fluctuating response even when optimum geometries are employed in the transmitting and receiving coils. An alternative approach to the problem is to provide separation between overburden and subsurface conductors by virtue of differences in conductivity and propagation delays through the overburden. This technique is used in the INPUT system, which utilizes the fact that the pulse transient response from overburden in general has a much shorter decay time than the transient response from underlying conductive mineralization. By the use of delayed sampling of the secondary field following the primary pulse, it is possible to gate out all or at least the greater part of overburden response and to detect only the transient field from underlying conductive bodies. In practice, this has proved quite effective, and the freedom from the effects of overburden experienced in the delayed channels of the INPUT system has been the main reason for its popularity.

In systems that detect only the quadrature component of the secondary field at two frequencies, separation of overburden effects from the underlying response of good conductors is not achieved, either by optimum minimal coupling with horizontal conductive sheets or by well-resolved conductivity separation. It is not uncommon, therefore, to encounter a substantial overburden response in both quadrature channels, a fact which has tended to mitigate against the use of these systems in areas characterized by extensive conductive overburden.

There remains a continuing requirement in airborne EM systems for increased penetration, particularly as relatively near surface ore bodies are discovered and the average depth of undiscovered deposits becomes greater. Many areas which have been surveyed with shallow penetration systems will require resurveying as deeper penetration methods are developed. Research is required in methods for reducing not only the instrumental noise associated with aircraft motions, but also geological noise. As instrument sensitivity is increased, the effects of overburden conductivity will become more significant. Therefore future trends are likely to be in the direction of increasing discrimination and resolution by the use of multiple frequencies in combination with the measurement of orthogonal components of secondary field response. Such methods, when used in conjunction with appropriate mathematical models and computerized data handling, will provide the exploration geologist with feasible targets under conditions with which present-day methods are unable to cope. A similar situation has occurred in oil exploration, where recent improvements in seismic techniques have opened up previously inaccessible areas.

0-201-04245-2

An additional application of airborne EM systems utilizing inductive fields is for resistivity mapping in connection with groundwater exploration. Some encouraging results have been obtained, and an expanded use of inductive field as well as plane wave systems for groundwater and engineering problems seems likely (Collett, 1969).

## III. VERY-LOW-FREQUENCY (VLF) RADIO FIELD METHODS

The geophysical electromagnetic systems normally referred to as EM methods are inductive field techniques in that they rely on the generation of alternating or pulsating magnetic fields by circulating currents in a transmitting loop. The oscillating fields normally employed are in the audio-frequency range, and consequently true propagation of a radiated field from the transmitting loop does not occur since the loop dimensions are a small fraction of a wavelength. Airborne systems have recently been developed which make use of propagated fields from radio transmitters. The primary field is radiated essentially as a plane wave, and behaves in a very different fashion from the inductive fields used in the so-called airborne EM systems. In the latter, the localized nature of the dipolar fields infers that eddy currents are confined to the immediate vicinity of the survey aircraft, and the decrease of detectability with respect to distance from the aircraft tends to follow an approximately inverse cube relationship for large targets and an even more rapid rate for small targets. In the case of a plane wave, however, very large geological structures are excited by the widespread field, and the falloff of secondary fields from such features occurs at a much lower rate. The net result is that there is a far greater sensitivity to geological structure in radiated plane wave systems.

Another important difference between propagated and inductive fields is that the energy carried in a propagated wave is divided between the electric and magnetic field components, whereas in inductive fields of the type generated by geophysical transmitters the electric component is insignificant. In practical airborne systems utilizing plane waves, the electric field component can be ignored and geophysical measurements made on the magnetic components only, or the electric field can be utilized as a phase reference as well as for other applications (Barringer and McNeill, 1968, 1969, 1970).

Most radio geophysical systems in use today take advantage of the VLF radio stations operated by various government agencies as their signal source. These stations radiate energy at the megawatt range and serve such purposes as communication with submerged submarines and transmission of time signals. Frequencies lie in the vicinity of 20 kHz, and skin depths of penetration into the earth (which are shown in Table 8.2) vary from 50 to 500 ft, depending on surface conductivity. The propagation characteristics of VLF signals are such that they can be usefully employed at distances of 3000 miles from the transmitter. Indeed, successful geophysical surveys have been carried out at ranges in excess of 3000 miles, although some restrictions may occur in the early hours of the morning because of fading of the signal at these extreme ranges.

0-201-04245-2

**TABLE 8.2**
**Penetration Depths[a] (18 kHz Frequency)**

| | Skin Depth (meters) | | |
|---|---|---|---|
| P *(ohm-meters)* | E = *5* | E = *15* | E = *80* |
| 10,000 | 383 | 400 | 542 |
| 1,000 | 119 | 120[b] | 124[c] |
| 100 | 37.6 | 37.6 | 37.6 |
| 10 | 11.9 | 11.9 | 11.9 |
| 1 | 3.75 | 3.75 | 3.75[d] |

[a] $P$ = resistivity; $E$ = dielectric constant.
[b] Typical value for rocky soil.
[c] Typical value for fresh water.
[d] Typical value for salt water.

Instrumental approaches to the utilization of VLF signals have included measurement of the absolute field strength and the tilt angle of the magnetic component perpendicular to the direction of propagation (Paal, 1965), and measurement of the tilt angles of the in-phase and out-of-phase components in a direction perpendicular to the direction of propagation, using a horizontal-axis coil as a phase reference (Paterson, 1970). Another method, known as RADIOPHASE,* employs the vertical electric field as a phase reference against which to measure the in-phase and out-of-phase components of the horizontal magnetic field detected with a horizontal-axis coil (Barringer and McNeill, 1968, 1969). The vertical electric field was chosen as a phase reference because calculations have shown that it is relatively phase stable against changes in underlying terrain conductivity. The results of modeling with this system have indicated that the amplitudes of the in-phase and out-of-phase magnetic components are indicative of conductivity contrast between conductors and the surrounding medium, and differ importantly from in-phase/out-of-phase responses in conventional airborne EM systems.

Experience with the RADIOPHASE system has indicated that the flow of currents in the ground associated with the propagation of VLF radio fields over the surface is highly sensitive to certain types of geological structure. Thus, variations in conductivity associated with stress metamorphism and shearing can provide in contoured RADIOPHASE maps patterns and trends which may differ considerably from the trends indicated on aeromagnetic maps of the same area. In general the contours on aeromagnetic maps tend to follow zones of equal susceptibility and hence are related to lithological layering within the rocks, whereas conductivity lineations may be associated with foliation, compression, and shearing, and may on occasion cut right across the aeromagnetic trend lines associated with lithology and geologic strike. Thus the structural picture that is obtained from VLF mapping is generally complementary to that provided by aeromagnetic maps.

*Trademark.

0-201-04245-2

Practical experience with RADIOPHASE in zones of deep tropical weathering has indicated that, although capability is not sufficient to penetrate the zone of weathering, structural patterns are nevertheless obtained which clearly relate to the major structures known to be present, apparently indicating that conductivity patterns may be retained in the weathered zone in relict form. An example of this phenomenon has been noted in surveys of the Fiji Islands flown at a terrain clearance of 1000 ft and a traverse interval of 1 mile. The VLF station used for this survey was located at the Northwest Cape in Australia, at a range of 3600 miles; and despite the depth of weathering, the extreme ruggedness of the topography, the width of the traverse intervals, and the range of the station the survey results tend to correlate well with the known major structural directions of the region, which are associated with the great ocean floor trench features surrounding the Fiji Islands. Such results were obtained, however, only after correction for terrain scattering effects. The portion of the Fiji survey shown in Figure 8.6 illustrates the evidence of structures paralleling the Tonga trench direction and cutting across the mountain ranges. The corresponding aeromagnetic map of the area is shown in Figure 8.7; it will be noted that the structural and geological information on each map is different from, and quite complementary to, that on the other.

In the RADIOPHASE system it has been found possible to use the amplitude of the vertical electric field component as an index of scattering since it is strongly responsive to terrain scattering and is only slightly affected by changes in ground conductivity. By employing computerized data-handling methods, it is possible to apply a topographic correction factor to the component of the horizontal magnetic field that is in phase with the vertical electric field, in order to minimize topographic scattering errors. Such corrections appear to be almost essential if VLF airborne surveys are to be carried out with any dependability over mountainous or rugged regions.

The systems described above make use of the magnetic field components of the VLF signals. A new method, termed E-PHASE,* has been developed by the writer and his colleagues; it differs from the others in that it operates entirely on the electric field components. It can be shown that when a radio wave propagates over a homogeneous earth the horizontal currents generated at the earth-air interface are phase shifted by 45° with respect to the propagating field over a very wide range of resistivities and frequencies. These currents flow horizontally in the direction of propagation and are associated with a horizontal electric field component which is also phase shifted by 45° with respect to the vertical electric field. Hence the resultant electric field is tilted slightly forward in the direction of propagation and is elliptically polarized. The horizontal electric field amplitude is related to the square root of the resistivity of the underlying terrain, and the tilt varies at VLF frequencies from a few minutes of arc over highly conductive terrain to 1° or more over resistive ground. For a given field strength, therefore, measurement of the horizontal component is sufficient to define the resistivity of underlying homogeneous earth.

*Trademark.

0-201-04245-2

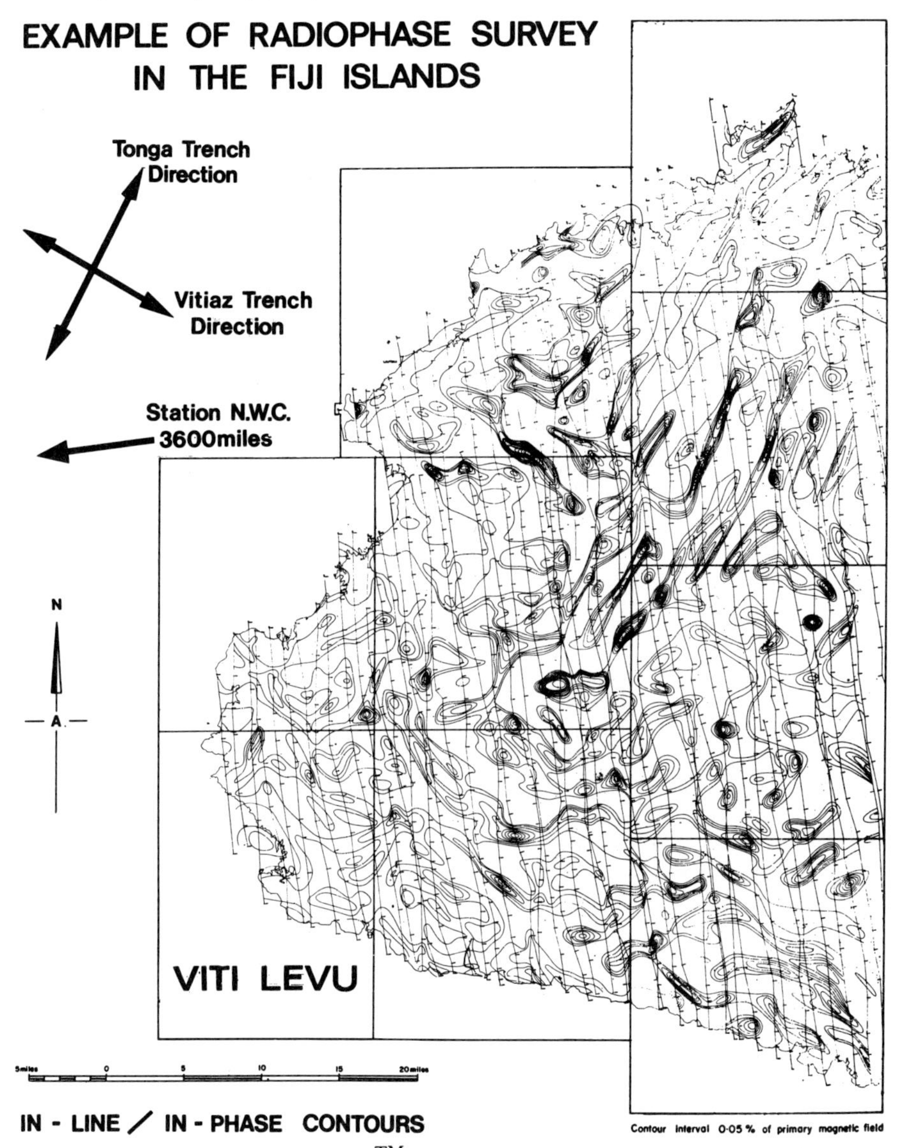


**Fig. 8.6.** A typical RADIOPHASE™ Survey in the Fiji Islands indicating parallelism between Radiophase lineament structures and major structural directions in the Southwest Pacific. (BRL, Toronto)

The problem encountered in measuring this component is that it is extremely small in relation to the vertical field strength and hence the antenna must be precisely horizontal. In the E-PHASE system, this difficulty is overcome by measuring only that component in the horizontal antenna which is 90° out of phase with the signal detected in a vertical antenna. This concept makes the system relatively immune to minor misorientations of the horizontal antenna since the receiver is in-

0-201-04245-2

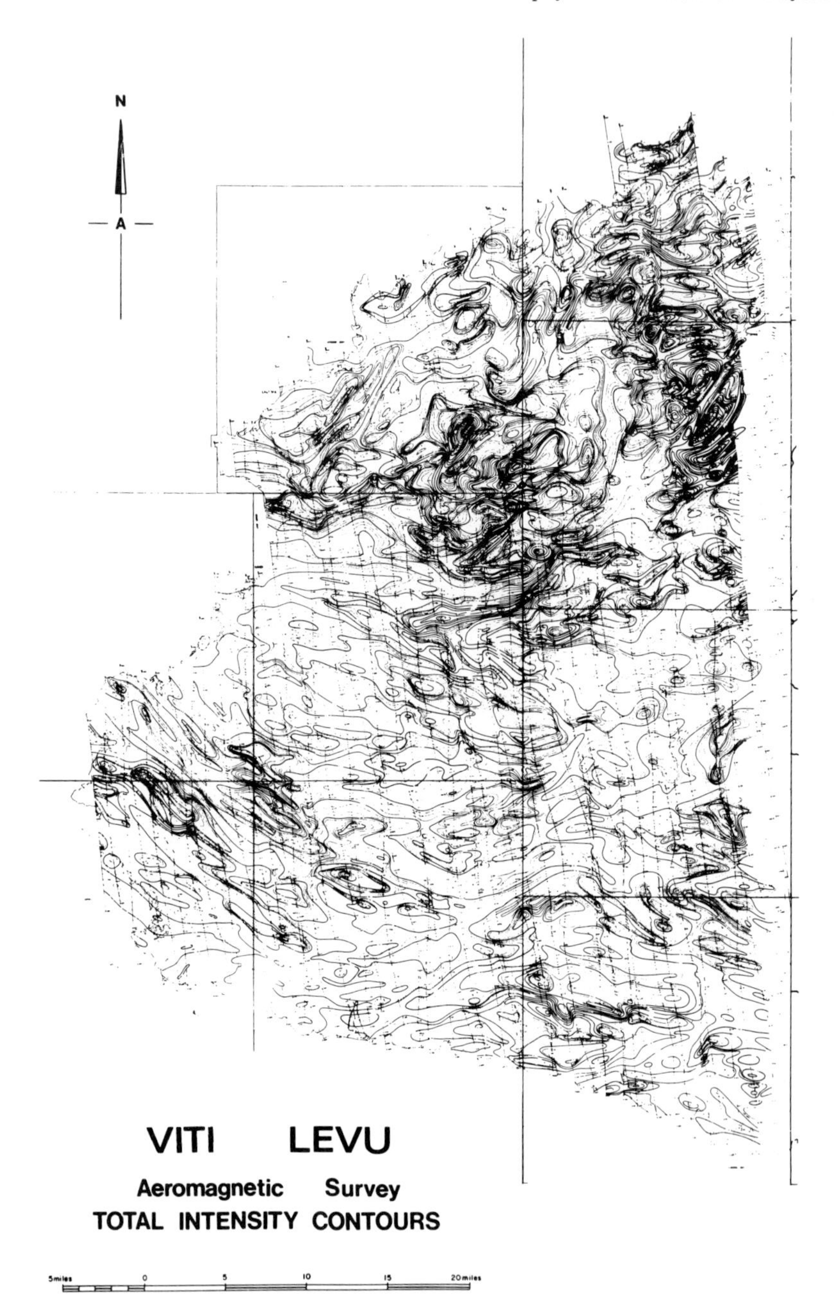


**Fig. 8.7.** Aeromagnetic map of area in the Fiji Islands corresponding to the RADIOPHASE survey in Figure 8.6. This map illustrates differences in geological correlation between RADIOPHASE and aeromagnetic surveys. (BRL, Toronto)

0-201-04245-2

sensitive to components of the vertical field picked up on the tilted horizontal antenna. Furthermore, since the phase shift of the horizontal field is known to be at 45° to the vertical field for a homogeneous earth, it is possible to derive a good approximation to the total horizontal field strength solely from knowledge of the quadrature component.

A typical helicopter E-PHASE installation is shown in Figure 8.8. Vertical and horizontal antennas are mounted on a nose boom and are gyrostabilized to maintain their approximate orientation during helicopter movements. Traverses are flown perpendicularly to the propagation direction so that the horizontal antenna lies in the propagation direction. When high-resolution data are required, a close interval between traverses is employed (as low as $^{1}/_{10}$ mile), these traverses being flown at a typical terrain clearance of 200 ft. When regional reconnaissance information only is sought, traverse intervals of as much as 1 mile can be employed, at flying heights of up to 1000 ft.

Resistivity maps produced by E-PHASE are contoured in ohm-meters, indicating apparent resistivity, and are not dissimilar from resistivity maps produced by conventional ground methods. In most cases the approximations are sufficiently accurate to produce a map which correlates well against ground resistivity surveys carried out with a comparable, fixed electrode spacing. A typical E-PHASE aeroresistivity map flown over a region of Pleistocene sands, gravels, and clays is shown in Figure 8.9.

**Fig. 8.8.** Stabilized E-PHASE boom mounted on helicopter showing vertical and horizontal electric dipole antennas. (BRL, Toronto)

0-201-04245-2

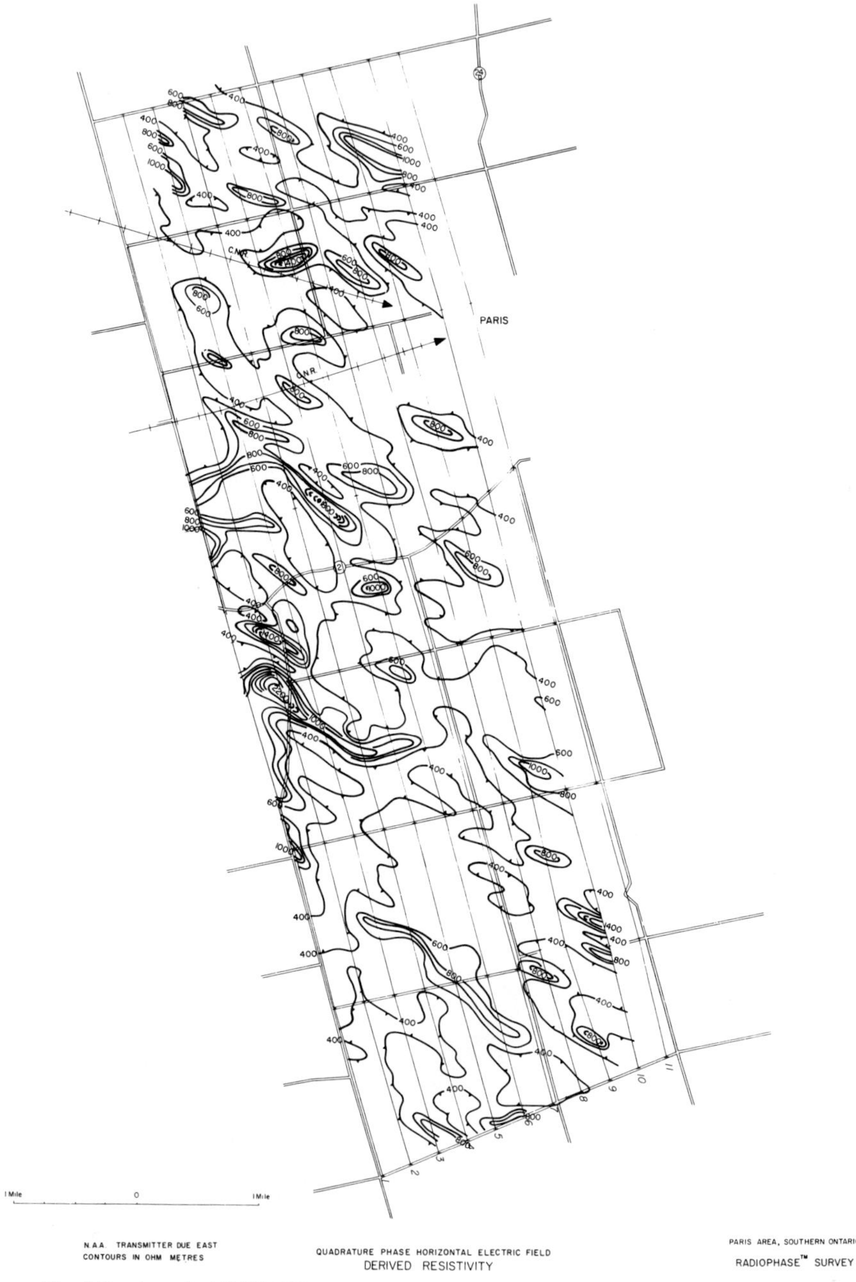


**Fig. 8.9.** A typical E-PHASE survey with apparent resistivity contoured in ohm-meters illustrating high resistivity zones over glacial areas carrying sands and gravels. (BRL, Toronto)

0-201-04245-2

One advantage of E-PHASE is that it achieves good discrimination between highly resistive materials. This is the precise converse of normal airborne EM detection systems, which operate with magnetic field components and generally show little response over resistive terrain. It is apparent, therefore, that the E-PHASE method has different areas of application which warrant careful investigation.

In considering these applications it is worth looking at some of the parameters which affect the conductivity of terrain. In general, earth materials such as gravels and sand which have a very low clay mineral content are relatively resistive unless they are saturated with saline waters (Wilcox, 1944). Conversely, materials which have a high clay content are normally relatively conductive as a result of ion-exchange effects (Keller, 1969). Frozen ground such as that occurring in permafrost zones tends to be much more resistive than thawed ground because of the restriction on the mobility of ions (Ogilvy, 1969). Conversely, heated ground such as that occurring in the vicinity of geothermal areas is likely to be more conductive than the surrounding cooler ground (Studt, 1968). Alteration originating from hydrothermal activity or weathering can produce marked increases in conductivity due to the associated rises in clay content. Typical examples are kaolin areas within granites, and the montmorillonite-rich weathered zone over kimberlite diamond pipes (Gerryts, 1969). Excessive salinity associated with sea water invasion or the development of salt pan areas in semiarid regions also can cause large increases in conductivity. Even higher conductivities occur in mineralized zones where sulfides act as semiconductors.

When all these conductivity associations are taken into account, it can be seen that the aeroresistivity maps produced by E-PHASE have a number of significant applications, including the search for gravel deposits, exploration for kaolin and diamond-bearing kimberlites, the mapping of ice lenses in discontinuous permafrost zones, exploration for geothermal areas, the mapping of sedimentary strata where subtle facies changes are accompanied by resistivity changes, the mapping of Pleistocene geology for water resource purposes, and hematite and manganese exploration, as well as several other uses, some of which have considerable economic importance.

The techniques described in this section have involved radio transmissions in the VLF range. It is possible, however, to operate at higher frequencies in order to restrict the measurements to shallower penetrations. For some types of engineering studies, such as the development of gravel resources, the use of two frequencies, one in the VLF and the other in the broadcast band frequency range (0.5–1.5 MHz) appears to have potential advantages. Two such frequencies can provide information on layering, and at the time of writing tests are commencing on simultaneous mapping at 0.5 MHz and 18 kHz. Another application of higher frequencies is the mapping of pollution in water where it is necessary to measure the resistivity of a relatively thin layer of water and to avoid penetration of the electromagnetic waves to the bottom. Providing that the skin depth of penetration is less than the depth of the water, it is possible to make fairly sensitive measurements of variations in water resistivity and hence to map the distribution of pollutant ion concentrations.

0-201-04245-2

There is some possibility that transmitters will be placed in operation in the ELF frequency range at the lower end of the audio-frequency spectrum. If this proves to be the case, these transmissions will be of great value for relatively deep penetration geophysical studies using some of the techniques described above. Dual-frequency electric field component mapping at VLF and ELF would make it possible to produce residual maps of deep resistivity with computed suppression of overburden and weathering "noise."

## IV. NATURAL ELECTROMAGNETIC FIELD SYSTEMS

Section III referred to the measurement of geophysical parameters by mapping distortions in man-made electromagnetic plane waves. These signals are generally easy to work with because of their relatively high field strengths and predictable transmission characteristics. There are, however, other plane wave signals propagated over the surface of the earth which originate from natural sources and which can be used for geophysical remote sensing purposes. These signals, known as "sferics," are radiated as oscillatory transients from lightning discharges on a more or less continuous basis during electrical storms.

Currents as large as 20,000 A flow during these discharges, and the long ionized air paths through which the currents flow form large vertical electric dipoles and are thus extremely powerful radiating antennas. The frequency spectrum of signals generated by the discharges ranges from a few hertz up into the megahertz range, and the propagation losses are so small at the subaudio end of the frequency spectrum that signals from a single event can be detected over the entire surface of the earth. To date, however, the use of these natural oscillating fields for airborne geophysical measurements has been confined to frequencies lying between 100 and 500 Hz, where attenuation limits detection to ranges of a few thousand miles from the source. When operating at these frequencies, surveys can be carried out during the summer season only when thunderstorm activity is located within a few thousand miles of the survey area. During the winter the signals are either absent or so attenuated as to be of no value. Such seasonal variations have tended to inhibit widespread application of this approach. If the very low audio and subaudio frequencies could be used, year-round operation would be feasible, but unfortunately there are major problems in detecting these low frequencies in an airborne system because of various sources of noise which are extremely difficult to suppress.

Virtually the only commercial airborne system which has operated using natural fields is termed AFMAG; it is based on measurement of the tilt of the polarization ellipse of the radiated magnetic field components of sferic signals at 150 and 500 Hz (Ward, 1958, 1959, etc.; Sutherland, 1969). Two coils are flown, mounted orthogonally in a "bird" with their axes in line with the flight direction and at 45° to the horizontal. One coil is used as a phase reference for the other coil, and the relative amplitudes of identical phase components of signals on each coil are compared. Flight traverse directions are chosen as far as possible so as to be perpendicu-

0-201-04245-2

lar to the geological strike of the region, and the method provides a record of the tilt angle of the polarization ellipse of the natural field along each flight line. Anomalies occur when the tilt angles change from forward dips to reverse dips along the traverse lines. The crossover points where the field is momentarily vertical are identified during interpretation, and line-to-line correlation of these crossovers is indicative of the strike of conductive zones associated with geological structures, conductive horizons containing graphite or sulfides, and so forth.

The AFMAG method is capable of considerable penetration because of the low frequencies employed and the nature of plane waves, as explained in Section III. Skin depths depend on ground resistivities and lie in the range between 500 and 5000 ft, which is adequate to penetrate most overburden and weathering. Sensitivity to large geologic structures is good since the rate of falloff of anomalies with depth or aircraft altitude is very small as compared with the values for active EM systems. However, the records are much noisier than those obtained with VLF plane wave systems because of the fact that the signals are only semicontinuous and may often be arriving more or less simultaneously from two or more storm centers in very different directions. This can cause marked fluctuations in the dip angle of the audio-frequency magnetic field associated with individual structures since these structures have different coupling for each of the incoming wave directions. Substantial improvements are possible, however, with more sophisticated signal processing.

The very deep penetration of ELF natural electromagnetic fields provides them with great potential for subsurface geological remote sensing. Only a fraction of this potential has so far been exploited, however, because of the instrumental problems associated with detecting these fields and the complexities of interpretation. The measurements made to date have been of a very simple form, and the use of electric field components has so far been ignored. The possibilities for producing multiple-frequency resistivity surveys using methods related to the electric field VLF technique already described are attractive and merit development. Many applications in water resource, geothermal, and engineering studies lie ahead for an airborne system capable of deep resistivity mapping. A new system, termed TELTRAN, is currently under development which adapts some of the concepts of the RADIOPHASE and E-PHASE systems to sferic detection and analysis (Barringer, 1969).

## V. GAMMA-RAY SPECTROMETRY

The airborne gamma-ray spectrometer ranks as an important low-altitude geological remote sensing tool. Naturally occurring radioactive elements in the earth's crust, including potassium 40 and the radioactive decay products of uranium and thorium, vary widely and often diagnostically in their relative distributions. Originally, airborne gamma-ray methods were confined to exploration activities for uranium; however, the potential of gamma-radiation measurements as a geological mapping aid is receiving increasing recognition (Darnley, 1968; Boltneva et al., 1964;

0-201-04245-2

Foote, 1969; Johnson, 1970). Considerable effort has been expended in recent years on the development and improvement of airborne gamma-ray spectrometers, and a variety of commercial instruments which incorporate gamma-ray energy discriminators of both the threshold and window types are available. In general, the type of spectrometer in which energy-discriminating windows are precisely located to select the energy photopeaks from the different naturally occurring radioactive elements has improved discrimination when compared with the threshold types. Nevertheless, both classes can be used with considerable effect as an aid in geological mapping, and even a simple total-count system which does not separate potassium from the uranium and thorium daughters can be most useful. An example of a total-count radiometric map from a survey flown over Liberia is shown in Figure 8.10,* in which the wealth of geological structure and information is self-evident.

The airborne gamma-ray spectrometer can in a sense be referred to as a remote sensing geochemical tool capable of defining certain chemical parameters. A study of Table 8.3, which shows the typical potassium, uranium, and thorium distributions for various rock types, gives an indication of the usefulness of this device in discriminating lithological units.

**TABLE 8.3**
**Potassium 40, Thorium, and Uranium in Igneous and Sedimentary Rocks (Parts per Million) (from Guillou, 1964)**

| | *Igneous Rocks* | | *Sedimentary Rocks* | | |
|---|---|---|---|---|---|
| | *Basaltic* | *Granitic* | *Shales* | *Sandstones* | *Carbonates* |
| Potassium 40[a] | 0.8 | 3.0 | 2.7 | 1.1 | 0.3 |
| Average range | 0.2–2.0 | 2.0–6.0 | 1.6–4.2 | 0.7–3.8 | 0.0–2.0 |
| Thorium | 4.0 | 12.0 | 12.0 | 1.7 | 1.7 |
| Average range | 0.5–10.0 | 1.0–25.0 | 8.0–18.0 | 0.7–2.0 | 0.1–7.0 |
| Uranium | 1.0 | 3.0 | 3.7 | 0.5 | 2.2 |
| Average range | 0.2–4.0 | 1.0–7.0 | 1.5–5.5 | 0.2–0.6 | 0.1–9.0 |

[a]Chemical potassium contains 0.0119% $^{40}K$.

Radioactivity maps tend to be complementary to aeromagnetic maps since they can reflect a great deal of detail in granitic or sedimentary areas where magnetic susceptibilities may be low and the geological information from magnetics may be meager. The penetration of gamma rays is very limited with the result that airborne measurements are directly related only to the top foot or less of soil. Maps tend to express surface conditions, therefore, and are heavily modified by the presence of surface water. Nevertheless, the effective penetration obtained is often surprising because of the fact that surface soils, even when of glacial origin, tend to reflect the chemistry of underlying rocks. In glaciated areas it is well known that larger boulders can travel great distances; however, the fine material, which represents the bulk

*See color insert.

0-201-04245-2

of the overburden, has often been translocated by only a few hundred feet or less and the gamma-radiation emission remains indicative of the underlying geology (Popenoe, 1962). This expression of underlying terrain, which has been noted in countless surveys, parallels the pseudo-deep penetration which can occur with VLF methods in deeply weathered residual overburden. The combined and complementary use of gamma-radiation mapping, VLF methods, and airborne magnetometry is extremely attractive, therefore, as a general geological mapping and exploration aid.

With reference to direct economic applications, the most obvious one lies in the detection of uranium ores (Pemberton, 1969). In this case the use of spectrometric methods rather than total-count radiation detection is of considerable importance since quite large radioactive anomalies can sometimes occur because of the presence of highly potassic rock outcrops, accumulations of boulders, and so forth. The use of a discriminating gamma-ray spectrometer incorporating subtraction circuits which provide real-time correction for the contribution of thorium in the uranium window is of value in providing immediate identification of uranium-rich sources. The use of a spectrometer instead of a total radioactivity instrument also makes it possible to estimate uranium-to-thorium ratios, which can provide a basis for priority selection of anomalies for follow-up prospecting. There is some debate on the relative merits of real-time computing of corrected values for uranium, thorium, and potassium, and of recording the outputs of the discriminating windows on magnetic tape and subsequently computing corrected values during data compilation. The latter method is more precise but involves more complex equipment and more sophisticated data handling. Although the future trend will undoubtedly be toward such methods, the simpler approach remains popular among those looking for uranium.

A less obvious economic application of gamma-ray spectrometry is in the detection of potassium alteration halos around mineralized areas and the identification of anomalous radioactive element ratios (Moxham and Alcarez, 1965). Variations in such ratios have been reported in association with some types of mineralization (Bennett, 1970), and this approach appears to have merit providing that sufficiently high-quality equipment is used for the surveys. Exceptionally good signal-to-noise ratios are required if good element ratio data are to be obtained, since there is a substantial degradation in signal-to-noise ratios when element ratios are derived. Large crystal volumes are necessary in order to acquire good statistics, and much of the airborne equipment in general use is inadequate for ratio applications.

## VI. AIR-SAMPLING TECHNIQUES

The possibility of detecting ores by sensing vapors being released to the atmosphere appears to have been suggested first by Williston for the case of mercury vapor (Williston, 1964, 1968). Williston published a profile in 1964 on the presence of atmospheric mercury over a mercury prospect, but does not appear to have reported any further work on air sampling over mineralization since that time. Other

0-201-04245-2

studies have been carried out by workers at the U.S. Geological Survey (McCarthy et al., 1969), who have constructed experimental airborne equipment for amalgamating on gold filters the mercury vapor contained in large air samples. This technique has been limited by having a time resolution of ½ min at best, but has nevertheless successfully demonstrated that atmospheric mercury anomalies can occur not only over mercury deposits but also over gold and porphyry coppers.

The writer and his colleagues have been involved in problems of mercury geochemistry and the development of mercury detection instrumentation since 1963. An analytical mercury detection technique that has been used extensively for soils analysis (Barringer, 1966) has been adapted for airborne survey applications (Figure 8.11) (Barringer, 1970a, b). Tests have indicated that in some cases excess concentrations of mercury up to 10 times background can be obtained over some mineral deposits. The atmospheric mercury detection method appears, therefore, to have promise as an aid to mineral exploration, particularly when used in conjunction with complementary geophysical methods. Much remains to be learned, however, regarding the meteorological and geochemical controls of atmospheric dispersions of mercury.

Atmospheric sampling techniques for remote sensing need not be confined to vapor detection since there is a considerable dispersion of aerosols and particulate matter in the atmosphere, which to some extent represents a sample of the underlying terrain. Weiss has carried out experimentation in the use of particulate sampling as a prospecting method and claims to have obtained encouraging results (Weiss, 1967, 1969). Generally the atmosphere contains at least 10 $\mu g\ m^{-3}$ of particulate matter, and analysis of these particulates shows that a wide range of elements is present. In urban areas industrial pollution and smog cause very large increases in aerosol and particulate backgrounds, and obviously contamination in the vicinity of populated areas is likely to eliminate the usefulness of atmospheric sampling methods other than for pollution studies. However, in sparsely populated regions under appropriate meteorological conditions the particulate content of the atmosphere at low altitudes will inevitably reflect the composition of underlying surficial inorganic and organic materials.

Development of atmospheric geochemical sampling methods for earth science purposes has so far received very little attention. It is apparent, however, that these methods are likely to receive more careful study in the future as a potential aid in the exploration of large areas of the earth's surface where detailed prospecting has so far been very limited.

Operation of atmospheric detection systems is normally carried out at terrain clearances of substantially less than 500 ft and typically in the region of 200 ft. Concentrations of vapors and particulates in the atmosphere decrease markedly with increasing altitude, particularly in the case of anomalous dispersions arising from point sources. The need for a rapid response of a few seconds for instrumentation flown at low altitude is particularly important, therefore, when localized sources are being sought. When looking for regional geochemical patterns in the

0-201-04245-2

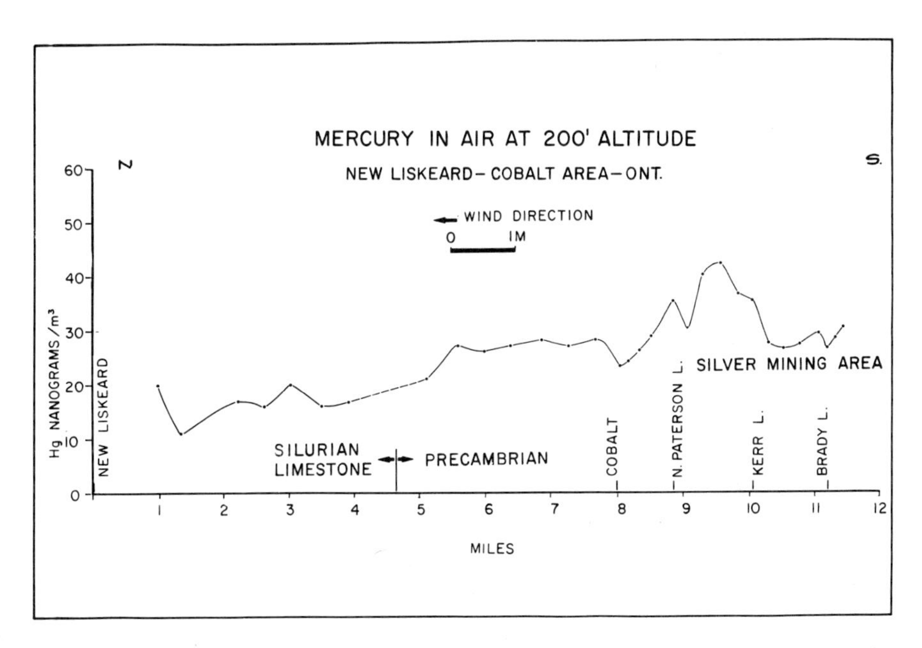


**Fig. 8.11.** Mercury atmospheric vapor detection system mounted on the outside of a helicopter and a profile of atmospheric concentrations over the Cobalt silver mining area Ontario, Canada. (BRL, Toronto)

0-201-04245-2

underlying terrain, however, there are fewer constraints, and instruments with slower response operated at high flying altitudes yield satisfactory results.

## VII. FRAUNHOFER LINE DISCRIMINATOR

An ingenious method of detecting luminescence in daylight by noting changes in the apparent depth of Fraunhofer lines in the solar spectrum has been investigated by the U.S. Geological Survey (Hemphill and Stoertz, 1969). Fraunhofer lines are dark lines in the solar spectrum caused by selective absorption of light by gases in the relatively cool upper part of the solar atmosphere. Widths range from less than $^1/_{10}$ Å to several angstroms, and the central intensity of some lines is less than 10% of the adjacent continuum. The lines are sharpest, deepest, and most numerous in the near-ultraviolet, visible, and near-infrared regions of the electromagnetic spectrum.

Successful use of the Fraunhofer line depth method was first reported by Kozyrev (1956), who detected luminescence in the ray systems of certain lunar craters. Other astronomers have also reported positive results using this method, and many, like Kozyrev, use the H and K lines of calcium in the violet because the relatively broad widths of these lines permit better spectral definition than narrower Fraunhofer lines at longer wavelengths.

The feasibility of using the Fraunhofer line depth method to detect luminescing earth materials was indicated in experiments conducted at the IIT Research Institute, Chicago, Illinois, using the following Fraunhofer lines: $D_2$, the sodium line at 5890 Å; G, a hydrogen line at 4340 Å; and H and K, two calcium lines at 3968 Å and 3934 Å, respectively. When luminescence from samples of calcite, colemanite, and a phosphate rock was observed using the Fraunhofer line depth method, the results were sufficiently encouraging to induce the U.S.G.S. to develop an airborne Fraunhofer line discriminator. Instrumentation was constructed for the U.S.G.S. by the Perkin-Elmer Corporation.

The Fraunhofer line depth method involves observing a selected Fraunhofer line in the solar spectrum and measuring the ratio of the central intensity of the line to a convenient point on the continuum a few angstroms distant. This ratio is compared with a conjugate spectrum reflected from a material that is suspected to luminesce. Both ratios normally are identical, but luminescence is indicated when the reflected ratio exceeds the solar ratio. Reflectivity differences between the central intensity of the Fraunhofer line and the adjacent continuum can generally be ignored because variation of reflectivity with wavelength is negligible for most materials over spectral ranges of only a few angstroms.

The critical components of the Fraunhofer line detector (FLD) are two narrow-band Fabry-Perot etalon filters. Each etalon consists of two highly reflective dielectric films separated by a glass plate 80 $\mu$ thick, parallel, and flat to 0.1 wavelength. These filters have a half-power band pass of less than 1 Å with peak transmission greater than 50%.

0-201-04245-2

In operation, light enters the FLD at the top and the bottom. The light entering at the top is sunlight which has been diffusely reflected from a Lambertian surface. The light which enters from the bottom is sunlight reflected from the target. One of the Fabry-Perot filters is tuned to the central wavelength of the Fraunhofer line, and the other filter to a convenient point on the solar continuum adjacent to the line. Low-noise photomultipliers are located behind each filter, and a series of lenses, beam splitters, and choppers permits the instrument to "look" alternately at the ground and upward at the sky through both filters, thereby monitoring the ratio of the central intensity of the Fraunhofer line to the continuum in each look. An analog computer, housed in a control console, determines and compares the ratios acquired in the solar look and the ground look and converts any variation in the ratios into a signal proportional to the intensity of luminescence in the ground target. This signal is produced in real time and may be displayed on an oscilloscope, plotted on a strip chart, or recorded on magnetic tape.

The principal application of the airborne FLD equipment to date has been in studying the distributions of rhodamine dye in water. Rhodamine dye is a water-soluble luminescent dye used by marine geologists and hydrologists to monitor current dynamics in rivers, estuaries, and coastal waters. The luminescence emission peak of rhodamine dye is near 5800 Å, and the tests carried out in the detection of dye by the U.S.G.S. were made with the FLD unit operating on the $D_2$ Fraunhofer line of sodium at 5890 Å, with a reference at 5892 Å. Tank tests carried out in daylight with the equipment indicated that concentrations of well under 5 ppb of rhodamine dye could be detected. Further shipboard tests confirmed this type of results, the FLD unit being suspended from a davit over the side of the ship and the luminescence intensity recorded as the ship moved slowly through a cloud of rhodamine dye.

In 1969 helicopter tests were conducted in the San Francisco area, the unit being mounted on the outside of a helicopter in a position where it could receive light from below and reference light from above. Shore-based tracking radar was used to plot the precise flight path of the helicopter and the location of dye clouds, which were placed in the water by dropping plastic bags containing the dye solution. These tests were successful and indicated that sensitivities of 0.1-0.2 ppb could be achieved. Further tests of altitude range indicated that at a ceiling altitude of 5000 ft the dye remained clearly detectable.

Apart from water pollution studies there are many other potential applications of the FLD. These include pollution measurements where the pollutant exhibits fluorescent properties. Substances such as lignin sulfonate, a by-product of paper manufacturing, exhibit fluorescence, as well as all crude oils. Tests carried out by the U.S.G.S. on crude oil spills offshore from the coast of Santa Barbara, California, were unsuccessful, probably because of an inability to use the optimum Fraunhofer line for the crude oil luminescence. Calculations and laboratory tests have indicated, however, that an improved FLD unit having a larger light throughput and an optimum selection of Fraunhofer lines should be capable of detecting oil spills.

0-201-04245-2

Some types of atmospheric luminescence may also be detectable with Fraunhofer line methods, particularly with reference to carcinogenic hydrocarbons. The problem is, however, a fairly sophisticated one which may require measurement at several Fraunhofer line wavelengths in order to separate other types of fluorescence from those due to dayglow and other atmospheric sources.

An interesting potential application of Fraunhofer line detection is in the monitoring of chlorophyll luminescence. The absorption and emission of light by chlorophyll is intimately related to photosynthesis, and the luminescence of plants increases when they are stressed. Detecting this fluorescence with an FLD unit may present serious problems, but if this can be done a new method of measuring plant stress will become available. Such plant stress can be related to plant disease or the presence of toxic materials in the soil nutrients.

## VIII. OPTICAL CORRELATION REMOTE SENSORS

Photographic and line-scanning optical techniques are mainly concerned with producing images at certain specific wavelengths. The principal emphasis on information derived in these systems involves spatial characteristics, and the bit density of data is tied to spatial resolution. However, it is possible to consider optical remote sensing systems which provide highly detailed spectral information and relatively low spatial resolution. These methods are based on the fact that radiant energy at optical wavelengths reflected or emitted from the earth's surface or from the surface of objects carries a spectral signature which may provide considerable information regarding the nature of the surface or of the gases lying between that surface and the detector. In some cases the spectral information may be carried as a very weak modulation of the total optical flux, and special methods are required in order to identify the spectral components present.

The technique of correlation spectroscopy (Barringer, 1965, 1966, 1968; Williams and Kolitz, 1968; Davies, 1970) is one such approach to the problem; it is based on the concept of correlating the spectra of radiation received in real time against a stored reference. A typical correlation spectrometer is shown in Figure 8.12; this particular instrument employs interchangeable dual telescopic foreoptics having difference acceptance angles for different types of application. The light received by the foreoptics passes through the entrance slit of a spectrometer, is dispersed by a grating, and is focused onto a pattern of exit slits which are arranged so as to correlate with the spectra being detected. These exit slits are etched onto an optical disk which rotates at high speed and carries on it a series of slit patterns sequentially located in the exit plane of the spectrometer. In its simplest application the spinning disk correlator can carry two different patterns, one set of slits being located in positions which correspond to the peaks of the spectrum being detected and the other set placed in positions which correlate against the various minima of the same spectrum. This technique of correlation provides maximum sensitivity and supercedes an earlier method in which the spectrum was vibrated across a single cor-

0-201-04245-2

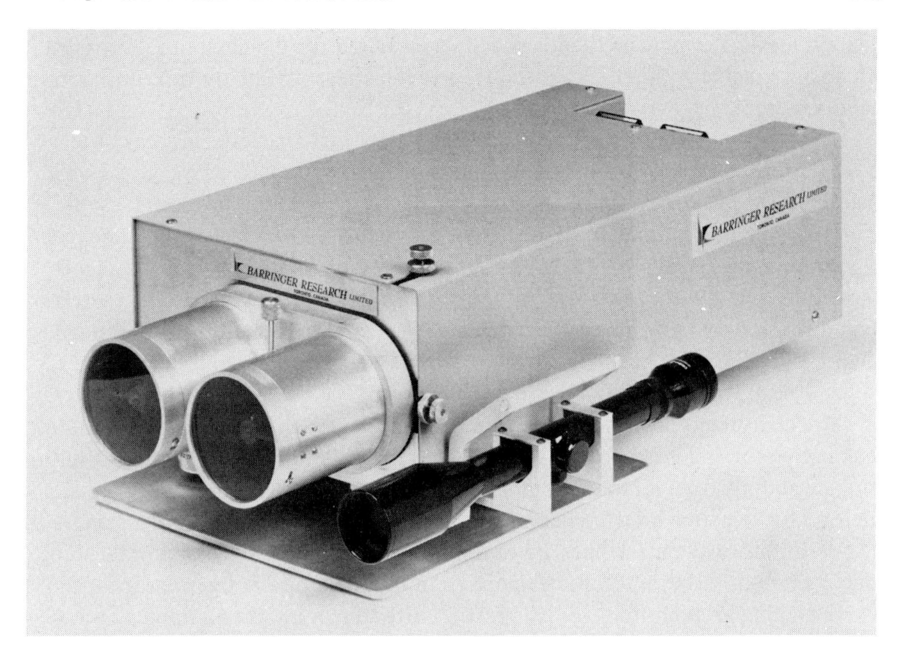

**Fig. 8.12.** Monitoring pollutant concentrations across an airfield with remote sensing dual gas monitoring correlation spectrometer. (BRL, Toronto)

0-201-04245-2

relation mask. The new method is equally well suited for the detection of spectra which have regularly spaced lines and those which have irregularly spaced lines or band structure.

The signal-processing portion of the correlation spectrometer consists of a photosensitive detector behind the spinning disk, an automatic gain control which maintains the average output of the photodetector at constant level, and a synchronous detector which senses in the output of the photodetector the presence of a pulsating component that matches the frequency of rotation of the disk. Thus there is no signal from the system in the absence of spectra matching the correlation disk. However, in the presence of a small spectral modulation representing less than 1 part in 10,000 of the total light intensity reaching the photodetector, a significant output is obtained.

The instrument illustrated in Figure 8.12 employs eight patterns on its rotating disk correlator and is capable of simultaneous detection of two gases such as sulfur dioxide and nitrogen dioxide. Four slit patterns are used for each gas, as part of a method for rejecting interference caused by Fraunhofer lines in the solar spectrum. The positions of the mask lines are optimized by computer analysis in order to provide maximum rejection of interference from fine structure in the solar spectrum.

Surveying air pollution from aircraft is currently an important application of the instrument in the passive mode of operation. The equipment has been installed in various types of aircraft; and when flown above the inversion layer, looking vertically down at the terrain, it produces a continuous profile of the total vertical burden of selected pollutants such as sulfur dioxide and nitrogen dioxide (Moffat and Barringer, 1969). A typical example of this type of survey was carried out for the U.S. Department of Health, Education, and Welfare over Chattanooga, Tennessee; the result is shown in Figure 8.13. Large concentrations of nitrogen dioxide gas over the city can be noted in this illustration. Further application of the instrument in the passive mode includes traversing along highways in trucks with the instrument looking vertically upward. This is an effective and inexpensive means of producing vertical burden profiles of air pollution; and if perimeter traverses are made around areas such as oil refineries, it is possible to calculate the mass production in tonnages per hour of a given pollutant, providing that measurements of wind velocity are available.

Tests carried out for the National Aeronautics and Space Administration of the United States have indicated the potential application of the correlation spectrometer for monitoring air pollution from orbiting satellites. A traverse was made in 1969 across Chicago with two spectrometers mounted in a high-altitude balloon recording profiles of $NO_2$ and $SO_2$. The significance of the test was that it established the feasibility of remotely sensing vertical burdens of these gases over Chicago from an altitude of 120,000 ft through the greater part of the earth's atmosphere and ozonosphere, using the UV and blue-visible spectra (Moffat and Barringer, 1969).

0-201-04245-2

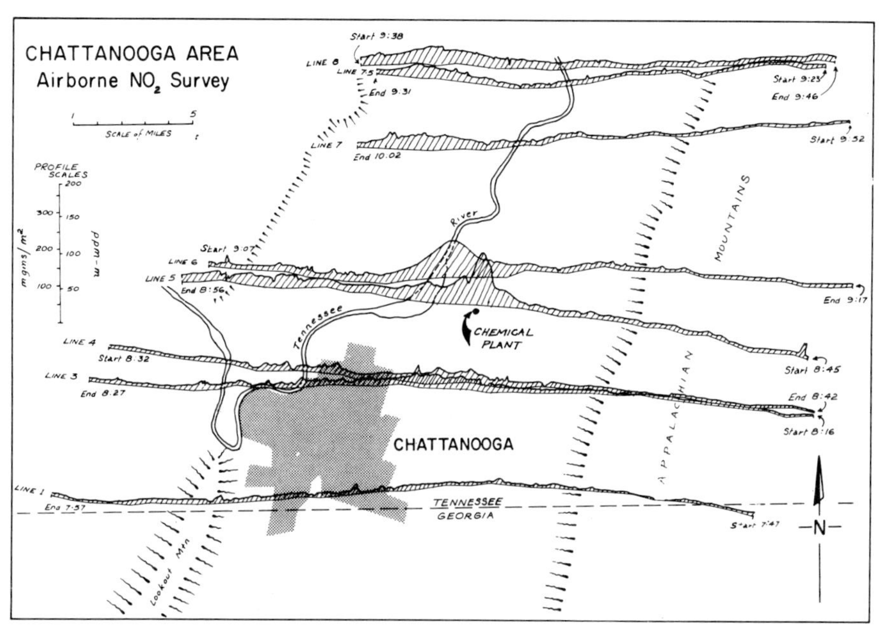


**Fig. 8.13.** Airborne profiles across Chattanooga, Tennessee, showing the remotely sensed vertical burden of nitrogen dioxide over an industrial area. (BRL, Toronto. HEW)

The correlation spectrometer has a number of other applications in the passive mode, apart from pollution monitoring. Experimental surveys have been carried out over coastline areas to measure the distribution of iodine vapor over kelp beds. Anomalous concentrations of iodine occur over such beds, and it is expected that the iodine detection method will have application also in defining general areas of biological activity in the oceans. This is due to the fact that many marine organisms act as concentrators for iodine (Goldschmidt, 1958), and those which live near the sea surface are believed to liberate some of this iodine to the atmosphere in elemental form (Kylin, 1929, 1931). There is economic significance to this fact, since plankton are the first stage in the food cycle for fish and the presence of anomalous iodine concentrations over the ocean could be an indicator of primary fish food sources and hence of fish themselves. This method may supplement another remote sensing technique under development, namely, the detection of chlorophyll in the oceans by optical methods.

Recent tests over an oil field area in California have also indicated that anomalous concentrations of iodine vapor can occur in the atmosphere overlying oil field source rocks. This is attributed to the marine biological associations of oil and the well-known occurrence of concentrations of iodine in oil field brines (Collins and Egleson, 1967).

0-201-04245-2

Numerous other potential applications of remote sensing correlation spectrometry in the passive mode include the surveillance of forest fires, the detection of gases associated with military targets, measurements of water vapor distribution, meteorological and atmospheric studies, and the study of planetary atmospheres.

The optical correlation spectrometer can also be used in an active mode with the spectrometer looking at a distant chopped light source or inspecting an area illuminated by the modulated light from a specially designed searchlight (Newcomb and Millan, 1970). In the active mode the instrument is capable of measuring average gas concentrations along a beam of light, and promising tests have been made of the feasibility of monitoring the pollution across airports by this method. Measurements have also been made along beams of light directed diagonally across freeways just above the traffic, so that it becomes possible to continuously monitor the pollution levels encountered by persons traveling on the highways.

Much of the gas detection work undertaken in correlation spectroscopy was originally carried out in the ultraviolet and visible portions of the spectrum, but increasing attention is now being paid to infrared studies. A very broad range of gases has characteristic infrared absorption spectra, some of which extend into the thermal infrared, a fact which makes it possible to detect these gases during the hours of darkness. Correlation interferometers have also been constructed which are capable of detecting gases in the infrared at high sensitivity (Davies, 1970). These interferometers are based on the Michelson principle, in which rotation of a thick compensation plate generates Fourier transforms of the spectra being detected. This plate is synchronized with a correlation disk carrying a stored memory of the interferogram of the spectrum being detected. The device offers considerable promise for achieving exceptionally high sensitivity in the infrared since it features an extremely large light throughput.

0-201-04245-2

Chapter 9

# Sensors for Spacecraft

*J. Lintz and D. S. Simonett*

## I. INTRODUCTION

Although both oblique and vertical aerial photographs are familiar to most people, there was a dramatic novelty to the first colored photographs taken from space. Made by one of the Project Gemini astronauts with an unauthorized hand-held camera, these early pictures were especially noteworthy for their synoptic views (all of peninsular India in two single frames, for example) and for the value of the color rendition. They whetted the appetite of earth scientists to move to operational space sensing. As early as 1962 or 1963 preliminary consideration was given to total global photographic coverage.

In the early 1960's what later became the National Aeronautics and Space Administration's Earth Resources Program was given impetus by the formation of instrument development teams funded by Dr. Peter Badgley of NASA's lunar program. Composed of university, government, and industry scientists (electrical engineers, physicists, and geologists) these teams had as their function to address the problems of developing instruments for lunar sensing. The teams were instrument oriented–radar, passive microwave, thermal infrared, photography–and ground test sites and laboratory experiments were, in the first instance, selected as surrogates for the lunar surface. Field studies were begun with volcanic sites in desert areas (Pisgah crater and Mono craters in California), and laboratory analysis commenced with simulated lunar surface materials, including expanded vermiculite, dry sand, and unweathered volcanics. By 1966, under the guidance of Dr. Badgley, the program was shifted from a lunar- to an earth-sensing orientation, and the teams were expanded to include a wide variety of earth scientists, ranging from agriculturalists, foresters, geographers, and hydrologists to oceanographers. In parallel with the instrument development teams NASA enlisted the support of federal agencies with earth resource research, management, and development responsibilities and funded research in remote sensing in the agencies and in additional university and industrial groups. These user agencies–U.S. Department of Agriculture (Agriculture, Rangeland, and Forestry), U.S. Department of Interior (Geology, Hydrology, Geography, and Land Management), and Naval Oceanographic Office–chose additional test sites, widely scattered throughout the United States, of interest to the various resource disciplines. Finally, NASA sponsored the development of remote sensing overseas and flew missions, especially in Mexico and Brazil, in recognition of the potential value of space sensing to the developing world.

0-201-04245-2

*Remote Sensing of Environment*, Joseph Lintz, Jr. and David S. Simonett (eds.) ISBN 0-201-04245-2

By the mid-1960's this fabric of a community of scientists, instrument development specialists, resource development agencies, and NASA was welded into a strong earth-sensing experimental program. As the early results from these experiments came in, as the potential practical as well as scientific value of earth sensing became more fully appreciated through the photographic experiments on Apollo 9, and as the use of satellites for meteorology, navigation and telecommunication expanded and became increasingly successful, planning for the first unmanned satellite specifically to address earth resource problems began to take form in 1967 and 1968. In 1969 the Earth Resources Technology Satellite Program was initiated. Worldwide participation was invited in 1970. The launch of the first Earth Resources Technology Satellite (ERTS-1, renamed LANDSAT 1) took place in July 1972.

## II. LANDSAT SATELLITES

In mid-1975 LANDSAT 1 was still providing useful information over North America, but its capabilities to tape record data collected outside the range of recording stations had degenerated to a point where it was useful only over North America. LANDSAT 2 was launched in January 1975, 9 days out of phase with LANDSAT 1, and is expected to be in full operation for approximately 20 months, with data restricted to North America thereafter. LANDSAT C, approved for construction and launching, is expected to be launched during 1978. (The NASA scheme for assigning letters and numbers to satellites is confusing to the uninitiated but is actually quite simple. Planned satellites are assigned letters; once launched and in operation satellites are designated by numbers thereafter.)

The LANDSAT program is experimental and was intended to establish the value of relatively coarse-resolution, large-area, synoptic, reflective multispectral imagery for earth resource analysis. In order to have the widest form of scrutiny and analysis, NASA sponsored several hundred investigators, including almost 40 foreign countries, and funded about one half of the domestic experiments. Experiments were funded in all of the following areas: (*a*) agricultural production, (*b*) water resources management, (*c*) rangeland management, (*d*) forestry management, (*e*) land use and urban and regional planning and management, (*f*) environmental conservation and management, (*g*) geologic survey and mineral resource management, (*h*) marine resource, oceanography, and coastal engineering, (*i*) cartographic and thematic mapping applications, and (*j*) disaster warning and relief.

Components of the LANDSAT System (see Figure 9.1) include (1) solar illumination, (2) modified by passage through the atmosphere, (3) variably reflected from the ground surface, (4) again modified by atmospheric absorption and scattering, and (5) sensed by the spacecraft sensor; (6) power for the sensor, for satellite altitude and orbital control, and for telemetry to the earth is provided by solar cells linked to nickel-cadmium batteries; (7) the sensed signal is tape recorded for later telemetry or is directly processed and telemetered to earth (in line of sight to +5°

0-201-04245-2

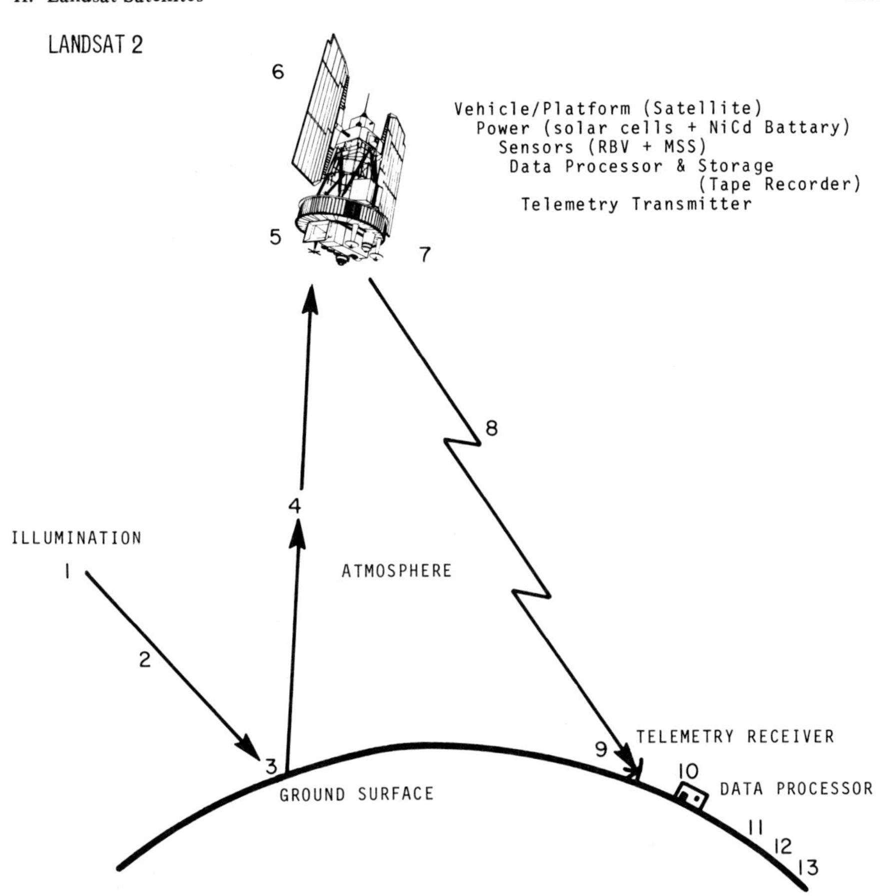


**Fig. 9.1.** Image data acquisition, telemetry, processing, distribution, and analysis flow chart for LANDSAT. 1: Solar illumination; 2: solar illumination modified by passage through the atmosphere; 3: reflection from the ground; 4: propagation through the atmosphere of reflected energy; 5: signal sensing at the satellite; 6: power supply from solar cells; 7: signal processing (e.g., digitizing, tape recording, and transmission); 8: propagation through the atmosphere; 9: telemetry received by ground receiving station; 10: signal processing for geometric, radiometric, and other corrections; 11: transmission to data bank; 12: distribution and public access facility; 13: analysis by users.

of the horizon); (8) the telemetered signal is propagated through the atmosphere as a digital signal; (9) the telemetered signal is received at a ground antenna; (10) the received signal is processed in a variety of ways, (11) stored in a data bank, (12) made available to the general public in a distribution and public access facility, and (13) analyzed. Discussion of these components is deliberately unbalanced in the following presentation with emphasis on the aspects compatible with this book.

LANDSAT is a medium-sized orbiter with the capability to achieve close orientation with respect to the earth and to modify its orbit slightly upon command

0-201-04245-2

from mission control. Solar panels unfold to power the sensors, attitude controls, and communications instruments. The orbit of LANDSATs 1 and 2 was carefully selected after lengthy consultation with the scientific community. Until the orbit was fully established, few parameters of the final products could be known. The highest priorities were given to (1) useful life span of a full 12 months to allow monitoring of a complete growing cycle in the temperate zones; (2) a near-polar orbit to cover most of the globe's land areas: the selected near-circular orbit (99° inclination) yields data to within 9° of the geographic poles; and (3) a constant local sun time of about 9:42 A.M. (originally intended for 9:30, but the LANDSAT 1 launch was late). Thus the satellite is sun synchronous and at a relatively less cloudy time of day, with adequate solar illumination for agricultural sensing at the Northern Hemisphere summer solstice. When these requirements were met, an orbital height of approximately 900 km was established and a sensor spatial resolution of 90-150 m. The actual spatial resolution varies, however, with the band and the scene contrast. By and large, these parameters are more favorable for rangeland, hydrologic, and oceanographic uses than for agriculture, cartography, geology, forestry, geography, land-use, and urban studies. All these disciplines, however, have obtained much valuable information even at the 100-m resolution.

The orbit allows the satellite to circle the earth 14 times each day as the earth revolves beneath it. The orbital parameters are such that the segment of the earth being sensed each day is adjacent to the previous day's coverage and partially overlaps it. The fact that 18 days (251 revolutions) are required to complete the cycle means that each segment is covered once every 18 days or approximately 20 times per year (Figures 9.2 and 9.3).

Because of its altitude and narrow instrument field of view (11½°) LANDSAT imagery is near-orthographic: relief displacement per 1000 m is of the same order as the resolution, namely, 100 m. The sun-synchronous nature means that adjacent areas may be seen on successive days with virtually no change in either solar illumination angle or azimuth, leading under cloud-free conditions to mosaicking of very large areas with essentially the same sun angle and direction. The higher the latitude, the greater is the seasonal contrast in solar illumination angle and azimuth at 9:42 A.M. (LANDSAT 1). The within-season changes are smallest, as would be expected, at the solstices, and greatest at the equinoxes. The ranges are as follows for latitude 30° N:

| *Season* | *Solar Elevation (deg)* | *Azimuth (deg)* |
|---|---|---|
| Summer solstice (June 22) | 60 | 90 |
| Equinoxes (March 21, September 22) | 45 | 125 |
| Winter solstice (December 22) | 28 | 145 |

The payload of the satellite has two sensor systems plus ancillary instrumentation. The first system is a return beam vidicon camera (RBV) with capacity to generate data in three channels. The selected channels were (in micrometers) 0.475-0.575 (blue-green), 0.580-0.680 (green-yellow), and 0.698-0.830 (red and

0-201-04245-2

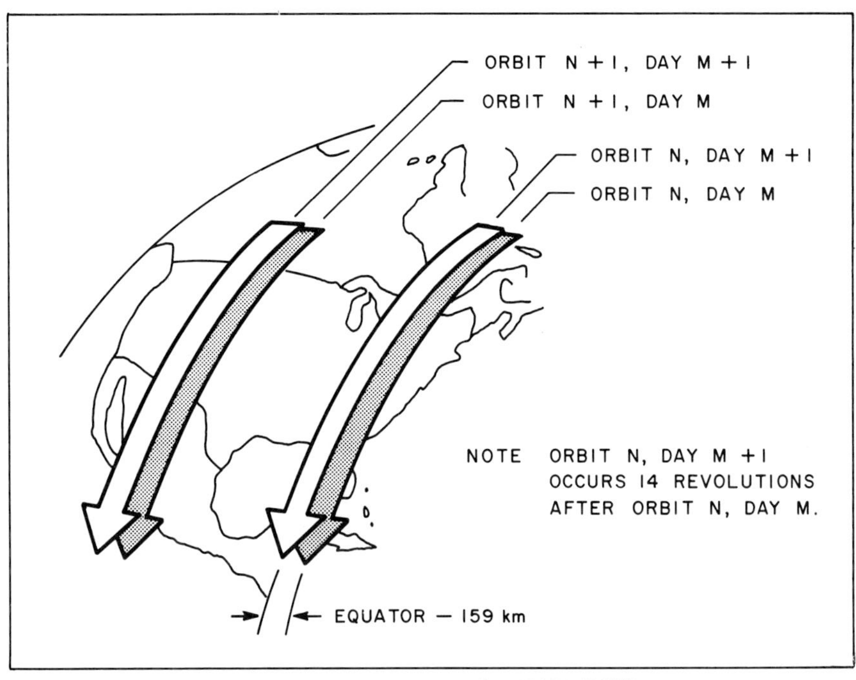


**Fig. 9.2.** LANDSAT ground coverage pattern. After NASA (1972).

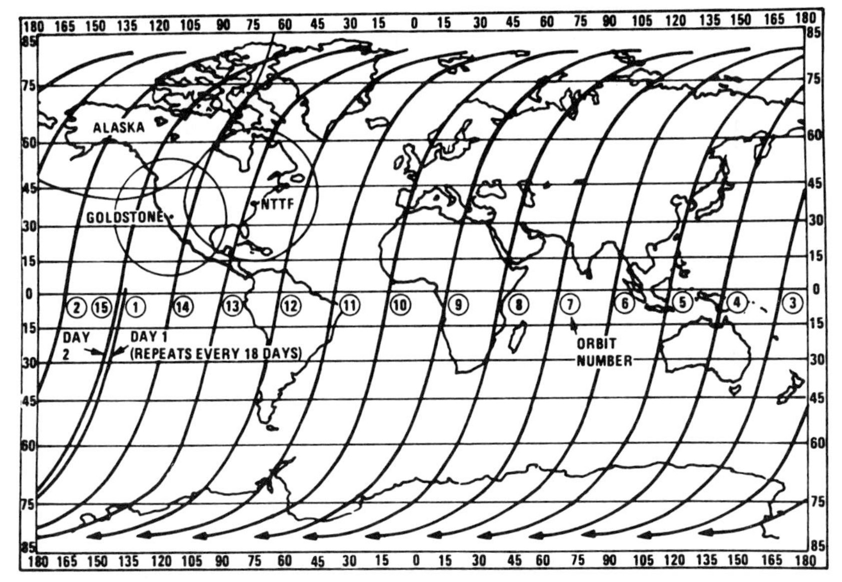


**Fig. 9.3.** Typical LANDSAT ground trace for 1 day (only southbound passes shown). After NASA (1972).

0-201-04245-2

infrared). When the satellite was over North America, the output of the cameras was scanned and directly telemetered to the ground, and at other times was recorded for transmission. The RBV had a field of view 185 km square, and the three channels were prefocused for 100% overlap. The RBV system generated some data of excellent quality in the early orbits of LANDSAT 1. However, it was found to be using more electrical energy than had been programmed and accordingly was shut down early in the life of the satellite.

The second sensor system is a four-channel multispectral scanning system (MSS) (Figure 9.4). The scanning technique employed an oscillating plane mirror. The scanning length was 185 km at approximately right angles to the orbit trajectory. The four channels selected for LANDSATS 1 and 2 were (in micrometers) 0.5-0.6 (green), 0.6-0.7 (yellow), 0.7-0.8 (red and near-infrared, and 0.8-1.1 (near-infrared). An additional thermal infrared channel (10.4-12.6) with a spatial resolution of 240 m is scheduled for inclusion on LANDSAT C. The MSS and RBV systems are sensing at similar wavelengths and should be expected to create nearly identical data via the two differing techniques. Although NASA had anticipated that the RBV would prove superior for cartographic and other uses requiring higher geometric fidelity, the early shutdown of the RBV on LANDSAT 1 interrupted this experiment. (In LANDSAT 2, both instruments are functioning properly.) In practice the performance of the MSS proved better than anticipated with respect to geo-

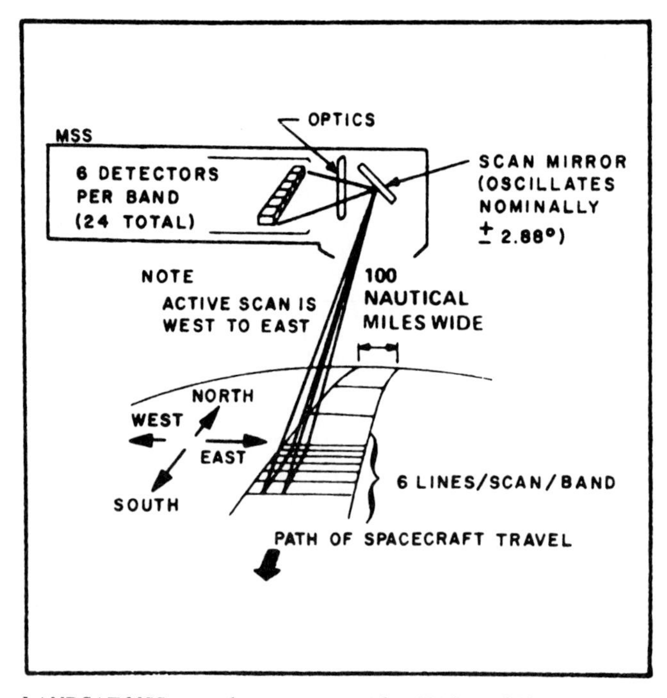


**Fig. 9.4.** LANDSAT MSS ground scan pattern. After NASA (1972).

0-201-04245-2

metric fidelity, and this superiority, coupled with its inherent advantages, has made it a star performer.

The properties of the MSS which make it such a valuable instrument are as follows:

1. Identical geometry in each channel, achieved by beam splitting, leading to close geometric congruence (band-to-band registration within one picture element or pixel). This greatly facilitates computer pattern recognition studies and the production of three-color image combinations, using three channels or ratios of channels.
2. High radiometric fidelity—a characteristic typical of multispectral scanners but not of either RBV or cameras.
3. Wide range in recorded gray-scale values (wide dynamic range), considerably greater than that available in black and white photography in the same bands. This wide dynamic range is divided into 256 equal gray-scale steps, a number far in excess of the corresponding value for film recording. The digital tapes therefore include many data lost when film recording is used.
4. Digital data record in *x-y* format, readily and directly usable for computer processing. The digital tapes are thus referred to as CCT's (computer-compatible tapes).

Both the RBV and the MSS lap sideways sufficiently for stereoscopic viewing, especially valuable for structural geology studies. In addition, the RBV has 10% forelap. Sidelap varies as the cosine of the latitude and is designed for 14% at the equator, leading to 26% at 30°, 34% at 40°, and 85% at 80° latitude. Thus the higher latitudes not only have available a greater stereoscopy (most satisfactory about latitude 50°) but also have in effect double viewing on successive days, on 18-day centers, thus increasing the likelihood of cloud-free coverage. A number of experiment sites were deliberately chosen in the overlap zones for this reason.

The ancillary payload includes a tape recorder and telemetering instrumentation. The tape recorder is used to record data collected outside of North America, and it dumps each time it comes within range of the North American telemetry receiving stations. In LANDSAT 1 it exceeded its life expectancy by 100% but eventually failed. Telemetry is accomplished with S-band instrumentation. United States ground stations are located at Fairbanks, Alaska; Goldstone, California; and Beltsville, Maryland. In mid-1975 ground stations were also installed in Canada and Brazil, and additional stations were under construction, or contracted for, in Zaire, Iran, and Italy, and under consideration in Saudi Arabia, Canada (a second one), and Australia. Venezuela had considered and rejected a ground station at this time because of the excessive cloud cover over the country.

Communications channels provided in addition to the S-band for data telemetry include command functions, satellite monitoring (tracking), and receivers for the data collection platforms, which send up "ground truth" information of especial value in hydrologic studies (Figure 9.5). More details on these data collection platforms can be found in Chapter 12.

0-201-04245-2

**Fig. 9.5.** Data collection system for transmitting point instrument measurements and calibration to LANDSAT. This information is then retransmitted to central ground stations. After NASA (1972).

The NASA Earth Resources Program established a large data processing facility at Goddard Space Flight Center. Here all telemetered data are received, either directly or via Fairbanks or Goldstone. Bulk processing involves the creation of master negatives from the data on magnetic tape. From these master negatives large quantities of copies can be run off. For bulk processing NASA offers its principal investigators a variety of products, including 70-mm negatives and positives and 240-mm positive and paper prints. Color-combined formats at 240 mm include film positives and paper prints. Additionally, NASA offers duplicate magnetic tapes for those interested in doing their own processing. All these products may be purchased also at the Sioux Falls (EROS) data center of the U.S. Geological Survey.

A second and more costly category of LANDSAT data requiring substantial additional processing is labeled precision processing. Here, select frames of bulk-processed data are further corrected geometrically for both the MSS and the RBV and radiometrically for the RBV. Ground control points are used in this process.

0-201-04245-2

Precision process formats include 240-mm materials and magnetic tapes. The photographic outputs include negatives, positives, and paper prints for black and white (individual channels) plus transparencies and prints for the color-combined materials.

The bulk-processed images remove a number of sources of geometric errors, such as scale variations due to satellite altitude variations, internal systematic distortions, dynamic attitude and velocity variations, image skew arising from earth rotation during scanning, and a variety of other mechanical scanning defects. The residual root-mean-square (rms) errors in latitude and longitude of point location are roughly 775 m for the RBV and 750 m for the MSS.

The equivalent figures for precision processing are about 100 m for the RBV and 245 m for the MSS.

Internal geometric accuracies (within-band, within-scene) and registration accuracies (band-to-band) are as follows:

| *Approximate Internal Geometric Accuracies (m)* | *Band-to-Band Registration Accuracies (m)* |
|---|---|
| MSS: bulk 300; precision 50 | MSS: bulk 159; precision 154 |
| RBV: bulk 70; precision 50 | RBV: bulk 336; precision 118 |

Radiometric accuracies, unlike geometric accuracies, which do not differ much between the RBV and MSS, are strikingly different for the two instruments. The net radiometric errors are as follows:

**Net Radiometric Errors (%)**

| *Instrument* | *Processing* | *Film* | *Computer Tape* |
|---|---|---|---|
| RBV | Bulk | 10 | 30 |
| RBV | Precision | 8 | 7 |
| MSS | Bulk | 6 | 2 |
| MSS | Precision | 6 | 6 |

From the above tables it is clear that by far the highest radiometric accuracies are obtained with bulk-processed MSS, computer-compatible tapes, and that the bulk MSS band-to-band registration accuracies are acceptably close to the precision MSS and RBV values. Only for internal geometric accuracies and latitude and longitude point locations are the MSS bulk products notably inferior to the precision products. More recent improved bulk MSS processing has in fact reduced both the latter sources of error, so that now the bulk MSS CCT's are clearly the preferred mode of data format and use.

The information telemetered to the satellite from the data collection platforms and retelemetered to Goddard is also available on magnetic tapes, on punch cards, and on computer printouts. Additional details of the LANDSAT satellites and their instruments may be found in the *ERTS-1 Data User's Handbook* (NASA, 1972).

0-201-04245-2

## III. SKYLAB

The manned satellite SKYLAB, which finished its mission in early 1974, carried a mixed complement of sensors for resource evaluation, including a six-band multispectral camera (Table 9.1), a high-resolution camera (Table 9.2), a 13-channel scanner which covered the visible, near-, mid-, and thermal infrared regions (Table 9.3), a single-channel microwave radiometer (1.43 GHz), and a radar altimeter-scatterometer-radiometer (Figure 3.2). The quantity of data obtained by SKYLAB, although substantial, was constrained by the amount of film and magnetic tape which could be carried to and from the satellite by the astronauts and by conflicts with other experiments on board. SKYLAB potentially could have obtained data over one-half of Earth between latitudes 50° N and S. In fact much less was acquired with the multispectral and high-resolution cameras, coverage being confined to small areas where experiments were underway outside the United States, except for Latin America (about one-third covered) and the United States itself (fully covered).

Correspondingly small areas were covered with the multispectral scanner. Severe noise problems in most bands, malperformance of some detectors, and a difficult and costly working format for the S-192 multispectral scanner significantly reduced the value of this sensor. On the other hand, the superb photography obtained with both camera systems has already demonstrated the great value of high-resolution space photography in conjunction with LANDSAT data. Further details regarding the SKYLAB instruments and performance and numerous examples of color im-

**TABLE 9.1**
**SKYLAB S-190, a Multispectral Photographic Camera**
**Lenses–six f/2.8 15.24 cm. Focal length (21.2° FOV)**
**Coverage–163 km square (26,585 km²)**

| Film | Spectral Coverage (μm) | Estimated Ground Resolution *[a] for Low-Contrast Targets, 1.6:1 (m)[b] | Visual Edge Match **[a] Estimated at 6:1 Contrast Ratio Negatives: Original (m)[b] | Visual Edge Match **[a] Estimated at 6:1 Contrast Ratio Negatives: Duplicate (m)[b] |
|---|---|---|---|---|
| B & W IR (EK2424) | 0.7–0.8 | 73–79 | 54 | 64 |
| B & W IR (EK2424) | 0.8–0.9 | 73–79 | 59 | 70 |
| Color IR (EK2443) | 0.5–0.88 | 73–79 | 52 | 70 |
| High-resolution color (SO-356) | 0.4–0.7 | 40–46 | 26 | 37 |
| B & W Pan-X (SO-022) | 0.6–0.8 | 38–38 | 27 | 31 |
| B & W Pan-X (SO-022) | 0.5–0.6 | 40–46 | 30 | 31 |

[a] Source: *NASA (1974), **Kenney (1975).
[b] Resolution in meters per line pair.

0-201-04245-2

agery and analyses may be found in *SKYLAB Earth Resources Data Catalog* (NASA, 1974).

**TABLE 9.2**
**SKYLAB S-190B, Earth Terrain Camera Configuration**
**Lens–f/4 Focal length 45.72 cm.**
**Coverage–109 km square (11,950 km$^2$)**

| *Film* | *Spectral Coverage (μm)* | *Nadir Estimated Ground Resolution*[a] for Low-Contrast Targets, 1.6:1 (m)[b]* | *Nadir Visual Image Evaluation Resolution**[a] Estimated at 2:1 Contrast Ratio Duplicate Negatives (m)[b]* |
|---|---|---|---|
| High-resolution color (SO-242) | 0.4–0.7 | 21 | 18–26 |
| B & W high-definition (EK3414) | 0.5–0.7 | 17 | 14 |
| Color IR (EK3443) | 0.5–0.88 | 30 | N.A. |
| High-resolution color IR (SO-131; SL-4) | 0.5–0.88 | 23 | N.A. |

[a] Source: *NASA (1974); **Oldham (1974).
[b] Resolution in meters per line pair.

**TABLE 9.3**
**SKYLAB S-192, Multispectral Scanner Configuration**
**IFOV–79.3 m square ground coverage. Swath width–68.5 km.**

| *Band* | *Description* | *Spectral Range[a]* |
|---|---|---|
| 1 | Violet | 0.41– 0.46 |
| 2 | Violet-blue | 0.46– 0.51 |
| 3 | Blue-green | 0.52– 0.56 |
| 4 | Green-yellow | 0.56– 0.61 |
| 5 | Orange-red | 0.62– 0.67 |
| 6 | Red | 0.68– 0.76 |
| 7 | Near-infrared | 0.78– 0.88 |
| 8 | Near-infrared | 0.98– 1.08 |
| 9 | Near-infrared | 1.09– 1.19 |
| 10 | Mid-infrared | 1.20– 1.30 |
| 11 | Mid-infrared | 1.55– 1.75 |
| 12 | Mid-infrared | 2.10– 2.35 |
| 13 | Thermal infrared | 10.2 –12.5 |

[a] Source: NASA (1974).

0-201-04245-2

## IV. LANDSAT D

A more advanced multispectral scanner known as the thematic mapper is under consideration for LANDSAT D, another sun-synchronous satellite (11 A.M.), tentatively scheduled for launch in 1980. This satellite was originally named the EOS (Earth-Orbiting Satellite).

The thematic mapper was the concern of a meeting called by NASA in April 1975 to consider the spectral resolution and band location, number of channels, radiometric and spatial resolution, swath width, geometric accuracy, and other parameters of the mapper (Harnage and Landgrebe, 1975). In comparison to LANDSATs 1 and 2, which had four broad channels, and LANDSAT C, which adds a fifth thermal IR channel, the thematic mapper on LANDSAT D will have six and possibly seven channels, more sharply defined and providing higher radiometric and spatial resolution. It will also, as noted above, have a later (11 A.M.) orbit.

The first LANDSAT was decidedly an early experimental satellite, and the band selection and orbital parameters involved a series of compromises, principally between agricultural, hydrologic, and geologic uses. The thematic mapper in the LANDSAT D satellite, on the other hand, would be optimized for near-noon agricultural use, with band locations, widths, radiometric resolutions, and spatial resolutions approximately as given in Table 9.4. A swath width the same as that of the other LANDSATs (185 km) is planned for the thematic mapper.

In addition to the thematic mapper, other instruments suggested for the EOS satellite include a high-resolution pointable imager (10-m resolution, 40-km swath

**TABLE 9.4**
**Expected Configuration of the LANDSAT D (EOS) Thematic Mapper (Multispectral Scanner) for Agricultural Targets[a]**

| *Band[b] (μm)* | *Dynamic Range (range from minimum to maximum reflectance)* | *Sensitivity (radiometric[c] resolution)* | *Spatial Resolution (m)* |
|---|---|---|---|
| 0.52–0.60 | 4–58% | 0.5% | 30–40 |
| 0.63–0.69 | 4-53% | 0.5% | 30–40 |
| 0.74–0.80 | 2-75% | 0.5% | 30–40 |
| 0.80–0.91 | 2-75% | 0.5% | 30–40 |
| 1.55–1.75 | 2-50% | 1.0% | 30–40 |
| 10.4 –12.5 | 270–330°K | 0.5°K | 120[d] |

[a] Source: Harnage and Landgrebe (1975).

[b] Consideration may be given to a possible seventh band from 0.45 to 0.52 μm.

[c] Expressed as $NE\,\Delta\rho$ at 13% reflectance for total system in all visible and near-IR bands for the range of reflections associated with vegetation problems. Here $NE\,\Delta\rho$ is defined as the smallest system-detectable difference in reflectance with a signal-to-noise ratio of 1. There is an $NE\,\Delta T$ of 0.5° K at 300° K for the total system, including atmospheric attenuation.

[d] Probably a 3× or 5× multiple of resolution in the thermal and other bands for convenient centering and registration.

0-201-04245-2

width, and four bands the same as those of LANDSAT 1), an atmospheric aerosol and scattering sensor, and an X- and L-band synthetic-aperture radar. In mid-1975 it did not appear likely that these additional sensors would be carried on LANDSAT D. At that time some consideration was still being given to having two thematic mappers and/or two satellites at 9-day intervals, and possibly a multispectral scanner of the present LANDSAT type as well.

Before considering other sensors, it is appropriate to discuss the thematic mapper characteristics further, since this instrument is likely to be more useful for agricultural sensing in developing countries (with small-field agriculture) than the earlier mappers of the LANDSAT series. The radiometric resolution will be significantly improved over the capabilities of the earlier LANDSAT satellites, thus potentially leading to slightly higher crop identification accuracies from this factor alone than has been achievable with LANDSATs 1 and 2. The substantial improvement in spatial resolution–if actually achieved by the instrument designers–will also lead to improved crop identification accuracies, since the number of edge (field edge) pixels will be correspondingly reduced. (Identification is always better with interior pixels of a field, which contain only the crop in question. Edge pixels may include coverage of several abutting different crops, pastures, and trees; see Figure 9.6.) Finally, the bands are more sharply defined in LANDSAT D than in LANDSATs 1 and 2 in order to emphasize more clearly fundamental *single* energy-matter interactions. In this respect most of the LANDSAT 1 and 2 channels have overlapped several energy-matter interaction zones.

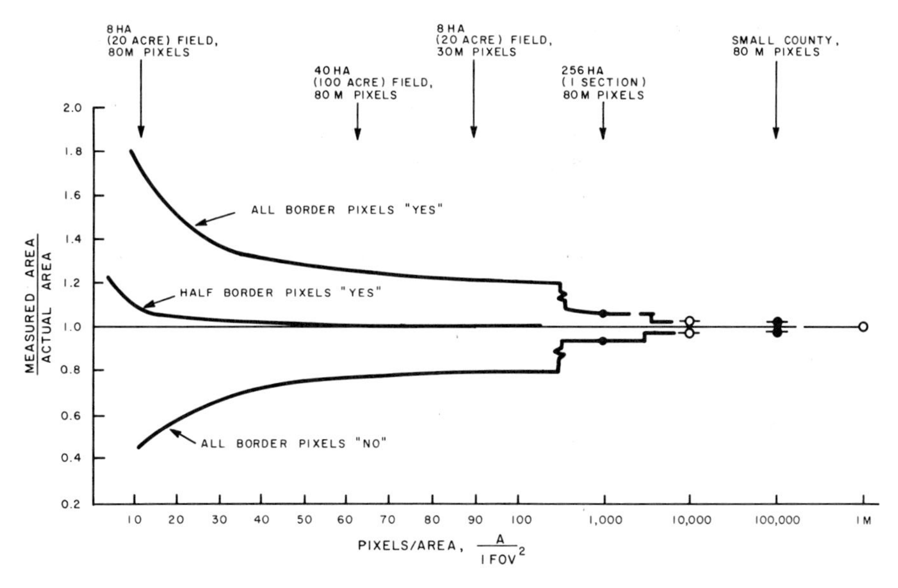


**Fig. 9.6.** Graph of area measurement errors in relation to various agricultural field sizes and picture element (MSS pixels) sizes. After General Electric (1974).

0-201-04245-2

**A. Band 0.52-0.60 μm**

This band has consistently proved useful in many agricultural applications with aircraft data, as reported in studies by Coggeshall and Hoffer (1973), Coggeshall et al. (1974), and Thomson et al. (1974), and in the Corn Blight Watch Experiment (1974). The same references apply to the following bands as well. The 0.52-0.60 μm band carefully straddles the so-called green hump (See Figure 3.2) lying between the two chlorophyll absorption bands between 0.42-0.45 and 0.64-0.67 μm. In addition to being an important band for detecting green vegetation, it is also useful in studies of coastal waters and in identification in geologic and urban-area studies.

**B. Band 0.63-0.69 μm**

This is a very important band carefully straddling the chlorophyll absorption region. Many studies show it to be the most important single band in ADP classification accuracy of green vegetation with full ground cover. The 0.63-0.69 μm band gives very high contrast in many scenes and therefore is also one of the best bands for soil boundary and geologic boundary discrimination. In addition, it is useful in water body and in urban land use studies where the urban areas have moderate to high tree cover.

**C. Band 0.74-0.80 μm**

Inclusion of this band stems from a number of studies with LANDSAT 1 (Rouse et al., 1974), using ratios of bands 5 and 6 (0.6-0.7 and 0.7-0.8 μm). These band ratios are notably sensitive to the amounts of green biomass and moisture in the vegetation and may be used to assess range feed condition. The 0.74-0.80 μm region was tentatively suggested at a Purdue University meeting as being more suitable for this purpose than the 0.70-0.80 μm. Other studies by Hoffer and his coworkers at Purdue University also support the possible use of this band (Hoffer and Fleming, 1975). Further study is needed, however, to confirm the selection of the 0.74-0.80 μm band.

**D. Band 0.80-0.91 μm**

This is the flat area of high-plant near-infrared reflectance and ranks second to the 0.63-0.69 μm region in discrimination of active green vegetation. Unlike the LANDSAT band 7, which continues to 1.1 μm, this has a long-wavelength cutoff at 0.91 μm to avoid the effects of the minor water vapor absorption band at 0.925 μm. In addition to contributing strongly to identification accuracies for growing crops, the 0.80-0.91 μm band is valuable for soil-crop contrast for water-land boundary delineation. It tends, however, to be of less value for within-class (i.e., within-water, within-soil, and within-ocean) discrimination than the 0.63-0.69 μm region.

0-201-04245-2

### E. Band 1.55–1.75 μm

This is known to be an important band for crop green matter identification accuracy and is potentially very important for crop stress detection, since it is sensitive to the water contained in the crop, for moisture estimation in fire-prone wildland vegetation, for surface soil moisture detection shortly after rainfall, and for land-water delineation.

### F. Band 10.4–12.5 μm

The importance of the thermal channel for vegetation classification is well documented. (For aircraft, see the Corn Blight Watch Experiment, 1974, and Thomson et al., 1974; for SKYLAB, see Biehl and Silva, 1975). The 10.4–12.5 μm band is also likely to be useful in crop stress studies, even though the 11 A.M. local sun time passage of the satellite is not optimal for stress detection, for which a 2–3 hr lag would be better. The band also will have some use in estimation of water at some depth in the soil profile, through coupling and differential lags of dry or moist soils. With high-quality detectors this is potentially a most valuable band because of its location, separation from other bands, the fact that emission, not reflection, is measured, and the sensitivity of vegetation of different height to the atmospheric thermal gradient from 0 to 10 m, coupled with the low thermal mass of vegetation. For further information on these topics reference should be made to Thomson et al. (1974), Norwood (1974), and Harnage and Landgrebe (1975).

## V. HEAT CAPACITY MAPPING MISSION SATELLITE (HCMM)

The Heat Capacity Mapping Mission Satellite, carrying a near-IR (0.8–1.1 μm) and a thermal infrared (10.5–12.5 μm) channel will be launched in 1977 or 1978, mainly to investigate the feasibility of identifying surface soil and rock composition (principally in arid and semiarid areas), surface and subsurface soil moisture content, and other resource characteristics through the use of heat capacity and thermal inertia. The greatest value of the HCMM is likely to be for geologic, hydrologic, and agricultural uses on a broad scale, since the spatial resolution will be 500 m. It will be sun synchronous with a 1:30 P.M. local sun time passage over the United States and a 2:30 A.M. nighttime passage (International Astronautical Federation, 1974). These passage times represent a compromise between technical and orbital requirements and are as close as possible to the optimum times of 1:30 P.M. and 4:30 A.M., the average maximum and minimum, respectively, of the daily surface soil temperature heating curve.

The HCMM is essentially similar to the high-resolution surface composition mapping radiometer carried on the meteorological satellite NIMBUS 5, with 0.5–1.1 and 10.5–12.5 μm bands.

The function of the satellite will be to measure the maximum and minimum of the daily heating and cooling curve and the albedo for a given location, and through

0-201-04245-2

maintenance of a daily log of meteorological data (radiation, temperature, rainfall) and soil water data (moisture content and various depth to, say, 50 cm) to develop calibrations for extrapolation to other areas. These data should enable inferences to be made regarding surface rock and soil composition (sand, clay, bare solid rock of acid as opposed to basic composition, etc.) and variations in contained soil water content and should assist in mapping large-scale thermal effluents, estimating plant transpiration and plant stress, and monitoring snow-melt for runoff prediction.

## VI. OCEAN DYNAMICS MEASURING SATELLITE (SEASAT A)

Also due for probable launch in 1978 is SEASAT A, the first of a potential series of ocean dynamics-measuring satellites. SEASAT A will be a long-lived (800-km orbit) satellite with a repeat time of 36 hr (108° inclination), thereby alternating between day and night ocean coverage of a given site at 36-hr intervals. The satellite will continue and expand radar and microwave experiments initiated on SKYLAB and will examine the feasibility of all-weather monitoring of the dynamics of major ocean currents, sea-ice movements, sea states, and tsunami sensing. In addition, monitoring of ocean color, as a guide to potential fish productivity, will be examined.

The instruments to be used in this ambitious program, of interest to the less developed as well as the developed countries, are given in Table 9.5.

**TABLE 9.5**
**SEASAT A Instrument Complement[a]**

| *Instrument* | *Wavelengths* | *IFOV, Nadir Spatial Resolution* | *Swath (km)* |
|---|---|---|---|
| Coherent imaging radar (CIR)—direct measurement of large waves and coastal areas | 22 cm | 25 m | 100 |
| Scanning multifrequency microwave radiometer (SMMR) | Five bands between 6.6 and 0.8 cm | 16–100 km | 1000 |
| Visible and infrared radiometer (VIRR)—ocean color and temperature | 0.52–0.73 and 10.5–12.5 μm | 2–5 km | 1900 |
| Compressed pulse radar altimeter (CPRA) | To measure altitude between spacecraft and ocean to ±10 cm rms. | | |
| Microwave wind scatterometer | To measure surface wind velocity by sensing capillary waves; expected to be accurate to ±2 m at velocities of 25 m/sec. | | |

[a] See International Astronautical Federation, 1974.

0-201-04245-2

## VII. SPACE SHUTTLE

SPACE SHUTTLE, expected to be operational by 1980, will be a manned space aircraft almost as large as a Boeing 707. It will carry an extraordinarily diverse array of instruments as needed for different experiments or operational missions. Launches will be made roughly every 30 days, and missions will last from 7 to 14 days. A long-lived space laboratory, to which the shuttle will join, forms part of the SPACE SHUTTLE system. Potential applications in support of integrated earth resources space sensing, employing unmanned satellites, meteorological satellites, and communications satellites, run the full gamut of agriculture, geology, energy and minerals, forestry, land use, and ocean and hydrologic use studies. The orbits can be tailored to the individual applications and experiments of prime concern during an individual mission.

Figure 9.7, reproduced from a recent study by General Electric Corporation (1974), shows the various types of spacecraft platforms needed for sensing in the SPACE SHUTTLE area and the roles of SPACE SHUTTLE as well as the other platforms, in meeting experimental or operational data-gathering needs. Careful study of this figure shows the *systems* thinking about resource problems increasingly needed in remote sensing programs; it is rare for a single spacecraft or satellite

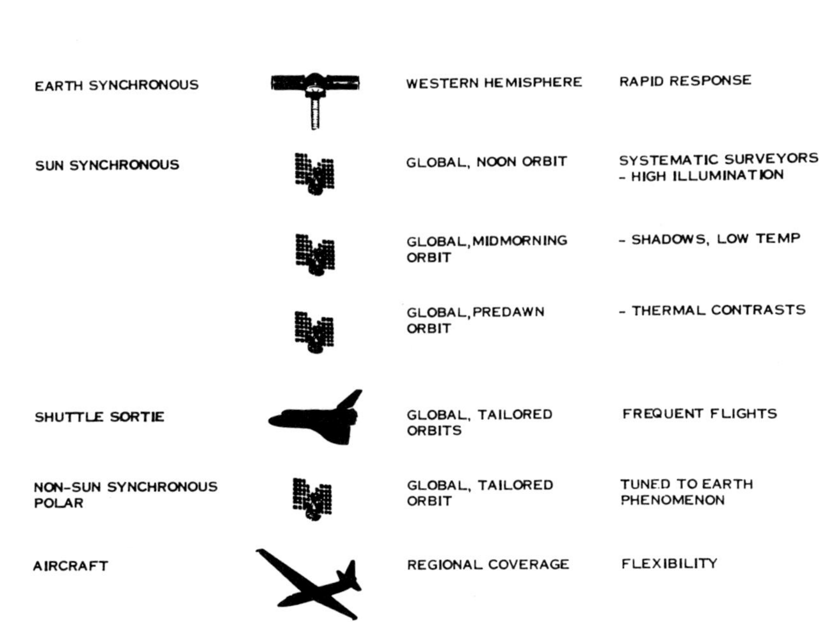


**Fig. 9.7.** The seven basic platform types and their characteristics. After General Electric (1974).

0-201-04245-2

to answer all needs for a given resource discipline or use. The *combination* of *sensors* and platform varies from use to use, as indicated in Figure 9.7.

Experiments to be carried out on SPACE SHUTTLE will lead to the development of improved unmanned satellites, since space-hardened equipment is not needed to the same degree as with the unmanned systems. In addition a variety of special-purpose missions will be run. These will fill a niche in an integrated resource evaluation system, either by carrying one or two bulky instruments essential for the experiment in question or by carrying a complete suite of instruments. Examples of possible instruments and their expected spatial resolutions are given in Figure 9.8, reproduced from a study by TRW/EarthSat (1973).

The general advantages of SPACE SHUTTLE as part of an integrated resource development program include:

- Use of non-space-hardened equipment.
- Function as test bed for techniques, concepts, and instruments.
- Combination of large, heavy instruments.
- Expanded capability for mission planning and phasing.

Specific advantages arising from the numerous and complex instruments which will be carried include:

- Addition of a new dimension to multistage sampling.
- Greater variety of observation cycles.
- Early retrieval in film form.
- Calibration of unmanned spacecraft.

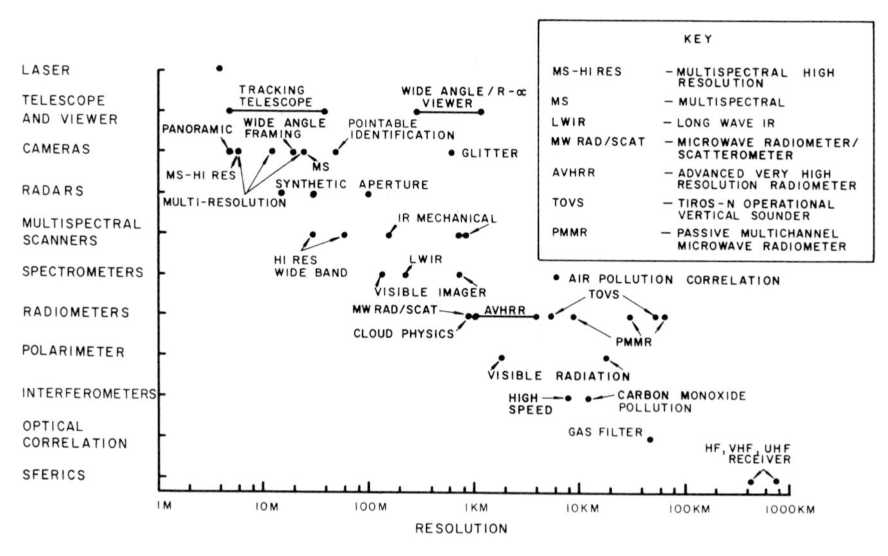


**Fig. 9.8.** Examples of possible instruments and spatial resolutions for SPACE SHUTTLE (given for the nadir and 400-km altitude). After TRW/EarthSat (1973).

0-201-04245-2

- Rapid response to the need for high-quality information.
- Capability for complex experiments requiring the combined use of various instruments.

For more information see TRW/EarthSat (1973) and General Electric (1974).

## VIII. METEOROLOGICAL SATELLITES

The existing meteorological satellites contain sensors of use to earth scientists, and the same will be true of planned ones. In addition, the frequency of coverage—twice daily or more, coupled with progressively improving spatial resolution of some sensors to 0.5 km or better, will make the meteorological satellites increasingly important for Earth survey programs. The meteorological data will be useful for a wide array of atmospheric inputs, atmospheric calibration and correction of other sensors, various modeling problems in hydrology and agronomy, and ground observations on surface and topsoil moisture.

There are currently six operating meteorological satellites: the NOAA 2, NOAA 3, SMS/GOES, NIMBUS 5, DAPP, and the ATS-F systems, all intended for operation through 1976. The sensors carried by these satellites, although of relatively coarse spatial resolution, have high radiometric resolution which make them suitable for use with higher-spatial-resolution LANDSAT data. In snow studies, for example, frequent coarse-resolution meteorological data are of high value in modeling snow melting. LANDSAT data provide infrequent but relatively high-spatial-resolution calibration to improve the interpretation of the frequent but coarse meteorological data.

NOAA 2 and NOAA 3, now operating, provide worldwide coverage of 0.5-0.7 $\mu$m at 3.6 km and of 10-12 $\mu$m at 7.2 km. Selected coverage is obtained at 0.9-km resolution in the same bands for cloud cover by day or night, surface temperature by day or night, and daytime albedo (Bale et al., 1974).

The SMSs (Synchronous Meteorological Satellites), launched May 1974 and February 1975, orbit Earth at 36,000 km, which is synchronous with Earth's rotation at the equator. Thus continuous coverage of any point within the field of view may be obtained. These satellites are planned for a 2-year life. The SMS has two channels: a visible (0.55-0.70 $\mu$m) with 0.9-km resolution, and a thermal (10.5-12.6 $\mu$m) channel with 8-km resolution. It also has a data collection system capable of handling 10,000 data points. The GOES (Geostationary Operational Environmental Satellite) Program will provide two satellites: one at 100° W and another at 60° W, both on the equator. GOES A is scheduled for launch in October 1975 with a 3-year life. Sensors include visible- and thermal-infrared-sensitive instruments with spatial resolutions of 0.9 km and 3.6 km, respectively. Continuous day and night coverage of the United States is planned for cloud cover and surface temperature, with daytime coverage for surface albedo. These will be of importance in detailed energy budget studies for hydrologic and agronomic water and energy budget models.

0-201-04245-2

NIMBUS 5, a NASA research satellite launched in December 1972 and currently in sun-synchronous polar orbit, carries a temperature-humidity radiometer (THIR), visible and thermal infrared scanners with 7.2-km spatial resolution, an electrically scanned microwave radiometer (ESMR) of 25-km resolution, and a microwave vertical atmospheric sounder (1.55 cm and 23.4-km spatial resolution). These sensors permit observation of cloud cover and surface temperature (including microwave temperatures) day or night and will detect cloud situations in which the clouds either are precipitating or contain large amounts of water droplets. In mid-1975 only the ESMR was still functioning.

The DAPP satellites operated by the U.S. Air Force observe four times daily in polar sun-synchronous orbit (6 A.M., noon, 6 P.M., and midnight local sun time) in both visible (albedo) and thermal infrared wave lengths. Spatial resolution is, respectively, 0.9 and 3.6 km.

Applications Technology Satellite-F (ATS-F) is an experimental geostationary satellite containing two very-high-resolution solid-state cameras for the visible and thermal regions. The instruments are designed primarily for coarse-scale, hemispheric, cloud, and weather motion studies with 22-km resolution in the visible and 17-km in the infrared. Focusing, however, can be achieved for 3-km resolution for smaller areas.

NIMBUS 6, launched in June 1975, carries a radiometer of 3.2-km resolution to measure Earth radiation budget. A 37-GHz radiometer is used for recording data for liquid-water vapor measurements when clouds are not present. Also, a vertical microwave sounder operates through nonprecipitating clouds.

For NIMBUS G (1979) a spectrometer is planned for ocean color imaging, plus a radiometer and scatterometer for sea state data. Other passive microwave sensors may also be aboard, but details have not yet been announced. Sensors to detect stratospheric pollution are planned.

TIROS N (January 1979) is expected to contain a sensor package with four channels of data and high spatial resolutions in both the visible and infrared sectors. It will also include an atmospheric temperature sounder and possibly a radiation budget instrument.

SEOS (Synchronous Earth-Observing Satellite) is a 1981–1985 satellite which will be placed in earth-synchronous (36,000-km) orbit. It will sense the entire hemisphere, but high-powered telescopes will provide local areas of high resolution with long dwell times, repeated looks, and a useful life of 24 months. Instrumentation for SEOS includes solid-state cameras, new-generation telemetry, and spectrometers for pollution detection. The satellite is intended principally for studying short-lived phenomena of meteorological hazards (e.g., extensive flooding and violent storms) and features such as estuarine dynamics and major fires.

In mid-1975 several additional satellites similar to SMS/GOES were under construction in Europe, Japan, and the U.S.S.R. LANDSAT-like satellites were under consideration by Canada, Germany, the U.S.S.R., France, and Japan, and various feasibility studies were underway for European Space Agency satellites carrying imaging radar, passive microwave and other sensors, and for a Canadian sea-ice monitoring satellite (International Astronautical Federation, 1974).

0-201-04245-2

## IX. AIRCRAFT PLATFORMS

To support the spacecraft program of the 1975–1985 period, continued, expanded, complementary, and perhaps even competitive aircraft sensing will be required. Although no single platform can meet all requirements, aircraft are the most versatile of the various platforms (Figure 9.7), encompassing helicopters, light aircraft capable of operating to 5000 m, heavy aircraft operating to 10,000 m such as the P3-A, C130 B-52, and Caravelle Jet, and high-altitude aircraft operating to 20,000 m such as the U-2, RB57F, and the new Leartek Harv, now (1975) under design by Gates-Learjet for commercial marketing in 1979. The high-altitude aircraft are the most competitive with certain phases of space sensing in that coverage of normal U.S.G.S. 7½-min quadrangle maps areas may be provided with single-image swath coverage in the east-west direction and 1-m ground resolution (Leartek Corporation, 1975).

Unless systems of very high spatial resolution are carried as part of future spacecraft complements, however, the role of aircraft will be secure in providing the multistage high-resolution data needed for a balanced sampling and calibrated system. Testing of potential space systems may also be carried out in part on aircraft. In addition, there are numerous sensors which cannot function effectively when carried on spacecraft. These include many geophysical instruments, ultraviolet devices, and to some degree laser and certain passive microwave devices. Generally, also, the higher the aircraft flies the larger is the area required for cost-effective survey, a problem at present when many special, single-purpose, and limited-area surveys are typically contracted for commercially. This situation may change as recognition of the various advantages of high-altitude surveying stimulate rethinking of image acquisition policy and practice by federal, state, commercial, and foreign users.

## X. SOME IMPLICATIONS OF FUTURE SATELLITE PROGRAMS

There are two important implications of an expanding, continuing, diverse array of satellites and instruments through the 1980's as described above. First, the opportunity for a large-systems approach to resource monitoring, using various mixes of satellites and instruments as outlined in Chapter 3, will be substantially increased, although the mix will fall short of a theoretical optimum. There will also be sharp improvements and reduction in the cost of ground processing and in the design and implementation of geobase information systems which merge satellite and supporting data.

Second, there will be parallel problems of handling numerous complex, principally digital, forms of data. For the technologically advanced countries and institutions these will be seen as opportunities, but for lower levels of government (state and local governments in the United States) and for the developing world this high technology will be viewed as much in terms of problems and threats as of advantages.

0-201-04245-2

# Color Illustrations

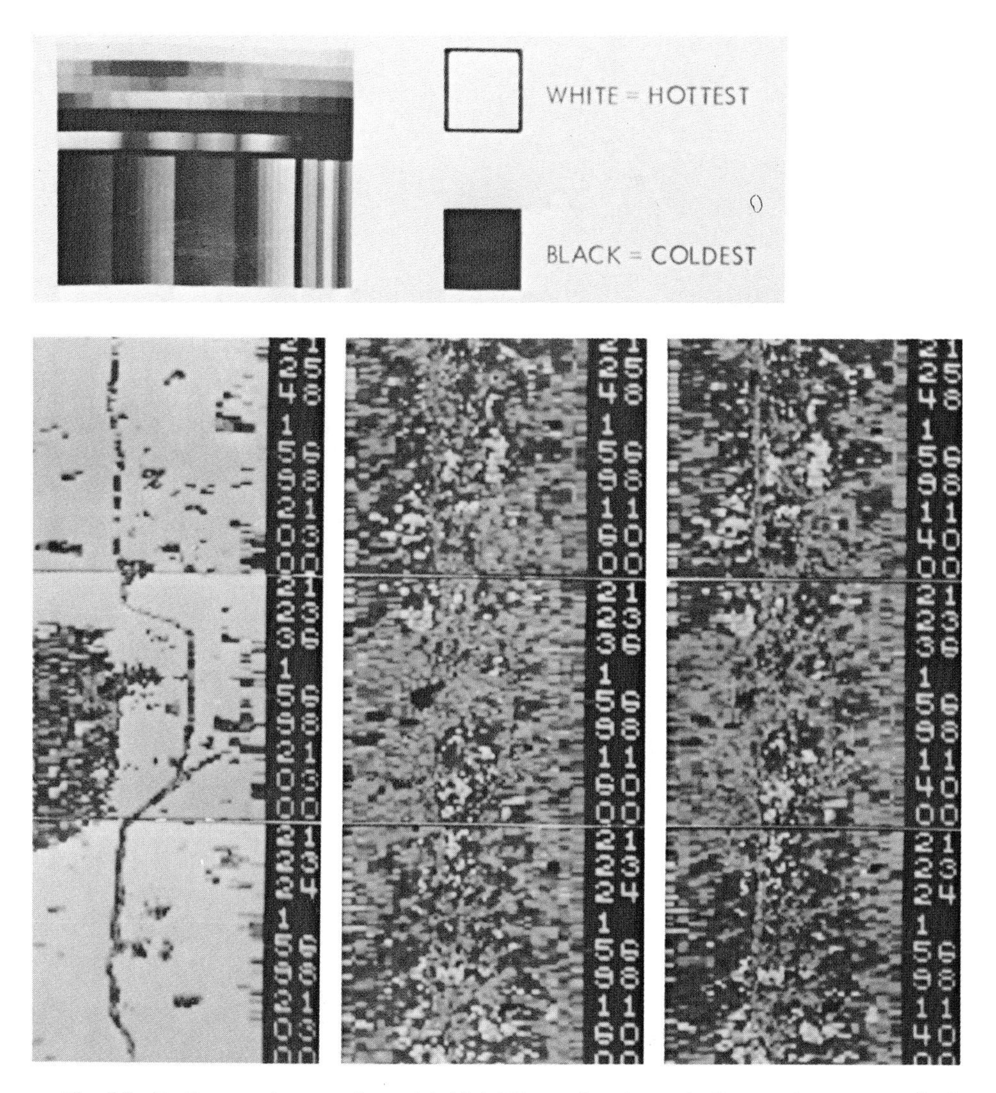


**Fig. 6.8.** Radiometer image, taken with 19.4-GHz passive electronically scanning system, of All American Canal along the United States-Mexico border. Radiometrically cooler area in Frame 21:23:36 corresponds to the city of Mexicali, Mexico. Calexico, California, lies between Mexicali and the curve in the All American Canal. Calexico appears radiometrically warmer than its sister city in Mexico. Courtesy of Aerojet-General Corporation.

ISBN 0-201-04245-2

**Fig. 8.10.** Combined radiometric and aeromagnetic survey over Liberia illustrating differential patterns of geologic response with the two systems. Courtesy Lockwood Surveys Limited, Toronto, Canada.

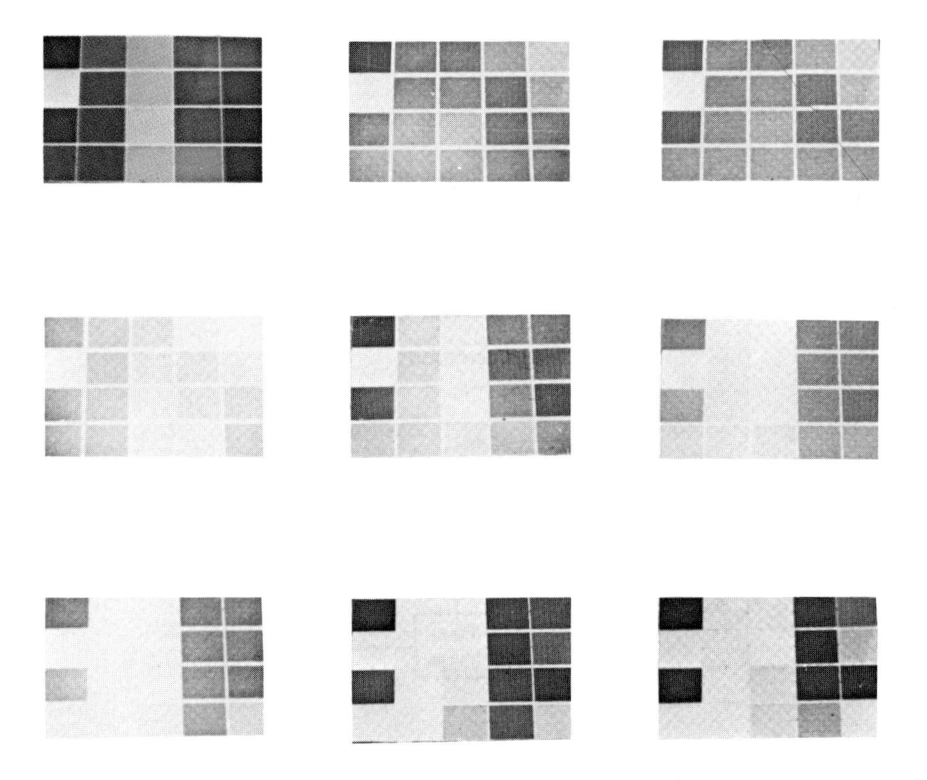

**Fig. 10.4.** Multispectral photography of color cards.

**Fig. 10.15.** Simulated color infrared of Lake Texoma frame.

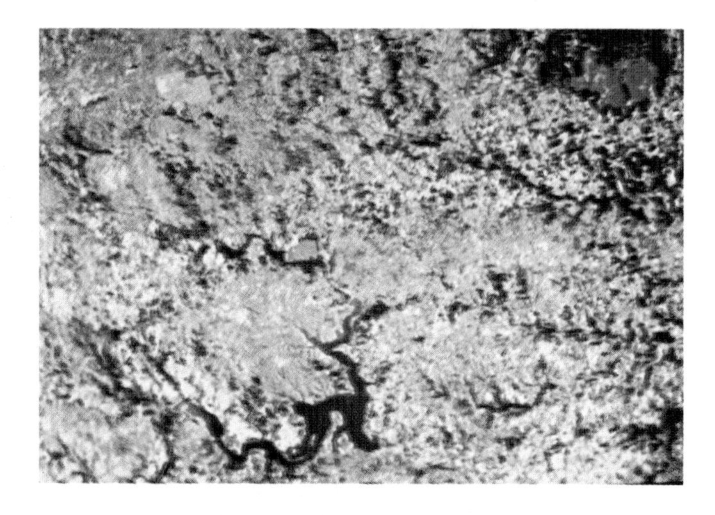

**Fig. 10.16.** Color-coded classification results of Lake Texoma frame.

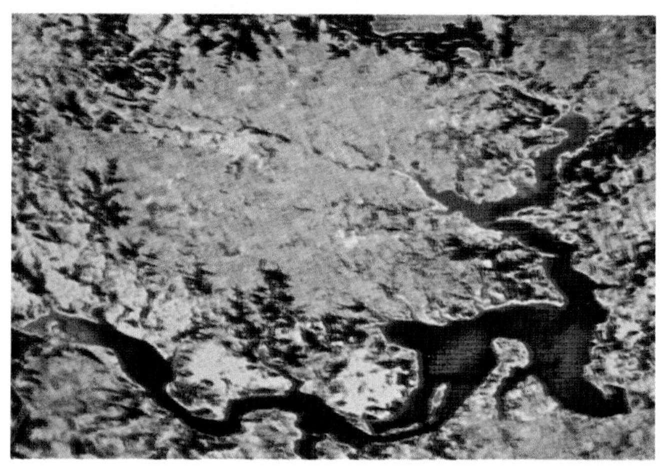

**Fig. 10.17.** Color-coded classification results of Lake Texoma.

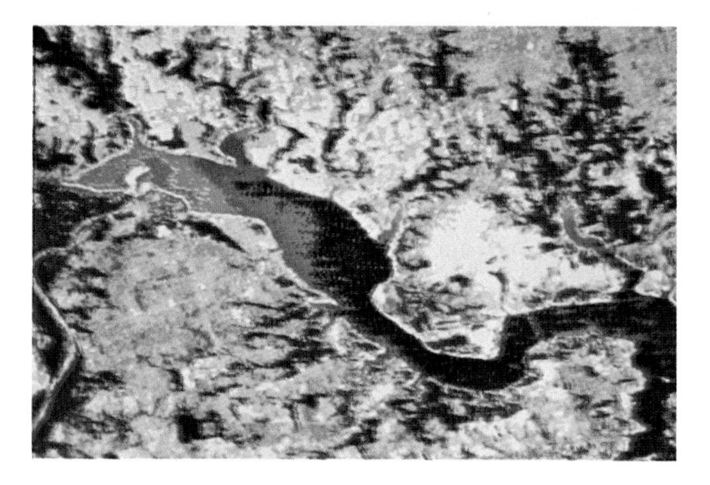

**Fig. 10.18.** Color-coded classification results of the west end of Lake Texoma.

**Fig. 10.19.** Low-altitude view of west end of Lake Texoma.

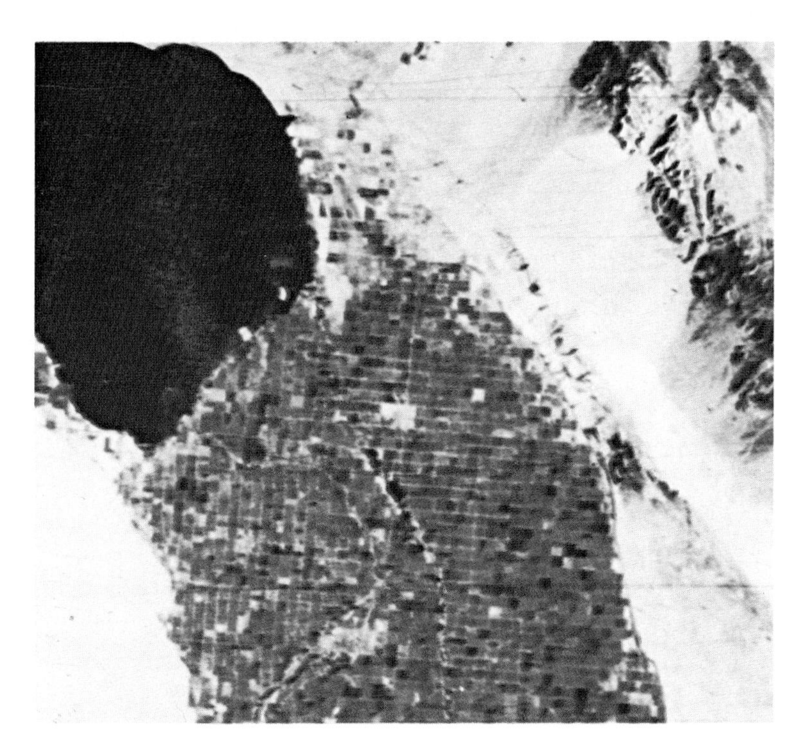

**Fig. 14.3.** Agricultural land-use map and corresponding infrared photograph of the Imperial Valley and Salton Sea Region. Courtesy of Department of Geography, University of California, Riverside.

**Figs. 16.7a. and 16.7b.** Stereo pair taken simultaneously from two aircraft.

**Fig. 16.11.** Infrared photograph of citrus groves near Mesa, Arizona.

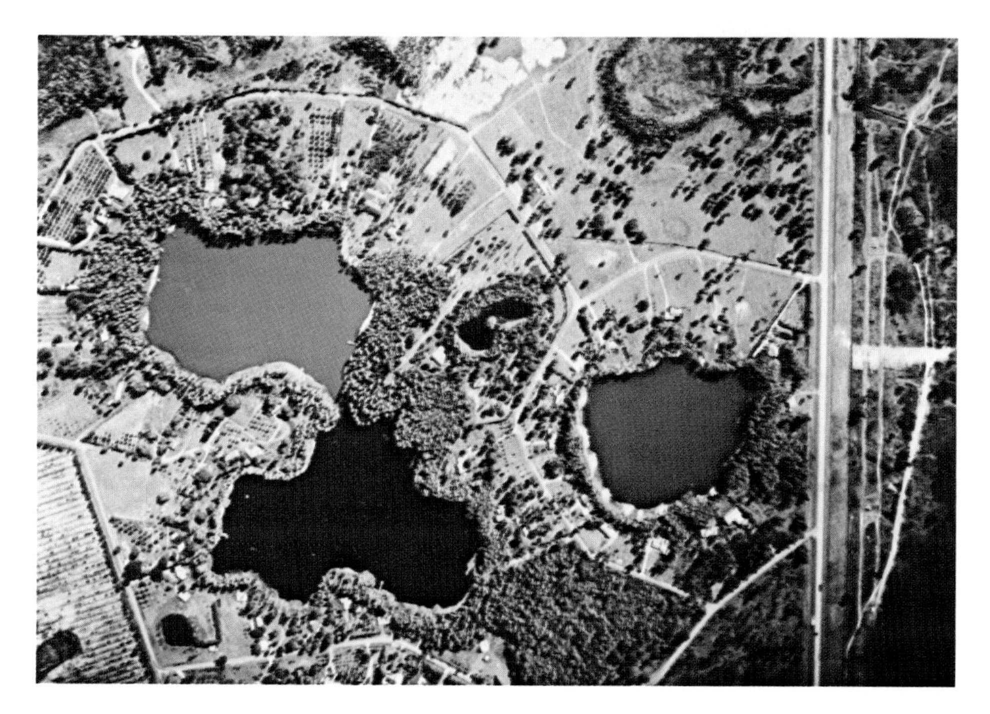

**Fig. 16.12.** Florida lakes, showing distinct tonal changes due to chemical and biologic conditions.

**Fig. 16.13.** Dye markers set by helicopter in Bolinas Lagoon, California. Earlier markers are diffused.

**Fig. 16.14.** Dyes set in Colorado River by helicopter. Photo taken concurrently by aircraft flying overhead.

**Fig. 16.18.** Continuous dye sources set in the Colorado River, near Needles, California.

Part III

# Remote Sensing Supporting Functions: Data Processing, Ground Truth, and Mission Planning

0-201-04245-2

## INTRODUCTION

Remote Sensing has its dramatic and interesting components; the acquisition of data and the presentation of results in the form of analyses and interpretations. Some other events, less conspicuous, are also critical to the success of a remote sensing project. In Part III we examine some of the behind-the-scenes activities which are vital to completion of the project.

Raw data acquired in remote sensing comes in many forms: photographs, magnetic tapes, strip charts from nonimaging devices, and field notes from "ground truth" teams. The combination is unwieldy. Data processing is the means by which order is created from this chaos. In its simplest terms it involves changing scales to some common one, reformatting for uniformity, and, for magnetic tape data, cleaning the data by the elimination of noise to enhance the signal. Also in this category of operations are relatively simple geometric rectifications of data to restore true angular relationships recorded with some distortion, such as that accompanying most scanners at the outer margins. These activities, both photo laboratory and machine processing (computer), have as their objective the presentation of data in a format convenient for the analyst. These operations are primarily mechanical.

A step beyond these mechanical activities, data processing crosses the line to become an integral part of the analysis. In this realm the data are actually manipulated to obtain some insights about subtle data which otherwise would evade attention. Examples of this type of activity includes additive and subtractive color techniques, density slicing, and ratioing procedures. These operations, which involve the use of computers or special-purpose analytical instruments, are among the more exciting developments in remote sensing.

At a third level there are programs for automated interpretations in which computer discrimination and decisions are accomplished on large-volume projects. This type of activity has been demonstrated often, particularly for areas such as (1) differentiation of mixed field crops; (2) rangeland identifications for management; (3) solutions of some hydrologic problems, especially lake inventory and flooding situations; and (4) broad-scale land-use mapping.

Ground truth is a vital link in a remote sensing system. It introduces reality to the system by stating what actually is being sensed. This appears redundant, but in essence what is required in ground truth is detailed spot data which the interpreter can extrapolate into adjacent areas from the remotely sensed data. Thus ground truth is an important calibration device. The acquisition of ground truth data is a varied endeavor, as it takes many forms, varying with the objective and the discipline orientation of the remote sensing mission.

0-201-04245-2

*Remote Sensing of Environment*, Joseph Lintz, Jr. and David S. Simonett (eds.) ISBN 0-201-04245-2

Another important behind-the-scenes activity in the remote sensing system is mission planning and operation. Although this appears to be merely a coordinating task, in reality it is here that the full research and flight design is developed and the activities of all personnel involved are integrated. Primarily it is concerned with the acquisition of the data. Once the mission operation plan is adopted, it should be difficult, but not impossible, to amend. Contingency plans ("Go to plan B") become the available options in the event of adverse meteorological conditions, instrument or logistical failure, and other perils in mission execution.

0-201-04245-2

# Chapter 10

# Machine Processing of Remotely Acquired Data

*D. A. Landgrebe**

## I. INTRODUCTION

Machine processing of remotely sensed data may be carried out in any portion of the spectrum and with data initially in any format. It is easiest, however, with data that are internally consistent spatially, and for images it is demonstrably far more satisfactory with data which *ab initio* are in machine-readable format.

It is for these reasons that data (images) obtained with the multispectral scanner are the most thoroughly analyzed examples of machine-processed data. The multispectral scanner forms the basis upon which this chapter is developed. In regard to the application of the multispectral scanner, we are interested in employing the device in such a way as to make optimal use of variations in the spectral, spatial, and temporal characteristics and in the energy fields of the environment.

Figure 10.1 is a diagram of a satellite survey system, the first elements of which comprise a sensor system and provide for on-board data processing, including the merging of sensor calibration data and the geographic location of the area images.

The data next are transported back to earth for analysis and processing via telemetry, as with the LANDSAT or through direct package return, as with SKYLAB. Usually the data are then preprocessed before final processing with one or more data reduction algorithms. At this point the data may be merged with ancillary information, either previously space-derived or obtained by ground observations.

An important part of the system which must not be overlooked is the last block in Figure 10.1, information consumption: there is no reason to go through the whole exercise if the information produced will not be used. In the case of an earth resource information system, this last portion can prove to be the most challenging to design and organize, since the many potential consumers of these data are not accustomed to receiving them from a space system and may indeed know very little about its information-providing capabilities.

It is necessary to understand thoroughly the portion of the system preceding the sensor, particularly that involving energy exchange in a natural environment

*The author is with the Laboratory for Applications of Remote Sensing (LARS) and the Department of Electrical Engineering, Purdue University. During the time when this material was written, LARS was sponsored by NASA under Grant NGL 15-005-112. Grateful appreciation is due for this support.

0-201-04245-2

*Remote Sensing of Environment*, Joseph Lintz, Jr. and David S. Simonett (eds.) ISBN 0-201-04245-2

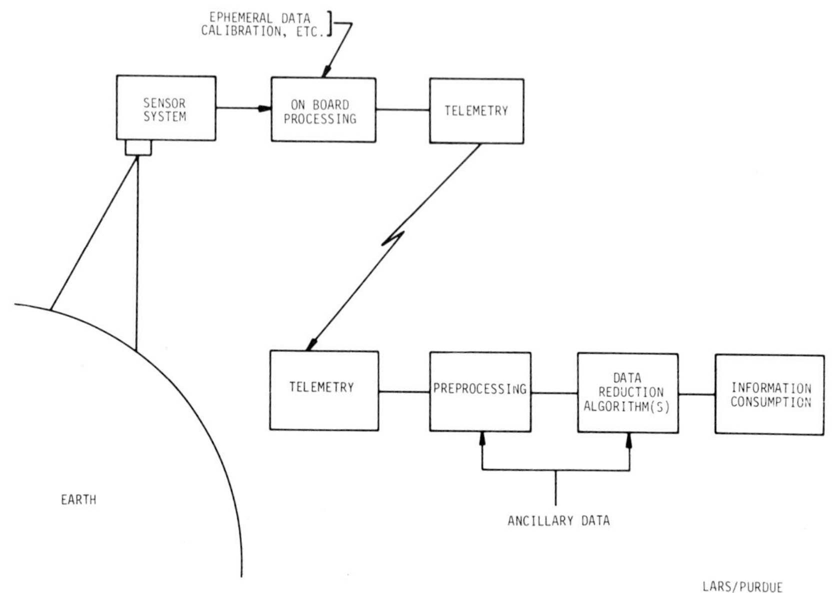


**Fig. 10.1.** Organization of an earth survey system.

(Figure 10.2). It is possible to detect the presence of vegetation by measuring its reflected and emitted radiation. One must understand, however, the many variables involved. Although the sun provides relatively constant illumination above the atmosphere, the radiation reflected from the earth's surface depends on the condition of the atmosphere, the existence of surrounding objects, and the angle between the sun and the earth's surface, as well as the angle between the earth's surface and the point of observation. Even more important is the variation in the vegetation itself. It is possible to deal with these variables in several ways, as discussed later.

In summary, one may derive information about the earth's surface and the condition of its resources by measuring the spectral, spatial, and temporal variations of the reflected and emitted energy derived from a given area. The measurements may then be related to specific classes of materials. To do so, however, requires an adequate understanding of the materials to be sensed. Finally, for the information to be useful, precise knowledge is needed about how and by whom it will be employed, and how accurate it must be.

## II. THE TWO MAJOR BRANCHES OF REMOTE SENSING

Remote sensing has two major stems originating from two different technologies. These two systems are based on (1) image orientation, and (2) numerical orientation.

An example of an image-oriented system might consist simply of an aerial camera and a photo interpreter. Photographic film is used to measure the spatial

0-201-04245-2

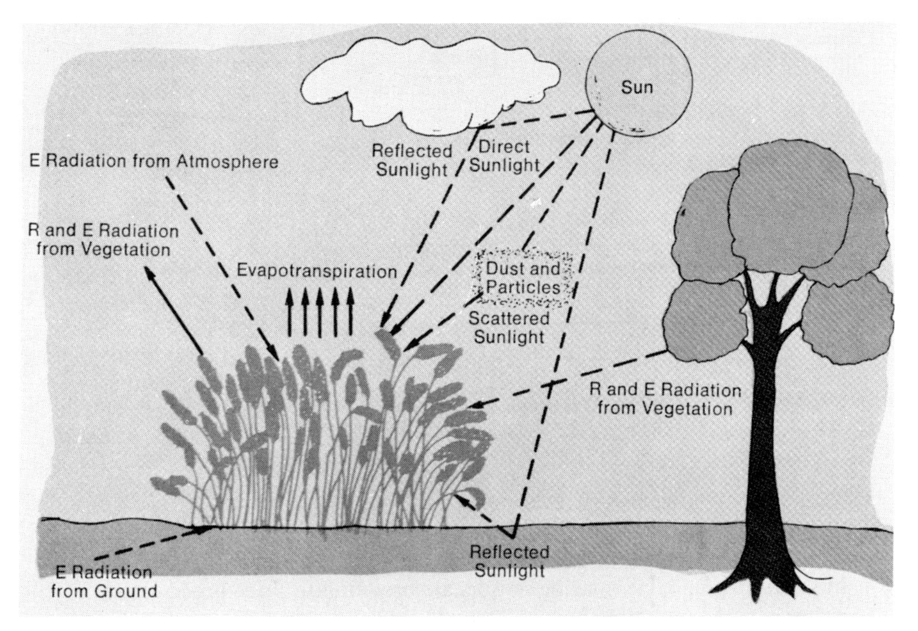


**Fig. 10.2.** Reflected (R) and emitted (E) radiation energy exchange in a natural environment.

variations of the electromagnetic fields, and the photo interpreter relates these to classes of surface cover. Numerically oriented systems, on the other hand, tend to involve computers for data analysis. Although the photo interpreter and the computer, respectively, tend to be the identifying components in the two systems, it is an oversimplification to say that there is a unique relationship in each case. This becomes clearer upon further examination.

The two systems (Figure 10.3) both need a sensor and some preprocessing; however, the distinction between them is seen by noting the locations of the "form image" block in the two diagrams. In the image-oriented type, it is a step in the data stream and must *precede* the analysis. A numerically oriented system, on the other hand, need not necessarily form an image. If it does—and in earth resources work these systems usually do—this may occur at the side of the data stream, as shown in Figure 10.3. Images may be used to monitor the system and perhaps to do some special analysis. An image is, of course, the most efficient way to convey a large amount of information to a human operator. Thus both systems use images, but in different ways.

## III. SENSOR TYPES AS RELATED TO SYSTEM TYPES

In designing an information-gathering system, the sensor as well as the means of analysis must be well mated to the type of system orientation. Some examples of imaging spaceborne sensors are next reviewed to lend emphasis to this observation.

0-201-04245-2

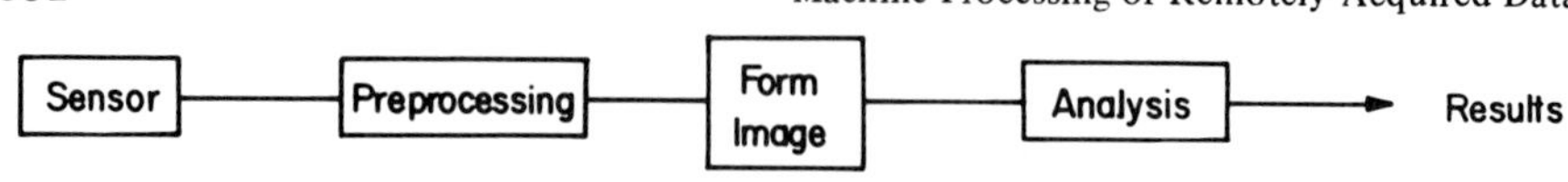


Image Orientation

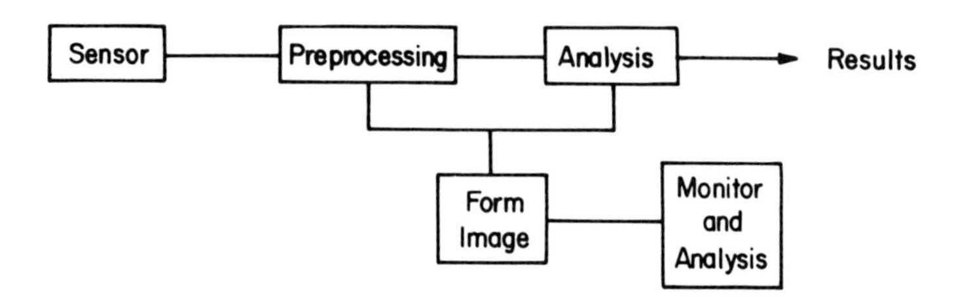


Numerical Orientation

**Fig. 10.3.** Organizations of image and numerically oriented systems.

In the following table imaging sensors are broken into three broad classes: photograph, television, and scanner. The table also provides examples of the advantages and disadvantages of each type.

| *Sensor Type* | *Example Advantage* | *Example Disadvantage* |
|---|---|---|
| Photography | Spatial resolution | Data return |
| Television | Size/weight | Spectral range |
| Scanner | Spectral range | Mechanical complexity |

Types of imaging space sensors

Photographs have the great advantage of very high spatial resolution, but to maintain this resolution film must be returned to the ground. Also, photographs can be obtained only in the visible and part of the near-infrared spectrum.

Television has the advantage that the signal is electrical and thus is immediately ready to be transmitted back to the earth; storage of the data, however, is not inherently present in the system in a permanent form, as in photographs. Thus one is faced, not with the task of carrying along a large quantity of film, but instead with the problems as well as the advantages of telemetry. One may view these advantages as involving either size and weight or efficiency, in that a satellite may be operated for a very long time with a single servicing. Television sensors are, however, restricted to much the same spectral range as photographs.

Scanners can be built to operate over the entire optical wavelength range. They can also provide a greater photometric dynamic range than either television or film, and can present all channels in exact geometric and spatial congruence, thus facilitating both image and computer comparisons.

The advantages and disadvantages mentioned must be viewed as examples only, since in any specific instance the advantages and disadvantages will depend on the

0-201-04245-2

precise details of the system. General statements are also difficult in regard to the types of sensors that are best for image-oriented and numerically oriented systems. There is a clear tendency to favor photography for image-oriented systems because of its high spatial resolution capability. Multiband scanners tend to be preferred for numerically oriented systems, since they make available greater spectral and dynamic ranges and have internally consistent spatial registration between channels.

The technology for pictorially oriented systems is well developed. Sensors best suited to this type of system have long been in use, as have appropriate analysis techniques. This type of system also has the advantage of being easily acceptable to the layman or neophyte, a feature important in the earth resources field, with its many new data users. Similarly, it is well suited for producing subjective information and is especially adapted to circumstances in which the classes to be identified in analysis cannot be precisely selected in advance. Thus man with his superior intelligence is (or can be) deeply involved in the analysis activity. Pictorially oriented systems also offer the possibility of being relatively simple and inexpensive.

In the case of numerically oriented systems, the technology is much newer and not nearly as well developed, although very rapid progress is being made. Because the various steps involved tend to be more abstract, they are likely to be less readily understandable to the layman. This type of system is best suited for producing objective information, and surveys covering large areas are certainly possible. In general, numerically oriented systems tend to be complex.

In summary, the state of the art is that there are two general branches of remote sensing; this duality exists primarily for historical reasons and because of the fact that these technologies began at different times. One type is based on imagery; therefore a key goal of an intermediate portion of the system is the generation of high-quality imagery. In the other case imagery is less important and indeed may not be necessary at all. It is not appropriate to view these two types of systems as being competitive, since they have different capabilities and are useful under different circumstances.

Numerically oriented systems and a particular type of data analysis useful for them will now be examined.

## IV. THE MULTISPECTRAL APPROACH AND PATTERN RECOGNITION

How does one begin the task of devising a technique for analyzing large quantities of remotely sensed earth observational data? Certainly one must make optimum use not only of man's abilities but also of modern computer devices. This consideration strongly influenced the route that the technology has taken. Although much basic research effort has been expended, few practical methods have been unearthed for the machine analysis of data as complex as earth observational imagery on the basis of *spatial* variations in the scene. Thus, if much of the routine and repetitive aspects of analysis are to be successfully turned over to a machine so that low-cost, high throughput can be obtained, a fundamentally simpler approach is needed. Basing the analysis primarily on *temporal* variations in the scene also caused concern in

0-201-04245-2

early work, since in this case no information based solely on observations taken at a single time could be derived.

Fortunately, the third of the three possibilities, *spectral* variations, showed, in feasibility studies, promise for machine analysis, and has become the basis of what is now called the *multispectral* approach. The route taken by the numerical branch of remote sensing has been to use spectral variations as fundamental to the analysis, and later to add spatial and temporal information as circumstances require and permit.

An initial understanding of what is meant by the term "multispectral approach" may be obtained by examining Figure 10.4.* Shown in the upper left of the figure is a reproduction of a conventional color photograph of a set of color cards. The rest of the figure shows photographs of the same color cards taken with black and white film and several different filters. The pass band of each filter is indicated beneath the particular color and card set. For example, in the 0.62–0.66 μm band, which is in the red portion of the visible spectrum, the red cards appear white in the black and white photograph, indicating a high response or a large amount of red light energy being reflected. In essence the multispectral approach amounts to identifying any color by noting the set of gray-scale values produced on the black and white photographs for that particular color rectangle.

As a further example of the approach, Figure 10.5 shows images of an agricultural scene taken in three different portions of the spectrum (LARS, 1967a). In the three bands alfalfa has responses that are dark, light, and dark, respectively, whereas

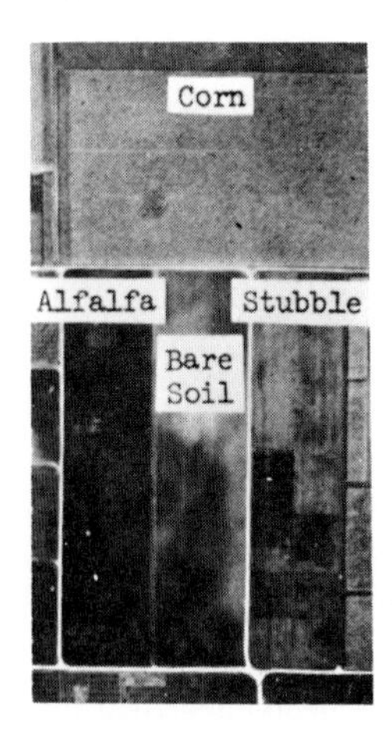

**Fig. 10.5.** Multispectral responses of corn, alfalfa, stubble, and bare soil.

*See color insert.

0-201-04245-2

bare soil is gray, dark, and white. Thus alfalfa can be discriminated from bare soil by identifying the fields that are, in order, dark, light, and dark in these three spectral bands.

To see how a numerically oriented system may be based on this approach, consider Figure 10.6. Shown at the left is a graph of relative response (reflectance) as a function of wavelength for green vegetation, soil, and water. Let two wavelengths, marked $\lambda_1$ and $\lambda_2$, be selected. Shown on the right of this figure are the data for the three materials specified at these two wavelengths, plotted with respect to one another. It is apparent that two materials whose responses as a function of wavelength are different will lie in different portions of the two-dimensional space.* When this occurs, one speaks of the materials involved as having unique spectral signatures. This concept will be pursued further presently; however, the concept of spectral signature is a relative one—the interpreter cannot know that vegetation has

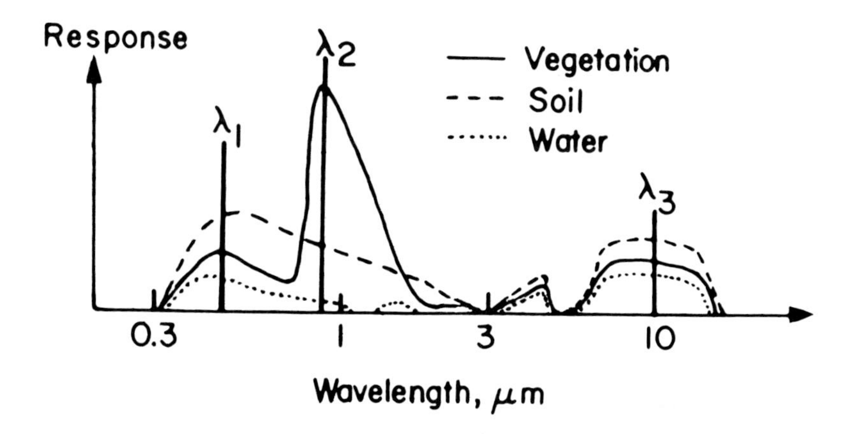


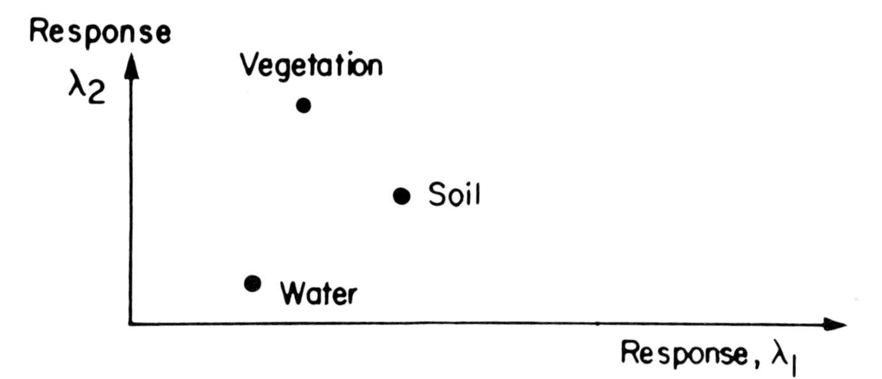


Fig. 10.6. Spectral data in two-dimensional feature space.

*This space is referred to as feature space.

0-201-04245-2

a unique spectral signature, for example, until he sees the plots resulting from the spectral responses of other materials present within the scene to be analyzed.

Note also that a larger number of bands can be employed. The response of $\lambda_3$ could be used, and the data plotted in three dimensions. Indeed, four or more dimensions have meaning and utility, even though an actual plot of the data is not possible.

So far no temporal or spatial information has been involved, only spectral. Temporal information can be utilized, however, in several ways. Time is always a parameter of the spectral response of surface materials. As an example, the problem of discriminating between soybeans and corn is examined in Figure 10.7. Under cultivation, these two plants have approximately 140-day growing cycles. Figure 10.7 illustrates how the two-dimensional response plot might look for fields of these two species with time as a parameter. Upon planting and for some period thereafter, fields of soybeans and corn would merely appear to be bare soil from an observation platform above them. Eventually, though, both plants would emerge from the soil and in time develop a canopy of green vegetation, mature to a brownish dry vegetation, and diminish. Thus, as viewed from above, the fields of soybeans and corn would, in fact, always be mixtures of green vegetation and soil. However, in addition to the vegetations of the two plants having slightly different responses as a function of wavelength, the growing cycles and plant geometries are different; thus the mixture parameters provide an even more obvious difference between the two plants than the spectral response difference in the plant leaves themselves. This

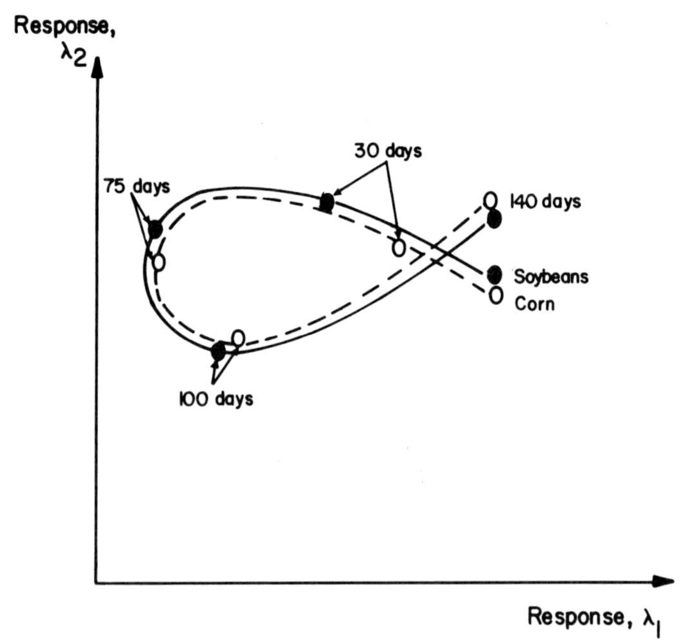


Fig. 10.7. Temporal change in two-dimensional feature space.

0-201-04245-2

is the implication in Figure 10.7, as shown by the rather large difference between the plants 30 days from planting date (partial canopy), as compared to the situation at 75 days (full canopy). Thus one way in which temporal information is used is in determining the optimum time at which to conduct a survey of given materials.

A second use of temporal information is perhaps less obvious. Consider the situation of Figure 10.7 at the 75-day and 100-day points. In this case the difference between the two plants is relatively slight. However, if these data are replotted in four-dimensional space, with responses at $\lambda_1$ and $\lambda_2$ and 75 days as dimensions 1 and 2 and responses at $\lambda_1$ and $\lambda_2$ and 100 days as dimensions 3 and 4, the small differences at the two times can often be made to augment one another.

A third use of temporal information involves change detection. In many earth resources problems it is necessary to have an accurate historical record of the changes taking place in a scene as a function of time.

Let us consider how one may devise a procedure for analyzing multispectral data (LARS, 1967a; Fu et al., 1969). In the process, one further facet of the multispectral approach must be taken into account. The radiation from all soybean fields will not have precisely the same spectral response, since all will not have the same planting date, soil preparation, moisture conditions, and so on. Moreover, response variations within a class may be expected of any earth surface cover. The extent of response differences of this type certainly has an effect on the existence of a spectral signature, that is, on the degree of separability of one material from another. Consider, for example, a scene composed of soybeans, corn, and wheat fields; if five samples of each material are drawn, the two-dimensional response patterns might be as shown in Figure 10.8, indicating that some variability exists within the three classes. Suppose now that an unknown point is drawn from the scene and plotted, as indicated by the point marked *u*.

In this case the design of an analysis system comes down to partitioning this two-dimensional feature space in some fashion, so that each such possible unknown point is uniquely associated with one of the classes. The engineering and statistical literature abounds with algorithms or procedures by which this can be done. (See, for instance, Nilsson, 1965, and Nagy, 1968.) To illustrate the concept, one very simple example is shown in Figure 10.9. In this case the conditional centroid or center point of each class is first determined. Next the locus of a point equidistant from these three centroids is plotted and results in three segments of straight lines, as shown.* These lines form, in effect, decision boundaries. In this example the unknown point *u* would be associated with the class soybeans as a result of its location with respect to the decision boundaries.

In very simple situations where data from the various classes are quite well separated in this feature space, an even simpler technique, called level slicing, can be used. The term "level-slicing" came into use when analysts of black and white photography began to identify (Figure 10.5) certain materials from others in a scene by

*When more than two dimensions (spectral bands) are being used, this locus will produce a surface rather than a line.

0-201-04245-2

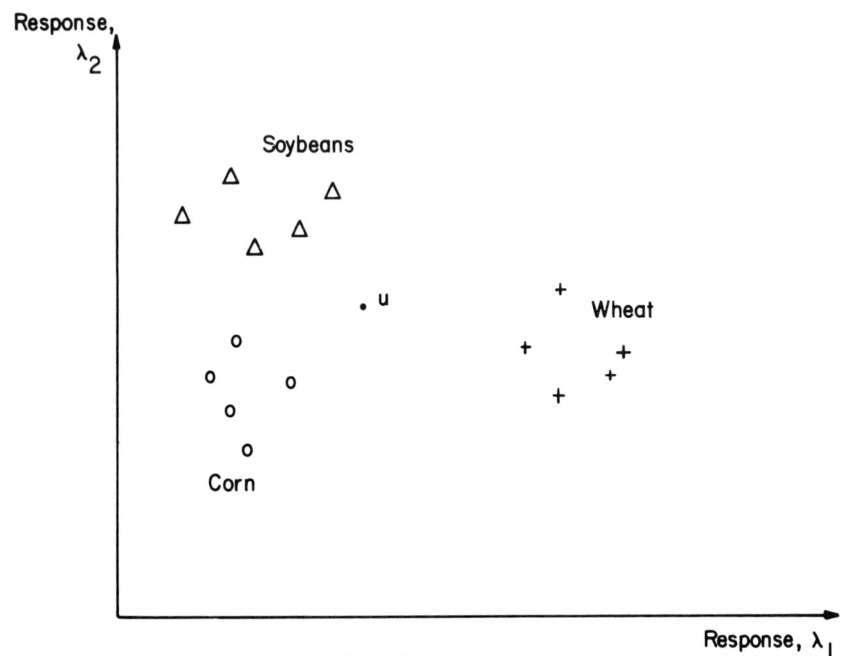


Fig. 10.8. Samples in two-dimensional feature space.

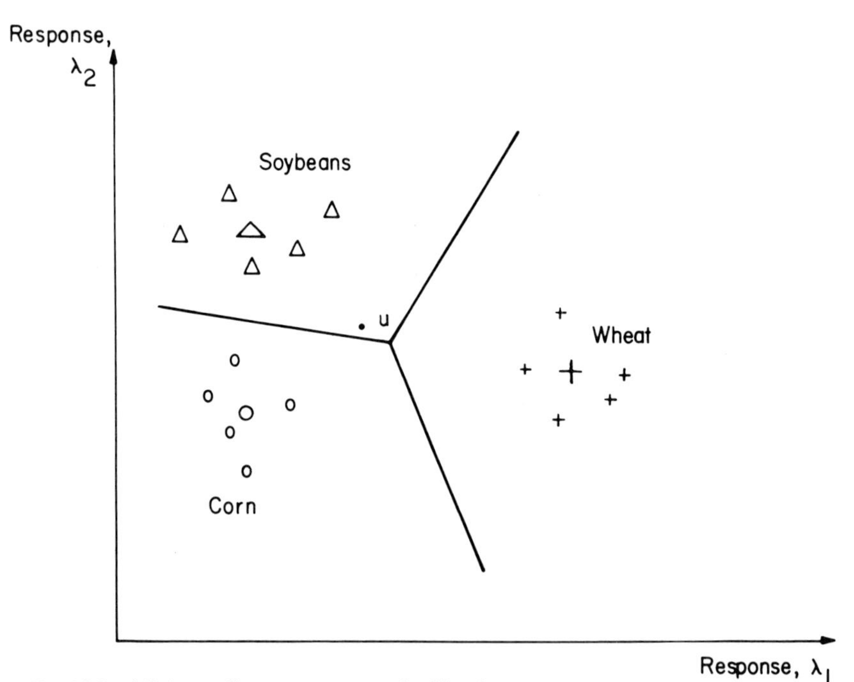


Fig. 10.9. Minimum distance to means classification.

0-201-04245-2

their range of gray levels. Thus, by identifying the areas on the film that had this range or slice of levels, one could locate all of the regions containing a particular material.

This simple concept can be immediately extended to the multiband case, where one looks for areas in which the data fall in one range in the first band, in a second range in band 2, and so on. From the feature space viewpoint this method identifies data points that fall in a horizontally or vertically oriented rectangle.

For many analysis problems of practical interest, a somewhat more sophisticated procedure allowing for greater generality in the location and form of the decision boundaries is required. One such algorithm, the Gaussian maximum-likelihood classifier, has been especially well studied for this purpose. In this case the initial samples of each class are used to estimate not only the mean value for the class, but also its covariance matrix. This matrix shows not only the variance present in data from each spectral band but also the degree of correlation between bands. Under the assumption that the data from each class have a Gaussian (or bell-shaped) distribution, the mean value and covariance matrix completely define the class distribution; for the two-spectral-band case they might appear as shown in Figure 10.10*a.* A given data point is then assigned to a class, according to which class Gaussian density function as shown in Figure 10.10*b* and are in general segments of second-order curves, that is, parabolas, hyperbolas, ellipses, and circles, with straight lines as a degenerate case.

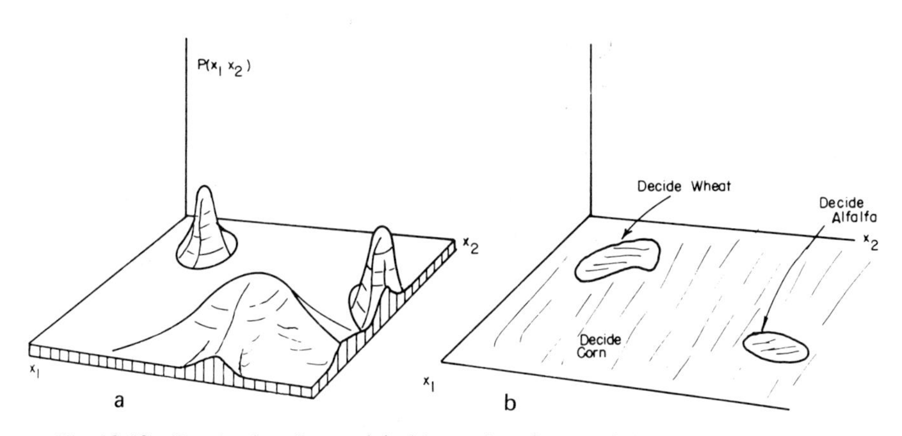


**Fig. 10.10.** Density functions and decision regions for a statistical approach.

This technique of analysis is referred to as pattern recognition. Many even more sophisticated procedures result in both linear and nonlinear decision boundaries; however, the use of a few initial samples to determine these boundaries is common to a large number of such procedures. The initial samples are referred to as training samples, and the general category of classifiers in which training samples are used in this way are termed supervised classifiers.

0-201-04245-2

## V. THE MULTISPECTRAL SCANNER AS A DATA SOURCE

Up to this point, the implication has been that color or multispectral photographs are the source for generating data for this analysis. Although cameras may be used, it is more appropriate to employ a multispectral scanner. Figure 10.11 shows a scanner mounted in an aircraft.

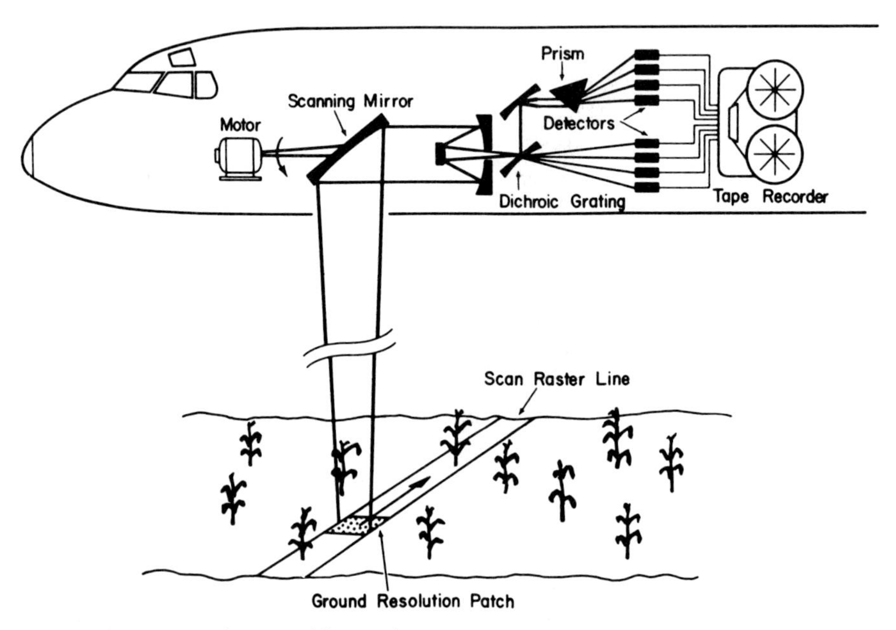


**Fig. 10.11.** An airborne multispectral scanner.

Essentially the scanner is a multiband spectrometer whose instantaneous field of view is swept across the scene by a motor-driven scanning mirror. At a given instant the device gathers energy from a single resolution element. The energy from this element passes through appropriate optics and may, in the visible portion of the spectrum, be directed through a prism. The prism spreads out the energy according to the portion of the spectrum involved; detectors are located at the output of the prism. The output of the detectors can then be recorded on magnetic tape or transmitted directly to the ground. Gratings are commonly used as dispersive devices for the infrared portion of the spectrum.

A most important property of this type of system is that all energy from a given scene element in all parts of a spectrum passes through the same optical aperture. Thus, by simultaneously sampling the outputs of all detectors, one in effect determines the response as a function of wavelength for the scene element in view at that instant and achieves spatial congruence in all channels.

The mirror scans the scene across the field of view transverse to the direction of platform motion, and the motion of the platform (aircraft or spacecraft) provides

0-201-04245-2

the appropriate motion in the other dimension, so that in time every element in the scene has been in the instantaneous field of view of the instrument.

## VI. AN ILLUSTRATIVE EXAMPLE

One example of the use of a scanner with pattern recognition procedures involved the classification of a 1-mile by 4-mile area into crop classes. Four-dimensional data (four spectral bands) were used, and the classification scheme was Gaussian maximum-likelihood discrimination (Fu, Landgrebe, and Phillips, 1969). A conventional air photo of the flightline is shown in Figure 10.12 in which the correct classification of each field has been added to the photo by hand. The symbols on the air photo and their associated classes are as follows: S, soybeans; C, corn; O, oats; W, wheat; A, alfalfa; T, timothy; RC, red clover; R, rye; SUDAN, sudan grass; P, pasture; DA, diverted acres; and H, hay.

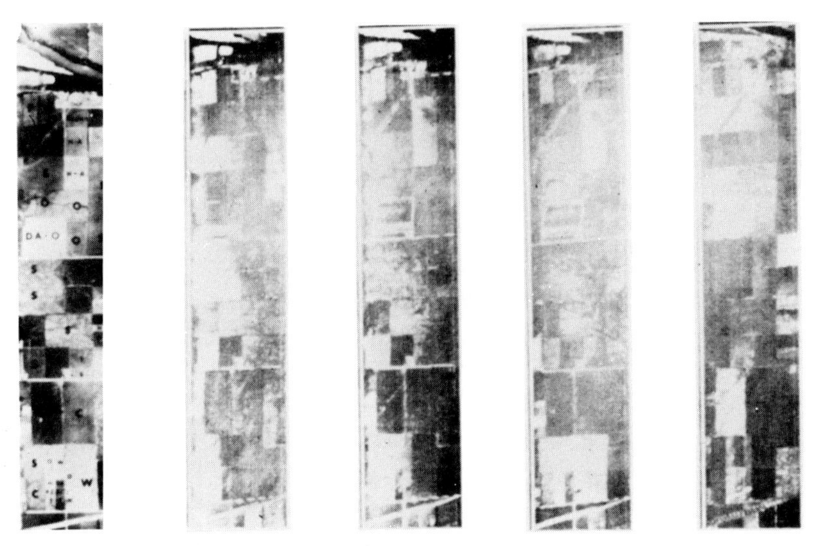

**Fig. 10.12.** Data in four wavelengths.

Figure 10.13 also shows the results of the classification. Two simple classes are shown. All points of the scene classified as row crops (either corn or soybeans) are indicated in the center of the figure. On the right are all points classified as cereal grains (either wheat or oats).

A quantitative evaluation of the accuracy was conducted by designating for tabulation the correct classes of a large number of fields in the flight line. The result of this tabulation is shown in Figure 10.14. It is seen that for all classes the percentage of correct identification was above 80%. A fuller analysis of accuracy

*This example was originally prepared by Professor Roger Hoffer of LARS/Purdue.

0-201-04245-2

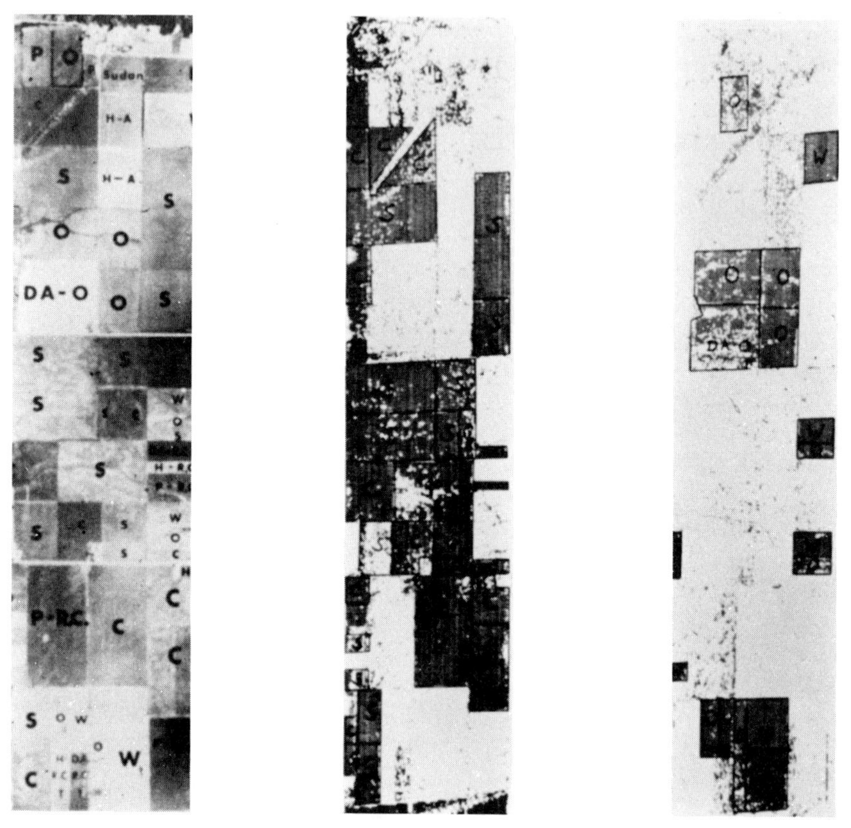


Fig. 10.13. Spectral pattern recognition of row crops and cereal grains.

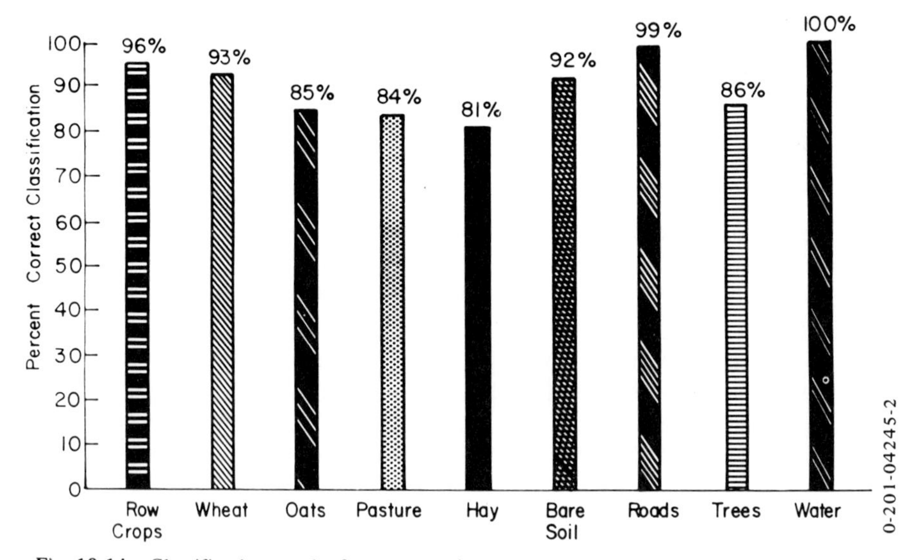


Fig. 10.14. Classification results for test samples.

0-201-04245-2

would include errors of omission and of commission, as well as separability of corn and soybeans in the row crop class. Inspection of Figure 10.14 illustrates these points, and the figure should be referred to also during the following discussion on class separability.

The same procedures using both aircraft and satellite data have been utilized for a wide number of classification tasks in addition to crop species identification. Some of these are as follows: tests of agricultural and engineering soils, involving mapping and delineating soil types, including mineral content, organic content, and moisture content; geologic feature mapping; water quality mapping and mensuration, using both reflective and emissive spectra; forest cover identification and tree species delineation; and division into geographic and land-use mapping categories.

## VII. SOME PROCEDURAL DETAILS IN THE USE OF PATTERN RECOGNITION

With the basic concept of pattern recognition in mind, some further details on how it may be applied are in order. One of the most important of these details is the definition of the classes into which the data are to be categorized.

Two conditions that a class must meet in order to be useful are to be *separable* from all others and to be of *informational value.* It does no good to define a class whose spectral response is not sufficiently distinct from the responses of other classes. Nor is it of use to define and accurately map a class that excites little interest. It will be seen presently that one may name classes of informational value and then check their separability, or vice versa.

A second matter is determining when a class actually becomes defined. In an agricultural survey, for example, simply designating a class as soybeans does not define it precisely enough. For instance, what percent ground cover is required before a given resolution element should have its classification changed from bare soil to soybeans? What percent of a resolution element may be covered with weeds? The fact is that a class becomes precisely defined *only* by the training samples to be used for it. Therefore an important step in the procedure is the selection of training samples that are sufficiently typical of the whole class in question.

One must also recognize that the definition of a class is always a relative matter, that is, it is relative to the other classes used in the same classification. It is well known in this regard that the broader the classes, the more accurately will identification be made. The effect of the decision boundaries is to divide up the feature space (see Figure 10.8) into nonoverlapping regions, depending on the location of the class training sets in relation to one another.

As a result every point in the space automatically becomes associated with one of the named classes. It is necessary, therefore, that the list of classes be exhaustive, so that there is a logical class to which every point in the scene to be analyzed can be assigned.

Because of these factors the selection of training samples is especially important. There are two approaches to obtaining training data; we shall refer to them here as the signature bank approach and the extrapolation mode.

0-201-04245-2

Using the signature bank approach, the researcher first decides on a list of appropriate classes and then draws from a signature bank previously collected data on the classes of material identical to those selected. This approach has a considerable amount of esthetic appeal. Presumably one could accumulate a very large bank of data from typical classes and thereafter always have training data available for any situation without further effort.

However, such an approach places stringent constraints on the sensor system, since absolute measurements of scene radiance are necessary if they are to be referenced to a future data-gathering mission. Furthermore, the extent to which detailed and sophisticated classes can be utilized is limited by the ability to determine and adjust for the instantaneous values of all other variable parameters, such as the condition of the atmosphere, the sun and view angle, possible seasonal variations in the vegetation, and the natural statistical distribution of the data for various classes. In short, although such a procedure is possible, it will result in more stringent requirements on the sensor system and in considerable data preprocessing in order to achieve this maximum utility. Alternatively, it must be restricted to cases in which only relatively simple classes are necessary.

The extrapolation mode, on the other hand, has somewhat different characteristics. In this case, training data are obtained by locating within the data to be analyzed specific examples of each of the classes that will be utilized. The classification procedure, therefore, amounts to an extrapolation from points of known classification within the scene to the remaining portions of the data. This approach has the advantage of requiring less exactness in the calibration capability of the sensor system and in the knowledge of the other experimental variables, since only variation of these factors within the data-gathering mission, and not from mission to mission, must be accounted for. On the other hand, it has the disadvantage of requiring some knowledge about the scene to be analyzed before the analysis can proceed. In the case of populated or accessible areas, this knowledge is usually obtained from ground observations. For inaccessible and/or unpopulated areas, it could perhaps come also from a very limited, low-altitude aircraft mission or from direct manual interpretation of imagery. The relative cost of this additional information may be modest or substantial, depending on the particular circumstances. The extrapolation mode was used in both the preceding example and the one to follow.

An example in which a small training set was used to classify a very large area will serve to illustrate further these details and procedures. Data for this experiment were collected by the multispectral scanner system (MSS) of the LANDSAT I. From its orbital altitude of approximately 900 km this sensor has an instantaneous field of view of some 80 m on the ground. The all-digital system transmits data to the ground in frames of imagery covering a square area 185 km on a side. The data presented in image form would result in an image made up of 2300 scan lines with 3350 samples per scan line. The sensor provides spatially registered data in four spectral bands from 0.5 to 1.1 $\mu$m. Thus a single data set (frame) consists of approximately 7.5 million four-dimensional points.

0-201-04245-2

In the previous example it was seen that pattern recognition techniques may be used effectively on data gathered by an airborne system; however, they are even more ideally suited to satellite-gathered data, since it is possible to collect a tremendous number of data in a very short time (in the case of LANDSAT-1, 23 sec!), thus holding constant many significant and important variables during the data-gathering activity. In other words, if a specific problem requires the collection of a large number of data, pattern recognition techniques that function efficiently and cost-effectively only on very large quantities of data are ideally suited, from the standpoint of data throughput, for the analysis test.

The particular frame of data used for the present example is the first frame gathered by the LANDSAT-1 MSS system after its launch (Frame ID 1002-16312). It was gathered on July 25, 1972, and is of the Red River Valley areas of Texas and Oklahoma. The frame is centered on a point 15 miles southeast of Durant, Oklahoma, and approximately 5 miles north of the Red River. A simulated color infrared photograph made from the frame is shown in Figure 10.15.*

A maximum-likelihood Gaussian classification scheme was trained, using 17 classes from this data set. The data were then classified into these 17 spectral classes. Figure 10.16* shows a display of the classification results in image form. The detailed analysis results themselves, that is, the identifying class number associated with each point plus the likelihood value for it, are stored on magnetic tape. Figure 10.16 presents these quantitative results in a form suitable for qualitative evaluation. This figure was constructed by associating each class or group of classes with an individual color. It is emphasized that Figure 10.16 is not an enhanced image; rather, it illustrates a means of displaying a large volume of quantitative results in qualitative image form.

The 17 classes are presented here in terms of 11 colors. This grouping was necessary because the human eye cannot discriminate between many more than about 10 or 11 colors. The colors and their corresponding estimated cover types are as follows:

| *Category* | *Spectral Class* | *Color Code* | *Cover Type* |
|---|---|---|---|
| 1 | 2, 3, 4, 5, 6, 7, 8, 12, 13 | Yellow, tan, light green, brown | Various classes of rangelands and pastures |
| 2 | 1, 17 | White, light gray | Sandy or bare soils, light vegetation, agricultural fields with sparse canopy |
| 3 | 9, 10 | Dark green | Forests and woodlots |
| 4 | 14, 15, 16 | Dark blue, blue gray, aqua | Water (3 subclasses) |
| 5 | 11 | Dark purple | Atoka Reservoir |

*See color insert.

0-201-04245-2

There remains in this experimental situation the problem of establishing the accuracy with which the classification algorithm performs. This is not a simple task, because of the very large area and number of classes involved. In this case low-altitude air photography and ground observations obtained 5 days after the satellite pass were used for verification. Figure 10.17* shows a more detailed classification of a region near Lake Texoma. This classification employed 18 spectral classes in four broad land-use groups. The color code used is as follows:

| *Category* | *Spectral Class* | *Color Code* | *Cover Type* |
|---|---|---|---|
| 1 | 12, 13, 16, 17, 18 | Light blue, dark blue, blue gray, medium blue | Water (5 classes) |
| 2 | 1 | White | Sandy and light bare soil, light-colored dry vegetation, agricultural fields with low crop coverage |
| 3 | 7, 14 | Red | Agricultural fields with high crop coverage |
| 4 | 2, 3, 4, 5, 6, 15 | Yellow, light brown, light tan | Pasture, rangeland |
| 5 | 8, 9 | Light green | Forest, sparse canopy |
| 6 | 10, 11 | Dark green | Forest, more dense canopy |

Figure 10.18* shows an enlargement of the west end of Lake Texoma. The Red River is seen winding through a forested area beginning in the southwest corner of the frame and entering Lake Texoma near the westernmost end. A delta has been formed at the point of entry and has been placed in the class indicated by white (sandy and light bare soil). Subcategories of water are indicated on the color-coded classification by various shades of blue. There is a small inlet on the north shore directly north of the delta. Figure 10.19* shows an oblique low-altitude air photograph taken near this inlet. The data and the gradations in water turbidity are apparent in the air photograph.

The instantaneous field of view of the multispectral scanner is approximately 80 m. Nevertheless, an ordinary two-lane bridge crossing Lake Texoma in Figure 10.18 was correctly classified on the basis of the classes available to the pattern recognition algorithm.

A study of a number of low-altitude photographs in comparison with the color-coded classification results suggests that the broad land-use categories shown in Figures 10.16, 10.17, and 10.18 are indeed well classified. If comparable results can be obtained elsewhere, one has a capability for deriving a land-use map of a region almost immediately. Viewing Figure 10.16, one could quickly draw boundaries between areas having various types of agricultural, range, forest, and other land uses. In addition, the intrinsically quantitative nature of the approach allows one

*See color insert.

0-201-04245-2

immediately to estimate the acreage in each land use by counting the sample points assigned to each of the classes.

A more complete discussion of the analysis of this frame is available in the literature (Landgrebe et al., 1972). Other demonstrations of the use of satellite imagery (Anuta and MacDonald, 1971; Anuta et al., 1971) already have established the accuracy possible with machine analysis for land-use mapping purposes; future work will no doubt corroborate this conclusion. We will concentrate here, therefore, on the procedures and computational algorithms necessary to achieve these results.

At present the use of various classification algorithms for this purpose is fairly well understood; however, the training of an algorithm is still a time-consuming process requiring a trained and sophisticated analyst. Earlier it was pointed out that the chosen classes must simultaneously satisfy two criteria in order to be valid. They must be *separable,* a restriction imposed by the reflectance properties of the scene, and they must have *informational value,* a restriction imposed by the intended use of the analysis results.

Furthermore, as was previously pointed out, the classes are not really defined until the training data or statistics describing them exist. The difficulty of determining 18 sets of four-dimensional mean vectors and covariance matrices can be readily envisioned, particularly if it were necessary to identify the sample points used to estimate these statistics manually. Over the last few years research has been directed toward machine-aided methods for this process. One such procedure involves the use of a type of classifier, termed an unsupervised classifier, that does not utilize training samples. Assume, for example, that one has some two-dimensional data (as shown in Figure 10.20); assume also that one knows there are three classes of material in these data but that the correct associations of the individual points with the three classes are unknown. The approach is to assume initially that the three classes are separable and to check this hypothesis subsequently.

Algorithms (computational procedures) are available (Ball, 1965; Friedman and Rubon, 1967; Haralick and Kelly, 1969) which will automatically associate a group of such points with an arbitrary number of mode centers or cluster points. These procedures, known as clustering techniques, can be used to divide the data, and the results of applying such a procedure (Wacker and Landgrebe, 1970) might be as shown in Figure 10.21. There remains, then, the matter of checking to be sure that all the points assigned to a single cluster belong to the same class of material. In other words, the method automatically establishes classes that are separable but do not necessarily have informational value. Thus, in comparing supervised versus nonsupervised classifiers, it is accurate to say that in the supervised case one names classes of informational value and then checks to see whether these classes are separable, whereas in the nonsupervised case the reverse is true.

Figure 10.22 shows the results of applying such a clustering technique to some multispectral data. The algorithm was instructed to form five clusters. A comparison of the clustering results with the data in image form shows that the clusters indeed were associated with individual fields. Cluster 4, for example, was associated

0-201-04245-2

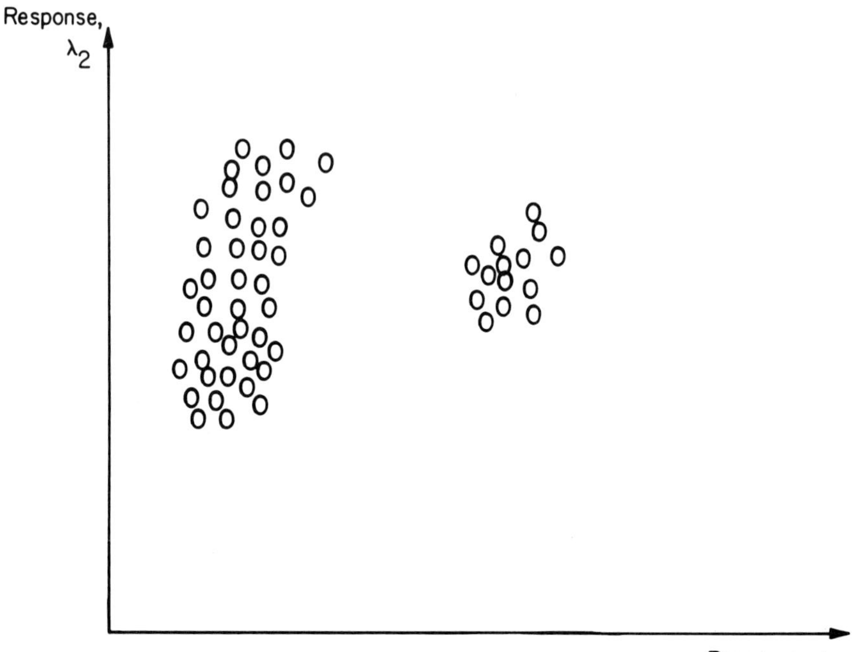


**Fig. 10.20.** Samples in two-dimensional feature space.

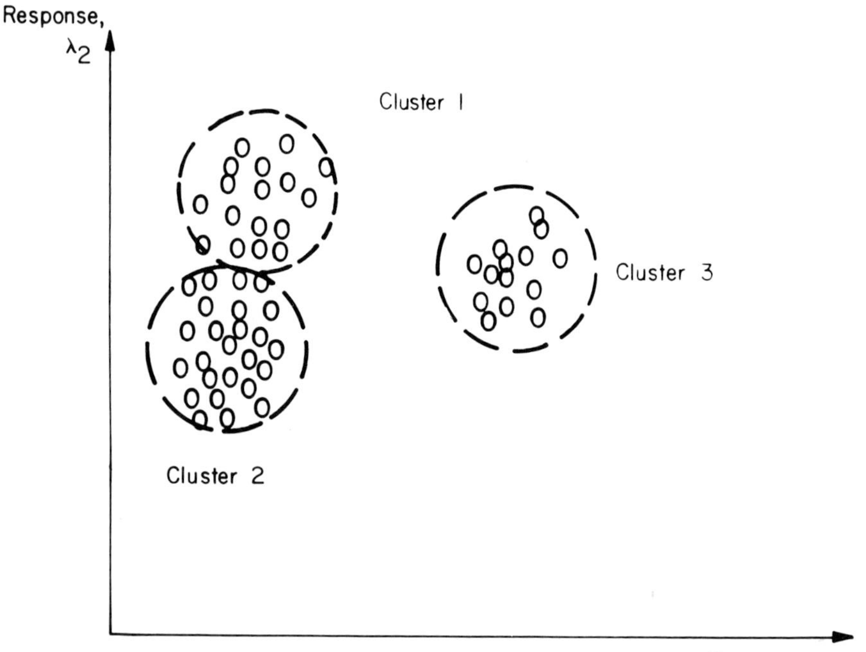


**Fig. 10.21.** Clustering in two-dimensional feature space.

0-201-04245-2

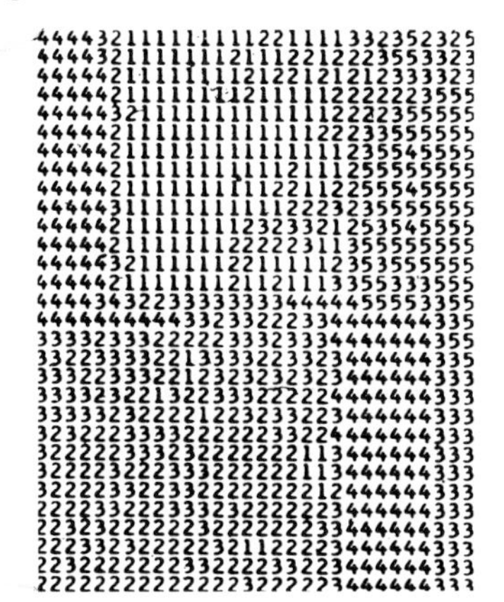

**Fig. 10.22.** Clustered data using four spectral bands.

with fields in the upper left and lower right, clusters 2 and 3 were associated with the field in the lower left, and so on. Such a technique is useful to speed the training phase of the classifier by aiding the human operator to obtain points grouped according to the class from which they originate. The statistics of each cluster point can be immediately computed from the cluster results, so that decision boundaries are quickly established. The operator is thus relieved of the necessity of locating and separating individual resolution points or fields for training each class.

The value of such a procedure is even greater in cases where large groups of points associated with the same class and located contiguously to one another are not present as they were in Figure 10.22. Figure 10.23 shows the result of clustering data for a soils mapping classification. Here it would be more difficult to select samples associated with specific soil types. As a result of the clustering, the operator has only to associate the soil type with each cluster point, and training samples are immediately available for further processing.

Such clustering techniques are very useful in the training phase of utilization of a supervised classifier. The specific steps to be followed in training a classifier are

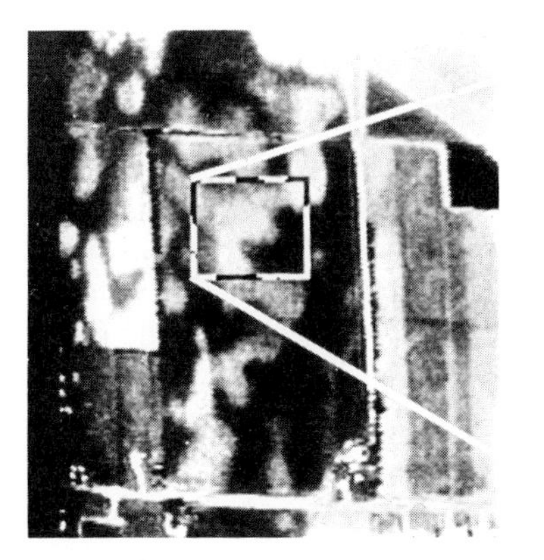

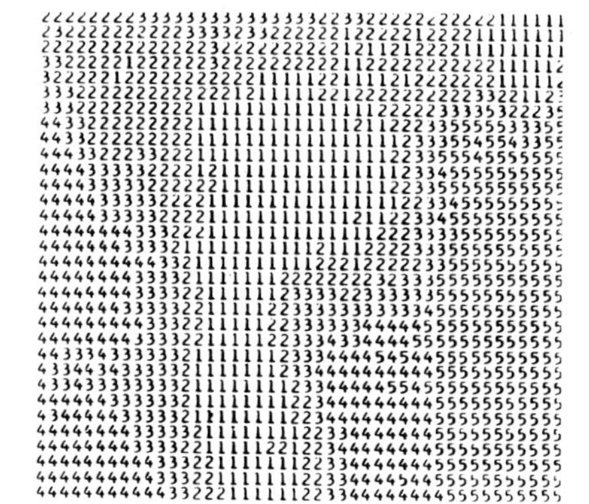

**Fig. 10.23.** Clustered data using four spectral bands.

0-201-04245-2

dependent not only on the data set but also on the user's informational needs. The following steps, which are similar to the ones used in the LANDSAT data analysis example, are fairly typical.

1. Make available to the clustering algorithm every *m*th sample of every *n*th scan line. It is ordinarily not necessary to cluster every point; indeed, to do so unnecessarily wastes computer resources. The values of $m$ and $n$ needed in each situation are a matter of judgment and depend on the availability of computing resources and the manner in which the classes distribute themselves over the data set to be analyzed.
2. Examine the results of the clustering to establish first that each cluster is sufficiently separable from the others and second that the clusters are associated with some class of material of interest.
3. Manually select such additional training sets as may be needed to treat special situations.
4. From this point the statistics of each class may be computed from appropriate clusters and the classification.

## VIII. THE SPEED AND COST OF DATA PROCESSING

Let us return to the question of processing speed and economics. It was mentioned earlier that, in order to deal with the large volume of data, special care must be exercised to choose a method offering great throughput capability. It was also pointed out that the aspects of simplicity and processing of a parallel nature contribute in this direction. Perhaps it is now more apparent why the multispectral approach is valuable in this respect. All of the data relevant to a single resolution element on the ground are collected and become available for processing at the same instant of time. Thus, rather than having to process several different images (e.g., from several different spectral bands), one need only process a single vector at a time. The mathematics and algorithms of multivariant analysis are thus immediately available, and implementations of this mathematics in parallel processor form are known and well understood.

Implementations of the classification algorithm used in the above examples have been made to-date on general-purpose digital computers, such as the one used to generate these examples, and special-purpose analog processors.* Some work has begun in examining the operational implementation of the processing algorithms which have been studied under research circumstances. Both software techniques, such as "table lookup" implementations of classifiers (Eppler et al., 1971a), and advanced hardware techniques (Preston, 1972; Bouknight et al., 1972) could produce reductions of several orders of magnitude in operational processing costs over those of general-purpose digital computers and the highly flexible software neces-

*The SPARC processor of the Environmental Research Institute of Michigan is an outstanding example of a special-purpose processor.

0-201-04245-2

sary in the research environment. Results on quantitative comparisons between various processing approaches (Joseph et al., 1969) and their costs are not yet abundant, however.

## IX. THE USE OF SPATIAL INFORMATION

So far we have discussed the use of spectral and temporal variations to derive information from measurements of electromagnetic energy arriving at the sensor. It is also possible to utilize spatial information within the multispectral approach to further increase the amount and accuracy of information that can be derived. One approach is to use a so-called per field classifier (Landgrebe, 1969; Wacker, 1972; Wacker and Landgrebe, 1972). In essence, the use of spatial information in this approach is based on the fact that points in near vicinity to one another are likely to be members of the same class. Consider, for example, the situation as shown in Figure 10.22. Here one might be willing to say, "I don't know what all the points in cluster 4 are, but I am willing to say that they are all members of the same class. What is this class?" Thus in this case one sees a situation in which a set of points rather than an individual point is available for a single classification. In essence, then, the mathematics of the situation permits one to use this set of points to estimate the statistical distribution function of the points. This estimated distribution can be compared with the distribution function of the points, and then with the distribution of each training set to decide on the proper classification.* Thus one is comparing a point set to a set of distributions, rather than comparing a single (vector) point to a set of distributions. As may be seen in Wacker's (1972) study, a generally higher classification accuracy is thereby achieved. One does have the preliminary problem, though, of grouping all points into point sets. This may be accomplished either by a boundary drawing algorithm (Wacker and Landgrebe, 1970) or through the use of clustering itself as shown in Figure 10.23.

## X. DATA PREPROCESSING STEPS

Now that the portion of the system involving data analysis (see Figure 10.1) has been discussed, it is appropriate to return briefly to the preprocessing section of the system. Depending on the type of information needed and therefore the type of analysis to be used, a large number of possible data preparation or preprocessing steps prove necessary or helpful. The list below enumerates some of them.

CALIBRATION

- Radiometric (intensity) manipulation.
- Geometric manipulation.

*The mean and variance of this estimated distribution correspond roughly to the tone and texture used by the human photo interpreter.

0-201-04245-2

ENHANCEMENT

- Spatial frequency operations.
- Multivariate transformations.
- Convolutional filtering.

MULTI-IMAGE OPERATIONS

- Interimage addition, subtraction, etc.
- Interimage registration.

DATA PRESENTATION

Generally, on board the sensor platform, as previously mentioned, various types of calibration information are derived. Hence a possible preprocessing step is to apply this calibration information to the data. This may take the form of radiometric calibration, in which corrections are made for variations in the atmospheric transmission, system gain, sensor aging, and the like. A second type of calibration is associated with the geometry of the image. Usually it is necessary to make intraimage corrections of both a relative and an absolute type. For example, in the LANDSAT-1 MSS images a type of skew distortion arises because the earth is rotating beneath the satellite during the period of time that a single frame is being sensed. Therefore a correction must be applied on an intraimage basis in order that individual resolution elements will have the proper locations with respect to one another.

On the other hand, it may be necessary for storage and retrieval purposes to establish the location of each resolution element relative to an earth-oriented coordinate system on a more absolute basis. An example of such an operation is the so-called scene correction or precision image processing done with the LANDSAT-1 imagery.

Another example of a calibration type of preprocessing is the angle correction process. The amount of radiation that is reflected from a given scene area is dependent on both the angle from which it is illuminated and the angle from which it is viewed (Kriegler et al., 1969). This problem has long been known to the field of photo interpretation in terms of the so-called image "hot spot." It is possible to process the data in such a way that the hot spot seems to have disappeared. Unfortunately, since the angle effect depends not only on the illumination angle and view angle but also on the scene material on the ground, it is generally not possible to process the data to a point of radiometric correctness. In short, a suitable approach to this problem from a radiometric standpoint is not known at this time.

In recent years a considerable amount of work has been done in the image enhancement area. Spatial frequency operations have been widely studied for such purposes as enhancing boundaries in the imagery, removing low-frequency shading effects, correcting for distortions introduced by the data transmission system, and removing single-frequency coherent noise (Billingsley, 1970). Enhancement by carrying out a multivariant transformation on multispectral imagery has also been examined.

0-201-04245-2

Multi-image operations are also sometimes necessary and desirable. One of the most important multi-image operations is that of achieving registration between two images collected over the same scene at different times or in different portions of the electromagnetic spectrum. This problem has great importance and has been extensively studied. Techniques today tend to fall into two broad classes. The first type involves primarily optical techniques developed principally in the field of photogrammetry, using image projection techniques and ancillary data derived from system characteristics and operation. The second approach to the problem differs in that registration is achieved generally through point-by-point processing and utilizing information derived from the data themselves. Two-dimensional image correlation is a common approach in this latter case (Anuta, 1969, 1970; Lillestrand, 1972).

Once image registration has been achieved, access is gained to temporal information by means of techniques described earlier. In addition, interimage manipulation for the purpose of noise minimization, for example, by adding two images of the same area gathered at nearly the same time, and for highlighting certain types of changes in the imagery through the ratioing of registered images may prove useful (Kriegler et al, 1969).

A very important additional type of data preprocessing is data compression. Data compression may be desirable to minimize the data volume, either in terms of the necessity for data transmission through a given link or in terms of reducing the data storage and retrieval problem. Relatively simple compression techniques based on both spectral and spatial redundancy appear to be possible at this time. Compression ratios of 5 or 10 to 1 without loss of essential information in the data appear within reach (Ready et al., 1971; Ready, 1972).

A most important area and one receiving considerable attention at this time is that of data display. Since the number of data to be dealt with is typically very large, methods by which to view them are most important in both pictorially and numerically oriented systems. Various types of viewers and image display systems, including those involving color in various ways, are being constructed, marketed, and used, especially in the pictorially oriented field. Though of perhaps less central importance to the numerically oriented field, image data display systems are utilized as effective means by which to monitor processing system performance and to interact with it. The very difficult operation of merging ancillary data with the data stream is often best done in this way. In addition, various types of printers and plotters are utilized to present results in map form for the user's purposes.

0-201-04245-2

Chapter 11

# Comparisons of Qualitative and Quantitative Image Analysis

*Earl J. Hajic and David S. Simonett*

## I. INTRODUCTION

In the analysis of remote sensor images, both qualitative and quantitative data and information may be obtained. The principal aim in *qualitative image analysis* is the distinction or discrimination between kinds of things (materials or phenomena). *Quantitative analysis* is concerned with the measurement of properties of the image and with probabilistic statements, such as error rates, associated with identification.

As in all such philosophical pairs, or dyads, which express two related aspects of an entity, it may be difficult to see where one ends and the other begins. As a general rule, human methods of image analysis tend to be qualitative, descriptive, and judgmental, and to carry unknown components of preceptual and training biases, as well as both systematic and random error. Quantitative analysis tends more to be associated with machine- (usually computer-) aided procedures. In these, attempts are made to numerically assess bias and systematic and random errors. To state these differences, however, is to realize that they are merely two positions in a continuum: computer studies which have no error analysis or in which the error analysis itself turns out to be flawed are *ipso facto* qualitative; manual studies which follow precise, repeatable decision rules and which assess interpreter error are quantitative.

Another aspect of human, qualitative image analysis is that it is the preferred or at least typical method for many professionals, in spite of the unknown errors associated with the results.

- Despite the countless lineament studies by geologists there are only a few which critically assess differences between interpreters and on the part of a single interpreter as a function of the type and magnitude of lineaments.
- Plant ecologists publish maps which have no accompanying statement on problems of categorization or of extrapolation from scattered ground observations, and which draw on much unstructured accessory material in undefined ways.
- Geographers produce regional and urban analyses without any or with little standardization of categories, a phenomenon also characteristic of the planning community: error analyses are rare indeed of land-use maps!

In this respect, then, quantitative image analysis is an *evolving* aspect of image analysis as a whole. Virtually all professions are increasingly concerned with the

*Remote Sensing of Environment*, Joseph Lintz, Jr. and David S. Simonett (eds.) ISBN 0-201-04245-2




0-201-04245-2

quantitative aspects of their fields; image analysis shares the common situation of being in a state of transition. The quantitative aspects include the identification, without ambiguity, of points, lines, and areas and the development of probabilistic statements about the points, lines, and areas. Error-free identifications are usually infeasible, and probabilistic statements are required concerning the accuracy of such identifications. Along with the *identification* or *probable identification* of an object, we are concerned with *enumeration* or counting the numbers of things in classes (points, lines, areas—e.g., houses, roads, and crops of different types), and with *mensuration,* the determination of lengths, area, and volumes.

A measure of the differences between man and machine for identification, enumeration, and mensuration is given in Table 11.1. The first column in this table

**TABLE 11.1**
**Comparison between Manual and Machine Methods of Identification, Enumeration, and Mensuration for Single and Multiple Images with Respect to Accuracy and Speed**

**One-Image Accuracy 1/1 Speed**

| | *Manual* | *Complex Analog* | *Digital Computer* | *Hybrid Interactive* |
|---|---|---|---|---|
| Identification | | | | |
| Points | 1/1 | 0.4/10 | 0.7/10 | 1/5 |
| Lines | 1/1 | 0.4/10 | 0.7/10 | 1.5/5 |
| Areas | 1/1 | 0.5/10 | 0.8/10 | 1.5/5 |
| Enumeration/Mensuration | | | | |
| Points | 1/1 | 0.4/10 | 0.7/10 | 1/5 |
| Lines | 1/1 | 0.4/10 | 0.7/10 | 1.5/5 |
| Areas | 1/1 | 0.5/10 | 0.8/10 | 1.5/5 |

**Multiple Images (Four or More) in Computer-Compatible Congruent Digital Registration**

| | *Manual* | *Complex Analog* | *Digital Computer*[a] | *Hybrid Interactive* |
|---|---|---|---|---|
| Identification | | | | |
| Points | 1/1 | 0.8/100 | 1.2/100 | 1.5/50 |
| Lines | 1/1 | 0.8/100 | 1.2/100 | 1.5/50 |
| Areas | 1/1 | 0.8/100 | 1.2/100 | 1.5/100 |
| Enumeration/Mensuration | | | | |
| Points | 1/1 | 0.8/100 | 1.2/100 | 1.5/50 |
| Lines | 1/1 | 0.8/100 | 1.2/100 | 1.5/50 |
| Areas | 1/1 | 0.8/100 | 1.2/100 | 1.5/50 |

[a] Future digital processing using parallel processors, microprocessors, and more efficient programs will work revolutionary changes in the speed and relative cost of digital processing. As an indication of this, a number of LARS (Laboratory for Applications of Remote Sensing) programs have recently been converted by Robert Ray (University of Illinois) for operation on the ILLIAC IV, a very large parallel processor. Factors of increased speed of 1400- and 6000-fold are reported for certain image-processing algorithms.

0-201-04245-2

rates the relative skill and speed of man as unity in achieving a given objective when he uses very simple analog equipment such as hand lenses, stereoscopes, measuring devices, and desk calculators. The other three columns treat sequentially (1) complex analog, (2) digital computer, and (3) fully interactive hybrid systems involving man + analog + digital components. For each of these methods of performing quantitative analysis, relative skill and accuracy are estimated in comparison to manual methods. As a general rule, when working with single images and engaged in identification, enumeration, and mensuration, man is as skillful—indeed frequently more skillful—than complex analog or digital systems. The advantage of the machines lies less in accuracy than in speed. In moving from single to multiple images and to statistical analysis, however, the advantages shift very rapidly from man to machine. Complex pattern recognition studies with large numbers of images taken at different times and in many spectral bands are feasible only with digital or analog computers.

The rough judgments on accuracy and speed given in Table 11.1 should not be used to make decisions between systems; quantitative analysis of images is still too fluid. Rather, they indicate that there are many ways of developing quantitative data from images and that for some tasks (and for some developing countries and some small offices) a high-labor, low-technology method of analysis may be almost as satisfactory (and certainly more cost effective) than a more sophisticated technique.

Examples of the equipment used for the three general forms of machine-aided processing and decision-making are as follows.

*Complex Analog.* These machines include the Digicol ($I^2$S Corporation) and the Interpretation Systems Incorporated (ISI) systems.

*Digital Computer.* Any second-generation computer equivalent to an IBM 360, 50, 65, or 75 series, can form the core of a digital system. Output may be in a variety of forms, depending on the supporting equipment. Recently developed minicomputers can also handle the volume of data involved.

*Interactive Hybrid Computer.* Commercial systems include the M-DAS system (Bendix Corporation), the Image 100 (General Electric), and the IDIMS (ESL Incorporated) systems, all of which use color television displays and minicomputers, plus some hard-wired logic. A number of universities, research institutes, and government laboratories in the United States have very versatile, large interactive hybrid systems used in substantial remote sensing programs. Prominent among these are Purdue University, The University of Kansas, The University of California, Pennsylvania State University, NASA Laboratories at Johnson Space Center (Houston), Jet Propulsion Laboratory (Pasadena), and The Environmental Research Institute of Michigan (Ann Arbor).

0-201-04245-2

## II. IMAGE ANALYSIS PROCEDURES

Image interpretation involves both the recognition of patterns and the discrimination of boundaries between the pattern entities in an image. Experienced scientists use contextual clues and inference to recognize different classes and patterns in an image of geologic structures, plant communities, and other features. The image clues include spatial relationships of tone, texture, and object size, shape, and geometry, as well as the spectral information in black and white or color images. Both field and interpretation experience, and the methods of particular scientific disciplines, are used to establish the image entities and their boundaries.

An interpreter, in qualitatively recognizing patterns in images, processes and compacts large numbers of data to extract useful information and to classify the patterns into meaningful classes. The "pattern recognition" of quantitative analysis, however, is statistical in character, and the "pattern" is a vector made up of a number of measurements. Pattern recognition algorithms (to a more or less acceptable degree) partition the statistical space so that each region of it can be assigned to a class of patterns.

For either scientist or computer, pattern recognition also involves the comparison of information derived from new data with that taken from known sample patterns or paradigms. A professional develops a paradigm, drawing upon long experience in fieldwork and image interpretation. Computers initially draw upon a much smaller pool of sample patterns. Nevertheless, just as with man, the computer program adaptively improves its knowledge of the sample characteristics through use of learning algorithms. Either way, the man or the machine arrives at a decision about an unknown pattern through comparison with known patterns.

The concepts given above can be extended to include the various ways of enhancing (or preprocessing) imagery, as well as the extraction and the analysis of qualitative "patterns" in the information extracted from the image.

In the following pages comparisons will be made between qualitative and quantitative and manual and machine procedures for enhancing, processing, rendering decisions on, and recording the results of image analyses. Comparisons and contrasts of methods and of preprocessing algorithms will be discussed so as to make clear the relations between different methods of qualitative and quantitative analysis.

The steps through which manual and machine analyses tend to go are clearly parallel, although one is called "image interpretation" and the other statistical "pattern recognition." The main steps are as follows:

- Obtaining input images.
- Image preprocessing.
- Feature extraction.
- Choosing and applying decision and classification rules.
- Displaying classification output.

0-201-04245-2

All the preceding steps are shared by both qualitative and quantitative analysis. A final step is confined to quantitative analysis:

- Error analysis of results.

*Input images* come in many formats, geometrics, and wavelengths. These include black and white photographs, color photographs, digitized photographs, multispectral scanner computer-compatible tapes, analog television images, side-looking radar images, and line trace radiometry.

*Image preprocessing* includes the removal of distortions, both geometric and radiometric, translation, rotation, change of scale, noise removal, image sharpening, gray-scale level slicing, filtering, color combination, image subtraction, image ratioing, and mathematical transformation (such as the Fourier and Hadamard transforms). Preprocessing is carried out to make the data more easily interpretable for human or machine analysis and more convenient as to format and more tractable in later analysis, to reduce its dimensionality, to emphasize special features, and to decrease systematic noise, thereby restoring degraded images.

*Feature extraction* involves the determination of features through the context of the problem or through statistical manipulation of the data. Contextual features are derived from the specialized intrinsically "human" interpretation, field, and laboratory experiences of scientists. Their professional judgments are based on convergence of evidence and on a great variety of clues in the imagery.

Statistical feature extraction employs many forms of multivariate analysis, including discriminant function analysis, clustering, maximum likelihood, and Bayes decision rules.

It is in the feature extraction phase that a real distinction may be drawn between the scientist using context and convergence-of-evidence clues and the computer implementation of statistical algorithms. Since the procedures used are so dissimilar, it should not be surprising that the two systems produce very different results on occasion. What is remarkable is the degree of similarity found in some cases.

In feature detection a small sample of the image data is analyzed to extract the "essence" of the context or of the statistical "patterns" in the images. This essence is employed to make decisions and classify the remaining data.

In *decision and classification*, features present in the data are grouped into appropriate classes, and decisions are made either qualitatively or quantitatively regarding the class into which a given item will be placed. A typical human process is that followed, for instance, by a plant ecologist who recognizes a dozen plant communities in an image, prepares a legend (feature selection), and then proceeds to map the distribution of the communities, mentally classifying and deciding in the process. A typical computer process involving LANDSAT multispectral scanner data after feature extraction (band selection on multidate tapes) is to use a Bayes decision rule with the selected channels. This is employed with the remaining data.

*Classification outputs* range from hand-prepared overlays for images to displays on color television, alphanumeric output from a line printer, color maps obtained

0-201-04245-2

from computer-driven scanners, and tabular arrays organized by interpreters or printed by a computer.

*Error analysis* involves verification of the output through some appropriate sampling process and includes sensitivity analyses.

## III. IMAGE FORMAT TYPES

The principal image formats are (1) those of point-perspective, continuous, and whole-scene instantaneous imaging systems (i.e., cameras), and (2) those of sampled images, which are frequently line perspective and noninstantaneous, and which include a great variety of devices employing discrete elements: vidicons, optical-mechanical scanners, real and synthetic-aperture radars, image orthicon tubes, and solid-state cameras.

Mapping cameras have the desirable quality (from their central-perspective projection of true stereoscopy) of enabling the construction of very accurate maps. This is not true of sampled systems, which are all (more or less) plagued by radiometric and, particularly, geometric distortions. The types of images employed will therefore have a considerable influence on the amount and types of preprocessing required.

A number of the physical characteristics and limitations of the various sensors with respect to image fidelity will now be examined, along with the several advantages and disadvantages of image sampling.

The eventual association of a particular gray tone of a picture element (pixel) with a particular feature (e.g., a crop) is in part a function of the characteristics of the sensor and of all the data manipulations which affect image fidelity. The significance of image fidelity may be understood by first considering the nature of the data from a conventional camera. Any optical lens images a point of light (e.g., reflected from a tiny portion of a crop) as a blur. This blur, called a *circle of confusion,* is due to two factors:

- The effects of the diffraction of light. This is a function of the size of the camera lens opening or aperture and establishes the theoretical limit of the resolving power (or resolution). Resolution is a concept of separating a repeated pattern (e.g., alternate black and white bars) into its components. Limits are reached when the separated components merge together to an unacceptable degree.
- The effects of aberration, which depend on lens quality. This effect can be minimized.

Thus a hypothetical pinpoint object becomes a disk or circle of confusion on the film. The more out of focus it is, the larger the disk diameter. The importance of considering this (Gibson, 1970) is that the *forms of detail* on a photograph/image are recognized by an interpreter (in decreasing order of importance) via (*a*) edges or lines, (*b*) "point lights," and (*c*) tonal gradations. These forms of detail when placed in context (e.g., spatial orientation vis-à-vis a background of all other

0-201-04245-2

objects) are part of the process of human recognition of objects. Aspects (*a*) and (*c*) are essentially self-evident. An example of (*b*) is that texture as a structural element in a photograph is indicated by *separated* point lights. It is important to note that most of the development of contemporary processing of features for automatic object recognition is in inverse order to this list. The reason is that processing- or decision-making algorithms generally become more complex, and probe increasingly more universal and complex domains, in this sequence: (1) gray tones, (2) textures, (3) edges, and (4) context. This brings home the difference between human and machine analysis, in which man relies heavily on context, whereas the machine at this time is principally confined to a lower order of analysis—that is, of tones.

Since human analysis is done via *forms of detail*, the detail must be *separated* and *sharp* in order for adequate recognition to take place. The rendition of "fine detail" governs the sharpness of a photograph as subjectively experienced. Detail separation is not, however, equivalent to sharpness. This seeming contradiction can be clarified by noting that picture sharpness (acutance) is due more to the sharpness of the grain clumps on a photographic print than of the grain size. If the grain clumps have soft, indistinguishable edges, a picture will lack crispness no matter how fine the grain. Acutance is measured by determining the sharpness of edges in an image (Gibson, 1970).

We can also note the significance of resolution to human photo interpreters. The human eye is compatible with a resolution of about 8 lines/mm (i.e., alternating black and white bars each 0.0625 mm wide). Thus a photographic image with a resolution of 80 line pairs/mm can usefully be enlarged by no more than a factor of 10 for use by a human photo interpreter, unless details need to be mapped. It is for this reason (extraction of detail in mapping) that the LANDSAT multispectral scanner 70 mm film product with an original image scale of 1 : 3,369,000 and a resolution of 34 line pairs/mm may reasonably be enlarged to a scale of 1 : 250,000, an enlargement of 13.5X.

Improved methods of measuring resolution now determine a modulation transfer function for a given equipment or image-processing technique. This is in effect a measure of the relative response of a system as it views increasingly finer spatial separation in the line pairs (i.e., tests at successively higher numbers of line pairs per millimeter). A useful, alternative way of thinking about these physical aspects involves the space frequency domain. By examining the spatial frequency distribution of variations in the gray level in an image, the effects of various image-processing methods can be evaluated. Haralick (1973) provides an excellent glossary for the further understanding of "modulation transfer function" and many other remote sensing terms used in image pattern recognition.

## IV. PREPROCESSING

Preprocessing may be carried out using photographic, electronic, and digital methods to achieve roughly comparable effects, but with equipment ranging widely in cost, accuracy, and repeatability.

0-201-04245-2

Preprocessing has the following goals:

1. To eliminate geometric or radiometric distortions in the image, thus improving any quantitative manipulations to be performed,
2. to emphasize or enhance data, and
3. to restore certain features not present in the original,

thereby enabling qualitative judgments to be made or improving qualitative judgments concerning the data.

## A. Image Geometry Preprocessing

Images of a variety of scales and geometries must be reconciled so that they may be overlaid for comparison. If an analyst lacks suitable equipment for high-quality rectification, he may either tolerate data having a variety of scales and distortions or develop laborious manual procedures so that he can analyze the material effectively.

It is common, for example, to have aerial photographs of an area of different dates, scales, and quality. The correction of geometric errors can involve the determination and implementation of a transformation, $T$, which maps a distorted image, $f(x, y)$, into a corrected image, $f(x', y')$, namely, $f(x', y') = f(x, y)T$ for all $x, y, x'$, and $y'$. Conceptually this correction could encompass system geometries, empirical functions, scale changes, and variable map projections. Sometimes a projective transformation (variably stretched or compressed) is approximated on a subimage basis by a linear transformation, for example, that which would correct the yaw effect (as opposed to the pitch or roll) of a sensor platform. In this case the transform, $T$, would be as follows (Kingston and LaGarde, 1972):

$$T = \begin{vmatrix} a_{11} & a_{12} & a_{13} \\ a_{21} & a_{22} & a_{23} \\ a_{31} & a_{32} & a_{33} \end{vmatrix} \qquad \text{for the subimage,}$$

and the corrected grid positions, $x', y'$, would be determined from

$$x' = \frac{a_{21} + a_{22}x + a_{23}y}{a_{11} + a_{12}x + a_{13}y},$$

$$y' = \frac{a_{31} + a_{32}x + a_{33}y}{a_{11} + a_{12}x + a_{13}y}.$$

The $T$ matrix would be determined by knowing at least four noncolinear points in each coordinate system. Additionally, since a pixel (picture element, a discrete sample) at grid position $x', y'$ may not coincide with an input pixel at $x, y$, radiometric correction is also used. For example, pixel $x', y'$ may take on a gray value derived from the neighbors of $x, y$ and $x + \Delta x, y + \Delta y$.

0-201-04245-2

For most investigators it would be too expensive—if a large area is involved—to attempt low-altitude geometric congruencing of multidate, multiscale imagery. The near orthography and narrow field of view from space plus the large area covered in one image make rectification feasible to obtain congruency.

When several missions are to be merged for analysis by digital computer and pixel decisions are made with statistical pattern recognition methods, the images or data must be in point geometric congruence. Such is the case for a single flight with a multispectral scanner, but multidate images must be brought to point congruence by matching images to a reference image, map, or orthophoto. This is very difficult, or even infeasible, for images of different type or date unless the following requirements are substantially met:

1. the area imaged is flat or, at most, gently undulating, and/or
2. the images are obtained with very narrow field-of-view systems, and/or
3. the images are obtained from very-high-flying aircraft or spacecraft (in which case even moderate relief may be tolerated; the image displacement on LANDSAT imagery of a mountain 1000 m high is less than a resolution cell across at the edge of the imagery), and/or
4. the images are produced by framing cameras, and orthophotos can be produced.

An important advantage of satellite imagery is the fact that time-differing images may be brought to congruence with much greater ease and at distinctly less cost per unit area than aircraft images of any type.

If exact pixel correspondence is not required, simple optical (and moderately expensive electronic and digital) congruence may readily be obtained. Steiner et al. (1972) and Bakis et al. (1971) discuss digital congruencing by various methods. The *Manual of Photogrammetry* (American Society of Photogrammetry) is a standard reference for optical and electronic rectification systems.

## B. Radiometric Correction and Normalization of Images

Radiometric correction and normalization eliminates or reduces the influence of image distortions involving scene radiance variations* such as falloff in light intensity $I$ away from the center ($\theta = 0°$) of a photographic lens ($I = \cos^4 \theta$), cross-field variations in lighting, systematic variations in scene illumination and developing and printing procedures, radar antenna patterns, scan line variations arising from differences in viewing angle, and faulty equipment biases.

Incomplete radiometric correction of an image is called *normalization.* Not all methods achieve near correction equally well. Some work well with a single image. Others are required for normalization of two or more images of the same scene. Scene radiance differences in mountains are not removed by normalization.

Typical methods for normalization include the following.

*See Chapter 2, "Effects of the Atmosphere on Remote Sensing."

0-201-04245-2

- Cameras equipped with an antivignetting lens filter, which varies in density as $I' = 1/\cos^4 \theta$. Compensation for improperly used filters can be obtained during printing.
- Ratioing, $R$, of the light received ($a, b, c, \ldots$) on different photographic layers on a color film or between channels on a multispectral scanner image (Brooner and Simonett, 1971; Maurer, 1971; Crane, 1971; Vincent, 1973; Smedes et al., 1971). There are many forms of ratioing. Common ones include these three:

$$(1)\ R_1 = \frac{a}{a+b+c}, \quad R_2 = \frac{b}{a+b+c}, \quad R_3 = \frac{c}{a+b+c};$$

$$(2)\ R = \frac{a}{b};$$

$$(3)\ R = \frac{a-b}{a+b}.$$

  Ratioing is a very important procedure, not only for normalization but also for the emphasis of special features such as variations in rock chemistry in desert areas (Vincent, 1973).
- Scan normalization, to reduce fore- and back-lighting variations in images and radiometric distortions (from changing viewing angle in multispectral scanners). This may be achieved by adjusting the received energy as a function of pixel position on a radar image (Simonett et al., 1967). For example, unsharp (defocused), low-density positive images may be sandwiched with a negative image to correct the final positive. Digital processing algorithms are employed by many investigators, including Crane (1971) and others at ERIM, Swain and associates at Purdue (Hoffer and Goodrick, 1971), Haralick and colleagues at Kansas (Haralick et al., 1973), and Billingsley at Jet Propulsion Laboratory (Goetz et al., 1973).
- Another normalization procedure involves removing the effects of an intervening atmosphere by subtracting the darkest object in a scene from all others (Crane, 1971). Equal-interval quantizing (Brooner et al., 1971) helps normalize images subject to along-track atmospheric variability.

Although atmospheric effects are discussed at length in Chapter 2, some summary comments are appropriate here. The intervening atmosphere has an impact on the radiant energy eventually sensed from the remote object and its context/background. This occurs because the atmosphere affects the transmission of radiant energy to the object and of the reflected energy. The atmosphere does this via processes of absorption, scattering, and irradiation. The energy received at the sensor is thus a sensitive function of both the geometry of the sun and the sensor vis-à-vis the object sensed, as well as the measurement wavelengths and the object

0-201-04245-2

and background signatures. This ensemble of effects operates to selectively modify signatures and contrasts of (and between) objects of interest and their backgrounds as a function of the geometry.

The intent is thus to selectively correct for these effects. Complete correction implies understanding and modeling all the significant processes of the interaction of electromagnetic energy with the transmission media and the sensed objects. Since models of necessity incorporate some idealized assumptions (e.g., homogeneous, stratified layers), radiometric corrections are incomplete. Thus alternative methods are also used. Some are based on use of the received energy itself as reference information for radiometric normalization.

The relationship between the radiance of the target/object as measured by the sensor and some significant physical effects is given by (Turner and Spencer, 1972)

$$L(h) = \frac{\rho^t E T(h)}{\pi} + L_p(h)\,,$$

where $L(h)$ = total radiance of the target as measured by the sensor,

$\rho^t$ = effective diffuse reflectance of the target/object material,

$E$ = irradiance onto the target,

$T(h)$ = atmospheric transmittance from the target to the sensor,

$L_p(h)$ = path radiance generated between the target and the sensor,

$h$ = altitude of the sensor.

Turner and Spencer note that $\rho^t$ is a pure object attribute only if it is diffusely reflecting. It usually is not, however, for it varies with the illumination and viewing geometry. Therefore it is necessary to specify the mean background albedo of the local terrain (usually via ground truth) with reflectances weighted relative to their respective areas. These effects are incorporated into an algorithm which additionally includes standard meteorological data such as station pressure at the surface and horizontal visual range.

The latter two data are used to calculate the standard Rayleigh optical depth for a particular wavelength, $\lambda$. It is then corrected for the effects of station pressure. An aerosol optical depth is also determined as a function of $\lambda$ and the visual range. The sum of these enables the determination of atmospheric transmittance.

The basic correction algorithm becomes a rather extensive computer model, because radiative transfer theory, still developing after 70 years, has significant mathematical intricacies even with simplifying assumptions. If coherent light-source remote sensing becomes of future importance, the model will increase in difficulty.

An alternative to this correction algorithm are processes of image normalization. These are used to obtain the normalized form of the spectral signature of an object.

0-201-04245-2

Haralick (1975) has summarized this process. He uses the previous equation in simpler form:

$$\rho = \frac{R_r - b}{R_s T},$$

where $\rho$ = object reflectivity,

$R_r$ = radiance received at the sensor,

$R_s$ = source irradiance,

$b$ = backscatter and path irradiance,

$T$ = atmospheric transmissivity.

Here it is assumed that $R_s T$ and $b$ are approximately constant for adjacent relatively narrow spectral bands. Then for $\lambda_1$ and $\lambda_2$, respectively,

$$\rho_1 = \frac{R_{r_1} - b_1}{R_{s_1} T_1} \quad \text{and} \quad \rho_2 = \frac{R_{r_2} - b_2}{R_{s_2} T_2};$$

and, if $b_1 \approx b_2$ and $Rs_1 T_1 \approx Rs_2 T_2$,

$$\frac{\rho_1}{\rho_2} = \frac{R_{r_1} - b_1}{R_{r_2} - b_1}.$$

If, additionally, the backscatter is assumed to be negligible, the ratio of the object reflectances is equal to the ratio of the received energies at the detector:

$$\frac{\rho_1}{\rho_2} = \frac{R_{r_1}}{R_{r_2}}.$$

Thus preprocessing the image with this normalizing algorithm will provide the corrections for variable geometries, irradiances, and attenuations.

It is to be noted that in a progressive sequence of normalization of the $i$th channel by the ($i$th + 1) channel the normalization of the last channel by itself constitutes a loss of one dimension (or degree of freedom). This, however, is considered a nominal cost for the normalization improvement.

If the backscatter component cannot be dismissed but can at least be assumed to be equal for each channel, an additional step of differencing adjacent channels before ratioing eliminates the backscatter variable in the normalization process:

$$\rho_1 - \rho_2 = (R_{r_1} - b_1) - (R_{r_2} - b_1) = R_{r_1} - R_{r_2},$$

$$\rho_2 - \rho_3 = (R_{r_2} - b_2) - (R_{r_3} - b_2) = R_{r_2} - R_{r_3};$$

0-201-04245-2

and the ratio of the differences is

$$\frac{\rho_1 - \rho_2}{\rho_2 - \rho_3} = \frac{R_{r_1} - R_{r_2}}{R_{r_2} - R_{r_3}},$$

that is, the ratio of the differences in target reflectance signatures for neighboring channels is equal to the ratio of the differences of the received energies.

### C. Image Enhancement Procedures

The main image enhancement procedures are density slicing, edge enhancement, change detection through image addition, subtraction and averaging, and contrast stretching. These can be achieved by use of positive and negative sandwiches of images, and by use of optical color combiners or by electronic or digital processing.

#### *1. Density Slicing*

A slice of the gray scale may be selected to emphasize a given feature (e.g., water on a near-infrared black and white photograph). Through exposure control one may photographically remove gray tones lighter than those for water. Photographic equidensity slices may be obtained using Agfa contour film (Ranz and Schneider, 1970).

Density slicing can be carried out using an electronic system. Transparencies are cathode-ray-tube (CRT) scanned to convert image densities to voltages. Voltage levels may then be selected and displayed on a CRT or printed on a film writer. Frazee et al. (1971) used density slicing and a color display in a test mapping of soil conditions. Helbig (1972) obtained equidensity slicing in color photographs in each color layer with a Vario-chromograph color scanner.

With a digital system any level or levels of gray may be selected for output on a line printer, on a flat-bed or drum plotter, or directly onto a photographic film writer (see Figure 10.6). The level of quantization employed in digitization, however, markedly affects the fineness of slicing and can introduce quantization noise or "spurious contouring" in an image.

#### *2. Edge Enhancement*

Edge enhancement is an important preprocessing step in that it can be used to "sharpen" an image by restoration of high-frequency components, and to emphasize certain preferred edges or all edges to aid interpretation. For example (Rosenfeld, 1969), to sharpen an image in one direction one may take the derivative of that image (i.e., the rate of *change* of gray level) or the absolute value of the derivative in the desired direction, for example, $|df/dy|$. To sharpen the image in all directions the highest directional derivative would be used. It is equal to

0-201-04245-2

$$\left[\left(\frac{\partial f}{\partial x}\right)^2 + \left(\frac{\partial f}{\partial y}\right)^2\right]^{1/2}$$

where $x$ and $y$ are orthogonal directions along the image, and $f$ is the gray-level function. For a digitized image a first-order approximation could be obtained by operating on the gray-level difference between adjacent row and column pixels via

$$[(f_{i+1,j} - f_{ij})^2 + (f_{i,j+1} - f_{ij})^2]^{1/2}.$$

Photographic edge enhancement is straightforward. Positive and negative images are sandwiched and slightly displaced with respect to one another. Edges normal to the displacement direction will be emphasized as white lines at the leading edge and black lines at the trailing edge in the resulting photograph exposed through the sandwich: this is a directional derivative. Note that edges of equal sharpness parallel to the displacement direction will not be emphasized. To emphasize all edges independently of direction requires off-axis illumination and exposure of the sandwich on a rotating disk. Another procedure is to sandwich an unsharp positive and a sharp negative and print through both. Electronic equivalents of these procedures include differentiation of an image along the scan lines of a CRT (the directional derivative) or the use of a spiral scan to obtain directional invariance.

The Fourier transform of an image may be obtained both optically and digitally. In the former case a collimated light source or laser is used to filter out all but the high-frequency components, and then some directions reconstituting the image are masked in, thereby emphasizing certain edges or locations.

To describe this mathematically (see Rosenfeld, 1969), let $f(x, y)$ be the picture function of the image in the $x$ and $y$ directions. The Fourier transform $F(p, q)$, where $p$ and $q$ are spatial frequencies in the $x$ and $y$ directions, is

$$F(p, q) = \int\int f(x, y) e^{2\pi j(px+qy)}\, dx\, dy,$$

where $j$ is $\sqrt{-1}$. Then, if the filtering of spatial frequencies is accomplished by an appropriate mask, $f(g)$, this masking operation is equivalent to the mathematical process of multiplication of $F(p, q)$ by $F(g)$, which is the Fourier transform of the mask. The inverse Fourier transform of $F(p, q) \cdot F(g)$ then describes the filtered image.

In the digital implementation digitizing takes place in a sequential raster scan comparable to that of a CRT, giving a preferred orientation from the scanning direction. Scan line noise is first removed; then either directional derivative edge enhancement or this-direction-invariant enhancement may proceed. Functions which are neutral with respect to orientation, such as a rotation-invariant second-derivative operator (i.e., a Laplacian) or a Fourier transform, are used to obtain enhancement of all edges.

0-201-04245-2

A simple example (Bakis et al., 1971) of a three-point digital filter, $f(z)$, modifying the input gray level, $z$, is

$$f(z) = a_1 z + a_0 + a_1 \frac{1}{z},$$

where $z = e^{at}$, and $t$ is the sampling period. Then by appropriate choice of the filter amplitudes, $a_0$ and $a_1$, this function can approximate an emphasis of the higher sampling frequencies which is useful for image smear removal.

Although all these procedures are helpful in image analysis, they must be used with caution, for they only enhance edges. Human evaluation is required after the enhancement. Plant ecologists, geologists, and others find edge-enhancement images both valuable and potentially very misleading if not used in conjunction with other materials in boundary delineation, lineament analysis, and similar procedures.

### 3. *Change Detection*

Detection of change from one image to another seeks to emphasize only relevant change and ignore extraneous change or noise. Since most changes are in fact irrelevant for a particular study (e.g., soils wetted by irrigation or patchy rainfall when one is interested in crop condition), change detection turns out to be much less easy to implement on a selective basis than its intrinsic attractiveness would suggest. Thus change detection algorithms give ambiguous results, requiring much interpreter analysis. Typical procedures used in change detection are flicker (switching between two images at 10 frames/sec) and image subtraction (sandwiching positive and negative images of different dates). Change detection has been used by investigators working with LANDSAT data, as reported in Chapter 3.

### 4. *Contrast Stretching*

Contrast stretching is used to make subtle gray-scale differences, not readily detectable with the naked eye, more obvious for interpretation. Any image of low contrast or any portion of the gray scale of an image may be enhanced this way.

To improve contrast, new negatives and positives may be prepared by processing with high-gamma film, use of a $\log_e$ (of the gray scale) processer, or histogram equalization in digital processing. For LANDSAT frames histogram equalization may be carried out before color combination to enhance the color contrast.

### 5. *Image Smoothing*

Image smoothing is occasionally useful. It is achieved by defocusing a camera, CRT blurring of spot size, or digital reduction of variance (Brown, 1972). A simple, nonlinear, digital smoothing function used by Holloway (1958) is as follows. Let

0-201-04245-2

the pixel (position gray value) $E$ be surrounded by neighbors $A$–$I$:

$$\begin{matrix} A & B & C \\ D & E & F \\ G & H & I; \end{matrix}$$

then replace $E$ by $E = {}^1/_4E + {}^1/_8(B + D + F + H) + {}^1/_{16}(A + C + G + I)$.

### 6. *Other Options*

Other options include *thresholding* (Steiner et al., 1972), *optical* and *electronic color combination* of images (Peterson et al., 1969; Steiner et al., 1972; Estes and Senger, 1971; Lent and Thorley, 1969), and *digital image sharpening* with Laplacian operators (Steiner et al., 1972). *Quantification of textures* may use auto-correlation functions, nearest-neighbor algorithms and dichotomous keys, Fourier transforms, variance of the signal, number of edges per unit distance, and micro-densitometer traces (either raw or after analyses such as power spectral analysis) (Kaizer, 1955; Haralick and Anderson, 1971; Brown, 1971; Brooner et al., 1971; Simonett and Brown, 1965).

## V. FEATURE EXTRACTION

Feature extraction takes place in both qualitative and quantitative measurements, mapping, and subsequent analyses. The steps involved in each case are (1) measurements, (2) determination of the characteristics of the variables, (3) isolation of significant variables via an analysis logic, and (4) reduction of the number of channels or data sources (i.e., collapsing to the "essence" of the data structure by eliminating highly correlated data).

In descriptive and qualitative procedures these steps are shown in such processes as field surveys, the development of legends, comparison between ground observation and image appearance, preparation of photographic keys, and the search for correlations between easily observed photographic features and important secondary features of greater interest. The problem is basically one of seeking order in the data.

The similarity between qualitative photo interpretation and the extraction of the essence of data through quantitative feature extraction is described by Steiner (1970) as follows:

> New observations, not contained in the original data set, can be allocated in exactly the same way, by making use of . . . discriminant functions. It should be noted, therefore, that automated aerial or space surveys do not differ from the basic principles of conventional photo interpretation, i.e., the selection of a sample of subjects of known identity, the establishment of a key based on this sample, and the application of the key information to classify all other

0-201-04245-2

> objects in the study area whose identity is not known as yet. For the mathematical solution the discriminant functions or any other appropriate functions represent the key, and the comparison of the classification of sample subjects as produced by the analysis with the true identity of these subjects provides a statistical estimate for the overall accuracy of the survey.

The search for the "essence" of the data structure thus arises simply from the need to be parsimonious.

Although quantitative analyses actually involve simpler decision rules than those used by the human analyst, the associated mathematics may seem forbidding. Hence we shall progress briefly through several examples.

## A. Quantitative Measurements of Image Gray Scale

Quantitative gray-scale data in photography may be obtained with microdensitometers and then be transferred (after analog-to-digital conversion) to magnetic tape for digital processing. Spot and scanning microdensitometers are usually equipped with color filters for use with color photography to obtain congruent, full-field registry of the three frequency bands of a color film. Point densitometry has been used by Maurer (1971) to compile statistics suitable for feature extraction.

Feature extraction may consist merely of quantifying the within-class variance from a densitometer trace. Pestrong (1969) used the variance of traces across images in coastal wetlands on multiband photographs to separate plant communities, and to profile channel depth in inaccessible sites.

Isodensitracing with scanning microdensitometers is a common practice. Myers (1969) temperature-contoured water and soils. while James and Burgess (1970) produced bathymetric maps and thickness maps of sediment plumes and outfall sites based on isodensity measurements.

Microdensitometry is widely employed by foresters. Ciesla et al. (1972) used the cyan layer of false color photography to test for the significance among three plots sprayed against the forest tent caterpillar.

As indicated by these initial examples, measurements give way to a search for structure or order.

## B. Statistical Structure and Order

The problem of recognizing order in apparent chaos has been evident since man first sought to measure and classify. The statistical methods used in quantitative feature extraction seek to find such order in multichannel data too complex for easy manual analysis. The search for order by means of statistical methods employs different tentative mathematical structures and is a function of the researcher and his perception of the problem and results. Statistical and probabilistic inquiries provide significant analysis procedures with a relatively common language for communication. To help in this process, histograms, covariance matrices, means, standard deviations, and scatter diagrams are commonly prepared from pixel gray-scale data.

0-201-04245-2

It is thus useful for us to differentiate between methods of feature extraction that are nonparametric and those that are parametric. The former elect the flexibility of "initial" nonspecificity and will be discussed shortly. The latter assume that sample data are part of a population that can be described by classical statistical distributions.

### *1. Parametric Structures*

One of the most commonly invoked parametric structures for a set of pixel gray values is the normal (or Gaussian) density function described by:

$$f(n) = \frac{1}{\sqrt{2\pi}\,\sigma} \exp\left[-\frac{1}{2}\left(\frac{x - \mu}{\sigma}\right)^2\right]$$

Herein two features may be extracted as indicative of the one-dimensional gray data in $x$. These are the mean, $\mu$, and standard deviation, $\sigma$; that is, the training data are assumed to be representative of a true population described by that statistical structure. The reasons for the lack of complete conformity are assumed to be due to the limited set of empirical data available and to the "noisiness" (random uncertainties) of the measurement process and the effects of confounding (i.e., other, previously unrecognized) variables.

It is generally considered that different kinds of measurements improve one's ability to adequately describe dominant, characteristic features. Thus multispectral measurements of remotely sensed objects are usually made. The extension of the mathematical structure to accommodate the multidimensional ($m$) measurement vector, $\overline{X}$, is often accomplished via the multinormal, joint probability density function. It is described (Morrison, 1967) by

$$f(x_1, x_2, \ldots, x_n) = \frac{1}{(2\pi)^{n/2}\sigma_1 \ldots \sigma_n} \exp\left[-\frac{1}{2}\sum_{i=1}^{n}\left(\frac{x_i - u_i}{\sigma_i}\right)^2\right].$$

In the convenient form of matrix algebra a vector $X$ is represented by a column matrix:

$$X = \begin{bmatrix} x_1 \\ x_2 \\ \cdot \\ \cdot \\ \cdot \\ x_n \end{bmatrix}$$

or by its transpose, a row matrix:

$$X^T = [x_1, x_2, \ldots, x_n];$$

0-201-04245-2

thus the expression is

$$f(\overline{X}) = \frac{1}{(2\pi)^{n/2}\,|\Sigma|^{1/2}} \exp\left[-\frac{1}{2}(\overline{X} - \overline{\mu})^T\, \Sigma^{-1}\, (\overline{X} - \overline{\mu})\right].$$

Here the means, $\overline{\mu}$, are also expressed as a vector, and the variance has been generalized to a diagonal matrix, $\Sigma$. The square root of the determinant of $\Sigma$ has assumed the role of the previous, univariate scale factor, $\sigma$. The $ii$th element of the $\Sigma$ matrix is the variance, $\sigma_{ii}$, the $ij$th element, $\sigma_{ij}$ (row $i$, column $j$), is the covariance of the $i$th and $j$th components. If all the covariances are 0, the $n$ components of $\overline{X}$ are independently distributed.

Usually, however, it is found that an $n$-dimensional feature measurement (e.g., the vector $\overline{X}$) is such that measurements $x_i$ and $x_j$ may covary in some fashion. Thus the generalization of the variance to multidimensional variates incorporates the covariance of the elements $x_i$ and $x_j$ of a random vector. It is the product moment of those variates about their respective means, that is,

$$\text{cov}\,(x_i, x_j) = E[(x_i - \mu_i)\,(x_j - \mu_j)],$$

where $E[\ ]$ is the expected value of the product, and $\mu_i$ and $\mu_j$ are the respective means of $x_i$ and $x_j$. This product moment will tend to be large when the $x_i$ and $x_j$ fluctuate together (directly or inversely). Thus the process of feature extraction seeks to explicate a reduced set of orthogonal (uncorrelated) vectors which carry the full significance of the initial, partially redundant correlations.

One such mathematical procedure is the principal components method, which is an "accounting of" the total variance that characterizes the data. It is useful to reflect briefly why it is informative to do such accounting and to understand the simpler one-dimensional basis of the more sophisticated forms of variance analysis.

Variance is basically a measure of uncertainty in describing the characteristics of an incompletely determined "population" (e.g., the spectral reflectance of a remotely sensed crop). If the complete set of remotely sensed data at one wavelength for this one crop was divided into several categories or groups, each would provide a separate variance estimate. If, furthermore, we determined two separate variance estimates for the between-group and within-group characteristics, their ratio, $F$, would be a measure of the homogeneity of the groups. This ratio would be about 1 for similar groups. If, instead, we measured the spectral reflectance at different wavelengths, $\lambda$, for the same crop and placed the data from wavelength 1 in category 1, etc., a significantly high $F$ ratio would indicate that $\lambda_1$ and $\lambda_2$ measurements characterized different features of the crop.

The variance ratio, $F$, is defined by

$$F = \frac{s_b^{\,2}}{s_w^{\,2}},$$

0-201-04245-2

where $s_b{}^2$ and $s_w{}^2$ are the estimates of the population variance, $\sigma^2$, from the between-group and within-group statistical characteristics, via

$$s_b{}^2 = \frac{m_j \sum_{j=1}^{c} (\mu_j - \mu)^2}{c - 1}, \qquad s_w{}^2 = \frac{\sum_{j=1}^{c} \sum_{i=1}^{m_j} (x_{ij} - \mu_j)^2}{m_j - c},$$

where $m_j$ = number of data samples per group or category $j$,

$c$ = number of categories,

$\mu_j$ = mean of $j$th category,

$\mu$ = population mean,

$x_{ij}$ = data sample.

As noted, if factors other than randomness are involved in the data, the variance ratio may depart significantly from unity. These concepts are then extended to encompass multidimensional analysis, often in the extended context of confounding data.

Confounding variables are often part of a training sample because they cannot be avoided. For example, a remote sensor gathering training data from one crop on days of variable cloud cover (incompletely corrected for by radiometric normalization) constitutes part of the source of confounding variables. In the analysis of variance, correlation analysis is frequently used to mitigate some of the influence of confounding variables.

The identification of the statistically independent features of an $n$-dimensional vector often uses the overall covariance matrix, $\Sigma$. Michael and Lin (1973) define a between-category-to-within-category variance ratio, $BWR$, by

$$BWR = \log \frac{|\Sigma|}{|W|} = \log \frac{|B + W|}{|W|},$$

where $B$ is the average between-class covariance matrix, and $W$ is the average within-class covariance matrix. Then

$$W = \sum_{j=1}^{c} P(c_j) W_j$$

and

$$W_j = \frac{1}{m_j - 1} \sum_{i=1}^{m_j} (x_{ij} - \bar{\mu}_j)(x_{ij} - \bar{\mu})^T,$$

0-201-04245-2

where $P(c_j)$ = a priori probability of class $j$,

$m_j$ = number of patterns in class $j$,

$x_{ij}$ = $i$th pattern vector from class $j$,

$\bar{\mu}_j$ = estimated mean vector from class $j$,

$c$ = number of classes,

$\bar{\mu}$ = overall mean vector, and

$$B = \sum_{j=1}^{c} P(c_j)(\bar{\mu}_j - \bar{\mu})(\bar{\mu}_j - \bar{\mu})^T.$$

Michael and Lin point out that the ratio $BWR$ is also related to eigenvalue $\lambda_i$ by

$$BWR = \sum_{i=1}^{d} \log(1 + \lambda_i) = \log \left| \frac{W + B}{W} \right|,$$

where the eigenvalues are determined from a characteristic equation,

$$B\mu_i = \lambda_i W \mu_i.$$

The eigenvalues are significant since they are an exact measure of the between-class to within-class variance ratio measured along the appropriate eigenvector, $\mu_i$. This variance ratio "can be understood as a kind of signal-to-noise ratio with the signal represented by the average distance between the classes (the numerator) and the noise by the average distance within each class (the denominator). The larger the ratio, the better separated the classes should be." A comparative and excellent summary of similarly useful "distance measurements" is given by Haralick (1975).

Ready et al. (1971) show the utility of eigenvalues, $\lambda_i$, for data compression by determining them from the characteristic equation

$$\Sigma\mu_i = \lambda_i \mu_i$$

for $i = 1, 2, \ldots, n', \ldots, n$, where $n'$ is the $n'$ largest eigenvalues (and hence the significant extracted features) of the covariance matrix $\Sigma$, and $n$ is the original dimensionality (i.e., number of features) of the multivariate vector $X$. They note that for highly correlated data the eigenvalues decrease rapidly; the resultant data compression ratio is $n/n'$. Compression ratios as high as 3 : 1 for spectrally correlated data have been used with negligible classification degradation errors on these extracted features, as compared to using the original $n$ features. This method also leads to the principal components representation (cf. Hotelling, 1933). The extraction of a typical decision rule under the assumption of a multivariate normal distribution is discussed later.

0-201-04245-2

### 2. *Nonparametric Structures*

In some instances, the assumption of a multivariate probability density function to describe the nature of the $n$-dimensional feature data is too restrictive. Then the use of nonparametric methods is appropriate. In this approach a selected set of pixel gray values called *training data* is used to determine distribution-free decision rules. Haralick (1975) summarizes this procedure as follows. The first step is to estimate the set of conditional probabilities,

$$\{P(j \mid i) \mid j \in c, i \in \bar{X}\},$$

that is, the probability that category $j$ is present, given that feature $i$, of all features $\bar{X}$, has been measured and under the further condition that $j$ is a subset of all the categories of interest, $c$. These conditional probabilities can be estimated by the proportion of data points having true category identification $j$ in the subset of the training data whose measurements generate feature $i$. Then, depending on whether or not the a priori probabilities of the various categories are used, the decision rule is either a simple Bayes rule or a maximum likelihood rule, respectively. These rules are briefly discussed later.

## C. Discriminant Functions

There are many ways in which the essence of a problem can be distilled. In photo interpretation it may be the discovery that at one time of the year complete discrimination between two entities can be made, using a single channel. At other times of the year even three or more channels may be unsatisfactory. Similarly, with a multispectral scanner we may discover that to discriminate one crop from all others may require as few as two channels and that the addition of extra channels gives little improvement (see, e.g., Biehl and Silva, 1975).

This may be illustrated in a simple, probabilistic description. Assume that three channels are independent in identification capability (usually this is not the case, but the assumption does not invalidate this example). Let the respective crop identification probabilities be

$$P(1) = .8, \quad P(2) = .5, \quad P(3) = .3.$$

Then the combined probability of identification of various combinations (wherein the identification is adequately made by one or more of the combined channels) is found from

$$P(A, B) = P(A) + P(B) - P(A)P(B)$$

or

$$P(1, 2) = .9, \quad P(2, 3) = .66, \quad P(1, 3) = .86, \quad P(1, 2, 3) = .93.$$

0-201-04245-2

Thus the addition of channel 3 to channels 1 and 2 does not significantly increase the combined probability of crop identification. In actual evaluations attention must also be paid to the effect that combining has on false identification probabilities (errors of commission).

The diversity of discriminant function types may now be summarized.

In pairwise tests of the difference between means Morrison (1967) used:

$$t = \frac{|\mu_1 - \mu_2|}{\hat{\sigma}},$$

where $\mu_1$ and $\mu_2$ are the sample means of groups 1 and 2, and $\hat{\sigma}$ is the estimated standard error of the difference between two means. Included in the assumption is that the two population variances are equal. Estimated via the pooled variance, $\hat{\sigma}$ is found to be

$$\hat{\sigma} = \left( \frac{n_1 s_1{}^2 + n_2 s_2{}^2}{n_1 + n_2 - 2} \right)^{1/2} \left( \frac{n_1 + n_2}{n_1 n_2} \right)^{1/2},$$

where $s_1{}^2$ and $s_2{}^2$ are the sample variances, and $n_1$ and $n_2$ are the sample sizes.

LeSchack (1970) prepared a power spectrum analysis of forest types (waveforms were treated as a time series, though the measurements themselves were in the context of amplitude, as a function of distance.) The Fourier transform of this series provides terms related to the power-frequency spectrum. In this manner various power spectra may be compared.

Biehl and Silva (1975) scanned color and multiband films from SKYLAB over Indiana to carry out maximum-likelihood land-use identifications. The conditional probability of various film color densities, $x_i$, caused by various land-use categories, $c_j$, is $P(x_i \mid c_j)$. Under the maximum-likelihood rule, given the feature or data, $x_i$, the assignment is made to the land-use category, $c_j$, satisfying the inequality $P(x_i \mid c_j) > P(x_i \mid c_k)$ for $k = 1$ to the total number of categories. Biehl and Silva (1975) and Anuta et al. (1971) also spatially registered multiband images and then used divergence analysis to measure the pairwise separability between classes.

### 1. *Pairwise Separability*

The measurement of pairwise separability may be briefly treated by the following steps (Fu et al., 1969). Consider a feature measurement vector, $\overline{X} = [x_1, x_2, \ldots, x_n]$, with $n$ features, which is characteristic of class $C_i$. Assume that the conditional probability, $P(\overline{X} \mid C_i)$ (i.e., that vector $\overline{X}$ is measured when $C_i$ is sensed), is distributed as a multivariate Gaussian density function (see Figure 11 of Chapter 10). Let a linear discriminant function, $g_i(X)$, be defined as

0-201-04245-2

$$g_i(\overline{X}) = \overline{B}_i{}^T\overline{X} - C_i,$$

where $B_i = [b_1, b_2, \ldots, b_n]$,

$T$ = the transpose,

$C_i$ = a constant.

Thus

$$g_i(\overline{X}) = b_1x_1 + b_2x_2 + \ldots + b_nx_n - C_i.$$

Similarly define another discriminant function, $g_j(X)$, relative to category $C_j$. Then let $P_{ij}(e)$, the pairwise probability of misclassification of the classes $C_i$ and $C_j$, be defined as

$$P_{ij}(e) = \tfrac{1}{2}P[g_i(\overline{X}) > g_j(\overline{X}) \mid C_j] + \tfrac{1}{2}P[g_i(\overline{X}) < g_j(\overline{X}) \mid C_i].$$

Thus the error probability is the weighted sum of the probabilities that the discriminant function $g_i(\overline{X})$ is greater than $g_j(\overline{X})$ when $C_j$ is present or that it is less than $g_j(\overline{X})$ when $C_i$ is present. A measure of the separability between the two classes, $d_{ij}$, is then defined by relating the difference of these two discriminants to the mean and covariance matrices of the feature vector thus:

$\overline{B}_{ij} = \overline{B}_i - \overline{B}_j$,

$\overline{M}_i, \overline{M}_j$ = mean feature measurement vector matrices for $\overline{X}$ when sensing $C_i$ and $C_j$,

$\overline{K}_i, \overline{K}_j$ = covariance matrices of $\overline{X}$ when sensing $C_i$ and $C_j$.

Then the measure is

$$d_{ij} = \frac{\overline{B}_{ij}{}^T(\overline{M}_i - \overline{M}_j)}{(\overline{B}_{ij}{}^T\overline{K}_i\overline{B}_{ij})^{1/2} + (\overline{B}_{ij}{}^T\overline{K}_j\overline{B}_{ij})^{1/2}}.$$

For multiclass separation, the measure $d_{ij}$ is optimized by maximizing the expected separability for all pairs of classes.

Because of the excessive computation with the maximum-likelihood ratio test on each data sample, many alternative procedures are being tested. Crane and Richardson (1972) note that, although it is possible to reduce the number of data through sampling in the classification step (see the problems of sampling mentioned later), it is also possible to save time via the nature of the decision rule itself. The decision rule in common use is

0-201-04245-2

> . . . based on two assumptions: (1) that the data are Gaussian (normal); and (2) that training data for each class adequately represent the entire class. With these two assumptions the maximum-likelihood decision rule becomes a quadratic rule. This rule requires a large number of multiplications for each decision, especially when many channels of data and many classes are used.

Crane and Richardson found that a linear decision rule compared favorably in accuracy and was 50 times faster than the maximum-likelihood function.

> In choosing a subset of channels for processing, the linear method is especially promising. The reason is quite simple: there are many subsets of channels that provide near-optimum recognition performance for the training data. Thus, even though linear methods might occasionally choose a different subset than would the quadratic, the difference in performance is likely to be negligible.

The quadratic decision rule was to choose an $i$ that minimized the function

$$[\overline{X} - \overline{M}_i]^T \overline{K}_i^{-1} [\overline{X} - \overline{M}_i] + \ln |\overline{K}_i|,$$

and thus category $C_i$ was sensed. Here, as before,

$\overline{X}$ = feature measurement vector,

$\overline{M}_i$ = mean feature vector of $\overline{X}$ when sensing $C_i$,

$\overline{K}_i^{-1}$ = inverse covariance matrix of $\overline{X}$ when sensing $C_i$.

As a comparison, the linear rule used the function

$$[\overline{X} - \overline{M}_i]^T \overline{K}_0^{-1} [\overline{X} - \overline{M}_i] + \overline{X}^T F_i + D_i,$$

where $\overline{K}_0$ is a modified equal covariance matrix used in all calculations. Since the quadratic portion of this function is common to all processes, it is eliminated from the decision process. The constants $F_i$ and $D_i$ were found by minimizing the difference between the quadratic and linear rules, weighted by $P(\overline{X} \mid C_i)$.

In contrast, the use of the elliptical boundary condition model (Richardson et al., 1971) considers that the distribution of accidental errors associated with experimental measurements may be described with elliptical curves. The decision rule is summarized as follows: Classify measurement vector $\overline{X}$ as belonging to category $C_k$ if

$$\sum_{j=1}^{n} [\overline{X}(j) - \overline{M}_k(j)]^2 < \sum_{j=1}^{n} [\overline{X}(j) - \overline{M}_i(j)]^2$$

for all $i = k$. Here

0-201-04245-2

$\overline{M}_i$ = the $(m-1)$-mean measurement vector of $\overline{X}$, calculated from $m-1$ categories, excluding category $C_k$,

$\overline{M}_k$ = the $k$th-mean measurement vector, calculated for $\overline{X}$ from category $C_k$,

$n$ = the number of features.

The potential function, $P$, is related to classical electrostatic potential descriptions. A slightly modified function for feature extraction defines it as $P = 1/(1 + ad^2)$, where $a$ is an empirical coefficient, and $d$ is any convenient way of measuring the distance from feature point $\overline{X}$ to each of the points in category $C_i$.*

### 2. *Bayesian Approaches*

Whatever the method used, there remains the perennial problem of using equal or weighted a priori probabilities for the category presence. Equal probability assumes equal likelihood of all items in the prediction set. Weighted probability assumes that the distribution of items within the prediction set is the same as that in the training set (or in some already established data, such as county statistics). Each approach has major problems associated with it. The Bayesian decision rule is very commonly applied with LANDSAT data because the use of a priori weighting leads to better results than using maximum likelihood with equal probability for all classes. The simple Bayes decision rule may be related to the maximum-likelihood decision process in the following way.

Let $P(C_j)$ and $P(C_k)$ be the a priori probabilities of categories $C_j$ and $C_k$, respectively. These are frequently determined from training, historical, or other data. The fundamental Bayes equation relates the probability of a category $C_i$, given the measurement vector $\overline{X}$, to the a priori probability of category $C_i$ and the conditional probability that $\overline{X}$ will result when $C_i$ is sensed, in the following way:

$$P(C_i \mid \overline{X}) = \frac{P(C_i)P(\overline{X} \mid C_i)}{\Sigma_i P(C_i)P(\overline{X} \mid C_i)},$$

where the denominator describes the probability of $\overline{X}$. Then the Bayes rule is to accept category identification $C_i$, given feature measurement $\overline{X}$, when

$$P(C_i \mid \overline{X}) > P(C_j \mid \overline{X}) \qquad \text{for } j = 1\text{-}k \text{ categories, excluding } i.$$

*See also Bond and Atkinson (1972) (linear discriminant functions), Su et al. (1972) (a composite sequential clustering algorithm), Richardson et al. (1971) (an elliptical boundary condition model), and Eppler et al. (1971a). Additional discussions on extracting features include those on linear discriminant functions (Steiner, 1970; Steiner and Maurer, 1969), polynomial discriminant functions, potential functions (the $n$-dimensional potential exerted by a point in space), and nonparametric partitioning. Also see Fu (1971), Steiner et al. (1972), and Haralick (1975).

0-201-04245-2

Examples include the studies of Brooner and Simonett (1971) and Brooner et al. (1971), who used point densitometry on color images of crops in eastern Kansas and the Imperial Valley of California, respectively. The Bayesian approach followed normalization of the images.

### D. The Use of Training Sets: Multispectral Scanner

Feature extraction with the multispectral scanner uses training sets (for "supervised classification") or clustering (for "unsupervised classification"). A training set is a small, identified sample. It is used to generate statistics to efficiently implement decision rules. The value of feature selection depends on the quality of the sampling.

One of the few papers investigating the relationship between size of training sets and accuracy of results is that by Roth and Baumgardner (1971). They examined the relation between soil color and organic matter, employing the coefficient of determination $r^2$ for one channel and for six channels, for sample sizes of 1, 9, 25, and 144. The best single channel improved from an $r^2$ of 0.42 to 0.46, the best six from 0.61 to 0.72, in going from 1 to 144 cases. Very limited regional sampling would be typical for this type of solution; however, most situations require much higher sampling rates.

A related study (Basu and Odell, 1974) showed the effect of intraclass correlation in training samples on the predicted error rates for test data. They point out that, more frequently than not, training samples are rather dependent, at best equicorrelated. In the latter case an unbiased maximum-likelihood estimate is made for a modified covariance matrix $[\overline{K} - \overline{R}]$ from

$$\frac{\sum_{i=1}^{n} (X_i - \overline{X})(X_i - \overline{X})^T}{n - 1}$$

where $X_1, \ldots, X_n$ are equicorrelated samples. The analysis is followed by a simple example illustrating the effects of correction and of noncorrection on the predicted error rates. The effect of correction is to predict more closely the higher-order errors.

The general question of sampling rates over large areas is of considerable importance now that many workers are using LANDSAT imagery. Before the launch of LANDSAT Hoffer and Goodrick (1971) showed that for aggregated classes (bare soil, water, green vegetation) training samples selected from a very small area could be extrapolated over considerable distances (with 97% accuracy for soil, 100% for water, and 99% for green vegetation). Since time is a valuable discriminant function (Steiner, 1970), this suggested that multidate LANDSAT imagery and gross categories could achieve good discrimination over large distances. In general this has proved correct in extending the application over large areas. Unfortunately, though,

0-201-04245-2

such gross categories fail to address the central problem of specific identification (e.g., wheat versus all other forms of green vegetation).

A paper by Higgins and Deutsch (1972) on the effects of picture operations in the Fourier domain is important in that it shows the types of errors introduced by sampling (in the classification step) in different areas. The more complex and more finely divided a given scene (e.g., fields in southeast Asia), the more difficulty is encountered in row and column sampling to reduce data volume. Such sampling is a common procedure with the multispectral scanner. The paper indicates some of the problems encountered as one moves from one environment to another in terms of the likely accuracy of results, and shows that spatial domain sampling introduces both high- and low-frequency noise, depending on the area, into the data.

## E. The Use of Clustering Algorithms

In areas for which training sets are not available, feature extraction can be derived by using a clustering algorithm. Alternatively, clustering may precede the selection of training samples as well. This is similar to photo interpretation, in which an interpreter places lines around regions that appear photographically similar. The process of mapping natural landscapes is a manual clustering routine with single photographs or images (see Figures 22–24 of Chapter 10).

Turner (1972) notes that "despite the existence now of a large body of literature on cluster analysis there is no unanimity on a universally based criterion although there is some consensus on the best criteria for certain types of data." When data can be considered a sample from a continuous multivariate normal distribution, Turner points out that the generalized variance (the determinant of the pooled, within-group variance-covariance matrix) is widely used. Data sampled from discrete distributions or mixed data from both distribution types lead to measures based on a Euclidian distance metric, which may be defined as $\sum_{n=1}^{N} (x_{in} - \mu_j)^2$ for each $i$ and $j$. Turner (1972), following Tryon and Bailey (1970), set up his trial group centroids and preceded via iterative condensation on centroids, each point being assigned to the group to which it had the smallest Euclidian distance. The process is repeated iteratively until no change occurs.

A thoughtful analysis of the procedures and dilemmas in supervised and unsupervised approaches to feature extraction has been given by Schell (1972). Both work well with small data sets, but the cost of processing is substantial for large areas. Schell employed a "spatial-spectral clustering classification philosophy," using a one-pass clustering procedure akin to that by Nagy et al. (1971). He used a distance function measure (maximum of the absolute values of the differences between the spectral components) with a spatial contiguity constraint. Initially he weighted the spatial portion of the analysis more strongly than the spectral portion, using a two-dimensional spatial correlation function.

0-201-04245-2

A valuable review of the problems of clustering and training sets in feature selection has been given by Nagy et al. (1971), who also developed clustering with spatial contiguity constraints; this is similar to procedures in use for some years by geographers. Nagy and his colleagues observed problems of overlapping of statistical and real classes like those observed by Haralick et al. (1969).

Fu (1971) and Steiner et al. (1972) discuss in some detail the various transformations used in clustering. The most commonly used clustering measures are those of Euclidian distance, similarity ratios, normalized correlations, minimum distance to the mean, minimum of least squares, iterative determination of centroid locations, principal components analysis, canonical discriminant functions, and separability measures.

The process of clustering begins with full exposure to the data available and seeks to determine logial, consistent, or "natural" groups in the data. After clusters have been formed, the relationships to the environment are sought. Generally the cluster is a hybrid—a mixture of sensed objects each having an associated measurement vector (which may be statistically described by a covariance matrix). Thus the clustered data are often a weighted combination, the weighting being the natural result of the composite object and the method of measurement.

A mundane (but graphic) example of this would be the measurement vector of an orange grove—more specifically, a component of the radiation reflectance from the leaves, the fruit and/or blossoms, the branches and trunk, and the surrounding soil. To carry the illutration further, if the spectral reflectance characteristics of lemon trees are essentially the same—with only the lemon providing a different return from that of the orange—it is apparent that these singular differences may be dominated by the more extensive, similar measurement vectors of background soil, leaves, shadows, etc. Thus the cluster would be grouped by measurements which are essentially devoid of fruit difference. After the clustering process the groups may become identified as "citrus," or even "green vegetation" if many additional crops were also included in it.

Generally, therefore, clustered data are contaminated with spurious responses that are not part of the information sought. To describe the cluster may require a category so general as to deprive one of significant environmental perception. Thus there is a tradeoff between the probability of errors associated with the decision rule (derived after the clustering process) identifying the cluster and the amount of information in that cluster identity (or level of generality). One method of clustering measurement vectors performs the aggregation by minimizing the sum of the "distances" between the actual or eventual center of the cluster. Though, as noted above, there are many clustering algorithms, there are certain common features, which we may now introduce, in most mathematical algorithms.

Assume an initial cluster center as $\mu_j$, where $j = 1\text{-}c$ potential clusters, and scattered measurement data $x_i$ in one-dimensional space (i.e., the data are pertinent to a single variable). Let $i$ vary from 1 to $n$, the total set of data. Then an elementary clustering process may successively compare the distances

0-201-04245-2

$$|x_i - \mu_1|, \quad |x_i - \mu_2|, \quad \ldots, |x_i - \mu_c|$$

for $i$ varying from 1 to $n$. (Since only distance and not direction is important here, the absolute value is used). Then cluster assignments are made on the basis of associating each point to the closest center.

To encompass the more general case of an $n$-dimensional vector (an $n$-tuple), the Euclidian distance is used. A relative distance, here the squared distance, saves computer analysis time. Some variants of the ways to test potential cluster centers may or may not be effective and efficient. Here "effective" refers to the final validity/cohesiveness of a cluster(based on some criterion), and "efficient" refers to an economical algorithm.

One method combines sequential variance analysis with means clustering (Su, 1972). The output of the first process serves as an input to the second, and the weaknesses of each one are thereby somewhat mitigated. The procedure is to take a small ($M$) sample of $K$-dimensional vectors and calculate a variance-related measure:

$$\Delta x_i^2 = \sum_{k=1}^{K} (x_{ik} - \mu_{ik})^2,$$

where $\mu_{ik}$ is a mean value, and $i$ varies from 1 to $M$.

The maximum $\Delta x_i^2$ is found from this sample, normalized, and compared with a threshold, $T$, by means of

$$\frac{\max \Delta x_i^2}{\mu_{ik}} \leqslant T.$$

If this inequality is satisfied, the $M$ samples are considered to form a new population. The process is continued until some cluster limit is reached. Tests are made to aggregate existing clusters by combining those with closest distances. Mergers require an updating of statistical descriptions. By iterating the process, all the data are eventually examined. These clusters are then inputed to a $K$-means clustering algorithm, whereby two maximally separated clusters are identified. Then pairs of distances to each of the remainders are evaluated. The minima of each pair are aggregated, and the maximum is designed as the third cluster. This procedure is iterated to a set number. The remaining clusters are tested for merging on the basis of minimum distance measures and as a way of ensuring cluster means dispersion throughout the measurement space.

Decision boundaries are initially set up orthogonally to the line joining the clusters. From an initial midpoint position these are shifted toward the cluster with the smaller variance, the shift being proportional to the variance ratio. This balances probabilistic type I and type II errors, which are errors of omission and commission, respectively. After the clustering process, "ground truth" is used to identify them. It must be recognized, however, that different clusters will be produced de-

0-201-04245-2

pending on the size of the area examined. In short, there are serious unresolved questions on indeterminancy when either supervised or unsupervised classification is used.

## F. Image Texture Analysis

It was observed earlier that the order of value for human interpretation of image was context, edges, texture, and gray scale, and that this was reversed in terms of present computer capability. In fact, overwhelmingly the greatest use of (and effort in) further recognition has been concentrated in point, gray-scale identification procedures. Over the last 5 years there has been significant research in texture analysis, but hardly any operational application. The more complex areas of edge and context analysis remain virtually the exclusive domain of the interpreter.

Analog texture analysis is commonly carried out using densitometer line scanning and recording signal variance. Examples are seen in the study by Nunnally (1969a) on grouping natural landscape units in radar images of Tennessee and in a paper by Morain and Simonett (1967), where texture differences on radar images were assessed in plant community mapping through analysis of class probability density functions.

In the future, digital texture analysis will undoubtedly increase markedly. Important papers in this area are those by Ramapriyam (1972) and Palgen (1970). Therein spatial frequency analysis of multispectral data employs the discrete Fourier transform.

Because signal variance is related to texture in the image, most measures use some aspect of signal variance. One of the earliest of these is contrast frequency (Rosenfeld, 1962), the average number of times a specified departure from the average density occurs in a selected portion of a picture. This will be recognized as one simplification of the spatial probability density function.

Matrices of spatial gray-tone dependence have been used by Haralick (1971) for analysis of texture. The elements of the matrix are the relevant frequencies with which selected gray tones occur in neighboring pixels of defined angular relationships and distances. Haralick defined scalar functions of these matrices, including the angular second moment inverse difference, the angular second moment, and the correlation between neighboring gray tones. Since these features are direction dependent, they are not invariant, but may be made invariant by summing over four directions to perform feature extraction. Other techniques used by Hord and Gramenopoulos (1971) are those of thresholded derivative densities and geometric analysis of the signal-to-noise ratio in the Fourier transform plane. These techniques were refined by them (1973) through the prior regionalization, or partitioning, of the format and were followed by application of the algorithms to samples from each region.

Woolnough (1972) used microdensitometry to evaluate the textural components of muskeg and obtained 92% identification with a simple texture measure derived from the microdensitometer data. Maurer (1971) measured the textures of

0-201-04245-2

crop fields, using a precise scanning microscope photometer, and employed a nearest-neighbor probability training and prediction procedure. Another important technique recently introduced involves taking a power spectrum of a picture, color-coding different frequency components, and recombining them into a color photograph. This procedure was developed by Andrews et al. (1972) and is a method of visual enhancement of texture characteristics of terrain.

More recent papers on tone/texture features are those by Haralick and Shanmugan (1973) and Haralick and Bosely (1973). Gramenopoulos (1973) has employed spatial and spectral components of LANDSAT images (using the Fourier transform) to improve identification accuracies.

## VI. DECISION AND CLASSIFICATION

Feature extraction takes place on a training set which is a subset of the data. This subset may be a sample, chosen according to some well-established sample design. Much more commonly, however, the subset is chosen for convenience (along a road), in areas free of clouds on the imagery (areal biasing) and in homogeneous, large, contiguous entities (biasing with respect to categories, and data quality). The study by Von Steen and Wigton (1973) on LANDSAT data is one of the few LANDSAT studies which attempts proper sample design in *both* selection of training sets and assessment of prediction accuracy. Gradually more studies are showing attention to these concerns.

The decision and classification step uses the decision rules for discriminating between categories, derived in feature extraction, to place new data in an appropriate category; instead, they may be thresholded into a nonidentified class.

Qualitative analysis employs judgmental decision rules. Rules and procedures are built up for handling overlapping categories. In land-use mapping, for example, the development of decision rules is iterative. First classifications are preceded by designation of the categories to be mapped. As mapping progresses, many situations are noted in which modifications of the decision rules are necessary. These in turn may require additional codification or even elimination or splitting of categories.

In vegetation (plant community) and geological mapping, decision rules are usually not formally structured. Most qualitative interpretation with remote sensing imagery is in fact quite informal. A later interpreter can never exactly reconstitute the boundaries and classes designated.

Formal reconciliation of soil or geological units and boundaries at county map-sheet boundaries is now built into soil and geological survey procedures in the United States. If it were not, the classes and boundaries would differ enough to prompt such questions as, "What causes the frequency of fracturing changes at the border (county, state, country?)" or "Why are the same rocks in your area of different age from those in mine?" "Reconciliation" does not necessarily mean "more truthful"; it implies only that some adjustment or compromise has been reached.

An example of the different perceptions of boundaries is shown in Figures 11.1 and 11.2, where several persons mapping an area in central Australia have per-

0-201-04245-2

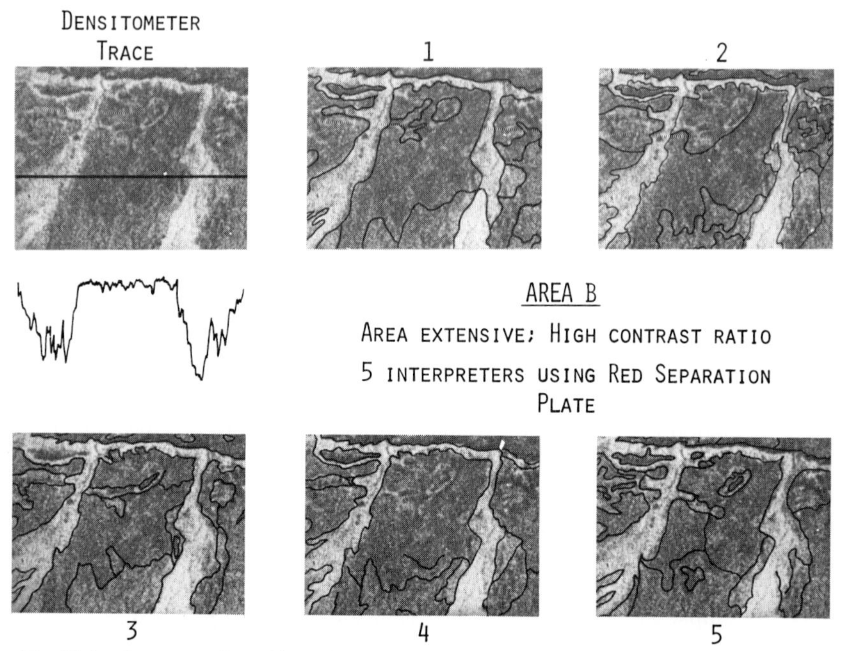


**Fig. 11.1.** Interpretation of boundaries by five interpreters on space photography of an area (Alice Springs, Australia) where extensive entities are separated by high-contrast boundaries.

ceived boundaries in different fashions, depending on the contrast between entities and their degrees of spatial complexity. As new situations are encountered, some observers erect a new class composed of a complex mixture of two or more entities, whereas others prefer to maintain discrete categories. Thus continuing problems of lumping and splitting occur when qualitative judgments are made.

Qualitative judgments of these kinds, though intrinsically fuzzy, could be formalized if the effort was warranted. Most workers either do not believe this to be so or are not comfortable with the self-discipline needed for codification.

On the other hand, quantitative procedures demand structured rules for computer implementation of decisions on classification or thresholding. Thresholding is used to produce a discard class into which all nonidentified items may be placed. With no thresholding all pixels in a scene are assigned to an established class. However, not all items in an area are desirable in a classification. (The smaller the number of classes employed, the less the computer charges. Hence there are real, practical advantages in having a large "all other" or thresholded category.) No thresholding causes a large number of errors of commission, whereas severe thresholding produces many errors of omission (failure to classify points which should be classified). The level of thresholding therefore dictates the proportion of errors of omission and commission. Depending on the problem at hand, it may be desirable to have no thresholding, severe thresholding, or some intermediate level. The choice

0-201-04245-2

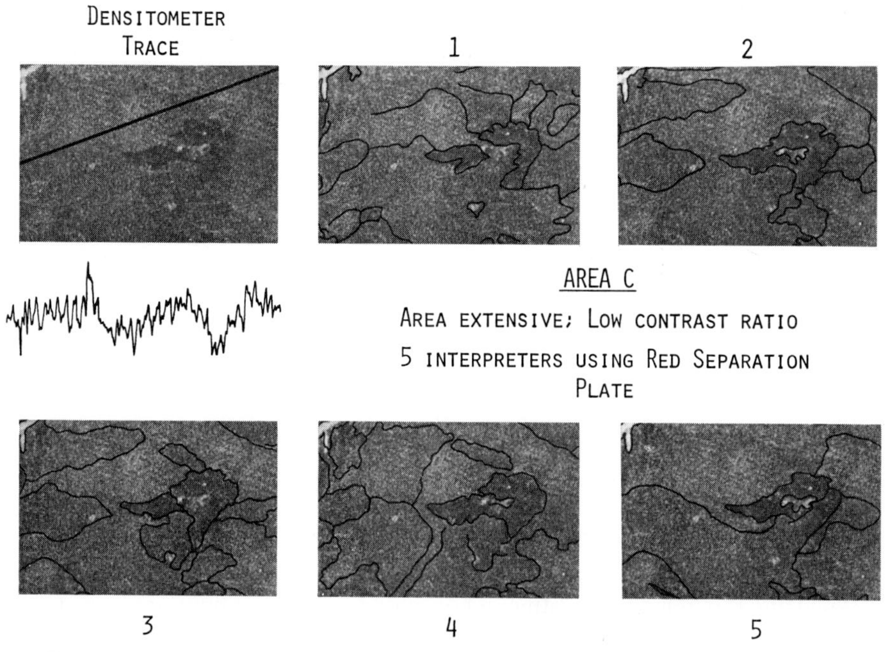


**Fig. 11.2.** Interpretation of boundaries by five interpreters on space photography of an area (Alice Springs, Australia) where extensive entities are separated by low-contrast ratios.

of thresholding level is dictated by the requirements of the problem, the penalties associated with type I or type II errors, and the distribution in statistical space of the data.

The decision rules employed with the thresholding are the codification of the location of the boundary values in *n*-dimensional space between one class and another. To reduce the cost of classifying very large data sets, simplified procedures may be used: table lookup or a linear approximation, which is simpler and less demanding of computer time than the maximum-likelihood function.

## VII. CLASSIFICATION OUTPUT

The four principal formats for presenting results are tables, graphs, *x-y* format, including images and maps, and perspective projection of three-dimensional surfaces. All may be produced either with the computer or manually, and some may be produced with analog equipment.

### A. Tables

A very important table format for remote sensing is a confusion matrix giving correct identifications on the diagonals, errors of commission on one axis, and errors

0-201-04245-2

on omission on the other. These may be derived by point-by-point tabulation of photo interpretation results through overlaying an interpreted map over a previous map and then using a point-counting grid. Computer tabulation of pixels is the digital equivalent. Numerous other tables may be prepared.

## B. Graphs

Histograms of the frequency of occurrence of a single identified category in the area may be prepared manually. Any comparable graphical plot may be easily produced with the computer or readily displayed on a CRT. It can be photographed to give a permanent record.

## C. *X-Y* Format

An interpreted map is probably the most common form of *x-y* output. This can be either a qualitative, a mixed qualitative-quantitative, or a quantitative map based on tabulation of statistics.

The list of maps is as endless as the ingenuity of man to devise: single-value and multiple-category maps; isarithmic (contouring continuous functions) maps; dasymmetric (discrete identification of noncontinuous distributions) maps; maps based on the use of absolute values, percentages, ratios, and changes from one time to another; and point and line, as well as area, formats.

Monochromatic and color television displays are common methods of presenting the final output map, from either an analog or a digital identification system. An alphanumeric output from a digital system is widely employed, using a line printer. Several such outputs may be photographed as separation plates and later combined photographically, or by printing, to give a color map. Some installations may have film writeout devices, which use discrete gray steps to map with considerable accuracy the various categories. The geometric fidelity of film writeout devices is now so high that a hybrid digital analog system can produce very accurate output maps. Classification may be used to produce three-color separation plates directly for color printing, using computer classification and a film writeout device.

The *Manual of Remote Sensing,* recently published by the American Society of Photogrammetry (R. G. Reeves, *Ed.,* 1975) contains many examples of output types, and discussions of the advantages and disadvantages of each may also be found in Tomlinson (1972, pp. 891–1123), as well as in the *Manual.*

## D. Perspective Projection

Three-dimensional surfaces may be effectively rendered and perceived for some forms of data. For example, the quantitative contours which describe a variable over a mapped region can be computer graphed to provide a perspective drawing, and a variety of computer programs exist. Appropriate masking of hidden contours enhances this perception, although this is often the most difficult part of the program. Often several vantage points are used for the rendering; this improves the data assimilation.

0-201-04245-2

## VIII. VERIFICATION OF RESULTS

Verification of results is frequently ignored in qualitative analysis, and improperly carried out in so-called quantitative analysis in remote sensing. In many qualitative studies by biologists and social and earth scientists, errors become known only because other workers find them. Even such a simple step as candidly assessing the weaknesses of one's own work is rare.

Verification of "quantitative" studies is now beginning to be approached with proper sample design. In almost all studies made before the last 2 years not only were there biases in the selection of training sets and cluster groups, but also there were considerable biases in sets used to test the accuracy of prediction. Among these biases were those of selection along roads, selection of large, homogeneous entities, failure to use a systematic or properly random method of test site selection, and concentration on the central pixels of an entity. Some early studies even used the training set for verification! Recent studies have approached the question of assessing accuracies much more rigorously. Unfortunately very few of these studies have been published at this writing or are in readily accessible journals. Relevant papers are those by Biehl and Silva (1975), Hord and Brooner (1976), Hoffer and Fleming (1975), and Simonett et al. (1976).

## IX. EFFECTIVE SENSING OF INFORMATION OF VALUE– SOME CONCLUDING REMARKS

Assessing the comparative merits of the many alternative clustering and feature extraction methods is very difficult, costly, and time consuming and consequently is inadequately handled. Some unifying theme and comparison method is essential. Let us briefly review what one seeks in remote sensing. There is information of varying value to be gained from sensing the earth, and methods of sensing and processing data which vary in effectiveness are available. Since the information of highest value is not always the easiest or more effectively obtained, a decision maker or researcher balances the desire to identify categories of high information value against the question of what kind of information can be effectively sensed and identified.

The somewhat analogous counterparts are the approachcs of feature extraction and clustering. The focus of the former is to ascertain effective ways of sensing the information that has been designated as valuable. Clustering, in contrast, seeks ways of aggregating data that have been sensed and then interprets the information content and value.

Information value and sensing effectiveness are thus significant variables for a unifying/evaluative framework. Usually only intuitive or subjective appraisals of "information of value" are possible. Thus "I'd prefer to know the acreage of cotton rather than of wasteland" implies some type of valuing. Methods of quantitative assessment of information value are practically nonexistent. As much time may be spent by researchers in identifying areas of limited economic value as regions or

0-201-04245-2

categories of high economic value. An approach to assessing value may be to equate such value with the penalty of not knowing the information. This may sound like circular reasoning; however, it does have merit, as well as being linked to quantitative ways of describing identification effectiveness.

As previously noted in statistical and/or probabilistic decision-making instances, descriptions of error (here, in identification) are called type I and type II errors. These are, respectively, the error of nonidentification (omission) and the error of false identification (commission). Often the rationale for establishing decision boundaries is the minimization of the sum of these errors, with little thought given to the possible different consequences of type I and type II errors. Alternatively, one may choose to associate different penalties with each type of error.

A simple equation for expressing the composite of these errors and penalties and for incorporating available a priori knowledge is the Bayes risk formulation. This provides comparative evaluations of sensing/identification effectiveness. The approach usually is to minimize this risk.

In a simple form, the Bayes rule is defined as follows:

$$R = P(1 - P_I)C_{\bar{I}} + (1 - P)P_f C_f,$$

where $R$ = average Bayes risk,

$P$ = a priori probability of a category of interest,

$P_I$ = probability of category identification (note that $1 - P_I$ = probability of type I error),

$C_I$ = cost of missed identification,

$P_f$ = probability of a false identification, a type II error,

$C_f$ = cost of a type II error.

The results of this formulation are consistent with an information-theoretic approach which determines the product of information value and sensing effectiveness, *Ie*. The latter may be evaluated as

$$Ie = \frac{I\,(\text{noisy identification})}{I\,(\text{noiseless identification})},$$

or

$$Ie = \frac{H(X) - H(X/Y)}{H(X)},$$

where $H(X)$ is the input entropy (or information rate), and $H(X/Y)$ is the equivocation or noisiness of the sensing and identification process. The denominator provides a normalization by comparing the feature extraction/clustering method and decision rules with ones that are hypothetically noiseless.

0-201-04245-2

There is, of course, considerable variation in quantifying the significance of errors of each type. Currently the results of most decision rules are compared on the basis of the sum of types I and II errors. This is analogous to considering that

$$PC_{\bar{I}} = (1 - P)C_f$$

in the Bayes risk formulation.

Note also that even apparently equal performance (i.e., equal sums of errors) may often in reality describe different performance. For example, sometimes conditions are different for the tests on training versus prediction data.

Generally, the "moment of truth" comes when the decision logic is committed to the analysis of new data. This step is still a cautious one in that most "new data" have some a priori ensemble of ground truth. This is as the case would be during the early stages of technology development. The extent to which a man-machine decision is accepted as valid may be described by some probabilistic or qualitative estimate of confidence in such procedures. Some remote sensing currently is regarded askance because it fails to provide data within the confidence estimates desired.

A final dimension of this evaluative framework is the associated costs, which include those of image acquisition, processing, feature extraction, clustering, and decision rule optimization. System costs thus require evaluation in a Bayesian framework. Such analyses have barely begun in remote sensing.

0-201-04245-2

Chapter 12

# Ground-Truth and Mission Operations

*Joseph Lintz, Peter A. Brennan, and Peter E. Chapman*

## I. GROUND TRUTH FOR REMOTE SENSING

### A. Definition

The concept of "ground truth" as applied to remote sensing has never been clearly defined and has grown haphazardly over the years, becoming progressively more diffuse in the process. This has led some to suggest eliminating the phrase, but this suggestion has not been adopted because, with all of its attendant weaknesses, "ground truth" carries useful connotations and possesses the great asset of being short and catchy.

"Ground truth" is defined here as the parameters actually creating the signal received by the remote sensor. Ground truth parameters are the items which the sensor actually "sees" or "records" through the atmosphere and ground truth is the physical reality that is being sensed remotely. Some ground truth parameters may be in a state of flux, changing constantly; others may be exceedingly stable, exhibiting little change over long periods of time. For example, agricultural, geographic, and oceanographic features will vary over short periods of time, whereas geologic features will show minimal or no change over long periods.

All remote sensing is weather dependent, and short-term meteorologic conditions greatly influence the radiant energy detected by sensors. The solar energy flux received is constantly varying because of sun angle, atmospheric absorption, scattering, and surface cloud masking. Thus the recording of atmospheric phenomena is definitely a part of ground truth. Figure 12.1 illustrates this experimental approach.

### B. Purpose and Objectives

The principal reason why "ground truth" means so many different things to so many people is that their requirements for ground truth vary greatly, with varying sensor packages and experiments. The investigator seeking to determine the role of surface roughness on the backscattering of microwave radiation is in an entirely different position from the citrus association which wishes coverage of its members' 50,000 acres of citrus groves in a disease detection program.

These examples indicate the two prime divisions of ground truth users. The first group is involved in understanding the full capacities and limitations of various sensors, whereas the second group wishes only to apply the sensors to its immediate

*Remote Sensing of Environment*, Joseph Lintz, Jr. and David S. Simonett (eds.) ISBN 0-201-04245-2


0-201-04245-2

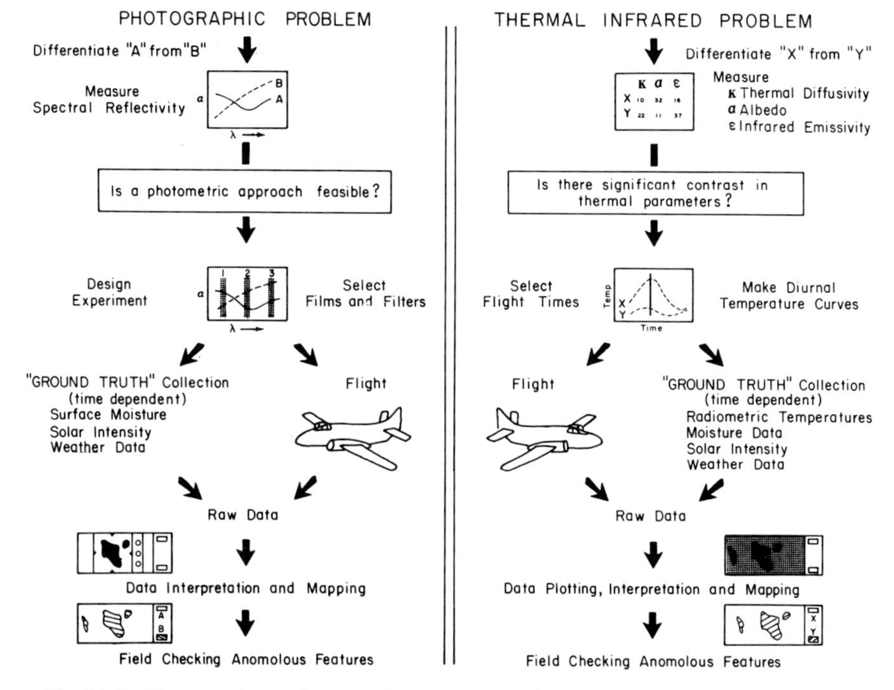


**Fig. 12.1.** The experimental approach to remote sensing involves the proposal and execution of a problem which fully identifies instrumental capabilities.

problems without necessarily understanding how or why the sensor produces its resulting signal. The first group is concerned with devising experiments to test the sensors so that it can attain its long-range goal of learning the relative importance of each of the several parameters that combine to produce the total signal that is sensed. The second group is operationally oriented; it can be equated with the majority of motorists who say that they need not know technically how an automobile functions in order to drive.

The research and development groups see their experimental role continuing for one or two decades. At some future time, depending on how much funding has been secured, we will have attained sufficient background and understanding of what the sensors sense. Then we will have the confidence to decrease our efforts in this direction, unless, of course, new sensors utilizing additional principles have evolved.

Our lives are confined to an environment that we individually see with our own eyes in the 0.4–0.72 μm range. Just recently through instrumentation we have gained the opportunity to "see" at wavelengths from 0.29 μm to as great as tens of centimeters. At these wavelengths our familiar 0.4–0.72 μm world takes on totally new dimensions and aspects very unfamiliar to most of us. The research-oriented remote sensing scientist is faced with a huge volume of ungathered data that need

0-201-04245-2

to be collected, sorted, and utilized as a research tool. The volume is still not fully known, and new awareness will create needs for additional basic and calibration data. An illustration will emphasize this point. Until infrared scanners became available, in the military and then in the civilian sectors, we had little knowledge of or even interest in the thermal regimes of plants, soils, rocks, and water. Today the thermal regimes of all natural bodies are of interest. Many data are being accumulated and processed to permit necessary future comparisons and generalizations.

The ground truth requirements of operationally oriented remote sensing programs are less rigorous. Those involved in such programs are not faced with the need to design experiments to check theories and extend principles. Most of their activities involve a small number of the parameters which affect the total signal. Their ground truth operations will generally occur after data have been remotely sensed and will be limited to areas of anomalous data. This type of ground truth essentially constitutes a "field check" and can be undertaken only after the existence of anomalous data has been identified by the sensors.

In this connection GEMINI photographs of the Midland-Odessa region of Texas provide an excellent illustration (NASA photos S-65-34702-7). These photographs of a semiarid region displayed a large semiparabolic shape, spread across one or two county-sized areas, for which no immediate explanation was available. Within a few days after the availability of the film, investigation of weather reports and interviews with ranchers showed that severe local thundershowers had occurred during the night before the photography. "Ground truth" informed the photographic interpreter that the dark-colored anomaly was the path of a rainstorm through this otherwise dry area. In this illustration, the remote sensing researcher would now be aware that the parameters affecting the signal of the anomaly involved changes in (1) spectral reflectivity, (2) albedo, and (3) moisture content, which in turn affects the (*a*) thermal diffusivity and (*b*) dielectric constant.

Both research and applied remote sensing can and should exist side by side, and from such coexistence valuable knowledge will be gained. New imagery and photography will raise problems of interest to the research groups, while the ever-increasing lore of the research groups will provide greater facility for the applications-oriented group to solve its problems and potentially reduce its needs for postoperative field checks. Contract aerial photography for minerals exploration and for agricultural objectives is now performed routinely with minimal field checks, whereas 20 years ago the field check played a larger role because of the smaller body of experience and hence low confidence in the sensor record.

## C. Ground Truth Requirements

Ground truth requirements vary widely with the individual sensors and specific objectives to be achieved. A research effort to assist in the calibration of a new sensor requires much more thorough and complicated preparation and execution than an applications-oriented mission with a limited objective. A light aircraft with one or two sensors requires less in the way of ground truth than a heavy four-

0-201-04245-2

motored aircraft with a complement of sensors examining terrain features in seven or eight segments between the ultraviolet and longer microwave regions of the electromagnetic spectrum.

To an important extent ground truth is a function of the wavelengths being sensed. The ground truth requirements for the microwave region are not the same as those for the ultraviolet region, and the intermediate infrared and visible segments possess still other requirements. Many variations are of degree rather than kind, for the emphasis shifts with the wavelength.

In addition there are time-critical versus time-stable measurements. Thus, in some disciplines, especially geology, it is possible to ascertain time-stable quantitative data independently of the overflight, and once these determinations have been made they are valid for all subsequent overflights of the target area. Conversely, time-critical data must be monitored with each overflight.

Frequently in the research-oriented overflight the target area will be chosen by the ground truth team. If the experimental objective is to assess the effects of roughness on microwave scatterometer data, an experiment might be designed around a geological test site where surface roughnesses of several orders of magnitude were available for comparison as a result of differential weathering of several rock types. Instead, it would be possible to use an agricultural area with adjacent fields of one crop at varying growth stages, or one might consider an oceanographic test site with several flights over ocean waves of varying amplitude (sea state). Selection of the actual test site from among these possibilities would involve several tradeoffs, such as the advantages in the oceanographic experiment of eliminating such variables as albedo and thermal diffusivity at the cost of repeating flights over several days as the sea state varied. In multisensor flights over an agricultural test site, it would be necessary to account for variations in albedo and probably thermal diffusivity, and the geologic test site would be likely to offer similar advantages and disadvantages.

Once a commitment is made, the details for even a simple experiment may be surprisingly complicated. Consider, for example, the application of multispectral photography to a relatively simple geologic problem, the discrimination of rock unit *A* from *B*. Let us assume that originally *A* and *B* were somewhat similar in mode of origin and physical characteristics, but that at some intermediate time between their origins and the present they were subjected to selective hydrothermal alteration from an underlying pluton. Thus only one unit, *A*, might carry sufficient mineral values to be of economic importance, whereas *B* would be barren.

The ground truth program for such an experiment, Figure 12.2, could be expected to comprise the following steps.

1. Representative exposures are located.
2. Two variables, (*a*) albedo and (*b*) spectral reflectivity, are determined; this can in part be done independently of an overflight.
3. The monitoring activities required during each overflight are performed; these include determination of (*a*) solar intensities, (*b*) surface content moisture,

0-201-04245-2

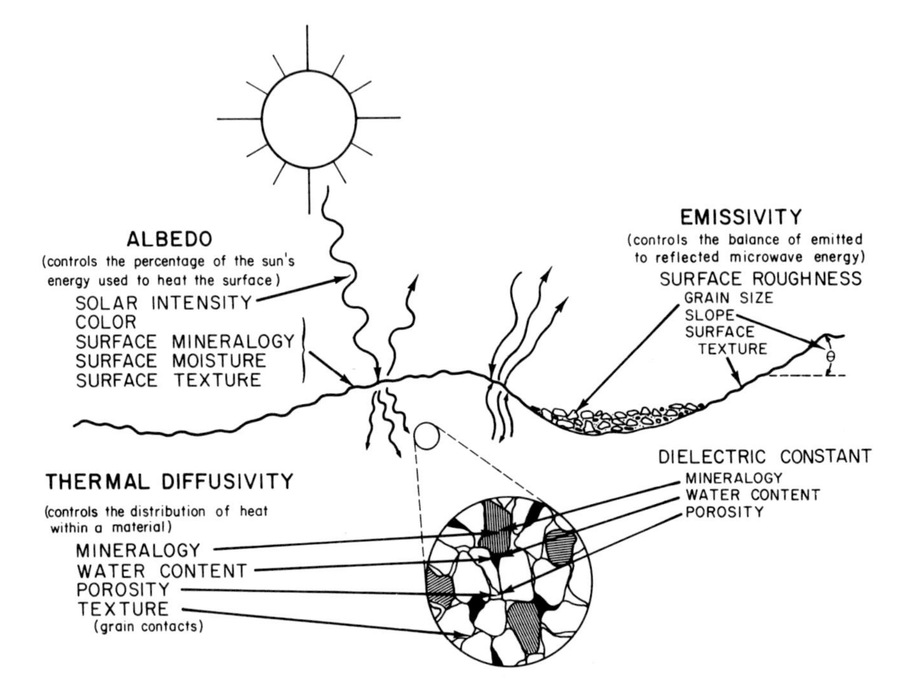


**Fig. 12.2.** Factors affecting the electromagnetic signals reflected or emitted by the earth. The relative importance of each factor is a function of the wavelength utilized. Detection and measurement of each parameter are the objectives of research-oriented ground truth operations. Although the illustration is geological, the same parameters apply to all disciplines.

(*c*) meteorologic conditions, and (*d*) directional illumination, reflection, and emissivity.

4. Rigorous darkroom procedures must be followed to ascertain that the film emulsions have been properly developed. Since multispectral photography generally employs black and white films, this requirement is not as severe as when color emulsions are involved.
5. Data presentation modes can and do vary, and probably trial and error experimentation may be followed with reconstituted color techniques or false color enhancement procedures, which will precede interpretation of the data.
6. Lastly, it may well develop from the data that some anomalous condition has been overlooked in both the preflight and overflight activities on the test site, so that a postflight field check is required.

The experiment outlined above is somewhat typical of a rational approach to multispectral photography, which is one of the less expensive and more rewarding remote sensing techniques and one which is becoming more widely available. If an advanced research aircraft were to fly the mission and simultaneously generate data in the ultraviolet, thermal infrared, and microwave segments of the spectrum, the ground truth task would become correspondingly more complicated.

0-201-04245-2

With the importance of certain specific ground truth measurements identified in the above experiment, let us look at the variety of determinations frequently made in establishing ground truth to facilitate understanding the sensors and the parameters that affect them. They are numerous, and to clarify the problem of handling so many entities the discussion has been organized in terms of the wavelengths being used.

### 1. *Ultraviolet Segment*

Investigation has not progressed as rapidly in the ultraviolet segment as in other regions. Few commercial operators offer data from UV sensors. Ultraviolet radiation is approximately 10% of the total solar radiance and is subject to both absorption by the atmosphere in the lower wavelengths and to heavy dispersion at all wavelengths. Since blackbody UV emission requires minimal temperatures of 6000°K, no terrestrial materials are self-emissive.

Important ground truth measurements for UV experiments will involve principally meteorologic investigations and reflectance studies. Three categories of objects yield luminescence in the UV region: certain minerals, some marine organisms, and some dyes.

Table 12.1 is a list of minerals which fluoresce or yield luminescence when stimulated by an UV source. These substances should be detectable by solar stimulation. Preliminary studies by the U.S. Geological Survey at Mono Lake, California, reveal that sensors have detected luminescent salts about Paoha Island. Ground determinations involve the collection of mineral specimens and the generation of reflectance curves for the UV segment of the spectrum, as well as the monitoring of atmospheric conditions at the time of overflight.

Many marine animals emit luminescence in the UV regions which should be capable of detection with sensitive detectors such as the Fraunhofer line discriminator (FLD). Common marine organisms exhibiting this phenomenon, sometimes referred to as "phosphorescence," are divisible into two groups: (1) larger vertebrates inhabiting totally dark waters below 100 fathoms, and (2) the Protista and other invertebrates of tropic surficial waters. The polychaete worm *Odontosyllis* secretes luciferin and luciferase in conjunction with its reproductive activities. Admixing these two enzymes produces luminous clouds. In addition, the bodies of the females glow green to attract the males, which glow continuously, and in addition emit bright flashes. Many marine dinoflagellates, especially *Pyrodirium* and *Gonyaulax,* luminesce at night, as does the "red tide" created by *Noctiluca.* Not all "red tides," however, cause nocturnal luminescence. Ground truth determinations for these marine forms involve identifications of the collected fauna by marine zoologists and can be performed within several days of the overflight with time limitations imposed by the life cycles of the faunal constituents. Meteorological data are required at the time of the overflight.

The third principal UV detection problem involves certain dyes such as fluorescein and rhodamine. In this case, it might be claimed that the process of dispensing

0-201-04245-2

**TABLE 12.1[a]**

**(*a*) Minerals with Strong Luminescence**

| | | |
|---|---|---|
| Anglesite | Hackmanite | Powellite |
| Anothoclase | Halite | Priceite |
| Aragonite | Hanksite | Rubellite |
| Autunite | Hexagonite | Ruby |
| Barite | Howellite | Sassolite |
| Benitoite | Hydromagnesite | Scheelite |
| Borax | Hydrozincite | Sapphire |
| Calcite | Inyoite | Sodalite |
| Calcium larsenate | Kunzite = pink spodumene | Sphalerite (wurtzite impurity?) |
| Celestite | Manganapatite | Spinel |
| Colemanite | Mayerhofferite | Strontianite |
| Copalite | Nasonite | Trona |
| Crocolite | Opal | Uranium salts (16 in addition to autunite) |
| Cuproscheelite | Ozocerite | Wernerite |
| Dumorturite | Pearls (not artificial) | Willemite |
| Elaterite | Pectolite | Witherite |
| Epsomite | Petroleum | Zircon |
| Fluorite | Phosgenite | |
| Glauberite | | |
| Gyrolite | | |

**(*b*) Minerals Which Do Not Luminesce, with Local Exceptions**

| | | |
|---|---|---|
| Agate | Calamine | Gypsum |
| Albite | Chalcedony | Lepidolite |
| Alunite | Clinohedrite | Quartz |
| Amazonite | Curtisite | Terlinguaite |
| Amethyst | Diamond | Thaumasite |
| Apatite | Diaspore | Topaz |
| Axinite | Dolomite | Wavellite |
| Bauxite | Emerald | Wallastonite |
| Beryl | | |

Luminescence in these minerals is probably caused by impurities.

[a] Data from O. C. Smith, *Identification and Qualitative Chemical Analysis of Minerals,* Van Nostrand, Princeton, N.J., 1953.

the dye and recording its initial location in critical areas for monitoring its dispersal by natural systems is a form of ground truth. Certainly testing to determine optimum dyes for remote sensing would be part of the general area of "ground truth."

### *2. The Visible Spectrum*

In this narrow segment of the spectrum more data have been collected concerning what the camera sees than in any other segment. Man is clearly more at home here than in any other wavelengths in the electromagnetic spectrum, and ground truth requirements are better established for the visible region than for any other spectral region. Nevertheless, there is a demonstrable need for quantification of data fre-

0-201-04245-2

quently handled on an empirical basis. An excellent example is the requirement for careful, fully defined spectral reflectance data for a variety of common objects: soils, plants, water bodies, bare rocks, lichens, snow, ice, deserts, etc. Many of these requirements are multiple, as, for example, wet and dry conditions in soils, stressed and unstressed conditions in plants. This catalog of data is very important for automatic data processing operations.

The four most important measurements to be obtained in association with remote sensing in the visible segment of the spectrum include (1) spectral reflectivity, (2) moisture content, (3) meteorologic conditions during overflight, and (4) total solar radiation. The relationships between some of these factors is obvious. Thus albedo will vary with solar radiation—a meadow may be a bright, fresh green on a clear sunny day and appear grayish and dull during a storm or on an overcast day. Meteorologic conditions affect total solar radiation too. Moisture content likewise affects spectral reflectance, as the MERCURY illustration cited has already shown.

The details of the quality and quantity of data to be collected for each of these areas will vary with the problem. Field spectral measurements may require multidate determinations with high-resolution spectrometers for, say, crop stress studies, whereas for a geologic study fewer measurements and simpler, coarser spectral data may be adequate.

### 3. *Infrared Segment*

The two most important ground truth parameters requiring measurement in this section of the spectrum (1.5-14 μm) are the albedo and the thermal diffusivity. The albedo is the sum of all the spectral reflectances over the visible and infrared band widths and therefore an indication of total solar illumination (heating). The second most important item is the thermal diffusivity of the objects under study. Diffusivity is the ability of a substance to transmit thermal energy; it differs from thermal conductivity by taking into account differing specific heats and densities.

Of less significance than the above two parameters is a third, the infrared emissivity. Emissivity is defined as the ratio of energy radiated at the surface of a material to that radiated by a blackbody at the same temperature. Thus a substance with an emissivity of less than 1.0 at a given temperature appears to an IR radiometer to have a lower temperature. Emissivity varies as a function of wavelength and is strongly affected by the microroughness of the surface (which with increasing roughness increases the surface area). Emissivity can be measured in the field or in the laboratory by means of an emissivity chamber as described by Buettner et al. (1965). Measurements of emissivity require much calibration and care.

Data summarizing these three parameters—albedo, thermal diffusivity, and emissivity—can be obtained without respect to the contribution of each by means of IR radiometers measuring over a diurnal cycle. The observed temperatures equal the sum of the three parameters and are of value for those performing a less rigorous type of ground truth measurement.

The last of the ground truth parameters important in the IR is the moisture contents of the surficial and subsurface materials. Moisture directly affects the al-

0-201-04245-2

bedo of a substance and its thermal diffusivity. At the same time moisture alters the passage of heat to the interior away from the surface. Thus time-varying data can be used as a guide to vertical, within-soil variations in moisture content.

### 4. *Microwave Region*

With longer wavelengths the relative signifiance of the most important parameters shifts so that microwave body reflectivity and emissivity are more important than thermal diffusivity. The microwave portion of the spectrum is very broad, being approximately 10,000 times as wide as the IR sector, which in itself is 200 times as wide as the visible region. With the great range of wavelengths in the microwave sector, it is dangerous to make generalizations for the entire sector. While albedo (reflectivity) is the single most important parameter in the microwave sector, the shorter wavelengths, such as the K band, are more highly reflective than longer wavelengths such as the L band.

Emissivity, throughout the microwave sector, is a very complicated factor. It is composed of two principal components, namely, the dielectric constant and surface roughness, and these are further subdivided as follows: *dielectric constant*– (*a*) moisture content and (*b*) layering; *surface roughness*–(*a*) slope and (*b*) particle size. The complex dielectric constant is a measure of a material's ability to attenuate an electric field passing through it, and it is frequency dependent. Its measured values typically vary between 70–80 (for salt-free water) and 3–4 for ice, 2.5 for dry loamy soil, and 1-2 for freshly fallen snow. Actual measurement of the dielectric constant is difficult, and relatively few laboratories are equipped to do this. However, the dielectric constant is affected most strongly by the moisture content of the sample; and since this may vary widely in near-surface conditions, it is the moisture content that governs the dielectric constant. Fortunately measurement of moisture content is relatively simple and can be performed with unsophisticated equipment.

The surface roughness of dense material and the particle size of particulate matter are very important subparameters of emissivity. Particularly is this true when the scale of roughness or particle size approaches some fraction of the wavelength being used, such as $^1/_2$, $^1/_3$, or $^1/_4$, and also at multiples of the wavelength up to about 10. Thus we are concerned with roughness ranging in size between 0.25 and 10.0 mm for a 1.0-mm wavelength, and for a 13-cm system we would be concerned with roughness varying from about 4 up to 130 cm.

The angle of slope is also important in the microwave region, especially since the scatterometer and SLAR are active systems which measure the backscatter of emitted energy. Still, slope also affects passive microwave radiometers. Usually slope is of sufficient magnitude that it can be measured without special tools and can be stated in conventional terms.

The last subparameter affecting emissivity is layering, and it influences the penetrative effects at microwave wavelengths. For active systems the penetration is the approximate point below which $1/e$ of the backscattering occurs. For passive systems it is the point below which only $1/e$ of the energy is originating. Hetero-

0-201-04245-2

geneous materials, such as soils, sedimentary rocks, and oil slicks, which exhibit layering characteristics commonly show their characteristics in their emissivity measurements.

The final parameter affecting the microwave systems is the thermal diffusivity factor, which controls the distributions of temperature with depth.

### D. Instrumentation and Equipment for Ground Truth Studies and Missions

The instrumentation and equipment selected for ground truth activities will depend on the interest, objectives, and resources of the investigator. For applications-oriented studies relatively simple and few instruments may be required, whereas a major mission in which many sensors are tested and calibrated simultaneously will necessitate considerable investment, planning, and quantities of equipment.

Ground truth requirements were introduced above as a function of the spectral regions, and the same pattern will be followed in discussing instrumentation. Since some instruments are utilized in more than one spectral region, they will not be re-described on repeat appearances unless modifications are required for adequate performance in the longer wavelength. Upon completion of the spectral review, meteorologic instrumentation and ancillary ground equipment common to all types of instrumentation will be discussed briefly.

By and large, manufacturers' names have been omitted, especially in competitive situations. When a specific manufacturer is mentioned, our experience has been either that the field is dominated by that manufacturer, or that his product is totally superior and other merchandise is essentially noncompetitive.

#### *1. Ultraviolet Sector*

The single most important instrumentation for ultraviolet studies is a spectral radiometer (spectrometer), filtered to measure the UV sector. In this way reflectance curves for UV radiation can be developed for numerous substances. Fieldwork is difficult to perform, as the scattering of the direct UV energy with the reflected energy contaminates the readings. Generally the instrumentation must be kept very close to the object to reduce such scattering. In the laboratory conditions can be controlled to permit the recording of reliable reflectance curves.

The addition of an ultraviolet source to the UV spectral radiometer permits the work to be carried on at night and gives a measure of the induced luminescence of various objects. This is helpful in the study of various minerals and of natural petroleums.

As modifiers for these two principal instruments, it is desirable to undertake surface moisture readings and to record the total quantity of solar energy being received in the UV sector. Moisture measurements can be made in either of two ways.

In the first a gamma neutron detector is applied directly to various soils, rocks, particulate matter, or plant matter. This is a direct-reading instrument weighing about 20 kg.

0-201-04245-2

A second method of determining soil moisture requires a much simpler array of equipment. With sample tins of uniform volume and a simple beam or platform balance, the percentage of moisture can be ascertained by weighing a sample as it comes from the field, drying it in an oven or over a hot plate, reweighing, and making a simple calculation. The usual sample weighs 100–150 g before heating, which is performed at approximately 100°F. This is a time-critical measurement and should be made at the time of overflight.

Lastly, solar radiation arriving at the earth's surface is measured with a spectrometer with the direct incidence pointed toward the zenith. This, too, is always time critical.

### 2. *The Visible Sector*

Instrumentation for the three principal parameters are the same as that described above for ultraviolet data. Filters for the spectral regions will be changed to accommodate the appropriate sector for the spectrometers. Once again, moisture and solar radiation are modifiers of the spectral reflectance curves which are the prime data.

### 3. *Infrared Sector*

In the infrared sector the total sum of the reflectance curves, or albedo, is more important than spectral reflectance curves. Albedo is more important because it controls the quantity of heat absorbed or reflected during the sunlight portion of the diurnal cycle. Reflectivity may be calculated by comparing the reflected energy with either the sky or a standard such as a white card or magnesium oxide powder. For wavelengths of less than 5 $\mu$m both spectral solar intensity curves and albedo measurements are important, but above 5 $\mu$m albedo becomes more important (see also Chapter 2).

Thermal diffusivity is measured rather easily in the laboratory. A standard plug or core is placed in a holder, and a known temperature is placed in contact with one end of the core. The time required for the heat increase to be felt at the opposite end of the core is a measure of the diffusivity. Cores measuring ½ in. in diameter by 1 in. in length are frequently used.

The IR emissivity chamber designed by Buettner et al. (1965) has been modified by P. E. Chapman to permit field readings. Basically the system for field readings utilizes a Barnes PRT-4 infrared radiometer attached to a chamber which is capable of measuring the reflected component $(1 - \epsilon)$ of the IR signal by varying the environmental (sky) temperature.

The emissivity is influenced especially by surface roughness, particularly in the range of ¼–10 $\lambda$. Attempts to collect data for surfaces such as rocks with roughnesses in this range have been frustrating. The most promising approach appears to be either to diamond-saw samples of approximately 2 cm$^2$ area for study, or to saw cross sections for viewing at right angles. In either case microscopes with mechani-

0-201-04245-2

cal stages are utilized. With the cross section the view is parallel to the topography, and the amplitude of the microrelief appears in profile. With the 2 X 2 cm slab, one looks down and measures the microareas as a function of slant direction and dip. Angles can be measured by placing the specimen on a universal stage without the upper hemisphere and examining it with reflected light. Microscale roughness measurements of soils and especially vegetation are correspondingly more complex and uncertain as to value.

Infrared radiometers are used for performing ground truth measurements in the IR region. Essentially this instrument measures the sum of the imputs of the albedo, thermal diffusivity, and emissivity. The preferred IR radiometer has a chopped temperature-stabilized detector. We have been thoroughly satisfied with several models of the Barnes Engineering Company. The PRT-4 is a laboratory model that has been strengthened for field service. In the field it is handicapped by the requirement for 100 V ac power, which can be supplied either by an inverter-equipped 12 V dc automobile battery with a life of approximately 6–8 hours, or by a portable generator, which must be frequency controlled to $\pm 1/2$ Hz. The PRT-5 model is battery operated and gives very fine results, but is heavy and bulky. The PRT-10 model, as well as similar models from other manufacturers, is a lightweight battery model, nonchopped and non-temperature-stabilized. It is capable of very fine performance in the hands of those who recognize its limitations, but field use requires frequent recalibration against standards.

As in the UV and visible sectors, the measurement of solar radiation is highly desirable and is time critical. It can be effectively performed for this sector by a properly calibrated silicon solar cell.

We have used thermistors often in connection with IR ground truth. In most cases they appear to give readings fairly consistent with those of the Barnes radiometers for surficial material. Since direct penetration of IR radiation is of the order of a wavelength, the depths of potential penetration are too small to be of concern. On the other hand, thermistors located at various depths can profile the daily thermal wave and thereby show the coupling between deeper and shallower layers of interest and value in the IR region. Surficial thermistor data can be hand recorded from various meters, fed onto a strip chart recorder, or placed on magnetic tape. The U.S. Geological Survey has developed an interesting mechanical sequencer which permits the operation of numerous thermistors in rotation through one meter or recorder—a technique which seems desirable.

### *4. Microwave Region*

The prime requirements in the microwave area are for data concerning albedo, microwave emissivity, and thermal diffusivity. The dielectric constant is a very important variable of the microwave emissivity. Because of the penetration capabilities of the longer wavelengths it becomes necessary also to look in detail at layering phenomena. Albedo is measured in the same fashion as for the shorter segments of the spectrum and poses no special problems.

0-201-04245-2

In microwave emissivity, surface roughness as well as water content plays an important role and should be measured for an understanding of its contribution to the emissivity parameter. Again, we are concerned with roughness in the magnitude of $1/4$ to $10\ \lambda$, which for the microwave region is conveniently handled by readily available and inexpensive equipment. For surface roughness in the millimeter range a carpenter's tool known variously as a contour marker or contour gauge, and under various trade names such as "Formit," is especially convenient. It consists of a row of wires approximately 1 mm in diameter, held in place by friction between two plates. When the row ends are pressed against an irregular surface, the individual wires will adjust to reproduce the irregular contour form. Transferrance can be accomplished with minimal delay, the surface roughness being recorded as a pencil line or spray-painted on paper.

Another mode of collecting surface roughness data especially valuable for particulate matter involves the formation of a latex mold of an area approximately 25 X 25 cm; once in the laboratory, a cast made from the mold will reproduce the field conditions and permit study and analysis of the amplitude and frequency of the roughness. Photogrammetric methods using vertical stereophotographs can also be employed.

The dielectric constant is a property possessed by matter that affects its emissivity and is especially important in the microwave wavelengths. The dielectric constant is difficult and expensive to measure in the laboratory. Measurement is done by placing a quantity of material in a tuned cavity in a coaxial cable. Microwave energy is transmitted through the cable, and the cavity is tuned until a "null" in the voltage standing wave ratio is found. By knowing the frequency of the microwave source and the size of the cavity at which the "null" was recorded, the dielectric constant may be calculated. A variation of this technique, used at Jet Propulsion Laboratories and the University of Nevada, Reno, employs a microwave of fixed wavelength. The radiometer is directed vertically at a sheet of aluminum ($\epsilon = 0$) on which an even layer of material is spread until a "null" in the microwave signal is recorded. By measuring the depth of the material over the plate and knowing the wavelength of the radiometer, the dielectric constant may be calculated. The advantage of this technique is that radiometers used in other ground truth measurements may be employed.

Slope measurements are important to the understanding of the microwave signal, as they strongly influence backscatter intensities in the active systems and exhibit variation of intensities in the passive systems. The magnitude of the slope measurements permits formulation of the slope characteristics in the field by estimate, noting directions and gradients, or in the laboratory from either aerial photography or, if available with sufficient quality of detail, topographic maps.

Ground truth for layering, which also contributes to the total microwave emissivity signal, is simple in concept but difficult in the practice of systematic measurement and quantification. The requirement for the third dimension arises with the greater penetration/emission capabilities of the longer wavelengths.

0-201-04245-2

Layering may occur in several terrain features: snow over ice over dense ice over soil, oil slicks over water, microstratigraphic relationships within soils or sedimentary rocks, sedimentary veneers over igneous layers, etc. The determination of ground truth for this phenomenon entails the exposure of the cross section and the measurement of the various data, especially the dielectric constant and moisture content for each horizon or layer. The surface roughness of each interface should also receive consideration.

The last component of the microwave emissivity data is moisture, which affects principally the dielectric constant. Techniques of determination of the moisture content for the shorter wavelengths are applicable here, but there is a need for data from below the surface as well.

The last of the major parameters affecting the microwave signal is thermal diffusivity, which is significant because the longer-wavelength energy has greater penetration power and it is important to know the temperature distribution over the entire interval. Previously described equipment is adequate for microwave ground truth studies.

A measurement of the total solar energy, an index of heating, is most helpful in understanding microwave data. Equipment previously described is adequate to perform this task as well, but it must be remembered that readings should be commenced a day or so before an overflight because the deeper penetration of thermal energy is being measured and meteorologic conditions vary from day to day.

Lastly, it is desirable to produce a thermal profile for microwave radiometer surveys. This can be best done by using thermistors, which can be attached to stakes and gently forced to the desired depth. Usual depths are about 0.1, 0.2, 0.5, and 1.0 m. With very-long-wavelength microwave systems and in very dry particulate matter, however, it may be necessary to embed thermistors for a considerably greater depth.

As was the case for the IR sector, there exist relatively small portable radiometers for microwave use. One microwave radiometer operating in the L band weighs approximately 10 kg and can be carried in a suitcase-sized container (less antenna). It gives a brightness temperature reading which integrates the albedo, emissivity, and thermal diffusivity components. Although of limited use, it is especially convenient in postflight confirmation of anomalous remote sensing data.

### *5. Meteorological Equipment*

As noted in the requirements for ground truth and in Chapter 2, meteorologic influences in remote sensing are very important. It is necessary to document the attenuation of the energy data by meteorologic phenomena. Ideally it is desirable to construct a profile between the sensor and its objective. By means of kites, tethered ballons, and radiosondes this has been accomplished for high-altitude flights. With low-altitude flights surficial meteorologic conditions alone are measured, and it is assumed either that these are representative of the sensor-

0-201-04245-2

ground air column or that simple, standard depth corrections can be made for the column.

Instruments required for weather inputs for sensor calibration include thermometers for ambient temperature, spectral hygrometers and sling psychrometers for water content, and anemometers and wind direction indicators for wind strength and direction. Determination of the aerosol content of the atmosphere is complicated and represents an area in which more experimentation is required. Lastly, the total cloud cover must be noted; it is usually estimated by observation but should be recorded on 35-mm film with a fish-eye lens on a vertically oriented camera.

With the exception of the aerosol content, which has yet to be fully measured, none of these readings poses a major problem; all must, nevertheless, be made to fully interpret the data collected on the mission.

### 6. *Ancillary Ground Truth Equipment*

To conduct a major test site experiment or to calibrate the instruments for the LANDSAT satellites the ancillary equipment required is appreciable and bulky. Since test sites are not always conveniently located, frequently much logistic support is required. The scope of the ancillary equipment is such that it can be conveniently discussed under several headings.

*a. Transportation.* Ground transport can involve nearly everything that moves, from walking technicians and animals to cars and trucks, boats, and even helicopters. Some test sites are rather extensive, and there is a need to get around from one segment to another, sometimes quickly and always easily. For this purpose anything that moves is acceptable, and one should attempt to think broadly, omitting nothing, when deciding on proper transportation.

The necessity or even the desirability of instrumenting vans for ground truth is debatable. The advantage of the van is that it centralizes operations, permits automatic recording of data, and allows a greater array of instrumentation than would otherwise be possible. The principal drawback is a loss of flexibility. In all probability the van will require a heavy truck chassis or heavy-duty trailer, thereby becoming restricted in range. Such equipment cannot venture very far from roads. Also, air conditioning is required, and it is possible to escalate equipment and instrumentation almost endlessly, and not necessarily with good value for money spent. Instruments to measure second- or third-order effects may well be as costly or even costlier than those needed for first-order effects!

Vans are very desirable for microwave ground truth research because the equipment, especially the antennae, is likely to be bulky and heavy. "Cherry pickers" or high rangers have been used by several researchers to enable them to get 3-10 m above the terrain surface and look down. Vans become a necessity when the degree of ground truth sophistication includes telemetering techniques from land-based stations. Obviously, however, much can be done without vans, and the added flexibility is very valuable.

0-201-04245-2

*b. Recorders.* Here is a serious problem which is highly important to the success of the ground truth program. Conceivably large quantities of data are being generated on the ground, perhaps some of it for months before the overflight, much of it in conjunction with the monitoring of the time-critical factors at overflight. Occasionally postflight ground data are required. There are three types of usable systems. The most primitive, but under some conditions still the most desirable, is the observer-reader with a notebook. Advance preparation of a chart or matrix that can be filled in at the proper boxes will, it is hoped, ensure that all readings are made at the proper time(s) and that what is written down is accurate. More sophisticated is the recording of data continuously by machines, either strip chart recorders or tape recorders with magnetic tape. Most strip charts are limited in the number of parameters they can measure, and although multichannel tape recorders of large capacity are commercially available, they are heavy, bulky, expensive in terms of energy consumption, and often highly temperamental. Final choice will depend on many tradeoffs: the complexity of the entire operation, its location, its duration, and perhaps numerous other factors. Demands for portability will be important. Large, complex multichannel recorders may be used at one or two fixed sites, supplemented by small portable units recording fewer data for distant sites.

*c. Communications.* There are three principal problems in communications, Figure 12.3. The first involves internal communications from one part of the test site to another. Ground teams may be deployed over areas from a few to hundreds

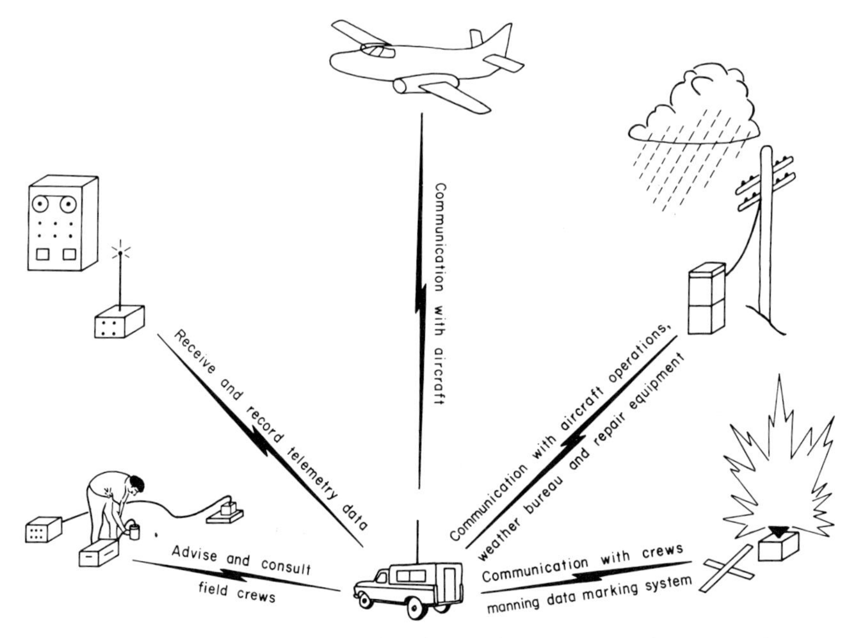


**Fig. 12.3.** Communications for ground truth operations are vital to the success of multisensor aircraft or spacecraft missions.

0-201-04245-2

of square kilometers in size. For this purpose citizens' band walkie-talkies are nearly indispensable. They are highly helpful too during times of overflight: persons far removed from the ground chief may notice phenomena requiring his attention, or some data may require transmission to aircraft or spacecraft. Second, if the test site is isolated, single-sideband transceivers are of great importance. They are vital for receiving weather reports and for ordering supplies that may require air dropping to the ground party. Especially they are required to maintain coordination between the ground truth party and the aircraft in the days approaching the overflight, when aircraft schedules are subject to slippages or short-notice changes because of system malfunction or unexpected weather changes. The third problem concerns the communication between air and ground, which is of great importance. This type of communication permits the ground team to know which sensors are critical on each pass down the flight line, and the team may make minor changes in the placement of some of its devices on successive passes as part of the experiment design. When instrument malfunction occurs on board the aircraft, the ground team knows of it and learns whether the malfunction is correctable in flight or whether the particular sensor involved is "dead" for the rest of the mission. Sometimes scattered cloud cover makes it desirable to rerun a flight line; and when the aircraft has completed its operation, a cheery good-bye and breakoff are greatly preferable to a silent departure, especially if the planned operation has had to be altered or is incomplete. There are several instances of aircraft having difficulty locating the precise test run because of errors in visual reckoning, and the air-ground radio link has saved the day as the ground crew talked the aircraft into the correct position and lined it up on the flight line. The usual communication zone used is in the VHF range, which is commendable for line-of-sight transmission. All communicators must possess a Federal Communications Commission radiotelephone operator's license, and chatter is forbidden, with conversations restricted 100% to business. Allocation of a proper channel should be requested in areas near busy airports, but is less important in remote areas.

*d. Power Sources.* Although acceptable ground truth can be performed in many instances without recourse to large quantities of power, sooner or later some type of power becomes necessary on the test site. For some pieces of equipment, such as some models of Barnes radiometers and certain tape recorders, the energy requirements are specific with tight constraints. Battery-powered operations possess flexibility and mobility, but they suffer in that batteries are relatively heavy for the quantity of power provided and have the annoying habit of running down at critical times. Hence the use of electrical generators is recommended. For many purposes a medium-size gasoline-driven system available in hardware stores is desirable; for others a larger-capacity machine with frequency stabilizers and greater reserves is to be preferred. Once the decision to use generators is made, it is usually wiser to go oversize, perhaps 10 KW, because there are always additional items that can operate conveniently if the power is present. Perhaps the optimum mix is a generator system for the base camp with battery sources on the outlying nodes. Batteries can be recharged as required.

0-201-04245-2

*e. Flight-Line-Indicating Equipment.* As early as 1963 it was recognized that ground truth teams would mark the flight lines to ensure that the aircraft flew over the proper stretch of land or water. Equipment for this function is relatively simple. For nighttime operations large, blinking flashlights can be used effectively. Not only can they show the pilot the beginning and the terminus of each flight line, but also they can indicate to the pilot directions for corrections, if required. A revolving beacon similar to those used on the tops of emergency highway vehicles and equipped with an amber globe has been used with success. This is preferred if the test site normally has other lights which may confuse the pilot. These beacons operate on either 6 or 12 V dc.

For daytime operation mirrors are the first choice if the sun is shining. Although any mirror more than 100 $cm^2$ in area is satisfactory, inexpensive heliographs are made of heavier glass and have a port for easier sighting of the target. Smoke flares or bombs have not proved worthy of the trouble in most situations on land. A pile of old automobile tires ignited with gasoline would provide quantities of smoke (and pollution complaints!) that might be visible from 15 to 20 km away when the aircraft attempted to line up for the data collection run. Smaller quantities of dark smoke would also be most helpful for hydrologic missions over ice or snow fields. For the oceanographic test site dye markers in the water are frequently of great help.

*f. Telemetering Equipment.* Here is an area that has lagged insofar as environmental studies have been concerned, mainly because of expense. Telemetering is useful to coordinate quickly the ground truth data and the remotely sensed data. It has the advantage of concentrating the data in one place, and there is less likelihood of their being lost through accident or misfortune in coming out of a remote test site. Most likely, however, the telemetered data will be raw data and may well require additional operations before they become usable, and these additional operations are generally performed by the collectors of the raw data. The principal remote sensing operations telemetering data are those using the DCS (data collection system) on LANDSAT. Plans for monitoring vast stretches of oceans by permanently anchored buoys include the use of telemetry systems to transmit a variety of sea and meteorological data to communications satellites, which would rebroadcast the data to shore installations for recording and analysis. The attractiveness of this plan is that low-powered VHF transmitters requiring very little energy and thus servicing no oftener than once a year can be used for the telemetry. Telemetry will undoubtedly become increasingly more common with the transition to unmanned environmental satellites.

*g. Air-Droppable Ground Truth Stations.* The military remote sensing specialists have been interested in the development of ground truth packages which could be dropped by remote sensing aircraft. This has obvious advantages when, for military reasons, a ground truth team is unlikely to be in a position to operate. The idea has not yet been discussed widely in civilian circles, and with current operations there is little need for this type of activity. Nor is the transition to remote sensing satellites

0-201-04245-2

likely to bring about a need for this type of activity in the civilian sector. The data generated by an air-dropped ground station are usually telemetered back to the aircraft which dropped them. Like parachute flares for night photography, droppable ground truth stations are classified as highly expendable.

*h. Ground Documentation.* Equipping ground truth teams with small field cameras is highly recommended. They are then able to record on film a variety of features present on the test site, both time critical and time stable, for future study and analysis. Already mentioned is the utilization of a fish-eye lens for recording cloud cover. A variety of films, lenses, and filters is desirable. Some black and white film will be used, but so will normal color film. Ektachrome infrared is commercially available in 35-mm format. Recommended filters include the following Wratten numbers: 1a (haze), 12 (Ektachrome IR), 25, 47, 58, and 87 (black and white IR).

*i. Data Markers.* For nonimaging systems which present their data in line or graph formats, the use of some type of "spiking" device is invaluable. Without the imaging capability of the majority of sensors, one is geographically lost in the interpretation of the data. Ideally, the nonimaged data are stored on tape and are recorded simultaneously with some imaging system such as an infrared scanner. Frequently, however, the look angle of a nonimaging device differs from the look angle of the imaging devices. This introduces uncertainties of correlation which hinder data interpretation. Unless some distinctive localized item lies along the flight line, and preferably one such feature at or near each end of the flight line (a stream or a lake, for example), it is nearly impossible to tell what the nonimaging instrument is looking at while the imagers are looking somewhere else. The problem may be eliminated by use of boresighted cameras, but this is not always feasible. Since the nonimaging devices are generally in the microwave sector (passive microwave radiometers and the scatterometers) a ½-sec blast from a microwave transmitter on the ground will artificially "spike" the data and provide certain correlation between the nonimaged data and the aircraft's geographic location. An inexpensive though less effective marker of active microwave systems is a three-cornered radar reflector, made by lining three corners of a carton with aluminum foil.

*j. Miscellany.* Various maps, charts, and pre-existing aerial photographs are of great assistance in preparing the test site for ground truth studies in connection with instrument calibration. Here belong such items as (1) caged beetles hung in the trees of a forest to attract wild beetles to the area for a controlled experiment in sensing beetle infestation damage, (2) dyes used in dispersal studies in water and to trace flow patterns, and (3) insect traps and other special items needed for particular experiments.

*k. Expendables.* A host of small supplies commonly used in test site operations must also be considered: sample containers of all kinds, jars for sea water or pol-

0-201-04245-2

luted water, small tins for moisture-content-determination specimens, cloth sacks for rock and boulder specimens, books for leaves and other segments of floral anatomy.

There are also flags and markers to recover certain positions on the ground when no other means of marking are available. Often large marker crosses 5 to 8 m in size are placed on the ground as points of reference in aerial photographs (a form of spiking), for rapid determination of the scale of the imaged data, and for daylight aircraft guidance. A number of types of resolution targets are used by the military and other photographic experimenters, not only to provide a scale for photography but also to ascertain resolution and color balance.

## E. Ground Truth Techniques

Although the range and scope of ground truth studies can vary enormously, some insight into the preparation of a ground truth test site for a complicated experiment is of interest. The actual selection of a test site for research activities is in itself an interesting study. Experiment parameters and objectives are fully identified before a search is begun for terrain where the experiment might be accomplished. Frequently weather patterns are critical to the acceptance or rejection of a proposed test site. Problems of incompatible demands by different investigators at a single site are of great importance. For the forester, range manager, or agriculturalist areas of bare rock and soil normally represent undesirable factors that render a given site less valuable than one with a variety of plant communities. Conversely, for the geologist the problem is to find an area of limited vegetation and/or great structural or lithologic diversity. Assuming that the best compromise site has been identified and is accepted by all concerned, what does one do, how does one begin?

The preliminary studies that led to the adoption of the site have already defined the major parameters—the types of crops, marine plankton, rock units, and so on that are of interest. The schedule then leads to the establishment of a center or headquarters at some central or nearby location. The design of the flight lines is one of the early decisions, with special attention paid to providing the greatest or least contrast of differing units or parameters for the experiment. Once this is performed, the time-stable data gathering can commence. These are measurements which essentially are invariables and when measured once do not require remeasurement.

Measurements of this type appear to be best covered by nodes, which may be defined as local areas of concentration of readings. The ground truth specialist must constantly be aware of the resolution cell or area of integration at any instant of the sensor instrumentation for which he is working. His task is to provide measurements which represent averages over the area of such a resolution cell. This may necessitate short traverses over the area with various instruments and certainly requires thinking in terms of rather large spatial arrays at all times, especially with satellite data.

0-201-04245-2

It is probably wise to measure the time-critical parameters during this phase of the work in order to prepare a base line or point of departure for the time-critical readings to be taken concurrently with the overflight. These are primarily weather controlled and include, of course, moisture content, solar flux measurements, and meteorological data.

Thus phase I can be accomplished as the schedule permits with relatively few personnel. These people should be highly trained and possess a thorough understanding of all ground truth work as well as a good knowledge of the sensors and the signals they measure. They will be called upon to make many decisions as their work progresses, and these decisions can strengthen the experiment if they are correctly made. Conversely, incorrect ground observations can nullify an expensive aircraft mission. For experiments in disciplines with fewer time-stable parameters than geology, the work of phase I will be correspondingly reduced. Much phase I analysis is performed in the laboratory on specimens brought in from the field.

During phase I the plan for the monitoring of the overflight is evolved in as much detail as possible: the deployment of men to each test node, the dispersal of equipment, and the procedures to be followed are all enumerated and developed so that, if necessary, alert and qualified technicians unfamiliar with the details of phase I can move into the nodes and quickly perform effective roles in their new environment.

Phase II is the time of maximum manpower needs. The setting up and testing of the various instruments at the widely spaced nodes, the integrating of equipment or procedures where node teams must cooperate, and the beginning of accumulation of time-critical data, usually starting some 24–36 hr before the scheduled overflight and increasing in tempo and numbers as the aircraft or spacecraft approaches the area, and the generating of the final set of data penecontemporaneous with the remote sensing mission are the principal events.

Reduction of data is customarily performed back at the home base and may require several weeks. During this time the field gear is retested in preparation for the next mission, repaired or modified as required, and then stored.

The ground truth data, flight data, and remotely sensed data are eventually coordinated during analysis. This is best accomplished by a team consisting of scientists and engineers who are sensor oriented and fully understand the theory of their instruments, along with ground truth team leaders who have measured the necessary parameters of their disciplines to provide the factual knowledge of what actually transpired during the mission at the time of data collection.

## F. Ground Truth for Space Sensing of Environment

The LANDSAT program had launches in 1973 and 1975; details of the operations and instruments are presented in Chapter 9. LANDSAT C is expected to be launched before 1980 and, it is hoped, will be able to sense in the thermal infrared sector.

What sort of ground truth requirements do these systems present? For LANDSATs 1 and 2, operating solely within the visible sector, the requirements

0-201-04245-2

were not especially complex. Time-stabilized data needs were minimal; more often it was time-critical data that were needed. One of the most interesting phenomena was the wide variety of ground truth collected. Each principal investigator (there were over 100) was able to determine how much and what he wished to do with respect to ground truth. The majority actually did very little in the way of ground truth, studying their LANDSAT data empirically. Others took pains to attempt to understand the moderating influences either of the atmosphere or of their test site characteristics.

As a part of the LANDSAT program for the purposes of assisting in the collection of ground truth, NASA developed the data collection system utilizing data collection platforms (DCP's). The DCP's provided eight channels for telemetering signals to the satellite for retransmission to the LANDSAT data bank. Principal investigators were free to select the data to be transmitted by their DCP's. The DCP's also possessed identification codes. Examples of data transmitted included meteorologic, soil moisture, and surface water velocity measurements. With the 18-day cyclicity of the LANDSAT orbiters, the DCP's provided a very welcome technique for acquiring ground truth data and had the special advantage of getting these data into the central data bank rapidly and accurately. Several hundred DCP's were used.

LANDSAT C is expected to provide a capability to sense in the thermal infrared. The increased spectral range will require a broader range of ground truth than has been acquired to date for LANDSATs 1 and 2. As a minimum infrared radiometers and subsurface thermister profiles will be required, and the data will have to be collected over a longer period of time because of the time-lag buildups in the emissivity field.

The SKYLAB satellite (1973–1974) was manned by astronauts who boarded in shifts. It carried out a very large number of experiments in many fields of space research. Relatively few were concerned with earth resources and environment. However, it must be noted that in conjunction with the SKYLAB overflights (200 km) some simultaneous aircraft flights were made, not only to obtain duplicate ground data, but also to assist in establishing the nature of the energy profile by attempting to demonstrate variations in the quality of the data from the two platforms, as well as by direct meteorological sampling. Thus "ground truth" achieved a new dimension.

## II. MISSION OPERATIONS

### A. Introduction

0-201-04245-2

Mission operations for remote sensing programs may be carried out on the ground, by aircraft, and by spacecraft. Usually ground-based sensing is regarded as a portion of ground truth. It will not be considered in this segment of the chapter.

For aircraft operations there is, once again, such a wide range of variables that generalizations are difficult. In a broad sense, all aerial photography is now a part of remote sensing activity. The collection of aerial photography, whether it be black and white, color, or color infrared, has progressed to a point where these rather run-of-the-mill operations are widely performed.

The addition of the nonvisible sector sensor, however, may change the parameters of flight to a major degree. The most common nonvisual sector sensor is the infrared scanner. To make most effective use of this instrument, radiation should be measured at several times during the diurnal cycle. This involves a flight at the coldest part of the day, immediately before sunrise, and sets into motion several factors not part of the standard daylight mission. Pilot navigation at night differs considerably from daylight navigation because customary reference points may not be visible.

The objective of the mission influences both the way in which it is flown and the degree of need for ground truth data. Many missions are performed to acquire relative, not absolute, data on thermal stream pollution or atmospheric pollution, to provide SLAR examination of terrains consistently obscured by clouds, and to accomplish a variety of other purposes. All of these operations are valid, provided that the sensors operate, the plane is correctly located, and the data generated meet the objectives of the mission.

The remote sensing mission is the focal point of the entire research program. It is the one unifying activity which is of vital concern to every individual involved in any way with the project. Since it represents the greatest concentration of expenditure in the shortest time, it is a highly critical operation. Nevertheless, it usually must be performed on a tight schedule with a large number of variables in position to force last-minute changes. Tremendous flexibility in the thinking and actions of all involved in the conduct of the mission is required. The situation is far simpler with the space satellite, which, once launched, proceeds unchangingly across the heavens, carrying out its assigned tasks as programmed without regard to weather constraints below it or the inadequacies of the human segments participating.

## B. Mission Planning

Mission planning for some less complex operations may be very simple indeed. In some cases the mission request leaves all details to the operator because of insufficient knowledge of the parameters or techniques of generating remotely sensed data. The U.S. Forest Service could conceivably order infrared coverage of a forest fire to see through the smoke and ascertain quickly the extent of the problems, leaving the details of height, direction, number, and position of flight lines to the mission director.

Usually when more than one sensor is to be utilized, mission planning must be carefully organized to optimize instrument use and aircraft time. Flying aircraft, even single-engine light ones, are expensive, and inadequate performance constitutes gross waste. Invariably the pilot or captain can do a better job if he understands the

0-201-04245-2

sensor capabilities and limitations, and at the same time he best knows the capabilities and limitations of his aircraft. As the number of sensors employed increases, the complexity of the mission also increases. The size of the aircraft and the cost of the operation rise disproportionately. Planning a complex mission of many sensors and numerous flight lines over one test site can become a complicated affair, calling for extremely close interface between the aircraft commander, the sensor systems commander representing the sensor operators, the experimenter, who may represent the sensor design team or who may well be discipline oriented, and the ground truth leader, who may also be an experimenter in his own right or who may be in the loop between the sensor design team and the mission director. The list of points to be agreed upon is lengthy for a full mission with all sensors.

Mission operations brings into existence another new set of data which become part of the experiment package. These are the flight data, involving such parameters as flight speed, direction, drift, altitude, time, yaw, pitch, and roll, radio-frequency interference between on-board sensor systems or between sensor and aircraft systems, including navigation, and precise aircraft position. These are all true data that have bearing on the final analysis and understanding of the sensor data, and they can vary considerably from the planned mission.

Planning a multisensor mission will logically involve the following criteria for each pass: (1) time, (2) location, (3) altitude, (4) direction of flight lines, (5) definition of minimal acceptable weather conditions, (6) number of passes required for each flight line, allowing for multiadjustments of sensors, if required, and (7) selection of sensors to be operated on each pass, together with precise definition of all sensor options, such as look angle for passive microwave radiometers, polarity for microwave systems, and film-filter combinations for cameras.

Commonly all of these factors can be arranged in a satisfactory sequence and fitted together to provide a workable field program. To prevent misunderstandings and preclude forgetfulness the final program requires early writing and distribution to all principals. An important segment of this working plan is a provision for ground truth monitoring activities with the following details enumerated: (8) location of the ground truth collectors within the test site with occupation of test nodes noted, (9) position of aircraft signalling devices, whether lights, mirrors, or other means, (10) presence and intent to use data markers (if the sensor technician is unaware of the intent to spike data, he may catch the $^{1}/_{2}$-sec energy burst and believe his sensor is malfunctioning), and (11) the frequency on which ground-air communications are to be established.

Frequently it is desirable to establish alternative plans to permit maximum usage of time away from base by the aircraft, or alternative modes of accomplishing the mission. These plans take into account the probabilities of weather vagaries and address themselves to instrument and aircraft idiosyncracies, for such aberrations are always present in the merging of several complicated and delicate scientific systems into a flying laboratory. Although we strive for the highest reliability, we must provide for minor failure to attain it.

0-201-04245-2

## C. Mission Execution

The ideal mission execution follows the prescribed plan without deviation, enjoying optimal weather, full reliability of each and every sensor and of each aircraft component, and 100% efficiency of all personnel, including aircraft support crews, flight crews, and ground truth teams. Such missions are rare. Usually there are changes and modifications of the plan. Improvisation is the key to preventing degradation and resultant failure. For example, on a July 1968 mission a heavy sensor-laden aircraft was delayed 15–20 min on takeoff by a malfunction of the prime sensor for the flight. If it could not be repaired, there was no reason to fly the mission and it would be canceled. But the difficulty appeared minor, and the sensor operator considered the instrument repairable. It was, but the 15-min delay allowed clouds to build up to generous proportions. Although the flight was completed with all passes flown and the sensors performed creditably, the shadow of the clouds on the terrain tended to degrade the data, something which had been avoided in the mission planning session several weeks earlier. At that time account had been taken of the fact that a buildup of clouds occurred each day between noon and 1:00 P.M. local time.

Essentially this was a minor problem; the mission was flown and valuable data were gathered, although not as valuable as if the sky had been 100% cloud free. The serious problems are those in which the severity of the degraded conditions approaches the limit of tolerance for the particular parameter involved.

Weather is usually the most common degradation problem. Changes developing on short notice may cause abandonment of a plan on even shorter notice. Occasionally it is possible to reverse the planned order of execution of a multi-test-site mission and proceed to another test site that still enjoys clear weather. On the other hand, some special missions are flown only with the arrival of poor weather to test certain sensors. All in all, the greatest amount of flexibility is required of all hands if the venture is to be a success.

One problem that can cause minor difficulties in flying the flight line with sensors generating data is electronic activity. Inadvertent inputs from telephone microwave networks or comsat satellites have been known to impinge on the data generated by remote sensor aircraft. The presence of a number of highly sensitive systems relatively crowded together within the confines of an aircraft sometimes generates radio-frequency interference between sensors or between sensors and aircraft systems. This is usually eliminated after a long and patient, sometimes a very trying and difficult, search. A certain master tape recorder showed a cyclic aberration that was eventually traced to the activities of the camera intervalometer. Utilization of ground-air communications during a data run should be avoided except in emergency conditions. Diodes in sensors are extremely sensitive devices affected by the stray energy of unrelated systems, and infrared imagery has sometimes been slightly distorted as a result of communications exchanged during data runs.

0-201-04245-2

Field processing of the day's data is highly desirable, but usually not possible. A large aircraft can have a small darkroom squeezed into the aft end, but unfortunately there are not many large remote sensing aircraft. The ability to process raw data, either in a darkroom or on computers, will reveal early enough the necessity for reruns of data where there has been partial degradation or failure. For oceanographic missions in particular, where research vessels are scheduled a year in advance, the ability of an aircraft to repeat a substandard run will very probably save a year for the experiment.

0-201-04245-2

Part IV

# Remote Sensing Applications: Analysis, Interpretation, and Resource Management

0-201-04245-2

## INTRODUCTION

Remote sensing specialists are divisible into two categories: those who are instrument oriented and those who are applications oriented. The former are usually physicists and electrical engineers and constitute a small minority. The latter are earth science discipline oriented and comprise a large majority. Their interest focuses on the applications of remote sensing to solve problems. To them remote sensing is a useful tool which shows constant improvement. The earth science disciplines are those indicated in the chapter headings of Part IV. At first glance they appear to be quite diverse, but underlying the diversity are common factors.

First, these applications-oriented scientists are wedded to their disciplines. Remote sensing is a subsequently acquired tool which enables them to increase their mastery of their spheres of interest by providing data that have not previously been available.

Second, these earth scientists go about their interpretations pretty much the same way whether they are oceanographers or geologists. They seek to establish the "big picture," using small-scale data, and then they close in on smaller-scale phenomena requiring larger-scale data or increased spatial resolution.

Third, this group tends to avoid nonimaged data. They show a distinct preference for data with an *x-y* dimension, and three-dimensional data are even more desirable. They eschew strip chart data, principally because of the extra step required to correlate them with ground features. This dislike has brought pressure for sensor design to contain an integral imaging capability.

Fourth, the patterns of mental gymnastics in the interpretation step, once the analysis is done, are very similar. Strong reliability is placed on the discipline-oriented fund of knowledge each has previously acquired, and the newly analyzed data then are accommodated in these specialists' reasoning—either in full acceptance with concepts previously known, or with a slight change or modification of these concepts.

Fifth, there tends to be a multidisciplinary teaming in many cases, often involving collaboration between the instrument-oriented and the applications-oriented specialists. These joint efforts provide excellent papers, as what is difficult for one author may be within the ready competence of the other. The bibliography provides many illustrations of such collaboration.

Sixth, many of the data generated by remote sensing are of value and of interest to several disciplines. LANDSAT images have something for almost everyone in every frame. This would be expected in any small-scale view of 34,000 km$^2$. However, larger-scale data also are interesting and valuable to scientists of differing disciplines.

*Remote Sensing of Environment*, Joseph Lintz, Jr. and David S. Simonett (eds.) ISBN 0-201-04245-2


0-201-04245-2

Seventh, in resource management no manager can afford to take a narrow approach to his specific problem. He must give consideration as to how his problem interfaces with the general ecosystem. Accordingly resource managers are achieving a broader outlook with respect to their tasks. Remote sensing is a tool available to these managers which facilitates this broader perspective. With increasing frequency multiple uses are being programed into remote sensing, from definition of objectives through mission planning, data acquisition, analysis, and interpretation.

Eighth, the EROS facility of the U.S. Geological Survey at Sioux Falls, South Dakota, is a prototype of a center of activity which brings together a cross section of diversely oriented users. Cross fertilization of ideas may occur from the random grouping of visitors and transient scientists. Several nations (Brazil, India, Iran, Canada, for example) have established (or are in the process of establishing) similar centers, and other, smaller nations are considering sharing in facilities designed to serve regions.

These are some of the themes which are interwoven in the chapters of Part IV. Some are explicit; others form a subdued leitmotif. All, however, are present.

During the past decade the application of remote sensors to a variety of problems in many disciplines has lead to the evolution of a considerable body of knowledge and experience in this aspect of remote sensing. The contributors have used some of these experiences in their applications-oriented papers, but at best they could only select a few illustrations which sample the much greater number of illustrations available. Volume 2 of the *Manual of Remote Sensing* (R. G. Reeves, Ed., 1975) is completely devoted to applications in remote sensing and presents a greater number of illustrations than we do.

0-201-04245-2

Chapter 13

# Remote Sensing of Cultivated and Natural Vegetation: Cropland and Forest Land

*David S. Simonett*

## I. INTRODUCTION

Among the principal concerns in the remote sensing of crops, forests, natural vegetation, and soils are census taking, stress detection, and change monitoring for agriculture and forestry, and entity detection and boundary delineation for natural vegetation and for soils. In census taking we are interested in the identification of crops and tree species, the determination of acreages, and, ultimately, the estimation of yields by crop type and timber type. Stress detection deals mainly with the spread of plant disease and of undesirable insects, as well as the beginning and spread of forest fires. In essence, then, we are concerned with what is where, how much there is of it, and what the state of its health may be.

In the mapping of natural vegetation and of soils, we are concerned mainly with discrimination between classes of natural vegetation and soil types, with quantifying the nature of within-class variation, and with the detection of boundaries between classes which correlate with those observed on the ground. Again, we are concerned with the deceptively simple questions of what is where, how much there is of it, and what its state or condition is. As always, it is easier to ask these questions than to provide clear and unambiguous answers.

There are no simple answers to these questions because of the complex manner in which natural and cultivated vegetation is mixed both in space and in time as well as at various scales. In analyzing these various relations we also need to look carefully at the multispectral concept, along with the use of time and resolution as discriminants. In addition, we need to consider the meanings attached to the delineation of classes that occurs in any hierarchical or orderly system of classification, both in respect to the sensor and to the objects being sensed.

In this chapter key papers are reviewed in agriculture and forestry which address themselves in a substantive manner to questions of identification accuracy for the categories, regions, and conditions under study. The greatest attention is given to papers which are quantitative rather than descriptive, and to those which address themselves to the larger questions and problems in each area. Running as recurring threads throughout this discussion are continuing references to problems related to natural and cultural complexes in space, time, and scale, to questions on the multispectral concept, to time-domain and resolution-related sensing, to the

*Remote Sensing of Environment*, Joseph Lintz, Jr. and David S. Simonett (eds.) ISBN 0-201-04245-2


0-201-04245-2

systems concept of joint spacecraft, aircraft, and ground sampling, and to classification problems. These concerns are common to all areas in remote sensing and were introduced in Chapter 3. Review articles which touch upon these matters are those of Colwell (1968a), Simonett (1968a), Myers and Allen (1968), Driscoll and Francis (1970), the International Union of Forest Research Organizations (1971), and Hall et al. (1974).

## II. AGRICULTURE

Census taking of crops and pastures and forest inventory are as complex as the numerous forms of censuses and inventories themselves. There is no single problem through which agricultural census taking is addressed, nor by the same token is there a single remote sensing method which may be used for problem solving. Some problems, such as the raisin-lay surveys described by Colwell (1970), may be handled adequately with conventional black and white photography at low altitudes, whereas other surveys may be so substantially improved through the use of false color photography at critical times of the year that this is the preferred method. Minnich et al. (1969), for example, have suggested that infrared Ektachrome photography is so superior for mapping natural vegetation communities in the San Bernardino Mountains that they strongly recommend its use in comparable environments in preference to conventional black and white or color photography. They have, in fact, also used IR Ektachrome in studies on Santa Cruz Island, California (Orme et al., 1971).

Other problems may be amenable to multistage sampling techniques using spacecraft and aircraft imagery and ground sampling, as has been documented by Langley (1969), Langley et al. (1974), Nicholls et al. (1973), and Heller (1975) for forest applications. For forest fire detection and monitoring thermal IR scanners are essential (Hirsch, 1968, 1971). Because of inclement weather conditions radar may be needed in other circumstances (Simonett et al., 1969c). Finally, either certain problems or the requirements of semiautomated and automated surveys may suggest the use of multispectral scanners (Fu et al., 1969). In short, it is important to recognize that remote sensing problems in agriculture and other resource evaluation must be studied in a systems context, in which the problem is viewed (1) in space and time as well as on different scales; (2) under different constraints and acceptable accuracies; (3) with alternative data collection strategies; and (4) by competitive remote sensing systems.

Few investigators at the present time can afford to purchase complex instruments such as radars and multispectral scanners, which rarely cost, with their accessory equipment, less than $150,000; the price of more advanced instruments may exceed $1 million. Thus only a few such instruments are currently collecting data. The availability since mid-1972 of LANDSAT multispectral data has substantially reduced the emphasis on aircraft multispectral data.

0-201-04245-2

## A. Crop Identification with Time-Sequential Photography

Two models which represent considerable extremes in conceptual basis and instrument complexity are available for use with LANDSAT satellite images. In one, time is employed as the prime basis for discrimination; in the other, a number of multispectral channels serve this purpose. With aircraft, there is a tendency to select one model or the other on the basis of cost. Time is used (i.e., selected dates for obtaining information) when it is known that crop phenologic differences are such that major contrasts exist between species and hence improved discrimination will be feasible at the times in question. All that may be tested is a single channel or perhaps color or false color film. Schepis (1968) has discussed time-lapse remote sensing in agriculture, drawing his inspiration primarily from earlier work by Brunnschweiler (1957). He used stereoscopic coverage at 1-month intervals throughout the year. Since the photographs were obtained at a scale of 1 : 12,000 (1 : 20,000 would be almost as good), excellent stereoscopy was available to aid in identification. Steiner (1969) also showed that stereoscopic height was a better discriminant of crops in Switzerland than color photography during the peak of the growing season. Schepis worked with a 120-km strip covering 3200 fields (30,000 acres) which were identified and evaluated for crop yield forecasting. Interpretation keys were developed for yield forecasts of corn, hay, and oats, and yields were estimated in four classes: excellent, good, fair, and poor. Production figures in each category were obtained from cooperating farmers, and a stereoscopic examination of the different fields with panchromatic photography established standard keys for estimating yields.

The same principle but with much reduced use of stereoscopy was employed by Pettinger (1969) in his study analyzing earth resources on sequential high-altitude multiband photography. Concurrently with the acquisition of multiband space photography by the Apollo 9 astronauts in March 1969, aerial photographs were obtained in Arizona and California from an altitude of approximately 21 km of areas covered in the space photographs. In introducing the study Pettinger noted that, since Colwell et al. (1969) had shown that many crops can be identified as consistently on space photographs as on high-altitude photographs, certain valid inferences regarding the interpretability of sequential space photographs could be made through the interpretation of sequential high-altitude space photographs. Unlike Schepis' (1968) study, in which extensive use was made of both stereoscopy and high resolution to determine small-scale features of the agricultural landscape, Pettinger was unable to use features such as texture, harvest patterns, plant height, and plant spacing to facilitate identification and was confined to photographic tone. As Lent and Pettinger point out in Pettinger (1969), a major reason for obtaining sequential photography is to detect changes in earth resource features. Changes in images are susceptible to visual interpretation, to automated analysis, and to various enhancement techniques.

Steiner (1969) used densitometry and stereoheight of crops with aerial photo-

0-201-04245-2

graphs in Switzerland to discriminate between 11 different crop types. Using the best 3 variables, he achieved an overall identification accuracy of crops of 90%; with all 13 variables, 100%.

Brunnschweiler (1957) used 1 : 12,000 scale photographs in Switzerland to study time-sequential data. He found that the size, shape, and boundaries of fields could best be determined in May and August, the peak months of image contrast. Individual crops, on the other hand, were easiest to dsicriminate in June or July, and May proved to be the best month for interpreting soil condition. Additional studies using time-sequential photographs have been carried out by Goodman (1959, 1964).

Pettinger and his co-workers found that, although there were very few cases in which one crop had a unique tone signature at a single date, it was possible to improve interpreter identification when photography during the growing season was available and that, when a variety of films and filter combinations were compared, the best results were obtained with IR Ektachrome. In the Phoenix area they used photographs obtained on March 12, April 23, and May 21, and found that it was feasible to identify almost without ambiguity only dry bare soil on March 12, cotton on April 23, and mature barley on May 21 from a group of crops consisting of the following units: March 12–dry bare soil, moist bare soil, sugar beets, weeds, cut alfalfa, mature alfalfa, and barley; April 23–cotton, sugar beets, weeds, cut alfalfa, mature alfalfa, and barley; and May 21–moist bare soil, sugar beets, wheat, weeds, cut alfalfa, mature alfalfa, and barley. All other crops showed very substantial overlap, and some were virtually coincident with one another. The same situation was obtained with a different sequence of crops on August 5 and September 30, when considerable overlap was again observed.

These data, obtained with panchromatic-25A photography, are shown in Figure 13.1 and indicate the caution with which one must approach each remote sensing concept. The use of time as a discriminant is intuitively most attractive, yet as these results indicate it is no panacea if only a single band or even color photography is used.

Crop calendars were developed for both the Phoenix and the Imperial valley areas and were demonstrated as useful for describing patterns of crop sequences. In addition, the calendar enabled a prediction of the optimum date for identification of the major crop types found in the test areas.

Pettinger and his co-workers found it infeasible to prepare photo-interpretation keys for crop identification which used the color tones of color IR film as a dependable characteristic. The problem of calibration of color photography is well known, and for all except the most highly controlled experimental situations it is very nearly intractable in an operational situation.

Some authors have argued that color composite images prepared from multiband photographs by either optical or electronic combination techniques may reduce the variability of color images taken on different dates (Yost and Wenderoth, 1968; Colwell et al., 1969). However, Pettinger points out that, unless the results

0-201-04245-2

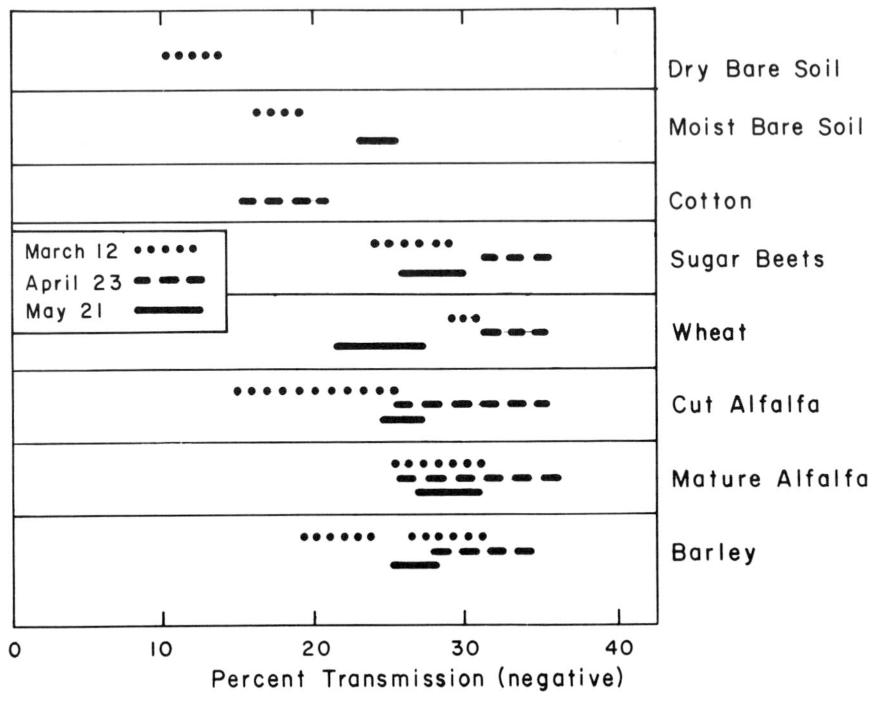


**Fig. 13.1.** Overlap in the reflectance regions of various crops, on three dates, using panchromatic-25A high-altitude photography near Phoenix, Arizona. After Pettinger (1969).

are viewed directly on a screen and interpretations are made in that fashion, the same problem will apply to color calibration of color photographs derived from screen combinations as occurs with conventional color or false color photography.

Comparison of the panchromatic-25A and IR Ektachrome photographs showed, for March 12, 56 and 67%, respectively, of correct identifications by photo interpreters; for April 23, 59 and 73%; and for May 21, 71 and 77%. Concurrent interpretation of IR Ektachrome for all three dates gave 84% correct identification, an average improvement of nearly 12% over the other three dates.

From the results obtained by Schepis (1968), Steiner (1969), and Pettinger (1969) one may conclude that the accuracy obtainable with multidate photography is largely a product of the ability to use high-resolution information, along with variations in the crop assemblages in the area under consideration.

In some circumstances, however, variations in tone alone provide an adequate basis for distinction. For example, Byrne and Munday (1972) used multidate photography in studying nonwestern agriculture in the Bahamas. They observed the impact of shifting agriculture on subtropical vegetation, as related to photo density. Photo density increased as the proportion of pioneer species declined and the average height of the regrowth increased. Simonett et al. (1970) had previously observed the same phenomenon on clearing-induced landslides in New Guinea.

0-201-04245-2

## B. Crop Identification Using Spacecraft Photography

Four studies using space photography before the launch of LANDSAT have been chosen for discussion because the range in problems and approaches is representative. The first (by Colwell et al., 1969) involved the use of a small group of interpreters who were trained on sample fields in the area and then predicted for the balance of the area. Their results were summed and formed the basis for estimates of interpretability of individual crop types. The second (Wiegand et al., 1969) utilized film transmittance values obtained with an isodensitracer and multiple filters and computer analysis of the results. The third (Simonett, 1970) involved the use of a pair of interpreters matching colors to National Bureau of Standards color chips on the basis of a 16-ha decision cell grid laid over the space photography. The fourth (Anuta and MacDonald, 1971) employed scanned, digitized multispectral Apollo imagery and maximum-likelihood computer analyses.

By far the most comprehensive and wide ranging of these reports is the one by Colwell and his associates, to which 11 authors contributed. Studies on crop discrimination based on training samples for interpreters near Phoenix, Arizona, showed that the percentage of correct identifications on a per-field basis lay in the mid-40's for red-sensitive film (pan-25A filter) and for black and white IR film (used with an 89B filter), whereas for color IR photography a 65% correct identification was achieved.

The results of the individual crop identifications for a color IR photograph are given in Table 13.1, along with a comparison of interpretation results based on a high-altitude-aircraft color IR photograph obtained synchronously with the Apollo 9 photographs. From Table 13.1 it is seen that the percentage of correct identification is identical for the spacecraft and high-altitude photography. (The same was also true of the red and IR films.) Table 13.1 also shows the nature of the errors of omission and commission in identifying crop categories by four interpreters. The numbers in the body of the table indicate the cumulative number of fields identified by the four interpreters while the numbers in boldface diagonal blocks indicate the number of fields identified correctly. Of a total of 92 fields of recently cut alfalfa, 84 were correctly identified on the Apollo 9 photograph; 8 were missed (errors of omission); and 22 fields of bare ground and 10 of sugar beets were incorrectly identified as cut alfalfa. All told, 124 fields were identified by the interpreters as alfalfa instead of 92, for a commission error of 40%. Notice also that the bare soil categories (dry bare soil and moist bare soil), while capable of moderate confusion between one another, were separable with very high accuracy from all other categories. These results are not unexpected with color IR photography, which is very sensitive to the presence of growing vegetation. In areas of bare ground, dark blue to pale green tones are commonly found, and these contrast sharply with the delicate pinks and reds of growing vegetation. It is important also to note that the separation between bare dry soil and bare moist soil was surprisingly good, indicating that acceptable probabilistically based estimates of the course of irrigation may be derived from this sample of color IR space

0-201-04245-2

**TABLE 13.1**
**Test Results Comparing the Interpretability of Agricultural Crop Types on an Infrared Ektachrome Space Photo with Those on an Infrared Ektachrome High-Altitude Aerial Photo, near Phoenix, Arizona[a] (after Colwell et al., 1969)**

*Image 6: Infrared Ektachrome, Apollo 9*
*March 12, 1969*

| Photo Interpreter's Results | Ground Truth[b] B | Am | Ac | SB | W | BSm | BSd | Total Seen by P.I. | Commission Error (%) |
|---|---|---|---|---|---|---|---|---|---|
| B | **66** | 21 | 1 | 4 | | | | 92 | 26 |
| Am | 31 | **33** | 2 | 3 | 2 | | | 71 | 38 |
| Ac | 22 | 5 | **84** | 10 | | | 3 | 124 | 40 |
| SB | 5 | 3 | 2 | **6** | | | | 16 | 10 |
| W | 7 | 2 | | 1 | **2** | | | 12 | 10 |
| BSm | | | | | | **43** | 14 | 57 | 14 |
| BSd | 1 | | 3 | | | 9 | **51** | 64 | 13 |
| Total Fields | 132 | 64 | 92 | 24 | 4 | 52 | 68 | 436 | |
| Incorrect | 66 | 31 | 8 | 18 | 2 | 9 | 17 | | 151 |

Total percentage of correct identification: 65%

*Image 7: Infrared Ektachrome, High Flight*
*March 12, 1969*

| Photo Interpreter's Results | Ground Truth B | Am | Ac | SB | W | BSm | BSd | Total Seen by P.I. | Commission Error (%) |
|---|---|---|---|---|---|---|---|---|---|
| B | **43** | 17 | 1 | | 4 | | | 65 | 22 |
| Am | 38 | **36** | 6 | 2 | | | | 82 | 46 |
| Ac | 24 | 3 | **72** | 10 | | | | 109 | 37 |
| SB | 24 | 5 | 2 | **11** | | | | 42 | 31 |
| W | 2 | 3 | | 1 | **0** | | | 6 | 6 |
| BSm | | | | | | **39** | | 39 | 0 |
| BSd | 1 | | 9 | | | 13 | **67** | 90 | 23 |
| Total Fields | 132 | 64 | 90 | 24 | 4 | 52 | 67 | 433 | |
| Incorrect | 89 | 28 | 18 | 13 | 4 | 13 | 0 | | 152 |

Total percentage of correct identification: 64%

[a] Numbers in body of tables indicate the cumulative number of fields identified by four interpreters. Numbers in boldfaced diagonal boxes indicate the number of fields identified correctly.

[b] B = barley; Am = mature alfalfa; Ac = cut alfalfa; SB = sugar beets; W = wheat; BSm = moist bare soil; BSd = dry bare soil.

0-201-04245-2

photography. In fact, the earliest results with LANDSAT very nearly repeated this experiment in the Central Valley of California. Notice also that the continuous cover crops—barley, alfalfa (mature), sugar beets, and wheat—were not consistently differentiated from one another, but the groups were easily separated from cut alfalfa and bare soil. As it happens, mid-March, the time of the Apollo 9 photography, is known to be among the more difficult times to achieve crop discrimination.

Broadly similar results were obtained in the Imperial Valley by Colwell et al. (1969), with, however, a decrease to around 50% (from 65%), using color IR photography for crop identification. As was true in the Phoenix area, there was essentially no difference between the high-altitude aircraft and the spacecraft photography. These results suggested very strongly in 1970 that, once one eliminated really high resolution and stereoscopy from the decision matrix, it would be possible to move to very high altitudes, including space photography, for crop identification when the area studied has very large fields. LANDSAT has in fact shown this to be true. The qualification of "very large fields" is important, for with modest spacecraft resolution acceptable results can be obtained only when fields are at least 8–16 ha in size. A further qualification in regard to the results in the Imperial Valley should be added: the tests were carried out over a restricted range of crops out of the total available in the valley. If a larger array of crops had been used, the identifications would have deteriorated appreciably because of substantial overlap in the probability density functions for the larger set. A further factor which must be borne in mind in interpreting these results is the partitioning of the percentage of correct identifications between crop classes, and the various proportions of the area in each crop class. In the case in question the correct answers came substantially from the categories of bare soil and barley, for which identifications were of the order of 80 and 60%, respectively, using IR Ektachrome SO-65 Apollo photography. As it happens, barley and bare soil together constitute more than 50% of the area evaluated by the interpreters. Thus the average of 50% correct identification is somewhat misleading because it depends in part on the proportion of bare soil and barley present in the sample.

In both the Phoenix, Arizona, and the Imperial Valley tests it is important to note that the best overall results were obtained with color IR photography. Various forms of enhancement, both optical and electronic, using the Forest Research Laboratory optical combiner and Philco-Ford electronic combiner, did enable certain categories to be readily separated. However, the overall performance of these methods of combination was not as good as the straight color IR interpretation. In 1975 we are still a considerable distance from being able to use such devices effectively as a *consistent* aid to interpretation. For examples of valuable results with such devices see Frazee et al. (1971), who studied soil properties in North Dakota using electronic enhancement.

Wiegand et al. (1969) studied soil conditions and crop discrimination from the Apollo 9 SO-65 experiment for a small area of the Imperial Valley, California. The 90-square-kilometer area studied contained 303 fields, for which ground truth data

0-201-04245-2

was furnished by the University of Michigan. An isodensitracing reproduced at a scale of 1 : 30,000 was obtained of the tiny ($2 \times 6$ mm) area on the space photograph, using a variety of filters. This was then run through a computer signature selection procedure based on minimum distance to the mean obtained from the same data set. It is important to note that the mean values were derived from the same data set used for prediction and consequently did not represent a true test of prediction, for which separate training and prediction sets should be used. Wiegand and his coworkers sorted fields into crop species and soil condition categories that were sufficiently replicated so that 50 usable fields resulted, including 9 sugar beet, 10 alfalfa, 11 bare soil, 9 barley, and 11 salt flats. The small number of replications also warrants some caution in interpreting these results. The overall identification was 54% correct, partitioned among the crops in the following manner: all 11 salt flats were properly identified, as well as 6 of 9 sugar beet, 3 of 10 alfalfa, 1 of 11 bare soil, and 2 of 9 barley fields.

Wiegand et al. point out that "a severe limitation is put on photographic analysis in the narrow wavelength interval available for exploitation. Variability in ground cover and height in the plants in the test fields exceeded the differences between the species." They also point out that the results of Richardson et al. (1969) indicate the photographic wavelengths are not very useful for species differentiation. The Richardson paper, read at the Sixth Symposium on Remote Sensing at the University of Michigan, showed that within the wavelength interval 500–2500 nm wavelengths between 1.3 and 2.4 $\mu$ were the most useful in species discrimination, although the information in this region was derived primarily from plant cell size and secondarily from species (Gausmann et al., 1969).

Another study, in a 200-square-kilometer area northwest of Dallas-Fort Worth, Texas, was reported by Simonett (1970). Land use in the area was based on a very simple three-use system consisting of bare ground, pasture, and small grains (barley and wheat). A small training area based on these three categories was established, and then two experienced interpreters made predictions over the rest of the area. Predictions were made on the basis of 16-ha cells, a 16-ha grid being laid over an enlargement of the space photograph. Each cell was then identified as to which of three groups corresponding to the training set it matched. Colors were checked against the National Bureau of Standards ICC/NBS system of color chips. Pasture was detected 80% of the time, bare earth 46%, and small grains 77%. The overall identification accuracy was 72%. Unfortunately, a very high commission error depreciated the value of these detections. It is noteworthy that the bare ground category was not identified with nearly the reliability obtained in March for the data in the Phoenix area reported by Colwell et al. (1969). (See also Wiegand et al., 1969, who had even poorer results with the bare earth category.) Finally, in areas of smaller fields, further to the east, identification deteriorated so badly that the interpreters found they were guessing rather than making probabilistic interpretations (see Simonett, 1970).

Multispectral photography from Apollo 9 was used by Anuta and MacDonald (1971) over the SO-65 test site in the Imperial Valley, California, for a pattern

0-201-04245-2

recognition crop identification study. Three bands covered the following regions: 0.47–0.61 $\mu$m (3400 Panatomic X), 0.68–0.89 $\mu$m (50-246 black and white IR), 0.59–0.715 $\mu$m (3400 Panatomic S). The fourth film was SO-180 Ektachrome color IR (0.51–0.89 $\mu$m).

The four duplicate transparencies were scanned with an Optronics rotating drum microdensitometer with 2200 $\times$ 2200 resolution elements to the film frame, and with 160 linear gray-scale steps. Figure 13.2 shows portion of the frame, and Figure 13.3 gives the alphanumeric printout of the same area. The box encloses the study area.

The sample line printer gray-shade map was prepared by sampling 3% of the picture elements (pixels) (every sixth row, every sixth column). Figure 13.4 shows the center of the study area, as a full pixel gray-shade map. Ground truth in this area in the vicinity of Brawley, California, was obtained by University of Michigan (Institute of Science and Technology, now the Environmental Research Institute of Michigan: ERIM) personnel. The ground resolution of 60 m is very similar to that obtained with LANDSAT. Symbols added to Figure 13.4 indicate surface cover as follows: A—alfalfa, B-barley, BS—bare soil, SB—sugar beets, SF—salt flats, W—water.

After scanning, the four transparencies were spatially registered, and the data were then used with the Purdue University maximum-likelihood classifier to produce the alphanumeric classified output shown in Figure 13.5. The classes are those listed above, but with barley, alfalfa, and sugar beets grouped as green vegetation.

**Fig. 13.2.** Portion of an Apollo 9 frame over the SO-65 test site in the Imperial Valley, California. After Anuta and MacDonald (1971).

0-201-04245-2

**Fig. 13.3.** Alphanumeric printout of gray scale of a black and white image of the site seen in Figure 13.2: 3% field sample. After Anuta and MacDonald (1971).

The accuracies of identification of the individual crop types are as follows, comparing multiband and multiemulsion results on a per-field, rather than per-pixel, basis. Note that there are sometimes significant differences, suggesting that slight shifts in wavelengths can markedly affect accuracies. Some green crops were identified with only 32% accuracy (sugar beets) and others with 71% (barley), while alfalfa, bare soil, salt flats, and water were discriminated with accuracies of 40, 62, 88, and 99%, respectively, with multiband data. These results are about the same as those generally obtained by manual interpretation. The results on color separations of color IR data were, respectively, barley 54%, alfalfa 36%, sugar beets 43%, bare soil 58%, salt flats 93%, water 96%. The overall result was 57% (cf. 59% for the multiband data) (see Anuta and MacDonald, 1971).

From these results, it is clear that field size, crop assemblage, and time of year all play some part in influencing crop discrimination capability using space photographs. All the fields studied in these tests were quite large, and consequently the mixing together of several categories within a resolution cell was kept to a minimum. As has been shown by Simonett and Coiner (1971), such mixing is a critical matter in regard to identification when fields are small and the resolution of the space photos is modest. In short, the environment plays no small part in strongly influencing the percentage accuracy obtained. Thus, in situations where the environment collapses to a simple structure (Dallas-Fort Worth), a high overall accuracy (72%) was obtained, even though the decisions were based on 16-ha cells.

0-201-04245-2

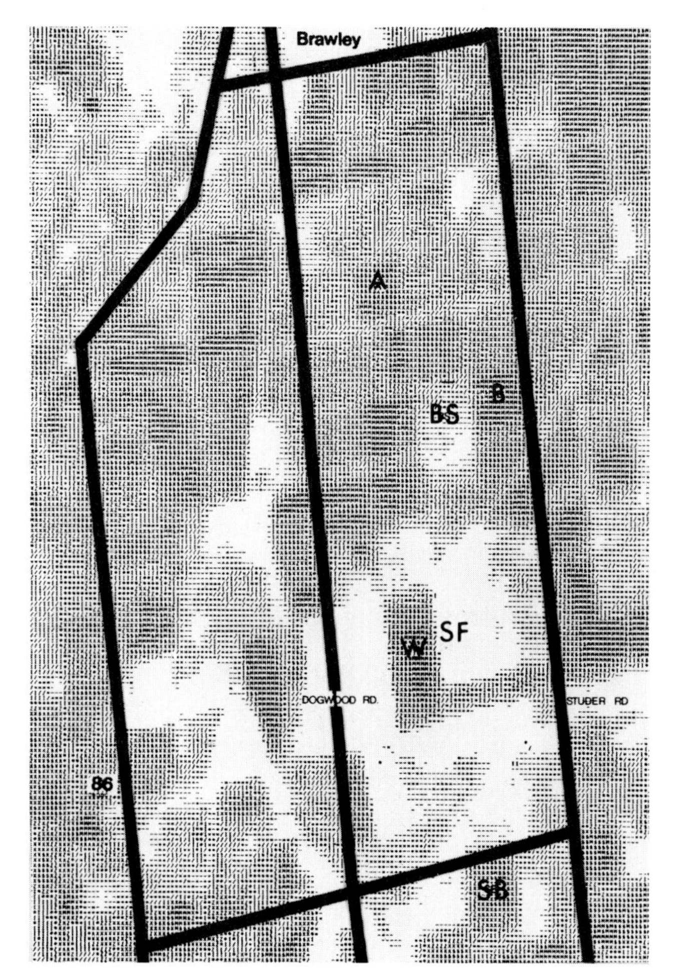


**Fig. 13.4.** Full pixel printout of the center of the boxed area in Figure 13.3. After Anuta and MacDonald (1971).

The qualification made earlier in connection with the proportion of the sample which falls into one or two of a large number of classes is also important, and it is to be expected that values for correct identification below 50% will be very common in complex environments when single photographs are used (see Imperial Valley results). Identification accuracies as low as 35% may be obtained with color IR photography, as has been shown for the Lawrence, Kansas, area in several instances (Simonett, 1970; Brooner and Simonett, 1971). This critical appraisal, however, should not serve to discount the value of space color photography, for further experimentation will undoubtedly reveal situations and areas for which space photographs obtained at the appropriate time of the year will prove accept-

0-201-04245-2

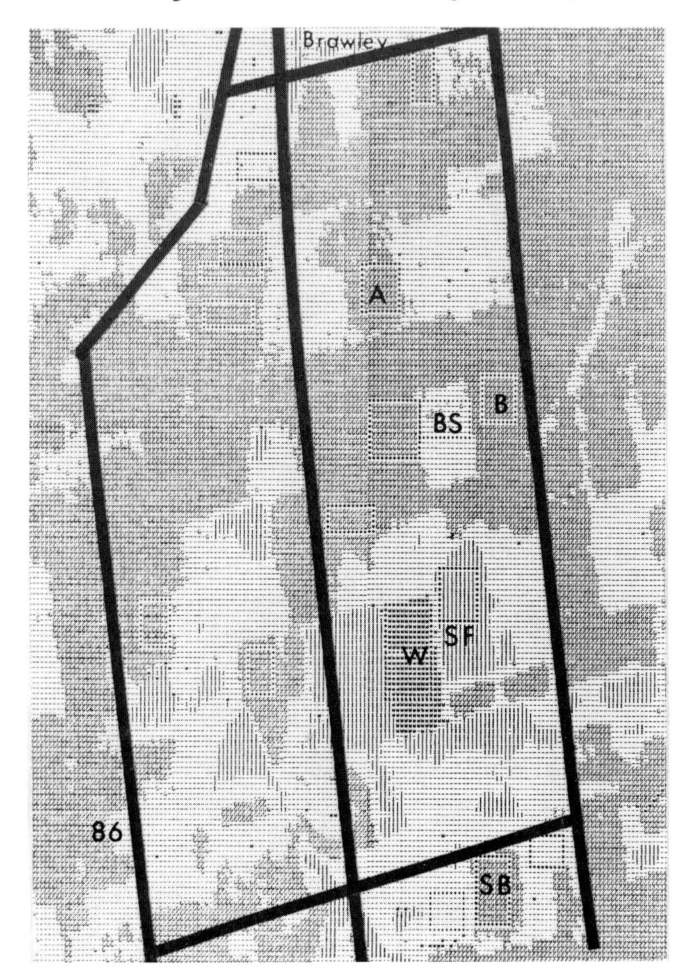


**Fig. 13.5.** Classified output of the area shown in Figure 13.4. Maximum-likelihood classifier used. After Anuta and MacDonald (1971).

ably accurate for particular crop discrimination. On the other hand, it should lead one to view with caution the extrpolation of particular results, in single situations, to other environments.

## C. Studies with Multispectral Scanners

The use of thermal IR scanners in the study of crops and their conditions of health has been carried out by a wide spectrum of investigators. On the other hand, work with aircraft multispectral scanners has been dominated by workers in the Laboratory for Applications of Remote Sensing at Purdue University and at the ERIM. The early research of the Purdue group appears in its annual reports, published as

0-201-04245-2

research bulletins of the Agricultural Experiment Station. More recent publications have appeared in the periodical literature.

An important early paper was that by Fu et al. (1969). They applied a maximum-likelihood classification rule to different crop classes in a series of classification experiments. On the basis of one experiment they considered that, using data from a 12-channel scanner covering an area between 0.40 and 1.0 μm, they could recognize with very high accuracy both bare ground and areas of green vegetation down to as little as 20% ground cover. They pointed out, "It is expected that it may be possible not only to map green vegetation but, in addition, to show on a map the percent of the ground cover," and illustrated this with three computer printouts, using different thresholds for green vegetation.

In another experiment they tackled a nine-class crop classification problem. The nine classes were oats, corn, soybeans, alfalfa, red clover, rye, bare soil, and two types of wheat. Training on one set of samples and predicting for another in the same flight run (the fields were all close together), they obtained the following percentages of correct identifications: soybeans 96%, corn 91%, wheat 99%, oats 92%, rye 97%, red clover 91%, alfalfa 88%, soil 99%—overall performance 94%. These results are so far superior to those obtained in the photographic (interpreter) investigations reported earlier that it would appear at first sight that multispectral scanner-computer analysis would supplant competitor systems. This judgment is in part correct, for all the training and prediction results obtained in the same flight line in the studies at Purdue have shown broadly comparable quality. However, these results also must be approached with some caution as to their interpretation, and the extent to which we may extrapolate from them to other areas.

Fu et al. (1969) approached this problem with "a more severe but important test on the generalizing capabilities of the classifier"; that is, he tested the classifier on different flight lines some miles apart. Table 13.2 shows the results of this experiment. It will be seen that the percentage of recognition ranged from 87% for wheat to a low of 58% for soybeans. The overall percentage of recognition was 71%. The two flight lines were only 5 km apart and perhaps had a comparable degree of agricultural variability to those encountered in the space photos studied. Notice that the results dropped to a level not notably better than those obtained with just color IR photography and human interpretation in the earlier examples studied in the western and southern United States.

Numerous factors influence the quality of the integrated spectral reflectance from crops. (Integrated reflectance is the composite return from the crop and background, mixed within the resolution cell at the particular solar angles, viewing angles of the remote sensing instrument, solar illumination quality, and so on; see Colwell, 1974). These factors impose on training and prediction sets severe requirements for data comparability.

Two quite dissimilar strategies have been used to surmount this problem. The first involves the use of training and prediction sets obtained at the same time. The ground truth for the training set is established close to the time of aircraft or

0-201-04245-2

**TABLE 13.2**
**Training on One Set of Data and Predicting on Another Set of Data Several Miles Distant in Tippecanoe County, Indiana, Using a 12-Channel Multispectral Scanner (after Fu et al., 1969)**

| *Class* | *Percent Recognition*[a] | *Number of Samples* | *Number of Samples Classified into:* Oats | Soybeans | Red Clover | Corn | Rye | Alfalfa | Water | Wheat | Bare Soil | Thresholded |
|---|---|---|---|---|---|---|---|---|---|---|---|---|
| Oats | 80.4 | 4,170 | 3351 | 183 | 240 | 83 | 88 | 61 | 0 | 76 | 0 | 88 |
| Soybeans | 58.2 | 9,295 | 19 | 5409 | 2 | 2990 | 0 | 0 | 0 | 7 | 411 | 457 |
| Corn | 58.8 | 6,941 | 6 | 2669 | 12 | 4084 | 1 | 0 | 0 | 61 | 6 | 102 |
| Wheat | 87.2 | 2,901 | 224 | 2 | 0 | 4 | 120 | 0 | 0 | 2529 | 0 | 22 |
| Total | 71.1 | 23,307 | 3600 | 8263 | 254 | 7161 | 209 | 61 | 0 | 2673 | 417 | 669 |

Overall performance = 65.1
Average performance by class = 71.2

Channels used: all 12

[a] Quantitative classification results of test fields in flight line C2, based on training samples from flight line C1.

0-201-04245-2

spacecraft passage. The other approach is to build up an extensive library of remote sensing "signatures" obtained under a great variety of circumstances and locations. These library data sets, coupled with the use of appropriate statistical and atmospheric models, provide the basis for prediction. Each approach rests on a conceptual basis which warrants continuing close examination. Thus the use of concurrently obtained training and prediction sets is based on the implicit assumption that the training set is a viable subsample of the prediction set itself. The degree to which this assumption is violated in space, in time, and at various scales of instrument resolution has not been examined, and this represents a problem of elementary crop geography to which far too little attention has been given. In a similar manner the use of "library signatures" rests on an implicit assumption of the validity and practical utility of the multispectral concept in agricultural applications.

The results using training and prediction sets drawn from areas several miles apart reported by Fu et al. (1969), as well as the comments made earlier in regard to extrapolation over long distances from training sets, acquire special point when considered in the light of studies during the last several years, culminating in the Large Area Crop Inventory Experiment (LACIE) now in progress by the National Aeronautics and Space Administration, the National Oceanographic and Atmospheric Administration, and the U.S. Department of Agriculture. Representative papers are those by Marshall et al. (1969), Earing and Ginsberg (1969), Brooner and Simonett (1971), Anuta and MacDonald (1971), Bauer et al. (1971), MacDonald et al. (1972), Draeger et al. (1971), and Hall et al. (1974). The study by Marshall et al. was carried out in Tippecanoe County, Indiana, and employed data for May and June derived from a single aircraft flight line only a few kilometers long, viewed from several altitudes. A total of 20 fields of wheat was investigated in this area, using a training set subsample and a prediction set subsample from the single flight line. It is uncertain from the paper whether 12 channels in the visible and near-infrared region or all 19 channels covering the area 0.32–14 μm were used. The percentage accuracy of determination of wheat against a background of other crops and bare ground lay in the range 70–80%, depending on the time of year, altitude, and number of samples used in the training sets. The authors make the point that increasing the number of samples in the training sets would improve identification, and this is an important qualification on the relatively modest results they obtained. Notice, however, that the results are broadly comparable to those obtained by Fu et al. (1969) when the latter trained on a set some 5 kilometers distant from the prediction set.

In the companion paper by Earing and Ginsberg a model for the selection and prediction of the appropriate channels to be used at seven stages of growth in the Midwest was developed. The model embodied knowledge of the life cycle and percent ground cover of wheat and all other crops capable of being confused with wheat at all seven stages of this cycle, extending throughout the growing season. Based on a library of crop signatures (reflectance statistics) developed over a span of years at the University of Michigan, three models were used for the determina-

0-201-04245-2

tion of statistics of the apparent radiance of a scene. "These models statistically describe the incident irradiance and its angular distribution; the reflectance characteristics of the scene; and the transmission of the atmospheric path between the scene and the sensor." From these data and previously derived computerized modeling and pattern recognition techniques developed at the University of Michigan, calculations of the probability of detection and of false alarms for wheat against the weighted background of all other backgrounds were developed. Table 13.3 represents the results of the Earing–Ginsberg calculations in channel selection. Different channels were found to be more important at different stages in the growth cycle of wheat. It is important to recognize the impossibility of extrapolating from this table to other situations. Identification results in actual prediction situations do not equal those obtained from training and predicting on the same data. This is the reason for the remarkable accuracies in the Earing–Ginsberg paper. In the last few years the "library signature" approach has been substantially discounted. It remains to be seen, however, whether, as our knowledge improves, there may be a continuing role for some applications of this approach.

**TABLE 13.3**
**Channel Selection for Highest Probability of Detection and Fewest False Alarms (Training and Predicting on Same Data) (after Earing and Ginsberg, 1969)**

| *Stage* | *Target* | *Best Channels* | *Probability of Detection (%)* | *Probability of False Alarms (%)* |
|---|---|---|---|---|
| 1 (September) | Plowed | 9 | 96.86 | 8.34 |
| | (wet soil) | 7, 8, 9 | 99.66 | 0.88 |
| | | All | 99.96 | 0.10 |
| 2 (October) | Planted | 9 | 81.45 | 25.69 |
| | (dry soil) | 5, 7, 8 | 98.10 | 4.75 |
| | | All | 99.18 | 1.12 |
| 3 (October-November) | Seedling | 8 | 79.33 | 55.19 |
| | (wheat = 0.35)* | 6, 7, 8 | 99.33 | 1.26 |
| | (dry soil = 0.65)* | All | 99.59 | 0.77 |
| 4 (March-April) | Seedling | 9 | 89.36 | 37.96 |
| | (wheat = 0.35) | 6, 7, 8 | 98.48 | 4.71 |
| | (dry soil = 0.65) | All | 99.50 | 1.13 |
| 5 (May) | Mature | 9 | 93.65 | 20.58 |
| | (wheat = 0.80) | 6, 7, 8 | 99.94 | 0.13 |
| | (dry soil = 0.20) | All | 99.98 | 0.05 |
| 6 (June-July) | Ripe | 1 | 65.84 | 9.80 |
| | (wheat = 0.80) | 1, 2, 3 | 93.25 | 1.66 |
| | (dry soil = 0.20) | All | 97.34 | 1.24 |
| 7 (July) | Stubble | 1 | 62.54 | 10.89 |
| | (wheat = 0.40) | 1, 2, 9 | 91.05 | 3.16 |
| | (dry soil = 0.60) | All | 97.05 | 1.87 |

*Proportions respectively of wheat and bare ground in October-November. Comparable units are used for the succeeding months.

0-201-04245-2

In an early comparison in eastern Kansas of equal probability and a priori weighting (maximum-likelihood and Bayesian decision rules) Brooner and Simonett (1971) warned of the modest results to be expected from training and predicting in areas where there would be substantial confusion between crops. Their predictions of low midsummer accuracies in complex cropping situations were, in fact, substantiated in a later LANDSAT study by Von Steen and Wigton (1973). Anuta and MacDonald (1971) obtained somewhat better results in a study using digitized space photography of the Imperial Valley, California, where large-field irrigated agriculture is intrinsically simpler at certain times of the year.

Bauer et al. (1971) found in their studies of southern corn leaf blight that digital processing gave more accurate results than human interpretation, a conclusion of considerable operational significance for later crop studies. The corn blight design study by MacDonald et al. (1972), like that by Bauer et al. (1971), had considerable influence on all later large-area crop analyses. The experience gained in the cooperation between the U.S.D.A., NASA, Purdue University, and ERIM was invaluable and provided the first operational test of the problems in setting up major remote sensing procedures. The study by Draeger et al. (1971) was important in that it showed that there were some situations and environments where prior stratification of crop assemblages was not valuable, a conclusion again borne out in later studies with LANDSAT by Wiegand et al. (1973).

The study by Hall et al. (1974) discussed earlier raised a series of important questions, most of them not fully resolved as of this date (mid-1975). Clearly, large remote sensing analyses of complex, large-area natural and cultural systems, such as agriculture, involve far more than taking the results of a single successful experiment and extrapolating. The environment is too complex for so simple a procedure. A large-systems analysis of the interacting natural and cultural systems is, in fact, required. This will be a major continuing task requiring the active cooperation of numerous disciplines.

The results obtained in a remote sensing experiment, as well as the meaning to be attached to them, must be subjected to very careful scrutiny. Remote sensing is advancing so rapidly on so many fronts and such a significant number of valuable data are being obtained that there is a tendency to be uncritical of results from outside one's area of immediate professional competence. Students from the natural sciences with only a modest background in statistics frequently tend to be overwhelmed by pattern recognition terminology used in conjunction with multispectral scanner studies. This analysis has shown, however, that on careful scrutiny the overall context of an investigation (as well as the details of the statistical techniques used) can be seen to have a significant influence on the quality of the results.

### D. Crop Discrimination with Radar Imagery

0-201-04245-2

Because information on crops must be obtained on a timely basis the cloud-penetrating value of radar is potentially significant for agriculture, notably in areas where extensive warm frontal cloudiness occurs during the growing season, or where cumuliform build-up almost every day is experienced. The agricultural areas of western Europe and portions of southeast Asia provide examples of these two situations. Agricultural studies to date have been conducted with a K-band multiple-polarization radar in a test area near Garden City, Kansas (Simonett et al., 1967; Haralick et al., 1970; Morain, 1971), and with an X-band single-polarization system near Phoenix, Arizona (Meyer, 1967), an X-band system at Garden City and Lawrence, Kansas (Morain, 1971; Morain and Coiner, 1970), and a 4–8 GHz broadband system also in Kansas (Ulaby et al., 1972). The results, though scanty, offer enough encouragement to extrapolate to other radars and other test areas.

At the Garden City, Kansas, site the K-band radar images have been evaluated for the effect of crop type, cover, moisture, and other parameters on radar return with the following results:

1. Statistical analysis (analysis of variance, analysis of covariance, and multiple correlation) indicates that the type of crop is the most important variable influencing radar return.
2. Allowing for differences in crop type, other major influences on radar return are crop height, percentage of the ground covered by vegetation, and row direction (bare ground and low cover).
3. Crop moisture, which shows a weak relationship to radar return, requires further study as a variable.
4. Soil moisture, even from bare ground, shows small influence on radar return during a single month. Unpublished theoretical studies at the University of Kansas indicate that as a contributor to radar return, soil moisture will be overwhelmed at short wavelengths by surface roughness effects, but that at longer wavelengths its contribution will be more significant.

Numerical taxonomic clustering techniques have been applied to the September 1965 and July 1966 multiple-polarization data to determine "natural" clusters in the data (Schwarz and Caspall, 1968). Other studies employing contingent probability and Bayesian decision theory (Haralick et al., 1970) and adaptive pattern recognition (Haralick and Kelly, 1969) have also been made for the determination of "natural" sets within the data and for crop discrimination training and extrapolation to adjacent area. In another study Haralick (1967) developed a chaining procedure using likelihood functions, which avoids the excessive computation required for the multiple-linkage method employed in numerical taxonomy.

As an example of the level of discrimination which may be achieved with K-band radar under very favorable conditions, Figure 13.6 shows the September 1965 data on HH and HV polarization K-band radar. Fields of sugar beets, corn, grain sorghum, wheat, alfalfa, and bare ground were examined. The $y$ axis is the like-density HH polarization; the $x$ axis, the cross-density, HV polarization. Notice that the hyperplanes erected lie at about 45°, implying that both polarizations are contributing information.

0-201-04245-2

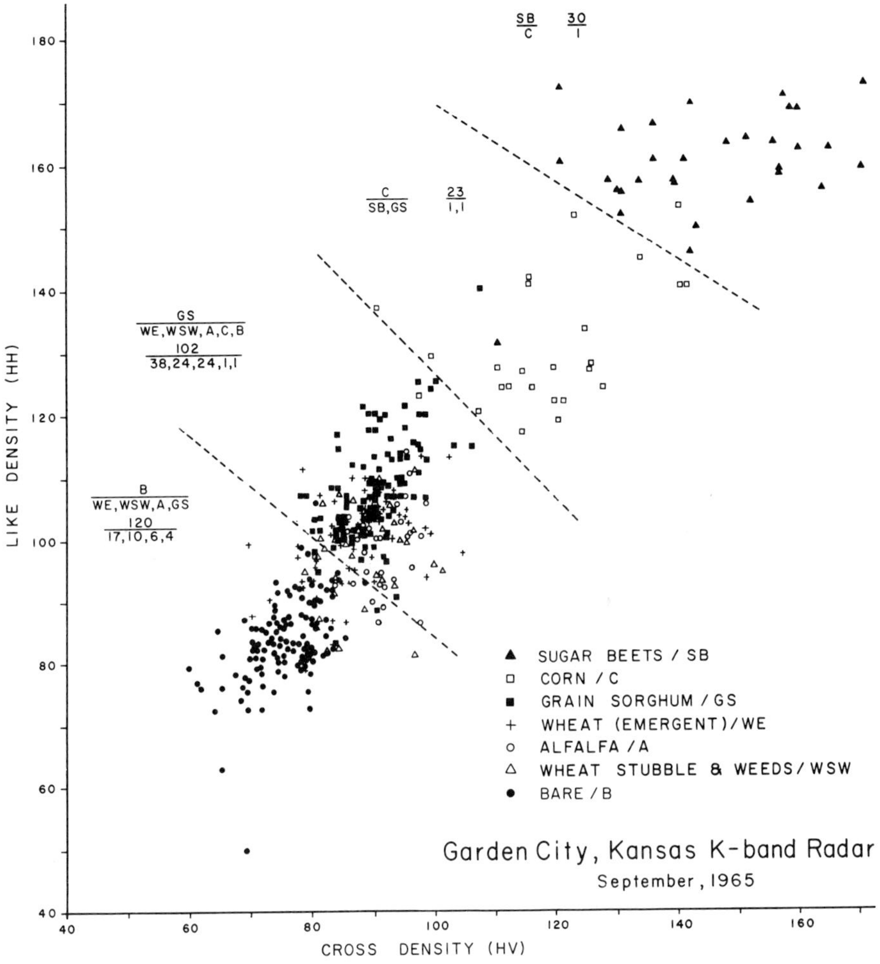

Garden City, Kansas K-band Radar
September, 1965

In the upper group there are 30 fields of sugar beets and 1 field of corn, suggesting that if this relationship had held in adjacent areas it would form a useful basis for extrapolation and discrimination. The next lowest region has 23 fields of corn, 1 field of sugar beets, and 1 field of grain sorghum. The lowermost group has 120 fields of bare ground; 17 of emergent wheat, in which the wheat occupies less than 5% of the total area—for all intents and purposes this is really bare ground; 10 of wheat stubble and weeds, in which the wheat stubble is dry and makes little contribution to the return; 6 of alfalfa; and 4 of grain sorghum.

It will be seen that with the exception of alfalfa and grain sorghum the lowermost category represents bare ground—or is, practically speaking, bare ground. Consequently, the lowermost group also looks reasonably encouraging for projection by extrapolation to adjacent areas.

0-201-04245-2

The next to lowest group contains emergent wheat, wheat stubble and weeds, alfalfa, corn, bare ground, and grain sorghum, though grain sorghum predominates. It is obvious that the two-polarization single-frequency radar achieves inadequate discrimination in this region. However, the uppermost portion is dominated by grain sorghum, and a subset might be discriminated which could serve as a basis for the prediction for grain sorghum in the lower portion of the mixture. Radar imagery at a different time optimized for the detection and discrimination of wheat (say, April) would aid in improving predictions in this area.

It is important to emphasize that this discussion of Figure 13.6 is not based on training and predicting with different data sets. Rather, it shows only that reasonable grounds exist for carrying out training and predicting on different sets in the future.

In portion of the study by Haralick et al. (1970), however, training and prediction were based on different data sets. These results form a better basis for assessment. Their tables of prediction accuracy must be used with care, however, for they are misleading without qualifications. Table 13.4 shows the form of the tables. The quoted "accuracies" of classification hold only in the context of "how the sensor saw the crop group," not for individual crops. In part, then, the high accuracies quoted are a statistical artifact of grouping.

A quantitative study has also been made by Meyer (1967), involving calibration of crops in the Phoenix, Arizona, area. He employed the AN/APS-73 (XH-4) synthetic-aperture X-band radar system and found the following:

1. Radar return data from crops could be calibrated and replicated to within about ±1.0 dB.
2. The condition of cotton, sorghum, wheat, and lettuce (usually height or density of vegetation) had a significant effect on the backscatter coefficient ($\sigma_0$) at the time of measurement. Crops within which separation could *not* be made by condition during the experiment included alfalfa, grapes, citrus fruits, barley, sugar beets, cabbage, onions, carrots, potatoes, and roses. Further work is needed to find the basis for these differences.
3. The average size of clods on bare ground had a significant effect on backscattered returns, increasing up to clods about the wavelength of the radar (circa 3.0 cm).

In another study (Simpson, 1969) with $K_a$ band radar, corn fields in New England could be readily discriminated with high accuracy (>90%) from pasture in the same vicinity.

Ulaby et al. (1972), extended work with radar to broad-band analysis in the 4–8 GHz (3.75–7.5 cm) region. They observed that multiple radar channels significantly improved crop identification accuracies over single-channel data (from 35 to 65%). Thus multispectral radar sensing for crop monitoring appears feasible.

Similar results, established for X (3 cm)- and L (23 cm)-band multiple-polarization radar for natural vegetation and forest (Drake and Shuchman, 1974) and for urban land-use categories (Bryan, 1975), also strongly support the possibility of multispectral crop discrimination.

0-201-04245-2

0-201-04245-2

**TABLE 13.4**
**Radar Clustering for Crops: Six-Category Breakdown, Using Multiple-Linkage Cluster Analysis Based on the Euclidean Distance Coefficient (after Haralick et al., 1970)**

| *Category* | *(a) Bare Ground and Wheat* | *($b_1$) Grain Sorghum and Wheat* | *($b_2$) Grain Sorghum and Alfalfa* | *($b_3$) Grain Sorghum* | *(c) Corn* | *(d) Sugar Beets* |
|---|---|---|---|---|---|---|
| Bare ground | 45 | 1 | | | | |
| Grain sorghum | 1 | 28 | 24 | 37 | 1 | |
| Weeds and wheat stubble | 2 | 6 | 8 | | | |
| Fall wheat | 9 | 18 | 4 | | | |
| Alfalfa | 1 | 6 | 16 | 2 | | |
| Sudan | 1 | | | 2 | | |
| Corn | | | | | 5 | |
| Pasture | 2 | | | | | |
| Sugar beets | | | | | 1 | 11 |
| Number correct/Total | 54/61 | 46/59 | 40/52 | 37/41 | 5/7 | 11/11 |
| Percent correct | 89% | 78% | 77% | 90% | 71% | 100% |

To properly evaluate the radar return from crops it will be necessary to test multifrequency and multipolarization systems sequentially through a number of growing seasons. It will then be feasible to determine the nature of both the time dependencies of the data and their internal consistency from year to year, as well as to assess the spatial and temporal bounds on the use of ground truth training sets for extrapolation.

## E. Interaction of Solar Radiation with Plant Canopies

Along with empirical studies on crop discrimination, fundamental research is being carried out at a number of institutions on the interaction of solar radiation with a plant canopy. These studies have become increasingly necessary as laboratory spectrometer measurements on single leaves have proved inadequate in many cases and thoroughly misleading in others. As Myers and Allen (1968) point out:

> . . . Reflectance from a crop canopy is a difficult measurement because (*a*) $O_2$ and $CO_2$ and water vapor absorption reduce incoming solar radiation in certain wavelength bands; (*b*) illumination from the sun varies in intensity with numerous conditions; (*c*) radiance from field crops is affected by crop geometry, background soil reflectance, and other factors; and (*d*) the intensity of the sun peaks at about .5 micron, falling off rapidly at shorter and longer wavelengths.

They continue:

> . . . Once light strikes a crop canopy, reflectance, tranmittance, scattering, and absorptance all influence disposition of the incident energy. Recent model and field studies by Myers and Heilman (1969) have shown that near infrared spectrophotometer studies of single leaves can be very misleading for predicting reflectance from crops. This is due to near infrared light being transmitted through the top of the crop canopy, changes in light quality within a canopy, multiple internal reflections within the canopy, and reinforcement of reflections from the top of the canopy.

C. E. Olson (personal communication) also found substantial differences between species in this regard, leathery leaves such as those of some oaks or rhododendron giving a much higher percentage reflectance from a single leaf than, say, a cotton leaf. There is a continuing need for spectral measurements obtained from aircraft using short- and long-wavelength radiometer-spectrometers of a type developed by R. J. P. Lyon of Stanford University (Lyon, 1967).

Other important studies on this topic include those by Suits (1972), who developed a model of the directional reflectance of a vegetative canopy, including the use of the solar hotspot as a means of observing two-layer canopies; by Colwell (1974), who examined the factors influencing vegetation canopy reflectance in terms of eight interacting sets of variables; and by Kriegler et al. (1972), who developed procedures for adaptive multispectral pattern recognition of crops. Kriegler and his colleagues pointed out that some aspects of scene variability could be compensated for by applying corrections for scene look angle, changing solar

0-201-04245-2

elevation, aircraft drift, and changing scene conditions. They found that their corrections led to improved performance at little cost and allowed more direct interaction of modeling and processing. Many of these corrections are now routinely applied in using LANDSAT data.

Smith and Oliver (1972) developed Monte Carlo simulation of plant canopy models of composite scene spectroradiance in the 0.4–1.05 μm region. DeBoer et al. (1974) examined spectral signature variations in relation to crop stage of growth in the range 0.4–2.3 μm and tested crop vigor and biomass estimation in the chlorophyll absorption band and the near-infrared region. They also estimated biomass at 0.87 and 1260 μm.

In a very careful experiment Pearson and Miller (1972) had earlier used band ratioing for estimation of standing crop biomass in the short-grass prairie. Accuracies, compared to handclipping, were better than 95%. They ratioed bands in the 0.68- and 0.78-μm regions, comparable to those within the LANDSAT and SKYLAB multispectral scanner systems.

A recent cooperative study by the U.S.D.A. and NASA is the Crop Identification Technology Assessment for Remote Sensing (CITARS) Experiment. It is based principally on the use of LANDSAT data. The questions posed in this experiment are very fundamental ones (see Hall et al., 1974):

- How does corn, soybean, and wheat identification vary with time during the growing season?
- How does crop identification performance (CIP) vary among different geographic locations having different soils, weather, management practices, crop distributions, and field sizes?
- Can statistics acquired from one time or location be used to identify crops at other locations and/or times?
- How much variation in CIP is observed among different data analysis techniques?
- Does use of multitemporal data increase CIP?
- Does use of radiometric preprocessing extend the application of training statistics and/or increase CIP?
- How much variation in CIP results from varying the selection of training sets?
- Does rotation or registration of LANDSAT data affect classification performance?

### F. Remote Sensing Studies on Crop Disease and Insect Infestation

The damage done each year to agricultural crops by pathogens and insects has been estimated as equal to $3.7 billion and $3.8 billion, respectively (National Research Council, 1970), in the United States alone. The magnitude of the losses and the degree of localization of particular pathogens and insects are such that a broad-scale attack is warranted, including techniques as diverse as genetic control and remote sensing. The role for remote sensing which most observers see most clearly is that of early warning and detection of the hazards, so that prompt control action may be taken to reduce the spread and the impact away from the infestation

0-201-04245-2

points. Colwell et al. (1969) list the most important agricultural candidate problems for remote sensing; these include, in cropped areas, (1) control of leaf and steam rust on wheat and oats; (2) barbary bush eradication; (3) cereal leaf beetle eradication; (4) imported fire ant eradication; (5) soybean cyst nematode control; (6) control of burrowing nematode disease on citrus; (7) phoney peach and peach mosaic eradication; and (8) control of various root-rot conditions on agricultural crops and timber stands. Each one of these control and eradication problems has its own peculiar remote sensing constraints and requirements. It is certain that in addition to studies on major diseases there will be a wide flurry of activity in the testing of remote sensing for disease control for a large number of crops. Recently, for example, studies have begun on sugar beet diseases in Texas in association with Texas A & M University.

Black and white IR and color Ektachrome IR photographs have been used for detecting both blights and the depredation of insects, nematodes, and viruses (Colwell, 1970). Although the efficiency of these films in detecting disease varies from crop to crop and circumstance to circumstance, and the explanation is certainly more complex than early suggestions (Knipling, 1970), positive evidence from a variety of crops indicates that in selected cases these films do enable previsual detection of the advance of disease and insect attack and that they warrant very wide testing. For crops for which IR photography proves unsuitable there remains wide-scale testing of the multispectral scanner as an aid to detection. There is also considerable interest in the agricultural community in the possible application of multispectral scanner data derived from spacecraft for the monitoring and prevention of the spread of blights of major crops such as wheat stem rust.

The best-known and most successful use of IR and color IR photography is that on potato blight detection in England and in Maine (Brenchley and Dadd, 1962; Manzer and Cooper, 1967). Manzer and Cooper note that "periodic survey flights over potato fields sprayed by aircraft can, via infrared photography, show the aerial applicator and the grower when and where added fungicidal protection may be needed. Moreover, the ability to obtain current information on disease incidence over a large area is useful not only in plant protection work but also in the making of reliable estimates of total crop production."

A variety of explanations have been suggested for the marked reduction of IR reflectance associated with higher incidence of disease and insect attack. These range from collapse of the spongy mesophyll structure through changes of the internal refractive indexes in the leaves to a stunting of growth and consequently the presence of more bare soil in the affected areas in the photographs. A number of investigators (see Hoffer, 1967) have found that there is no change in infrared reflectance when measured on the plant and that the explanation in these cases appears to be the general reduction in crop size and the greater area of soil exposed.

Color IR photography is now used relatively routinely in surveys of citrus orchards in Florida affected with root spot and other diseases and in studying the incidence of late potato blight. Also, research continues on the photographic effects of a variety of stem and leaf rusts of wheat and other small grains.

0-201-04245-2

In the corn blight watch experiment carried out in the early 1970's (see Macdonald et al., 1972; Bauer et al., 1971; Kumar and Silva, 1974), it was found that *early* detection of corn blight was of only modest accuracy because blighting took place on the lower leaves first. However, as blighting progressed, up to five distinct levels were separable with a 12-channel scanner operating in the range from 0.4 to 11.7 μm The thermal IR channel was important in this separation. Also the machine classifications were of higher accuracy than those of photo-interpretation teams using color IR photography. The results using the average transformed divergence measure in machine processing were particularly encouraging in that, despite much scene variability, the separability of blight classes was very high. Confusions were between classes, not with other crops.

Jackson et al. (1974) found that computer processing of scanned color IR imagery, coupled with careful calibration of the photography, made possible the discrimination of corn aphid infestations. This suggests that direct use of a multispectral scanner would also be applicable to such infestations.

Two studies on coconuts dealing with detection of fertilizer response (Dakshinamurti and Summanwar, 1972) and previsual detection of coconut wilt (Dakshinamurti et al., 1971) obtained positive results with color IR film. For coconut wilt, photography from a helicopter of the crown—invisible from the ground—showed that both conventional black and white and false color imagery detected the presence of wilt, but not necessarily in the early stages. In the fertilizer trials only nitrogen showed a significant response when crown centers were monitored with false color from an altitude of 300 m.

Other studies of interest are those by Frazee et al. (1971) on detecting soil hardpans, low spots, and other yield-affecting features on imagery; Gausman et al. (1969a, b, 1970) on the relationship between soil salinity and cotton leaf reflectance in Texas; Gausman (1974) on cellular characteristics of vegetation in relation to near-infrared reflectance of stressed plants; Gausman et al. (1974) on iron deficiencies in sorghum in relation to leaf area index, reflectance, and chlorophyll concentration; Meyer and Chiang (1971) on simulated insect defoliations (grasshopper and army worm) with multiband photography, in which timing was shown to be critical for detectivity; DeCarolis et al. (1974) on thermal IR detection of "Giallume," a yellows-type disease in European rice fields, successfully contrasted with healthy plants in early-afternoon flights over *heavy* stands of rice; and Younes et al. (1974) on spectral reflectance studies of mineral-deficient corn plants in Egypt, using laboratory spectral measurements and false color photography (with a 25 filter); the near-infrared region showed previsual symptoms of phosphorous and potassium shortages particularly well. Also, nitrogen-deficient plants could be distinguished from those with deficiencies of phosphorous or potassium or from control plants by band ratioing in the 0.55–0.65 μm region. All the preceding studies showed mixed results, but with a sufficient level of success both to warrant continued work and to hold out hope that detection and monitoring of the spread of disease and insect attack, as well as the observation of soil deficiency effects on crops, will in due course become feasible.

0-201-04245-2

## III. FORESTRY

In the remote sensing of forest resources there is a need for a variety of resolutions both to cover the range of problems involved and to integrate both detailed and broad-scale knowledge into broad management planning. The necessity for a variety of scales of observations suggests that mixed supportive systems involving spacecraft and high-, medium-, and low-altitude aircraft remote sensor data, coupled with ground sampling of forest environments, may be combined into a rational package. In a series of studies from 1969 to 1975 this has been demonstrated as feasible by Langley (1969), Langley et al. (1975), Nichols (1973), and Heller (1975).

Timber inventory, for example, requires detailed information on species composition, which can be determined in many cases only with ground truth very-low-altitude aircraft photography (necessary to obtain fine resolution, large scale, and excellent stereoscopy). At the same time this very detailed localized information needs to be extrapolated over much larger areas. For such large-area extrapolation the very detailed resolution required for point samples is neither necessary nor reasonable in terms of cost, acquisition, or processing. Consequently, many timber inventories of the future will combine high-altitude and space photography with very-low-altitude photographs and ground observations along the lines used by Langley and others cited above. Since the use of high-resolution black and white photography is a standard part of forest survey procedures, attention will be confined here mainly to the use of the newer films and to high-altitude and spacecraft studies.

There is now considerable evidence that the most useful single general-purpose film in studying forests is IR aerial Ektachrome. Table 13.5, based on extended study by Colwell and his associates in California and elsewhere, shows that at scales of 1 : 10,000, 1 : 30,000, and 1 : 1,000,000 IR Ektachrome is clearly the most suitable film for a variety of forestry-related studies. This conclusion is supported by numerous other published and unpublished investigations, several of which are reported in Johnson (1969). Anson (1966) reached the same conclusion from studies in North Carolina, as did Douglass et al. (1974) from studies of black spruce in subboreal areas. As already mentioned, Minnich et al. (1969) were so impressed with the superior quality of color IR photography in the mapping of natural plant communities in the San Bernardino Mountains of California that they recommended its exclusive use in preference to other films.

Colwell (1970) notes that:

> . . . The photo-interpretation approach to tree species identification is much the same as that already discussed for crop type identification. However, in two notable respects the problem is more difficult for the forester. . . . A given piece of farmland is usually occupied by a single type of crop, and its photo identification is facilitated by explaining the uniform mass effect that this single type of crop produces on the aerial photographs. In contrast, a given piece of forest land is often occupied by a random mixture of many tree species, making both the photo identification of the component species

0-201-04245-2

0-201-04245-2

**TABLE 13.5**
**Feasibility of Identifying Various Natural Resource Features on Aerial and Space Photography, Based on Studies Conducted at NASA Test Sites in California and Arizona (after Colwell, 1968a, b)**

| | *Photographic Scale and Film-Filter Combination* | | | | | | | | | | | |
|---|---|---|---|---|---|---|---|---|---|---|---|---|
| | *Scale 1:10,000* | | | | *Scale 1:30,000* | | | | *Scale 1:1,000,000* | | | |
| *Type of Resource Feature: Vegetation Resources* | *Pan-12* | *IR-89B* | *Ekta-HF-1* | *IR Ekta 12* | *Pan-12* | *IR-89B* | *Ekta-HF-1* | *IR Ekta 12* | *Pan-12* | *IR-89B* | *Ekta-HF-1* | *IR Ekta 12* |
| Vegetated or not | ++[a] | ++ | ++ | ++ | ++ | ++ | ++ | ++ | + | + | + | ++ |
| Wild or cultivated | ++ | ++ | ++ | ++ | ++ | ++ | ++ | ++ | + | + | + | ++ |
| Fields with crops | ++ | ++ | ++ | ++ | ++ | ++ | ++ | ++ | + | + | ++ | ++ |
| Fallow fields | ++ | ++ | ++ | ++ | ++ | ++ | ++ | ++ | – | + | + | + |
| Mature conifers | ++ | ++ | ++ | ++ | + | + | ++ | ++ | – | + | – | + |
| Mature hardwoods | + | ++ | ++ | ++ | – | + | + | ++ | – | + | – | ++ |
| Open brushfields | ++ | + | ++ | ++ | ++ | + | ++ | ++ | + | + | + | + |
| Riparian vegetation | + | ++ | ++ | ++ | + | + | + | ++ | + | + | + | ++ |
| Aquatic vegetation | + | ++ | ++ | ++ | + | ++ | ++ | ++ | – | + | + | + |
| Meadow or grassland | ++ | ++ | ++ | ++ | + | + | + | ++ | + | + | + | + |
| Sparse or drying vegetation less than 1 m high | – | – | + | ++ | – | – | – | + | – | – | – | + |
| Vegetation not yet in leaf | – | – | + | ++ | –– | –– | + | ++ | –– | –– | –– | –– |
| Herbaceous vegetation in standing water | – | ++ | + | ++ | – | + | + | ++ | – | – | – | + |
| Sprayed brushfields | + | + | ++ | ++ | + | – | ++ | ++ | – | – | – | – |
| Dead or dying vegetation more than 1 m high | – | + | + | ++ | – | – | + | ++ | – | – | + | + |
| Snags or other downed timber | + | –– | ++ | ++ | + | –– | + | ++ | – | – | – | – |
| Burned areas | + | ++ | + | ++ | – | + | + | ++ | + | + | + | ++ |
| Windrowed brush | ++ | ++ | ++ | ++ | ++ | ++ | ++ | ++ | – | – | – | – |

[a] Legend: ++ generally and easily identifiable, but often requiring very close study
– inconsistently identifiable
–– unidentifiable

> and a classification of the timber into meaningful categories most difficult. Most crop plants on a farm have been purposely spaced so as to receive full sunlight. Consequently, these plants are uniformly exposed to the aerial view and exhibit a remarkably uniform appearance on aerial photographs. In contrast, many plants the forester wishes to identify as to species or type from study of aerial photos are in the "forest understory"; since they are obscured to varying degrees by overtopping trees, their appearances also are quite variable. This makes species identification difficult. . . . Topography, stand density, elevation range, and the size and configuration of dominant trees in the stand are more easily interpreted from aerial photos than are species. A forester can often establish a high correlation between these factors and species composition. Nevertheless, a great deal of confirmatory field checking is required.

Considerable evidence is at hand that, despite the higher cost, color and false color photography in forest surveys will repay the extra investment in many instances. Commonly, it is found that photo interpreters make more certain identifications more quickly using color and false color photography than with conventional panchromatic minus-blue film. Thus there is a tradeoff between the increased information and faster laboratory processing of data and the higher acquisition costs and duplication costs in the laboratory of the color film. Although in some instances tests have shown insufficient difference to warrant the higher cost, there is no doubt that the next decade will see a very much greater use of color and false color photography in the study of forest areas. As investigators concerned with the monitoring of forest conditions, logging, and growth become increasingly aware of the multiple options available to them and the diverse constraints of their several environments, experimental aircraft flights will be much more widely used before beginning operational forest photography.

At the Sixth International Symposium on Remote Sensing at the University of Michigan in October 1969 three papers reported the use of multistage sampling techniques (Heller and Wear, 1969; Wert, 1969; Langley, 1969). That by Langley will be reviewed here, and the others will be discussed in Section III.C on disease, insect, and pollution effects on forest stands. The abstract in Langley's paper describes very succinctly the multistage sampling technique and is quoted here in full:

> A new multistage sampling technique, with wide application in earth resources surveys using remote sensing, has been developed and tested on several occasions. For the first time a complete theory is available with the capability of utilizing information from sample imagery of increasingly finer resolution simultaneously. First-stage samples are selected at random from space or aircraft imagery with the probability proportional to a prior prediction as to the relative resource quantity contained in the population units. Increasingly higher resolution imagery is obtained as subsamples within subsequent stages, again with probability proportional to the prediction made at the appropriate stage. Finally, sampling is undertaken on the ground to obtain the necessary ground truth data. These ground measurements are expanded through the system to obtain estimates that are valid over the entire area of interest. The method has been proven to yield unbiased estimates,

0-201-04245-2

> and furthermore the sampling errors have been solely on the accuracy of the prediction made at each stage. Consequently, the estimates are free of the sampling errors customarily rising by virtue of the inherent variations existing in the raw population units. Allocation formulas have been developed for optimally allocating survey funds to minimize the sampling errors for a given fixed cost of the survey.

The importance of Langley's study can hardly be overemphasized, for it formally quantified techniques which had been intuitive and qualitative. It therefore marked a break from a strictly qualitative tradition and was the first of a series of such sampling strategies now being developed in response to various mixes of problems and space and aircraft photography of various resolutions. It represented a breakthrough in finding a rational way for incorporating spacecraft photography as part of an orderly system of resource evaluation. Langley evaluates probability sampling (probability proportional to size), introduced in 1943 by Hanson and Hurwitz, as having considerable advantages over simple sampling with equal probabilities. He discusses the alternatives and points out that he extended the general theory of probability sampling to include multistage sampling with or without stratification, with arbitrary probabilities of selection at all stages. He adds:

> . . . By the same token, there are many ways in which the basic multistage probability sampling design can be applied to a forest inventory using space and aircraft photography. If a large area is being covered, a sample of space photographs may be drawn on which further subsampling is carried out. In probability sampling, the entire population of space photographs must be examined to determine the detection probabilities. If this is not feasible, a simple random sample of space photographs may be drawn with equal probability provided the spatial distribution of the forest population is fairly uniform or random but not clustered. If extensive clustering is present, stratification or cluster sampling should be considered. A technique we used to guard against clustering in one of the mortality surveys was to select random clusters or systematically arranged photographs. The probability of selecting the cluster may be proportional to the number of photos or land area contained in each.

Langley's study employed a sequence of events which finished with obtaining only 10 ground plots totaling 2.4 ha out of the 2.4 million ha covered by the Apollo 9 photographs in Louisiana, Mississippi, and Arkansas. From this sequence an estimate of 63 million gross $m^3$ of timber with an estimated sampling error of 13% was obtained. It began with a partitioning of the space photographs by means of a square grid, into 6.4 × 6.4 km primary sample units. When the 6.4 × 6.4 km grid was laid over the Apollo 9 IR color photographs, pine and hardwood forests were identified. A sample of five 6.4 × 6.4 km squares in the Mississippi River Valley was selected from the Apollo 9 photographs, two from the pine and three from the hardwood areas. The probabilities of selecting the squares were proportional to the area of forest contained in the squares as determined by visual interpretation. The assumption was that on areas of that size (greater than 4000 ha) forest area and timber volume would be highly correlated. In the next sample stage the selected 6.4 × 6.4 km squares were rephotographed at a scale of 1:60,000, using Polaroid

0-201-04245-2

film. This was quickly analyzed on the plane, and while still airborne a strip grid was superimposed on the Polaroid photographs. Out of these, two sample strips were selected for more detailed photography from each square with a probability proportional to predicted volume as determined by examining the Polaroid photographs. As Langley points out, proportionately more samples were taken in areas containing more timber volume than in those containing less timber volume; consequently the survey is weighted toward the more valuable timber lands. These in turn were later subsampled on the ground. Langley's estimating procedure for each step was as follows:

> To obtain timber volume estimates applicable to the entire survey area, the measured tree volumes are expanded back through the sampling formula. In the Apollo 9 study, the estimate for each stratum was
>
> $$v = \frac{1}{m} \sum^{m} \frac{1}{p_i n_i} \sum^{n_i} \frac{1}{p_j} \cdot \frac{A_j}{a_c} \cdot \frac{1}{p_p t_p} \sum^{t_p} \frac{v_k}{p_k}$$
>
> in which
>
> $v_k$ is the measured volume of the $k$th sample tree on a selected ground plot,
>
> $p_k$ is the probability of selecting the $k$th sample tree,
>
> $p_p$ is the probability of selecting the $p$th plot from the cluster of plots delineated on the 1:2000 scale 70-mm photos in a strip,
>
> $p_j$ is the probability of selecting the $j$th sample strip in a selected 6.4 × 6.4 km square area,
>
> $p_i$ is the probability of selecting the $i$th sample square,
>
> $a_c$ is the area covered by the cluster of 1:2000 scale 70-mm photographs within a strip,
>
> $A_j$ is the total area of the $j$th sample strip,
>
> $t_p$ is the number of sample trees measured on the $p$th plot,
>
> $n_i$ is the number of sample strips in the $i$th 6.4 × 6.4 km square, and
>
> $m$ is the number of 6.4 × 6.4 km squares.

This study has been described at more than usual length because it led the way to later space satellite (LANDSAT), aircraft, and ground multistage, very-cost-effective timber inventory (Langley et al., 1974; Nichols et al., 1973).

A multiauthored 1971 book-length publication by Section 25 of the International Union of Forest Research Organizations, entitled *Application of Remote Sensors in Forestry,* is a useful compendium, and though now somewhat dated contains excellent reference lists and thoughtful review articles on a dozen or more topics, ranging from foliage reflection and the use of multiband photography to automated mapping of forest communities. A number of authors noted the value

0-201-04245-2

of very-large-scale photographs in forestry studies. During the last 5 years there has been increasing use of color and false color photographs at scales as large as 1 : 800, as well as of scales of up to 1 : 120,000 from high-altitude aircraft. Krumpe et al. (1971), for example, found in Appalachian mixed forests that crown characteristics were critical for species identification: high-resolution, large-scale photographs were therefore required. They also found that color of trees was an unacceptable basis for automatic mapping because within-species variability in the fall or winter was so great as to largely overlap the color ranges of many other species.

Smith and Chiam (1970) also found that high-resolution, large-scale aerial photographs were essential in refining estimates of stand density, through use of crown width ratios in separating tree stocking from density.

Ulliman and Meyer (1971) examined the feasibility of forest-cover-type interpretation with small- and medium-scale photography (1 : 15,840; 1 : 31,680, 1 : 63,360; 1 : 90,000) and found that, although interpreter accuracy was greater with the larger scales, there was no significant difference in the classification of forest cover types on the various scales. Interpreters working at 1 : 31,680 considered they could do as well as with 1 : 15,840. Those working with the two smaller scales, however, were conscious of loss of information and preferred the larger scales. For certain uses, though, the small-scale photos were acceptable.

## A. Use of Radar in Forest Studies

A number of studies on the effect of natural vegetation on radar returns have been made with multipolarization K-band radar in several climatic and topographic environments in the United States, and in all areas some influence of vegetation on radar returns was observed (Morain and Simonett, 1966; Simonett and Morain, 1967; Peterson et al., 1969). However, not all forest areas show the same degree of influence, and the area at Horsefly Mountain, Oregon, investigated by Simonett, Morain, and Peterson in the studies cited above, shows a more-than-normal influence. The results at Horsefly Mountain are, in a sense, biased because of this effect and should be viewed as an indication of best possible discriminations likely, rather than as being typical.

By analyzing the texture and return in the like- and cross-polarized return at Horsefly Mountain it has proved possible to delineate various types of vegetation. The vegetation types in this area consist of ponderosa pine forest, juniper woodland, white fir forest, hardwood forest, sagebrush on basalt rubble, chaparral shrub, grassland, and recently burned areas which are almost entirely vegetation-free. In comparing the two radar images it is clear that the forest-nonforest boundaries are enhanced on the cross-polarized images through emphasis of the image texture, increased gray-scale contrast, and sharpened acutance of the edge of the forest-sagebrush contact.

Broad vegetation structures are easily separated on this multipolarized imagery. Forest, shrub land, and grassland are readily distinguished. However, within-class

0-201-04245-2

distinctions are difficult to infeasible. For example, white fir forest is impossible to discriminate from ponderosa pine of the same age and height characteristics. On the other hand, variations within ponderosa pines related to the degree to which the forest has been cut over, the presence of relic emergent tall pines, and the size and pattern of regrowth are distinguishable on the radar imagery. Cutover ponderosa pine, for instance, has a very marked radar image texture on the HH image which normally permits its rapid disctinction from surrounding vegetation types. In general, it has a coarse texture and a wider range of point gray-scale values than adjacent vegetation types.

Image texture is generally quite fine in nonforest areas, and inspection shows much the same gray-scale values for the various vegetation types, although there are subtle differences in texture and tone between grasslands and shrub lands.

On like-polarized imagery (HH), burned patches are not easily detected, whereas on cross-polarized imagery they are much more readily seen. Mesic grassland sites give very low radar returns and are easily distinguished. Dry grassland gives returns which may occasionally be confused with sagebrush.

Radar imagery of natural plant communities is very sensitive to variations in their densities. Thus Morain and Simonett (1966) found that variations in sagebrush density were readily detected in the Escalante Valley, Utah, with radar imagery. Similarly, Peterson et al. (1969) showed that color combination multipolarization radar imagery of Horsefly Mountain (the color combinations were produced with the IDECS system, an electronic enhancement system developed at the University of Kansas) was sensitive to density and height differences within the ponderosa pine community as well as in sagebrush communities.

The methods used in these studies for the interpretation of vegetation have included analysis of tone and texture, as well as the use of the IDECS system. The techniques also include tricolor image combinations, the generation of probability density functions, and the employment of a data sensor to enable distinctions to be made between vegetation types which the eye normally sees as much the same. Despite the sophistication of the techniques applied to this study area, machine (IDECS) training and predicting with adjacent aerial photographs has produced very modest results (Peterson et al., 1969)—a not unexpected situation, for the degree of variability in natural environments is considerable.

The results of these studies indicate that radar may have a role in the following areas: (1) preparation of small-scale, regional, or reconnaissance maps of vegetation types, particularly where there are pronounced structural differences between plant communities; (2) delimiting vegetation zones that vary with elevation; (3) tracing burn patterns from previous forest fires; (4) determining the altitude of the timber line; (5) identification of species by inference in areas characterized by monospecific stands; (6) possible discrimination of structural subtypes in cutover, burned, and regrowth forest; (7) deriving estimates of vegetation density in sparsely vegetated areas; and (8) supplementing high-altitude, low-resolution photography in which texture differences related to vegetation are weakly expressed.

0-201-04245-2

Images produced at long wavelengths and high resolutions at a variety of wavelengths are only now beginning to be studied. Laboratory studies have previously suggested that a combination of short and long wavelengths together with multipolarization and high resolution would permit more significant discrimination between natural plant communities, including structural subunits, than has been obtained with single-frequency radars. That this possibility will prove feasible in dense mixed forest environments has been suggested by the observations of Drake and Shuchman (1974) with an X (3 cm)- and L (23 cm)-band, dual-polarization system in forest areas in the southeastern United States.

### B. Forest Studies with Multispectral Scanners

Since early 1972 there has been relatively little work with aircraft multispectral scanners in forestry, as the attention of most workers in remote sensing in forestry has been focused on LANDSAT satellite investigations. Since most forestry investigations with the multispectral scanner did not begin until the late 1960's, this potentially promising area has been somewhat short-circuited, and only a few papers are available, notably those by Driscoll (1974)—a succinct review, Driscoll and Spencer (1972), Smedes et al. (1971), Olson et al. (1971), Weber and Polcyn (1972), Olson (1972), Rohde (1972), and Rohde and Olson (1970, 1972). Roughly half these papers deal with the detection of disease and stress and will be discussed later.

Driscoll (1974) notes that aircraft-borne multispectral scanners "are expensive. They are very temperamental, requiring highly trained electronic engineers and technicians to operate and maintain them. In addition, personnel skilled in computer-assisted analysis techniques are normally required in data processing." He also notes that the only operational programs involving nonphotographic imaging systems are for wildfire detection and surveillance (Hirsch, 1971) and for water pollution detection (Sherz, 1971).

As was noted in Section II on radar mapping of plant communities, the multispectral scanner has been of mixed value: useful in general community discrimination, but less successful in identifying specific subcommunities or individual species. Rohde (1972) found that the multispectral scanner performed consistently better than manual interpretation with black and white IR photography. Conifers were separated from broad-leaved trees as accurately as in any photographic analysis, and separation among conifers was more successful than in any photography. Particularly successful was the discrimination between pine and spruce. Discrimination between hardwoods was of the level of 80% or better. Driscoll and Spencer (1972), on the other hand, found that multispectral data obtained from 900 m in Colorado (in the Manitou Forest) was useful in general community mapping, but did not compare favorably with photography in mapping specific within-community systems. They also found that intensive preprocessing was required to eliminate bidirectional reflectance effects arising from variation in

0-201-04245-2

scanner viewing angle and plant community geometry. They further noted that categories were established in the classification which were *not* vegetation groups, being based on amount of bare soil and litter (useful in forest fire potential analyses). Sunlit and shadowed areas also posed problems in computer processing.

Broadly comparable results were obtained in Yellowstone Park by Smedes et al. (1971), and in the Black Hills (South Dakota) and Georgia.

Broad-scale natural resource and watershed mapping with the multispectral scanner appears more promising, and in fact with LANDSAT data has been demonstrated to be of use in updating maps and detecting changes. Other areas of potential value include the estimation of foliar moisture content (species independent) from the near-infrared reflectance (Olson et al., 1971).

## C. Remote Sensor Studies on Disease, Insects, and Pollution in Forests

Among the more damaging diseases and insects affecting forest stands are root rots, gypsy moth infestation on hardwoods and some pines, woolly aphid infestations on conifers, loss from pine bark beetles, spruce budworm infestation, conifer limb and stem rust, hardwood dieback, and dwarf mistletoe on conifers (Colwell et al., 1970). In a study by Heller and Wear (1969) it was shown that interpretations from color aerial photography are a reliable method to detect and evaluate ponderosa pine (*Pinus ponderosa* Laws) mortality from the Black Hills bark beetle (*Dendroctrones ponderosae* Hopk.). They used a multistage probability sampling design comparable to that employed by Langley. Visual observation, sample strip color photography at a scale of 1 : 8000, and ground study were used to derive estimates in the Black Hills National Forest with a sampling error of only 12%. They found that further improvements appeared possible by correct seasonal timing and by joint use of photography taken at scales of about 1 : 30,000 and 1 : 8000. The most intensive stage of the probability sampling was done on the ground to obtain tree counts, tree volumes, and beetle population trends. Heller and Wear concluded that their suggested sampling method should be accurate and inexpensive and should provide estimates of the timber killed and the trend of the epidemic, thus making it possible to experiment with sprays and other control methods. If a combined multistage sampling and control system can be developed in the Black Hills National Forest, it will set the stage for widespread experimentation with comparable systems, for the pine bark beetle is one of the most destructive agents in coniferous forests. It carries a blue stain fungus on its body as it bores into the cambium layer of the tree. Heller and Wear comment that "the combination of thousands of beetles invading the growing source of the tree and the growth of the fungus through the water-carrying tissues usually spells death to the tree—and during an epidemic, many thousands."

The results of the probability sampling survey among the pine bark infestation were very encouraging indeed. The coefficient of variation was only 12% on the estimates made, a value "well below the 25% coefficient of variation considered

0-201-04245-2

acceptable on similar surveys made in the past. The entire survey including analysis was made in 10 calendar days. The total cost was $3400–about 0.2¢ per acre for the total area of 1.35 million acres" (Heller and Wear, 1969).

One additional conclusion worth mentioning is that during the 2 years of testing the authors could discover no advantage in using IR color film instead of conventional color film. They noted that more commission errors were made with the false color film and that in no case could either film serve as a previsual sensor. When the pine foliage began to change color, even very subtly, the change was as visible on one film as on the other.

Ciesla et al. (1967) investigated color and false color photography for detecting outbreaks of the southern pine beetle. The results were broadly comparable to those found by Heller and Wear, in that no previsual symptoms were available with the color IR photography. The two types were equally satisfactory for detecting diseased trees which had changed color or lost leaves. Since the southern pine beetle (*Dendroctones frondris* Zimm.) is one of the most destructive of the forest insects affecting southern yellow pines, the inability to detect any previsual symptoms with IR Ektachrome is unfortunate but does not invalidate the use of color photography in such surveys. Ciesla et al. concluded that Ektachrome IR aerofilm was superior to Anschrome D-200 film, not because of its ability to obtain previsual symptoms, but rather because of its ability to discriminate between pines and hardwoods and to penetrate haze when used in combination with a Wratten 12 filter. Their identifications were made within the 90% confidence limits.

These early studies have been followed by a spate of publications. Especially valuable is that by Murtha (1972), who developed a guide to air photo interpretation of forest damage in Canada. He gives a thorough review of the use of conventional panchromatic, black and white IR film, and color and false color in the detection of forest damage. In addition he has developed a detailed key for the various forms of damage as displayed on the different types of film with large- (1 : 3000), medium- (1 : 3000–1 : 9000), and small-scale photography (1 : 9000+). To develop the key a very detailed classification of damage types first had to be developed, and thorough testing of the various physical and color changes in foliage and limbs was required. Heavily illustrated with black and white, color, and false color examples, Murtha's publication is an outstanding professional paper and belongs on every forester's bookshelf. Table 13.6 compares, for hardwoods and conifers, various types of normal and damaged trees, limbs, and foliage as seen on color and false color small-scale photography.

Douglass et al. (1974), in a series of careful comparisons, studied forest tree disease detection and vegetation classification in the subboreal forest region, using large- and medium-scale photography (1 : 6000 and 1 : 15,840). They found that (1) conventional color was best in the fall, though at other times of the year color IR was superior; (2) no spectral difference between diseased and healthy trees survived the reduction in scale from 1 : 6000 to 1 : 15,840; (3) optical combination, masking, and density slicing "did not yield any additional data or extend the threshold of detection" of hypoxylon canker in aspen; and (4) infected black

0-201-04245-2

**TABLE 13.6**
**Comparison of Normal Color of an Object with Color in False Color Photograph (after Murtha, 1972)**

| *Object* | *Normal Color* | *False Color Appearance* |
|---|---|---|
| Most hardwood foliage | Green | Magenta |
| Most coniferous foliage | Green | Magenta to dark-blue magenta |
| Young conifers[a] | Green | Magenta |
| Old conifers[a] | Green | Dark-blue magenta |
| Plant foliage | Light green | Pink |
| Sick foliage | Yellow | Mauve |
| Hardwood foliage | Autumn yellow | White |
| Larch foliage | Autumn yellow | Mauve |
| Hardwood foliage | Autumn red | Bright yellow |
| Dead, dry foliage | Straw-yellow to red-brown | Yellow to yellow-green |
| Defoliated branches | Gray to brown-black | Green, blue-green, blue |
| Defoliated branches with exfoliated bark | Whitish | Silvery, silvery green |
| Wet branches with exfoliated bark | Dark gray | Green, blue-green, blue |

[a] As they appear on small-scale photos (1 : 10,000 and smaller).

spruce was detected about equally well with *all* film-filter combinations. These authors state that the "costs of conducting these aerial surveys are low, ranging from $0.005 to $0.019 per acre and are not prohibitive." They also conclude that "aerial photographic surveys are more versatile than operation recorders because they can be conducted over any outbreak area, regardless of the conditions of the terrain."

In addition to the effects of disease and insects, pollution from city smog and industrial emissions is increasingly of concern in portions of the United States. Miller and Parmeter (1963) found significant ozone injury to the foliage of *Pinus ponderosa* in the Los Angeles basin. Wert (1969) notes that "about 100,000 acres of pine forest stands near Los Angeles, California, have been showing varying degrees of damage from oxidant air pollution" and that high-value recreation sites in the mountainous areas surrounding Los Angeles are threatened with the loss of dying pine trees. Similarly, he states that in the southeastern and eastern United States "steam-generating power plants burning high-sulfur content coal emit varying quantities of sulfur dioxide gas. These emissions have been found to be detrimental to the health and vigor of forest trees in the vicinity of these power plants."

Wert used multistage sampling techniques employing ground visits with simultaneously exposed 1 : 8000 and 1 : 1584 scale color photography. The larger-scale photos were exposed in triplets along the flight line at 10-sec intervals, starting at random after beginning the small-scale coverage. Randomly sampled triplets were then ground-inspected to provide a basis for the extrapolation from known diseased

0-201-04245-2

trees across the sample, following broadly the same strategy as used by Langley and by Heller and Wear, reported earlier. The survey showed that about 1,300,000 trees (of the order of 25% of the total forest pine stand) were affected by air pollution. This estimate was obtained at a cost of $3265 or about 3¢ per acre.

The first results of the use of optical mechanical scanner data for the detection of diseased or infested trees are now beginning to become available, and, as might be expected, wavelengths outside the photographic region are proving useful. Thus, for example, Heller et al. (1969) found that 15-channel optical mechanical scanner data collected by the University of Michigan aircraft in May 1968 showed very interesting results. Ten channels of the spectrometer data were processed to produce four color-coded film strips, each representative of one of four tree condition classes. The authors found that individual spectral bands could discriminate old-kill and faded trees, infested with the Black Hills pine beetle. However, all 10 channels were needed to detect previsually any infested trees. Of all the channels inspected, the 2–2.6 $\mu$ channel, in their judgment, "should lead to a significant improvement in identifying nonfaded infested trees if it could be added to the multispectral analysis." If this channel proves operationally useful in this area and other ones, it will make possible much earlier detection and hence the use of more effective control methods.

The suggestion that the 2–2.6 $\mu$ reflected IR channel may prove operationally useful is a welcome conclusion, for, as noted earlier, the photographic region has not poved suitable for previsual detection, nor had the thermal IR in earlier studies by Heller et al. (1967). These authors found that tree crowns within a forest canopy could not be distinguished on thermal imagery as to whether they were healthy or dying. Previsual detection with 8–14 $\mu$ imagery, however, is not beyond the realm of possibility, for Heller et al. found in mid-June in South Dakota that "the foliage temperatures of dying trees were 6 to 8°C higher than healthy at 1000 hours; the difference was slightly less at 1400 hours (4 to 6°C)." Detectors in optical mechanical scanners are capable of discriminating such temperature differences; however, the resolution cell at safe aircraft operating altitudes in hilly regions is usually larger in area than the dying tree crowns, which are surrounded by cooler healthy trees. Heller et al. also found in this study of Black Hills beetle infestation of *Pinus ponderosa* that radiometer readings showed consistently higher temperatures from infested trees which subsequently died than from those which did not.

Olson et al. (1969) have studied the changes in morphology and physiology of trees under stress, using remote sensors. With sugar maple (*Acer saccharum*) they found that remote detection of plant damage from salt accumulation was most likely at wavelengths between "1.43 and 2.19 micrometers." They also found differences in reflectance values for foliage on watered and on unwatered yellow poplar seedlings which agreed with earlier results reported by Weber and Olson (1967): "Leaves which reached full size before watering ceased showed a slight increase in reflectance at most wavelengths when the plants were subject to severe moisture stress, whereas leaves unfolded under severe moisture stress were less reflective at all wavelengths." Weber and Olson concluded that "in all cases the level

0-201-04245-2

of water stress at the time of leaf formation and development appeared to exert a greater influence on foliar reflectance in the near- and mid-IR region than did the level of water stress at the time the reflectance measurements were made."

In a continuation of these stress experiments Rohde and Olson (1970) girdled oak, red maple, and balsam poplar and found that girdled poplars under very severe stress were not detected on thermal IR imagery. Although oak and red maple were detected during the day, this was only because their locations were known. No nighttime imagery made detection possible.

Thus it appears to remain impossible to predict, from previsual thermal symptoms with airborne imagery, the location of low-vigor and even of severely stressed trees. Whether higher resolution and/or ratioing techniques will help remains to be seen.

Olson et al. (1969) also studied reflectance measurements for foliage from red and white pines (*Pinus resinosa* and *Pinus strobus*) infected with the root-and-heart-rotting fungus *Fomes annosus.* These measurements showed that the 0.6–0.66, 0.78–1.78, and 1.2–2.6 μm bands might provide data leading to earlier detection of this disease. However, by 1972 Olson reported that detection was proving more difficult, and that the ratio of bands 1.0–1.4 μm and 1.5–1.8 μm enabled detection of openings in canopies near infected areas (in this respect see also Weber and Wear, 1970). At the same time Olson noted that, in detecting stress in *Fomes*-infected canopies, multiscale photography (1 : 6000, 1 : 12,000, 1 : 24,000, and 1 : 40,000) and multidate (four dates) tests revealed that only the three best interpreters, working with the best images (1 : 6000) at the best dates (June and July, at the beginning of the growing season), did well enough in interpretation accuracy to warrant recommending a photosystem rather than ground surveillance.

The development of spectrosignature indicators of root disease impacts on forest stands has been studied by Weber and Wear (1970), using, *inter alia,* multispectral scanner and especially thermal IR data. While their studies were still at a very early age, they concluded that "remote sensing in the thermal infrared region shows great promise for the early detection for root-rot infected trees." However, by 1970 enough data had been accumulated to invalidate this conclusion with respect to *Poria weirii* infestations of Douglas fir. However, they did find a possible signature indicator—the presence of ringworm-like forest openings, observed on conventional panchromatic photography. Clearly, much more work will be needed to stabilize results.

An attempt was made by Wear to identify areas of known disease on Apollo 9 space photos. Known infestations of *Elytroderma* needle cast disease (epidemic in the ponderosa pine forests northeast of San Diego), Ips beetles (near Prescott, Arizona), and heavy *Lophoderium* needle cast disease near Tularosa, New Mexico, were studied. Although the results were discouraging, Wear concluded, reasonably, that a single attempt with modest-resolution imagery does not constitute an adequate test of the possible detection of disease or insect-infested stands using spacecraft data.

0-201-04245-2

Finally, Weber and Polcyn (1972) used aircraft multispectral scanner data for forests in Michigan in the detection of stress from insect attack and root and leaf infestations. Although they considered the results very encouraging in detecting dead and dying trees (but not for previsual symptoms), with variable results by species and group (hardwoods, firs, and pines), Driscoll (1974) states, in referring to this experiment, that the results were "not sufficiently encouraging to establish an operational procedure now, or else the job can be done more efficiently using high-quality aerial photographs."

0-201-04245-2

Chapter 14

# Cultural and Landscape Interpretation

*Robert W. Peplies*

## I. INTRODUCTION

Traditionally the focus of geography has been to explain and comprehend the varying arrangements, relationships, and processes among phenomena which characterize and differentiate small and large regions of the surface of the earth. The ultimate purpose of the discipline is to understand the entire surface of the earth; this goal is unattainable, but it is an ambition toward which the discipline must strive. Remotely sensed data of the earth from spacecraft and aircraft platforms, although they do not supply all of the information needs of geography, should bring the discipline a step closer to this goal.

The roots of geography are set in man's concern to know about the characteristics of the environment of which he is a principal component. It also receives strength from the human desire to know about new regions—to explore unknown lands and their diverse parts and relationships. Thus geography derives its substance from man's sense of place, a compound of a sense of territoriality, physical direction, and distance (NAS-NCR, 1965). These basic and universal concerns are as paramount to geography in today's world of spacecraft trips to the moon as they were at the time of Columbus's voyages across the ocean.

In the past geographers have satisfied their quest for comprehension of the various segments of the surface of the earth through an empirical, inductive, and idiographic approach. The empirical aspect still is a major methodological feature of geographic research; it has not been replaced. But since the end of World War II some serious and successful efforts have been made to utilize a theoretical-deductive approach to solve spatial problems. A trend in the recent development of the discipline, therefore, has been to incorporate research procedures from which spatial principles and laws are products and from which scientific predictions of phenomena in space can be established.

Associated with theoretical-deductive methodology are rigorous mathematical and statistical constructs which make liberal uses of the technique of modeling and models, often borrowed from other disciplines. Geography, which has existed somewhat apart from some subject-matter fields such as biology, physics, and sociology—fields which have capitalized on the theoretical approach for a long time—has moved into a position which closely approximates these disciplines. The "new approach" in conjunction with the so-called classical procedures of geography has contributed to producing a sound and desirable balance of methodologies which are at the disposal of geographers to attack relevant spatial problems.

0-201-04245-2

*Remote Sensing of Environment*, Joseph Lintz, Jr. and David S. Simonett (eds.) ISBN 0-201-04245-2

Since the beginning of the discipline, geographers have used remote sensing techniques, that is, if one accepts the definition of remote sensing to be *reconnaissance at a distance.* Human sensor systems—eyes, ears, and nose—have been major sources of primary data about the surface of the earth. Over the years, ground techniques for observing and recording the traits of the earth have been refined. But the field approach is a building-block approach. It begins with the world at a large scale, a world which approaches a one-to-one relationship, and comprehension of an area larger than that within the field of vision necessarily demands manipulation of field data through expository, cartographic, and statistical techniques. To be sure, geographers anticipated with great delight the development of air photo interpretation techniques, and they in turn have contributed to the development of this field (Roscoe, 1960). Photographic systems on aircraft platforms have given geographers an opportunity to appreciate and interpret larger areas than they could through conventional field approaches. However, photogeography was concerned largely with object interpretation in relation to purposes of the discipline. Generally, it was not concerned with the physical and psychological (perception) processes and factors affecting the acquisition of photos; these, however, are topics of interest to geographers involved in remote sensing research.

Since the late 1950's the increased interest in remote sensing techniques by the general scientific community has coincided with the development of the theoretical approach in geography. This historical "accident" has done much to bring the profession into direct association with the new technique-field, for some geographers were trained to relate to sophisticated mathematical models and instrumentation designs which are fundamental to sensing things remotely. Because of the concomitant growth of both subfields of geography, several new research avenues have been opened and several old approaches have been revived in geography.

## II. RECENT HISTORY OF REMOTE SENSING IN GEOGRAPHY

Although a few geographers realized the potential importance and significance of remote sensing in geography before 1960, the big impetus of remote sensing to the discipline started in the 1960's. A conference sponsored by the National Aeronautics and Space Administration and the Office of Naval Research was held in Houston in 1965 to evaluate the possible geographic uses of this new technique-field (NAS-NCR, 1966). Although the particular emphasis of the conference was placed on the use of spacecraft imagery in geographic research, some attention was devoted to remote sensing techniques per se. The main work of the conference was accomplished by eight panels, each of which considered a subfield of geographic science as potential consumers of data from spacecraft and remote sensing. The panel subjects were as follows: Energy and Water Budgets; Geomorphology and Glaciology; Plant Cover and Soils; Resource Utilization; Settlement; Population and Historical Geography; Urban Data and Data Systems; Transportation Linkages; and Mapping, Imagery, and Data Processing.

0-201-04245-2

It was generally agreed to by the conferees that, before remote sensing from spacecraft systems could be accomplished, it would be necessary to test the different capabilities of sensor devices from aircraft. As a result several geographers engaged in feasibility studies to test remote sensing systems with respect to the information needs of the several subareas of geography. A test site program was established at various locations throughout the United States. Originally this program was sponsored by the ONR with funds from NASA. Later, the Geographic Applications Program was formed in the U.S. Department of Interior. This office is responsible for coordinating the geographic effort of the national aircraft and spacecraft program for the geographic profession.

Test sites were established at various locations throughout the United States, including Asheville in western North Carolina, Chicago, southern California (Imperial Valley and Los Angeles Basin), and Cascade Glacier in Washington. Flights were made over the test sites at low altitudes—5000 m and less—using largely photographic sensor systems. Color and color infrared film were the main film types used, although a nine-lens multiband system, thermal and radar scan systems, and a four-channel passive microwave radiometer were on-board. Later, higher-flying platforms were acquired by NASA and greater altitudes were achieved over the test sites, more sophisticated instruments were installed, and several subsequent improvements in sensors and associated data systems were made. In addition to the photographic missions, aircraft radar overflights were made over several test sites. These radar flights were made with a Westinghouse aircraft using an APQ-97 K-band multiple-polarization radar set. The radar and photographic imagery became the main research focus during the early stages of the remote sensing program.

As the program progressed and the number of geographic test sites increased, it became obvious that the geographic remote sensing research goals and aims needed to be refined and that the techniques required for this purpose should be consolidated. Thus, in the latter part of 1967 and the beginning of 1968, a large integrated test site mission was planned. This mission, known as NASA Mission 73, focused on southern California and was carried out during May 1968. It involved a hundred or more geographers and other scientists. Particular attention was devoted to the Imperial Valley, the Coachella Valley, the Salton Sea area, the Indio region, the Anza-Borrego Desert, and the Los Angeles Basin, and several aircraft with select instrumentation were utilized.

Although several scientific disciplines were involved, the greater portion of the attention was devoted to spatial problems (geographic) and the testing of the various types of equipment to be used to acquire spatial information about the area. Ground truth data were collected for such factors as urban housing quality, soil conditions, crop types, geomorphic features, land-use systems, and various other landscape components. These data were analyzed in conjunction with the aircraft remote sensor returns, and various reports were published. The significance of Mission 73 with respect to geography was that it sharpened the research direction which the various investigators and program coordinators needed to take. In particular it showed that certain avenues of research could produce significant, valu-

0-201-04245-2

able, and relevant results. Thus, from the Mission 73 program, greater attention was given to certain research projects with respect to remote sensing technology in geography.

More attention was devoted also to small-scale remote sensing returns from high-flying aircraft. In 1969 NASA's Manned Spacecraft Center, Earth Resources Program, acquired an RB-57 jet aircraft. This aircraft, equipped with a cluster of Hasselblad camera systems, two Wild RC-8 cameras, and other photographic systems, gave the geography program the capability of viewing the earth from hyperaltitude locations—on or about 20 km. In addition, since the advent of the program, researchers involved with remote sensing have devoted attention to the analysis of low-resolution spacecraft imagery returned from the various GEMINI and APOLLO missions. Special attention has been given to imagery of the southern parts of the United States, over which these spacecraft missions have flown.

Most of the remote sensing investigations carried out during the last 5 years were directed toward the uses of remote sensing with respect to basic geographic inquiries. An attempt is now being made by the Geographic Applications Program of the U.S. Department of Interior to provide new geographic analysis and prediction services for resource management, environmental quality control, and land planning. This applied geography program was designed so that it would be possible to conduct integrated urban and regional studies by the early 1970's, with special emphasis on areas with high environmental stress and rapid changes in resource use, where remote sensing can be an important source of data. The program implementation was accomplished by a series of research activities aimed at the development of new products and services for specific user groups with needs for geographic information and environmental prediction services. Examples of the products and services initiated by the Geographic Applications Program are the following: an atlas of urban and regional change; a geographic information service (maps, publications, computer graphics, and other media for transmitting information derived from satellite monitoring systems); a geography program input to a "state of the environment" annual report; an assessment of disasters, environmental hazards, and environmental quality as related to land-use development and population concentration; and integrated regional studies, i.e., analysis and prediction of urban and regional development and environmental change. This new program is interacting strongly with recent advances in geographic and environmental research which stress the spatial systems analysis approach necessary for understanding and predicting complex terrestrial relationships at different scales and resolutions.

## III. SCALE AND RESOLUTION

Several themes are especially characteristic of geographic research. Among others are the problems associated with scale and resolution. ["Scale" is defined as the mathematical relationship between objects on the earth and their representative features on maps and/or air photos (or other types of remotely sensed format).

0-201-04245-2

"Resolution" is generally considered to be the ability of the entire remote sensing system (e.g., lens, exposure, processing, and other factors) to resolve and define objects. Admittedly there are some variations and differences of opinion regarding the definition of "resolution" as it relates to the different types of remote sensing devices.] Although most disciplines focus on phenomena of various types at the best scale which is obtainable, geographers also have directed their attention to the problems associated with changes of information transfer which occur with increases or decreases in the scales of observation.

If one conceives of the scale of reality as extending from a world which is infinitely small—the world of electrons, atoms, and molecules—to the astronomer's world of the universe (Broeke, 1957), then the scale concerns of geography occupy a middle position. This position is centered on man and the earth, the macroworld, and ranges from one-to-one relationships to a magnitude of one-to-ten to the eighth power. The latter is a scale which considers the earth as a whole. To reduce the earth and its various relationships and component parts to a comprehensible size, it is necessary to condense reality. This process has been accomplished by various means, including cartographic techniques and, more recently, remote sensing from aircraft and spacecraft platforms.

Haggett (1965) suggests that scale enters into geographical research in three main ways: "in the problem of covering the earth surface; in the problem of linking results obtained at one scale to those obtained at another; and in standardizing information that is available only on mixed series of scales." He reduces these problems simply to the scale-coverage problem, the scale-linkage problem, and the scale-standardization problem. Each of these problem areas is directly related to the uses of remote sensing in geographical research.

With respect to the scale-coverage problem, Haggett mentions that if each researcher in the various geographic professions were responsible for only one segment of the land surface of the earth, each would have approximately 1930 square km for which to account. With an increased population and concomitantly an increased proportion of geographic researchers, the areal responsibility of each would be reduced. But the suggestion is somewhat facetious, for increased populations tend to increase the complexities of geographic problems. Nevertheless, the example does throw some light on the magnitude of the scale-coverage problem.

Solutions to the scale-coverage problem, according to Haggett, are divided into two groups—internal and external. Internal solutions are concerned with sampling procedures which are of indirect concern to the thesis of this chapter. These are particularly significant in the general applications of remote sensing with reference to "ground truth" or fieldwork procedures. External solutions are concerned with improving sources of information about the earth. Haggett notes that there has been a vast increase of information about the surface of the earth, but this increase has not been consistent, for the various segments of the earth surface and the information contrasts are extremely great. He predicts that satellite data from earth orbit, when made available on magnetic tape, will provide continuous recording of certain terrestrial information on a coordinate system and will greatly facilitate the

0-201-04245-2

scale-coverage problem, particularly with respect to the information contrast between the various segments of the earth.

Although scale and resolution are separate and distinct concepts, they are nevertheless related, especially with respect to linkage problems. Stone recognized a scale-related problem with respect to the methodology for handling rural settlement morphology and the settling process (Stone, 1968). He notes that it is important, if not mandatory, to consider these phenomena, as well as others, from multistage positions (i.e., at small, intermediate, and large scales), foı, as McCarthy et al. (1956) have observed, "every change in scale will bring about a statement of a new problem, and there is no basis for assuming that association existing at one scale will exist at another."

Linkage problems are as significant to the applications of remote sensing in geography as they are to geographic methodology. More often than not, the geographic linkage problems in remote sensing, however, are concerned with the resolution of the objects portrayed rather than the scale. Data on a film or imagery format are, in part, a product of the system's ability to detect objects on the surface of the earth. It is possible to change from larger scale to smaller scale or vice versa while retaining the same resolution, for example, if the imagery is enlarged or reduced or if the data bits from the imagery are placed on magnetic tape. Geographic analysis, however, can be no better than that of the smallest bit of data which the system is capable of detecting. Simonett and others at the University of Kansas (Simonett, 1968a) have observed, after carrying out rather intensive research, that a single resolution fitted to a very large area will not convey the same class of information for all regions. The size and spatial distribution of phenomena in an environment setting will substantially influence the nature of the information transferred, particularly if the resolution of the remote sensing system is close to the critical level. Thus their research indicates that information transfer is environmentally modulated.

With respect to the scale-standardization problem, Haggett mentions two problems. These results from the fact that much of the data, particularly census data, are released for areas rather than for points, and these areas vary in size and shape both between countries and within countries. Variations of this type are of little significance when the degree of geographic analysis is highly generalized, but if more refined statistical analysis is desired, the size and shape of the units become of some importance. Through remote sensing from low-, medium-, hyper-, and orbital-altitude platforms, some of the problems which are central and critical to scale standardization can be at least partially resolved. APOLLO 9 spacecraft photographs of northern Alabama are herewith used to illustrate the situation (Figure 14.1). The size and areal extent of forest cover in the various counties and their smaller minor civil divisions are indicated by census data (Table 14.1). However, these census data do not indicate whether the woodlands are concentrated in one or two (or more) large blocks or are distributed as small units among the several landholdings within the county area. Through the use of space and aerial photography this type of inquiry can be answered.

0-201-04245-2

0-201-04245-2

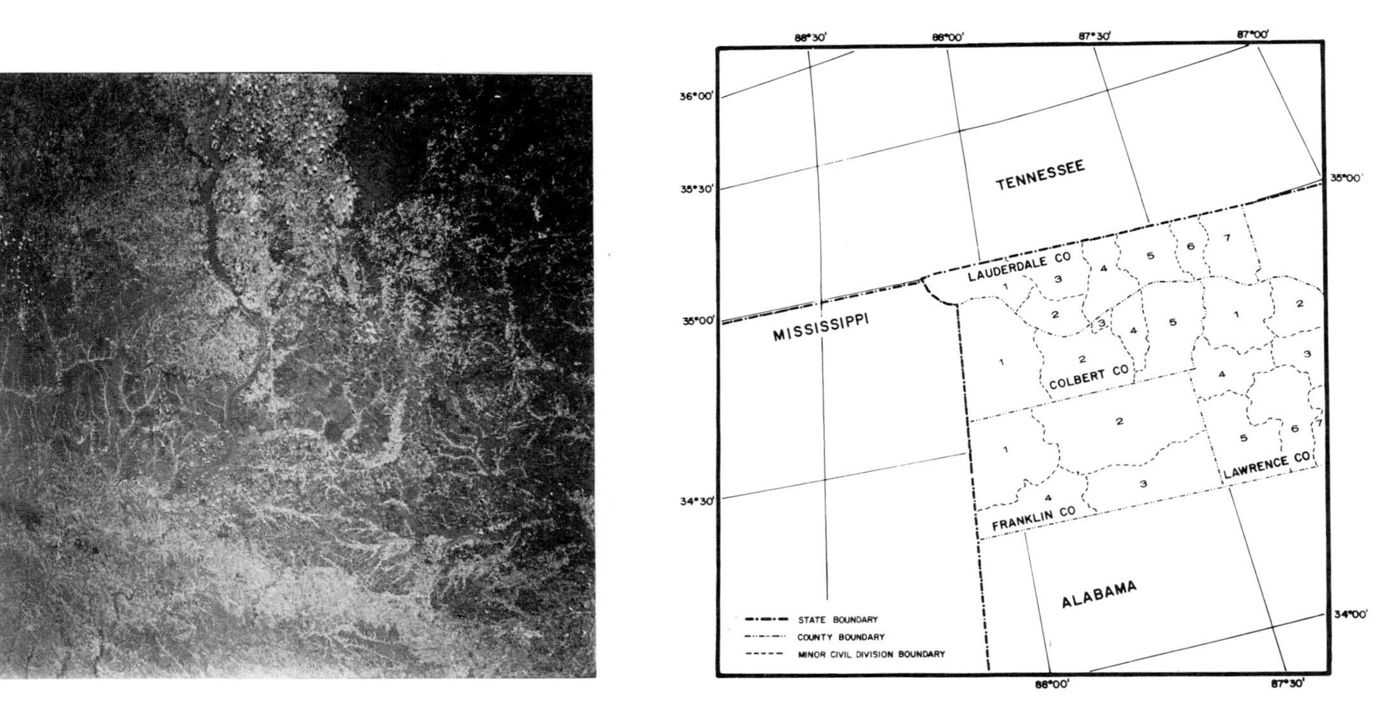


**Fig. 14.1.** Apollo 9 photography of segments of northern Alabama, northeastern Mississippi, and south-central Tennessee and corresponding map of the same area showing county and minor civil division boundary lines. Orientation between the photograph and the map can be accomplished by locating the course of the Tennessee River (in the center of photo), which corresponds to the locations of the boundary lines between Lauderdale, Colbert, and Lawrence counties.

**TABLE 14.1**
**Forest Cover in Colbert, Franklin, Lauderdale, and Lawrence Counties, Alabama[a]**

| *Counties and Minor Civil Divisions* | *Total Land Area (hectares)* | *Land in Farms (hectares)* | *Proportion of Land in Farms (%)* | *Woodland on Farms (hectares)* | *Total Forest Cover[b] (hectares)* |
|---|---|---|---|---|---|
| Colbert | 160,000 | 72,546 | 45.5 | 25,461 | 89,842 |
| 1 | | 19,456 | | 8,525 | |
| 2 | | 20,525 | | 10,193 | |
| 3 | | 90 | | 8 | |
| 4 | | 4,668 | | 1,116 | |
| 5 | | 26,187 | | 5,558 | |
| Franklin | 166,798 | 86,282 | 51.7 | 43,685 | 113,800 |
| 1 | | 18,949 | | 11,641 | |
| 2 | | 35,036 | | 15,063 | |
| 3 | | 19,613 | | 8,979 | |
| 4 | | 12,683 | | 8,002 | |
| Lauderdale | 178,195 | 100,551 | 56.4 | 28,375 | 74,140 |
| 1 | | 9,246 | | 6,461 | |
| 2 | | 18,462 | | 3,734 | |
| 3 | | 12,615 | | 3,806 | |
| 4 | | 15,772 | | 5,902 | |
| 5 | | 12,296 | | 3,464 | |
| 6 | | 10,681 | | 1,942 | |
| 7 | | 21,476 | | 3,468 | |
| Lawrence | 177,677 | 101,498 | 57.1 | 25,700 | 86,078 |
| 1 | | 24,092 | | 4,762 | |
| 2 | | 11,939 | | 4,980 | |
| 3 | | 12,134 | | 2,609 | |
| 4 | | 12,211 | | 3,391 | |
| 5 | | 12,065 | | 2,811 | |
| 6 | | 21,280 | | 5,115 | |
| 7 | | 7,854 | | 2,029 | |

[a] Source: 1964 Census of Agriculture, *Alabama Farm and Farm Characteristic Data, Available by Minor Civil Division,* unpublished data.

[b] Source: Tennessee Valley Authority, Forestry Division, Norris, Tennessee, unpublished data.

## IV. DIMENSIONAL CHANGE: NONPHOTOGRAPHIC AND MULTISPECTRAL ANALYSIS

As already mentioned, geographers have had a restricted view of the surface of the earth which has been determined largely by the human senses of sight, hearing, and smell. Of these three means of perceiving the surface of the earth, sight has been the most important, and therefore geographical reporting and analysis have been limited to the regions of the electromagnetic spectrum which are within the visual

0-201-04245-2

range. Because of remote sensing devices which operate beyond the visual range (e.g., infrared, ultraviolet, microwave), geographers now have the capability of extending their views of terrestrial processes, forms, and relationships. For example, the ability to detect phenomena within the infrared range, particularly the thermal region, permits a view of a "world" which was known to exist but could not be investigated except in a large-scale microclimatological or micrometeorological sense. Nonphotographic sensing from aircraft and spacecraft, therefore, increases the geographer's ability to "see" spatial phenomena beyond his human capabilities, thus providing more information about earth objects and processes with which he is familiar. It also presents whole new realms of inquiries about nonvisual processes and forms which make up the surface of the earth.

Remote sensor signature interpretation, especially with respect to spectral ranges outside of the one with which the geographer is most familiar, requires some understanding of the energy flow profiles—physical factors—associated with the sensor systems. It must also include some knowledge of the operations, characteristics, and capabilities of the remote sensing devices. With respect to the geographical analysis of air photographs it is enough to simply identify objects as they exist on the surface of the earth, because the objects as they are represented on the photos contain a great degree of familiarity to the photo interpreter. In other words, to extract spatial information from an air photo no real concern need be given to the energy flow and target interaction processes. To be sure, remote sensing returns other than photos can be interpreted in the same manner, but if maximum benefits are to be obtained from the imagery, the processes responsible for the image appearance should be understood to some degree (Colwell, 1963). For the geographer, therefore, interpretation of remote sensor imagery is not simply an extension of the procedures used for air photo interpretation. It necessitates at least some basic and elementary understanding of the electromagnetic spectrum, the energy flow profile, and the instrumentation used to detect electromagnetic (and other) force fields (Nunnally, 1969b).

Three approaches can be used for interpreting remote sensor imagery: (1) to consider the image as a direct representation of the earth object or process, that is, to recognize the object or process in terms of some identifying characteristic generally associated with the visual range; (2) to consider the image as a surrogate or proxy of some earth object or process; and (3) to consider the image as a direct representation of an earth object or process which is normally not detectable within the visual range of the electromagnetic spectrum. An example of these technique procedures is illustrated here in relation to the infrared thermal imagery of Brawley, California (Figure 14.2). The light fluffy objects which form the edges of the streets of Brawley can be interpreted as trees, a direct interpretation in relation to the visual range of the spectrum. But because of the environmental conditions (desert-like), they can be considered also as surrogates for sources of water, a substance which cannot be detected directly on this picture. The trees, therefore, serve as surrogate indicators of some type of irrigation process. Finally, because these features appear white on what is apparently a nighttime

0-201-04245-2

**Fig. 14.2.** Photography of infrared scan image of Browley, California. Source: D. S. Lowe, NSF Short Course in the Geographic Applications of Remote Sensing, University of Michigan, 1968.

infrared image, they serve to indicate that a greater degree of heat is present in and around the features than in the surrounding environment. These fluffy objects are the result of the fact that trees serve as heat sinks, a meteorological process invisible to photographic sensors or the human eye but detectable within the thermal IR range.

Another example that can be cited involves the tone and texture patterns which appear on the small-scale Apollo 9 color photograph of northern Alabama, reproduced here in black and white (see Figures 14.1 and 14.4). The stippled regional pattern located at point 5 in Figure 14.4 can be considered an aggregate of the direct representation of an assemblage of small land-use units. Information from fieldwork and air photo analysis has revealed that the pattern consists of small fields, woodlots, and farmstead units. As a surrogate device the textural and tonal pattern serves to indicate a region of small landholdings; furthermore it represents an area in which the majority of inhabitants are part-time farmers who commute to the nearby cities for work in urban occupations.

Images taken in different regions of the electromagnetic spectrum often are strikingly different for the same objects. A remote sensor interpreter can usually obtain more information from two sets of images obtained in different spectral bands than from one image in either band alone (Colwell, 1961). The multispectral concept provides the geographer with additional sources of data for the surface of

0-201-04245-2

the earth. Great difficulties arise, however, when one tries to cross regions, that is, to compare image characteristics in one band (e.g., visual) with those in another (e.g., radar). The resolutions, scales, and geometry of the presentations of the various remote sensor devices often are not alike, and thus spatial characteristics are not comparable. If the interpretation is carried out in relation to a system which records images in the same format for several parts of the spectrum, such as the University of Michigan's multispectral scanning 18-channel system, these difficulties are eliminated. Data from the Hasselblad multicamera system, which uses the same film size as well as the same types of lens systems, are also comparable. Experience has shown, however, that an image interpreter can integrate simultaneously only about three wavelength bands (Olson, 1968). This does not mean that the interpreter ignores the other bands, but in the process of interpreting certain specific or general targets he or she generally relies almost exclusively on three specific types of returns. Also, there is a considerable degree of redundancy of information when a system of more than four channels is used (Simonett, 1969a), though a large number of channels would be needed to obtain the *optimum* four for *each* land-use type.

## V. TEMPORAL CHARACTERISTICS IN GEOGRAPHY

Time, space, and the composition of materials and energy are the three great parameters that must be considered with respect to most scientific problems. The focus of geography is directed primarily toward space in time. Time, therefore, may be considered one of the subsystems of geographic landscapes. With respect to remote sensing, and particularly the interpretation of imagery, a temporal parameter should be taken into consideration (Prentice and Olson, 1968). The temporal parameter considers three characteristics: dynamics, periodicity, and durability. Dynamics involves the type of characteristic changes that occur to objects on the surface of the earth and the corresponding changes of signatures as they appear on imagery. Remote sensing devices, especially on spacecraft platforms, which operate over the same area on a regular basis can be used to note regional dynamic trends and directions of growth. Once changes are recorded, some of the other locational factors responsible for these changes can be identified. This methodological procedure then can be used to predict where changes may occur in other locations.

Some of the man-made and/or environmental changes which occur on the surface of the earth are permanent. The significant changes generally become a matter of historical records. However, some of the dynamics of the earth's surface which can be recorded by remote sensing signatures and images are cyclical—diurnal, seasonal, or annual. Because of the recurrence of these signatures from time to time, it is possible to catalog them in terms of periodicity and durability. The concept of periodicity refers to the moment in time when a remotely sensed signature appears on a sensor return. For example, water appears lighter than its surroundings at night and darker in the daytime on IR thermal imagery. This is a result of the

0-201-04245-2

fact that water is warmer than its surroundings at night and cooler than its surroundings in the daytime; thus, early afternoon and the time just before dawn, respectively, are periodicity points for these two types of signatures. Periodicity, thus, is a result of the interaction of electromagnetic radiation with targets in their environments at certain moments in time. The concept of durability is directly related to that of periodicity. Durability is the quality that causes a signature to have a characteristic appearance for an interval of time Water on an IR thermal image will, in general, appear light from evening to dawn and dark the rest of the day.

Man, for example, does things at certain periods of time (periodicity) and for certain lengths of time (durability). The products of his work appear at certain periods of time and remain for certain lengths of time. The natural environment reacts in a similar manner. Some cultural and natural phenomena of the environment are extremely exacting; they can be expected to appear at particular moments of time and/or remain for certain definite periods of time. Other phenomena per se are irregular in their timing but nevertheless obey statistical rules, for example, the planting, growth, and harvesting of crops (Steiner, 1970). The point can be made, therefore, especially with respect to geographical significance, that a property of remote sensor signatures includes certain temporal traits which are the result of the properties of the target in relation to the surrounding environment. Remote sensor signatures can be expected to change in time, to change at certain periods of time, and to remain for certain intervals of time.

## VI. GEOGRAPHIC RESEARCH IN REMOTE SENSING

As mentioned earlier, geographic research in remote sensing since the beginning of the 1960's has been concerned with testing and calibrating existing remote sensing systems with respect to several basic spatial research inquiries. To a large degree these research directions were determined by the competencies and interests of the investigators who engaged these questions. Out of this melange of research endeavors has developed a program which has specific objectives. The major geographic remote sensing research thrust has been in the areas of land use, urban studies, thematic mapping, transportation studies, geomorphology, energy budget, and the development of geographic remote sensing techniques. Since remote sensing experiments, in general, utilize a significant amount of equipment and personnel, in the form of the use of aircraft and interpretation equipment, it is uncommon to find these projects carried out by an individual researcher; most of them are done by research teams.

## VII. LAND-USE STUDIES

Land-use studies, particularly rural-land-use studies, have been the focus of attention of geographers for a long time. In many ways the study of man-environment interactions, a central theme of geography, receives its unity through land-use

0-201-04245-2

investigations. Although the term "land use" has been defined in different ways by different scientific investigators, it generally refers to the functional traits of an area with respect to the occupations or potential occupations of man and his culture. Land, therefore, may be conceived as the terrestrial space upon which life's economic, social, and physical interactions take place.

The problems concerned with land-use studies in remote sensing are of a threefold nature: (1) procedures for identifying land use from various types of remote sensor imagery, (2) classification and categorization, and (3) mapping land-use traits. These problems are not necessarily unique to remote sensing; on the contrary, they concerned geographers for many years before the advent of this technique. Nunnally and Witmer (1970) found that the most common problems associated with the interpretation of land-use data from remote sensing imagery and other sources are as follows: (1) to reconcile incompatible and inconsistent terminologies, and (2) to develop useful and comparable classification systems. After a thorough literature search, they reached the conclusion that there were not only wide diversities and discrepancies among land-use classifications, but also a great number of different types of land-use classes and definitions, each fitted or created to suit every possible seasonally and geographically distributed land-use type. In their opinion it is not practicable to create a land-use classification system which can be applied to all of the various types and scales of remote sensor imagery. Rather, what is needed, according to Nunnally and Witmer, is the establishment of a standard strategy for approaching the problems of developing land-use classifications. Such a strategy should permit each researcher the latitude to organize and classify land-use data in accordance with his particular needs and at the same time should maintain some degree of compatibility between the data organized by the researcher and those originating in other investigations. Similar conclusions concerning the procedure for land-use mapping from spacecraft imagery were reached by investigators at the University of Kansas (Schwarz et al., 1969). Schwarz and his colleagues showed that very few conventional land-use maps could be constructed from spacecraft photography. They noted that difficulties associated with land-use classification groupings were, in part, the result of technical aspects associated with the photography, but resulted primarily from the internal inconsistencies of existing land-use maps and classifications.Simonett (1969b) has observed that thematic maps (particularly land-use maps) reflect to a great degree the peculiarities of their locale and the idiosyncrasies of their cartographic compilers, in addition to the diversity of their data from different sources.

Nunnally and Witmer (1970) suggest that the strategy for land-use mapping should be based on three tenets: (1) interpret land use from remotely sensed imagery in as great detail as is possible with complete definitions of each category; (2) establish hierarchical categories by grouping similar or related uses; and (3) use a uniform point sampling technique for tabulating data for large areas. They tested their strategy with a group of 20 interpreters, divided into an experimental group and a control group of 10 interpreters each. The experimental group was instructed in the use of the strategy being tested, and the others were told to use any system

0-201-04245-2

or classification that they desired. Subsequent analysis of the results from the two groups produced the conclusions that the experimental group achieved the most accurate and detailed interpretations.

Thrower and Senger (1969) constructed a land-use map of the southwestern United States, following a procedure they developed for interpreting spacecraft photography. They used spacecraft color photography from GEMINI and APOLLO missions as their primary source of data. Land-use information was taken directly from the photographs and placed on acetate map overlays. Transparencies, magnifications, black and white prints (both rectified and unrectified), and a controlled GEMINI mosaic of the Southwest were used as supplemental sources of information to the color prints. These data were transferred onto acetate sheets at a map scale of 1:250,000, in order that the land-use information, being consolidated into more general categories, could be mapped directly onto existing U.S.G.S. 1:250,000 scale topographic maps. Imagery interpretation resulted in the delineation of a number of identifiable land-use phenomena and/or associations. Thrower and Senger noted that a basic dichotomy existed between the phenomena observed on the imagery and grouped the classes of imagery under two headings: (1) those that they could identify with a high degree of certainty; and (2) those that could be delimited only in terms of color and textural areas with no certainty of identification. With respect to both cases, identification and verification procedures were necessary to explain the nature of the phenomena. Published materials that related to the land use in the area were consulted, and large-scale aerial photographs were used to up-date the published materials. Unfortunately, however, most of the available large-scale air photos, 1:20,000, were old. The air photos were useful, however, in examining some of the more durable elements of the environment, particularly physical features. Field checks were made in several locations in the Southwest; in particular, an intensive investigation of a typical dry land region in southern Arizona was carried out.

Thrower and Senger spent considerable effort in developing a land-use classification scheme that would be amenable to the spacecraft photography of the southwestern United States. After examining the existing land classification systems of Dudley Stamp, F. J. Marschner, and Demetrios Christodoulou (Stamp, 1935; Marschner, 1958; Christodoulou, 1959), they constructed a system which represented categories of land use that could be interpreted directly from satellite imagery or infrared by association of the photography with a priori knowledge of man-land relationships of the area. Although the categories were very similar to those of Christodoulou's world land-use survey, several extensive modifications were made. Nine major categories and 10 subcategories were used: (1) transportation—roads, railroads, and airfields; (2) settlements; (3) croplands—irrigated and nonirrigated; (4) arboreal associations—forests (coniferous) and woodland; (5) extractive mining—mines, quarries, and oilfields; (6) grazing lands—unimproved; (7) water bodies—permanent and seasonal; (8) nonproductive land; and (9) noninterpretable.

0-201-04245-2

Rudd and Highsmith (1970) investigated the use of small-scale air photo mosaics as simulators of spacecraft photography for the specific purpose of mapping land use. The project was carried out using a 1:400,000 air photo mosaic of the northwestern part of Oregon. At first, they developed a very general classification of land uses. As the research progressed, these broad categories were subdivided on the basis of detailed examinations. For example, the forest category was separated into continuous well-developed forest, continuous poorly developed forest, cut-over; continuous poorly developed forest, burned-over; and dominantly forest but some agriculture. The agriculture signature was subdivided into cultivated, dominantly cropped, dominantly cultivated but with wooded tracts, and formerly cultivated or cleared (unimproved pasture, abandoned fields, etc.).

Land-use studies, particularly the identification and interpretation of crop types, were made from small-scale color infrared photography of the Imperial Valley, California, from the APOLLO 9 earth-orbiting spacecraft by researchers from the University of California at Riverside (Figure 14.3)* (Johnson et al., 1969). This study was done to show that worldwide agricultural land-use mapping can be accomplished from satellite color IR imagery. Although the researchers were able to define and classify areal units as small as 10 acres, they noted that the interpretation was dependent on prior knowledge of the region; the detection of identifying factors other than crop types in the imagery, direct and surrogates; and the careful selection of dates for the photography. According to the investigators, the principal factor permitting the identification of agricultural crops on small-scale imagery is the ability to detect and predict color variations among crop types. These variations, however, are seasonal; in addition, technical problems are introduced because of film quality control, variabilities at the time of exposure, the storage history of the film before processing, and the processing itself. Thus it is difficult for them to suggest the use of fully automatic interpretation procedures based on catalog programs for different crop types. Through the use of a semiautomatic processing system, however, it is possible to overcome the variations of crop image color. The system that these researchers suggested and investigated is based on location-specific data; that is, each field unit or cell on the image is examined individually to determine its color record and any other data present. By means of this approach, coupled with use of a computer, referral can be made to any previously stored data which exist within the system.

The research mentioned above on land-use characteristics, although based on small-scale imagery, focuses on discrete interpretation of large-scale phenomena—crop types, field patterns, housing, and settlements. Another approach to land-use mapping (and other types) is to delimit areal patterns from images which represent integrated physical and/or cultural features. This approach has been applied to small-scale spacecraft photography, small-scale aerial photography and mosaics, and small-scale imagery other than photography. Nunnally (1969a) in his investigation of radar imagery of the Asheville Basin in western North Carolina has observed that patterns could be delimited on the radar imagery which could be correlated with

*See color insert.

0-201-04245-2

known and observable variations of physical and economic phenomena. He showed that from radar imagery one can characterize a number of relatively distinct regions within the Asheville Basin. The small scale and limited resolution of the radar naturally prohibit interpretation of detailed regional variations, but if further analysis is required one need only refer to the large-scale photographic and other imagery of the area. Through this means one can quickly and accurately delimit significant regions from small-scale imagery at lower cost and in shorter time periods than through conventional methods. Although the technique can be applied to any type of landscape, physical or cultural, it has been subjectively observed that the majority of image characteristics and variations within populated region such as the Asheville Basin are the result of cultural morphological traits—land-use and settlement morphology.

A similar technique applied by MacPhail (1969) for delimiting resource areas in Chile utilizes maps and air photo indices at a scale of 1 : 100,000. Photographic patterns are formed from composite images of photo-identifiable features of the physical and cultural landscape. These photographic patterns, called photomorphic units, are homogeneous in character and recognizable in terms of their areal extent. Each unit image is a complex of a variety of tones which, according to the researcher, are the result of vegetation, crop types, soil moisture conditions, field patterns, drainage characteristics, and settlement morphology. The principal manner in which photomorphic units are described is qualitative rather than quantitative. The investigator is of the opinion, however, that the photomorphic area is a useful tool in regional studies, in that (1) photomorphic units can serve as references for further quantitative analysis; (2) the areas can be correlated and analyzed against existing thematic maps; and (3) they can serve as a basis for delimiting landscape or land type within a regional classification system.

MacPhail's approach has been used to analyze a spacecraft image of northern Alabama taken from APOLLO 9 on March 11, 1969 (Figure 14.1) (Peplies and Wilson, 1970). The research has tentatively shown that photomorphic "regions" can be identified from orbital-altitude photography, and that spacecraft photography has some significant advantages over the air photo map and/or air photo indices. For example, the lighting characteristic across the photography is basically the same; only a few control points are needed to develop a location-specific geographic grid, and tones and texture patterns can be assumed to be the same for the majority of the photographs. By means of the photomorphic technique, 11 different regions were identified in the northern Alabama-Tennessee Valley region (Figure 14.4). These regions were field checked, and the factors responsible for the identification of signatures associated with these areas were noted. These factors were basically the same as those that MacPhail found to be responsible for this photomorphic region. Even though the spacecraft image has certain advantages which cannot be obtained by any other means, some dangers are associated with its use. For example, regional delimitation based on texture and tone traits may not be indicative of significant regional situations, but instead may be related to temporal character-

0-201-04245-2

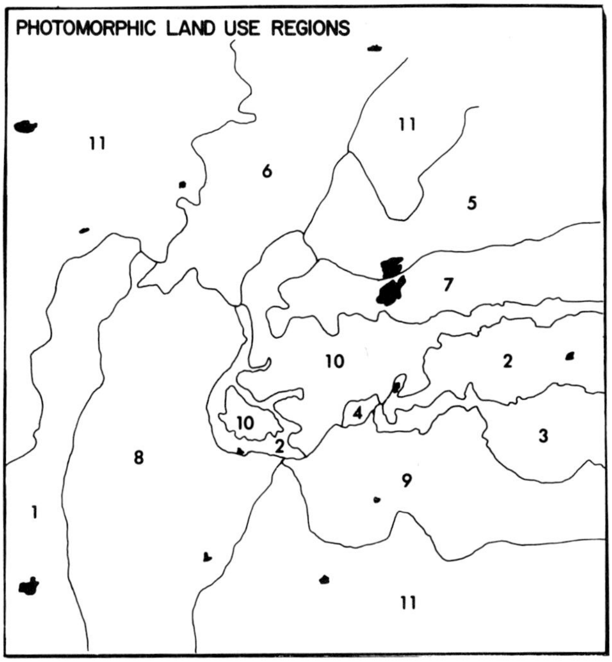

PHOTOMORPHIC LAND USE REGIONS

**Fig. 14.4.** Map of photomorphic regions of northern Alabama. These regions were delimited from APOLLO 9 photo of northern Alabama (See Figure 14.1). The regions were differentiated in terms of variation of textual and tonal properties of the images on the photograph. These assemblages of similar variations seem to be indicative of significant functional and regional traits of these various areas.

istics. To achieve repeatable results, areas under consideration should be thoroughly field-checked, and imagery should be obtained of the area on (at least) a seasonal basis.

In a different approach taken by Simonett et al. (1969b) with respect to using spacecraft imagery of Australia for resource mapping, boundary situations which separated meaningful and significant regions were examined. Through a comparison of boundaries uncovered in the field with those on spacecraft photographs, they showed that the smallest bit of space photographic data was related to qualitative changes in landscape characteristics, and the space boundaries were more readily apparent than the boundaries found on air photos or photomosaics. Discrimination between landscape units and different qualities that appear with similar tones on a photograph is somewhat difficult, however, and requires considerable field work before the imaged properties can be fully understood and appreciated.

One answer to the staggering problem of mapping and defining land use and other phenomena, especially on a global basis, is to use automatic interpretation procedures. Although some attempts were made in this direction before 1970, research had barely been initiated, and the state of the art was naive. The wide

0-201-04245-2

availability and standard format of LANDSAT 1 data substantially changed this picture, as may be seen in Chapter 3, devoted to LANDSAT results. Steiner (1969) presents a methodology for automatic photo identification of rural land-use types. He notes that vegetation (including crop) types can be described in terms of tonal, textural, and stereo height properties, and that it is easy to quantify the tonal characteristics (gray tones on black and white photography and color tones), as well as to extract height information by taking measurements with such instruments as a high-order stereo plotter. Tonal properties can be differentiated through the use of densitometers and colorimeters. Furthermore, according to Steiner, it is also possible to express texture numerically. His procedure for analyzing the information derived from photos involves discriminant analysis, that is, the optimum allocation of individual observations to statistical groups. Although Steiner's research has been limited to using large-scale multiband photography, his opinion that space photography and imagery can also be employed (Steiner, 1970), has been amply demonstrated in many LANDSAT 1 experiments since 1973.

Before the launching of LANDSAT 1, Dr. James Anderson and his co-workers (of the Geographic Applications Program, U.S.G.S.), in response to the need to develop a uniform legend for space and high-altitude aircraft data for land-use mapping, developed a classification which has been very widely applied, with slight modification on the basis of LANDSAT 1 experience. Their study, entitled "A Land Use Classification System for Use with Remote-Sensor Data" (Anderson et al., 1972), is hierarchical, with each minor categoric subdivision nested within the higher levels of classification. Thus there are 9 Level I categories—urban and built-up, agricultural, rangeland, forestland, water, nonforested wetland, barren land, tundra, and permanent snow and ice fields. These 9 Level I categories are further subdivided into 35 Level II categories. For example, within the urban and built-up Level I category they recognize 9 Level II classes—residential, commercial and services, industrial, extractive, major transport routes and areas, institutional, strip and clustered settlement, mixed, and open and other categories. It was anticipated that all Level I categories would be reliably obtained from LANDSAT 1 data, along with some but not all Level II and a few Level III categories.

Although this expectation has generally been fulfilled, there is increasing evidence that each environment has a different *mix* of Level II and Level III land-use categories, which may be reliably derived from LANDSAT data. In essence, as suggested earlier (Simonett et al., 1969a), the information extraction is environmentally modulated. Anderson et al. (1972) essentially anticipated this result and the parallel differences in emphasis for Levels III and IV in their classification of different portions of the United States. They stopped their classification essentially at Level II and left the finer levels for aircraft (as opposed to spacecraft) sensing and for state and other planning agencies to develop rather than attempting a national scheme. At this time it is still undecided whether a general Level III classification will be attempted at the national (U.S.) level.

An extended discussion of land-use studies with LANDSAT 1 data is given in Chapter 3.

0-201-04245-2

## VIII. DETECTION OF URBAN CHARACTERISTICS

As mentioned earlier, the geographic program in remote sensing has narrowed its research concerns to several themes, the greatest emphasis probably being placed on the detection of urban phenomena. The urban segment of the surface of the earth is undoubtedly undergoing the greatest amount of change at the present time. Urbanized areas are in a continual process of dynamic equilibrium which changes direction with each new input from the complex of environmental factors that compose the systems and subsystems of cities. A considerable proportion of the research concerns of geography in recent years has been directed toward understanding the spatial structure and behavior patterns of intra- and interurban areas.

The agglomeration and the behavior of the urban forms which compose an urbanized area may be perceived from a number of points of reference. Perhaps the most relevant "platform" from which to view the complexities of the geographical phenomena which constitute an area as a city or city-region is to consider such an area (or such areas) as a system which is the result of a mélange of social, economic, and physical processes and morphology. An understanding of the nature of urban areas from this viewpoint is dependent on the availability of data which describe the various subsystem processes of urbanized places. These data are needed not only to understand the spatial interrelationships which make up an urban place, but also to meet certain criteria, namely, timeliness, flexibility, compatibility, and reliability (Moore and Wellar, 1969).

Two general avenues of research into the nature of cities have been suggested. On one level is the analysis of cities in relation to each other–central place theory (macrosystems). The other level is concerned with the systems and subsystems which make up the individual city (microsystems). Remote sensing devices are capable of monitoring some of the morphological forms associated with both macro- and micro-urban systems. The scales and resolution levels to be used for the monitoring function, however, vary with the types of phenomena to be detected. The signatures of phenomena vary not only in terms of scales and resolution, but also in terms of temporal characteristics.

The spatial constructs and behavior patterns of cities and the relationships among cities have long been foci of attention from geographers, regional planners, and others. These topics have been studied from both the theoretical and the empirical points of reference (Lösch, 1967; Christaller, 1966; Berry and Pred, 1965). Although some attempts have been made to provide empirical evidence of the central place theories put forth by spatial scientists, these efforts have been made only in terms of conventional data collecting techniques or have been limited by constraints associated with small area studies. Remote sensing from hyperaltitude aircraft and spacecraft systems should provide a means for understanding the spatial interrelationships among cities and/or hinterlands. Understanding these interrelationships can have a profound effect on determining the direction of national and international policy with respect to local, regional, national, and international urban developments.

0-201-04245-2

It is possible to define sets of variables which reflect physical conditions, detectable via remote sensor systems, that permit the definition of the hierarchical positions of cities within local, regional, or larger systems of cities. Horton has suggested that among the needed variables, defined in the literature, are the following: (1) area and/or populations or urban places, (2) distances from a particular urban place to cities or equal or larger size, (3) number of major arterial highway links entering the urban place, (4) number of rail links entering the city, (5) number of all discernible commercial and/or industrial nodes apart from the CBD, and (6) area of the central business district (Frank Horton, University of Iowa, personal communication). These variables should provide a suitable starting point for analysis.

Holz et al. (1969), in a study for the AAG Commission on Geographic Applications of Remote Sensing, examined urban spatial structure based on remote sensing imagery. Specifically, their objective was to determine whether or not such phenomena as urban built-up area, highway and rail linkages, and distance to the nearest larger centers could be incorporated into the formal constructs designed to predict the structural characteristics of urban areas. Forty urban places within the Tennessee River Valley region were selected to test this approach. Data for urban places for two focal time periods—1953 and 1963—were chosen. These data could then be coordinated with the U.S. Census of Business for the same time frames. The following hypotheses based on central place concepts were formulated for testing: (1) that the population of an urban area is positively related to the number of links it has with other urban areas; (2) that the population of an urban area is positively related to the population of the nearest larger urban area; (3) that the population of an urban area is inversely related to the distance of the nearest larger area; and (4) that the population of an urban area is proportional to the observable area of occupied space of such a population. The outline of the built-up urban areas and the number of transportation links were counted on each aerial photograph or groups of photographs of the urban region. The population of the place, number of establishments for retail sales, and population of the nearest larger town or city were recorded from U.S. Census Reports on Business and Population. The distance to the nearest largest town was computed from state highway maps. The hypotheses were then tested, using a stepwise linear regression, and the investigators concluded that physical features visible from the imagery could be used to predict, quite accurately, various spatial characteristics (population, tertiary activities, and others) of urban areas.

High-resolution photography from orbiting spacecraft could be used to determine some of the morphological characteristics associated with urban places. However, research completed at the University of Kansas has shown that it is difficult to extract transportation linkages from color space photographs of the GEMINI and APOLLO scale and resolution types. Space photography of the Dallas-Fort Worth area proved to be insufficient for studying the complete transportation network in the area (Simonett et al., 1969c). To be sure, paved highways and federal interstate systems were easily detected because of their wide surface width,

0-201-04245-2

but as the class of road changes toward a smaller width, a threshold is reached at which only portions of the lengths of some primary and secondary roads are visible. Terrain, water bodies, linearity, contrasting land use, sun angles, and quality control present problems in terms of road identification.

Wellar and Tobler (Wellar, 1969; Tobler, 1968), utilizing a model developed by the Swedish geographer Nordbeck (Nordbeck, 1965) for relating population to areas of urbanization, found independently that spacecraft imagery could be utilized to check empirically some of the theoretical constructs which Nordbeck had developed. Nordbeck found that the built-up area of a settlement is proportional to the population raised to an exponent which is related to the culture (settlement packing) of the area. The theory has been tested experimentally by using data from GEMINI and APOLLO imagery of the south-central United States, as well as some foreign areas, namely, the Nile Delta region of Egypt. This approach seems to have considerable potential for countries where data collection procedures are not well developed.

At the present time remote sensing research on intraurban areas is directed toward such items as dwelling unit counts, housing quality, urban land use, and the limits of urbanization. These microstudies of urban morphology and behavioral processes are leading to the development and maintenance of an urban change detection system. It is nearly impossible to collect through conventional ground techniques the variety of needed social and economic data concerning households, businesses, and firms. Sufficient evidence now exists of a relationship between the physical morphology of the urban environment and the social and economic environment, and some components of the physical environment can be detected via remote sensing techniques. Remotely sensed data coupled with conventionally collected data can provide inputs into a useful and significant urban information system. Such an urban change detection system has been suggested by researchers at the University of Iowa (Dueker and Horton, 1970). A flow diagram of their suggested urban change detection system, which includes remotely sensed data, is illustrated in Figure 14.5. The problem in properly implementing an urban change detection system such as the one here indicated is that it would have to be updated on an annual basis, at least. This requirement obviously is predicated on the state of the art of the various components of the system and, naturally, on funding constraints.

Different levels of scale/resolution requirements among the various spatial phenomena have received attention in urban studies using remote sensors. Dwelling unit counts and housing quality determinations as conventionally carried out demand very-high-resolution systems, approximately in the order of 30 cm (Simonett, 1969b). However, recent studies with SKYLAB photographs show that part of this information can be obtained from much lower-resolution imagery. Similarly, although a high resolution (1 or 2 m) is required to define and delimit urban land use, lesser resolution can also be satisfactory for some purposes. Delimitation of urban versus nonurban areas requires lower resolution, but in relation to that needed for rural-land-use studies the value is still quite high—in the order of 10–20 m.

0-201-04245-2

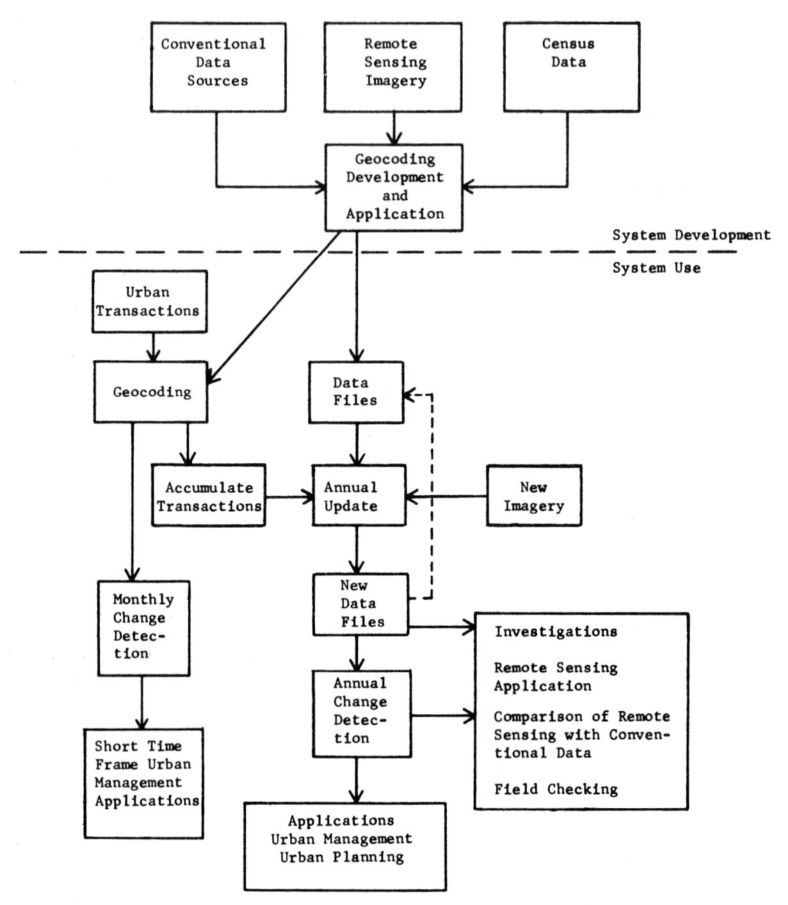


**Fig. 14.5.** Flow diagram of operational urban Change Detection System.

The densities and distribution of population within the various districts of urban places are significant data inputs for planning studies related to land-use changes and transportation. These demographic data are also necessary for human and cultural geographic studies. Although it is nearly impracticable, if not impossible, to count people from even large-scale imagery, it is feasible to estimate the number of dwelling units contained within an urban area (Binsell, 1967). Dwelling units, in turn, can serve as surrogates for population distributions and densities. Of some significance is the fact that "modification" and "calibration" are required of each urban environment (Simonett, 1969b). House or block population densities vary within a cultural region and, naturally, among different cultures.

In light of the urgency for federal, state, and local governments to acquire data on environmental conditions, it seems appropriate that several geographers have examined the use of remote sensing as a technique for collecting information data about urban quality. It seems particularly appropriate that areas of different

0-201-04245-2

housing classes can be differentiated through remote sensing techniques, especially in relation to the increased concern from urban areas which are potentially "volatile" because of social and economic needs. Conventional field enumeration techniques for collecting data on housing quality by federal, private, and municipal agencies are expensive and time consuming. Urban housing quality studies have been accomplished by several investigators using different kinds of remote sensing returns—large-scale multiband photography, small- and large-scale color, and small- and large-scale color infrared—particularly for Chicago and Los Angeles. All of these researchers have noted that remote sensing can provide housing data with a comparable degree of accuracy and a considerable savings of time and cost in relation to conventional techniques. Wellar (1968) adopted a procedure which used multiband aerial photography for determining areas of poor housing quality. Among other things he found that the most consistent perceptible indicators of low-quality housing that can be extracted from remote sensor returns are litter, garbage, wrecked derelict cars, rubbish, lack of landscaping, weeds, and locations which are in juxtaposition with hazards and nuisance factors emanating from such things as industrial plants and warehouses. In a study of Los Angeles, using small-scale color IR photography, Bowden (1968) was able to classify residential areas into four broad categories according to housing quality characteristics, employing such parameters as size and shape of houses, yard conditions, swimming pools, roof colors, numbers of automobiles, parking facilities, street patterns, and utility services.

Several urban land-use mapping studies have been made by urban geographers using large-, medium-, and small-scale aerial photography, as well as orbital photography, color and color IR photography, and high-resolution radar. In general, it has been found that color and color IR aerial photography can serve to map urban land uses with a greater degree of accuracy, in less time, and at lower cost than conventional ground survey techniques. In a study in Asheville, North Carolina, Hannah (1967) analyzed color and false color aerial photography for urban land-use mapping purposes. Field data were compared against the information obtained from photography. Of 1700 land-use parcels identified on photographs, only 26 were misinterpreted. The results obtained with photo interpretation techniques were compared with those of a consultant firm using ground survey techniques. It was estimated that by employing color IR photography the analysis for a town the size of Asheville (60,000 persons) would cost approximately $55,000, whereas to achieve the same results through field survey techniques would require $70,000.

Studies made using radar, particularly the APQ-97 multiple-polarization K-band set, indicated that gross categories of urban land use can be identified from images. Peterson (1968) found that 6 of 33 land-use areas in Lawrence, Kansas, were misinterpreted in the analysis of radar; however, she also noted that with improved application of techniques of analysis this proportion of error would probably be reduced. Moore (1968) used imagery from the same radar set to study lineal urban features, particularly transportation networks and associated major land uses, including manufacturing, commercial residential parks, and other open spaces. He

0-201-04245-2

concluded that the accuracy for land-use identification with such a system is not sufficiently detailed for planning purposes. Simonett (1969b), however, has commented that multifrequency and multipolarization radar with greater resolution than that obtained from the APQ-97 K-band set might considerably improve the prospect of obtaining such urban land-use data.

Geographers have always been interested in boundary interface problems. A significant amount of research attention has been devoted by geographers to "frontiers" of various sorts. In urban geography boundary separation problems are of a number of different types, for example, separation of different types of land use, separation between areas of different incomes, and separation of rural from urban lands. With respect to the last of these, it has been observed that monitoring interfaces between urbanized areas and nonurbanized areas is nearly impossible on a real-time basis using conventional techniques, but that timely data can be obtained from hyperaltitude photography or even space photography.

The location and associated problems of the growing edge of the city can be regarded from different points of reference. From one point it is important to know where new urban developments are taking place and where new urban services and facilities are needed. It is also important to note from the rural-agricultural point of view what lands can be expected to go out of agricultural production. Preliminary investigation of Asheville, North Carolina, seems to indicate that there are zones of change around an urban area that are affected by the growing edge of urbanism. The sequence of land-use changes from an outermost zone of purely agricultural lands to an inner zone of purely urban land use includes the following: a zone of changing agricultural occupance (e.g., dairy farms which become beef operations); a zone of speculation or idle land; a zone of part-time farming operations (i.e., "baby" farm operations); a zone of low-density subdivisions; and finally a zone of high-density subdivisions. Although the sequence of succession of agricultural to urban land use around each city will vary, in general concentric zones of transition exist around most American cities. Most features of these zones are detectable with high-resolution remote sensor systems, particularly photographic, from high-altitude and even orbital imagery.

## IX. THEMATIC MAPPING

Among the areas of remote sensing research in geography, land-use and urban studies have received the lion's share of attention. Nevertheless, varying degrees of interest have been directed toward other geographical topics as well; among these the uses of remote sensing from earth orbit for global and regional thematic mapping purposes have been of principal concern. Although thematic maps focus on some item of interest to geographers—soils, vegetation, land form features, land use—the most significant mapping problems are outside the realm of phenomena per se and are more closely associated with geographic methodology and techniques. One such problem is the resolution level necessary for thematic mapping

0-201-04245-2

purposes. Simonett (1969a) has indicated that a coarse resolution on the order of 90 or 120 m from spacecraft systems may not be acceptable for most problems in thematic mapping and that it may be necessary to use photography of higher and finer resolution (say, 30 m) to obtain data for most thematic mapping problems. This opinion has been advanced by many other investigators as well. Research carried out by a number of investigators, however, indicates that coarse-resolution photography taken at different times of the year can serve as a substitute for fine resolution to discriminate thematic mapping classes (Simonett, 1969b).

## X. ENVOY

In general, geography is without an "instrumentation tradition." Except for a few persons who have investigated large-scale meteorological processes, no significant attempt has been made in the discipline to develop a degree of competency whereby geographers can create the technology—instruments as well as procedures—that can be used to answer spatial questions. Too frequently geographers have had to depend on devices developed by others to manipulate spatial data. Moreover, when geographic data are put into a system which has not been designed to accept spatial data, the output information will always be suspect. Although no great advances have been made to provide for a scientific technology, small but nevertheless significant attempts are being made to design and construct geographic hardware models and appropriate software procedures (Morgan, 1967). To be sure, the move toward the theoretical-deductive approach in geography has encouraged the use of instrumentation techniques in the discipline. Also, some geographers engaged in remote sensing research have contributed to the development of various types of devices, techniques, and procedures useful for investigating spatial problems, such as film-filter combination studies (Pease and Bowden, 1969) and image-processing systems, for example, the IDECS (Image Discrimination Enhancement, Combination and Sampling) developed at the University of Kansas. From these modest beginnings it is expected that some significant advances will be made toward what might be called geographic remote sensing technology.

0-201-04245-2

Chapter 15

# Geologic Applications of Remote Sensing

*Floyd F. Sabins*

## I. INTRODUCTION

Remote sensing methods have great potential application to geologic exploration for fuel and mineral resources and to the study of environmental geology. Although we are still in the early stages of proving and establishing these applications, remote sensing surveys are gradually evolving from research and experimentation into routine operation. One advantage is that most geologists are acquainted with aerial photography interpretation, which is useful background for interpreting other remote sensor imagery.

The reference to black and white aerial photography recalls that this earliest form of remote sensing provided clues for the discovery of many valuable mineral deposits and oil fields. The typical black and white minus-blue photography employed for these studies utilizes only the narrow range of electromagnetic energy from 0.5 to 0.7 $\mu$m. Because so much valuable geologic information has been obtained from this narrow range of the electromagnetic spectrum, it is logical to anticipate comparable quantities of new data from hitherto unused portions of the spectrum, such as the infrared and microwave (radar) regions. If several types of imagery are obtained for an exploration area, the interpretations may reinforce each other and lend greater weight to the geologist's recommendations. In general, remote sensing should be regarded as a reconnaissance method to image large areas and to indicate promising sites for more detailed investigations. Many areas of the United States have already been explored in detail, and there may be little reason to resurvey them with reconnaissance methods unless a specific new application is involved. Many land areas of the world are relatively unexplored, however, and suitable remote sensing reconnaissance is a logical first step in an exploration program.

Perhaps because of the potential usefulness of the technology, geologists tend to expect a great deal of information from remote sensor imagery. The imaging remote sensors, however, do not have any depth penetration but only record the interaction between the earth's *surface* and electromagnetic energy of specified wavelengths. For a subsurface structure or mineral deposit to appear on the imagery there must be a surface manifestation of the target. Here again the experience of aerial photography has demonstrated that such targets may be expressed at the surface by anomalies of drainage, vegetation, color, and fracture pattern that are imaged at the wavelength of visible light. It is reasonable to expect that undetected surface anomalies exist that can best be imaged at other wavelengths.

0-201-04245-2

*Remote Sensing of Environment*, Joseph Lintz, Jr. and David S. Simonett (eds.) ISBN 0-201-04245-2

Geologists can readily learn enough about remote sensing technology to begin the interpretation of imagery. This basic understanding of the technology is essential in order to comprehend the significance of the imagery. Also one must understand the characteristic distortion of various types of imagery and be able to recognize imagery irregularities that could be mistaken for geologic anomalies. One problem in this regard is the shortage of experienced remote sensing instructors and the limited opportunities for training and experience. It is hoped that this publication will be a useful training aid. Most of the examples in this chapter have been used for several years in my evening seminar on remote sensing at the University of Southern California Geology Department and have proved very useful to students. I am grateful to the many individuals and organizations that supplied imagery and information. My employer, the Chevron Oil Field Research Company, has generously supported the effort that went into this report and has approved its publication. The revised manuscript was completed August 9, 1972.

## II. AERIAL PHOTOGRAPHY

### A. General

Geologic interpretation of conventional stereoscopic black and white air photographs has long been recognized as a basic tool for exploration and field mapping. Numerous excellent manuals on photogeology describe the methods and provide examples. One of the best illustrated and least expensive of these is the U.S. Geological Survey publication by Ray (1960). Because conventional aerial photography is so well described, the emphasis here will be on newer types and applications of photography.

Black and white stereoscopic coverage of most areas in the United States is available from government agencies at low cost. Because of the ready availability, modest cost, and proven values, conventional air photos are the standard against which other, newer forms of remote sensing must be judged. Wherever possible in this chapter, therefore, conventional photos have been provided for comparison with radar and thermal infrared imagery.

### B. Low-Sun-Angle Photography

Aerial photography is normally flown between 10:00 A.M. and 3:00 P.M., when the sun is at a high angle and shadows have a minimum extent. These conditions are desirable for topographic mapping, which requires unobscured terrain. For geologic mapping, however, other illumination may be useful. Hackman (1967a) photographed a topographic relief model at various angles of illumination and reached the following conclusions:

1. Photography with the sun 10° or less above the horizon reveals subtle differences in relief and textural pattern that are otherwise unrecognizable. This

0-201-04245-2

photography is more useful in areas of low relief than in areas of high relief, where shadows of large topographic features obscure too much of the area.
2. There is a tradeoff in low-sun-angle photography because tonal differences are less apparent. For this reason it is recommended that both high-angle and low-angle illumination photographs of the same area be employed.
3. If only one set of photos can be taken, those with a sun angle of 20–30° are the most satisfactory.

These concepts of illumination are illustrated with examples of aerial photography and satellite photography (Hackman, 1967a, Figures 6–8). Another example from the desert area of southern California was provided by Howard and Mercado (1970).

One factor that complicates the acquisition of low-sun-angle photos is the relatively limited hours in the morning and evening when the optimum illumination occurs. During these times the illumination values are changing rapidly, and proper camera exposures may be difficult to achieve. Clark (1971) and Lyon et al. (1970) compared low-sun-angle aerial photos with side-looking radar images.

## C. Color and Infrared Color Photography

Earlier chapters of this publication have described and compared these two types of film. Because of the expense of reproduction, published color examples are relatively sparse in contrast to black and white photos. Among the best published examples are those in the American Society of Photogrammetry *Manual of Color Aerial Photography,* edited by Smith (1968).

The eye can discriminate many more hues and shades of color than shades of gray; hence the information content of color photos is much higher than that of conventional black and white photos. Color differences may enable the geologist to recognize different rock units, soils, and zones of mineral alterations. From comparative evaluations it has been demonstrated that color and IR color are superior to black and white for all aspects of photo interpretation (Anson, 1966, 1970).

Infrared color photography was originally developed for camouflage detection. When the capability for early detection of plant disease was recognized (Colwell, 1956), the film came into use for agriculture and forestry surveys. More recently geologists have recognized that IR color photography has great potential application in their field.

Infrared color film is a three-layer color transparency medium in which the blue-imaging layer records reflected green light; the green-imaging layer, red light; and the red-imaging layer, near-IR radiation of wavelength from 0.7 to 0.9 $\mu$m. This portion of the IR spectrum (commonly called "photographic IR") has no thermal significance and comprises reflected solar radiation rather than heat radiated from the target. A yellow filter over the camera lens eliminates blue light and results in greater resolution on IR color film than on conventional color because blue light is preferentially scattered by the atmosphere. For this reason shadows are denser on

0-201-04245-2

IR color than on conventional color, where scattered blue light partially illuminates shadowed areas.

Vegetation is imaged in various shades of red on color IR because the maximum reflectance of plants is in the near-IR region. Plants have a wider range of reflectance variation in the near-IR than in the green regions; hence IR color film discriminates between various species and growth stages more clearly than conventional color (Colwell, 1956). These plant differences may indicate geologic conditions such as soil and rock type, moisture content, and in some cases trace element distribution. Water is typically blue or black on IR color and contrasts with the red-imaging vegetation. This is useful in drainage analysis, particularly in tropical and subtropical regions, where greenish water may be difficult to separate from vegetation on black and white or conventional color photographs. Surface moisture variations are more pronounced on IR color film. The only drawback is the false color aspect; however, geologic interpreters rapidly compensate for this and have no adjustment problems.

Geologists are coming to realize that for their purposes color IR is superior to black and white and color film. Paarma and Talvitie (1968) interpreted high-altitude IR color photography of an area in northern Finland that is covered by lakes, bogs, and glacial moraines. Fractures which control the occurrence of iron ore deposits are difficult to map in the field or on black and white photos. On the IR color photos, however, the fractures are indicated by moisture and vegetation differences and were confirmed by aeromagnetic surveys. Pressman (1968) shows simultaneous color and IR color photos of the Powder River Basin of Wyoming, flown at 750 and 3330 m. The low-altitude photos have similar quantities of data for geologic mapping. At the higher altitude, however, the IR color is superior because atmospheric scattering has degraded the resolution and contrast of the conventional color photos.

Suggestions for exposing IR color film have been provided by Fritz (1967), who also recommends color compensation filters to modify the color balance and enhance specific subjects. Pease and Bowden (1969) recommend auxiliary filters to improve the quality of IR color photos taken at high altitude or with pronounced atmospheric haze. The originals of IR color photos are in the form of transparencies, generally in rolls 70 mm wide. To produce film for direct stereoscopic viewing on a light table, the camera must be mounted in the aircraft so that the film-advance direction is parallel with the flight direction. Although color prints can be made from the transparencies, the cost is appreciable and the quality of the prints is generally inferior to that of the transparencies. One advantage of prints is that they can be used in the field, something which is difficult with transparencies. Meyer and Malkin (1969) recommend acetate overlays for annotating the transparencies; also the color contrast can be enhanced by suitable color filters placed over the lenses of the viewing stereoscope.

Unlike radar and IR scanner imagery, which can be flown by only a few operators, color and IR color photography can be flown by any number of competent aerial photography contractors throughout the United States. Sabins (1973) de-

0-201-04245-2

scribes a camera mount and reticle whereby a 70-mm reflex camera can be used in a light aircraft to obtain high-quality stereophotos. Surprisingly, the cost of IR color transparencies is competitive with that of new black and white photography. Another benefit is that interpretation is appreciably faster for IR color photography than for other forms. For all these reasons IR color photography is a significant remote sensing tool for the geologist. It should not be overlooked in the current enthusiasm for the newer forms of remote sensing.

## D. Multispectral Photography and Imagery

The simplest form of multispectral (also called multiband) data acquisition utilizes two or more cameras with different film and filter combinations that operate simultaneously. Ulliman et al. (1970) described a system of four 70-mm cameras with a common shutter release control to acquire simultaneous photography from aircraft. Another approach is to use a single camera body with multiple lenses and to subdivide the photographic spectrum into narrow wavelength regions with appropriate filters and black and white film. Molineux (1965) described such a camera with nine lenses and illustrated examples of the photography. It is difficult, however, for the interpreter to compare nine photos of the same target. Black and white multispectral photos of some geologic sites were shown by Cronin (1967). One approach to the interpretation problem is to project the narrow-band negatives with light sources of different colors to form false color images, which may enhance particular targets (Yost and Wenderoth, 1968; Orr, 1968). This has the advantage of combining up to four narrow spectral regions on a single presentation.

Multispectral cameras are limited to the spectral region that can be photographed directly on film. Holter (1970) described an optical-mechanical scanner for recording on magnetic tape 12 narrow spectral bands of imagery extending from the near-UV to the thermal IR regions. Smedes et al. (1971) prepared computerized maps of the ratios between various combinations of spectral bands of Yellowstone Park and were able to differentiate broad terrain types such as forest, bog, bedrock, talus, and water. Watson and Rowan (1971) computer-processed multispectral scanner data to discriminate rock types in the Arbuckle Mountains, Oklahoma. They also provided an objective discussion of the problems and limitations of this approach. Multispectral photography and imagery are still in the experimental stage but may have great potential for geologic applications.

## E. Satellite Photography

This discussion is based on earth photographs from space and excludes forms such as NIMBUS and TIROS imagery, which may show gross terrain features but generally is unsuitable for geologic interpretation. Merifield and Rammelkamp (1966) illustrated examples of TIROS imagery. Aside from early rocket photographs, space photography has been provided by the NASA MERCURY, GEMINI, and APOLLO flights, which were summarized and illustrated by Underwood (1968) and Lowman

0-201-04245-2

(1968, 1969). Most space photography has been in 70-mm format with color film. Table 15.1 lists numerous published examples of geologic applications of space photography. Reproductions of GEMINI and APOLLO photography are available from the Technology Applications Center, University of New Mexico at Albuquerque, New Mexico, which provides catalogs and price lists.

For some parts of the world space photography is the only coverage that is available. The main advantage for geology is that space photography provides a small-scale synoptic view of broad areas under uniform lighting conditions. Regional relationships and large but subtle structural features may be more readily apparent. This aspect is emphasized in the examples interpreted by Trollinger (1968) and Abdel-Gawad (1969).

In addition to extensive conventional color photography, APOLLO 9 collected multispectral photos with four 70-mm cameras in March 1969. The film and filter combinations employed are as follows:

**APOLLO 9 Multispectral Photography (SO 65)**

| *Band (NASA Designation)* | *Film and Filter* | *Mean Wavelength of Sensitivity* | *Nominal Band Pass* |
|---|---|---|---|
| A | Infrared color Wratten 15 | Green, red, and infrared | Total sensitivity of all dye layers 510–900 nm |
| B | Panchromatic Wratten 58B | 525 nm Green | 460–610 nm |
| C | Infrared Wratten 89B | 800 nm Infrared | 700–900 nm |
| D | Panchromatic Wratten 25A | 645 nm Red | 580–700 nm |

I have examined several sets of this photography to evaluate the geologic usefulness of the different spectral regions. Except for poorly exposed frames, the IR color is superior, as would be expected because of its broader spectral coverage. Surprisingly, no one of the narrow-band images is consistently superior even in the small sample considered. The areas examined ranged from desert to humid, and it may be that the type of terrain influences the selection of optimum wavelength.

Powell et al. (1970) analyzed APOLLO 9 multispectral photos of an area in east-central Alabama and detected previously unrecognized linear features cutting across the regional geologic strike. The fact that some of the lineaments are the sites of anomalous hydrologic conditions suggests that the lineaments may be major fracture zones. The authors compared the four types of multispectral photos for detecting various features and indicated that IR color was most useful and the blue-green band was least useful. The choice between the red or the IR narrow-band photos apparently depended on the type of feature being investigated.

0-201-04245-2

**TABLE 15.1**
**Geologic Interpretations of pre-LANDSAT Space Photography**

| *Locality* | *Geologic Information from Space Photos* | *Space Photos Utilized* | *Reference* |
|---|---|---|---|
| West Texas, southeast New Mexico | Regional structural trends in Permian Basin, drainage anomalies, unexplained tonal anomalies. | GEMINI IV; APOLLO 6 | Trollinger (1968) |
| East-central Alabama | Previously unrecognized lineaments are apparent. Anomalous hydrologic conditions along the lineaments suggest that they are fractures with little or no offset. | APOLLO 9 | Powell, Copeland, and Drahouzal (1970) |
| South Yemen | Excellent drainage detail. Differentiation of some rock types. Some major structural features. Foliation trends of metamorphic rocks in several areas. ". . . Should be regarded as supplementary to air photographs." | GEMINI IV, VII | Greenwood (1969) |
| Western Peru | Previously unrecognized lineaments are apparent. One lineament several hundred miles long is interpreted as a major fault. A variety of geologic features are identifiable in this poorly mapped, remote area. | GEMINI IX | MacKallor (1968) |
| Red Sea area | Offsets of two sets of prominent shear zones on opposite sides of Red Sea indicate major lateral displacement of the sea. The offsets were first noted on GEMINI photos and then substantiated from the literature. GEMINI photos show other fissure systems, some of which are related to oil fields along Gulf of Suez. | GEMINI IV, XI | Abdel-Gawad (1969) |
| Mali and Upper Volta | Space photo pairs were examined stereoscopically. Major lithologic contacts, volcanic features, fracture zones, and linears were distinguished. Notes that faults and folds less than 5 miles in extent tend to blend in with regional structural pattern. | APOLLO 6 | Muhm (1969) |

0-201-04245-2

0-201-04245-2

| | | | |
|---|---|---|---|
| Numerous areas | Large-format, high-quality color reproductions with various brief descriptions of topographic and geologic features. | Various GEMINI flights | Lowman (1968) |
| Pakistan | Space photograph emphasizes regional continuity of faults and folds that could be easily overlooked on conventional air photos. Some stratigraphic relationships can be provisionally inferred. | GEMINI V | Hemphill and Danil-chik (1968) |
| Southeast Arizona, southern California, Central America | Oblique aerial photos are compared with space photos to illustrate the advantages of both types of photography. The broad coverage provided by space photography is emphasized. | GEMINI | Colwell (1968) |
| Southern New Mexico, west Texas | In southern New Mexico, major faults are shown on space photos because of aligned valley segments, vegetation, and vertical offset of strata. Compares space and aerial photos for geologic purposes. | APOLLO 6, 7, 9 | Amsbury (1969) |
| Amadeus Basin, central Australia | Salt diapir, thrust fault, anticlinal ranges, unconformity, and a possible cryptovolcanic feature are evident, together with regional relationships. | GEMINI V | MacNaughton and Huckaba (1966) |
| Numerous areas | Large collection of color photos with brief geologic and geographic descriptions. Reproduction quality is, unfortunately, poor. | GEMINI III, IV, V | NASA (1967) |
| Numerous areas | Excellent large-format color reproductions, with brief descriptions. | GEMINI IV, V, VII XII, XIII; APOLLO 6 | Underwood (1968) |
| Arabia | Compares geologic interpretations of space photos with published data. There is good correlation, and in some areas the photos suggest revisions of publications. | GEMINI IV | van der Meer Mohr (1969) |
| Numerous areas | Geologic and sedimentologic interpretations are presented for numerous examples of space photography. | APOLLO 6; most GEMINI flights | Wobber (1969) |
| Africa, Pakistan, southwestern U.S. | In addition to photo interpretation examples there is a useful discussion of practical aspects of space photography. | GEMINI IV, VII, XI. XII | Lowman (1969) |

To provide even broader coverage the U.S. Geological Survey has prepared mosaics of western Peru and the southwestern United States from GEMINI photos. A second mosaic of the Southwest was prepared from APOLLO 6 photos (U.S. Geological Survey, 1970).

The unmanned LANDSAT launched in a near-polar orbit, provides repetitive telemetered multispectral photos that can be reconstituted into color IR. The manned SKYLAB earth-orbiting satellite has acquired additional multispectral photography and imagery. At the time this was written, imagery was not yet available from the satellites.

## III. RADAR IMAGERY

### A. General

Radar imagery is a form of remote sensing that is particularly useful for surface structural mapping, especially in areas where aerial photography is handicapped by poor weather or dense vegetation. Radar systems transmit pulses of microwave energy and measure the energy returned (backscattered) from the terrain. Radar is an "active" form of remote sensing because it supplies its own energy. Photography and thermal IR, on the other hand, are "passive" because they detect energy reflected or radiated from terrain, rather than providing a source of energy. Radar is classified on the basis of wavelength into K band (0.75–2.40 cm) and X band (2.40–3.75 cm), which are the commonly used imaging systems. There are longer wavelength bands (C, S, L, P), but these are not generally available as imaging systems.

Everyone is familiar with the rotating radar antennas at airports and defense installations that provide a 360° sweep of the surroundings. For airborne terrain mapping the antenna is rigidly mounted below the aircraft fuselage to transmit narrow pulses of energy at a right angle to the aircraft flight direction (Figure 15.1). This side-looking airborne radar system gives rise to the acronym SLAR.

### B. Terrain Response

The interaction between terrain and radar energy is complex, but for geologic interpretations the two most important terrain aspects are microrelief (roughness) and topography. For short-wavelength K- and X-band radar, surfaces with microrelief of less than a few millimeters are "smooth" and serve as specular (mirror-like) reflectors to radar energy. Playas, concrete, and quiet water are examples of surfaces that are specular reflectors. Surfaces with microrelief greater than a few millimeters are diffuse reflectors and backscatter the radar energy in various directions. The direction and intensity of backscattering are functions of roughness and orientation. Vegetation, lava flows, and gravel deposits are examples of diffuse surfaces. The importance of roughness has been analyzed by Rouse et al. (1966) and shown

0-201-04245-2

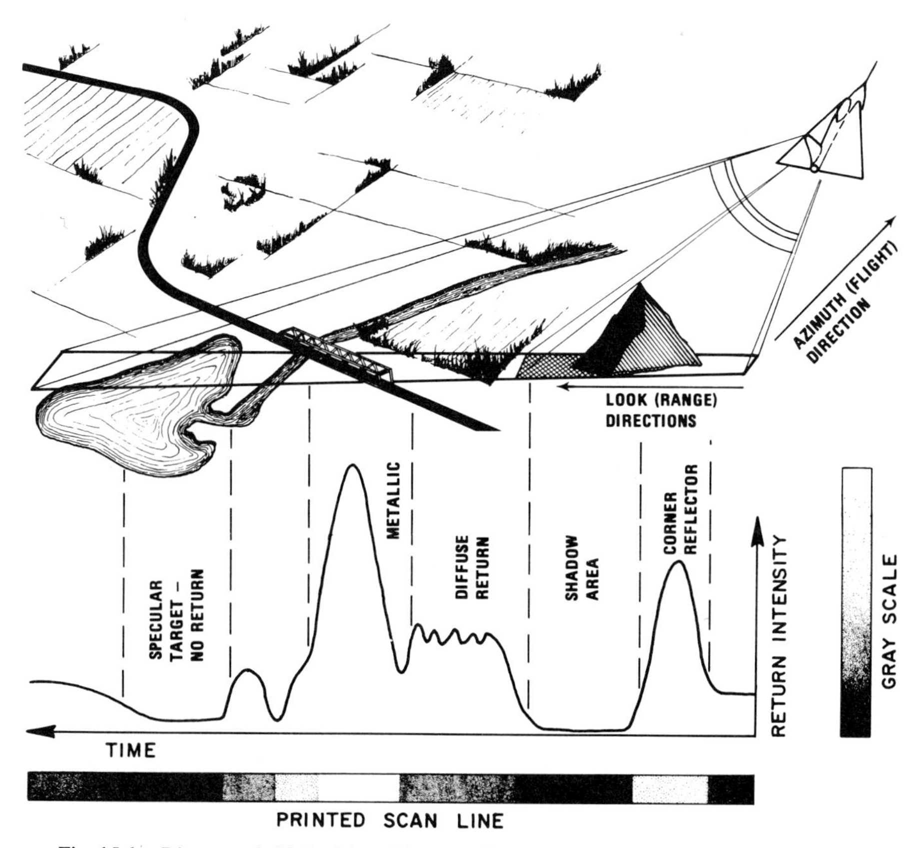


**Fig. 15.1.** Diagram of side-looking airborne radar mapper. From Sabins (1973, Figure 7).

diagrammatically by Rydstrom (1967). MacDonald and Waite (1971a) demonstrated that soil moisture variations may also influence radar imagery.

Topographic features such as ridges, valleys, and escarpments are the other major terrain element that influences the radar response. Figure 15.1 illustrates a single pulse of energy transmitted from the aircraft at a right angle to the flight direction. The narrow strip of terrain that is "illuminated" by the pulse reflects the energy back to the antenna, where the time and intensity of the returns are recorded. Variations in the two-way travel time determine the positions of the reflecting features, as shown on the plot of return intensity as a function of time in Figure 15.1. Variations in return intensity (or backscatter) are converted to a gray scale with white for highest-intensity and black for lowest-intensity return, and shades of gray for intermediate values. In Figure 15.1 the mountain face oriented toward the antenna is a large-scale corner reflector with a strong return. There is a shadow zone behind the mountain that is not illuminated and appears black on the image. Vegetation is a diffuse target that backscatters intermediate amounts of

0-201-04245-2

energy with variations depending on density, leaf and branch texture, and moisture content. Metallic objects such as bridges and railroads are conspicuous by their very bright returns. The lake is a specular surface that reflects most of the energy, the angle of reflection being equal and opposite to the angle of incidence; hence there is no return to the antenna. Details on radar imaging are given by Moore and Thomann (1971). A useful set of radar interpretation keys based on tone, texture, and topography is provided by Barr and Miles (1970).

The apparent similarity between radar imagery and low-sun-angle vertical aerial photography (Lyon et al., 1970; Clark, 1971) has already been noted. Figure 15.2 compares low-sun-angle photography with radar imagery of an area near San Clemente along the coast of southern California. Length and orientation of shadows are similar on both views, but the fault is more clearly expressed on the radar image. This comparison also illustrates the poorer spatial resolution of radar, which, however, does not hamper recognition of geologic structure. In addition to radar shadows, the following can also be noted on Figure 15.2*B:* bright signatures of ridges and scarps facing the antenna; dark specular signature of water with bright returns from surf zone; diffuse signature of vegetated slopes and valleys; bright returns from urban developments along coast; dark specular signature of a concrete freeway parallel with the coast.

In forested terrain, such as the Panama and Venezuela examples in Sections III.I and III.J, radar imagery commonly exhibits fractures and lineaments that are not apparent on aerial photography. This may lead to the mistaken conclusion that radar is penetrating vegetation or "stripping away the vegetation cover." Actually, the relatively short-wavelength energy employed in currently available systems interacts with vegetation and is backscattered by it. The following is my attempt at a more acceptable explanation for the higher content of geologic information on radar imagery of forested terrain.

Because of the lower spatial resolution (30–50 ft spot size) and small scale of radar imagery, individual trees are not resolved, thus diminishing the distracting details of vegetation cover. Just as some light penetrates even the densest forest canopy, so must some radar energy penetrate. Geologic features, which typically have dimensions of thousands of feet and strong preferred orientation, will produce a stronger return than the small, randomly oriented vegetation elements. The large spot size of the radar beam may be likened to a moving average filter which reduces small-scale random "noise," thereby enhancing larger "signals."

Because of the oblique "illumination" and scanning nature of the imagery, radar flight lines should be carefully oriented in relation to the structural "grain" of the survey area. As shown in Figure 15.3, structure lineaments oriented normal to the radar look direction are accentuated, but those parallel with the look direction are suppressed. Where there are two or more structure trends, the radar look direction may be oriented at an intermediate azimuth to all trends.

The angle of incidence, measured in the vertical plane, between the radar beam and the terrain is also important because it controls "illumination" and shadow

0-201-04245-2

A. AERIAL PHOTOMOSAIC FLOWN DEC. 1953. FAULT INDICATED BY ARROWS.

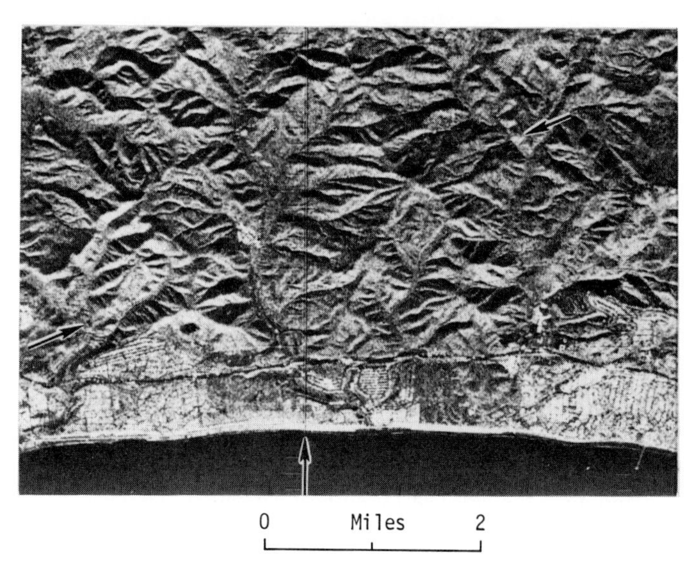

0 Miles 2

B. RADAR IMAGE FLOWN NOV. 1965. OBLIQUE RADAR ILLUMINATION CAUSES HIGHLIGHTS AND SHADOWS WHICH ENHANCE FAULT TRACE. LOOK DIRECTION SHOWN BY ARROW.

**Fig. 15.2.** Comparison of radar image and aerial photomosaic of fault near San Clemente, southern California. From Sabins (1973, Figure 8).

0-201-04245-2

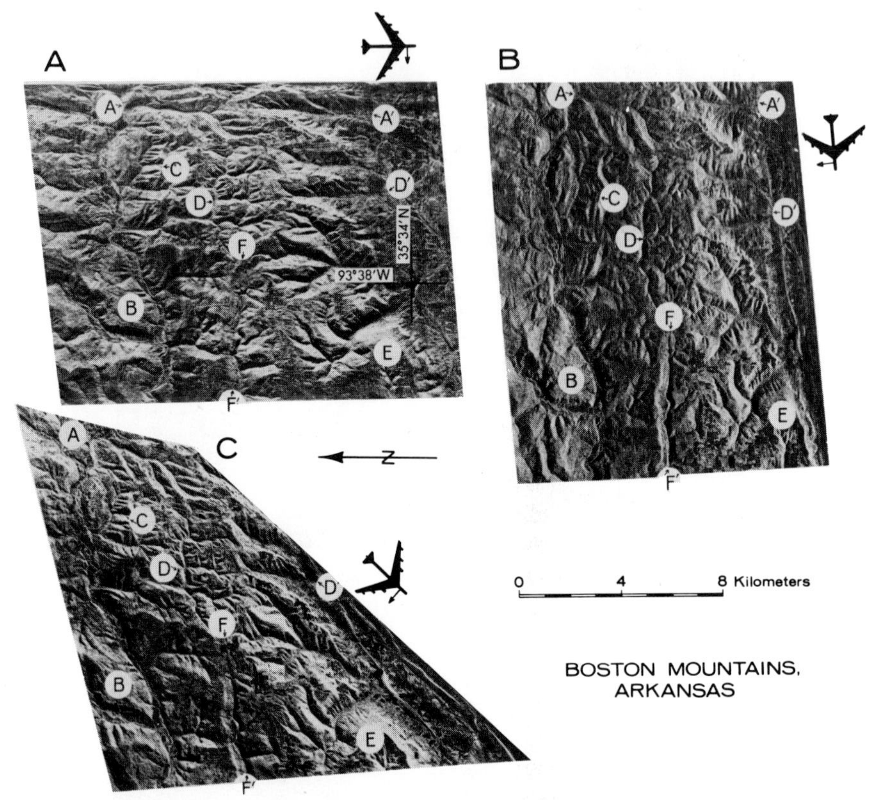


**Fig. 15.3.** Influence of radar look direction on detection of linears, Boston Mountains, Arkansas. In *A* the north-south linears are accentuated, whereas in *B* the east-west set is prominent. In *C* the look direction is oblique to both sets, which are equally apparent. From MacDonald (1969, Figure 27).

length. MacDonald and Waite (1971b) analyzed the optimum depression angles for five categories of terrain.

## C. Background and Status

The history of radar geology has been summarized by MacDonald (1969, p. 6), who notes that beginning in the late 1940's military personnel extracted terrain information from photographs of airborne radar scope displays. Scheps (1962) considers that Dr. Siegel at the University of Michigan carried out the first real radar geology in the late 1950's by estimating the composition and roughness of the lunar surface from radar responses. Geologic interpretations of radar scope photographs were published by Cameron (1964), but with the development of side-looking radar imagery the scope presentations are no longer used for geologic investigations. Among the earliest side-looking radar imagery to be declassified and published was the

0-201-04245-2

AN/APQ-56 system, which was flown by the U.S. Air Force in the 1950's and is now obsolete. Two antennas were employed to produce an image strip along either side of the flight path with a narrow blank strip directly below the aircraft. The resulting low-resolution, small-scale imagery provided wide regional coverage and was used for Arctic sea ice studies (Bradie, 1967) and for geologic investigations (Holmes, 1967; Wing and Dellwig, 1970). The AN/APQ-69 was another dual-antenna radar mapper that produced low-resolution, small-scale imagery. Dellwig et al. (1966) demonstrated its potential for recognizing regional structural lineaments in the Boston Mountains, Arkansas. Table 15.2 summarizes published geologic interpretations of radar imagery.

Few radar mappers are presently unclassified and available for civilian use; one is the AN/APQ-97 system operated by Westinghouse Electric Corporation, which produced much of the imagery in this chapter. In 1965 and 1966 Westinghouse, under contract to the U.S. Geological Survey and the Earth Resources Survey Program of NASA and through the cooperation of the U.S. Army Electronic Command, imaged a number of flight lines in the United States covering several hundred thousands square miles (Kover, 1968). Like other NASA remote sensing imagery, the radar strips are available for viewing at the Earth Resources Research Data Facility, Houston. In 1969 Westinghouse mounted the AN/APQ-97 mapper on a DC-6B aircraft and has since flown imagery for commercial and government contracts in a number of countries. Radar imagery is somewhat more expensive than aerial photography, but for many geologic problems the additional cost is justified.

The AN/APQ-97 mapper provides imagery in a like-polarized and a cross-polarized mode. The transmitted pulse is generally horizontally polarized, and the strongest terrain returns are likewise horizontally polarized. These are recorded as the parallel-polarized (HH) image. There are also weaker reflections at other polarization directions, including the vertical, which are recorded as the cross-polarized (HV) image. In some areas, such as Pisgah Crater, California, there are geologically significant differences between the two polarization modes (Dellwig and Moore, 1966), although the reasons are imperfectly understood. Details of multiple-polarization imagery are discussed by Ellermeier et al. (1966).

In 1971, the Aero Service Corporation entered the commercial radar mapping business with an X-band system built by Goodyear Aerospace Corporation and mounted on a Caravelle jet aircraft. The Aero Service system lacks dual-polarization capability but is a "synthetic-aperture" radar, which in practical terms means that it should have somewhat higher and more consistent spatial resolution than "real-aperture" systems. Aero Services has flown a large area for the Venezuela government, as well as nearly 4.5 million square km in the Amazon Basin for the Brazilian government.

It is difficult to make an objective comparison between Ka-band, real-aperture imagery and X-band, synthetic-aperture imagery because few areas have been imaged by both systems. One such area is Sabine Pass between Texas and Louisiana, which is shown in Figure 15.4. Even the images in this figure are not completely comparable, however. because they were flown at different times with different

0-201-04245-2

**TABLE 15.2**
**Geologic Interpretations of Radar Imagery**

| *Locality* | *Geologic Features on Radar Imagery* | *Remarks* | *Reference* |
|---|---|---|---|
| Quachita Mts., Oklahoma | Lineaments (previously unmapped) that cut across fenster boundary faults. | Future determination of time relationship between development of lineaments and thrusting would aid in evaluating various proposed structural interpretations of Ouachita Mts. | Dellwig et al. (1968) |
| Chase Co., Kansas | Two lineaments were recognized for first time on cross-polarized imagery, but are absent on like-polarized imagery or aerial photos. | Except for some stream alignments, there is no surface indication of the lineaments. One of the lineaments may be a surface expression of a known basement fault. | Dellwig et al. (1968) |
| Boston Mts. and Arkansas River Valley | Pronounced north-south lineament pattern was first recognized on small-scale imagery. | Existence of these radar lineaments was confirmed on imagery from two other radar systems. | Dellwig et al. (1966, 1968) |
| North-central Arkansas | Previously undescribed narrow zone of lineaments trends northwest for 110 miles on imagery. | Along the northwest and southwest projections of the lineament are mapped faults, giving a total extent of 675 km. Such a structural feature could be of major significance. | Dellwig et al. (1968) |
| Numerous localities | Faults, joint systems, and dip slopes. | Radar look direction influences detectability of these geologic features. Recommend minimum of two opposing look directions for geologic studies. | MacDonald et al. (1969) |
| Twin Buttes, southern Arizona | Outcrops of rhyodacite have conspicuous low return on cross-polarized image but are not revealed on like-polarized image or aerial photos. | There is no difference in topography or surface roughness between rhyodacite outcrops and surrounding alluvium. Major differences are rock composition (Cooper, 1966) and denser vegetation on rhyodacite (Gillerman, 1967). Cause of the radar expression is therefore undetermined. | Cooper (1966); Gillerman (1967) |

0-201-04245-2

0-201-04245-2

| | | | |
|---|---|---|---|
| Cane Spring area, northwest Arizona | A fault is more clearly imaged on radar than on air photo mosaic. Different types of alluvium are distinguished on radar. | Enhancement of fault on radar is due to suppression of distracting surface details on imagery. Oblique "illumination" on radar may also be a factor. | MacDonald et al. (1967) |
| Pisgah Crater, southern California | Faults, lava flows, alluvium, playa deposits, and other rocks in a desert environment are imaged in two polarization modes and in opposing look directions. | First geologic evaluation of like- and cross-polarized imagery, which shows strong contrasts for different rock units. The differences are attributed to variations in surface roughness; however, lithologic differences may also be important. | Dellwig and Moore (1966) |
| Eastern Panama | Radar mosaic portrays faults, fractures, folds, rock units, and drainage patterns with considerable detail. | Most of the area has never been covered by aerial photography because of prevailing poor weather. Imagery of areas in the U.S. is also illustrated. | MacDonald (1969) |
| Nevada and Alabama | Imagery has numerous linear features, some of which can be identified as faults. | Notes that chief value of remote sensor imagery is that it points out anomalous features for detailed study. | Reeves (1969) |
| Southeast Missouri | Linear feature on imagery is confirmed by field investigation to be a previously unrecognized major fault. | Radar linear proves to be part of major feature named "Roselle lineament." | Gillerman (1970) |
| Wyoming-Colorado | Ring dike complex and associated fractures in Precambrian terrain. | Another example of radar imagery used to recognize major geologic features. | Wing and Dellwig (1970) |
| Southwestern U.S. | Imagery of faults, topography, and recent sediments is interpreted. | Concludes that manner of illumination, surface roughness, and geometry of target are major factors controlling appearance of a natural target on radar imagery. | Rydstrom (1967) |
| Southern Utah | Fault that was only poorly discernible on aerial photography is clearly visible on imagery. Some differences in rock types more pronounced on imagery than on photography. | Lists the geologic capabilities and applications of imagery. Compares it with low-sun-angle aerial photography. | Hackman (1967b) |

**TABLE 15.2** *(continued)*

| *Locality* | *Geologic Features on Radar Imagery* | *Remarks* | *Reference* |
|---|---|---|---|
| Pennsylvania and Maryland | Imagery of portion of Appalachian Piedmont shows some geologic features that are expressed in slope and vegetation patterns. Geologic linears occur, but so do pseudo-geologic linears caused by tree lines and other features. | Expresses a worthwhile note of caution in the geologic interpretation of imagery. | Wise (1967) |
| Colorado-Wyoming | Small-scale imagery of Great Plains and adjacent Rocky Mountains is interpreted in terms of surficial materials. | Small scale of the imagery restricts it to a useful regional reconnaissance tool. | Holmes (1967) |
| Eastern Canada and Scotland | Major rock units and structures are discernible on the very-small-scale radar scope photos. | In contrast to side-looking imagery, this imagery is photographed from scope display of an airborne rotating antenna. Little has been published subsequently on this type of imagery. | Cameron (1964) |
| Northwest Arizona | Rock types and structures in San Francisco volcanic field are imaged. Variations in surface moisture content appear to be imaged. | Lithologic differences appear to be more distinct on cross-polarized image than on parallel-polarized image. | Schaber (1967) |
| New Mexico | Several prominent alignments (probably faults) and crests of domelike structures apparent on imagery are either not visible or poorly visible on air photo. | Small-scale radar image compared with high-altitude air photo. Return intensities of rock units on imagery were measured with microdensitometer. | Fischer *in* Levine (1966, pg. 1027-1030) |
| Arctic Ocean, north coast of Alaska | Arctic coastal plain; sea ice of various ages and thicknesses. | Sea ice types are identified by radar signatures. Repeat flights provide data on ice movement. | Bradie (1967) |

0-201-04245-2

0-201-04245-2

| | | | |
|---|---|---|---|
| Numerous areas | Imagery from many areas with emphasis on engineering geology. | Provides keys for interpreting imagery features. | Barr and Miles (1970) |
| Southwest and central Utah | Broad classes of rock types. | In this well-exposed region structures image equally well on radar and air photos. | Hilpert (1967) |
| Grand Canyon, Arizona | Fault pattern is expressed on imagery. | Radar lineaments indicate some previously unmapped faults. | Jefferis (1969) |
| Colombia, Panama, and U.S. | Coastal features, such as beach ridges, swamps, natural levees, and deltas, are interpreted. | Imagery is especially useful for coastal geomorphology in poor-weather areas. | MacDonald et al. (1971) |
| Southern Louisiana | Soil moisture variations are inferred from parallel- and cross-polarized imagery. | Vegetation cover must be dry or leafless. | MacDonald and Waite (1971) |
| Oregon coast | Rock types are recognized by distinctive radar signatures; faults and lineaments are imaged that are not identifiable on air photos. | Excellent example of advantages of radar in forested terrain. | Snaveley and Wagner (1966) |
| Eastern Panama | Structural and lithologic interpretation from radar mosaic. | This is a continuation to the northwest of earlier work by MacDonald (1969). | Wing (1971) |
| West Virginia | Fracture pattern on imagery is correlated with movement of thrust sheets. | Notes advantages of imagery for fracture analysis in this forested terrain. | Wing et al. (1970) |

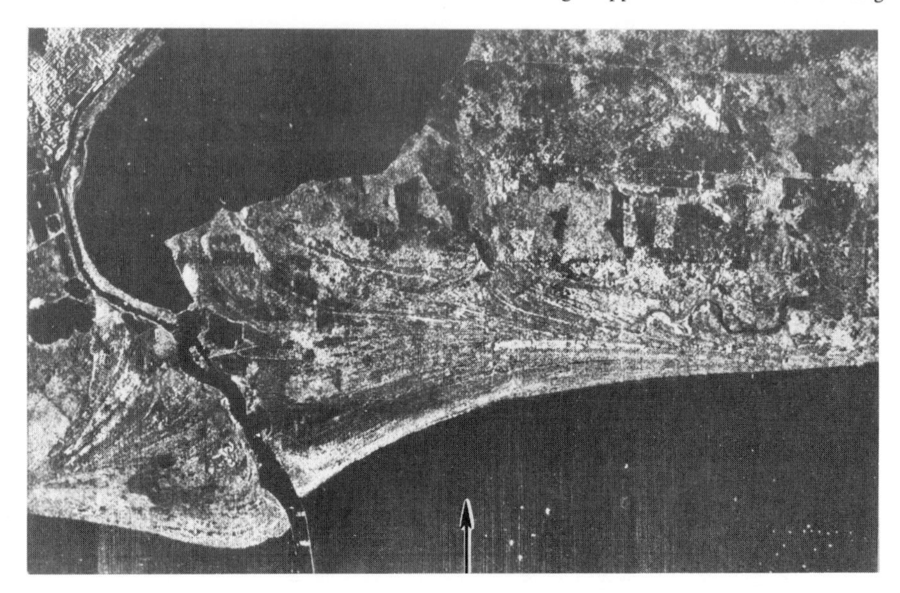

A. Synthetic aperture, X-band (2.40 - 3.75 cm) wavelength image flown by Goodyear Aerospace Corp.

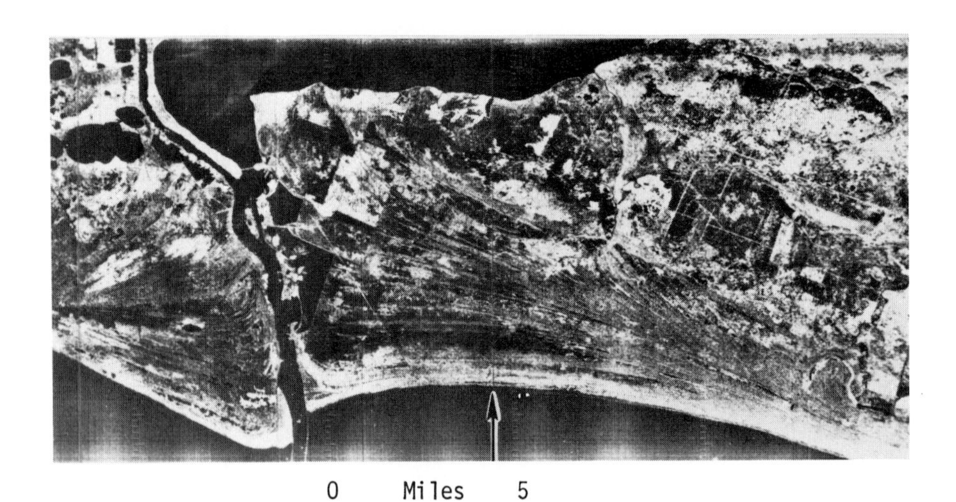


B. Real aperture, Ka-band (0.75 - 1.18 cm) wavelength image flown by Westinghouse Electric for NASA.

**Fig. 15.4.** Comparison of real- and synthetic-aperture radar imagery of Sabine Pass, Texas, and Louisiana. Look direction shown by arrows.

0-201-04245-2

surface conditions and different depression angles. Both systems produce imagery suitable for geologic interpretation; therefore selection of a system probably will be influenced by relative cost, availability, and operational considerations.

## D. Advantages of Radar Imagery for Geology

In discussing the advantages of radar imagery for geology, it is not implied that aerial photography, field geology, and geophysics will be supplanted by the newer method. Radar imagery does have some unique geologic capabilities, however, particularly for reconnaissance studies, which have been summarized by Pierson et al. (1965, p. 105).

1. *A generalizing tool.* A radar image strip may be several hundreds of miles long and a few tens of miles wide with minimal distortion. The interpreter can make major regional generalizations without magnification of the imagery. The imagery tolerates high magnifications, however, for detailed work. Aerial photomosaics of comparable scale are degraded by individual photo boundaries and tonal changes caused by varying lighting conditions. The Panama and Venezuela examples, discussed in Sections III.I and III.J, demonstrate that radar strips may be combined into a useful mosaic.
2. *Oblique "illumination."* The oblique "illumination" of terrain by the radar beam produces a highlight and shadow effect that enhances faults and fractures, as illustrated in Figure 15.2. Orientation of the radar flight line relative to the structural pattern is important because linear features trending parallel with the radar look direction may not be detectable, as shown by Figure 15.3. In areas of high relief shadows will obscure part of the terrain, and additional flight lines with opposite look direction will be needed to image the shadowed areas.
3. *Suppression of minor detail.* The advantage of radar imagery for suppressing details of vegetation cover because of small scale and low resolution have already been discussed. Minor cultural detail is likewise suppressed so that the interpreter is not distracted in his search for larger geologic features. This is illustrated in Figure 15.5, where the beach ridges are more distinct on the radar image than on the air photograph, which is "cluttered" with urban cultural features. Oblique illumination of the radar also aids recognition of the beach ridges.
4. *Detection of faint lineaments.* This is an outstanding attribute of radar imagery. Table 15.2 lists a number of previously unrecognized fractures, faults, and lineaments first detected on radar imagery. The Roselle lineament, described in Section III.H, is a good example. Several factors contribute to this capability, including the suppression of minor detail, oblique illumination, and continuous coverage without breaks. This combination enables the eye to recognize and integrate subtle changes in tone, texture, and geometry of the image, which may combine to form lineaments with structural significance.

0-201-04245-2

A. AERIAL PHOTOMOSAIC FLOWN DEC. 1953. URBAN DEVELOPMENT ALONG COAST OBSCURES TERRAIN FEATURES.

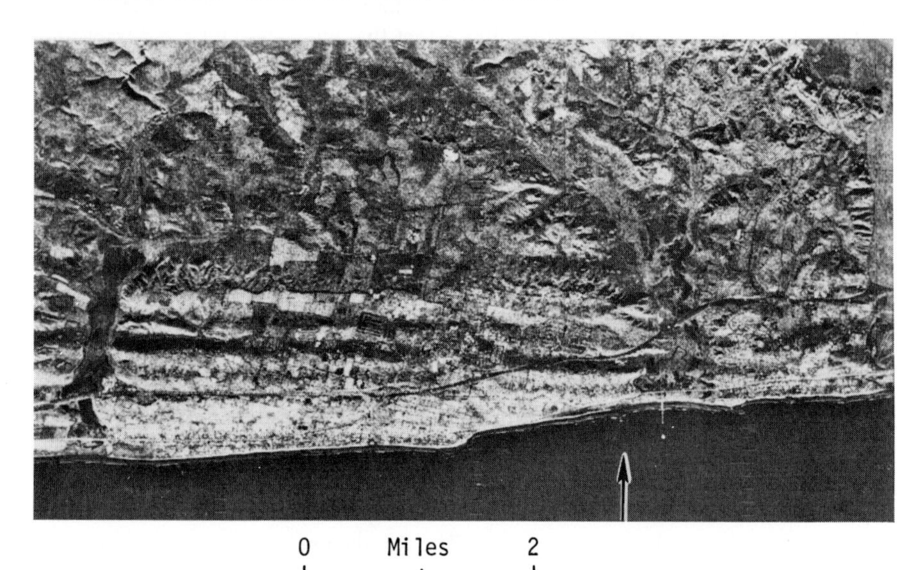


B. RADAR IMAGE FLOWN NOV. 1965. LOWER RESOLUTION SUPPRESSES URBAN "CLUTTER" AND ENHANCES BEACH RIDGES PARALLEL WITH COAST. OBLIQUE RADAR ILLUMINATION ALSO AIDS. LOOK DIRECTION SHOWN BY ARROW.

**Fig. 15.5.** Comparison of radar image and aerial photomosaic of beach ridges near Oceanside, southern California. From Sabins (1976, Figure F-21).

0-201-04245-2

5. *Lithologic variations.* Because of lower spatial resolution, this is not a leading capability, but in some instances lithologic differences are enhanced on radar imagery, particularly in comparing parallel- and cross-polarized imagery, as in the Pisgah Crater example (Section III.F). Most lithologic differences expressed on the imagery are probably due to differences in surface texture or roughness of the rocks rather than to compositional differences. This is shown in the example from Cane Springs, Arizona (Section III.G).
6. *All-weather capability.* Some parts of the world have never been satisfactorily photographed from the air because of prevailing poor weather. Except for active precipitation, radar imagery can be acquired under almost any weather conditions, thus providing information on hitherto poorly known regions.

## E. Stereoscopic Radar Imagery

Stereoscopic imagery may be obtained by spacing flight lines so that adjacent image strips overlap. As shown in the example of Figure 15.6, stereoscopic viewing is an advantage for radar interpretation. Unlike aerial photography, where stereoscopic coverage is obtained by forward overlap of successive frames, stereo radar is produced by side overlap. Hence stereo radar may require flight lines in addition to those needed for normal mosaic coverage.

**Fig. 15.6.** Stereographic pair of radar images from New Guinea. May be viewed in three dimensions with a pocket stereoscope. Look direction is toward the right. Courtesy of Westinghouse Electric.

0-201-04245-2

## F. Pisgah Crater, California

This area in the Mojave Desert 60 km east of Barstow illustrates differences between parallel- and cross-polarized imagery that were first described by Dellwig Moore (1966). As shown on the air photo (Figure 15.7) and the geologic map (Figure 15.8), there are two lava flows: the Pisgah basalt of Pleistocene to Recent age and the slightly older Sunshine basalt. Lavic Lake is a typical playa with a mud-cracked clay surface (Figure 15.9-1) that acts as a specular reflector and is dark on the radar (Figure 15.10). The bright line across the lake is a gravel road. Older bedrock outcrops and alluvial fans eroded from them comprise the rest of the area. Pisgah Crater is a favorite field-trip locality for my seminars in remote sensing at the University of Southern California; hence the air photo and imagery are indexed for ready reference. The comparison chart (Table 15.3) has proved useful for correlating signatures of specific features on different types of imagery; it can be expanded to include other imagery, such as color photography and IR scanner imagery when available.

The radar imagery in Figure 15.10 was flown with look direction to the southwest; additional flights were made with look direction to the northeast. Dellwig and Moore (1966) noted that differences in return intensity on different look-direction imagery were related to terrain slope angle. The HV imagery of Figure 15.10 has a series of parallel alternating lighter and darker bands along the upper margin, which

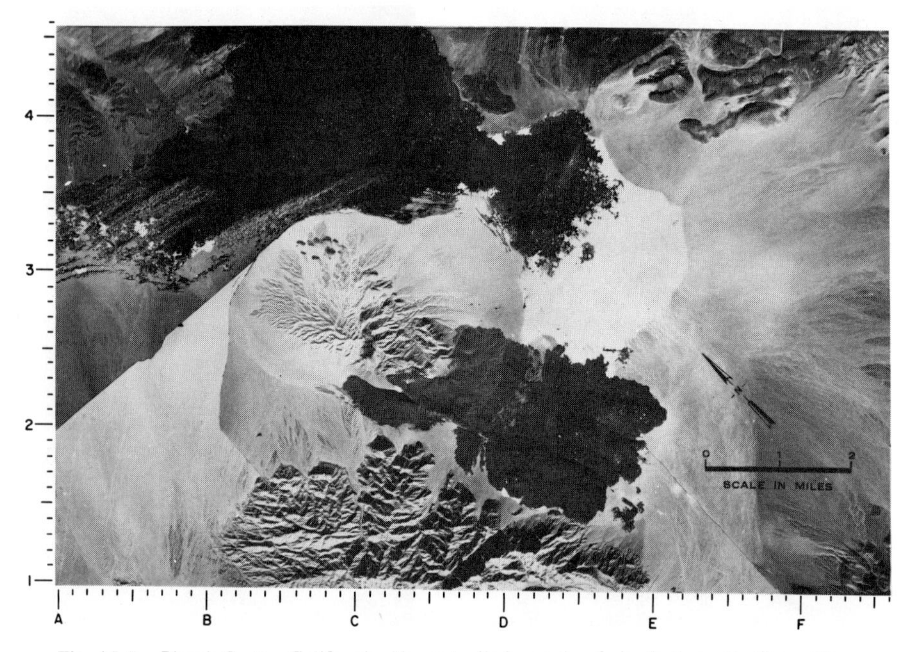


**Fig. 15.7.** Pisgah Crater, California. Uncontrolled mosaic of air photographs flown December 1954 by U.S. Army Map Service. Original scale 1:62,000.

0-201-04245-2

was toward the aircraft. These are "side lobe patterns" caused by the radar antenna and should be ignored.

Terrain has so many components to interact with radar energy (composition, texture, slope, moisture content, and vegetation) that it is presently impossible to state which factors control radar polarization response. In comparing parallel- and cross-polarized imagery, therefore, we generally settle for an empirical correlation with observed terrain features. These empirical differences may be valuable in pointing out significant geologic differences that are not apparent on any one form of imagery. The research by Ellermeier et al. (1966) suggests that more quantitative interpretation criteria may be available in the future.

The Pisgah and Sunshine basalt flows have a bright response on the HH imagery, but on HV the Sunshine flow is noticeably darker than the Pisgah flow. The two flows are similar petrographically and both have aa and pahoehoe phases (Figure 15.9). Desert varnish is best developed on the slightly older Sunshine flow and may be a factor. The aa and pahoehoe phases are not distinguished on the imagery of either flow. This may be due to the low resolution of the radar; also, both phases may appear equally rough at the wavelength of the radar. On the IR imagery discussed in Section IV the two phases of Pisgah basalt are distinguished.

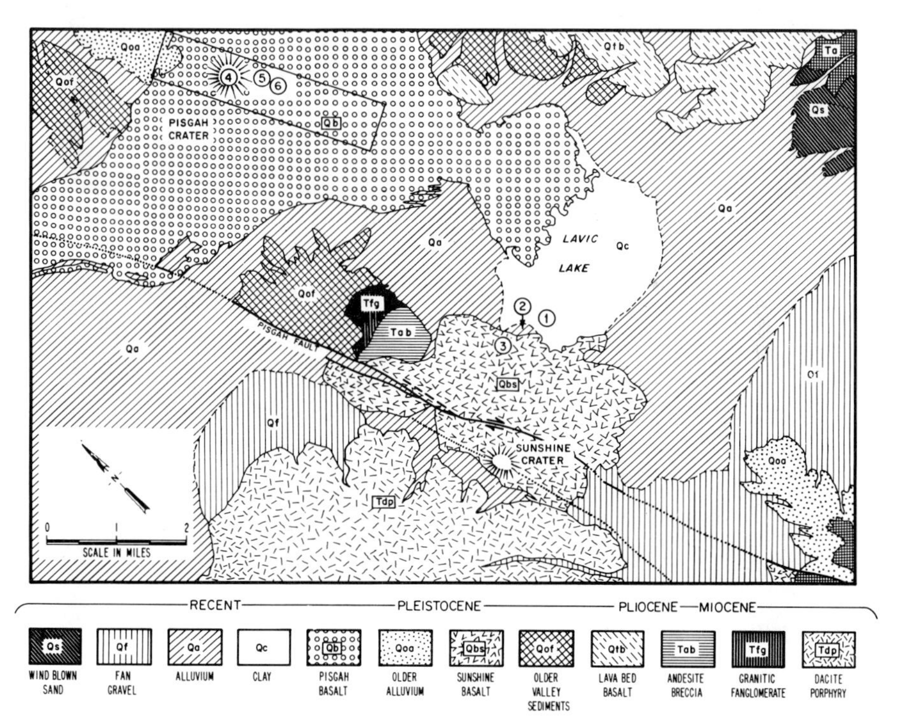


**Fig. 15.8.** Geologic map of Pisgah Crater area, California. From Dibblee (1966). Numbers 1–6 are localities of samples illustrated on Figure 15.9. Rectangle at Pisgah Crater is location of IR imagery on Figure 15.17.

0-201-04245-2

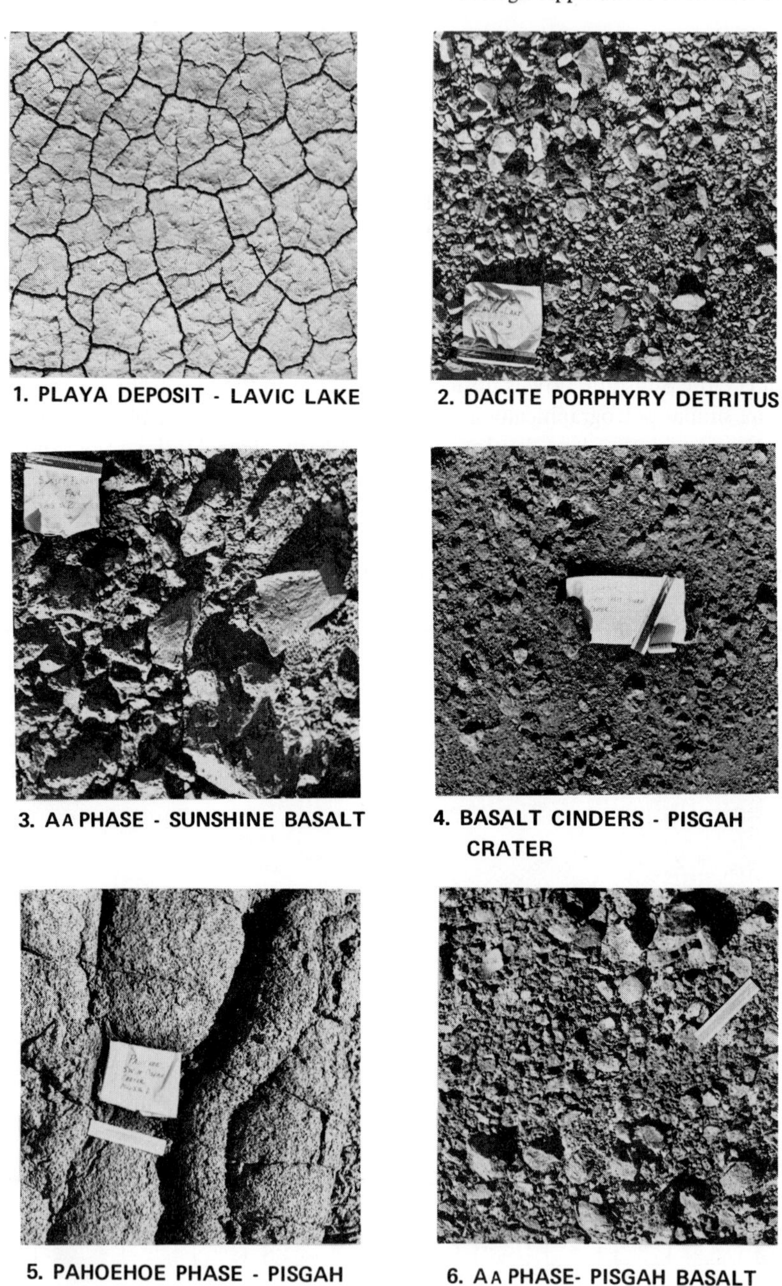


**Fig. 15.9.** Typical outcrops, Pisgah Crater area. Scale is 6 in. long. See Figure 15.8 for locations.

0-201-04245-2

0-201-04245-2

**TABLE 15.3**
**Comparison of Aerial Photography and Radar Imagery of Pisgah Crater Area**

(*Note:* In the "Feature" column, "Lo" indicates location; "Ex" indicates expression on the imagery.)

| *Feature* | | *Aerial Photography* | *Radar Imagery* | | *Field Observations* |
|---|---|---|---|---|---|
| | | | *HH* | *HV* | |
| Pisgah Crater | Lo. | 4.2, B.3 | 2.6, B.3 | Same | Basalt cinders and ash (Figure 15.9-4) |
| | Ex. | Black, not identifiable | Bright, with shadow | Same | |
| Pisgah basalt | Lo. | 3.3, A.0–3.5, D.5 | 2.2, A.6–2.0, D.1 | Same | Fresh basalt with pahoehoe and aa phases (Figure 15.9-5, 6) |
| | Ex. | Black | Bright | Bright | |
| Sunshine Crater | Lo. | 1.7, D.1 | 1.1, C.5 | Same | Basalt cinders and ash |
| | Ex. | Black | Bright, with shadow | Same | |
| Sunshine basalt | Lo. | 2.3, C.0–2.2, E.1 | 1.6, B.7–1.3, D.3 | Same | Basalt with pahoehoe and aa phases; some desert varnish (Figure 15.9-3) |
| | Ex. | Black | Bright | Intermediate | |
| Lavic Lake | Lo. | 3.0, D.0–3.0, E.1 | 1.7, C.5–2.0, D.4 | Same | Mud-cracked clay (Figure 15.9-1) |
| | Ex. | White | Dark | Dark | |
| Small alluvial fan | Lo. | 2.6, D.3 | 1.6, C.6 | Same | Cobbles and pebbles of dacite porphyry (Figure 15.9-2) |
| | Ex. | Gray | Gray | Bright | |
| Fan east of Lavic Lake | Lo. | 3.8, D.9 | 2.2, D.2 | Same | Predominantly pebbles and cobbles of basalt |
| | Ex. | Med. gray | Bright | Bright | |
| Fans south and west of lake | Lo. | 3.4, E.3–2.7, D.0 | 1.7, C.5–1.9, D.5 | Same | Predominantly pebbles and cobbles of andesite breccia |
| | Ex. | Med. to light gray | Dark gray | Dark gray | |
| "Distributary" | Lo. | 2.7, D.9 | 1.6, D.1 | Same | Shrubs and stabilized small sand dunes |
| | Ex. | Blends with lake | Bright | Med. gray | |

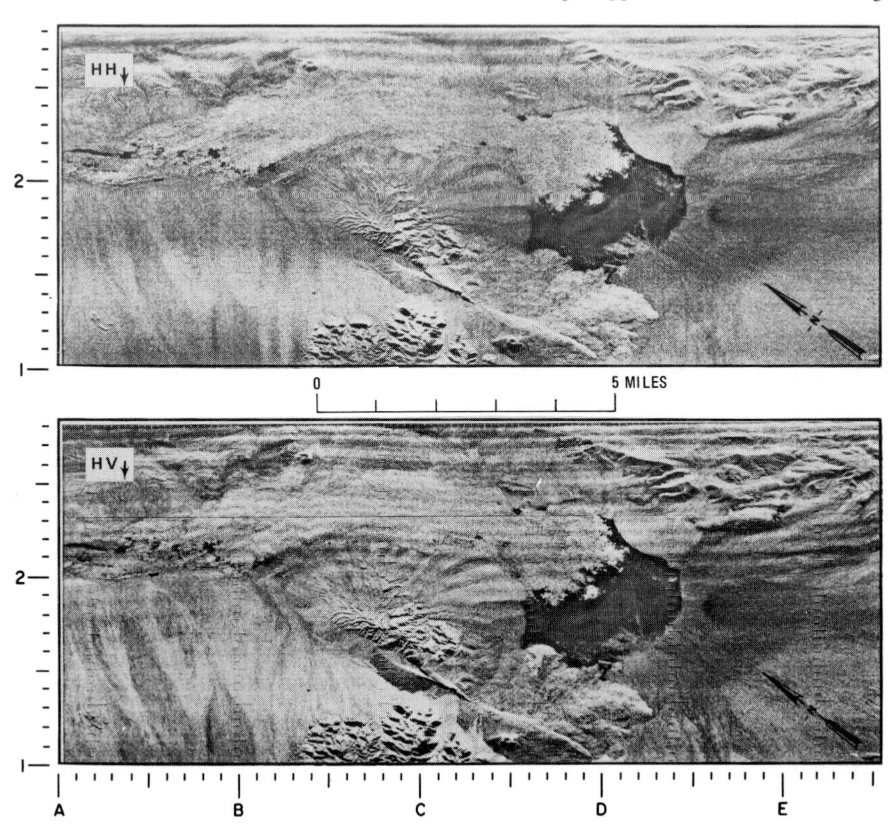


**Fig. 15.10.** Pisgah Crater, California. Like-polarized (HH) and cross-polarized (HV) radar imagery with look direction to the southwest. Imagery courtesy of NASA and the U.S. Geological Survey.

A small alluvial fan lies at the contact between Sunshine basalt and the west side of Lavic Lake. On HH imagery the fan has the same appearance as the basalt, but on HV imagery the fan is much brighter. The fan consists of cobbles and gravel of dacite porphyry (Figure 15.9-2) derived from outcrops farther west (Figure 15.8). Dacite alluvium between these outcrops and the Sunshine flow has the same radar expression as the fan (Figure 15.10, locality 1.3, C.1). Previous investigators suggested that the petrographic and textural differences between the dacite alluvium and the adjacent basalt caused the differences in HH and HV imagery. In the field, however, we have observed a greater density of vegetation on the fan than on the adjacent basalt and playa surfaces. We suggest that the vertical alignment of plant stems plays a dominant role in the polarization differences.

A large fan on the east side of Lavic Lake is uniformly bright on both polarizations. It contrasts with fans on the south and west margins which are darker on both polarizations. Cursory field examination showed that the fans are similar in

0-201-04245-2

slope, surface texture, and vegetation. The most obvious difference is that the bright-appearing fan consists predominantly of basalt detritus, whereas the darker fans are predominantly andesite detritus. In this example where surface roughness and vegetation are similar, compositional differences may be enhanced on the imagery.

The right-lateral Pisgah fault is clearly imaged by bright returns from the scarps facing toward the aircraft and shadows from the opposite-facing scarps. Dellwig (1969) compared C-band (6.73-cm wavelength) and P-band (70.00-cm) imagery of Pisgah Crater with the $K_a$-band imagery described here and noted some differences.

### G. Cane Springs, Arizona

MacDonald et al. (1967) compared radar imagery with aerial photography and prepared a geologic interpretation for this area in extreme northwestern Arizona (Figure 15.11). The westerly look direction of the radar causes a strong return from the prominent southeast-dipping hogback of resistant Kaibab limestone and casts a shadow on the northwest-facing escarpment. The low-return (dark) areas at *a* are compact, dry, fine-grained, limonite-stained sand overlying the upper unit of the Coconino sandstone. Locality *c* has a porous puffy texture and consists of fine-grained sand mixed with fragments of well-cemented sandstone, which produces an intermediate return on the imagery. Between the two *a* localities is an area of compact alluvium consisting of rock fragments of varied lithology with geometric irregularities that scatter the radar signal to produce a comparatively high-intensity bright return. These differences in texture are not as distinct on the air photo.

The Tertiary lava in the southern part of the radar imagery (locality *b*) has a uniform distinctive tone and is distinguished from the underlying Tertiary alluvium. On the air photo, however, the lava is marked by irregular light-toned areas where it is covered by a thin film of carbonate.

The northwest-trending fault at *d-d′* is more prominent on the radar imagery than on the photography. MacDonald et al. (1967, p. 77) suggest that the irregular distribution of vegetation along the fault has obscured its trace on the photo, whereas the lower resolution of the radar minimizes the vegetation and enhances the fault. The northeast-trending normal fault cutting the Kaibab hogback in the south-central part of the area is enhanced on the imagery by the radar shadow effect.

### H. Roselle Lineament, Missouri

Gillerman (1970) first identified and named the Roselle lineament from radar imagery (Figure 15.12, upper) as a "... well defined, remarkably straight linear feature extending from near Doe Run southward for over 20 miles to the edge of the imagery." The south part of the linear is partly defined by drainage alignments, but north of Roselle, although the lineament parallels Washita Creek, it lies

0-201-04245-2

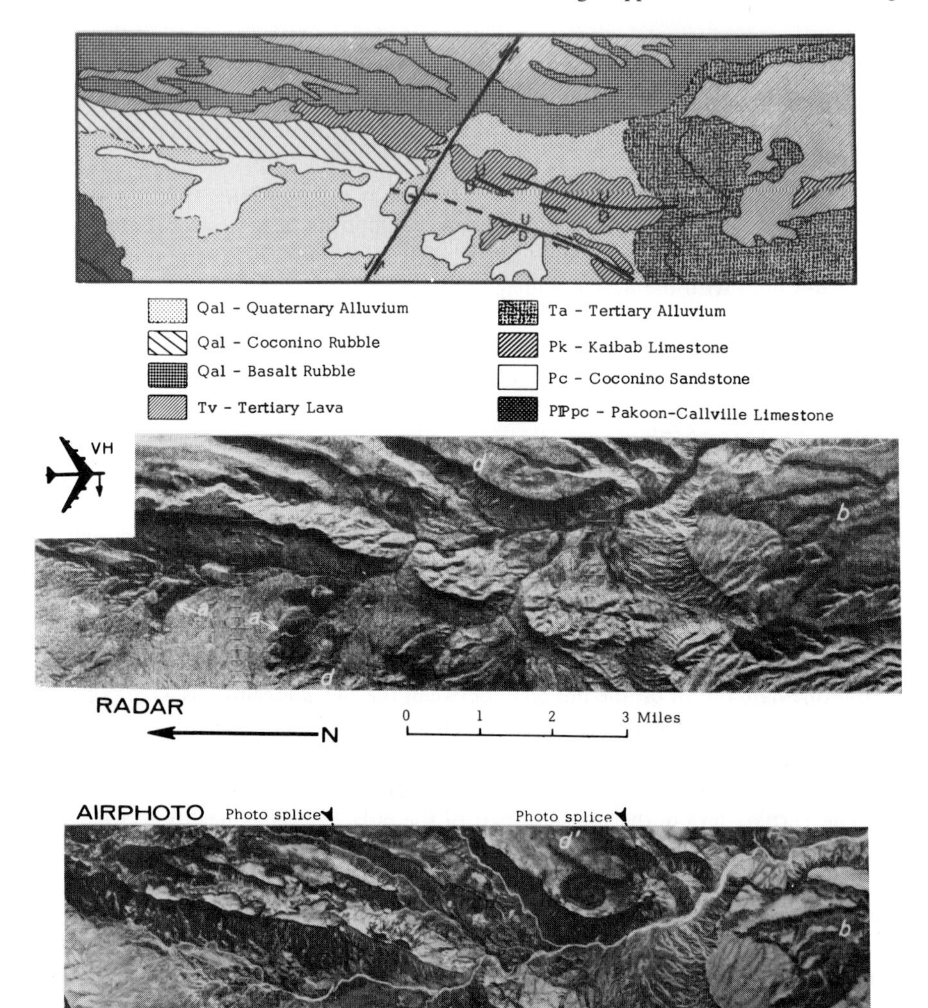


Fig. 15.11. Geologic map, radar image, and aerial photography of Cane Springs area, Arizona. From MacDonald, Brennan, and Dellwig (1967, Figures 4 and 5).

0.75 km west. Farther north there is no correlation with drainage. The lineament crosses Precambrian granite and volcanic rocks and Cambrian sedimentary rocks and is presumed to be a fault rather than a geologic boundary or trace of a bed or geologic formation (Gillerman, 1970, p. 975). Another previously unrecognized fault, called the Murphy Hill fault, occurs 1.6 km east of the Roselle fault and extends parallel with it for about 8 km. The linear expression of the Murphy Hill fault in the imagery is related to slope and drainage patterns. Neither fault is

0-201-04245-2

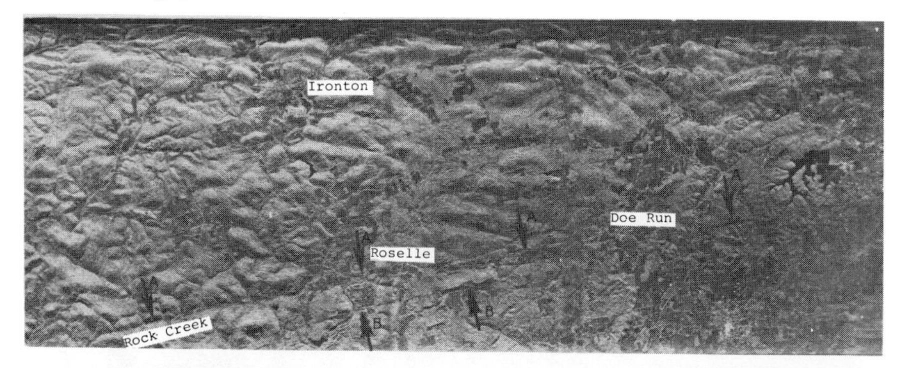

**Fig. 15.12.** Radar image (upper) and aerial photomosaic (lower) of area east of Ironton, Missouri, showing the Roselle fault, *A-A*, and the Murphy Hill fault, *B-B*. From Gillerman (1970, Figure 1).

apparent on the photo index sheet (Figure 15.12, lower) used for comparison. Although the photo sheet is not a controlled-quality mosaic, it does illustrate that subtle lineaments can be obscured by vegetation, culture, and photo boundaries, which are suppressed or eliminated on radar imagery.

In contrast to the strong relief, arid climate, and abundant outcrops of Pisgah Crater and Cane Springs, the Roselle lineament occurs in an area with subdued relief, humid climate, and few outcrops. These differences help demonstrate the versatility of radar for geologic purposes.

## I. Darien Province, Panama

In 1967 complete radar coverage was acquired of 17,000 km$^2$ in eastern Panama, where previous aerial photography missions had been unsuccessful because of perpetual cloud cover. The imagery was compiled into the mosaic of Figure 15.13, where the northeast-southwest flight direction is indicated by the alignment of bright parallel side-lobe patterns. In order to minimize the number of flight lines imagery was obtained on both the northeast- and southwest-bound flights; hence the look direction and shadow orientation are reversed in some parts of the mosaic.

0-201-04245-2

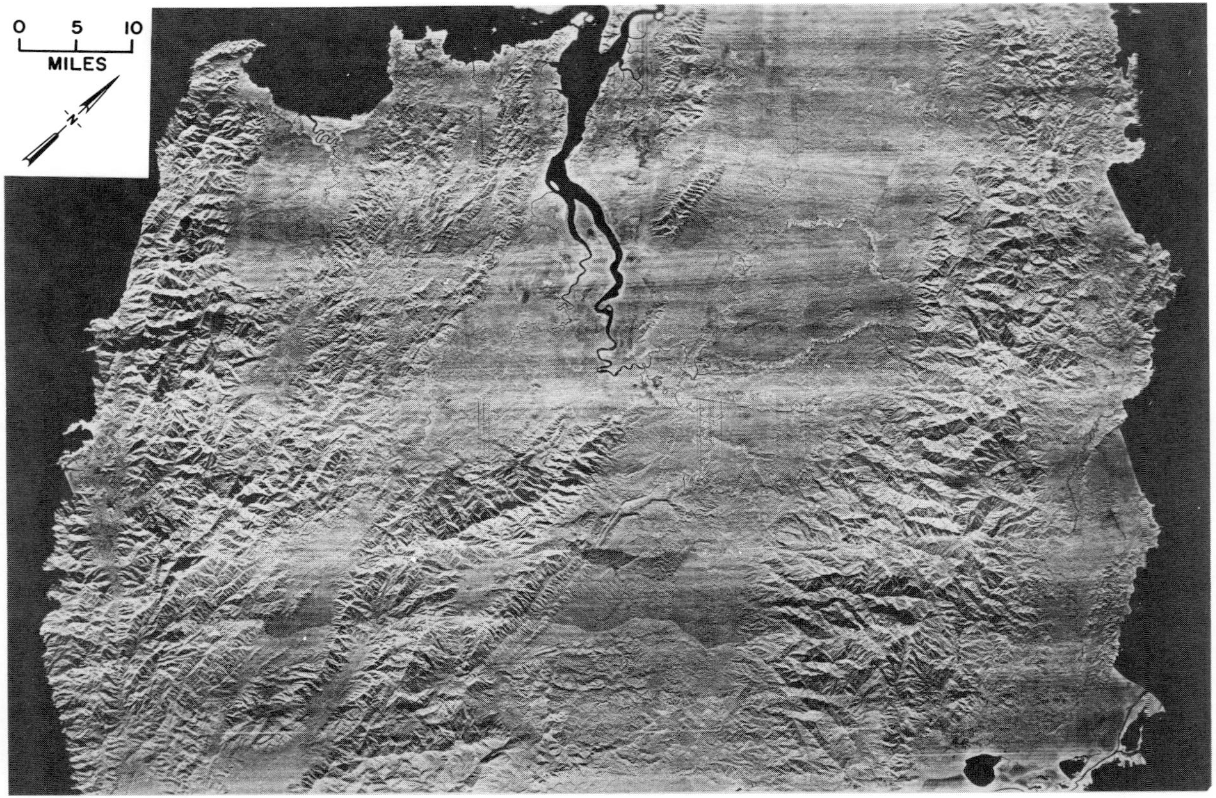

**Fig. 15.13.** Radar mosaic of Darien Province, Panama, flown by Westinghouse Corporation in 1967. Mosaic by Raytheon-Autometric. Courtesy of U.S. Army Engineer Topographic Laboratory.

0-201-04245-2

The original mosaic was compiled at a scale of 1:250,000 (1 in. = 4 mi/1 cm = 2.5 km). This scale was greatly reduced (approximately 1 in. = 14 mi/1 cm = 8.6 km) for Figure 15.13, but many details of this classic example are retained. Some boundaries between adjacent image strips in the lower part of the mosaic are shown by abrupt tonal changes along an irregular contact. MacDonald (1969, p. 19) points out that the mosaic contains errors in the form of offsets, duplications, and omissions that are inherent in using unrectified slant range imagery. The excellent report by MacDonald (1969) describing his geologic interpretation of the mosaic is the basis for this summary. The clear expression of topography and drainage patterns is the first striking impression gained from the mosaic. The outcrop patterns shown on the geologic interpretation map (Figure 15.14) were identified by their tone, texture, shape, and pattern on the imagery, together with the sparse geologic data available and limited field checking.

The pre-Tertiary (?) basement complex consists of altered basaltic and andesitic lavas, tuff, and agglomerate intruded by mafic rocks. The basement complex forms the cores of the mountain ranges and is recognized on the imagery by the rugged massive texture and lack of stratification. Differences in erosional maturity and intensity of deformation cause differences in radar expression of the basement complex. In some of the high-relief areas major joint systems produce angular, choppy, erosional remnants with jutting edges that dominate the topography. The interior mountain ranges in the northwest part of the area have more subdued relief, and fractures are represented mainly by joint systems and angular drainage patterns. Where joint control is minimal, the topography developed on the basement complex is more rounded and hummocky.

Sedimentary rocks are recognized by their stratified appearance and erosional land forms. Hogbacks and alternating ledge and slope topography formed on sedimentary rocks are clearly expressed in the south-central and north-central parts of the mosaic. Steeply dipping strata in high-relief terrain form prominent flatirons on the west side of the mountain range in the northern corner of the mosaic. Where independent field data can be correlated with the imagery, the interpretations can be extrapolated with greater confidence. For geologic reconnaissance mapping, it is generally more practical to recognize individual map units than to attempt a detailed lithologic interpretation.

Structure is the most useful geologic information that can be obtained from this and other radar imagery. Comparing MacDonald's geologic map (Figure 15.14) with the mosaic will demonstrate this fact. As reproduced here, the mosaic is somewhat degraded and reduced to a very small scale; consequently some of the mapped features may not be apparent in Figure 15.13. Also, MacDonald had access to additional imagery flown both parallel with and normal to the mosaic flights. A northwest-tending *en échelon* alignment of anticlines is immediately apparent in the center of the mosaic. A previously unmapped major fault that trends subparallel with the Pacific Coast is apparent on the northwest (left) side of the mosaic. This appears to be the continuation of a fault mapped in Colombia, giving a total length of approximately 300 km. East of the fault the regional fracture pattern on the

0-201-04245-2

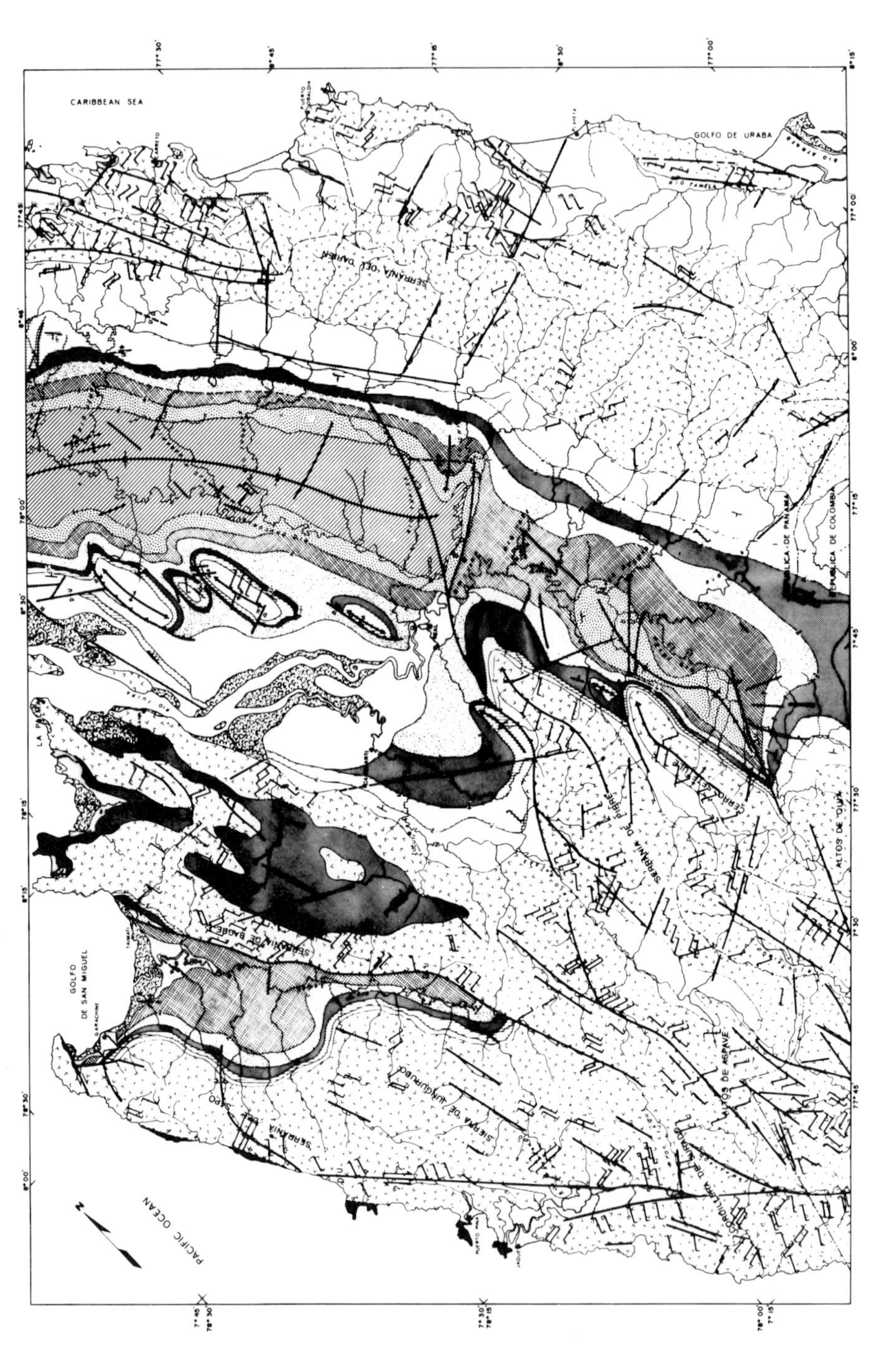
CARIBBEAN SEA
GOLFO DE URABA
SERRANIA DEL DARIEN
REPUBLICA DE PANAMA
REPUBLICA DE COLOMBIA
ALTOS DE QUIA
GOLFO DE SAN MIGUEL
ALTOS DE ASPAVE
PACIFIC OCEAN
N

0-201-04245-2

0-201-04245-2

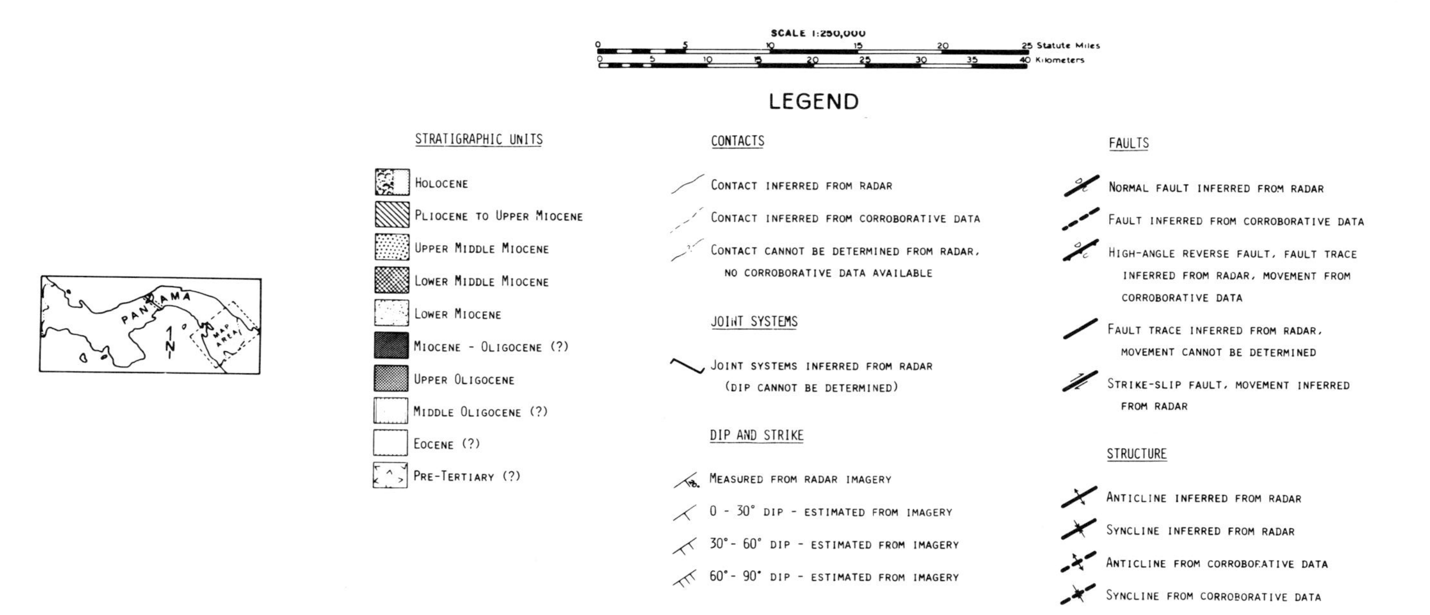


**Fig. 15.14.** Geologic interpretation of radar mosaic, Darien Province, Panama. From MacDonald (1969, Figure 3).

mosaic trends predominantly south, but on the west side there is an abrupt change to west-trending fractures. A shorter, but equally obvious, fault offsets Punta Garachine at the extreme west (upper left) part of the mosaic. Like the fault along the coast, this is presumed to have strike-slip displacement, although lateral offsets cannot be determined from the imagery. A major drainage anomaly occurs in the lowlands in the northern (upper right) part of the mosaic. The Tuquesa and Tupisu rivers diverge, suggesting that they are deflected by a subtle structural uplift.

A detailed structural interpretation is given in the report by MacDonald (1969), from which the preceding description was summarized. This example emphasizes the potential application of radar for geologic reconnaissance. Although eastern Panama is covered by tropical rain forest, much structural, stratigraphic, and geomorphic detail is readily apparent on the imagery.

The area between Darien Province and the Panama Canal Zone was later imaged by Westinghouse and interpreted by Wing (1970). As in the Darien example, a wealth of hitherto unobtainable structural information was extracted from the imagery. Wing, on the basis of his and MacDonald's work, prepared a stress analysis relating the structure of eastern Panama to postulated plate tectonics.

### J. Esmeralda Quadrangle, Venezuela

The mosaic of Figure 15.15 was originally compiled at a scale of 1:250,000 (1 in. = 4 mi/1 cm = 2.5 km) and is reproduced here at a greatly reduced scale (approximately 1 in. = 12 m/1 cm = 7.5 km). Despite this reduction much of the topographic and geologic detail of the original mosaic is retained. The Esmeralda Quadrangle is part of a larger project flown by Aero Service Corporation for the Venezuelan government. Flight lines were oriented so that the look direction was consistently westward, as shown by the radar shadows.

The Esmeralda Quadrangle is located in the Upper Orinoco Basin of the Guyana Highland physiographic province (Lopez, 1956, Figure 1). Much of the area consists of dissected granite plains that rise eastward with an elevation of about 240 m. In the west the spectacular mesa of Cerro Duida rises to an elevation of nearly 2400 m. The west-flowing Orinoco River, shown in the southwest part of the quadrangle, and its tributaries drain the area. In the extreme southwest part of the quadrangle, the south-flowing Rio Casiquiare diverts part of the Orinoco water to the Rio Negro, an example of unfinished stream piracy (López, 1956, p. 334).

What little is known about the regional geology of this remote area is here summarized from López (1956). The quadrangle is located near the center of the Guayana Shield, consisting of Precambrian rocks. Cerro Duida and smaller mesas in the area are erosional remnants of the nonmarine Roraima Series of Precambrian age, which consists of coarse conglomerate and sandstone in the lower part and fine sandstone in the upper part. The Roraima Series is underlain by a thick zone of weathered granite. López (1956, p. 334) notes that Cerro Duida is intensely folded, and the surface of this and other high sandstone mesas may be relics of an old peneplain.

0-201-04245-2

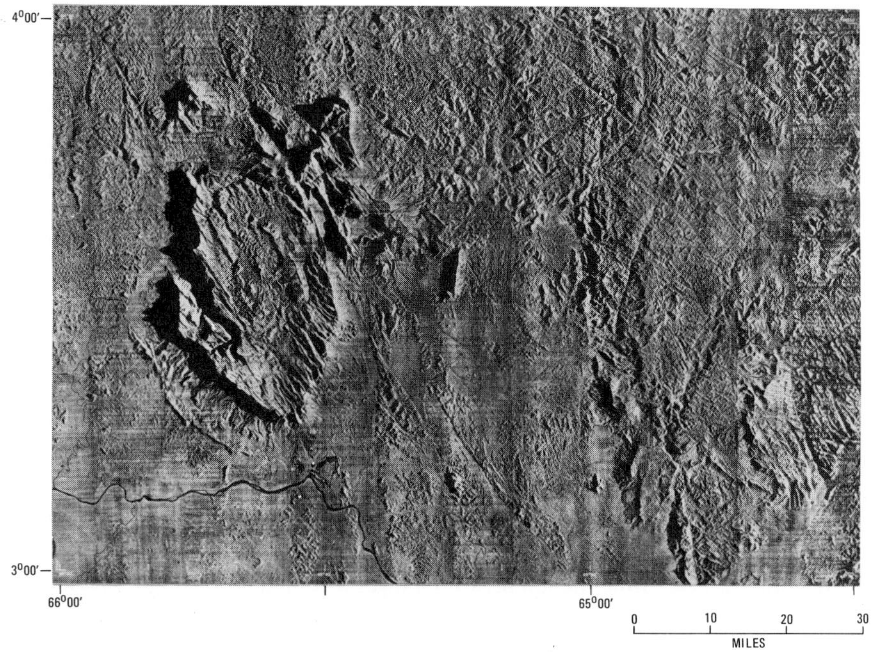

4°00′–
3°00′–
66°00′
65°00′
0 10 20 30
MILES

**Fig. 15.15.** Radar mosaic of Esmeralda Quadrangle, Venezuela. Flown in 1971 for Codesur, Venezuela government by Aero Service Corporation, using Goodyear synthetic aperture, X-band (3.12-cm) wavelength radar. Mosaic courtesy of Aero Service Corporation.

On the mosaic the relatively smooth texture of the sandstone outcrops contrasts with the rounded, "pebbly," fine texture of the crystalline rocks. Faults and fractures range in magnitude from the major NNW-trending lineament just east of Cerro Duida down to small fractures that control minor streams. The suggestion of topographic reversal along the trace of the major lineament suggests that it may have a strike-slip component of displacement. In addition to this NNW trend, NE and NW fractures are apparent. Cerro Duida appears to be a large synclinal remnant with the highest stratigraphic units preserved at the east-central part. The pronounced NNE linear grain of the mesa surface may reflect the folding mentioned by Lopez.

## IV. INFRARED SCANNER IMAGERY

### A. Equipment Considerations

Since the early 1960's geologists have been investigating the applications of thermal IR imagery. Early studies were handicapped by security restrictions, which have largely been surmounted in recent years. Early scanner systems were delicate and subject to malfunctions, particularly the detector and associated cryogenics. The

0-201-04245-2

development of trimetal detectors with liquid-nitrogen cooling and rugged simple scanners has increased reliability.

Short-wavelength imagery (3–5 μm) is acquired with indium antimonide detectors, and long-wavelength imagery (8–14 μm) with trimetal detectors or with doped germanium detectors, which are more difficult to maintain and operate. No definitive studies comparing imagery at the two different wavelengths are available. My personal experience in comparing limited samples of the two types is that the longer-wavelength imagery contains more geologic data, and this seems to be the opinion of most interpreters. Stingelin (1969), however, makes a theoretical case for the usefulness of the shorter-wavelength imagery.

On IR imagery, bright signatures represent relatively warm areas on the ground, and dark signatures are relatively cool. Variation in target emissivity, as well as temperature, influences the radiant temperatures imaged by the scanner, but for most interpretations "warmer" and "cooler" are adequate terms. The gray scale of the imagery is relative, and no quantitative temperature determinations can be made from conventional imagery. Stingelin (1969, Figure 7) illustrates a radiometer trace recorded simultaneously with imagery that provides quantitative temperature measurements along the center line of the imagery. This technique has potential application for calibrating the temperature ranges recorded on the imagery. Another approach to quantitative imagery, described by Van Lopik et al. (1968), incorporates blackbody temperature references within the scanner. Quantitative imagery is especially useful for hydrologic studies, because the uniform emissivity of water allows for more accurate temperature determinations. Radiometer profiles over land and water areas were reported by Lorenz (1968).

Until recently scanner imagery was recorded directly on film so that any operator error in estimating ground speed, in gain setting, or in processing the film could ruin a night's flying. Today, however, most scanners record the imagery on magnetic tape, which is played back in the laboratory to produce 70-mm film imagery. The tape may be replayed to obtain correct image scale, contrast, and density without reflying the project. The characteristic distortion of scanner imagery can also be rectilinearized during playback (Sabins, 1973). Some scanners record both short- and long-wavelength IR imagery, and the most sophisticated systems cover up to 18 wavelength bands from UV to IR.

### B. Time of Day

For most geologic purposes IR imagery is normally flown at night to minimize the effects of reflected solar radiation. As shown by the Pisgah Crater example (Section IV.H), daytime imagery portrays topography because of differential shadowing and exposure of slopes to the sun. At night these effects are lacking, and topographic expression, except for drainage patterns, is greatly reduced. Variations in surface lithology and moisture content are more apparent on nighttime imagery; this is another reason why most geologic imagery is flown at night.

0-201-04245-2

Imagery flown at different times of day in New York state has been illustrated by Stingelin (1969, Figure 1), who concludes that nighttime imagery is superior to daytime imagery for most applications. Rowan et al. (1970) believe that in the Arbuckle Mountains of Oklahoma predawn imagery is most useful in distinguishing rock types and for mapping faults or fracture zones. Daytime imagery of the Arbuckle Mountains displays much stratigraphic and structural detail, largely because of topographic expression. Wolfe (1971) compared day and night imagery of the Carrizo Plains, California, and noted that topography dominated the daytime imagery. Differences in rock type are more apparent on the nighttime imagery, although Wolfe (1971, p. 51) emphasized that the IR signatures are largely controlled by the properties of surface debris, not of bedrock.

Maximum thermal contrasts generally occur near sunset, but the radiant temperatures change as the night progresses. The most stable radiant temperatures occur in the predawn hours. If a number of lines are to be flown, more uniform results will be obtained from predawn flights. Local weather conditions, such as early morning ground fog, must also be considered. Imagery acquired shortly after heavy rainfall may be of little use because thermal effects of surface moisture variations may mask the more subtle geologic signatures.

### C. Flight Planning and Navigation

This summary is taken from the more complete discussion by Sabins (1973), to which the interested reader is referred for more details. To obtain a complete mosaic coverage of an area, a series of accurately spaced parallel lines must be flown. Navigating these lines is a difficult problem because existing navigation facilities are designed for point-to-point flights on radial courses rather than parallel lines. Formerly we employed ground personnel with spotlights for guidance, but this was difficult, particularly in remote areas without roads. A recently developed very-low-frequency radio navigation system (Sabins, 1973) greatly simplifies nighttime navigation problems and eliminates the need for ground personnel.

Flight elevation above terrain determines image scale and the spacing of adjacent flight lines. Sabins (1973) discusses the various considerations in selecting flight elevation. For general geologic reconnaissance with a 120°-field-of-view scanner and rectilinearized imagery we fly at 2000 m above average terrain and space flight lines 4.5 km apart. The resulting imagery, played back onto 70-mm film, has a scale of 1.0 cm to 1 km and covers a ground swath 6.0 km wide, or 3.0 km on either side of the flight path. The 4.5-km spacing provides 1.6 km of side lap between adjacent imagery strips, so that good mosaics can be compiled. Obviously other altitudes may be flown for other imagery requirements, but experience shows this flight program to be a good tradeoff between image quality and the economical covering of a large area with a minimum number of flight lines.

As with radar, the orientation of IR flight lines is important because linear geologic features trending parallel with the scan line pattern may be obscured. In

0-201-04245-2

general the angle between flight lines and known or suspected structural "grain" should be 45° or less.

### D. Imagery Irregularities

Geologic interpretation may be hampered or confused by a variety of imagery irregularities or defects. Again, this discussion is summarized from the more extensive presentation by Sabins (1973). Uncompensated aircraft roll (Figure 15.16*A*) occurs when the scanner roll compensation device is not operating properly. Straight lines, such as the road in this example, are distorted into a wavy pattern. In imagery of dipping sedimentary rock outcrops, this failure can produce patterns resembling plunging folds. Most scanners include a warning signal for roll compensation failure, which should be monitored by the operator. Turbulent air may cause aircraft roll in excess of the usual 10° limit on roll compensation; such turbulence should be noted in the flight log for the interpreter's information. Yaw (crab) is aircraft rotation about a vertical axis so that the aircraft longitudinal axis is not parallel with flight direction; it results from cross winds. Williams and Ory (1967, p. 1378) described the effects of yaw on imagery. These effects may be avoided by rotating the scanner mount in the horizontal plane to compensate for the yaw angle.

Although imagery flights should be made on clear nights with no surface winds, often flights must be made under less than ideal conditions. Clouds below the flight altitude produce the patchy warm and cool appearance shown in Figure 15.16*B*. Scattered rain showers result in a pattern of streaks parallel with the scan lines.

Surface winds produce characteristic "smears" and "streaks" on imagery. Wind smears (Figure 15.16*C*) are parallel, curving lines of alternating warmer and cooler signature that may extend over wide expanses of imagery. Wind streaks occur downwind from obstructions on flat terrain and typically appear as the warm (bright) plumes shown in Figure 15.16*D*. In this example the obstructions are clumps of trees, which image warm, and one distinct building with a cold-imaging metal roof in the lower right part of Figure 15.16*D*. The obvious solution to wind effects is to fly only on still nights, but in many areas surface winds persist for much of the year, so that their effects must simply be endured. It is important to avoid confusing wind patterns with geologic linears.

Transmissions from many aircraft radios may cause strong interference patterns on imagery. In the example of Figure 15.16*E*, the interference occurs as bands of electronic noise which obscure the underlying image pattern. Radio transmissions may also produce a wavy, moiré interference pattern. Electronic shielding of the scanner equipment may prevent this interference, but the simplest solution is to observe radio silence during image runs and communicate with the ground during turns and offset legs.

Cyclic repetition of discrete signal patterns (Figure 15.16*F*) is an annoying, but not serious, form of electronic interference. In this example the noise occurs as

0-201-04245-2

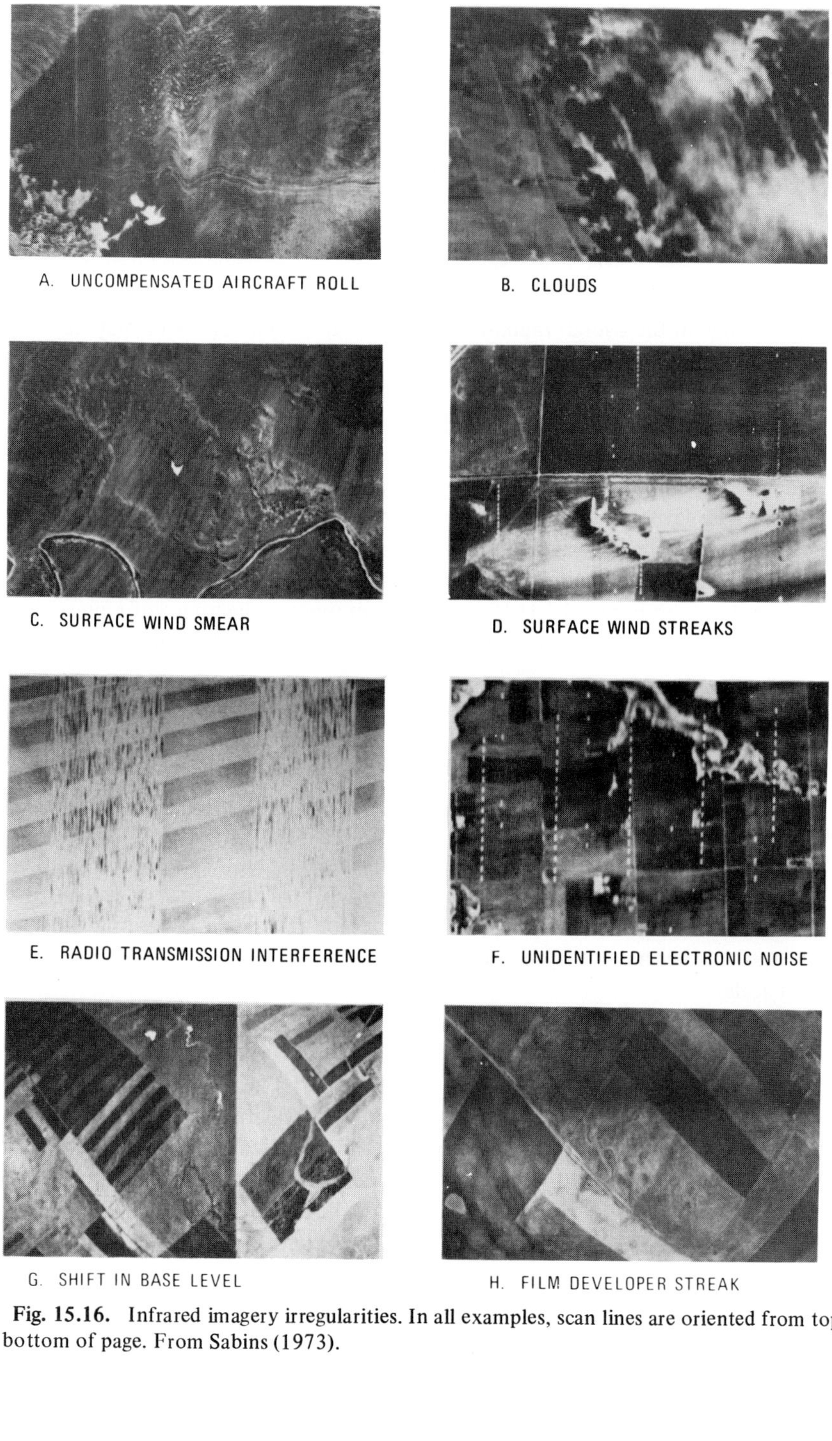

A. UNCOMPENSATED AIRCRAFT ROLL

B. CLOUDS

C. SURFACE WIND SMEAR

D. SURFACE WIND STREAKS

E. RADIO TRANSMISSION INTERFERENCE

F. UNIDENTIFIED ELECTRONIC NOISE

G. SHIFT IN BASE LEVEL

H. FILM DEVELOPER STREAK

**Fig. 15.16.** Infrared imagery irregularities. In all examples, scan lines are oriented from top to bottom of page. From Sabins (1973).

0-201-04245-2

positive (bright) dots, but it may also have a negative signature and occur as dashes. We have observed variations of this noise on imagery from a wide range of scanners, aircraft, and localities. It seems to occur sporadically and does not hinder image interpretation. It has been suggested, but not established, that outside sources such as air traffic radars may be responsible. Electronic shielding of the scanner installation may reduce the effect.

### E. Processing Effects

During image recording of a flight line, on either tape or film, there may be a progressive change in the overall radiant temperature level. The operator may have to shift the recording base level to stay within the optimum range, resulting in an abrupt change in image density, as shown in Figure 15.16*G*. With direct film recording, base level shifts may be partially compensated for by printing the denser and thinner negatives with different exposures. With magnetic tape, the compensation may be accomplished during playback by monitoring the signal level and adjusting for base level shifts to obtain a uniform film density.

The photographic development of the image film (either directly recorded or played back from tape) is another potential source of image irregularities. The developer streak on Figure 15.16*H* resulted from uneven alignment of a pressure roller in an automatic film processor, which was corrected before further use.

### F. Incorrect Image Scale

Image scale is determined by the elevation above terrain and the geometry of the scanner. To achieve correct scale in the flight direction, the recording film speed must be related to the aircraft ground speed and elevation ($V/h$ ratio). Scanner manufacturers supply tables relating $V/h$ to film speed, and average elevation is determined during flight planning. Incorrect ground speed information is the main source of scale errors, which cannot be corrected on direct film recording. Magnetic tapes, on the other hand, may be replayed with a corrected film speed to obtain correct scale.

### G. Geologic Interpretation

Ground data collected concurrently with the imagery flights are called "ground truth" and may be useful in imagery interpretation. Air temperature, relative humidity, wind speed, and direction can be recorded by the ground personnel at each control station. Brief notes on vegetation and surface moisture distribution may also be useful.

Except for volcanic and geothermal areas with high thermal contrasts, the most successful applications of IR imagery have been in arid and semiarid areas with little vegetation cover. In vegetated terrain the IR imagery simply records the strong thermal patterns of vegetation, which effectively mask any underlying geology. On

0-201-04245-2

nighttime imagery standing bodies of water are relatively warm because of the high thermal inertia of water. In contrast, moist areas appear very cool because of the evaporative cooling of absorbed water. In humid areas surface moisture variations, rather than geology, dominate the appearance of the imagery.

A tabulation of published geologic applications of IR imagery is given in Table 15.4.

### H. Pisgah Crater, California

Radar imagery of this California area was described earlier, and the location of IR imagery (Figure 15.17) is indexed on the geologic map of Figure 15.8. The imaged area consists of Recent flows of fine-grained porphyritic vesicular Pisgah basalt. There are aa phases, which are very rough (Figure 15.9-6), and pahoehoe phases, which are smoother and ropy, although highly vesicular (Figure 15.9-5). The Pisgah cinder cone rises about 250 ft above the basalt flows and consists of basaltic pumice, cinders, and ash (Figure 15.9-4).

There is greater detail on both the day and night IR imagery (Figure 15.18) than on the air photo of Figure 15.7; this is due partly to the larger scale of the imagery. Also, reproduction of the photo has degraded some of the tonal contrasts. On the original air photos and to the field observer, the pahoehoe basalt has a dark gray color owing to its smoother surface and higher light reflectivity; this contrasts with the jet black appearance of the aa basalt.

The most obvious contrast between the day and night imagery is the topographic detail on the daytime version, caused by differential solar heating of south-facing slopes and shadowing of north-facing slopes. Pisgah cone, for example, is obvious on the daytime imagery but obscure on the predawn. Gawarecki (1969, p. 35) noted that the cinder cone images very warm in the daytime but cools rapidly at sunset in contrast to the basalt flows, which do not change as rapidly. The cinders and pumice act as an insulator and cannot conduct the solar input much below the surface, causing high daytime temperatures. At sunset the surficial heat is rapidly reradiated back to the sky, so that the cinders appear cool.

Table 15.5 and the geologic interpretation map of Figure 15.17*A* were prepared to facilitate comparisons of specific features on the two images. Pahoehoe basalt is warmer on daytime imagery and cooler on predawn, whereas aa basalt has the opposite expression. During the day the rougher and somewhat darker aa phase absorbs more solar radiation and may conduct it deeper into the rock, whereas the smoother pahoehoe reflects some solar radiation and restricts absorbed heat to a shallower depth, resulting in a warmer daytime temperature. At night the pahoehoe cools more rapidly than the aa, which retains its heat longer and images warmer (Garawecki, 1969, p. 34). Some Pisgah flows have a mottled warm and cool appearance because the pahoehoe and aa surfaces are intermingled. Also, isolated patches of wind-blown sand produce cold spots on the day and night imagery.

The warm nighttime appearance of the fissures and collapsed lava tubes may be due to their behavior as blackbody cavities. Heat absorbed during the day is radi-

0-201-04245-2

**TABLE 15.4**
**Geologic Applications of Thermal IR Imagery**

| *Locality* | *Geologic Features on Imagery* | *Reference* |
|---|---|---|
| Yellowstone National Park | Hot springs and geysers are clearly indicated on day and night imagery flown in winter time. Previously unknown hot springs were recognized in remote areas. One thermal anomaly may be a surface indication of a subsurface geothermal heat source. | McLerran and Morgan (1964); Miller (1966); Pierce (1968) |
| Pennsylvania | On nighttime imagery, a sandstone outcrop images relatively cooler than shale outcrop, probably because of differences in vegetation cover. Water in springs and a swamp images warm. | Lattman (1963) |
| Michigan and California | Effects of water and plants on imagery are considered. | Blythe and Kurath (1967) |
| California | San Andreas fault is imaged by variations in soil moisture caused by water-barrier effect of fault. Stratigraphic units, stream offsets, and landslide topography are also identified on imagery and are useful for fault recognition. | Wallace and Moxham (1967) |
| Various areas | Compares air photos and imagery to illustrate the interpretation of imagery. | Cantrell (1964) |
| Imperial Valley, California | First published example of mosaic constructed from IR imagery. Compared with aerial photo mosaic. Causes of image distortion are discussed. | Williams and Ory (1967) |
| Goldfield, Nevada | Faults, fractures, and alteration zones are imaged. Some features were initially detected on imagery and later confirmed in the field. | Kilinc and Lyon (1970) |
| Imperial Valley, California | An obscure and previously unmapped anticline was recognized for first time on nighttime imagery. The outcrops on flanks of the structure are more readily distinguished by radiance differences on imagery than by reflectance differences on photos. | Sabins (1969) |

0-201-04245-2

0-201-04245-2

| | | |
|---|---|---|
| Coachella Valley, California | On nighttime imagery siltstone and sandstone outcrops are distinguished respectively by cooler and warmer signatures. A covered portion of San Andreas fault is imaged by cool anomalies caused by evaporative cooling of shallow ground water blocked by fault. | Sabins (1967) |
| Pend Orielle, Idaho | In this forested area with few outcrops, there is little correlation between the imagery and geologic features. Drainage pattern is well expressed. | Harrison (1968) |
| Various areas | Instructive comparisons of various types of imagery. | Stingelin (1969) |
| South Texas | Soil characteristics and imagery expression are correlated to determine that water content, texture, and porosity are most important factors influencing thermal expression. | Myers and Heilman (1969) |
| Various areas | Emphasizes importance of vegetation and moisture content on imagery appearance. | Blythe and Kurath (1968) |
| Southwest New Mexico | Geologic features of alluvium and igneous outcrops in desert terrain are correlated with imagery. Of particular interest are unexplained lineaments on imagery that may be fractures. | Pratt (1968) |
| Hawaii | Imagery of Kilauea caldera and rift zones depicts extent and relative intensity of abnormal thermal features. Structural details are also shown. Large springs issuing into the ocean were detected. | Fischer et al. (1964) |
| Mono Lake, California | Shoreline springs and seepage are detectable on imagery. | Lee (1969) |
| Arctic areas | Glacial features such as crevasses, meltwater, and lakes are imaged and compared with photos. | Poulin and Harwood (1966) |
| Surtsey, Iceland | Volcanic activity and related structural features were imaged by airborne and satellite scanners and compared with ground measurements. | Friedman and Williams (1968) |

**TABLE 15.4** *(continued)*

| *Locality* | *Geologic Features on Imagery* | *Reference* |
|---|---|---|
| Taal Volcano, Philippines | Imagery flown 2 weeks after an eruption showed relatively little thermal activity at the site, aside from hot water issuing from a new cinder cone. At other craters, imagery showed more activity than previously supposed. No thermal evidence of a postulated rift zone was found. | Moxham and Alcarez (1965) |
| Mt. Ranier, Washington | Some thermal anomalies are related to well-known fumaroles. Other anomalies may be related to older volcanic features. | Moxham et al. (1965a) |
| Pisgah Crater, California | Textural differences between lava flows are imaged, as are fissures and collapsed lava tubes. Features of a dry lake and associated alluvium are shown in more detail than on other types of imagery, probably because of moisture variations. | Gawarecki (1969) |
| Geysers area, California | In this geothermal area, imagery anomalies generally coincide with hydrothermally altered steaming ground. At the surface, the principal high-temperature zones are limited on the southwest by structural lineaments; their northeast extent is more diffuse. | Moxham (1968) |

0-201-04245-2

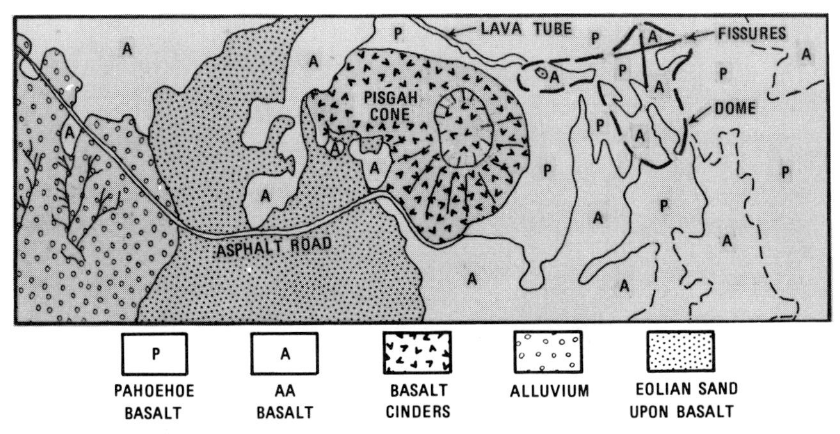


A. GEOLOGIC INTERPRETATION OF PREDAWN IMAGE.

B. PREDAWN IMAGE FLOWN FEB. 14,1965 AT 0629 HRS.

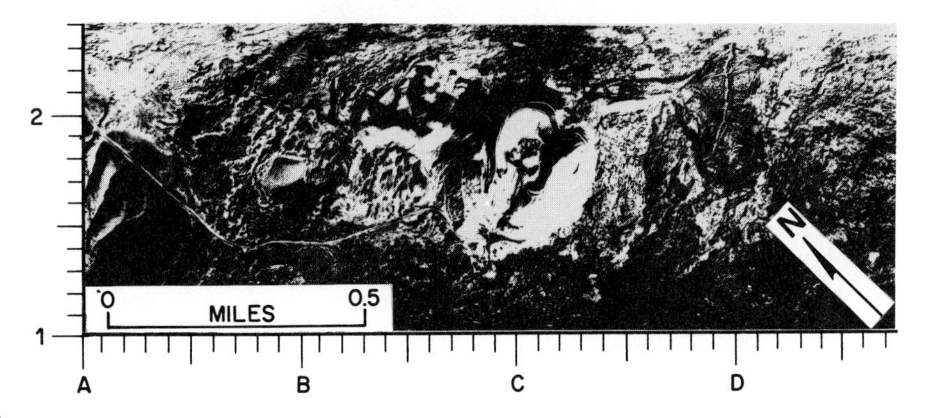


**Fig. 15.17.** Geologic interpretation of predawn and daytime thermal IR scanner imagery (8–14 μm), Pisgah Crater, California. Imagery courtesy of NASA Earth Resources Data Facility. See Figure 15.8 for location.

0-201-04245-2

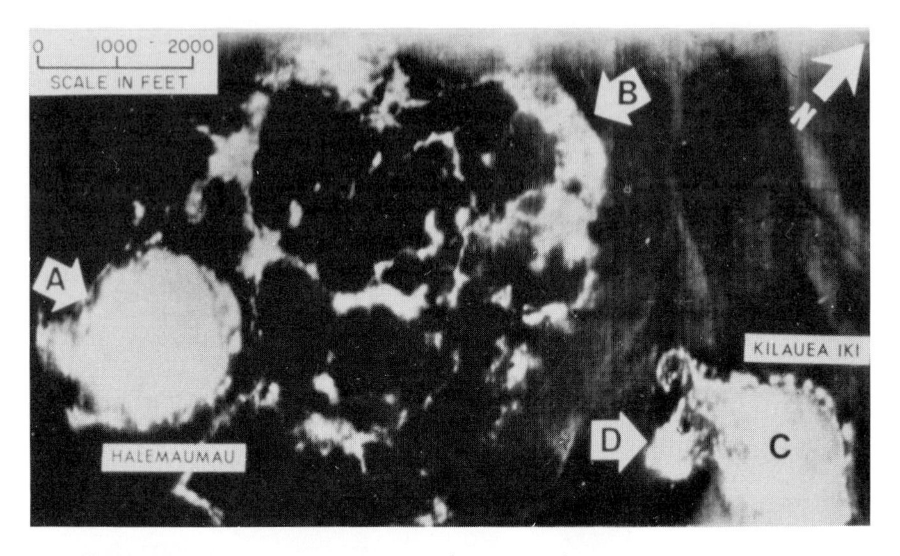


A. THERMAL IR IMAGERY (4.5–5.5) $\mu$m FLOWN JAN. 28, 1963 AT 0702 hrs; ALTITUDE 1800 METERS. SEE TEXT FOR EXPLANATION.

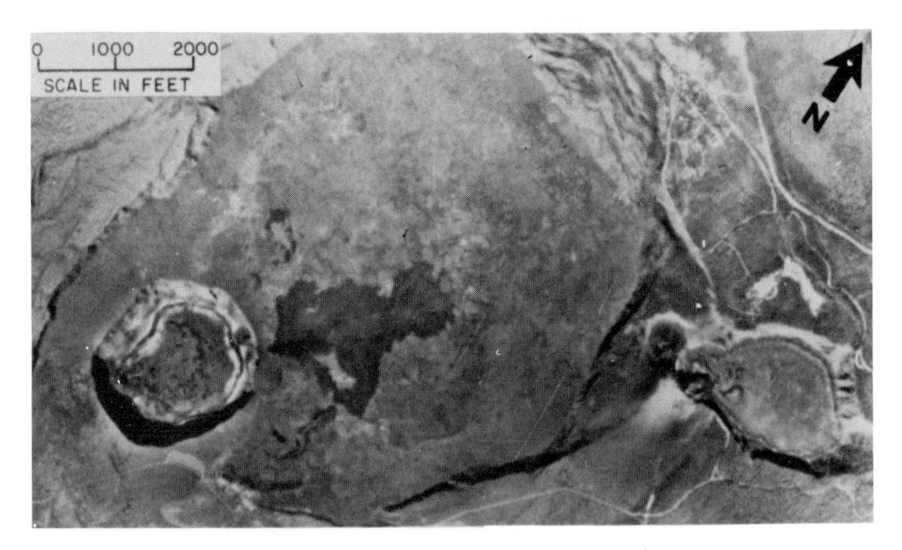


B. AERIAL PHOTOGRAPH

**Fig. 15.18.** Photography and IR imagery of caldera at summit of Kilauea Volcano, Hawaii. From Fischer and others (1964, Figures 8 and 11). Copyright 1964 by the American Association for the Advancement of Science.

0-201-04245-2

**TABLE 15.5**
**Interpretation of IR Imagery of Pisgah Crater Area**

| *Feature* | *Figure 15.17 Location* | *IR Imagery Expression* | |
|---|---|---|---|
| | | *Night Time* | *Day Time* |
| Pisgah cinder cone | 1.7, C.1 | Medium | Warm on sunlit side; cool on shaded side |
| Intersecting fissures in basalt | 2.2, D.0 | Warm | Warm on sunlit side; cool on shaded side |
| Collapsed lava tube | 2.4, B.9–2.2, C.2 | Warm | Cool |
| Aa basalt flow | 1.2, C.1 | Warm | Cool |
| Pahoehoe basalt flow | 1.7, C.5 | Cool | Warm |
| Alluvium | 1.5, A.2 | Cool | Cool |
| Wind-blown sand on basalt flow | 1.9, A.6 | Cool | Cool |
| Dry arroyos | 1.6, A.1 | Warm, possibly because of vegetation | Highlights and shadows on slopes |
| Metal-roofed buildings | 1.4, B.9 | Cold | Cold |
| Asphalt road | 1.4, A.7 | Cold | Warm |

ated outward at night; but rather than being lost immediately to the sky, some radiation is absorbed and reradiated from opposite walls of the cavities. Alluvium in the far northwest corner (lower left, Figure 15.17) of the imagery is cold both day and night, but the detailed warm expression of dry arroyos at night is noteworthy. This may be due to the blackbody behavior of the steepsided arroyos and to their concentration of vegetation. The diurnal radiometric expression of vegetation is discussed in the description of the Indio Hills locality in Section IV.K. In daytime only the sunlit slopes image warm. Wind-blown sand covering part of the basalt north of Pisgah cone is cool on both images. Buildings at the west foot of the cone are cold night and day because the metal roofs reflect the sky temperature, which is extremely low under cloudless conditions.

## I. Active Volcanism, Hawaii

After the discussion of extinct volcanic features at Pisgah, it is appropriate to consider the imagery of active volcanoes at Hawaii. This was described in the report by Fischer et al. (1964), from which the following discussion is summarized. The imagery and air photo of Figure 15.18 illustrate the caldera at the summit of Kilauea,

0-201-04245-2

a shield volcano on the island of Hawaii. The floor of the caldera consists of lava erupted in historic time, most recently in 1954. Steam issues from arcuate patterns of cracks on the caldera floor and adjacent areas. Within the caldera the crater of Halemaumau was formerly filled with liquid lava, but the crust is now solidified. Adjacent to the caldera on the east is Kilauea Iki, a crater filled by a lava lake during the 1959 eruption.

On the imagery some peripheral faults of the caldera form faint curved anomalies that are slightly warmer than the cooler background. One example appears to trend northwest from Kilauea Iki. Elsewhere the faults form more localized anomalies that correlate with steaming vents. The caldera is reticulated, with curvilinear elements of greatly varying intensity that also correspond in part to steaming fissures. One prominent subcircular feature (*B* in Figure 15.18) apparently corresponds to the buried margin of a sunken central basin that existed in the caldera during the nineteenth century. Other imagery shows that point *A* has the highest apparent temperature of the thermal anomalies at Kilauea. It is the vent and spatter cone of the July 1961 eruption into the floor of Halemaumau and is located where the southwest rift zone intersects the crater wall.

At Kilauea Iki an intense thermal anomaly occurs at the apex of the cinder cone (*D* in Figure 15.18) on the southwest flank of the crater immediately adjacent to the vent. Cinders on the crest of the cone are a bright yellow in contrast to dull gray on the flanks and base. This color contrast, evident in the air photo, is attributed to pneumatolytic alteration and deposition. The lava lake, formed during the 1959–60 eruption, is about 110 m deep; the solidified crust is now about 15 m thick. The molten lava at the base of the crust has a temperature of about 1065°C. A double row of vents bordering the lava lake and extending along the walls of Kilauea Iki is very warm on the image. These vents coincide with the peripheral fracture zone developed during back drainage of the lava. Although not distinct on this reproduction of Figure 15.18, there are differences in the apparent surface temperature of the lava crust. The central portion is hotter than adjacent areas, although there are no known corresponding compositional differences. There are differences, however, in the surface texture of the crust which relate to differences in its cooling history. It would be useful to compare temperature signatures of basalt types at Hawaii with their older counterparts at Pisgah.

In other Hawaiian areas repeated imagery flights have showed thermal changes with time; two such changes correlate with the location of subsequent eruptions, but the cause and effect relationship is uncertain. Imagery flights along the coast revealed 25 large springs issuing into the ocean (Fischer et al., 1966). The spring water images cold in contrast to the warmer ocean water.

Other areas of active volcanism that have been successfully imaged include Taal Volcano in the Philippines (Moxham and Alcarez, 1965) and Surtsey, Iceland (Friedman and Williams, 1968). The Friedman and Williams paper also includes imagery at Mount Aetna and Mount Vesuvius, together with a summary of other published and unpublished volcanic imagery.

0-201-04245-2

### J. Superstition Hills Fault and Anticline, California

This area on the west margin of the Imperial Valley was flown with 8–14 μm nighttime imagery by HRB-Singer Company in October 1963. A geologic interpretation by Sabins (1969) is summarized for the present description. The imagery and corresponding air photo are compared in Figure 15.19, and the geologic interpretation is shown in Figure 15.20. Aside from irrigated fields in the southern part, the area is very-low-relief desert with sparse vegetation. Scattered sand dunes are stabilized by mesquite bushes that image dark on the photo and bright (warm) on the imagery. Vegetation and sand are typically warm on this type of imagery. In the northern part of the image the characteristic scanner distortion of Imler Road, which is actually straight, can be noted. The road, which images warm, is surfaced with hard-packed sand. The faint trace along the power line on the imagery is actually a seldom-used access road. At the south edge, the Fillaree irrigation canal images warmer than the adjacent ground. Because of its high thermal inertia, water does not cool as rapidly as land after sundown and is warmer in the predawn hours. The thermal variations in the cultivated areas reflect different agricultural practices. The very-warm-imaging (bright) field north of the canal was probably flooded with water at the time to leach salts from the soil. The very-cool-imaging (dark) fields had probably been recently irrigated but had no standing water, so that evaporative cooling of the damp ground caused the cold pattern. Fields with intermediate temperatures probably had intermediate moisture content.

Bedrock in the area consists of nonresistant, brownish gray, lacustrine siltstone with thin interbeds of well-cemented brown sandstone. Flaggy pieces of the sandstone litter the surface where it crops out. Light-colored, nodular, thin layers within the siltstone help to define bedding trends in this otherwise monotonous sequence of late Tertiary age. Much of the area is covered by wind-blown sand, which supports most of the sparse vegetation. Numerous exposures of bedrock indicate that the sand rarely exceeds a few feet in thickness. The El Centro sheet of the geologic map of California, prepared by the California Division of Mines and Geology, shows that this area was covered by ancient Lake Coahuilla during Quaternary time. Except for some large boulders coated with travertine deposits, any Coahuilla deposits have been reworked by the wind.

On the imagery the bedrock outcrops are dark (cool), and the wind-blown sand is shown by lighter shades of gray (warm). The very warm Y-shaped image in the west-central part of the area represents a thick accumulation of wind-blown sand lodged against an earthen levee.

The major geologic feature is the east-plunging fold pattern in the center of the imagery, which had not previously been recognized. Fieldwork showed that this is an anticline in the Tertiary siltstone bedrock with an axial length of about 1.2 km and a width of 400 m. Had the imagery not indicated its presence, we could have walked across the anticline in the field without recognizing it, for there are no conspicuous lithologic or topographic patterns. The plunge of the anticline was defined

0-201-04245-2

A. VERTICAL AERIAL PHOTOGRAPH FLOWN MAY 5, 1953 FOR U.S. GEOLOGICAL SURVEY.

**Fig. 15.19.** Photography and IR imagery of Superstition Hills fault and anticline. From Sabins (1969, Plate 1)

0-201-04245-2

B. PREDAWN THERMAL IR SCANNER IMAGERY, FLOWN OCTOBER 1963 BY HRB-SINGER CO.

Fig. 15.19. (*continued*)

0-201-04245-2

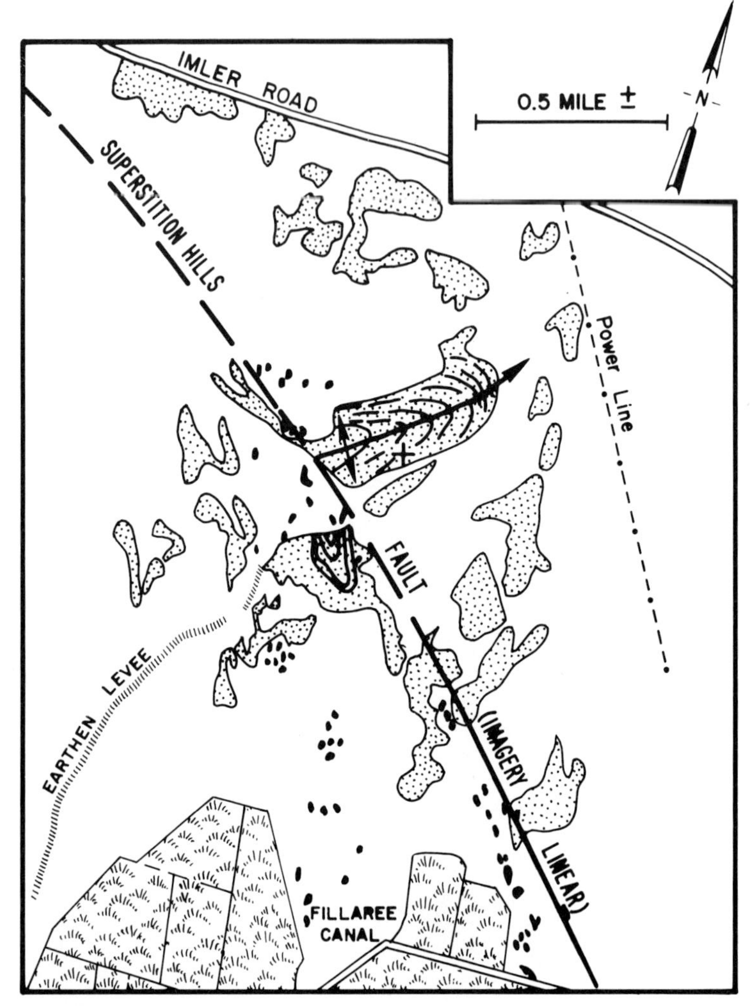


OUTCROPS OF DEFORMED TERTIARY LAKE DEPOSITS, SHOWING TRENDS OF BEDDING.

RECENT WIND BLOWN SAND COVER.

CULTIVATED AREAS.

STABILIZED SAND DUNES AND TUFA-COATED BOULDERS.

+ BRASS CAP MARKING SE CORNER SEC. 25, T.14S, R12E.

**Fig. 15.20.** Geologic interpretation of IR scanner imagery shown in Figure 15.19. Modified from Sabins (1969, Figure 5).

0-201-04245-2

by walking out the light-colored nodular beds and the resistant sandstones. Although structural attitudes are obscure in the siltstone, dips up to 45° were measured in the sandstones and the dip reversal across the fold axis was located. A broad, low ridge with up to 30 ft of relief coincides with the general area of the anticline.

On Figure 15.20 the anticline is shown as solid bedrock, but actually there are numerous thin patches of wind-blown sand, which account for the gray tones on the imagery. The dark-imaging core of the anticline consists of siltstone. The alternating light and dark pattern outlining the limbs of the fold appears to correlate with outcrops of sandstone (warm) and siltstone (cool), respectively. This was confirmed, although 4 years later (March 1968), by predawn measurements with portable radiometers. The sandstones generally measured 2°F warmer than the siltstones. The sand dunes, vegetation, and wind-blown sand were warmest, as indicated on the imagery. This relationship agrees qualitatively with the results of a similar survey made in the Indio Hills north of the Salton Sea (Sabins, 1967).

South of the anticline there is a small arcuate pattern of alternating dark and light bands. Field inspection showed this to be an exposure of gently dipping siltstone and sandstone bedrock with an outcrop pattern generally resembling that of the imagery. The pattern is more pronounced on the imagery than on the ground.

The west end of the anticline is truncated abruptly along a line coincident with the southeastward projection of the Superstition Hills fault, as shown on the El Centro sheet of the geologic map. The inferred trace of the fault is obscured by wind-blown sand, but siltstone outcrops in the immediate vicinity of the anticline are strongly deformed, suggesting drag folding. Along the projection of the fault in the southeast part of the image there is a subtle, but definite, southeast-trending linear marked by the sharp contact between cooler areas on the east and warmer ones on the west. The trend of the linear is parallel with, and about 160 m to the east of, the row of prominent warm sand dunes. In April 1968 an earthquake caused surface breaks along the trace of the Superstition Hills fault which were mapped by A. A. Grantz of the U.S. Geological Survey and M. Wyss of California Institute of Technology. Their unpublished field map (Grantz, personal communication, July 10, 1969) shows a break with less than 3 cm of right-lateral displacement that closely coincides with the linear on the imagery, which was flown 5 years before the earthquake. It appears that the imagery effectively pinpointed another important structural feature which is obscure on photographs and in the field.

The most recent large-scale aerial photography of the area was flown in May 1953 for the U.S. Geological Survey and predates the imagery by 10 years. The imagery itself was not available to us until 1967. Despite the elapsed time, it was possible to correlate features on the photography, on the imagery, and in the field with a high degree of confidence. This is probably due to the combination of low-relief terrain, in an arid, uniform climate, with little human activity, so that changes are minimal.

The most striking and obvious difference is the greater detail and contrast on the imagery than on the photography. In this area the range of nighttime thermal emittance variations is apparently greater than the range of daytime visible reflec-

0-201-04245-2

tance variations. Our own visual experience supports this statement, for anyone flying or walking over this country is impressed by its generally monotonous appearance. Although the lack of contrast on the aerial photography might be attributed to incorrect exposure or processing, inspection in the field shows that the photography is a faithful reproduction of the appearance of the terrain. Dark green mesquite bushes photograph black, and light-colored wind-blown sand photographs white, so the full dynamic range of the film was utilized. Under stereoscopic viewing, individual clumps of vegetation and minute details of drainage can be resolved; hence the photography is representative of the state of the art. On the basis of work in nearby areas, Howard and Mercado (1970) suggested that low-sun-angle photography might enhance the visible-light presentation; this is entirely possible. R. J. P. Lyon has successfully enhanced this anticline with photography taken with late afternoon light (Lintz, personal communication).

The geologists with whom I have consulted agree that during a normal stereoscopic study of the aerial photography they would not have noticed the anticline. After examining the infrared imagery, however, they were able to locate the anticline on the photography.

### K. Indio Hills, California

This area on the east side of the Coachella Valley was flown with nighttime imagery (8–14 $\mu$m) in August 1961 and described by Sabins (1967). The hills are a ridge of deformed clastic sedimentary rocks of late Tertiary age that trend southeast parallel with the San Andreas fault zone, which cuts the hills as shown on the geologic map of Figure 15.21. This arid terrain with little vegetation and excellent bedrock exposures is ideal for IR imagery. The imagery and corresponding air photos (Figure 15.22) cover the southern part of the hills. Alluvium surrounding the hills has a relatively cool and featureless expression on the imagery. Two types of bedrock are readily distinguished on the imagery. One type consists of poorly stratified, moderately to poorly consolidated conglomerate, sandstone, and siltstone that image with a relatively warm and uniform appearance. The outlying hill in the lower center of the imagery is a good example.The other bedrock type is the Palm Spring formation, consisting of well-stratified alternating beds of sandstone and siltstone up to 40 ft thick that weather to a ridge and slope topography formed by the more resistant sandstone and nonresistant siltstone. This well-stratified bedrock images with a distinctive pattern of alternating warm and cool bands. Careful correlation of the imagery, air photos, and field outcrops established that on the nighttime imagery the sandstones are warm and the siltstones cool.

Quantitative radiant temperature measurements were needed to verify this relationship, but portable field radiometers were not available until recently. Therefore large samples of the sandstone and siltstone were placed out of doors in Fullerton, California, where electric current was available for a laboratory radiometer operating in the 8–14 $\mu$m spectral range. During midday in October the siltstone

0-201-04245-2

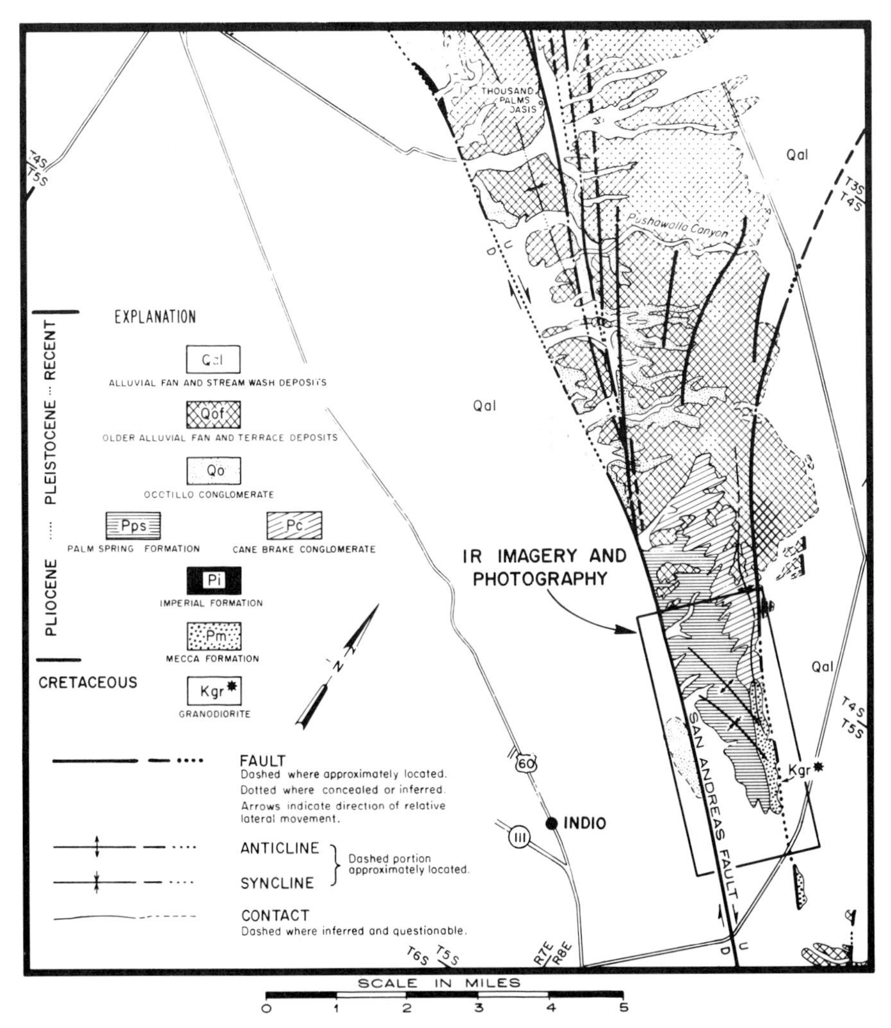


**Fig. 15.21.** Geologic map of south part of Indio Hills, Riverside County, California. Note location of IR imagery and photo of Figure 15.22. From Cummings (1964, Plate 3-B).

reached a maximum temperature of 37°C and the sandstone 35°C; both units cooled down in the afternoon, with the siltstone remaining a few degrees warmer. In early evening, however, there was a thermal crossover, for the siltstone became cooler than the sandstone. The latter condition agrees with the relative temperatures of sandstone and siltstone shown on the nighttime imagery. This experiment also demonstrated that the different topographic expressions of sandstone and siltstone outcrops do not control temperature variations, for both samples were equally exposed.

0-201-04245-2

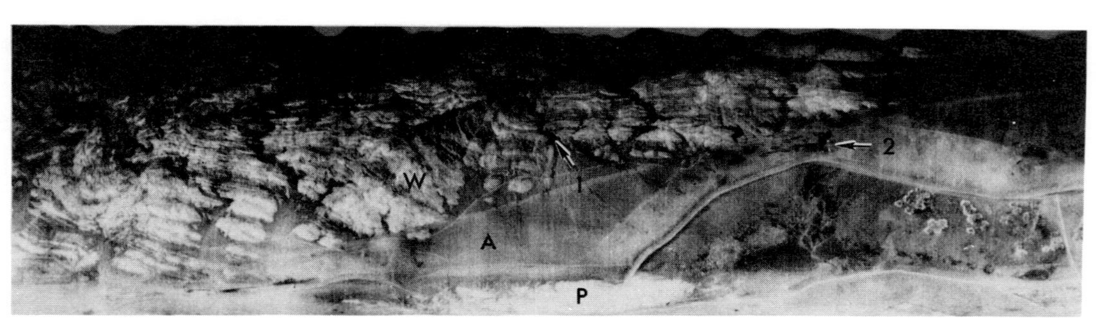


**Fig. 15.22.** Nighttime IR imagery (lower) and aerial photo (upper) of south part of Indio Hills, California. *A*–alluvium, *P*–poorly stratified bedrock, *W*–well-stratified bedrock. 1–locality for radiometric measurements of rocks, 2–locality for vegetation and soil measurements. From Sabins (1967, Figures 3 and 4).

0-201-04245-2

When a portable radiometer became available in January 1969, we made temperature measurements on eight pairs of sandstone and siltstone outcrops at locality 1, indicated in Figure 15.22. As shown in Figure 15.23, four sets of measurements were made: midnight, predawn, immediately postdawn, and daytime. In the predawn and early morning measurements the sandstones are warmer than the siltstones; when the canyon is fully illuminated by 8:35 A.M., however, the siltstones have warmed up much faster than the sandstones, a finding consistent with our earlier experiments. Figure 15.24 is a summary of the outcrop measurements. For each of the four times of measurement the average sandstone and siltstone temperatures are plotted. In the predawn measurements the sandstones are consistently warmer than the siltstones, as shown on the imagery. A thermal crossover occurs shortly after dawn with the siltstones becoming warmer than the sandstones. It is significant to note how rapidly the radiant temperatures of both rock types increase after daybreak.

The sparse vegetation images with a conspicuous warm pattern; for instance, between the canal and embankment at the lower right corner of the imagery there are irregular bright rings. These are large patches of desert shrubs that are green at the margin but dead in the center. In contrast to the warm nighttime expression, vegetation in the desert typically appears cool in daytime imagery. In order to evaluate this phenomenon quantitatively, a diurnal sequence of radiometer measurements were made at locality 2 on Figure 15.22. Here two large salt cedars and a third, smaller one imaged distinctly warmer than the surrounding bare alluvium. From noon onward, radiometric temperature measurements were made for the salt cedars plus three smaller shrubs and the values were averaged. For each observation period six soil temperature measurements were also averaged. The results are plotted in Figure 15.25, together with air temperature readings. This diagram shows that at night vegetation is consistently warmer than the soil, the maximum differential being 4°C. The reverse is true during the day, when the soil is much warmer than vegetation. It will be noted that the thermal crossovers occur within an hour's time both in the evening and in the morning.

The structural geology of the Indio Hills is also well expressed on the imagery. A series of plunging anticlines and synclines in the well-stratified Palm Spring formation is shown by the configuration of the warm and cool outcrop pattern. The most impressive structural feature is the San Andreas fault, which borders the west side of the Indio Hills. To the south it passes along the east side of the warm-imaging outlayer of poorly stratified Ocotillo conglomerate; farther south the fault trace is concealed by alluvium with no topographic expression. On the aerial photo the alluvium-covered trace of the fault is marked on the east side by denser vegetation that terminates abruptly at the trace (Figure 15.22). On the IR imagery the fault trace is clearly expressed by an alignment of very cool (dark) anomalies along the east side. The cool anomalies are not related to vegetation distribution, for we have already demonstrated that the vegetation is warm on this imagery, but are probably related to the barrier effect of the San Andreas fault on groundwater movement. In the spring of 1961, which was a few months before the IR survey,

0-201-04245-2

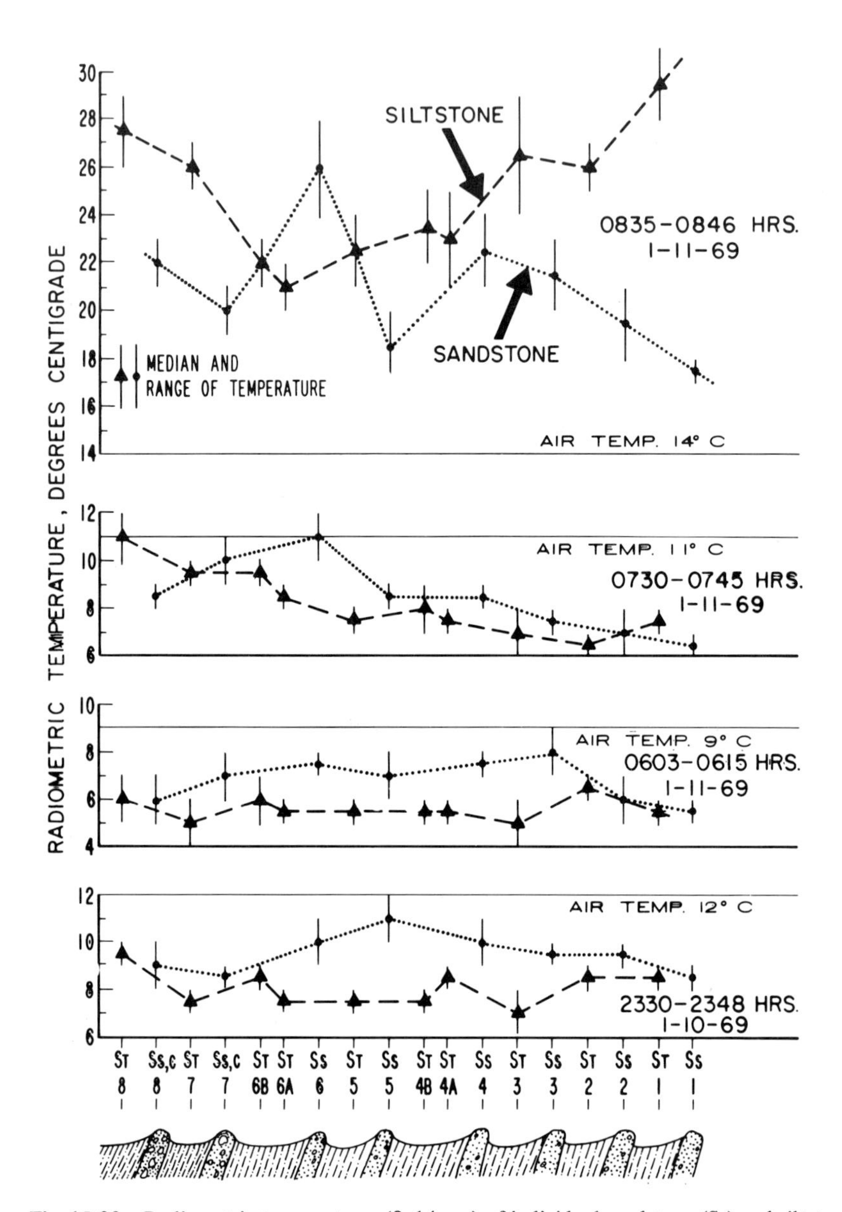


**Fig. 15.23.** Radiometric temperatures (8–14 μm) of individual sandstone (Ss) and siltstone (St) beds of Palm Spring formation, Indio Hills. Temperatures were measured at four different times of day and illustrate the thermal crossover of sandstone (dotted curve) and siltstone (dashed curve) near dawn. Measurement stations are shown on the diagrammatic stratigraphic section, which represents approximately 400 ft of beds. From locality 1 of Figure 15.22. From Sabins (1976, Figure E-18).

0-201-04245-2

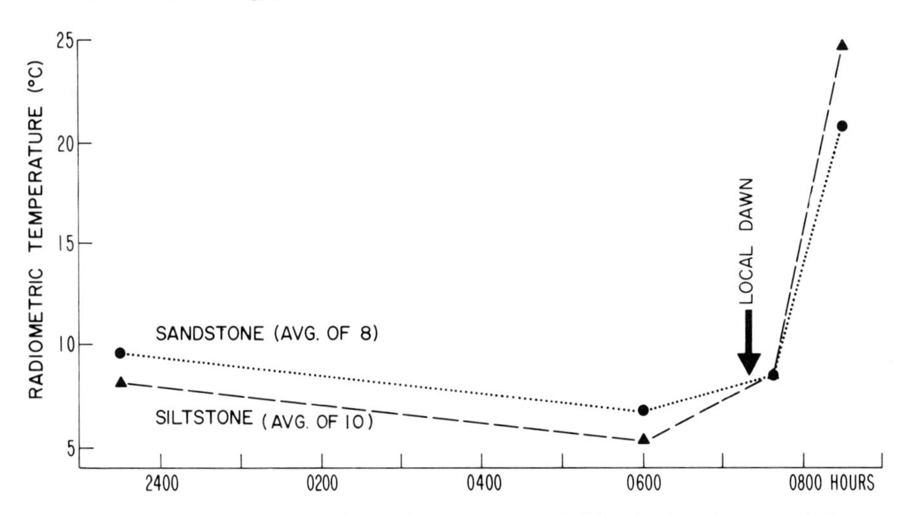


**Fig. 15.24.** Average diurnal radiometric temperatures (8–14 μm) of sandstones and siltstones of Palm Spring formation, Indio Hills, California. From locality 1 of Figure 15.22.

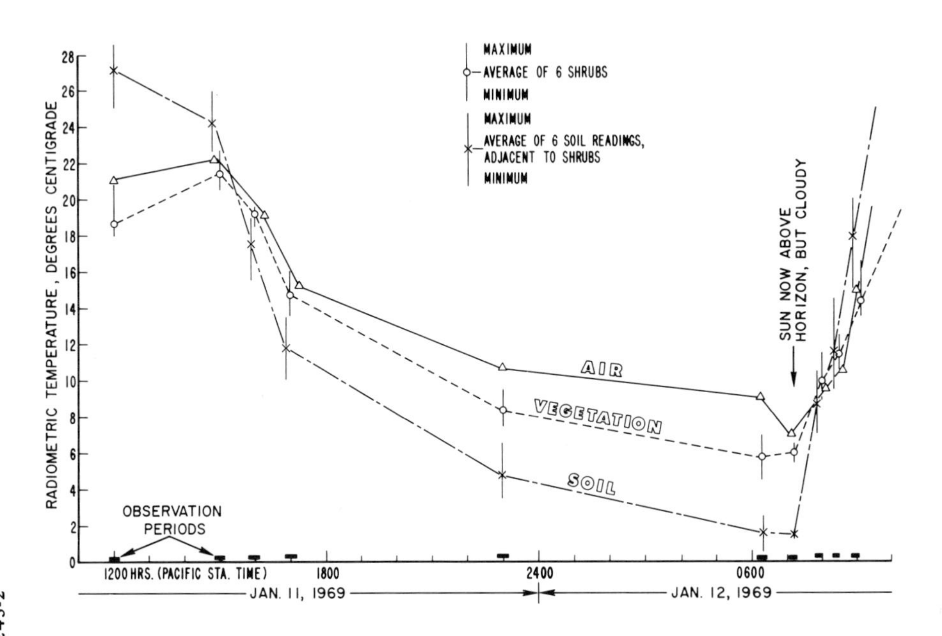


**Fig. 15.25.** Diurnal radiometric temperatures (8–14 μm) of vegetation and soil at Indio Hills. From locality 2 of Figure 15.22. From Sabins (1976, Figure E-19).

0-201-04245-2

Cummings (1964, p. 34) reported a 15 m difference in water table elevation across the fault in this vicinity. The shallower water table and higher moisture content on the east side of the fault could cause evaporative cooling. Similar anomalies appear on nighttime imagery of the San Andreas fault in the Carrizo plains approximately 320 km to the northwest (Wallace and Moxham, 1967, Figure 5). On other Indio Hills imagery (Sabins, 1967, Figure 2) the Mission Creek fault was also indicated by cold anomalies at Thousand Palms Oasis, where the fault serves as a groundwater barrier.

## L. Arbuckle Mountains, Oklahoma

In this area it is difficult or impossible on aerial photos to differentiate between limestone and dolomite outcrops of lower Paleozoic age. On predawn thermal IR imagery, however, Rowan et al. (1970) found sufficient thermal contrast between dolomite (warm signature) and limestone (cool signature) to distinguish these types.

In Figure 15.26*A* a sinuous band of warm-imaging dolomite outcrop trends generally north, bordered on the west by relatively cool limestone. On the southeast the dolomite is interbedded with limestone, producing an alternating warm

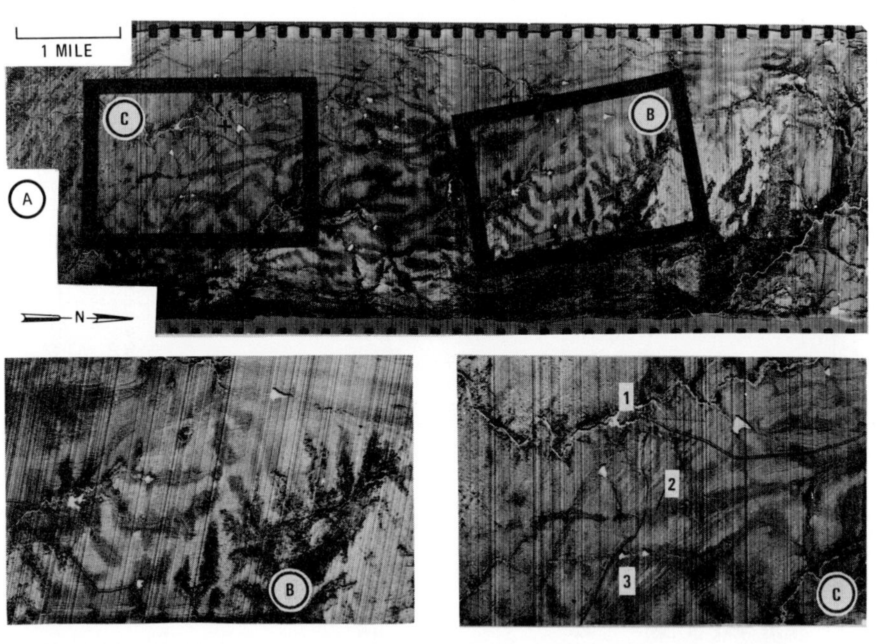


**Fig. 15.26.** Predawn thermal IR imagery (8–14 μm) of Tishomingo anticline area, Arbuckle Mountains, Oklahoma. A–north-trending sinuous belt of warm (bright) dolomite outcrop; locations of B and C are also shown. B–detail of interbedded limestone (cool) and dolomite (dark) at lower left; C–three cool-imaging linear fracture zones. All imagery from Rowan et al. (1970).

0-201-04245-2

and cool pattern shown in detail at the lower left of Figure 15.26*B*. This is also a good example of proper flight line orientation; if the scan lines were parallel with the outcrop strike, the rock type differentiation would be obliterated. Rowan et al. (1970) attribute the thermal contrast to a combination of albedo and thermal inertia characteristics typical of limestones and dolomites in many areas.

Fracture or fault zones not readily detectable on air photographs are shown on the imagery with cool, linear patterns (Figure 15.26*C*). In the field, Rowan and his colleagues found the fracture zones to be saturated with groundwater and attributed the cool signature to evaporative cooling.

### M. Airborne Infrared Emission Spectra

There are absorption bands in the energy spectra emitted from rocks and minerals that can be used to identify specific types. Lyon and Patterson (1969) described an airborne IR spectrometer operating in the 9–14 $\mu$ atmospheric window that records the emitted spectra for discrete patches of terrain. A computer program then identifies the unknown spectra by comparing them with a library of spectral signatures for known rocks. Test flights have successfully identified various igneous rocks in the Sierra Nevada and Pisgah Crater (Lyon, 1972). This is not an imaging system, but it has considerable potential for remote identification of rock and mineral composition.

## V. SUMMARY AND CONCLUSIONS

The preceding imagery examples were selected to illustrate the range of applications of the different types of remote sensing, which are summarized in Table 15.6. The first requirement for a successful geologic application of remote sensing is to select the optimum sensor or combination of sensors, based on the geologic objectives and the type of terrain. The nature of the geologic objective is vital; for instance, thermal IR or radar imagery could not be used in a search for surface color alteration associated with mineral deposits. On the other hand, a regional study of fracture patterns related to mineralization could utilize radar imagery. Thermal IR imagery may be effective for fault detection in arid country but has little geologic value in humid, forested terrain.

Aerial photography is the oldest form of remote sensing, but the geologic applications have by no means been exhausted. Infrared color photography in particular, and possibly multiband photography as well, have a wide range of potential applications. Of all the remote sensors, thermal IR imagery is most strongly influenced by surface environmental conditions, such as moisture, wind, and temperature. This fact, plus the requirement for nighttime flying, makes it more difficult to acquire satisfactory IR imagery, but the results may be worth the effort. Radar imagery has demonstrated potential for regional structural analysis, particularly in forested terrain. In recent years there has been much progress in making radar and IR im-

0-201-04245-2

**TABLE 15.6**
**Summary of Geologic Applications of Remote Sensor Imagery**

| *Method* | *Wavelength Region* | *Properties Imaged* | *Requirements for Optimum Geologic Imagery* | *Advantages for Geologic Interpretation* |
|---|---|---|---|---|
| IR color photography (aircraft) | 0.5–0.9 $\mu$ | Reflectance of green, red, and photographic IR | Daylght, clear weather | High resolution, good penetration of atmospheric haze. Good and versatile technique at large and small scales. Discriminates foliage types, drainage patterns, moisture variations. |
| Multiband photography (aircraft) | 0.4–0.9 $\mu$ | Reflectance of narrow spectral bands | Daylight, clear weather | Main potential is for enhancing subtle surface variations that are not apparent on photography. For maximum benefit, target and background spectra should be determined in order to select optimum spectral bands. |
| Satellite photography | 0.4–0.9 $\mu$ | Reflectance of visible and photographic IR | Daylight, clear weather | The above types of photography and conventional photography may be acquired. Main advantage is the repeating broad-area coverage with uniform lighting obtained on each frame. Useful for regional analyses. |
| Thermal IR imagery | 3–5 and 8–14 $\mu$ | Temperature and, to a lesser degree, emissivity | Nighttime, clear weather | Variations of rock type and moisture content may be more apparent than on other types of imagery. Optimum terrain is relatively dry with minimal vegetation. May detect thermal anomalies in volcanic and geothermal areas. |
| Radar imagery | Few centimeters | Surface roughness and, to a lesser degree, electrical properties | Day or night; rain is only interference | Low-resolution, small-scale imagery particularly suited for regional reconnaissance. Especially useful in forested terrain and areas of poor flying weather. Fracture patterns and lineaments are commonly enhanced. |

0-201-04245-2

agery more widely available. If this trend continues, this chapter can be rewritten in a few years with new examples and new applications.

Although most remote sensors form imagery using electromagnetic energy of wavelengths longer than light, there is no significant penetration of the earth's surface. For geologic features to be imaged, there must be a surface manifestation, which may be expressed in a manner not apparent to the eye. Temperature, emissivity, surface roughness, and electrical properties are detectable by suitable remote sensors under appropriate conditions. The major application of remote sensing for geologic exploration will be for airborne reconnaissance surveys to indicate areas of interest for more detailed and expensive surface studies.

0-201-04245-2

Chapter 16

# Remote Sensing for Water Resources

*Herbert E. Skibitzke*

## I. INFORMATION NEEDS IN HYDROLOGY

The type of information needed for hydrologic engineering now is not the same as was required 50–70 years ago. At the turn of the century it was enough to know the amount of water in a given section of a river system, or to know where groundwater could be found below the earth's surface. The development of the water resources in the West initiated a demand for types of information that had not been needed before. The design of reservoirs and water supply schemes for surface-water systems required data on the average amount of water available. In other cases, information on the flood peaks and the repetition interval between their occurrences was needed for the planning of public works.

As overdevelopment of surface waters occurred, the statistical information gathered in the traditional data collection program was no longer pertinent. For the western United States it has often been said that about 50 years would be needed to gather sufficient statistical information to determine the average long-term water supply at a given gaging station. During this period it should also be possible to determine the expected peak flow for future flood predictions. The measurement of stream flow along a river at a given point is a relatively simple engineering task. Thus it proved easy to locate a point where measurements could be made, to set up an adequate gaging system, and to man the station. The statistical data gathered should have been adequate for determining the long-term average water supply and flood peaks if sampled over a sufficiently long period of time. However, man-made developments along western rivers were occurring even as the measurements were being made. During the last 50 years such great changes have occurred that large rivers such as the Salt and the Santa Cruz rivers in Arizona no longer flow at their lower ends. The Colorado River has ceased to flow into the Gulf of Mexico. The whole surface flow system has been greatly modified by the works of man. Groundwater development along certain rivers has begun to dry up the river systems. The problem of interference between well development programs and rivers has become a significant issue. These factors render the statistical sample obtained of little value. And, of course, it would not be possible to go back and start another 50-year measurement system to determine what the flow would be at any given point because of these modifications by man.

Engineering data are needed today. The upper watershed use and watershed modifications have so completely changed the river flow systems that the statistical information derived earlier has no application to today's problems. It is now neces-

*Remote Sensing of Environment*, Joseph Lintz, Jr. and David S. Simonett (eds.) ISBN 0-201-04245-2


0-201-04245-2

sary to compute the effects of man on his environment in order to apply the data already collected. Water systems can no longer be analyzed from data acquired by statistical measurement at a single point; rather, data which allow evaluation of an entire region as it applies to the control of water supply are required. Such data could be used for computing the effects which man has wrought on the region and then would allow the application of this knowledge to changes that are projected for the water system. Measurements over the entire region would be required.

For a regional analysis, a large number of engineers would be needed to evaluate the environmental effects and their changes; but, among the other shortages that have occurred in the last 50 years, is a paucity of engineers. Salary costs have also risen sharply, so that it is highly desirable to reduce engineering staffs rather than to increase them. Yet to study the environmental impact of man on an area there would have to be sufficient manpower to cover the whole region. The application of remote sensing techniques can solve the problem of high engineering costs.

During the last two decades the ability to measure certain parameters from an airplane or a satellite has increased tremendously. Remote sensing techniques are now being applied to measurements concerning the dynamics of the environment. Environmental changes are being measured experimentally and are then being evaluated in terms of their effects on the hydrology of the region.

In most cases the data that will be most valuable in the next few decades will be those that document the changes induced by man over a region concurrently with the expansion of civilization. Of course, there are also important natural changes. Disasters such as hurricanes have violent effects on the environment. It is highly desirable to measure these effects and, to a certain degree, to determine quantitative parameters, so that damage caused by similar occurrences in the future can be evaluated. The day of point-measurement environmental engineering is gone in America. Possibly a few places in which exploratory work would be of interest may remain, but not in this country.

Evaluation of the effects of groundwater pumping will become more dependent on pumping patterns and water utilization in the area. Engineering analyses have already disclosed the nature of the aquifer, or the rock media, in which the water moves. The aquifer now becomes part of a piece of machinery, the machinery of water supply. Man is beginning to control where and how much water goes into the aquifer. He decides where and when it will be removed at a later time. More data will be required about the works of man. These, combined with the data already obtained on the nature of the rocks in which the water is flowing, can be used to analyze the water supply problem. Even today water levels in western regions depend more on the political question of how much agricultural development will occur in an area than on any natural factors. All these considerations lead the hydrologist to the conclusion that it will be necessary to collect vast amounts of information rapidly over large areas to obtain the answers that he needs. Moreover, such an operation must be automated because the engineering costs of even simple data collection by other means would be much too high.

0-201-04245-2

## II. A CONSIDERATION OF HYDROLOGIC PARAMETERS

It is evident, after analyzing current data needs in hydrology, that the application of remote sensing techniques for the purpose of exploration would have little value. Remote sensing is more important as an aid in the study of the dynamics of the environment and the changes that are being wrought on the environment by man. Pollution, for example, has become one of the biggest problems today in water supply and will be an increasing cause of concern in the years to come. The analysis of a pollution problem requires the determination of the chemical or biologic factors that are causing a water supply to be damaged in some way.

Sediment load and thermal properties affect the water supply in important and problematic ways. A modern power plant can dump into a lake or river sufficient quantities of hot water to completely upset the ecological balance that existed before the plant was built. It is obvious that sensors that will identify or measure the chemistry, the thermal nature, or the amount of sediment in the water would be valuable aids in determining the factors of pollution. All such parameters, of course, are measured in terms of some geometric function.

It is not of any great interest to know only the magnitude of the chemistry or sediment factors; the geometry of the region that is being invaded by these pollutants is a subject of much greater concern. The remote sensing device, then, would be an instrument that would depict, in a geometric sense, the pattern of contamination as it changes or as it is originally found. The rapid growth of algae as a by-product of chemical changes, for instance, is becoming an important source of pollution. In such a case it is important to be able to determine the geometric factor almost as a motion picture of the changing environment. Pollution problems must be attacked in this manner, and techniques for doing so will be developed in the coming years.

The use of water by man often takes the form of converting liquid water to water vapor through evapotranspiration processes. In many places in the world where water is in short supply, all of the available water is being converted by nature through one of these processes. In other words no liquid water is available. In this case the hydrologist's aim is to reroute the lost water through a new evapotranspiration process that is beneficial to man. It is necessary then to map the part of the region that is involved in the evapotranspiration effect. From these data the changes that man is planning can be analyzed.

Many of the changes being planned by man will require monitoring for many years to come. Infrared photography can provide the means for analysis of evapotranspiration changes over an entire region much more satisfactorily than can a point-measurement system.

Important aspects of groundwater systems can be analyzed by remote sensing, although in many cases it would appear unlikely that remote sensing techniques would be useful. However, as groundwater systems are developed in our country, data concerning the physical parameters at depth become known. These depth factors, such as permeability and storage coefficient, do not change. The factors

0-201-04245-2

that change in groundwater systems occur along the boundary, particularly the upper boundary, of the aquifer; examples are the evapotranspiration losses, the development of wells, and the seepage of water downward to the groundwater system from the irrigation of new farmlands.

Actually, in time man will make the groundwater system a part of the total water machine that he uses. The permeable aquifer will be a unit in the motion system that will occur in moving water from one part of the hydrologic cycle to another. The nature of the bounding limits of the aquifer, in many cases, can be measured easily by remote sensing techniques. However, groundwater boundaries are subject to changes. These changes, both as causes and as effects, are the features of the aquifer that will be important in the future.

The form of engineering analysis used by the hydrologist is an expansion of the data concerning a large region into the terms of a partial differential equation describing its physics. Inherent in the partial differential equation are certain parameters that describe the nature of the aquifer controlling the groundwater movement within it. To find a significant solution to a water problem in terms of this differential equation it is necessary to apply a boundary condition, such as the amount of water that is leaving by evapotranspiration or that is seeping downward from irrigation on the ground surface.

The description of the internal boundary of this same set of differential equations is affected by the drilling of wells in the region and the lowering of the water table near a well by pumping. This of course causes movement of water toward the well and eventually changes the water level throughout the region.

Surface boundary conditions can be measured from data collected by remote sensing techniques. As groundwater systems are developed, the discharge of groundwater from the limits of the boundary decreases. Phreatophytes begin to die, and springs to dry up. These occurrences can be seen clearly by using infrared photographic techniques. Groundwater systems, then, have many different parameters that will become of increasing interest to the community. These parameters can best be measured by remote sensing, unless very expensive field engineering techniques can be afforded.

The lakes of the world are another important hydrologic consideration. Remote sensing techniques will be a great aid in analyzing their dynamics. Large lakes, such as the Great Lakes system, are affected by several dynamic systems occurring simultaneously. A proper analysis would require measurements of all the parameters of the lake at one given instant; no method is available for making such massive synoptic measurements by surface operation. If measurements are to be made of the effects of dispersion, for example, or the regional movement of contaminants in a lake, at a given instant, it would be necessary to cover the entire lake by man engineering parties, making individual measurements simultaneously, something that would not be feasible. If, on the other hand, an overhead system that looked down and measured these parameters rapidly were used, the mechanics of a lake could be determined in terms that have not been possible to date. Of course, this again requires the conversion of some of the significant parameters of lake physics to

0-201-04245-2

parameters that can be measured by remote sensing instruments. Here too sensors are measuring parameters in terms of an overall geometry that allows the analysis of large-scale influences.

For large lakes the synoptic measurement techniques that are possible by remote sensing are probably of greater interest than any other single item. However, individual measurements of energy losses in a lake by devices that could be moved rapidly from place to place by helicopter are also of great interest. Thus the importance of individual measurements of such things as shear velocities in lakes, or the dying out of potential energy impulses into lake systems, that could be measured by devices carried within a helicopter must not be overlooked. Once again, a geometric pattern is being painted of some significant factor that can be seen by these devices.

Small lakes are also of interest. In many places in the United States, particularly in the North Dakota, South Dakota, and Nebraska regions, there are literally millions of lakes. If it were possible to study each of these lakes for 1 dollar, the total cost would run into millions of dollars. This amount of money is not and will not be available. However, these lakes have tremendous value, and it is becoming increasingly important to know their characteristics and the changes that are occurring in them.

The chemical nature of water, of course, often affects its color. The sediment in a lake can affect both its color and its infrared reflection; the amount of plant life also affects the IR reflection. The boundaries of lakes and their modifications as water level changes occur are simply geometric factors that almost any sensor "sees." The lake environment–in other words, the cities, towns, homes, and roadways around the lake–can easily be seen by aerial photography. It appears, then, that many of the data (as a matter of fact, almost all of them) that would be desired in regard to a small lake can be collected through aerial photographic techniques for a very few cents, making the study of small lakes economically feasible.

Lagoons and estuaries have some of the characteristics of a lake, although the oscillatory motion in them caused by ocean tides are of concern. These characteristics are properties that can be measured in many cases by natural tracers or by putting dyes and floats into the lake, lagoon, or estuary system. In certain parts of the world, such as Venice, it has become of great concern to determine the effects of huge industrial complexes on valuable artistic, cultural, and sociologic centers. Venice is being virtually destroyed. The causes of this process can be observed and measured by remote sensing systems. Without such measurements it will not be possible to regulate, plan, or design methods to prevent such tragedies from occurring.

Surface-water systems are controlled by the nature of their watersheds. As a result of the modifications that man is making in watersheds today, both the quantity and the quality of the water supply are being greatly changed. Floods are beginning to cut away the soils and ruin the watersheds. Computer techniques are available now for analyzing these effects, but the data required for the computer are not available. Much work is being done to develop remote sensors that will provide information concerning watershed modification.The factors involved, such as

0-201-04245-2

slope, roughness, and amount of plant cover, all lend themselves to one or more of the photographic or imagery techniques. Radar altimetry to produce valley profiles is being considered as a technique to analyze flood plains, in order to determine how high the water can rise above the ground surface. Many of these profiles will be required, some in areas that will no longer be accessible to ground engineering teams. Remote sensing devices can make the measurements rapidly and easily.

Today the glaciers of the world are being naturally modified. With few exceptions all glaciers in the Northern Hemisphere are receding, and receding at an increasing rate, resulting in important effects on the local environment, as well as the global water supply. The affected areas, of course, are vast and are not conducive to single engineering point measurement. Remote sensing techniques will furnish the required information. As the satellite sensing platforms become effective, they will be of immense value in analyzing the glacial changes that are occurring throughout the world. These changes are probably a better indicator of the effect of man on his environment, particularly in the atmosphere, than any other single factor. A study of the glacial changes may allow a prediction of the long-term effects of some of the developments that man is planning.

This discussion covers only a few of the parameters of possible interest in hydrologic considerations. Many more could be examined. As the field of remote sensing develops, an increasing number of engineers will begin to think in terms of converting their methods of regional measurements into the use of systems that can be measured by remote sensing techniques.

## III. MEASURING CONTAMINANT DISPERSION IN LAKES, BAYS, ESTUARIES, AND RIVERS

It would seem that a solution to mapping the spread of surface-water contaminants could be attained by making current meter measurements from which a vector field of water velocities could be plotted. The vector field would be the desired solution to determine the path taken by a contaminant. However, such a solution is not possible because the velocities can be exceedingly slow (a matter of a few feet a day, in many cases), and most current meters would not "see" these velocities.

If the concept of current meters were extended to include tracers of the contaminating elements that could be observed from the air, and a history of tracer movement were plotted as a function of time, the ability to analyze the spreading of a contaminant would be greatly enhanced. However, even this method requires a more extensive analysis than one would initially believe.

## IV. THE NATURE OF DISPERSIVE MOVEMENT

0-201-04245-2

Dispersive motion in the simplest sense can best be visualized by considering a circular spot of contaminant floating on surface water. If this water were moving, its movement could be followed by making repeated photographs at short intervals over a long period of time. From these photographs a history of the path of motion could be plotted. In addition, the length of each line segment for a fixed interval of time would give a vector field depicting the velocities. However, the spot of contaminant could be seen to be moving in other ways, such as spreading.

Consider a coordinate system that is described as moving with the center of the spot. The spread of the spot about its moving center could be described by an equation having the form

$$\frac{\partial c}{\partial t} = D\frac{\partial^2 c}{\partial x^2}, \tag{16.1}$$

where $c$ = concentration, $D$ = dispersive coefficient, $t$ = time, $x$ = distance along the flow direction.

This is the Fickian equation; however, the coefficient $D$ is many magnitudes greater than the $D$ describing diffusion processes at a molecular level. If the motion is extended to three dimensions, a physical picture of the measurements of polluting agents in surface-water systems can be attained. A measure of the changing size of the dye disk gives the magnitude and direction of $D$. A measure of the path of the disk provides the velocity of the water masses. Some other parameters of interest will be discussed later.

## V. TRACER TECHNIQUES

Two problems are encountered in the use of tracers. One is determining the type of tracer to be used; the other is placing the tracer in a satisfactory geometric pattern. To accomplish desired goals the quantity of dye required can be very significant.

Determination of the regional movements of quantities of contaminating material would require such vast quantities of dye that it would be entirely unrealistic to attempt this approach. Fortunately another method using IR (false color) film is available.

Vast discolorations occur in most bodies of water from sediments introduced from surface-water sources. Pulses of sediment are received from streams after local storms. The history of the movement of these natural tracers can be utilized to measure movement and dispersion in a particular water body. Infrared reflectance from a clean water body is very low. If sediment is in the water, however, the reflectance becomes quite high. Gradations of density of exposure on the film depend on concentration of sediment. This technique will become an important one. Much research is needed on movement in large bodies of water, and remote sensing techniques will probably provide the only means by which such movement can be determined.

The photographs of Figures 16.1, 16.2, and 16.3 show the appearance of natural tracers in lakes. Figure 16.4 is a photograph of Lake Erie, near Toledo, Ohio. If a mosaic of such photographs is studied, the movement of the contamina-

0-201-04245-2

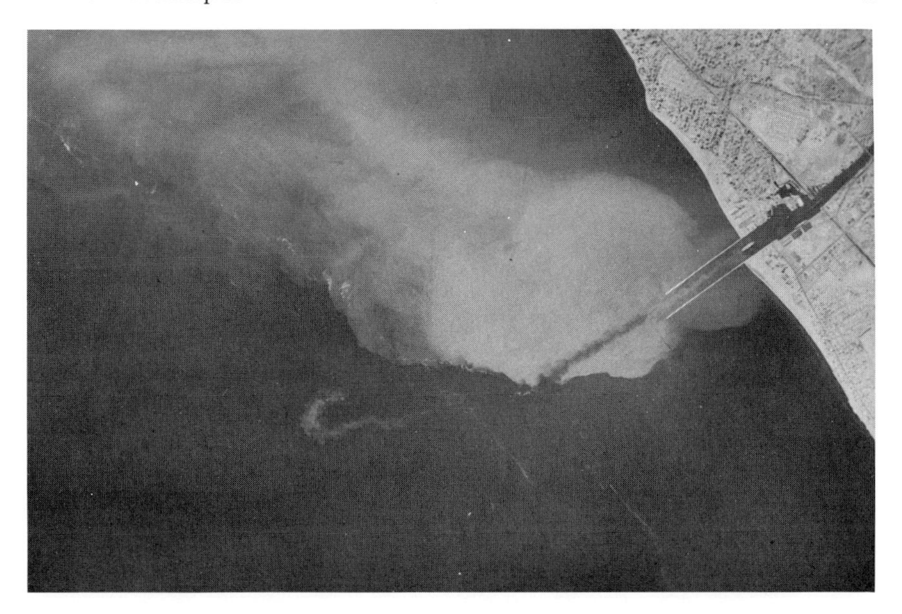

Fig. 16.1. Infrared photograph showing natural contaminants used as tracers.

Fig. 16.2. Infrared photograph showing natural contaminants used as tracers.

0-201-04245-2

tion of the lake surface gives the appearance of a weather map. When the mosaic is of a synoptic nature, it will indeed depict all motion in the lake since this is a representation of the bounding and free surface of the lake. From this boundary condi-

**Fig. 16.3.** Infrared photograph showing natural contaminants used as tracers.

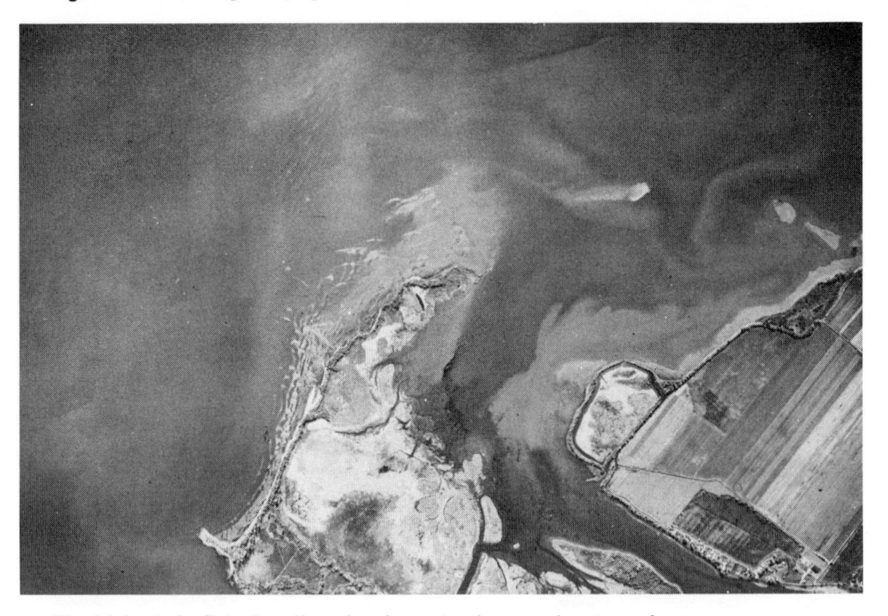

**Fig. 16.4.** Lake Erie shoreline, showing natural contaminants used as tracers.

tion the motion of the lake can be described in three dimensions. Figure 16.5 shows the movement of a vast surge of motion caused by hurricane Beulah along the Texas Gulf Coast.

0-201-04245-2

**Fig. 16.5.** Sediments entrained by violent action of hurricane Beulah along Texas coast.

The temperature of the water surface can also be used as a tracer. The pulses, or continual flow of water, of differing temperatures into a large body of water serve the purpose of a tracer, although there are complicating features, such as the loss of heat due to evaporation and radiation. A combination of the two approaches using sediment and temperature will become an important technique of interpretation. Examples of temperature discontinuities are shown in Figure 16.6.

## VI. CONTROLS ON LARGE DISPERSIVE SYSTEMS

The partial differential equation that describes the motion of any viscous liquid system is the Stokes-Navier equation. This may be written as follows:

$$\frac{\partial \overline{\mathbf{V}}}{\partial t} + (\overline{\mathbf{V}} x \bar{\nu}) = \nabla \bar{\mathbf{p}} + \overline{\mathbf{G}} + \nabla |\overline{\mathbf{V}}|^2 + \nabla \bar{\phi} + \mu \nabla^2 \overline{\mathbf{V}}. \tag{16.2}$$

These terms may be identified as follows: $\partial \overline{\mathbf{V}}/\partial t$ = local acceleration at a point, $(\overline{\mathbf{V}} x \bar{\nu})$ = gyroscopic term, $\overline{\mathbf{G}} + \nabla \bar{\mathbf{p}} + \nabla |\overline{\mathbf{V}}|^2$ = Bernoulli term, $\nabla \bar{\phi}$ = compressibility, $\mu \nabla^2 \overline{\mathbf{V}}$ = frictional force.

From a remote sensing standpoint each term has an interesting significance, with the possible exception of the first term, which must be measured as an effect secondary to the others. However, wave height and motions are controlled by this term.

0-201-04245-2

**Fig. 16.6.** Thermal imagery showing temperature discontinuities along shoreline.

0-201-04245-2

The term $(\overline{\mathbf{V}} \mathrm{x} \overline{\nu})$ is called the gyrostatic term because it responds to a force in the same general way as does a gyroscope. Gyroscopic motion occurs because the radius of curvature of the flow lines causes "gyroscopes" to occur as part of a continuum rather than as the discrete wheels so often considered. The vector is a cross product of velocity and vorticity and must therefore be perpendicular to both vectors. The outstanding effect of this term can be seen from wind streaks (long curves; see below) on water bodies. In the gyroscopic effect the water rotates about an axis parallel to the wind streak. Debris is collected where adjacent systems of motion are submerging. The actual motion pattern is characterized by long curves generally called wind streaks. The separation between the streaks is controlled by wind velocity, since this is the factor which sets the water in motion. The frictional force of the wind on the water surface rotates the water downward in a circular motion. The gyrostatic term, combined with the other terms, gives a lateral precession and the spiral motion results.

This motion system was studied extensively by Langmuir. The results are easily seen on color as well as IR photography. The motion is modified substantially by movement in shallow water, where other shearing forces begin to be significant. Water involved in the region having Langmuir motion can be at a different temperature from the water just below it.

The Bernoulli terms express the changes between kinetic and potential energy. Therefore they represent wave motion. The terms $\nabla \overline{\mathbf{p}}$ and $\overline{\mathbf{G}}$ define the potential energy, while the term $\nabla |\overline{\mathbf{V}}|^2$ expresses the kinetic energy.

Remote sensing techniques can detect the direction of wave motion, the spacing, and the amplitude. Matched stereo pairs of photographs taken at the same instant can show the amplitude of wave motion. The photography is accomplished by two aircraft, flying one behind the other at the desired stereo spacing. The two cameras are fired at precisely the same instant by radio control. When viewed through a stereoscope, the resulting photographs provide information for an analysis of the wavelength.

Stereo analysis of wave height allows the final friction term, $\mu \nabla^2 \overline{\mathbf{V}}$, to be analyzed. This component is represented in the decline of wave height because of friction.

The term $\nabla \overline{\phi}$ does not, in general, apply to studies of large water bodies because it represents the compressibility of the water. However, a similar term must be included when two-dimensional systems are considered. It becomes important when changes in the depth of a water body occur in shallow-water problems and represents the net change in volume per unit of surface area. Figure 16.7* is a stereo pair of photographs taken from two airplanes at the same instant. Figure 16.8 shows the change in the direction of wave motion as the depth of water in a lake declines.

These data can be applicable when it is possible to rewrite the Stokes equation into a macroscopic equation of the type

$$\nabla \cdot \overline{\mathbf{V}} = \ell \frac{\partial^2 h}{\partial x^2} + K \frac{\partial h}{\partial t} . \qquad (16.3)$$

*See color insert.

0-201-04245-2

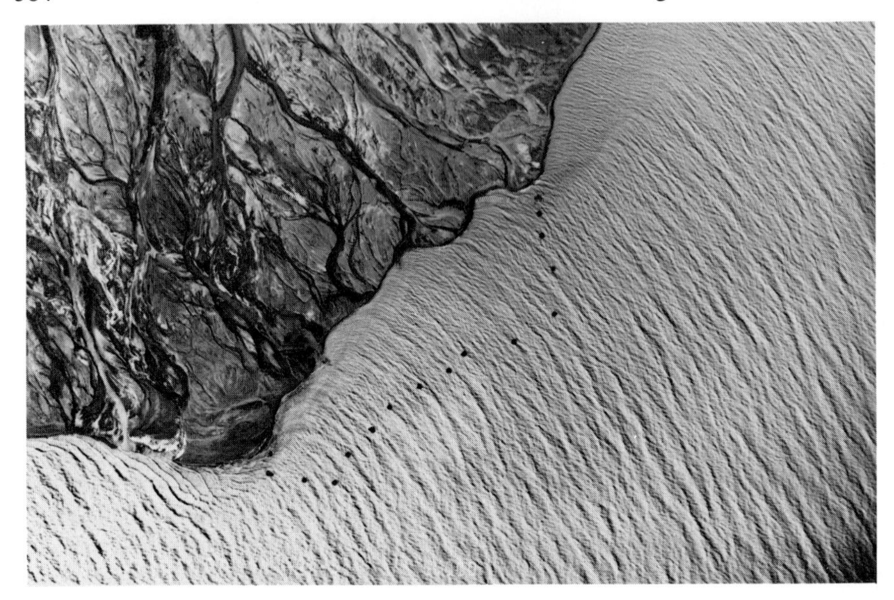

**Fig. 16.8.** Diffraction of wave patterns caused by shelf (dotted lines) along shoreline.

This equation considers a wave equation, where $\ell$ is the coefficient describing the ratio between the kinetic and potential energy terms of the Stokes equation. This is proportional to the wavelength and height. The term $K$ is the frictional component, measured by the wave height decay. The term $h$ is a potential term, and $\overline{\mathbf{V}}$, the velocity, is in turn a function of $h$. This type of expression can be solved by computer analysis. Therefore an expression such as Eq. (16.3) with its parameters measured by remote sensing can be used to compute bay and estuary movement.

Expression (16.3) also has a nonlinear term that can be measured by the lack of symmetry in wave motion. This term determines the steady-state underflow in a bay or estuary, when a part of the water body is contained in a smaller tidal basin. The tide can change the basin; however, if restrictions prevent the water from flowing out as fast as the tide goes out, this basin will allow water to run out of it even though the tide may be coming in. Remote sensing techniques could reveal the asymmetry as well as the barriers that are causing such delays in drainage.

## VII. REMOTE SENSING OF GROUNDWATER SYSTEMS

Groundwater systems characteristically defy direct observation. The water flows through rock formations that normally are not visible to the engineer. At best, only the bounding surfaces of the aquifer system can be seen. However, the bounding surface exerts a control on the entire region. The situation is similar to observing the boundaries of a lake and then determining the internal state between

0-201-04245-2

the boundary points. The internal state of a groundwater system is described by the equation

$$\nabla^2 h = \frac{S}{P} \frac{\partial h}{\partial t} + f(h), \tag{16.4}$$

where $h$ = hydraulic head, $S$ = storage coefficient, $P$ = permeability, $t$ = time, $f(h)$ = recovery system along the surface of the aquifer.

Of these parameters the storage coefficient and permeability are variables that are internal to the region. It is not possible to observe these parameters directly by any known remote sensing technique; however, much information is known about them in most cases. The permeability function is a tensor and has directivity. No engineering technique can be applied inside the aquifer to determine the preferred directions of motion. However, in shallow aquifers the sediments often are laid down so that they can be seen along the surface of the ground. A good example is the meander belt of the coastal areas, where the permeability of the old meander schemes is maintained in the direction in which the river was flowing at the time the meander belt was laid down. Thus internal factors such as permeability can often be determined.

Frequently some data concerning the storage coefficient ($S$) can be obtained by remote sensing measurements. Because $S$ is the amount of water stored in pore spaces and is the amount that can be drained to keep the aquifer working, it is an important consideration. The aquifer drains along the bounding surface of its upper region, and sometimes this surface is so near the ground that the amount of water in the pore spaces can be inferred from measurements made by remote sensing techniques. In the waterlogged areas typified by jungle regions, the water often is standing on the ground and the storage coefficient is unity in this region until such time as drainage occurs.

When the moisture is near enough to the ground that it is wetting the surface but not completely saturating it, this may be seen on color film. If research were completed to determine what colors are most affected by soil moisture, there is little doubt that color photography could be used to collect definitive data concerning the storage coefficients in many groundwater systems. The water standing on the ground, particularly in jungle regions, can best be observed by side-looking radar systems or by glint photography. The imagery from side-looking radar shown in Figure 16.9 shows clearly the water standing on the ground in a jungle region in Panama. Glint photography is taken looking into the sun and also clearly shows where water is standing.

The quantity $f(h)$, which describes the nature of the bounding surface, is difficult to measure. Loss of water, either by evaporation or plant usage, occurs on most of the aquifers in the world, at least at some points. Sometimes bounding surfaces are less obvious from the remote sensing standpoint, as in the case of the aquifers from the southern United States that empty into the Gulf of Mexico. However,

0-201-04245-2

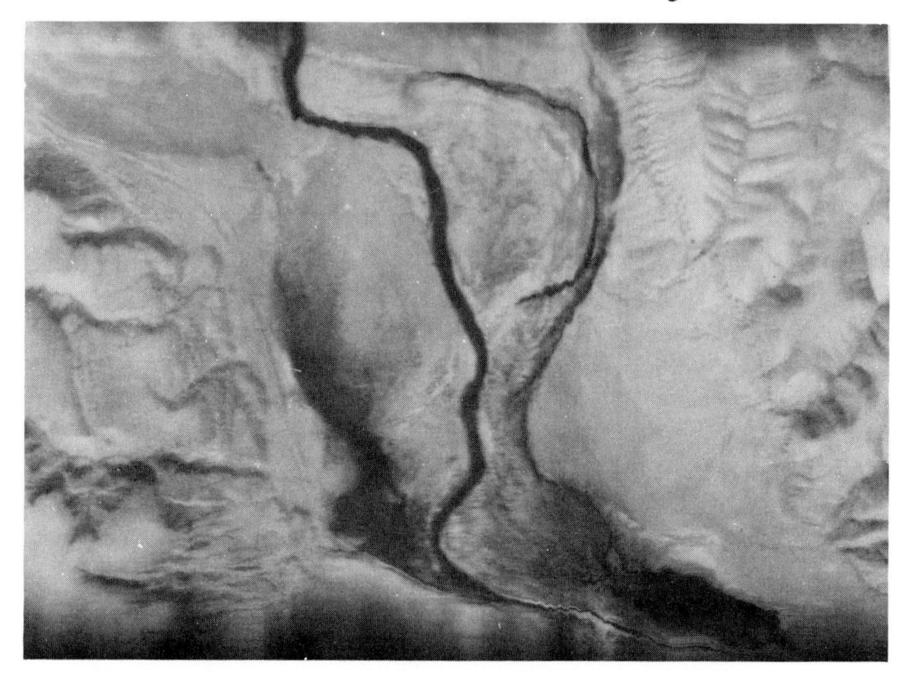

**Fig. 16.9.** Side-looking airborne radar imagery of jungle region in Panama.

when one of the bounding surfaces has water that is being utilized by phreatophytes, much information can be determined. R. C. Culler of the U.S. Geological Survey has conducted a research project for several years on the Gila River in Arizona to determine the water use by phreatophytes and the waterlogging that is taking place. When pumping begins, the water table is lowered and the water that was supporting the phreatophytes is reversed and used internally in the aquifer. The supply is recoverable and must be described by the function $f(h)$.

Evaporation sites, where the water is vaporized directly into the atmosphere from the surface of the water table, can be observed by remote sensing. A significant amount of water is involved in the evaporation process. The water development in the area would be enhanced if the evaporation or transpiration occurred where it could be utilized by useful plant life. The imagery of Figure 16.10 shows quite plainly the cool spots caused by the evaporation of water on the ground surface. Thus it becomes possible to map the regions where recovery will take place as the water table is lowered. The data for function $f(h)$ can be put into the computer system to evaluate the effects of water recovery.

In the same regions in which water is evaporating at the surface or being transpired by plant life, chemical concentration takes place. Chemical concentration is often severe enough that in time the usefulness of the land is completely lost. Places in northern Mexico near Yuma, Arizona, have been ruined in the last 10

0-201-04245-2

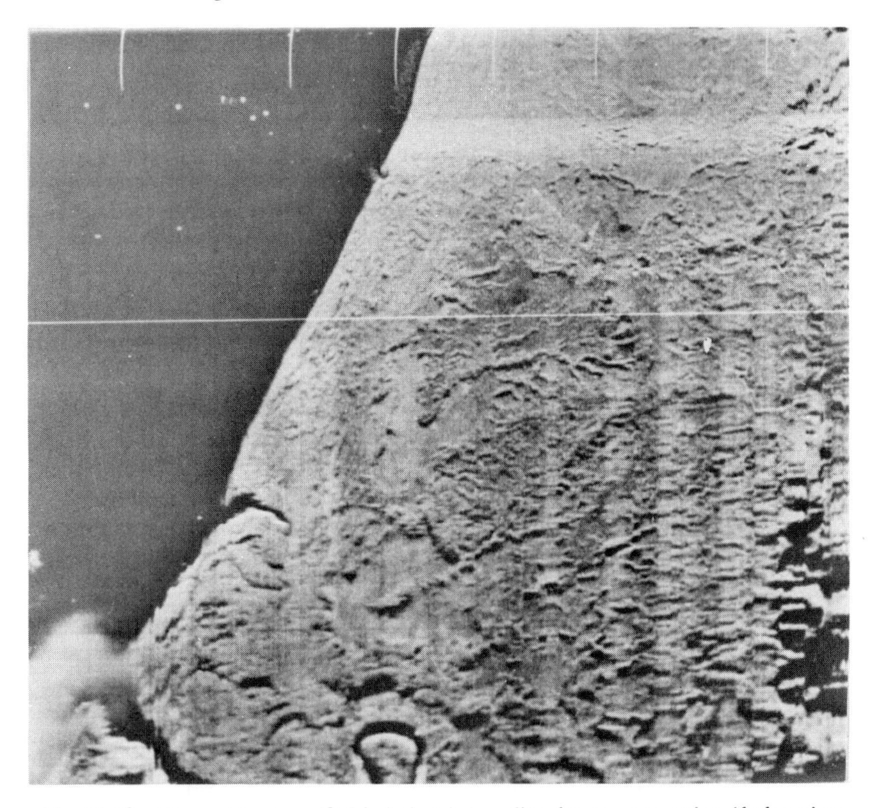

**Fig. 16.10.** Thermal imagery (9–14 $\mu$) showing cooling due to evaporation (dark region near stream).

years by the deposit of sodium on the land surface from evaporation of the groundwater supply. A similar situation exists in Pakistan. Remote sensing measurements and aerial photography show vividly the regions that are being destroyed in this way. The chemical effects of groundwater development can be seen plainly by remote sensing techniques.

Aquifers have other boundary conditions. Dynamic boundaries are caused by pumpage and drainage in the area and may be mapped most adequately by aerial photography. Also, if repeated measurements are made, pumpage in the area can be evaluated by the wetting of the agricultural area near the pumps, as illustrated in the photograph of Figure 16.11.*

Although groundwater measurements cannot be made directly by remote sensing techniques, many of the allied data required to compute the usefulness of groundwater can be acquired by remote sensing. In time these remote sensing techniques will be developed into quantitative methods able to aid in the analysis of groundwater development in any region.

*See color insert.

0-201-04245-2

## VIII. REMOTE SENSING OF LAKE SYSTEMS

In addition to large lakes, such as the Great Lakes, numerous small lakes in the United States should be studied. A severe engineering problem occurs because of the vast number of these lakes. As stated previously, states such as North Dakota and Minnesota actually have millions of small lakes with tremendous value as recreation sites, water supply sources, and breeding grounds for many of the water fowl in the United States. It would be very disastrous to dry these lakes up, yet in many regions they make agricultural operations difficult. On the one hand, then, the farmer would like to dry up the lakes; on the other, the fish and wildlife people would like to keep the lakes open, and the general community wants to retain them for recreational purposes. To survey these lakes and evaluate them to determine the most advantageous use of each is difficult because there are so many. Each lake is an individual entity. Each has its peculiar value to the people in the region. There is no ground-based way to study an average lake and draw important conclusions.

These lakes can be adequately surveyed, however, by aerial photography in color and IR color, producing most of the data—lake size, encroachment by natural vegetation, man's effect, chemical nature of the lake water, amount of sediment, year-to-year decline in the water level in each lake, and decline in the number of lakes in the area—that are needed. It would appear, then, that for a few tens of thousands of dollars (less than 10 cents per lake) most of the lakes in any given state could be surveyed by aerial photography. To attempt to make ground studies of a million lakes, however, even at a cost of 10 cents a lake, would represent an expenditure of 100,000 dollars; moreover it would not be possible to collect any worthwhile data concerning the lakes for such a relatively small expenditure. It appears, then, that remote sensing is the only technique that can be used to measure and evaluate for engineering purposes the numerous small lakes in certain areas. There is really no other way in which the problem can be attacked. This is, therefore, a very important application of remote sensing techniques that should be developed.

Figure 16.12* shows chemistry changes in several of the small lakes in a region in Florida. The encroachment of civilization around the lakes can also be seen. This is an example of how data concerning lakes can be obtained. If the lakes were mapped and classified by section, township, and range, it should then be possible to have a complete file of every range and township in the state, showing all of the lakes. Such surveying would produce most of the data necessary for planning for every lake in an entire state.

## IX. SEDIMENT CONTAMINATION STUDIES IN STREAMS AND BAYS BY REMOTE SENSING

The problem in streams and bays, as described earlier, is searching for the path of motion and determining the group velocity, as well as the dispersion parameters. The best way to obtain the data is through the use of dyes such as fluorescein or rhodamine. Today rhodamine B is used widely as the dye tracer. After a dye pat-

*See color insert.

0-201-04245-2

tern has been laid down within a bay, a lagoon, or a river, aerial photographs repeated at short intervals of time will disclose the streamlines in the area. Figure 16.13* shows a pattern of dye markers set in a lagoon in Bolinas Bay in northern California. Figure 16.14* shows a similar situation along the Colorado River near Needles, Arizona.

The original coordinate system is set by helicopter or airplane. Photographs are then taken at intervals of 5 min, 15 min, or 1 hr, depending on what is desired. A history of the movement of the dye pattern can be used to show the overall streamlines. As the circular dye paths change, either into ellipsoids or into exponential decay patterns, the dispersion parameters can be clearly seen. In the first case, quantitative information can be obtained by plotting the vectors that describe the movement occurring during the interval of time between photographs. This shows the overall path of motion. However, determination of the dispersion parameters requires quantitative information on the rate at which the dye is dispersing.

Quantitative data can be obtained from the photographs through densitometer analysis. In many cases, this may be practical and may solve the problem. However, the Fraunhofer line discriminator (FLD), which is described in Chapter 8, can be used in a helicopter to make quantitative measurements of the dispersion across the dye markers. To accomplish this goal a radar-helicopter-FLD system has been operated in several areas. Figure 16.15 shows the equipment being used in the

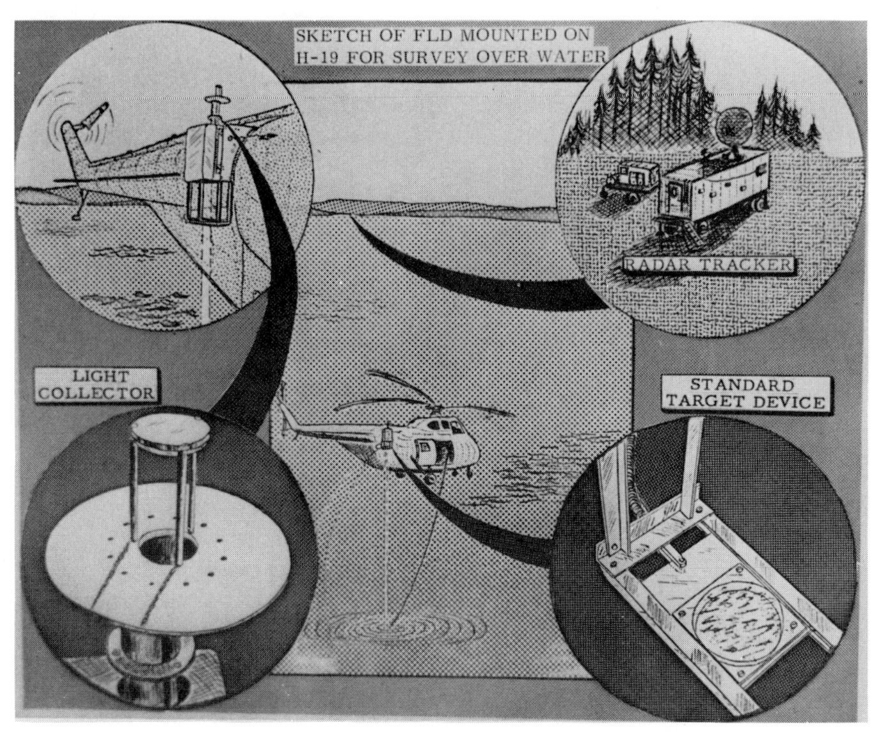


**Fig. 16.15.** Sketch showing Fraunhofer line discriminator mounted on H-19 helicopter and the radar unit used to track the path of the helicopter.

*See color insert.

0-201-04245-2

San Francisco Bay region. The equipment associated with the radar makes an *x-y* plot of the path of the helicopter as it follows the dye patterns, as shown in Figure 16.16. The dye patterns are detected by use of the FLD, which makes it possible to observe a very low level of dye contamination. As a matter of fact, it can measure levels of dye more dilute than the eye can see. Even more important, the data that are taken as the helicopter is flown across the dye pattern are quantitative and can be utilized to determine the dispersion factor. Information taken directly from the FLD has been plotted on probability paper to show the distribution of the dye concentration after a given period of time (Figure 16.17).

The combination of equipment described provides an effective technique whereby to study all contamination processes in lakes, rivers, lagoons, or other

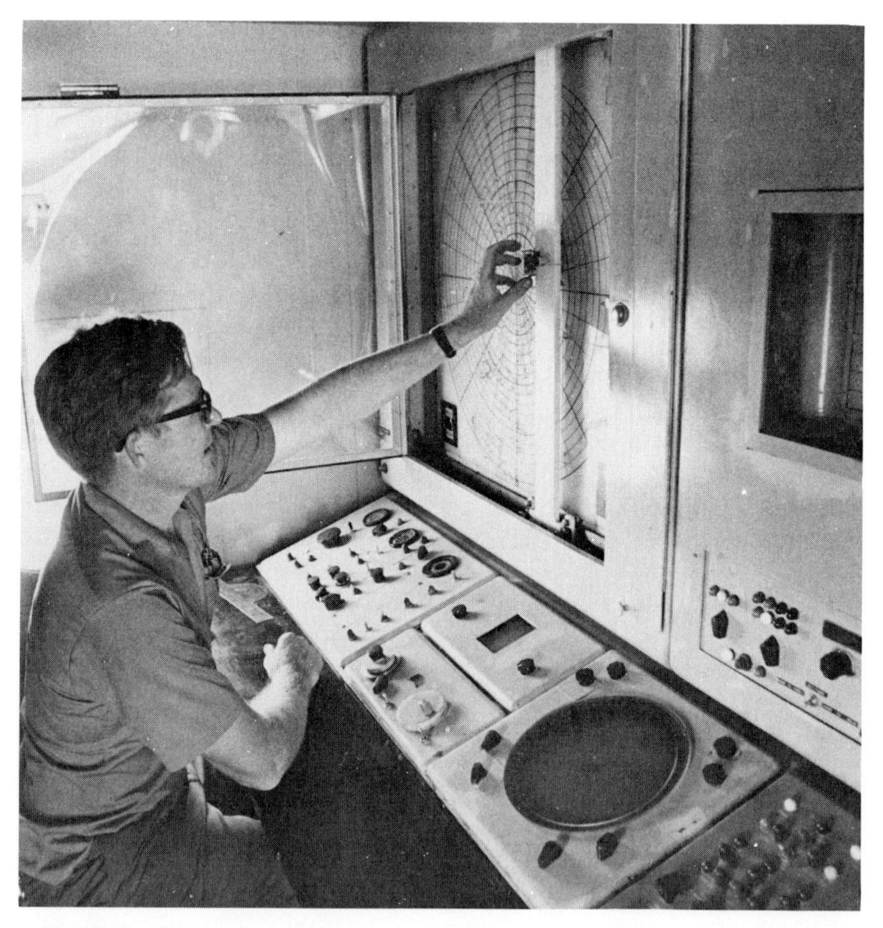

**Fig. 16.16.** The *x-y* plot of helicopter movement during dye studies is produced by radar tracking.

0-201-04245-2

bodies of water. It can also be used on a moving river to make surveys. Figure 16.14 shows a dye pattern that was set on the Colorado River, and Figure 16.18* shows the dye streaming in lines from a continuous supply source on the river. The data can subsequently be plotted in terms of time lines from aerial photographs taken at given intervals after the dye began to move. Again, to make a quantitative analysis of the spread of the dye, Fraunhofer line measurements were made across the dye path. The results are shown in Figures 16.17 and 16.18.

More research is being conducted in applying such techniques to the study of rivers and lagoons. At the moment this quantitative system of analysis looks like one of the best available for hydrologic studies.

## X. SUMMARY

This discussion does not by any means cover all of the uses of remote sensing in hydrology. Many techniques are being tried today, and the field is probably characterized more by ingenuity than as an exact science. However, as time passes and some of the techniques that are being utilized today are developed into quantitative forms of analysis, this field of endeavor will become more standardized than it is at the present time. One problem is that today very little money is being spent on research and development of techniques to measure hydrologic factors. This is most unfortunate, because of all the applications of remote sensing its use in the field of

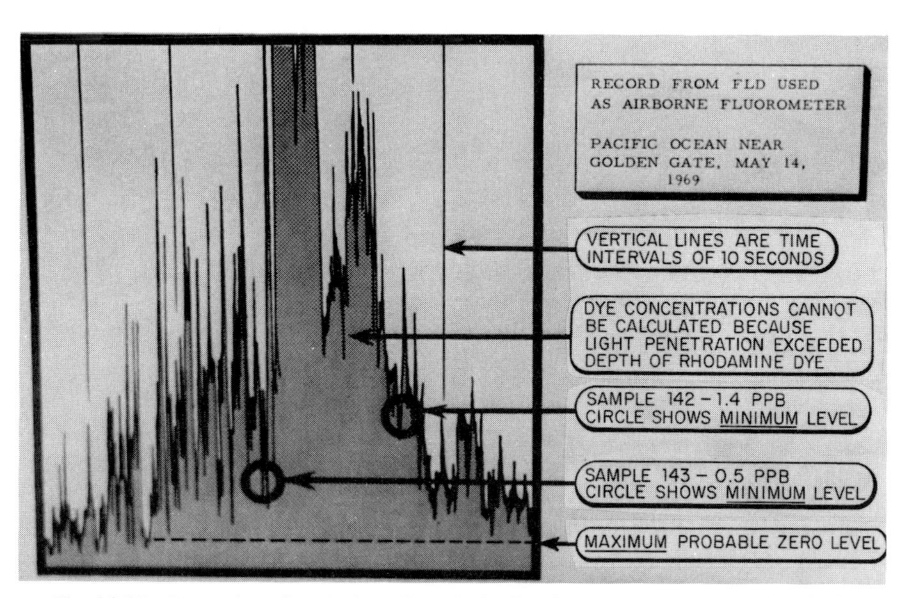


**Fig. 16.17.** Data taken directly from Fraunhofer line discriminator, showing distribution of dye concentration.

*See color insert.

0-201-04245-2

hydrology probably is the most promising and will be the most important to society in the years to come. Remote sensing will not be of much value for exploratory work in the United States. Instead it will be used to describe the dynamic systems that cause contamination and interfere with the utilization of resources. No tool could be more effective in this regard than remote sensing because measurements can be made without disturbing the area studied and without requesting permission from landowners. Particularly for governmental purposes, remote sensing in hydrology will become increasingly useful as the years go by.

0-201-04245-2

Chapter 17

# Remote Sensing in Oceanography

*Don Walsh*

## I. INTRODUCTION

Oceanography and atmospheric studies (meteorology) have common problems, though the densities of the fluid media are quite different. The interaction of the atmosphere and the oceans provides a global heat engine whereby the sun's energy fuels the ocean and atmospheric motion through the exchange of energy across the sea-air interface. Studies of the motions and properties of these fluid systems are complicated by long- and short-scale temporal variations. Thus to know and to apply knowledge of these motions requires a synoptic approach.

Each discipline faces the difficult problem of obtaining both synoptic coverage of vast areas and specific point-source quantitative data. For meteorology the advent of weather satellites provided a partial solution to this problem. Since 1960 the National Aeronautics and Space Administration has launched more than 24 weather stations (NASA, 1970), which have yielded insight into the possibility of looking at the oceans from spacecraft (NWRF, 1967). Earlier attempts in the late 1940's by the Woods Hole Oceanographic Institute (WHOI), employing aircraft, camera systems, and the then-new infrared radiation thermometer, did not find wide acceptance (Stommel et al., 1953).

Beginning about 1966, NASA's Earth Resources Survey Program sponsored the Spacecraft Oceanography Project (SPOC), which revived interest in the synoptic view of regarding the ocean at "arm's length" and making quantitative observations without physically touching the seas. The goals of remote sensor applications to oceanography were established in the SPOC objectives, which read, in part:

A. To identify, test and evaluate techniques which can be used on Earth spacecraft to provide meaningful and useful synoptic oceanographic data.
B. To establish the reliability of spacecraft-acquired oceanographic data by comparison with remotely sensed surface data, and in turn relate these data to surface and subsurface ocean phenomena.
C. To develop and test techniques of displaying space acquired data on a global basis consistent with conventional synoptic data.
D. To develop environmental forecasting techniques for dynamic ocean phenomena using space acquired data.

In order to lead into spacecraft studies, airborne experiments were begun by SPOC with NASA and U.S. Navy aircraft. Aircraft and spacecraft both yield synoptic views, while the point-source calibration (the aqueous equivalent of "ground truth") is conducted by either instrumented buoys or ships.

0-201-04245-2

*Remote Sensing of Environment*, Joseph Lintz, Jr. and David S. Simonett (eds.) ISBN 0-201-04245-2

Internal consistency of the data collected by point-source and synoptic collectors is essential in any remote sensing program. Past oceanographic surveys, based primarily on surface collecting, have been extremely variable in objectives and in philosophic outlook. Thus, although many data have been acquired, they are often difficult to relate to each other. Remote sensing applied to oceanographic research demands a particularly high order of internal consistency for the data, independent of its source, so that the various items can be welded into an efficient and scientifically desirable system for information and management studies of the ocean.

Finding solutions to oceanographic problems requires data collection over both long and short periods of time to accommodate long- and short-scale temporal variations in the processes under study. Frequently specific experiments have to be devised and programmed, which can be conducted only on short notice, in order to obtain information on very short-lived phenomena such as tsunamis.

Many of the needs for information about the oceans may be met most conveniently by remote sensing. Oil spills in the Gulf of Mexico have already been surveyed from aircraft with a variety of sensors, as have pollution from excessive siltation following hurricanes, and thermal and other pollution along industrialized shorelines. Not only does petroleum enter the oceans, but pesticides, biologic wastes, and plastic particles also are frequently encountered at sea and will undoubtedly become even more common in the future. Remote sensing from aircraft and possibly spacecraft gives promise of serving to identify and monitor these events, and from the knowledge thus obtained we can take action to change our patterns of dumping or to devise alternatives to this practice.

## II. BACKGROUND

The exact origin of oceanography by remote sensors is hard to define. A precise dividing line between where other remote sensing activities end and where oceanographic applications begin is impossible to draw. For example, aerial mapping can be considered oceanographic when it is conducted along coastal zones, and airborne weather observations become partially oceanographic over oceans. In general, however, it can be said that specific application of remote sensors to oceanography did not really become a viable technique until the 1960's.

Between World Wars I and II precision aerial mapping techniques evolved, and coastal zones received their share of attention. Amphibious operations in World War II required detailed beach data (bathymetry, wave data, beach contours, etc.).

The development of the infrared radiation thermometer increased the scope of remote sensing for the first time. In addition to the Woods Hole activity already noted, the Bureau of Sports Fisheries and Wildlife used this instrument on the Atlantic and Pacific coasts to map sea surface temperatures (Clark and Frank, 1963) and to study the schooling behavior and distribution of pelagic species, research which had been started earlier by commercial fishing interests (Squire, 1961).

0-201-04245-2

Rocketry (V-2, 1946–50; Viking, 1954–55) provided only limited views of the oceans. With the beginning of manned space flight in May 1961 there became available a long series of spectacular photographs of our planet, obtained by the astronauts using hand-held cameras. From the MERCURY, GEMINI, and APOLLO missions a gold mine of high-quality color photographs of the oceans and their coastal boundaries provided the oceanographer with a powerful bank of data, which are still being analyzed (Stevenson and Nelson, 1968).

Remote sensing for oceanographic applications became a firm commitment in August 1964, when a NASA-sponsored meeting for oceanographers was convened by the Woods Hole Oceanographic Institute. The proceedings of this meeting, *Oceanography from Space* (WHOI, 1965), were widely distributed and made all oceanographers aware that the troubling problem of the synoptic view was approaching solution. The next year the creation of SPOC within the Naval Oceanographic Office gave focus and structure to the effort to introduce remote sensors into the oceanographic programs.

One of the first problems that SPOC faced was to identify the sensors of most value to oceanography, in order to place them on some of the early Earth Resources Survey satellites. The ability to see the ocean surface clearly generally requires wavelengths of little interest to land-based scientists. On the other hand, limitations of the spacecraft demand commonality of remote sensors.

By 1980 an oceanographically oriented satellite, serving the specific needs of this community, should be in orbit. An accelerating factor is the International Decade of Ocean Exploration (IDOE), which began in 1970. The National Academy of Sciences Committee on Oceanography and the National Academy of Engineering's Committee on Ocean Engineering have jointly recommended that remote sensor oceanography be a part of IDOE (NAS/NAE, 1969). The 1970 report of the President of the United States to Congress on the subject of marine resources and engineering development (U.S., 1970) also brought remote sensors into the national IDOE effort, as well as identifying this area as one of the more promising new techniques in ocean sciences.

## III. PLATFORMS AND SENSORS

Although detailed information is presented elsewhere in this volume on sensors and, to some extent, on platforms, it is relevant here to review briefly the types that have proved useful for ocean studies and applications.

### A. Platforms

The platforms used for remote sensor oceanography range from fixed towers on the ocean floor to vehicles floating on the sea surface (ships and buoys) and spacecraft many thousands of miles from the earth.

0-201-04245-2

### 1. *Fixed Platforms*

In this category are towers and platforms that are fixed to the ocean floor. These are used in locations where long-term measurements are required and where placement of a mobile platform is impracticable. A typical example is the U.S. Navy's Argus Island, a tower on Plantagenet Bank 30 miles southwest of Bermuda. This has performed useful services as a test site for the calibration of remote sensors (Yotko and Zaitzeff, 1967). In addition, the Navy has three other towers located along the Gulf and Pacific coasts that can serve this purpose (OCEANAV, 1970).

### 2. *Floating Platforms*

Ships and buoys comprise this class. The first real applications of remote sensors to oceanography were made from ships, through the use of cameras and IR radiation thermometers to study surface characteristics. Many of the early applications involved the study of exchange across the air-sea interface. Floating platforms now provide useful "ground truth" correlation for airborne remote sensors during overflight periods.

### 3. *Balloons and Airships*

The modern hot-air balloon has provided modestly funded ocean operations with a compact, reliable means of placing the observer and his instruments above the ocean surface. In the mid-1960's the Bureau of Commercial Fisheries at La Jolla, California, experimented with hot-air balloons towed behind oceanographic vessels to study fish behavior. More fascinating, however, have been the applications of these devices in the search for seafloor archeological finds. Several sunken vessels have been found in shallow waters by observing their characteristic or anomalous patterns against the seafloor from manned balloons towed by the search vessel. Balloons also provide the capability for long-term investigation of a fixed area on the sea surface. A moored balloon can provide remote sensor measurements over periods much greater than may be obtained with aircraft or spacecraft, free from the biasing influence of a research ship.

The airship has been little used in oceanography, because there are so few of these blimps. In 1966, S. F. Singer of the University of Miami used the Goodyear blimp to make microwave radiometer measurements over Florida coastal waters.

In summary, only limited applications of this class of platforms have been made, but they can be very useful where low speed, endurance, and good payload capability are required.

### 4. *Aircraft*

This type of platform is by far the most active in application today. Table 17.1 provides a representative, but not all-inclusive, listing of aircraft used by various agencies to perform oceanographic-related work. Practical aerial survey work is being conducted on a regular basis, as well as sensor tests for spacecraft platforms.

0-201-04245-2

0-201-04245-2

TABLE 17.1
**Representative Remote Sensor Aircraft[a]**

| *Agency* | *Aircraft* | *Oceanographic Application(s)* | *Remarks* |
|---|---|---|---|
| U.S. Navy Oceanographic Office | Lockheed NC-121K (3) | Remote sensor support for ASWEPS (1). Birds Eye ice reconnaissance program (1). Project Magnet magnetic field surveys (1). | These 4 aircraft belong to the Oceanographic Air Survey Unit. |
| | Douglas NC-54R | Project Magnet magnetic field surveys. | |
| U.S. Naval Research Laboratory | Lockheed EC-121 | Sea surface mapping with 4-frequency radar system to detect sea ice and icebergs, wind fields, and slicks. | |
| U.S. Air Force | Lockheed RC-130<br>McDonnell-Douglas RF-4C | Coastal zone mapping by cameras for location of engineering construction materials (sands, gravels, etc.). RF-4C equipped with IR imager and SLAR | Program run by U.S. Army Corps of Engineers |
| | Convair JC-131 | Trisensor (visual, IR, and radar) mapping aircraft. Oceanographic information came from coastal and harbor overflights. | NASA-U.S. Army Corps of Engineers Geoscience Program. |
| U.S. Army | Grumman OV-1A | SLAR mapping of coastal areas for location of engineering construction materials. | For U.S. Army Corps of Engineers. |
| NASA Manned Spacecraft Center, Houston, Texas | Convair 240A<br>Lockheed C-130<br>Martin RB-57F | Multisensor aircraft, all used for Earth Resources Survey Program. | Sensor test and evaluation. |
| NASA Goddard Space Flight Center, Greenbelt, Maryland | Convair 990 | High-altitude microwave radiometry, using electronically scanned radiometer. "Bore-sight" camera for concurrent photography is also used. | Radiometer test and evaluation. A/c destroyed in crash 1973. |

**TABLE 17.1** *(continued)*

| *Agency* | *Aircraft* | *Oceanographic Application(s)* | *Remarks* |
|---|---|---|---|
| U.S. Coast Guard | Grumman UF-1G<br>Grumman UF-2G | Regular sea surface temperature mapping flights, using portable IRT's. | Flights made on Atlantic, Gulf, and Pacific coasts. |
| | Lockheed WC-130G | Ice reconnaissance, using radars, cameras, and microwave radiometers. | |
| U.S. Weather Bureau (ESSA) | Douglas DC-6 | Maritime weather investigations, including hurricanes. | Primarily a meteorology research aircraft. |
| Woods Hole Oceanographic Institution | Douglas C-54Q | General oceanographic support for WHOI programs requiring airborne platforms. | Used for early IRT work over Gulf Stream. |
| | Helio Courier | General work where extended range is not a factor and slow flight maneuverability is required. | |
| Scripps Institution of Oceanography (University of California) | Douglas DC-3 | Infrared imagery and IR two-wavelength radiometer measurements of temperature and heat flux at ocean surface. Also has photographic capability. | |
| University of Michigan | Douglas C-47 | Eight-channel spectrometer for measuring ocean color variations as insight into productivity, surface roughness, and sediment load. | Used by University of Michigan for general remote sensor studies. |
| University of Kansas | Beech C-45 | X-band scatterometer radar for wave and ice measurements. | Plane converted in 1970. |
| Philco-Ford Corporation | Cessna 260 | Metric and multispectral photography of ocean surface and coastal waters. | Aircraft leased for a series of specific studies. |

0-201-04245-2

0-201-04245-2

| | | | |
|---|---|---|---|
| Texas Instruments Company | North American B-25 | IRT's and infrared imagery. | Aircraft used for both sensor testing and contract surveys. |
| Space General Corporation | North American B-25 | Microwave radiometry with concurrent ("bore-sight") photography. | |
| Bendix Corporation | Beech D-18 | Multispectral scanner/imager with 9-channel coverage from UV to IR bands. Additional camera capability on 70-mm format. | Aircraft used for both sensor testing and contract surveys. |
| Westinghouse Corporation | Douglas DC-6B | Mapping radar for coastal applications in looking for engineering construction materials. | Program run by U.S. Army Corps of Engineers. |
| Grumman Aerospace Corporation | Grumman Gulfstream I | Multispectral and mapping cameras; IRT and IR imager. Magnetometer for geophysical surveys. Integrated data-recording system. | Operated by Grumman Ecosystems for contract work. |
| Nova University Physical Oceanographic Laboratory | Piper Twin Comanche | Through surface measurements, using air-dropped sensors and floats. For determination of upper-layer profiles of temperature, salinity, and circulation. Cameras provide additional capability of surface and shallow water coverage. | Used for the development and test of experimental sensors. |
| Geophysical Aircraft Service Company | Cessna 310G | Used for aircraft launch tests of an expendable bathythermograph (XBT), first developed for helicopters. At airspeed of 190 mph a 300-ft profile of temperature vs. depth was possible. | Joint experiment with Nova University, ONR, and NAVOCEANO. |
| Aero Service Corporation | Sud Caravelle | X-band SLAR plus photographic systems. | Coastal studies and coastal petroleum geology for extrapolation to offshore areas. |

[a] The aircraft listed in this table represent a cross section of types that have been used in a variety of oceanographic remote sensing projects. This is by no means a complete listing of either aircraft or projects, nor is it representative of all effort occurring at one point in time.

The utility of aircraft in oceanographic work is so well established that within a few years the major oceanographic laboratories and institutions will operate aircraft as regularly as their research ships. The variety of aircraft now in use permits the carrying of payloads ranging up to several tons to the very edge of the earth's atmosphere, in order to establish sensor feasibility for spacecraft applications.

### *5. Spacecraft*

Both manned and unmanned earth-orbiting spacecraft provide the advantage of rapid coverage of an area on a real-time basis. The advent of manned space flight demonstrated that visual acuity (human eye and camera) was much higher than expected (NASA, 1967, 1968). In the decade of the 1960's, 16 geodesy and 24 weather unmanned satellites were launched to contribute to man's knowledge of his planet. Although none was specifically ocean oriented, the data taken have been used in oceanographic studies. During the 1970's both manned and unmanned earth-orbiting laboratories will be applied to the study and assessment of both the resources and the pollution of this planet. The fundamental data taken by these craft will greatly enrich the basic sciences involved in these studies. By mid-decade specific ocean-oriented sensors should be in orbit, thus beginning the application of space technology to ocean sciences.

## B. Sensors

The wide variety of sensors applied to oceanographic studies is dependent on the platform, since the earth's atmosphere tends to penalize accuracy and resolution, as a function of altitude above the ocean's surface. In addition, sensors that are found to be suitable for space platforms must also be graded on their applicability to other earth studies, so that each instrument can provide the greatest quantity of data for the largest number of investigations. A "scientific fallout" of the advancements in sensor technology and applications will be greatly improved sensor combinations for other platforms, such as aircraft, ships, and fixed platforms. The listing of sensors below is provided to indicate the basic areas where oceanographic applications have been involved.

### *1. Passive Sensors*

*a. Ultraviolet Band.* The ultraviolet is not a band of primary interest in oceanography. Some multichannel spectrometers include UV together with visual and IR bands. The only UV sensor used for oceanographic work has been a NASA-converted IR imager, which has been employed with some success over the ocean (Walsh, 1969b).

*b. Visible Band.* This is the largest class of imagery, with respect to data now in existence, embracing primarily the use of cameras and photographic emulsions. Short of having a trained observer at the viewing location, photography is the most direct means of presenting data. Interpreting this type of information is less rigor-

0-201-04245-2

ous than dealing with the numerical values derived from other types of sensors, and the perception of key features is generally immediate.

*c. Infrared Spectrum.* This band can be selectively used, by choice of sensor frequency, to measure heat fluxes above, at, and beneath the sea-air interface. The band with the greatest utility is the 8–14 $\mu$ range, which corresponds to the zone of minimum atmospheric attenuation (the "atmospheric window"). The sensors used in the IR spectrum are of two basic types: the imager and the radiometer-spectrometer. The first type produces imagery of the sea surface mapped in thermal contours, while the second provides spot values of the apparent temperature of the surface. The use of a family of IR radiometers, operating on different frequencies, can provide the differential thermal measurements necessary for finding the energy flux across the air/water interface. The IR spectrum sensors are second only to photography for environmental remote sensing.

*d. Microwave Spectrum.* This spectrum covers the same general range as radars. Useful measurements possible here are temperature, surface brightness (index to wind and wave action), and possibly salinity. As this is a lower-frequency range, the sensors and their antennae tend to get larger with decreasing frequency. An advantage is that these sensors are not quite so blinded by atmospheric impedance, so that higher altitude and poor weather are not the serious problems they are with the visible and IR spectra. On the other hand, reduction of the data requires the consideration of many more factors than are involved with the higher-frequency bands. The sensors are of both the imager and radiometer types, and the use of multifrequency sensors provides the corrective data required to arrive at values of surface conditions.

*2. Active Sensors*

*a. Lasers.* Lasers provide a highly collimated energy source that can be directed at a spot on the earth's surface. Acting as an optical "radar," the laser can provide accurate altitude information for geophysical measurements (shape of the geoid, slope of the ocean surface) or for wave height measurements. Used as an energy source for stimulation of materials (oils, sediments, etc.) at the sea surface, it can generate distinctive signatures that permit identification of these materials.

*b. Radars.* Families of specialized radars are now employed in remote sensing to measure a variety of phenomena, including coastal features, river outflows, surface boundaries, fish schooling, and waves. The advantage of these devices is that they can be used on an all-weather basis under circumstances where passive sensors would be ineffective.

In addition to the types listed above, a large family of related sensor devices aid in oceanographic studies. The geophysical sensors, airborne magnetometers and gravimeters, provide basic global data on various phenomena. Variations of the

0-201-04245-2

magnetometer are used to locate and map geological features of economic interest, as well as serving as submarine detection devices for the U.S. Navy. Communications and navigation satellites provide a kind of reverse remote sensing in that they are precisely located in order to assist vehicles on the surface of the ocean in performing their functions. In addition there are under development capabilities whereby oceanographic data can be rapidly transmitted from surface platforms (fixed or floating) to the shore research laboratory via the orbital satellite. All of these devices are designed to speed the collection of ocean data so that true synoptic analysis is possible.

To help sort out the general types of sensors described in the preceding paragraphs Table 17.2 is provided. The sensors listed are not so much specific devices as they are representatives of a family of devices that have worked in the given spectral band. The wide range of resolution in some cases reflects the lower and upper altitude limits (aircraft to spacecraft) of the sensor platform.

**TABLE 17.2**
**Representative Oceanographic Remote Sensors**

| *Sensor* | *Spectral Band* | *Resolution (m/°C)* |
|---|---|---|
| Absorption spectrometer | 0.25–0.60 μ | 70–100/– |
| Ultraviolet imager | 0.29–0.40 μ | 5/– |
| Multispectral camera | 0.30–1.00 μ | 30–50/– |
| Metric camera | 0.30–1.00 μ | 20–30/– |
| Laser*[a] | 0.30–0.90 μ | 0.15–300/– |
| Image spectrophotometer | 0.40–0.70 μ | 700/– |
| Video visual band | 0.48–0.84 μ | 50–500/– |
| Infrared radiometer | 2.00–2.40 μ | 25/0.1 |
| Absorption spectrometer | 2.50–16.0 μ | 70–100/– |
| High-resolution IR radiometer | 3.40–4.20 μ | 9000/2.0 |
| Infrared radiometer-spectrometer | 8.00–14.0 μ | 30–300/0.1–0.8 |
| Infrared imager | 8.00–14.0 μ | 300–2000/0.5 |
| Microwave radiometer | 0.30–30.0 cm | 3–30,000/0.3–2.0 |
| Microwave imager | 3.0 cm | 300–10,000/0.3–0.7 |
| Scatterometer radar* | 4, 10, 21, 75 cm | 1–7000/– |
| Radar imager* | 1–19.3 cm | 30–1000/– |
| Magnetometer | 20–100k gamma | 1 γ |

[a] The asterisk indicates an active sensor.

## IV. OCEAN SCIENCES: THE DISCIPLINES

Oceanography is the application of four basic disciplines to the earth's hydrosphere. The biological, geological, and chemical sciences are distinct and clearly separated. Physical oceanography is divided by some authorities into two segments: (1) phys-

0-201-04245-2

ics of the oceans, and (2) meteorological oceanography, or the effect the air-water interface produces in the generation of global weather patterns, which play a primary role in atmospheric physics.

Each of the four basic disciplines related to oceanography has multiple problem areas, some of which show susceptibility of solution either by new data from remote sensors, or through new techniques of acquiring data of the type gathered in the past, also using remote sensors. Table 17.3 summarizes these problem areas. Topics pertinent to meteorological oceanography are included as subdivisions of A-6: Atmosphere-sea interaction.

**TABLE 17.3**

| | |
|---|---|
| A. Physical oceanography | B. Biological oceanography |
| 1. Sea surface temperature | 1. Chlorophyll assay |
| 2. Sea surface circulation | 2. Sea color indices |
| 3. Deep ocean circulations | 3. Fish oils |
| 4. Estuarine circulation | 4. Behavioral studies |
| 5. Water mass identification | 5. Temperature mapping |
| 6. Atmosphere-sea interaction | |
| a. Energy exchange | C. Geological oceanography |
| b. Filter influence of the atmosphere | 1. Sediment transport<br>2. Coastal bathymetry |
| c. Maritime weather | 3. Geophysical observations |
| 7. Sea state ( = surface roughness) measurement | |
| | D. Chemical oceanography |
| 8. Definition of the geoid | 1. Organic films and slicks |
| 9. Optical properties of seawater | 2. Salinity |
| 10. Oceanic fronts | 3. Chlorophyll measurement |
| 11. Salinity | 4. Gas exchange at the sea-air interface |
| 12. Surface slicks, foams, and films | |

Table 17.3 includes problems that are of concern to more than a single basic discipline, such as A-11 and D-2: Salinity; A-9 and B-2, having to do with the optical properties of seawater; A-6 and D-4, both of which are sea-air interface problems. Also, A-12, B-3, and D-1 are related. Since individual topics may be of interest to investigators in several disciplines, the following discussion is oriented toward sensor capabilities in order to eliminate duplications.

## A. Physical Oceanography

Physical oceanography can be simply defined as "descriptive oceanography." This is the branch of oceanography that deals with the motions (currents, waves, deep circulation, etc.) of oceans, energy exchanges at the boundaries, water mass descriptions, physical properties of seawater, and interaction of the oceans with their boundaries. Acoustics is related to this area, as the physical description of the oceans includes the propagation of acoustic energy within them.

0-201-04245-2

*1. Sea Surface Temperature (SST)*

This is perhaps the most vital parameter in physical oceanography. Knowledge of surface temperature leads to identification of water masses and currents and provides an index of energy exchange across the sea-air interface and an indication of biological activity. Fortunately this is one of the easier measurements to make with remote sensors. Early work with the infrared radiation thermometer (IRT) has provided sufficient experience (Stommel et al., 1953; Richardson, 1957; Clark and Frank, 1963) to put this remote sensor "on the line" as a standard instrument for aircraft and ship applications. A more advanced model, the airborne radiation thermometer (ART), is standard equipment in the Navy's antisubmarine warfare aircraft (Wilkerson, 1966) and since 1966 has been used regularly for surveys of the Gulf Stream (Pickett and Wilkerson, 1966).

The IRT and ART operate in the 8–15 $\mu$ range, where the earth's atmosphere provides the greatest transparency to energy radiated from the sea surface. This atmospheric window is closed by water vapor; and since the oceans average 60% cloud cover, an IR sensor located above the clouds will not function properly. For this reason most applications of these devices will probably continue to be made with aircraft rather than spacecraft (McAlister and McLeish, 1965).

The temperature measured by these IR radiometers is actually the "skin temperature" at the sea-air interface, where evaporation occurs. Skin temperature will generally be about 0.5°C cooler than the temperature 0.02 mm beneath the surface of the ocean (Boudreau, 1965). Since this is a relatively constant factor, the data can be adjusted to correct for this bias.

Through selection of flight altitudes below 300 m the effects of the earth's atmosphere can be minimized while still permitting good coverage of large areas. The IRT-ART data format is usually in the form of a strip chart requiring correlation to the earth's surface. This may be accomplished by plotting a time and flight trace, or using an IR scanning mapper which provides a photographic presentation of the sea surface in terms of thermal contours (Holter and Legault, 1965). This combination can permit very rapid and accurate mapping of SST over large regions. There has been some effort to combine the two instruments to provide imagery which has temperature data indicated on it.

Several sea surface phenomena can affect the accuracy of the SST measurements taken by these instruments. Slicks, films, wind, and waves are all agents of variation (McAlister, 1964; Walker, 1965; McAlister and McLeish, 1965). On the other hand, as will be discussed later, such variations also form the basis for sensing these features.

Despite the limitations noted above, the IRT and ART have a most promising future for the routine gathering of SST data. The low cost and small size of these units put them within the reach of the most modestly endowed investigator, their portability permits the use of almost any platform, and their reliability is good.

The value of even relatively crude SST data is demonstrated in the extensive work that has been done with weather satellites. Initially it was thought that the

0-201-04245-2

specialized design of the IR and video sensors would preclude application of the data to oceanographic studies. Inspection of the data proved quite the opposite (Wilkerson and Noble, 1969; La Violette and Seim, 1969; Warecke et al., 1969), however, and a new tool for ocean studies was developed. It was found that oceanic cloud patterns gave inference as to the SST patterns beneath them and that the IR sensors, operating over the open ocean, could provide a useful, though not precise, idea of the surface temperatures.

Microwave radiometers have provided a means of measuring SST that is not quite so sensitive to atmospheric interference, though a penalty is paid in having less accuracy than IR instruments afford. In addition, these microwave sensors require larger packages for antennas and electronics. Work by several investigators has shown that the real advantage in microwave sensing is that it can detect differences in materials that are at the same temperature and that the multifrequency approach can permit selective measurement of several parameters simultaneously (Mardon, 1965; McAlister and McLeish, 1965; Paris, 1969). The addition of an imaging capability in the microwave part of the spectrum provides the property for "mapping" surface characteristics over the multichannel bands. The insensitivity of the microwave band to atmospheric attenuation makes these radiometers prime candidates for the "all-purpose" requirements for spacecraft sensors (Sherman, 1969).

As man attempts to enlarge the harvest of foods from the sea, remote sensors will play an increasingly important role in this endeavor. Marine animals tend to be so selective in their temperature preferences that a small change in surface temperature can force changes in the normal temporal and geographic location of a particular population. For example, it has been shown that the passage of a hurricane can leave cold patches of considerable extent on the sea surface for many days afterwards (Leipper, 1967). Similarly, a cold-air outbreak from the land to the sea in the winter season can create a mass mortality of coastal marine life as a result of the upwelling initiated by the offshore winds (Gunter, 1940). By providing the capability to measure ocean temperatures accurately, remote sensing devices will make it possible to predict and perhaps ultimately control the migration of marine life.

### 2. *Sea Surface Circulation*

The surface circulation of the world ocean is primarily a result of prevailing wind patterns over the sea. The transfer of atmospheric motion energy to the sea surface sets up water transport mechanisms, which range from small-scale eddies to major ocean current systems. The wind-driven circulations are further modified by the earth's rotation, the shape of the ocean's boundaries, and interaction with other current systems. For the most part these circulations are major surface features on the world ocean and therefore are prime features for remote sensing.

The principal characteristics that indicate these currents are temperature, water color, surface roughness, salinity, and boundaries marked by slicks or films. In addition the thermal contrast between the moving current and the relatively static

0-201-04245-2

ocean will often manifest itself in distinctive cloud patterns along the boundary of the current.

The best-studied example of a surface current is the Gulf Stream, which is the largest surface current in the oceans. The Gulf Stream dominates the eastern North Atlantic from its origin in the Caribbean region and its flow up the East Coast of the United States to its expiration in the mid-North Atlantic. It was here that some of the first remote sensor work was done over a major ocean feature (Stommel et al., 1953; Clark and Frank, 1963; Picket and Wilkerson, 1966). More recently the Gulf Stream has become a favored "target area" for the analysis of photographic and radiometric data taken from space platforms. Photographic imagery from the GEMINI and APOLLO flights has provided unique coverage of portions of this ocean feature (NASA, 1967, 1968; Stevenson and Nelson, 1968). More prominent are the data taken by the imagers located on weather satellites, whereby it was first learned that meteorological sensors could have some oceanographic applications as well (LaViolette and Seim, 1969; Warecke et al., 1969). Perhaps the most dramatic illustrations came from the high-resolution IR radiometer on board the NIMBUS 2 weather satellite, whence television imagery was sent back to ground stations (Wilkerson, 1967a, b; Warecke et al., 1969). This imagery plainly showed the prominent features of the north wall of the Gulf Stream, where the temperature change can be from 15 to 25°C in a few hundred meters.

The weather satellites have also looked at other major currents, such as the Agulhas Current, the Brazil Current, and the Kuroshio Current. Their boundaries with other currents have established patterns of large thermal contrasts (National Council on Marine Resources and Engineering Development, 1967; Warnecke et al., 1969).

Many smaller-scale features, such as eddies, gyres, and coastal circulations, are also seen by remote sensors. Analysis of the GEMINI photography has demonstrated the value of relatively simple photography in looking at small sea surface circulation patterns (Stevenson and Nelson, 1968; Stevenson, 1969c). In addition, the weather satellites have been able to detect more prominent small-scale features of this sort where sharp boundaries or irregular shapes are involed (Wilkerson, 1967). Even surface patterns on lakes have been analyzed through these techniques (Barnett, 1966; Warecke et al., 1969).

Since ocean surface circulation features are generally large scale, the higher-altitude remote sensor platform is the preferred vehicle. High-altitude aircraft and spacecraft promote the capability to see the entire feature on a real-time basis. The primary sensors in this area are cameras, IR sensors, and microwave radiometers. Generally precision of measurement is not as important as the ability to track these circulations and define their boundary regions.

### *3. Deep Ocean Circulation*

These circulations result primarily from density differences and compensation for wind-driven transport (surface circulation). Very little evidence of these deep

0-201-04245-2

circulations are found on the sea surface, and modified remote sensing methods are required. Dr. W. S. Richardson of Nova University has developed a means of dropping from aircraft expendable buoys which descend to a programmed depth in the water before returning to the surface. By sighting each buoy's relative position and distance from the initial surface marker (dye or smoke) a current profile can be developed for that point in the ocean. Through the use of additional expendable, air-dropped sensors which can measure temperature and salinity as functions of depth and telemeter the information back to the aircraft, enough data can be assembled to develop accurate density profiles of the water column in the upper levels of the ocean. Despite the fact that such devices limit the remote sensor platform to low altitudes, their employment provides the first remote means of quickly and accurately measuring deep circulation over a large region in a brief period of time.

### 4. *Estuarine Circulation*

The estuarine zone is generally defined as the area where fresh water runoff from land meets the ocean. Within this area are found major gradients of temperature, salinity, water color, and chemical makeup. Because this zone has such large features associated with mixing and with man's use of the coastal region, it is a prime area of interest for remote sensor applications. Of particular interest are the regions where major rivers and harbors are located, as these tend to be places of high pollution probability. Unfortunately pollution does not exist only in the ocean waters, and often the remote sensor data are obscured by atmospheric pollutants which filter the data from the ocean surface (Arnold and Thompson, 1969). Some specific remote sensor work has been done in the delta areas of the Mississippi, Amazon, Colorado, and Columbia rivers (Walsh, 1969a; Stevenson and Nelson, 1968).

Estuarine areas may be used as test and calibration sites for remote sensors. The measured gradients here are significantly higher than any found over the open ocean. For this reason initial evaluation of a new sensor could be conducted over a target of great contrast, and the sensors in service could be calibrated near "home base." The logistics of supporting a test and evaluation site near the coast are much more reasonable than at some place over the open ocean, and it would be easier to monitor the tests with "ground truth" at such sites. The work in the Mississippi Delta with the SPOC aircraft from NASA, Houston, has demonstrated this advantage (Walsh, 1969b).

### 5. *Water Mass Identification*

In this context the term "water mass" defines bodies of water which can be considered homogeneous with respect to their surface characteristics. This definition differs slightly from the oceanographer's classic concept of a water mass as a body of water that can be traced throughout the world ocean from its source of origin

0-201-04245-2

until it loses its identity through mixing. The key parameters to tracing a water mass are its temperature and salinity versus depth. These factors provide the "signature" which permits tracing the mass in the ocean. In remote sensing, however, the investigator is limited primarily to surface characteristics and is therefore required to employ a simpler approach to the concept of water mass. In some cases surface characteristics of the ocean provide insight into internal motions. For example, bands of slicks, foam, and discolored water often are indicators of upwelling or internal waves (Ewing, 1950a, b, 1964; La Fond, 1959; Roll, 1965).

GEMINI color photographs have been used to locate specific water mass areas in the ocean because of the known correlation between different water masses and fisheries productivity. When the data interpreted from the photography were compared with those for known fisheries, the correlation was high (Stevenson, 1969b).

The use of dropped, expendable instrumentation from aircraft will permit the oceanographer to obtain profile data in the ocean's upper regions which will provide a basis for the direct determination of large water masses in a brief period of time.

### 6. *Sea-Air Interaction*

This area of ocean studies falls between oceanography and meteorology. The exchange of energy across this interface comprises the major part of the global heat engine, which is initially fueled by solar energy stored in the upper layers of the oceans. The release of this energy back to the atmosphere is the primary activating mechanism for the motions of the atmosphere. These motions, through wind stress at the sea-air interface, set up the global ocean circulation patterns on the sea surface. The point of this grossly oversimplified statement of a complicated process is that the marine scientist cannot divorce oceanography from marine meteorology, and likewise the meteorologist cannot overlook the oceans.

Of direct concern to understanding the sea-air interaction processes is measurement of the energy exchanges across this interface. This has been done successfully through the use of multiband radiometers which operate in the IR spectrum and permit differential measurement of temperature at and just below the sea surface (McAlister and McLeish, 1965; McAlister, 1969). The application of this technique to worldwide surveys of the world ocean can lead to meaningful revision of estimates of the ocean's role in the earth's heat budget. Ultimately this will require measurements to be made on time scales ranging from hours to years to develop true periodicity. At present numerical methods which will provide the format for the introduction of more precise remote sensor data are being applied to this problem (Wolff et al., 1969). These methods are used to obtain synoptic predictions of the conditions and processes of the surface layers (to a depth of about 400 m) of the ocean.

The most damaging weather known to man comes from maritime-generated storm systems which sweep across coastal regions to bring catastrophic destruction. More extensive and accurate knowledge of what factors initiate these weather sys-

0-201-04245-2

tems and a concept of how they are steered could improve prediction and perhaps lead to effective modification while they were in their formative stages. There is also the question of the role these mesoscale systems play in the transfer of energy from regions of surplus to those of deficit. Preliminary contributions to the solution of this problem have already been obtained from remote sensing.

*a. Energy Exchange.* Determination of the energy exchange at the sea-air interface must begin with measurement of incoming solar energy arriving at the edge of the earth's atmosphere. Satellite measurements have directly determined this value to be 2.00 $cal/cm^2 \cdot min$ (Byers, 1959), an increase of 9% over previously calculated values. Tropical regions receive the surplus energy, whereas arctic and subarctic regions have a deficit. From 70°N to 70°S latitude the oceans receive, on an average, $52 \times 10^{12}$ kW, but ocean temperatures do not increase over the long term (Sverdrup et al., 1942). The dispersion of this energy has been identified as 54% lost in evaporation (latent energy transfer to the atmosphere), 40% reradiated into the atmosphere, and 6% transported to other parts of the ocean by water circulation. This idealized situation is seldom encountered in nature, however, as much of the incoming solar energy is lost in the atmosphere because of scattering, absorption, and reflection. Clouds, with their highly reflective upper surfaces, constitute the major barrier to incoming solar energy. Snow and ice, to a lesser areal extent, are excellent reflectors. Weather satellites have already made it possible to improve the theoretical calculations of solar flux, and they also provide greater accuracy in determining the percentage of cloud cover. In addition, Earth Resources Survey satellites will offer better information on the snow and ice fields, insofar as cloud cover will allow.

*b. Filter Influence of the Atmosphere.* There is a basic difference between the atmospheric filtering effects over land and those over the sea. Man-made effects (smoke, pollutants, etc.), together with natural filters (e.g., dust, aerosols), change very rapidly over land areas and hence differ from the filtering effects (predominantly aerosols) found over the sea. The effects of the atmosphere on sensors will require minor modifications for oceanographic sensing, but the extent of the modifications is only now being determined.

*c. Maritime Weather.* The interests here range from the study of maritime-originated storms, which can affect shipping or coastal areas, to determination of the ways in which cloud patterns can serve as indicators of ocean surface characteristics. In the 1960's large cooperative expeditions were initiated to study air-sea energy processes in tropic and subtropic regions, where the most violent mesoscale maritime weather is generated. For example, the Barbados Oceanographic and Meteorologic Experiment (BOMEX) utilized several ships and aircraft in a simultaneous study in the Caribbean in mid-1969. This permitted some degree of synoptic data acquisition on a multidisciplinary basis.

0-201-04245-2

Cloud patterns, covering 60% of the ocean, are of interest if a correlation with sea state conditions can be ascertained. Initial studies in this area (Arnold, 1967; Stevenson, 1969a) show the desirability of continuing this line of investigation. The value of cloud pattern information is particularly apparent in looking at the APOLLO photography of the whole earth and the weather satellite imagery of it (Johnson, 1969). The relationship of cloud patterns to regions of warm and cold surface waters is quite apparent, and the major storm systems can be easily sighted. Through repetitive coverage static storm systems can readily be distinguished from moving systems.

### *7. Sea State Measurement*

In the most basic sense, sea state is the surface roughness of the ocean. This roughness is primarily created by wind stress at the sea-air interface, forming part of the energy exchange between ocean and atmosphere. Better understanding of this process on a worldwide basis will assist in measuring the energy flux across this interface. In addition, knowledge of the wave fields on the sea surface will lead to more precise predictions in regard to waves and marine weather, which are of major import to commercial shipping.

A basic problem in making good measurements of sea state is that the ocean's surface is a complex field of wave spectra which has its origin in wave trains converging from many different directions. The frequencies of these waves range from extremely-long-period tidal variations to very-short-period capillary waves. Thus for a good analysis of this field sufficient data to perform a statistical interpretation are needed.

The remote sensing techniques applied to sea state range from very basic visual observations of sun glitter on the surface to the use of several radars. Glitter pattern observations are based on the concept that the area of glitter increases with surface roughness to a point where whitecap and foam formation produce a "whiteout," in which the glitter patch is lost. This type of observation has been made from weather satellite imagery (Strong and McClain, 1969). Also it has been suggested that an observer in a manned spacecraft could make similar observations by means of very simple instrumentation, such as a battery-operated visible-light radiometer (Duntley, 1965). In this scheme the observer would be able to make qualitative estimates of sea state, surface wind velocity, and direction much as a mariner makes similar estimates on board ship at sea. The author has investigated the sea surface photography taken on the APOLLO 502 unmanned earth-orbital mission. The high quality of the photography permitted the measurement of long swells off the western coast of North Africa, where a wavelength of about 250 m was estimated.

The microwave sensor family will provide valuable tools for the measurement of sea state. The brightness temperature measured by these sensors is affected by the formation of foam, spray, and whitecaps in portions of the microwave band.

0-201-04245-2

With this information as a basis, work has been conducted to correlate surface roughness with sea state through the use of aircraft-mounted microwave radiometers (Williams, 1969; Sherman, 1969; Paris, 1969).

Since the ocean cloud cover averages about 60% at any given time, and since more precise roughness data are needed, the most promising sea state sensors are the active radars. The two primary radar sensors are the scatterometer and the wave profiler (Pierson, 1969). The scatterometer makes roughness measurements by comparing the intensity of the reflected energy to the energy transmitted by the unit. The greater the surface roughness, the greater is the scatter of the incident energy. The wave profiler, on the other hand, is a high-resolution radar altimeter that measures wave height profiles along the flight track. In both cases the movement of the platform provides wave field measurements rather than point measurements because of the high movement of the platform relative to the wave motion. Since 1963 the Department of Meteorology and Oceanography of New York University and the Center for Research in Engineering Science at the University of Kansas have conducted cooperative studies on the theories and applications of radar sensors to oceanography (Greenwood et al., 1967; Pierson, 1968, 1969; Moore and Pierson, 1965; Moore and Biggs, 1968; Chia, 1968).

The scatterometer radar has been used on board the NASA remote sensor aircraft (Badgley, 1969; Pierson, 1969), while the wave profiler has been employed by the U.S. Navy on its ASWEPS aircraft (Schule and Wilkerson, 1967). A comprarative sensor mission was flown by these two aircraft in the vicinity of Argus Island, near Bermuda, in 1966. The purpose of the experiment was to compare the radar sensors against one another and the fixed wave staff mounted on the Argus Island tower, which stands on Plantagenet Bank in 59 m of water (Beckner, 1968).

Two additional radar systems are being tested for sea state measurements. These are the imaging radar (SLAR) and the four-frequency radar. The imager is the conventional radar which provides imagery of the target area. Although these units do not have the precision of pulsed (altimetry) systems, they have shown a capability to detect wave patterns and major current boundaries over a small portion of the swath width (Moore and Biggs, 1968). The real value of these units will probably lie in mapping coastal zone processes, where a profiling of the land-sea interface is required.

The four-frequency unit is relatively new and was developed by the U.S. Naval Research Laboratory (NRL). It uses the spread of the Doppler spectrum as an index of significant wave height. Concurrent flight tests of this unit and the scatterometer radar are being conducted by NRL with its EC-121 aircraft to determine the relative effectiveness of these systems. One apparent problem with the scatterometer is a loss of signal at higher sea states, where the four-frequency unit may not have this problem.

The active radar sensors provide an all-weather, day-night capability to measure surface roughness over the oceans. In addition, some work has been done by the Naval Oceanographic Office on a laser wave profiler. An aircraft-mounted geodolite

0-201-04245-2

laser has been tested to obtain one-dimensional sea surface profiles. The Navy hopes to develop a mathematical model for two-dimensional spectra from the one-dimensional data (Noble et al., 1969).

Stereophotogrammetric methods were applied to aircraft-supported photography of the sea surface in the 1950's in early studies of wave spectra and energy distribution. The methods were laborious and could not be adapted to survey techniques or to high-altitude remote sensor platforms. The introduction of scatterometry techniques has made stereophotography obsolete for studies of surface roughness.

*8. Defining the Surface of the Geoid*

It is generally accepted that the earth's surface grossly resembles an ellipsoid of revolution. The surface of interest to the oceanographer is the geoid, that is, the level surface of equipotential where the gravity vector would be everywhere normal to the sea surface if all wave action ceased and the surface were perfectly calm. The ocean's surface deviates from this reference plane for several reasons of interest to the oceanographer. The most prominent causes of these deviations are tides, storm surges, wind-driven circulation, and baroclinicity.

However, on the global scale the surface of the geoid over the oceans is not a surface of revolution. Deviations from this theoretical shape are as great as 50–80 m on a scale of several thousand kilometers horizontal distance (Pierson, 1969). In addition, there are smaller "bumps and dips" associated with the great ocean trench systems, of which there are 20. For example, the Puerto Rican Trench has a depression of about 15–20 m over a horizontal distance of 3–4° of longitude (Von Arx, 1966).

The remote sensor required for these types of measurements is a precision active height finder mounted on an earth-orbital space platform. The key elements of the system will be the precision of the spacecraft orbit with reference to the center of the earth and the internal design of the sensor itself. A properly designed system should be capable of measuring within an error range of ±2 m (Pierson, 1969).

With accuracy in this range it will be possible to sense major tidal regimes. In the Bay of Fundy the spring tidal range can be as great as 17 m and hence can be tracked by spacecraft. Also, when storm surges and tsunamis are greater than 3–4 m, they can be detected by the radar altimeter before they arrive at the coastline.

The sensing of the slope of the ocean changes due to major surface current circulations will probably require improvement in the sensitivity of these altimeters. The western boundary currents (Gulf Stream, Kuroshio, Brazil, and Agulhas) feature the largest slopes, but even the most prominent of this family, the Gulf Stream, has only a dip of 1 m in a horizontal distance of 100 km along its greatest slope (Greenwood et al., 1967).

0-201-04245-2

Improvements in radar system design for altimetry and a better knowledge of the geoid surface of the oceans could lead to significant capability for remote sensing of large-scale ocean surface features.

### *9. Optical Properties of Seawater*

Primarily this category refers to the remote sensing properties of transparency and color. In the first case, optical transparency can lead to bathymetric mapping of coastal areas and analysis of beach deposits; in the second, color can provide indices for productivity, sediment transport, and surface wave action. This area has received a considerable amount of attention, since the camera continues to be one of the most popular remote sensors. Knowledge of the properties of the sea surface and upper layer in the visible light region (from near UV to near IR, 0.3–1.0 $\mu$) is vital to understanding this imagery.

Density slicing experiments have been conducted with the photographs taken during the GEMINI missions to determine bottom contours in selected coastal areas (Ross, 1968, 1969a). The experimental techniques involve reprocessing the imagery through selective filtering/masking to obtain a 0.010–0.025 $\mu$ band pass in the region from 0.4 to 0.6 $\mu$. Because of the selective attenuation of light with depth, this provides a capability for looking at the seafloor in shallow waters at fairly precise intervals. The results showed that six distinct contours could be mapped in water depths to 20 m and that in deeper waters agreement with the 20-, 40-, and 60-m contours was possible (Ross, 1968). This process may now be carried out with a number of commercially available electronic systems.

Normally the best band for maximum penetration of seawater is 0.425–0.550 $\mu$, the blue-green region; however, in coastal waters, where sediment and biological influences are more pronounced, the band tends to shift toward a yellow-green. Thus the transmission peak in the open ocean is at about 0.470 $\mu$, whereas in coastal waters it is at 0.560 $\mu$. One unique set of empirical data was taken on the penetration of light in seawater by using a deep-submersible research vehicle, the *Ben Franklin,* and a camera-equipped aircraft. The deck of the submersible was painted matte white as an optimum target, and it was then operated in the clear waters of the Gulf Stream off Florida at depths of 10, 15, and 25 m. The aircraft was equipped with a four-band multispectral camera system which covered the blue, green, red, and blue-green portions of the visible spectrum. In addition, the *Ben Franklin* was equipped with a spectroradiometer which looked upward to measure the downward irradiance incident on the deck of the submersible. This instrument scanned the visible spectrum from 0.35 to 0.75 $\mu$ in 0.025-$\mu$ increments. The results of the experiment showed that two pass bands, 0.46–0.51 $\mu$ and 0.51–0.56 $\mu$, provided the best information, based on penetration of seawater over most of the world ocean (Ross, 1969b).

The color of seawater is best evaluated by spectrophotometric techniques to equate radiance with key characteristics. In this way more precise signatures can

0-201-04245-2

be measured to determine finite features, such as water mass and major current boundaries, chlorophyll concentrations and distribution, sedimentary discharges from rivers, and oil pollution and other forms of environmental hazards on the oceans (Ramsey, 1968). The greater portion of work in this area has involved using ocean color as an index to biological productivity; this in turn can lead to direct benefits in greater food production from the sea (Yentsch, 1960; White, 1968).

A major problem area in establishing the utility of visible band sensors is the modification of the energy in this band by the atmosphere and the ocean (Duntley, 1963; Jerlov, 1964; Curcio et al., 1961). This involves concepts of attenuation, absorption, scattering, and reflection in these two fluid media and their interface.

A combination of cameras (relatively broad band) and multichannel spectrophotometers (narrow band) will provide the best sensor mix for detailing the ocean's surface in the visible spectrum. The data output from the latter instruments provides quantitative signature measurements and is therefore adaptable to computer processing (Ramsey, 1968).

As an inverse of the foregoing purposes of ocean transparency and color studies, the oceanographer utilizing remote sensors will be able to determine water transparency in many coastal areas where the bathymetry is known. Through the plotting of variations of the transparency factor additional insight into local processes may be gained.

As mentioned above, some sea color indices are indicators of biological activity. Since the dawn of history mariners have named bodies of water according to their coloration—for example, the Red Sea, the Black Sea, the Yellow Sea, and these characteristics are often aids to navigation (Ewing, 1969). In the instance of the infamous red tides occurring along many coasts the effect on marine life is negative. The small plankton of the tide deplete the oxygen in the upper layer of the ocean, and this can result in mass mortality of other life forms in the sea. Other colorations can be indicative of dead organic material or the presence of nutrients from either land runoff or upwelling. In each case investigations are required to correlate the color signature of the sea surface with the biological meaning of the colorant.

### *10. Oceanic Fronts*

These features are formed in a manner analogous to weather fronts. Two or more masses of water are brought into contact in such a way that there is a considerable difference between their properties (e.g., temperature, salinity, velocity, direction, and color) (Cromwell and Reid, 1956; Knauss, 1957; Ewing, 1964). Ocean fronts are of interest to the ocean scientist because the turbulent boundary provides a means of studying the energy and mass exchanges between the adjacent waters, as well as constituting a strong surface feature for the tracing of water masses.

0-201-04245-2

The oceanic front manifests itself usually as a strong demarcation line, which is detectable by almost all remote oceanographic sensors. Multisensor investigations off river mouths have shown that the frontal characteristic is prominent in UV, IR, and visual bands (Walsh, 1969a, b; Paris, 1968). From the economic point of view these zones can be important as locations where fish stocks tend to concentrate for feeding. This is especially true in coastal areas, where river discharge and upwelling are factors (Stevenson, 1969a, b).

Akin to oceanic fronts are boundary zones which include, in addition to oceanic fronts, other interfaces such as shore-sea, sea-air, and sea-seafloor. Work by fisheries oceanographers with the photography from the GEMINI and APOLLO spacecraft shows a definite correlation between these features and fisheries zones (Stevenson, 1969a, b; Lindner and Bailey, 1968).

In the shallow waters of the Bahamas it has been found possible to view the entire water column from sea surface to seafloor, a development which opens new possibilities for studying marine ecology (Conrod et al., 1968b). Although direct sighting of marine animals is not expected, the probability of their being in an area can be predicted by the conditions of the water column and the seafloor, along with observations of the marine botany.

### *11. Salinity*

Salinity is second only to temperature as a physical oceanography parameter. Knowledge of salinity, temperature, and depth permits the calculation of water mass identification, density profiles, and deep ocean circulation. To date, salinity measurements have yet to be achieved by remote sensing techniques, and considerable investigation toward this end will be required. It is hoped that microwave radiometers operating in multiple frequencies will eventually serve in this role (Paris, 1969; Texas A & M University, 1969; Edgerton et al., 1970).

### *12. Surface Slicks, Foam, and Films*

These phenomena are generally related to internal processes in the ocean. Slicks and films are associated with boundaries, fronts, and internal wave action (Dietz and La Fond, 1950; Ewing, 1950a, 1969b; La Fond, 1959; Knauss, 1957; Von Arx and Richardson, 1953). Wind action over the sea can also create these features by establishing upwelling, initiation of wave fields, and production of foam (Langmuir, 1938; Roll, 1965). For these reasons appearances of these phenomena can be used as indices for wind patterns over the sea and internal motions in the upper layer (Walsh, 1969a). Films and slicks of biological origin can provide evidence as to commercially valuable fish stocks, while those of mineral origin can lead to pollution monitoring and detection.

These features inhibit evaporation at the ocean's "skin" interface with the atmosphere and therefore appear warm to infrared sensors. In the microwave region the films and slicks increase the brightness radiometric temperature, also making

0-201-04245-2

these areas appear warmer. In the visual part of the spectrum glitter patterns on the sea surface generally indicate these features (Soules, 1969). As mentioned earlier, various types of radar sensors have been able to sense the larger concentrations of these features along oceanic boundary zones.

In addition to foams associated with low wind velocities and internal motions, there are the more common foams that come with high tides and seas. Several investigators are using this feature as a means of determining these wind and sea state conditions through the use of microwave sensors where brightness temperature is affected by foam generation (Paris, 1969; Williams, 1968; Edgerton et al., 1970).

### *13. Fish Oils*

It is known that when one fish stock feeds on another organic oils are liberated and form a molecular film on the sea surface. Such an oil film is detectable. However, without knowledge of its type, which then would have to be related to the proper predator, no real economic gain (i.e., fisheries information) is promoted. Thus both microwave and IR sensors can sense oil films, but identification of the type of oil requires more precise measurement.

The study of fish oil films in the laboratory has shown that there is a considerable difference between fresh and older slicks that have been exposed to air for several hours. These same studies show that in the 8–14 $\mu$ "atmospheric window" the optical absorption signatures between various oil types are very pronounced (Barringer, 1969). By use of correlation spectrometer techniques the fish oil type and approximate freshness can be determined. Investigation of the UV band from 0.3 to 0.4 $\mu$ has also shown strong absorption bands for a film approximately 0.025 mm thick. Unfortunately these data were for transmission rather than reflection in the UV band, and it appears that remote sensing of the reflected signature will not work in the UV range (Barringer, 1968). It was determined, however, that mineral oil and fish oil slicks can be differentiated by differences in their reflectivities. So far, the use of the IR band seems much more promising in these studies.

One area which is receiving investigation in the laboratory is the detection of fish oils and scales by stimulated fluorescence. It has been found that fish and crude oils are distinctly different and that the fish oils themselves can be discriminated as to type. Pulsed arc and laser light sources of stimulation are used. Tests have yet to be made at sea with live populations of fish, where several additional variables (multiconstituent makeup of ocean surface, foam, wave action, atmospheric interference with signal, etc.) will be encountered. If at-sea tests are successful, however, this type of active sensor could provide some improvement over past methods of surveying fish populations.

Unfortunately, the relationship between fish location and fish oil on the surface is anything but simple, and even a foolproof system of detecting fish oil will not invariably lead to fish detection. In brief, schools of fish do not emit oil at all times, nor do they stay in the same area where oil has been released by feeding or other processes.

0-201-04245-2

## B. Biological Oceanography

Remote sensor applications to the living resources of the sea will primarily involve indirect observation methods. In the past aircraft have been used for fish spotting and tracking, but from very low altitudes. The approach of the future will be to determine what data "indicators" from one or more sensors provide the conditions for a high probability that a certain fish stock will be located in a specified area.

Marine biology in this sense will range from detection of phytoplankton (the grazing grass of the sea) to behavioral studies of the largest marine life at the other end of the food chain (tuna, albacore, etc.). The remote sensor platform permits continuous and rapid coverage over target areas, plus the advantage that it is not coupled to the medium to scare away the objects of study (e.g., schooling fish, marine mammals).

In this section the emphasis will be on the sea surface characteristics that influence marine biota. Sea color indices, boundary zones, fish oils, and temperatures, topics which are all applicable to biologic oceanography, were discussed in Section IV.A.

### *1. Chlorophyll Assay*

Chlorophyll provides a basic index to productivity in the ocean. In waters with a high chlorophyll content the characteristic deep blue color of the ocean changes toward green, with the addition of the yellow substance associated with the photosynthetic pigments of phytoplankton (Yentsch, 1960). It has been shown that the concentration of this substance is directly linked to biological activity (Tyler, 1964). The spectral region of interest is in the visible band from 0.4 to 0.7 $\mu$, the greatest differences in chlorophyll concentration being seen at 0.4–0.5 $\mu$ (Ramsey, 1968). By way of reference the range of this concentration in the oceans, averaged over available data from the Pacific and the Atlantic, is 0.04–28.00 mg/m$^3$ for surface waters (maximum depth about 10 m). Furthermore, experimental data have shown that these surface measurements correlate with euphotic zone (total upper layer receiving sunlight) and primary productivity values (Lorenzen, 1966).

### *2. Behavioral Studies*

Behavioral studies of marine animals are important in order to determine how their activities are influenced by sea surface features that can be sensed by remote sensors. Almost every life form in the sea is represented in the top 40 m, even though many of them are capable of excursions into great depths. This statement is more accurate when applied to commercially important fish populations. Viable mapping of the valuable fisheries will depend on how well the sensor data can be related to the known behavior of a given species. These studies will involve not only direct examination of the habits of fishes but also past catch records of the stock for the fishery in question.

Some remote sensor behavioral studies may be possible from low-altitude platforms such as aircraft and airships. Large marine mammals and large schools of

0-201-04245-2

surface-feeding fish can be seen directly and their activity patterns noted without having the observer prejudice the study through his presence on a highly detectable ship platform. Aircraft have been used for some time for these purposes (Clark and Stone, 1965; Saur, 1965), employing both direct observation and IR band sensors (Walker, 1965). Useful ecological studies might be conducted with migrating species, such as whales and sea lions, when information on the size, route, and time of migration is desired.

## C. Geological Oceanography

Since remote sensor techniques are essentially ocean surface oriented, their applications to geological oceanography are limited. Measurement of sediment quantity and quality may be possible by color analysis with cameras and spectrophotometers. The configurations of coastlines and their changes due to man and nature can be mapped, and shallow-water bathymetry can be determined, depending on water clarity. Finally, in the geophysical realm airborne or spaceborne magnetometers and gravimeters can be used for determining the gross characteristics of the floor beneath the sea.

### *1. Sediment Transport*

The ocean acts as a transport medium for vast quantities of terrestrial materials. These are delivered to the ocean by water runoff from the land masses as they are modified by the actions of man and nature. By river, stream, and wind this material is dumped into the oceans at the rate of millions of metric tons a year. Since the heaviest fractions tend to be deposited only a short distance from where the material enters the oceans, it is in the coastal zones that the use of remote sensors can provide data on sediment volume, characteristics, and depositional areas (Emery, 1965; Pruitt, 1965; Phleger, 1965).

A multispectral approach with both active and passive sensors will be required to obtain useful measurements in estuarine and coastal zones. It is obvious that the same techniques can be applied equally well to the problems of pollution monitoring in these areas.

### *2. Coastal Bathymetry*

Coastal geological processes are reflected in the shape of the coastline and the contours of the offshore bathymetry. In many areas these features change considerably over short periods of time. A proper understanding of these changes and predictions of their effects can be best derived through the application of remote sensors.

Continuous "mapping" of coastal areas will provide not only data on changes but also a determination of the rate of change. This is particularly important in the estimation of what man-made changes might be required to arrest the process.

Remote sensing of coastal bathymetry requires "penetration" of the shallow coastal waters; this is accomplished by densitometric photographic techniques and

0-201-04245-2

by the use of certain active sensors, such as the laser. Other sensors (radars, metric photography, microwave radiometers, etc.) can provide measurement of the energy that is the prime mover in creating bathymetric changes (e.g., wave and sea action, coastal currents, weather). Imaging techniques can monitor and record the change in coastal configuration at the sea-land interface.

The photographic penetration techniques depend on knowledge of the light penetration in sea water, as mentioned in Section IV.A.9. The work with spacecraft photography from the GEMINI and APOLLO missions has shown that a "density slicing" technique can provide bathymetric data to depths of approximately 50 m (National Council on Marine Resources and Engineering Development, 1967; Ross, 1968). These results were obtained by reprocessing the photographic imagery after it was taken; a more advanced effort has employed aircraft using multispectral cameras, where the filtering is done at the time the imagery is made (Badgley, 1969). In these experiments NASA has employed six- and nine-lens cameras to determine the optimum filtration, band widths, and resolution with respect to space, weight, and power limitations for spacecraft. A four-lens camera was used in the Bahamas during the *Ben Franklin* tests mentioned earlier, which provided useful data on light penetration in coastal waters (Ross, 1969b). A final optical technique is the use of multispectral imagers such as the University of Michigan's 18-channel multispectral optical-mechanical scanner. Overwater flights with this system have shown that 12 of the channels cover the region of water transparency (0.4–10 $\mu$) and that the 0.55–0.58 $\mu$ band provides the greatest penetration (Polcyn and Noble, 1968). With any multiband technique proper choice of the bands, according to their selectivity of penetration of sea water, can lead to effective bathymetric mapping.

Recent experiments with lasers have demonstrated that active sensors can also be applied to bathymetric measurements. Calculations have shown that a unit with a peak power of $10^5$ W (operating on a wavelength selected for maximum penetration) can provide depth information to 40 m at a platform altitude of 300 m or to 1.5 m at about 333-km altitude (Polcyn and Noble, 1968). A pulsed neon laser has been built which operates at 0.54 $\mu$ and has given depth information to 9 m in turbid water near shore (Hickman and Hogg, 1969). Tentative aircraft flights with this system indicate that depth/bathymetric data can be obtained at from 30 to 36 m in clearer waters. The Naval Oceanographic Office has developed a unit called PLADS (pulsed light airborne depth sounder); this is a pulsed laser operating at 0.55 $\mu$. It will have a depth accuracy of ±0.5 m, or 5% of measured depth, whichever is greater, and a maximum useful depth range of about 45 m (Noble et al., 1969). The parameters stated for this use are for aircraft applications.

### *3. Geophysical Observations*

Although not a uniquely oceanographic application, the mapping of the earth's gravity and magnetic fields is an important activity over the 71% of the earth that is water covered. The data from these surveys have application in the improvement

0-201-04245-2

of navigation techniques and the location of subsurface structures which may have commercial value (ore bodies, petroleum-bearing formations, etc.).

### D. Chemical Oceanography

This area of oceanography deals with studies of seawater as a chemical solution and with this solution's interaction with the boundaries, interfaces, and living matter of oceans. The complex makeup of seawater and its high degree of spatial variability throughout the world ocean make the application of remote sensors to this discipline very attractive. Since almost every characteristic of the ocean is affected, in some way, by the chemistry of its waters, most of the techniques mentioned in earlier sections will have equal applications here. Some chemistry-related problem areas—for example, organic films and slicks, salinity, and chlorophyll measurement—were discussed in Sections IV.A and IV.B.

#### *1. Gas Exchange at the Sea-Air Interface*

Gases in the marine atmosphere result from processes of nature and from man's activities. Direct chemical precipitation of calcite (limestone) upon the seafloor occurs as a function of carbon dioxide in the atmosphere and the barometric pressure. Certain of these gases interact with the upper layers of the sea, and in some cases the sea supplies gases to the atmosphere. These gases have a selective attenuation effect in the atmosphere, which is related to their concentration. By using the multipath technique it should be possible to make differential measurements to obtain the concentrations of selected gases in the air column above the sea.

Studies in the absorption bands for the various gases are required to develop their distinctive signatures. At the present time, little work is being conducted in this area.

In addition to gases, vapors can be sensed by similar methods. By "vapors" is meant the substances which collect on the ocean surface and then, by virtue of their high pressures, produce discernible vapors in the atmosphere above the sea surface. Investigation into iodine vapor (Barringer, 1969) has shown that this substance is associated with marine surface life and that it can be sensed remotely. In addition, it has some relationship to concentrations of seaweed and its proliferation.

## V. OCEAN SCIENCES: THE APPLICATIONS

If the knowledge gained from remote sensors is to have practical utility, it must be applied to activities which have some economic relevance. For example, NASA's Earth Resources Survey Program is designed to achieve this end point rather than the development of data or knowledge for its own sake. The basic studies permit description of phenomena, construction of a theory to explain them, and finally a

0-201-04245-2

model with which prediction can be obtained. From this knowledge comes the fundamental inputs that permit conversion of academic work to economic work through remote sensing. In this section it will be demonstrated how a combination of the five oceanographic disciplines can be applied to remote sensor support of ocean operations.

### A. Maritime Weather Monitoring and Prediction

Maritime weather exercises two major influences on man's use of the oceans. In the coastal areas, where the highest population densities are located, the effects of wind and sea associated with storms can create catastrophic damage and considerable modification of the coastline. On the high seas storms can delay and damage shipping. In addition to the obvious costs of repairing storm damage to ships, the delays incurred when multimillion-dollar cargoes are unnecessarily kept at sea and away from their markets result in more subtle but none the less substantial costs. Such situations represent a stagnation of capital, and on an annual basis for the world's shipping the costs are staggering.

Although it is not yet possible to control maritime weather, it is possible to develop forecasts which will reduce hazard in both coastal areas and on the high seas to the absolute minimum. Good forecasting requires continuous monitoring of the world ocean to locate and track major weather systems, as well as the wind and wave fields generated by them. In previous sections the feasibility of meeting these requirements in remote sensor systems has been demonstrated. At present the earth-orbiting weather satellites provide a very coarse, but useful, real-time capability for tracking weather systems; and this, together with penetration flights of weather aircraft, has improved the path predictions for major marine weather systems (hurricanes, typhoons, etc.). Also, a program is in operation which uses computers and continuous weather inputs to provide optimum track routing to ships that request it. Unfortunately, many of the inputs to these systems come from climatological data (historical) and shipboard weather reports which have limited accuracy (uncalibrated sensors, untrained observers, and time delays). In the future specialized sensor platforms will fill an obvious need in these weak fundamental areas, permitting relatively refined surveys.

More basic work will be needed to understand what conditions are required to develop full-scale oceanic storm systems and what factors govern the path that they take. It appears possible that remote sensors could be used to develop probabilities of storm development and the putative path of such a system, once the critical parameters for generation are understood. Project Stormfury of the U.S. Weather Service, one of the many efforts in this area, is experimenting also with the possibilities of modification, both to reduce the intensity of storms and to steer them away from heavily populated areas. Although these efforts are still in the very basic stages, some preliminary results indicate possibilities that these goals may be reached.

0-201-04245-2

### B. Navigational Improvements

Since 1960 over 23 navigation satellites have been put into earth orbit (NASC, 1970). Through precise orbit, with respect to the center of the geoid, these "remote sensors" can provide both mariners and aviators with a navigational reference far more accurate than any used previously.

### C. Hydrography

Hydrography has been defined as "the description of the bucket that holds the oceans." Essentially this is the business of surveying, publishing, and distributing information required by the mariner. Because ships operate at the sea-air interface and are endangered by shallow waters, the features which may trouble them are likely to be those which can be mapped by remote sensors.

In preceding sections the means by which currents, wave fields, and bathymetry can be sensed were illustrated. However, more specific applications have been initiated for this field. The U.S. Naval Oceanographic Office utilized some of the early GEMINI photography to check photographic data against charted features. This was unique, being the first time it had been possible to get the features "at arm's length" in one piece of imagery. The photographs were scaled to the largest charted scale and then were sandwiched to compare contours. Rongelap Atoll in the Pacific Ocean was one of these areas. The superposition of chart (H.O. Chart 6029) and photography showed the chart to be in error in certain places. These were errors in survey, rather than changes in the configuration of the land. Another example was the Colorado Delta in the Gulf of California (Sea of Cortez), where the bathymetries of the delta as charted and as photographed were compared. Again there were differences, but in this case they were due to changes that had taken place in the delta because of sediment transport by the Colorado River since the last survey. Since this delta is on the orbital paths used by both GEMINI and APOLLO missions, additional comparison would show changes in the photography taken over the 5 years' duration (to date) of these two programs.

In situations where large changes in near-shore bathymetry can occur in brief time periods, the use of remote sensor techniques offers the fastest and most economical means of correcting charts. In areas subjected to maritime-generated storms such as hurricanes, it is not unusual to have entire small barrier islands along a coastline in a storm's path move some distance or even disappear. The time of the change will be less than a few days, and if this occurs near shipping lanes or channels there can be a hazard to navigation. The need for quick resurveys of storm-affected areas can be met by remote sensor platforms.

Throughout the world ocean there are areas that have been reported as being hazards to navigation because of the presence of shoal water. These have been reported by all manner of ships, and many of these sightings go back to the last century. In the interest of safety, however, they must remain on the charts until surveys can positively remove them. In the Gulf of Mexico alone there are 54 such

0-201-04245-2

doubtful soundings, according to the International Hydrographic Bureau. These are defined as locations where the reported water depth is thought to be 15 m or less. Because areas of this nature tend to be anomalous, characteristic features of the ocean in their vicinity should provide adequate indications of whether or not such a shoal exists. These features would be changes in water color, variations in wave field due to shoaling, localized anomalies in sea surface temperature, and anomalous cloud patterns (Noble et al., 1969). Tests have been conducted on sample areas using multispectral scanners in the visual band (Polcyn and Noble, 1968), and the results showed the feasibility of this type of survey in the clear waters off Florida. The addition of color photography, for the application of density slicing techniques, and a laser depth finder should provide the capability to check virtually all of these reported shoals in the world ocean. If an aircraft platform is used, a satellite navigation system will be necessary to ensure precise navigation.

In view of the number of these shoals still uninvestigated and the relatively small world population of oceanographic vessels, it appears that the use of remote sensor platforms is the only sensible way to resolve this hydrographic dilemma.

### D. Fisheries–Food from the Sea

By "fisheries" is meant the commercial application of marine biology to the business of deriving food from the sea. Fishing is man's oldest enterprise involving the sea, and its importance is increasing as the population on this planet increases. On the basis of the use of time, this activity is not as well organized as terrestrial food-producing operations. The fisherman spends 65% of his time searching, 10% catching, and the balance traveling to and from the fishing grounds. It can be seen that the biggest improvements should come in the 65% search time, and it is here that remote sensor capabilities can have the greatest impact (Chapman, 1969). As outlined in Section IV.B, multisensor applications have great potential for locating biological activity in the surface waters of the ocean. An additional task is to develop a system for rapid analysis of the data, developing forecasts for specific fisheries (i.e., tuna, cod, herring, etc.), and transmitting the information to the fisherman. The use of communications satellites and computers can help to keep the entire circuit "real time."

The range of fisheries prediction is certain to expand as more experience is obtained. At present it appears feasible to develop fairly short-term forecasts based on sea conditions which would affect an existing fish stock. But in the future, with better knowledge of fish behavior, it will be possible to forecast the seasonal potential for a given fishery (Chapman, 1969). There are certain measurable factors, such as water temperature, salinity, nutrient content, and turbidity, that affect the location and size of spawning grounds. Since these factors influence the size of the stock several seasons away, they could become the basis for long-term prediction methods. In addition, monitoring of ocean and atmosphere patterns for a given fisheries area on a continuous basis for comparison against climatological records of these patterns will yield insight into deviations from conditions which provide

0-201-04245-2

large catches. With forecasts available well in advance of fishing seasons, the fisherman will have the advantage of planning his investment in time and resources before he goes to sea.

To a lesser extent the sensor applications which will bring improvements in commercial fishing will also be applicable to sport fishing, although the economic return is not as attractive. The immediate gain to sport fishermen will be in fisheries where both commercial and sport interests participate.

Commercial fisheries involving sedentary species such as scallops and oysters can also benefit from remote sensor techniques. These resources are located in fairly fixed locations, and therefore the yield of such a fishery is extremely sensitive to factors which change the beneficial environment. Moreover, since the beds are located primarily in coastal areas, the additional factors of land runoff and pollution must be considered. Remote sensors provide the needed rapid survey capability to detect the onset of unfavorable changes of these types.

Although food from the sea is obtained primarily from animal species, the coastal areas also support a variety of ocean plant life which has some commercial value. For example, the *Macrocystis* kelp along the southern California coast is harvested regularly by both Mexican and U.S. interests. This is one of the fastest-growing plants in the world, and the derivatives of its processing are found in many food and medical products. Remote sensors can provide frequent surveys of the plant beds on a worldwide basis to monitor their growth and restoration after harvesting. It has been found that kelp is detectable by imaging radars and that the iodine vapor concentration at the sea surface is increased near seaweed beds. Such factors as these should permit accurate remote determination of the viability of a given marine plant resource. In addition, these marine plant beds are closely tied to the biological cycle, as many commercially important species spend part of their early growth cycle in such areas.

## E. Coastal Engineering

Coastal engineering involves the protection of the coastline against the modifying forces of nature, as well as man's efforts to rebuild it for his purposes. Wind, sea, and runoff from land all continually work to change the configuration of the coastline, while man modifies it to suit his requirements.

In preceding sections it has been shown that remote sensors can be used to detect most of the influences that result in coastal changes. From the detection and tracking of marine weather to the changes in near-shore bathymetry due to river deposition, the remote sensor platform can provide the continuous monitoring required to preserve the coastlines.

One major area of concern is the effects of pollution in the coastal zone. Much of the pollution is introduced into these waters by the discharge of wastes into harbors and rivers. The effluent plumes of offshore sewer outfalls also can be a source if proper water (sewage) treatment is lacking. Remote sensor techniques are

0-201-04245-2

ideal for routine monitoring of pollution and for study of its effects on the coastal zone as a whole. Since most pollutants have both thermal and color signatures, relatively simple sensor suits can be developed for use by local agencies in monitoring their coastal areas. More importantly the application of remote sensor techniques can provide accurate synoptic, periodic surveys of a coastal area before major changes are made. Thus harbor flushing characteristics, river outflow variations, and annual changes in estuarine zones could be determined to establish baselines and optimum locations for activities that would require dumping of effluent into the adjacent waters. Moreover, after such an activity was established, the same techniques could monitor it to ascertain compliance with legal standards.

Man's construction of coastal works (seawalls, landfill, jetties, harbors, etc.) requires great quantities of building materials. The coastal zone can provide much of the sands and gravels needed at costs more reasonable than those of terrestrial sources. Remote sensor surveys can help to locate and identify these resources in shallow waters.

In areas where the sea causes steady erosion of recreational sites, these beaches are being rebuilt by pumping the sand from offshore back onto the beach. This is a costly but necessary operation on many beaches if their recreational value is to be preserved. Remote sensor surveys could provide the continuous observations needed to determine the rate of erosion and the time when replenishment would be required.

### F. Coastal Hydrology

Many coastal regions, despite their proximity to the oceans, lack sufficient fresh water resources. A few years ago the U.S. Geological Survey studied Hawaiian volcanoes with an airborne infrared thermometer/imager. It was noted that on some of the imagery taken along the coast small cold-water streaks appeared where no surface stream existed. It was determined that these represented fresh water springs on the seafloor, where the terrestrial water table pressure forced the fresh water through the porous seabed. Fresh water is less dense, though colder, than seawater, and it rose to the surface, where the IR imager sensed it as an anomalous cold patch. Similar springs are known to exist off many coastal areas, and remote sensors could be employed to locate and map these resources (Kohout and Kolipenski, 1967). Although direct measurement of water volume is not feasible, it would be important to determine whether the flow was seasonally variable or constant at all times.

Another contribution to coastal hydrology would be the same type of regular coastal surveys that were mentioned in the preceding section. In this case the strength of coastal currents and their relationship to harbor and river outflows would be determining factors in locating saline-water-conversion plants. The intake for such plants should be relatively clean seawater, and the hot brines discharged from the process should be placed where they cannot damage the environment.

0-201-04245-2

### G. Sea Ice Prediction and Tracking

Sea ice has an economic impact in two ways. First, drifting ice poses a threat to shipping when it appears in heavily used lanes. Second, the increased interest in developing the resources of the arctic slopes has led to a consideration of using ice-strengthened ships to support these operations; knowledge of ice conditions will be vital to avoid costly damages and lost time.

Because darkness and cloud cover prevail over the polar regions a great deal of the time, direct eyeball or camera observations are difficult. Work with the remote sensors on board weather satellites has shown that the high-resolution IR imagers on the NIMBUS 1 and 2 satellites indicate that individual icebergs in the Arctic and Antarctic oceans can be tracked (Barnes et al., 1969). The Navy's Birds Eye flights and the Coast Guard's ice reconnaissance aircraft program both have applied a wide range of remote sensors to ice studies, which will contribute to both information on the regions and sensor development for their study.

The most promising sensors for remote surveys of the polar regions appear to be the radars and the radiometers (IR and microwave), which provide data even in adverse conditions of visibility (McLerran, 1969).

### H. Data Transmission

In the beginning of this chapter it was stated that one of the most powerful advantages of remote sensors in oceanographic work was the attainment of synopticity on a continuous basis. Related to this is an effort to tie together data taken at several points on the oceans at the same time. Groups of ships, aircraft surveys, data buoy systems, and various combinations of these elements all collect and store data, which usually are not processed at the site. Fixed data points on the ocean surface can serve a complementary purpose with remote sensors: to provide profile data (temperature, salinity, sound velocity, etc.) that can be correlated with surface characteristics seen by the remote sensor. For shorter-term studies such data collectors (i.e., buoys) can be dropped by the remote sensing aircraft itself.

The use of satellites to interrogate, receive, and transmit data from various data acquisition systems will be a vital step in achieving both synopticity and time compression in data processing. The National Aeronautics and Space Administration has a program for a system called IRLS (interrogation, recording, and locating system) which not only can tie together several data points but also can provide navigational information (Townsend, 1969). The ability to bring back real-time data from literally hundreds of points on the oceans to land stations via these satellites means that large-capacity computing equipment can be used to reduce and plot key data, thus permitting additional personnel to participate in expeditions by "remote control."

0-201-04245-2

# References


Abdel-Gawad, M. (1969). New Evidence of Transcurrent Movements in Red Sea and Area and Petroleum Implications, *Bull. of Amer. Assoc. of Petroleum Geologists.* 53, 1466–1479.

Abdel-Gawad, M., and Silverstein, J. (1973). ERTS Applications in Earthquake Research and Mineral Exploration in California, in *Symp. on Significant Results Obtained from the Earth Resources Tech. Satellite-1.* NASA SP-327, 1, 433–450. Goddard Space Flight Center, New Carrollton, Md.

Aldrich, R. C. (1975). Detecting Disturbances in a Forest Environment, *Photogram. Engin.* 41(1), 39–48.

Altshuler, E. E., Falcone, V. J., and Wulfsberg, K. N. (1968). Atmospheric Effects on Propagation at Millimeter Wavelengths, *IEEE Spectrum.* 5(7), 83–91.

Ambartsumyan, V. A. (1958). *Theoretical Astrophysics.* Pergamon Press, New York.

American Society of Photogrammetry (1960). *Manual of Photographic Interpretation.* George Banta Co., Menasha, Wis.

American Society of Photogrammetry (1968). *Manual of Color Aerial Photography.* George Banta Co., Menasha, Wis.

Amsbury, D. L. (1969). Geological Comparison of Spacecraft and Air-Craft Photographs of the Potrillo Mountains, Texas, in *Proc. of the Sixth Int'l. Symp. on Remote Sens. of Environ.* 493–515. Univ. of Mich., Ann Arbor.

Anderson, J. H., Shapiro, L., and Belon, A. E. (1973). Vegetative and Geological Mapping of the Western Seward Peninsula, Alaska, Based on ERTS-1 Imagery, in *Symp. on Significant Results Obtained from the Earth Resources Tech. Satellite-1.* NASA SP-327, 1, 67–76. Goddard Space Flight Center, New Carrollton, Md.

Anderson, J. R., Hardy, E. E., and Roach, J. T. (1972). A Land Use Classification System for Use with Remote-Sensor Data, U.S. Geol. Survey Circular 671, 1–16.

Anding, D., and Kauth, R. (1970). Estimation of Sea Surface Temperature from Space, *Remote Sens. of Environ.* 1, 217–220.

Andrews, H. C., Tescher, A. G., and Kruger, R. P. (1972). Image Processing by Digital Computer. *IEEE Spectrum.* 9(7), 20–32.

Anson, A. (1966). Color Photo Comparison, *Photogram. Engin.* 32(2), 286–297.

Anson, A. (1970). Color Aerial Photos in the Reconnaisance of Soils and Rocks, *Photogram. Engin.* 36(4), 343–354.

Anuta, P. E. (1969). Digital Registration of Multispectral Video Imagery, *Soc. of Photo-Optical Instrumentation Engineers.* 7(6), 168–175.

Anuta, P. E. (1970). Spatial Registration of Multispectral and Multitemporal Digital Imagery Using Fast Fourier Transform Techniques, *IEEE Trans. Geoscience Electronics.* GE8(4), 353–368.

Anuta, P. E., Kristof, S. J., Levandowski, D. W., Phillips, T. L., and MacDonald, R. B. (1971). Crop, Soil, and Geological Mapping from Digitized Multispectral Satellite Photography, in *Proc. of the Seventh Int'l. Symp. on Remote Sens. of Environ.* 1983–2016. Univ. of Mich., Ann Arbor.

Anuta, P. E., and MacDonald, R. B. (1971). Crop Surveys from Multiband Satellite Photography Using Digital Techniques, *Remote Sens. of Environ.* 2(1), 53–67.

Arnold, J. E. (1967). Reflections of the Sea Surface Temperature Field in Low Level Cloud Development: An Observational Air-Sea Interaction Study from Space, Ref. No. 67-4T. Dept. of Oceanography, Texas A & M Univ., College Station.

Arnold, J. E., and Thompson, A. H. (1969). Apparent Coastal Water Discolorations from Space–Oceanic and Atmospheric, in *The Oceans from Space* (P. C. Badgley, L. Miloy, and L. Childs, eds.). 82–91. Gulf Pub. Co., Houston, Tex.

Atroshenko, V. S., Feigelson, E. M., Glazove, K. S., and Malkevich, M. S. (1963). *Calculation of the Brightness of Light,* Part 2. Consultants Bureau, New York.

Avery, T. E., and Canning, J. (1974). Air Photo Measurement of New Zealand Pines, *Photogram. Engin.* 40(8), 957–959.

Badgley, P. C. (1965). Introductory Briefing to the Conference on the Feasibility Explorations from Aircraft, Manned Orbital, and Lunar Laboratories, in *Oceanography from Space* (G. C. Ewing, ed.). 1–17. Woods Hole Oceanographic Inst. Ref. No. 65-10, Massachusetts.

Badgley, P. C., and Vest, W. L. (1966). Orbital Remote Sensing and Natural Resources, *Photogram. Engin.* 32(5), 780–790.

Badgley, P. C., Childs, L., and Vest, W. L. (1967). The Application of Remote Sensing Instruments in Earth Resource Surveys, *Jour. of Geophysics.* 32(4), 583–601.

Badgley, P. C. (1969). Earth Resource Surveys from Space, in *The Oceans from Space* (P. C. Badgley, L. Miloy, and L. Childs, eds.). 1–28. Gulf Pub. Co., Houston, Texas.

Bakis, R., and Langley, P. G. (1971). A Method for Determining the Height of Forest Trees Automatically from Digitized 70 mm Color Aerial Photographs, in *Proc. of the Seventh Int'l Symp. on Remote Sens. of Environ.* 705–713. Univ. of Mich., Ann Arbor.

Bakis, R., Wesley, M. A., and Will, P. M. (1971). Digital Correction of Geometric and Radiometric Errors in ERTS Data, in *Proc. of the Seventh Int'l. Symp. on Remote Sens. of Environ.* 2, 1427–1436. Univ. of Mich., Ann Arbor.

Bale, J. B., and Bowden, L. W. (1973). Land Use in the Northern Coachella Valley, in *Symp. on Significant Results Obtained from the Earth Resources Tech. Satellite-1.* NASA SP-327, 1, 915–922. Goddard Space Flight Center, New Carrollton, Md.

Bale, J. B., Conte, D., Goehring, D., and Simonett, D. S. (Eds.) (1974). Remote Sensing Applications to Resource Management Problems in the Sahel, Technical Report, U.S. Dept. of State Contract No. AID afr-c-1058, Earth Satellite Corporation, Washington, D.C.

Bale, J. B., Goehring, D. R., Rohde, W. G., and Simonett, D. S. (1975). Evaluation of SKYLAB-EREP Data for Land Resource Management, NASA Contract No. NAS 9-13314. 1–20. Earth Satellite Corp., Washington, D.C.

Ball, G. H. (1965). Data Analysis in the Social Sciences: What about the Details? *IEEE Proc. Fall Joint Computer Conference.* 27(1), 533–560.

Barnes, J. C., Chang, D. T., and Willard, J. H. (1969). Use of Satellite High Resolution Infrared Imagery to Map Arctic Sea Ice. Report prepared for Naval Oceanographic Office Spacecraft Oceanography Project. Contract No. N62306-68-C-0276. Allied Research Associates, Concord, Mass.

Barnhardt, E. A., and Streete, J. L. (1970). A Method for Predicting Atmospheric Aerosol Scattering Coefficients in the Infrared, *Appl. Opt.* 9(6), 1337–1344.

Barr, D. J., and Miles, R. D. (1970). SLAR imagery and site selection, *Photogrammetric Engr.,* 36, 1155–1170.

0-201-04245-2

Barrett, A. H., and Chung, V. K. (1962). A Method for the Determination of High-Altitude Water Vapor Abundance from Ground-Based Microwave Observations, *Jour. of Geophys. Res.* 67(11), 4459–4466.

Barrett, T. P. (1966). Oceanographic Interpretation of Two Gemini 5 Photographs, Informal Manuscript Report No. 0-20-66. U.S. Naval Oceanographic Office, Washington, D.C.

Barrett, T. P., and Wilkerson, J. C. (1967). On the Generation of Ocean Wind Waves as Inferred from Airborne Radar Measurements of Fetch-Limited Spectra, *Jour. of Marine Res.* 25(3), 292–328.

Barringer, A. R. (1962). New Approach to Exploitation–The Input Airborne Electrical Pulse Prospecting System, *Mining Congress Jour.* 48, 49–52.

Barringer, A. R. (1965). Developments Towards the Remote Sensing of Vapours as an Airborne and Space Exploitation Tool, in *Proc. of the Third Int'l Symp. on Remote Sens. of Environ.* 297–292. Univ. of Mich., Ann Arbor.

Barringer, A. R. (1966). Interference-free Spectrometer for High-Sensitivity Mercury Analyses of Soils, Rocks, and Air, *Inst. of Mining and Metallurgy Trans.* (Sect. B: Appl. Earth Sciences). 72 B120-4.

Barringer, A. R. (1967). Detecting the Ocean's Food (and Pollutants) from Space, *Ocean Industry.* 2(5), 28–34.

Barringer, A. R. (1968). The Remote Sensing of Spectral Signatures Applied to Pollution Measurements and Fish Detection, in *Proc. of the Ninth Meeting of the Ad Hoc Spacecraft Oceanography Advisory Group.* Abstract. January. Spacecraft Oceanography Project, NAVOCEANO, Maryland.

Barringer, A. R., and McNeill, J. D. (1968). Radiophase–A New System of Conductivity Mapping, in *Proc. of the Fifth Int'l Symp. on Remote Sens. of Environ.* 157–167. Univ. of Mich., Ann Arbor.

Barringer, A. R., and McNeill, J. D. (1969). Recent Developments in Remote Sensing for Geophysical Applications, *Proc. of the Sixth Int'l Symp. on Remote Sens. of Environ.* 617–636. Univ. of Mich., Ann Arbor.

Barringer, A. R., Newbury, B. C., and Moffat, A. J. (1968). Surveillance of Air Pollution from Airborne and Space Platforms, in *Proc. of the Fifth Symp. on Remote Sens. of Environ.* 123–155. Univ. of Mich., Ann Arbor.

Barringer, A. R. (1969). Remote Sensing of Marine Effluvia, in *The Oceans from Space* (P. C. Badgley, L. Miloy, and L. Childs, eds.). 92–103. Gulf Pub. Co., Houston, Tex.

Barringer, A. R. (1970a). Optical Detection of Geochemical Anomalies in the Atmosphere, *CIM.* Special Vol. 11.

Barringer, A. R. (1970b). Remote Sensing Techniques for Mineral Discovery, in *Proc. of the Ninth Commonwealth Mining and Metallurgical Congress 1969,* Vol. 2, *Mining and Petroleum Geology.* 649–690. London, England.

Barringer, A. R., Davies, J. H., and Moffat, A. J. (1970). The Problems and Potential in Monitoring Pollution from Satellites, Paper No. 70-305. Earth Resources Observations and Information Systems Meeting, Amer. Inst. of Aeronautics and Astronautics, New York.

Barringer, A. R., and McNeill, J. D. (1970). The Airborne Radiophase System–A Review of Experience, unpublished paper. Canadian Inst. of Mining and Metallurgy, Toronto, April 1970.

Barringer, A. R., and McNeill, J. D. (1971). E-Phase™, A New Remote Sensing Technique for Resistivity Mapping, in *Proc. of the Seventh Int'l Symp. on Remote Sens. of Environ.* 1447. Univ. of Mich., Ann Arbor.

0-201-04245-2

Barringer, A. R., and Davies, J. H. (1972). Applications of Correlation Interferometry and Spectroscopy to Aerospace Monitoring of Earth Resources and Pollution, in *Proc. of the Seventh Int'l Aerospace Instrumentation Symp.* Cranfield Institute, England.

Barteneve, O. D. (1960). *Izv. Geophys. Ser.* 1237.

Barton, D. K. (1964). *Radar System Analysis.* Prentice-Hall, Englewood Cliffs, N.J.

Basu, J. P., and Odell, P. L. (1974). Effects of Intraclaus Correlation among Training Samples on the Misclassification Probabilities of Bayes' Procedure, *Pattern Recognition.* 6, 13–16.

Bates, C. C. (1953). Rational Theory of Delta Formation, *Bull. of Amer. Assoc. of Petroleum Geologists.* 37(9), 2119–2162.

Bauer, M. E., Swain, P. H., Mroczynski, R. P., Anuta, P. E., and McDonald, R. B. (1971). Detection of Southern Corn Leaf Blight by Remote Sensing Techniques, in *Proc. of the Seventh Int'l Symp. on Remote Sens. of Environ.* 693–704. Univ. of Mich., Ann Arbor.

Bauer, M. E. (1972). Remote Sensing as a Means of Detecting Crop Disease, NASA Grant No. NGL 15-005-112. Lab. for Applications of Remote Sensing, Information Note 010672. Purdue Univ., West Lafayette, Ind.

Baumann, R. L., and Winkler, L. (1955). Rocket Research Report No. XXI, Photography from the Viking 12 Rocket at Altitudes Ranging up to 143.5 Miles, U.S.N.R.L. Report R-4489. U.S. Naval Research Lab., Washington, D.C.

Baumgardner, M. F., Henderson, J. A., Jr., and LARS Staff. (1973). Mapping Soils, Crops, and Rangelands by Machine Analysis of Multitemporal ERTS-1 Data, in *Third Earth Resources Tech. Satellite Symp.* NASA SP-3517, 205–223. Goddard Space Flight Center, Washington, D.C.

Bechtold, I. C., Liggett, M. A., and Childs, J. F. (1973). Regional Tectonic Control of Tertiary Mineralization and Recent Faulting in the Southern Basin-Range Province, An Application of ERTS-1 Data, in *Symp. on Significant Results Obtained from the Earth Resources Tech. Satellite-1.* NASA SP-327, 1, 425–432. Goddard Space Flight Center, New Carrollton, Md.

Becker, A. (1969). Simulation of Time-Domain Airborne, Electromagnetic System Response, *Geophysics.* 34(5), 139–752.

Beckmann, P., and Spizzichino, A. (1963). *The Scattering of Electromagnetic Waves from Rough Surfaces.* Pergamon Press, Elmsford, N.Y.

Beckner, C. F., Jr. (1968). Comparisons of Remote Airborne Oceanographic Sensors, Tech. Report TR–204. U.S. Naval Oceanographic Office, Washington, D.C.

Beilock, M. M., Wilson, C., and Zaitzeff, E. (1969). Design Considerations for Aerospace Multispectral Scanning Systems, in *Symp. on Information Processing II.* 658-671. Purdue Univ., West Lafayette, Ind.

Bennett, R. (1970). Exploration for Hydrothermal Mineralization with Airborne Gamma-Ray Spectrometry, *CIM.* Special Vol. 11.

Benoit, A. (1968). Signal Attenuation Due to Neutral Oxygen and Water Vapor, Rain and Clouds, *Microwave Jour.* 11(11), 73.

Bergstrahl, T. A. (1947). Photography from the V-2 Rocket at Altitudes Ranging Up to 160 Kilometers, U.S.N.R.L. Report R-3083. U.S. Naval Research Lab., Washington, D.C.

Bernstein, D. A. (1974). Are Reforestation Surveys with Aerial Photographs Practical? *Photogram. Engin.* 40(1), 69–73.

Berry, B. J. L., and Pred, A. (1965). *Central Place Studies: A Bibliography of Theory and Applications,* Philadelphia Regional Science Research Institute.

Bhattacharyya, B. K. (1965). Two-Dimensional Harmonic Analysis as a Tool for Magnetic Interpretation, *Geophysics.* 30(5), 829–857.

0-201-04245-2

Bhattacharyya, B. K. (1966a). A Method for Computing the Total Magnetication Vector and the Dimensions of a Rectangular Block-Shaped Body from Magnetic Anomalies, *Geophysics.* 31(1), 74–96.

Bhattacharyya, B. K. (1966b). Continuous Spectrum of the Total Magnetic Field Anomaly Due to a Rectangular Prismatic Body, *Geophysics.* 31(1), 97–121.

Bhattacharyya, B. K., and Raychaudhuri, B. (1967). Aeromagnetic and Geological Interpretation of a Section of the Appalachian Belt in Canada, *Can. Jour. of Earth Sciences.* 4(6), 1015–1037.

Bhattacharyya, B. K. (1969a). Bicubic Spline Interpolation as a Method for Treatment of Potential Field Data, *Geophysics.* 34(3), 402–423.

Bhattacharyya, B. K. (1969b). Semiautomatic Methods of Interpretation of Magnetic Data, Mining, and Groundwater, *Geophysics.* 1967, 402–423.

Biehl, L. L., and Silva, L. F. (1975). Evaluation of SKYLAB Data for Land Use Planning, Paper AAS 74-145, Am. Astronautical Society, 1–38.

Bignell, K. J., Saiedy, F., and Sheppard, P. A. (1963). On the Atmospheric Infrared Continuum, *Jour. of Opt. Soc. of Amer.* 53, 466.

Bignell, K. J. (1970) The water-vapour infra-red continuum, *Royal Meteor. Soc. Quart. Jour.* 96, 390.

Billingsley, F. E. (1970). Applications of Digital Image Processing, *Appl. Opt.* 9(2), 289–299.

Binsell, R. (1967). Dwelling Unit Estimation from Aerial Photography, U.S.G.S. Contract 14-08-0001-10654. Geographic Applications Program, U.S. Geol. Survey, Washington, D.C.

Blanchard, B. J. (1973). Measuring Watershed Runoff Capability with ERTS Data, in *Third Earth Resources Tech. Satellite Symp.* NASA SP-351, 1, 1089–1098. Goddard Space Flight Center, Washington, D.C.

Blifford, I. H., Jr., and Ringer, L. D. (1969). The Size and Number Distribution of Aerosols in the Continental Troposphere, *Jour. of Atmos. Sciences.* 26(4), 716–725.

Blythe, R., and Kurath, E. (1967). Infrared and Water Vapor, *Photogram. Engin.* 33(7), 772–777.

Blythe, R., and Kurath, E. (1968). Infrared Images of Natural Subject, *Appl. Opt.* 7(9), 1769–1777.

Boles, William (1969). Scattering of Waves from a Rough Layer. Ph.D. Dissertation, CRES Tech. Report 118-19. Univ. of Kansas, Lawrence.

Boltneva, L. I., Vasilenko, V. N., Dmitriyev, A. V., Ionov, V. A., Kogan, R. M., Kuznetsova, Z. V., Nazarov, I. M., and Yagodovskiy, I. V. (1964). An Experiment for Studying the Radioactivity of Granitoid Intrusion by Airborne Gamma Spectrometry, *Akademeia Nauk SSSR Izv., Geophysics Ser.* 858–871. [*Acad. of Science of the USSR Bull. Geophysics Ser.* 520–527.]

Bond, A. D., and Atkinson, R. J. (1972). An Integrated Feature Selection and Supervised Learning Scheme for Fast Computer Classification of Multispectral Data, in *Remote Sensing of Earth Resources* (F. Shahrokhi, ed.). 1, 645–672. Univ. of Tenn., Tullahoma.

Boniwell, J. B. (1967). Some Recent Results with the INPUT[R] Airborne EM System, *CIM Trans.* 70, 60–66.

Boriani, A., Marino, C. M., and Sacchi, R. (1974). Geologic Features on ERTS-1 Images of a Test Area in the West-Central Alps, in *Proc. of a Symp. on European Earth Resource Satellite Experiment.* ESRO SP-100, 198–204. Frascati, Italy.

0-201-04245-2

Bortner, M. H., Davies, J. H., Dick, R., Grenda, R. N., and LeBel, P. J. (1971). Carbon Monoxide Pollution Experiment–A Solution to the CO Sink Anomaly, Joint Conference on Sensing of Environmental Pollutants. AIAA Paper No. 71-1120. Palo Alto, Calif.

Boudreau, D. R. (1965). Skin Temperature of the Sea as Determined by Radiometer, Ref. No. 65-15T. Dept. of Oceanography, Texas A & M Univ., College Station, Tex.

Bouknight, W. J., Denenberg, S. A., McIntyre, D. E., Randal, J. M., Saneh, A. H., and Slotnick, D. L. (1972). The Illiac IV System, *Proc. of IEEE.* 60(4), 369–388.

Bowden, L. W. (1968). Multi-Sensor Signatures of Urban Morphology, Function and Evolution, U.S.G.S. Contract 14-08-0001-10674. Geographic Applications Program, U.S. Geol. Survey, Washington, D.C.

Bradie, R. A. (1967). Slar Imagery for Sea Ice Studies, *Photogram. Engin.* 33(7), 763–766.

Bradshaw, M. D., and Byatt, W. J. (1967). *Introductory Engineering Field Theory.* Prentice-Hall, Englewood Cliffs, N.J.

Braithwaite, J. G. N. (1966). Dispersive Multispectral Scanning: A Feasibility Study, Final Report. Institute of Sci. and Tech. Contract No. 14-08-001-10053. Univ. of Mich., Ann Arbor.

Braithwaite, J. G. N., and Lowe, D. S. (1966). A Spectrum Matching Technique for Enhancing Image Contrast, *Appl. Opt.* 5(6), 893–906.

Brenchley, G. H., and Dadd, C. V. (1962). Potato Blight Recording by Aerial Photography, *NAAS Quart. Review.* 57, 1–21. London, England.

Brennan, P. A. (1968). Albedo Measurement at Mt. Lassen Test Site, Tech. Report NASA-CR-101784. Univ. of Nev., Reno.

Broeke, K. (1963). *Gateway to the Great Books* (R. M. Hutchins and M. Adler, eds.). 8(6), 597–644. Encyclopaedia Britannica, Chicago.

Brooner, W. G., Haralick, R. M., and Dinstein, Its'hak (1971). Spectral Parameters Affecting Automated Image Interpretation Using Bayestian Probability Techniques, in *Proc. of the Seventh Int'l. Symp. on Remote Sens. of Environ.* 3, 1929–1949. Univ. of Mich., Ann Arbor.

Brooner, W. G., and Simonett, D. S. (1971). Crop Discrimination with Color Infrared Photography: A Study in Douglas County, Kansas, *Remote Sens. of Environ.* 2(1), 21–35.

Brown, W. L., Polcyn, F. C., and Stewart, S. R. (1971). A Method for Calculating Water Depth, Attenuation Coefficients and Bottom Reflectance Characteristics, in *Proc. of the Seventh Int'l Symp. on Remote Sens. of Environ.* 1, 663–682. Univ. of Mich., Ann Arbor.

Brown, W. L. (1972). Reducing Variance in Remotely Sensed Multispectral Data–A Pragmatic Approach, in *Remote Sensing of Earth Resources* (F. Shahrokhi, ed.). 1, 525–537. Univ. of Tenn., Tullahoma.

Brown, W. M., and Porcello, L. J. (1969). An Introduction to Synthetic Aperture Radar, *IEEE Spectrum.* 6(9), 52–62.

Brunnschweiler, D. (1957). Seasonal Changes of the Agricultural Pattern: A Study in Comparative Airphoto Interpretation, *Photogram. Engin.* 23(5), 759–771.

Bryan, M. L. (1975). Interpretation of an Urban Scene Using Multi-Channel Radar Imagery, *Remote Sens. of Environ.* 4(1), 49–66.

Buettner, K. J., Kern, C. D., and Cronin, J. F. (1965). The Consequences of Terrestrial Surface Infrared Emissivity, in *Proc. of the Third Int'l. Symp. on Remote Sens. of Environ.* 549–561. Univ. of Mich., Ann Arbor.

0-201-04245-2

Bullrich, K. (1964a). *Advances in Geophysics* (H. E. Landsberg and J. van Mieghem, eds.). 10, 101–260. Academic Press, New York.

Bullrich, K. (1964b). Scattered Radiation in the Atmosphere and the Natural Aerosol, in *Advances in Geophysics* (H. E. Landsberg and J. van Mieghem, eds.). 10, 101–260. Academic Press, New York.

Unusual pagination arises from insertion of some recent key references.

0-201-04245-2

Byers, H. R. (1959). *General Meteorology,* McGraw-Hill, New York.
Byrne, R., and Munday, J. C., Jr. (1972). Photodensity and the Impact of Shifting Agriculture on Subtropical Vegetation: A Case Study in the Bahamas, in *Proc. of the Eighth Int'l Symp. on Remote Sens. of Environ.* 1311–1326. Environmental Research Inst. of Mich., Ann Arbor.

Cameron, H. L. (1964). Radar as a Surveying Instrument in Hydrology and Geology, in *Proc. of the Third Int'l Symp. on Remote Sens. of Environ.* 441–452. Univ. of Mich., Ann Arbor.
Cantrell, J. L. (1964). Infrared Geology, *Photogram. Engin.* 30(6), 916–922.
Carlson, N. L. (1967). Dielectric Constant of Vegetation at 8.5 GHz, Tech. Report 1903-5. Electro-Science Lab., Ohio State Univ., Columbus.
Cartier, W. O., McLaughlin, G. H., and Robinson, W. A. (1952). System of Airborne Conductor Measurements, *Official Gazette U.S. Patent Office.* 665(5), 1590. U.S. Patent 2,623,924.
Cassinis, R., Lechi, G. M., and Tonelli, A. M. (1974). Contribution of Space Platforms to a Ground and Airborne Remote Sensing Program over Active Italian Volcanoes, in *Proc. of a Symp. on European Earth Resources Satellite Experiments.* ESRO SP-100, 185–197. Frascati, Italy.
Catoe, C., Nordberg, W., Thaddeus, P., and Ling, G. (1967). Preliminary Results from Aircraft Flight Tests of an Electrically Scanning Microwave Radiometer, Tech. Report No. 622-67-352, NASA. Goddard Space Flight Center, Greenbelt, Md.
Chandrasekhar, S. (1960). *Radioactive Transfer.* Dover Publications, New York.
Chapman, P. E.,and Brennan, P. A. (1969). Ground Truth Acquisition International, *Remote Sens. Inst., First Annual IRSI Sym. Proc.* 2, 140–161.
Chapman, P. E., Brennan, P. A., and Chipp, E. R. (1970). Remote Sensing Evaluation of the Klondike Mining District, Tech. Report NASA-CR-129934, Contract No. NAS9-7779. Univ. of Nevada, Reno.
Chapman, R. C. (1970). Investigation of Scattering Characteristics of Surface with an Ellipsometer. Unpublished M.S. Thesis. San Diego State College, Calif.
Chapman, W. M. (1969). Implications of Space Research to Fishery Development, in *The Oceans from Space* (P. C. Badgley, L. Miloy, and L. Childs, eds.). 202–216. Gulf Pub. Co., Houston, Tex.
Chia, R. C. (1968). The Theory of Radar Scatter from the Ocean, Tech. Report 112-1. Center for Research, Inc., Univ. of Kansas, Lawrence.
Christaller, W. (1966). *Central Places in Southern Germany.* Prentice-Hall, Englewood Cliffs, N.J.
Christodoulou, D. (1959). *The Evolution of the Rural Land Use Pattern in Cyprus.* A. P. Taylor & Co., London.
Ciesla, W. M., Bell, J. C., and Curlin, J. W. (1967). Color Photos and the Southern Beetle, *Photogram. Engin.* 38(8), 883–888.
Ciesla, W. M., Drake, L. E., and Wilmore, D. H. (1972). Color Photos, Aerial Sprays and the Forest Tent Caterpillar, *Photogram. Engin.* 38(8), 867–873.
Clark, J., and Stone, R. B. (1965). Marine Biology and Remote Sensing, in *Oceanography from Space* (G. C. Ewing, ed.). 305–312. Woods Hole Oceanographic Inst. Ref. No. 65-10, Massachusetts.
Clark, J. R., and Frank, J. L. (1963). Infrared Measurement of Sea Surface Temperature, *Undersea Tech.* 20–23.
Clark, M. M. (1971). Comparison of SLAR Images and Small-Scale Low Sun Aerial Photographs, *Bull. Geol. Soc. Am.,* 82, 1735–1742.
Cloud, P., ed. (1969). *Resources and Man.* Nat'l Acad. of Sciences-Nat'l Research

0-201-04245-2

Council. W. H. Freeman and Co., San Francisco, Calif.
Coggeshall, M. E., and Hoffer, R. M. (1973). Basic Forest Cover Mapping Using Digitized Remote Sensor Data and ADP Techniques. M.S. Thesis, Purdue Univ., West Lafayette, Ind.

0-201-04245-2

Unusual pagination arises from insertion of some recent key references.

Coggeshall, M. E., Hoffer, R. M., and Berkebile, J. S. (1973). A Comparison Between Digitized Color Infrared Photography and Multispectral Scanner Data Using ADP Techniques, in *Proc. of the Fourth Biennial Workshop on Aerial Color Photography in the Plant Sciences.* 43–56. Univ. of Maine, Orono.
Cohn, M., Wentworth, F. I., and Wiltse, J. C. (1963). High Sensitivity 100 to 300 GHz Radiometers, *Proc. of IEEE.* 51(9), 1227–1232.
Cole, M. M., Owen-Jones, E. S., Custance, N. E. E., and Beaumont, T. E. (1974). Recognition and Interpretation of Spectral Signatures of Vegetation from Aircraft and Satellite Imagery in Western Queensland, Australia, in *Proc. of a Symp. on European Earth Resources Satellite Experiments.* ESRO SP-100, 243–287. Frascati, Italy.
Collett, L. S. (1969). Resistivity Mapping by Electromagnetic Methods, in *Mining and Groundwater Geophysics.* 1967, 615–625. Geol. Survey of Can. Econ. Geol. Report No. 26. Ottawa.
Collins, A. G., and Egleson, G. C. (1967). Iodide Abundance in Oilfield Brines in Oklahoma, *Science.* 156(3777), 934–935.
Collins, D. G., Blattner, W. G., Wells, M. B., and Horak, H. G. (1972). Backward Monte Carlo Calculations of the Polarization Characteristics of the Radiation Emerging from Spherical-Shell Atmospheres, *Appl. Opt.* 11(11), 2684–2696.
Collins, R. J., McCown, F. P., Stonis, L. P., Petzel, G., and Everett, J. R. (1974). An Evaluation of the Suitability of ERTS Data for the Purposes of Petroleum Exploration, Type III Final ERTS Report, Contract NAS-5-21735. Oklahoma City, Okla.
Colwell, J. E. (1974). Vegetation Canopy Reflectance, *Remote Sens. of Environ.* 3(3), 175–183.
Colwell, R. N. (1956). Determining the Prevalence of Certain Cereal Crop Diseases by Means of Aerial Photography, *Hilgardia.* 26(5), 223–283.
Colwell, R. N. (1961). Some Practical Applications of Multiband Spectral Reconnaissance, *Amer. Scientist.* 49(3), 9–36.
Colwell, R. N. (1963). Basic Matter and Energy Relationships Involved in Remote Reconnaissance, *Photogram. Engin.* 29(5), 761–799.
Colwell, R. N. (1964). Aerial Photography–A Valuable Sensor for the Scientist, *Amer. Scientist.* 52(3), 16–49.
Colwell, R. N. (1965). The Extraction of Data from Aerial Photographs by Human and Mechanical Means, *Photogrammetria.* 20, 211–228.
Colwell, R. N. (1968a). Determining the Usefulness of Space Photography for Natural Resource Inventory, in *Proc. of the Fifth Int'l Symp. on Remote Sens. of Environ.* 249–289. Univ. of Mich., Ann Arbor.
Colwell, R. N. (1968b). Remote Sensing of Natural Resources, *Scientific Amer.* 218(1), 54–69.
Colwell, R. N., and other personnel of FRSL (1969). An Evaluation of Earth Resources Using Apollo 9 Photography, Final Report, NASA Contract No. NAS 9-9348. Forestry Remote Sensing Lab., Univ. of Calif., Berkeley.
Colwell, R. N. (1970). Applications of Remote Sensing in Agriculture and Forestry, in *Remote Sensing* (Committee on Remote Sensing, eds.). 164–223.
Colwell, R. N. (1973). ERTS-1 Imagery and High Flight Photographs as Aids to Fire Hazards Appraisal at the NASA San Pablo Reservoir Test Site, in *Symp. on Significant Results Obtained from the Earth Resources Tech. Satellite-1.* NASA SP-327, 1, 145–157. Goddard Space Flight Center, New Carrollton, Md.
Condit, H. R. (1969). Spectral Reflectance of Soil and Sand, in *New Horizons in Aerial Color Photography.* Amer. Soc. of Photogrammetry. 3–15.

0-201-04245-2

Conrod, A. C., Boersma, A., and Kelly, M. G. (1968a). Investigation of Visible Region Instrumentation for Oceanographic Satellites, Report RE-31. Mass. Inst. of Tech., Experimental Astronomy Lab., Cambridge.

Conrod, A. C., Kelly, M. G., and Boersma, A. (1968b). Aerial Photography for Shallow Water Studies on the West Edge of the Bahama Banks, Report RE-42. Mass. Inst. of Tech., Experimental Astronomy Lab., Cambridge.

Conway, W. H., Bacinski, R. R., Brugma, F. C., and Falco., C. U. (1963). A Gradient Microwave Radiometer Flight Test Program, in *Proc. of the Second Int'l Symp. on Remote Sens. of Environ.* 145–175. Univ. of Mich., Ann Arbor.

Cooper, C. F., and Smith, F. M. (1966). Color Aerial Photography: Toy or Tool?, *Jour. of Forestry.* 64(6), 373–378.

Cooper, J. R. (1966). Geologic Evaluation–Radar Imagery of Twin Buttes Area, Arizona. Test Site 15. Tech. Letter NASA-28. Washington, D.C.

Cosgriff, R. L., Peake, W. H., and Taylor, R. C. (1959). Terrain Scattering Properties for Sensor System Design, in *Terrain Handbook* II. Engineering Experimental Station Bulletin 181, Ohio State Univ., Antenna Lab., Columbus, Ohio.

Coulson, K. L., Dave, J. V., and Sekera, Z. (1960). *Tables Related to Radiation Emerging from a Planetary Atmosphere with Rayleigh Scattering.* Univ. of Calif. Press, Los Angeles.

Coulson, K. L. (1966). Effects of Reflection Properties of Natural Surfaces in Aerial Reconnaissance, *Appl. Opt.* 5(6), 905–917.

Crane, R. B. (1971). Preprocessing Techniques to Reduce Atmospheric and Sensor Variability in Multispectral Scanner Data, in *Proc. of the Seventh Int'l Symp. on Remote Sens. of Environ.* 2, 1345–1355. Univ. of Mich., Ann Arbor.

Crane, R. B., and Richardson, W. (1972). Rapid Processing of Multispectral Scanner Data Using Linear Techniques, in *Remote Sensing of Earth Resources* (F. Shahrokhi, ed.). 1, 581–595. Univ. of Tenn., Tullahoma.

Cromwell, T., and Reid, J. L. (1956). A Study of Oceanic Fronts, *Tellus.* 8(1), 94–101.

Cronin, J. F. (1967). Terrestrial Multispectral Photography, Special Report No. 56. Air Force Cambridge Research Lab., Bedford, Mass.

Cummings, J. R. (1964). Coachella Valley Investigation, Calif. Dept. of Water Resources Bull. 108.

Curcio, J. A., Knestrick, G. L., and Corden, T. H. (1961). Atmospheric Scattering in the Visible and Infrared, U.S.N.R.L. Report 5567. U.S. Naval Research Lab., Washington, D.C.

Curran, R. J. (1972). Ocean Color Determination through a Scattering Atmosphere, *Appl. Opt.* 11(8), 1857–1866.

Cutrona, L. J. (1961). A High Resolution Radar Combat Surveillance System, *Inst. of Radio Engineers.* MIL-5, 127–131.

Dakshinamurti, C., Krishnamurthy, B., Summanwar, A. S., Shanta, P., and Pisharoty, P. (1971). Remote Sensing for Coconut Wilt, in *Proc. of the Seventh Int'l Symp. on Remote Sensing of Environ.* 25–30. Univ. of Mich., Ann Arbor.

Dakshinamurti, C., and Summanwar, A. S. (1972). Remote Sensing of Coconut Plants in Kerala (India), in *Proc. of the Eighth Int'l Symp. on Remote Sens. of Environ.* 1327–1331. Environmental Research Inst. of Mich., Ann Arbor.

Darnley, A. G. (1968). Helicopter tests with a gamma ray spectrometer, *Canadian Mining Jour.* 89(4), 104–106.

Darnley, A. G., Grasty, R. L., and Charbonneau, B. W. (1970). Mapping from the Air by Gamma-Ray Spectrometry, *CIM.* Special Vol. 11.

Dave, J. V. (1964). Importance of High Order Scattering in a Molecular Atmosphere, *Jour. of Opt. Soc. of Amer.* 54, 307–315.

0-201-04245-2

Dave, J. V., and Furukawa, P. M. (1966). *Scattered Radiation in the Ozone Absorption Bands at Selected Levels of a Terrestrial, Rayleigh Atmosphere.* Amer. Meteorological Soc., Boston.

Dave, J. V. (1969). Effect of Coarseness on the Integration Increment on the Calculation of the Radiation Scattered by Polydispersed Aerosols, *Appl. Opt.* 8(6), 1161–1182.

Davies, A. R., and McNeill, J. D. (1971). E-Phase™, A New Remote Sensing Technique for Resistivity Mapping, in *Proc. of the Seventh Int'l Symp. on Remote Sens. of Environ.* Univ. of Mich., Ann Arbor.

Davies, J. H. (1970). Correlation Spectroscopy, *Analytical Chem.* 42(6), 101A–112A.

Davies, J. H. (1972). Detection of Atmospheric Constituents, in *Proc. of the First Canadian Symp. on Remote Sens.* Ottawa, Ontario.

Davies, J. H., and McNeill, J. D. (1972). Airborne Remote Sensing of Resistivity Through the Use of E-Phase™ Techniques, in *Proc. of the First Canadian Symp. on Remote Sens.* Ottawa, Ontario.

Davis, E. E., and Parham, J. B. (1958). Modification and Utilization of AN/SSQ-41 Sonobuoys for the Collection of Volume Reverberation Data from Aircraft, Tech. Report TR-221. U.S. Naval Oceanographic Office, NAVOCEANO, Washington, D.C.

DeBoer, Th. A., Bunnik, N. J. J., van Kasteren, H. W. J., Uenk, D., Verhoef, W., and de Loor, G. P. (1974). Investigation into the Spectral Signature of Agricultural Crops During Their State of Growth, in *Proc. of the Ninth Int'l Symp. on Remote Sens. of Environ.* 1441–1455. Environmental Research Inst. of Mich., Ann Arbor.

DeCarolis, C., Baldi, G., Galli de Paratesi, S., and Lechi, G. M. (1974). Thermal Behaviour of Some Rice Fields Affected by a Yellows-Type Disease, in *Proc. of the Ninth Int'l Symp. on Remote Sens. of Environ.* 1161–1170. Environmental Research Inst. of Mich., Ann Arbor.

Dedman, E., Culver, J., and Janza, F. (1968). Computer Simulation of Remote Sensing by Microwave Radiometers, *Int'l Assoc. of Scientific Hydrology, Hydrology Symp.* Pub. 1 (2). Tucson, Ariz.

Deirmendjian, D. (1960). Atmospheric Extinction of Infra-red radiation, *Royal Meteor. Soc. Quart. Jour.* 86, 371.

Deirmendjian, D. (1964). Scattering and Polarization Properties of Water Clouds and Hazes in the Visible and Infrared, *Appl. Opt.* 3(2), 187–197.

Deirmendjian, D. (1969). *Electromagnetic Scattering on Spherical Polydispersions.* Elsevier, New York.

Dellwig, L. F., Kirk, J. N., and Walters, R. L. (1966). The Potential of Low-Resolution Radar Imagery in Regional Geologic Studies, *Jour. of Geophys. Res.* 71, 3995–4998.

Dellwig, L. F., and Moore, R. K. (1966). The Geological Value of Simultaneously Produced Like- and Cross-Polarized Radar Imagery, *Jour. of Geophys. Res.* 71(14), 3597–3601.

Dellwig, L. F., MacDonald, H. C., and Kirk, J. N. (1968). The Potential of Radar in Geological Exploration, in *Proc. of the Fifth Int'l Symp. on Remote Sens. of Environ.* 747–763. Univ. of Mich., Ann Arbor.

Dellwig, L. F. (1969). An Evaluation of Multifrequency Radar Imagery of the Pisgah Crater Area, California, *Modern Geol.* 1(1), 65–73.

Deshler, W. (1974). An Examination of the Extent of Fire in the Grassland and Savanna of Africa along the Southern Side of the Sahara, in *Proc. of the Ninth Int'l Symp. on Remote Sens. of Environ.* 23–30. Environmental Research Inst. of Mich., Ann Arbor.

0-201-04245-2

Dibblee, T. W. (1966). Geologic Map of the Lavic Quadrangle, San Bernardino County, California, U.S.G.S. Map L-472.

Dicke, R. H. (1946). The Measurement of Thermal Radiation at Microwave Frequencies, *Review of Scientific Instruments.* 17(7), 268–275.

Dicke, R. H., and Wittke, J. P. (1960). *Introduction to Quantum Mechanics.* Addison-Wesley, Reading, Mass.

Dietz, R. S., and La Fond, E. C. (1950). Natural Slicks on the Ocean, *Jour. of Marine Res.* 9(2), 69–76.

Dishler, J. J. (1967). Visual Sensor Systems in Space, *Trans. ICCC on Commun. Tech.* Com-15(6), 824–834.

Douglass, R. W., Meyer, M. P., and French, D. W. (1974). Remote Sensing Applications: Forest Tree Disease Detection and Vegetation Classification within the Sub-Boreal Forest Region, in *Proc. of the Ninth Int'l Symp. on Remote Sens. of Environ.* 1087. Environmental Research Inst. of Mich., Ann Arbor.

Dowsett, J. S. (1969). Geophysical Exploration Methods for Nickel, in *Mining and Groundwater Geophysics.* Geol. Survey of Can. Econ. Geol. Report No. 25. 310–321. Ottawa.

Doyle, F. J. (1969). Let Aircraft Make Earth Resource Surveys, a Rebuttal, *Jour. of Astronautics and Aeronautics.* 7(10), 78–79.

Draeger, W. C., Pettinger, L. R., and Benson, A. S. (1971). The Use of Small Scale Aerial Photography in a Regional Agricultural Survey, in *Proc. of the Seventh Int'l. Symp. on Remote Sens. of Environ.* 1205–1217. Univ. of Mich., Ann Arbor.

Draeger, W. C. (1973). Agricultural Applications of ERTS-1 Data, in *Symp. on Significant Results Obtained from the Earth Resources Tech. Satellite-1.* NASA SP-327, 1, 197–204. Goddard Space Flight Center, New Carrollton, Md.

Draeger, W. C., Nichols, J. D., Benson, A. S., Larrabee, D. G., Senkus, W. M., and Hay, C. M. (1973). Regional Agriculture Surveys Using ERTS-1 Data, in *Third Earth Resources Tech. Satellite Symp.* NASA SP-351, 1, 117–126. Goddard Space Flight Center, Washington, D.C.

Drahovzal, J. A., Neathery, T. L., and Wielchowsky, C. C. (1973). Significance of Selected Lineaments in Alabama, in *Third Earth Resources Tech. Satellite Symp.* NASA SP-351, 1, 897–918. Goddard Space Flight Center, Washington, D.C.

Drake, B., and Shuchman, R. A. (1974). Feasibility of Using Multiplexed Slar Imagery for Water Resources Management and Mapping Vegetation Communities, in *Proc. of the Ninth Int'l. Symp. on Remote Sens. of Environ.* 219–249. Environmental Research Inst. of Mich., Ann Arbor.

Driscoll, R. S., and Francis, R. E. (1970). Multistage, Multiseasonal and Multiband Imagery to Identify and Qualify Non-forest Vegetation Resources, NASA Contract No. R-09-038-002. Forestry Remote Sensing Lab., School of Forestry and Conservation, Univ. of Calif., Berkeley.

Driscoll, R. S., and Spencer, M. M. (1972). Multispectral Scanner Imagery for Plant Community Classification, in *Proc. of the Eighth Int'l Symp. on Remote Sens. of Environ.* 1259–1278. Environmental Research Inst. of Mich., Ann Arbor.

Driscoll, R. S. (1974). Use of Remote Sensing in Range and Forest Management, *Proc. Great Plains Agric. Council,* 111–133.

Driscoll. R. S., and Coleman, M. D. (1974). Color for Shrubs. *Photogram. Engin.* 40(4), 451–459.

Dueker, K. J., and Horton, F. E. (1971). Urban Change Detection System, *Proc. of the Seventh Int'l Symp. on Remote Sens. of Environ.* 1523–1536, Univ. of Mich., Ann Arbor.

0-201-04245-2

Duntley, S. Q. (1963). Light in the Sea, *Jour. of Opt. Soc. of Amer.* 53(2), 214-233.
Duntley, S. Q., Johnson, R. W., and Gordon, J. I. (1964). *Ground-Based Measurements of Earth-to-Space Beam Transmittance, Path Radiance, and Contrast Transmittance.* Wright-Patterson Air Force Base, Ohio.
Duntley, S. Q. (1965). Oceanography from Manned Satellites by Means of Visible Light, in *Oceanography from Space* (G. C. Ewing, ed.). 39–45. Woods Hole Oceanographic Inst. Ref. No. 65-10, Massachusetts.

Earing, D. L., and Ginsberg, I. W. (1969). A Spectral Discrimination Technique for Agricultural Applications, in *Proc. of the Sixth Int'l Symp. on Remote Sens. of Environ.* 21–32. Univ. of Mich., Ann Arbor.
Earth Satellite Corporation (1972). Aerial Multiband Wetland Mapping, *Photogram. Engin.* 38(12), 1188–1189.
Earth Satellite Corporation/Booz-Allen-Hamilton Applied Research Corporation (1974). *Earth Resources Survey Benefit-Cost Study.* U.S. Dept. of the Interior/Geol. Survey. Washington, D.C.
Eastman Kodak Company (1961). *Manual of Physical Properties of Kodak Aerial and Special Sensitized Materials.* Rochester, N.Y.
Eastman Kodak Co. (1970). *Wratten Filters,* Pub. No. B-3. Rochester, N.Y.
Ebtehadj, K., Ghazi, A., Barzegar, F., Boghrati, R., and Jazayeri, B. (1973). Tectonic Analysis of East and South-East Iran Using ERTS-1 Imagery, in *Third Earth Resources Tech. Satellite Symp.* NASA SP-351, 1, 783. Goddard Space Flight Center, Washington, D.C.
Edgerton, A. T., Mandl, R. M., Poe, G. A., Jenkins, J. E., Soltis, F., and Sakamoto, S. (1968). Passive Microwave Measurements of Snow, Soils, and Snow-Ice-Water Systems, Tech. Report 4, SGD 829-6, Contract No. 4767(00)NR387-033). Aerojet-General Corp., El Monte, Calif.
Edgerton, A. T., and Trexler, D. T. (1970). Oceanographic Applications of Remote Sensing with Passive Microwave Techniques, in *Proc. of the Sixth Int'l Symp. on Remote Sens. of Environ.* 767–789. Univ. of Mich., Ann Arbor.
Egan, W. G. (1970). Optical Stokes Parameters for Farm Crop Identification, *Remote Sens. of Environ.* 1(1), 165–180.
Eiden, R. (1966). The Elliptical Polarization of Light Scattered by a Volume of Atmospheric Air, *Appl. Opt.* 5(4), 569–575.
Ellermeier, R. D., Fung, A. K., and Simonett, D. S. (1966). Some Empirical and Theoretical Considerations of Multiple Polarization Radar Data, in *Proc. of the Fourth Int'l Symp. on Remote Sens. of Environ.* 657–670. Univ. of Mich., Ann Arbor.
El Shazly, E. M., Abdel-Hady, M. A., El Ghawaby, M. A., and El Kassas, I. A. (1973). Geologic Interpretation of ERTS-1 Satellite Images for West Aswan Area, Egypt, in *Third Earth Resources Tech. Satellite Symp.* NASA SP-351, 1, 919-942. Goddard Space Flight Center, Washington, D.C.
Elterman, L. (1968). *UV, Visible, and IR Attenuation for Altitudes to 50 km* [AFCRL-68-0153]. Air Force Cambridge Research Lab., Bedford, Mass.
Elterman, L. (1970). Relationships Between Vertical Attenuation and Surface Meteorological Range, *Appl. Opt.* 9(8), 1804–1810.
Emery, K. O. (1965). Recommendations of the Panel on Coastal Geography, in *Oceanography from Space* (G. C. Ewing, ed.), 413–417. Woods Hole Oceanographic Inst. Ref. No. 65-10, Massachusetts.

0-201-04245-2

Eppler, W. G., Helmke, C. A., and Evans, R. H. (1971a). Table Look-up Approach to Pattern Recognition, in *Proc. of the Seventh Int'l Symp. on Remote Sens. of Environ.* 1415–1427. Univ. of Mich., Ann Arbor.

Eppler, W. G., Loe, D. L., and Wilson, E. L. (1971b). Interactive Displays/Graphics Systems for Remote Sensor Data Analysis, in *Proc. of the Seventh Int'l Symp. on Remote Sens. of Environ.* 2, 1293–1306. Univ. of Mich., Ann Arbor.

Erb, R. B. (1973).The Utility of ERTS-1 Data for Applications in Agriculture and Forestry, in *Third Earth Resources Tech. Satellite Symp.* NASA SP-351, 1, 75–86. Goddard Space Flight Center, Washington, D.C.

Erskine, M. C., Everett, J. R., Haenggi, W. I., Kirwan, J. L., and Levin, S. B. (1973). Use of Small Scale Imagery in Geologic Exploration, in *Proc. of Second Latin Amer. Geologic Congress.* Caracas, Venezuela.

Erskine, M. C., and Levin, S. B. (1974). Geology and Mineral Prospecting, in *Remote Sensing Applications to Resource Management Problems in the Sahel* (J. B. Bale, D. Conte, D. Goehring, and D. S. Simonett, eds.). 101–122. U.S. Agency for Int'l. Development Contract No. AID afr-c-1058. Washington, D.C.

Estes, J. E., and Senger, L. W. (1971). An Electronic Multi-Image Processor, *Photogram. Engin.* 37(6), 577–586.

Estes, J. E., Jensen, J. R., Tinney, L. R., and Rector, M. (in press). Remote Sensing Input to Water Demand Modeling, in *First Annual Earth Resources Survey Symp.* L.B.J. Space Center, Houston, Tex.

Evans, R. D. (1955). *The Atomic Nucleus.* McGraw-Hill, New York.

Everett, J. R., and Petzel, G. (1974). An Evaluation of the Suitability of ERTS Data for the Purposes of Petroleum Exploration, in *Third Earth Resources Tech. Satellite Symp.* NASA SP-351, 2, 50–61. Goddard Space Flight Center, Washington, D.C.

Everett, J. R., and Russell, O. R. (1975). Evaluation of Aerial Remote Sensing Techniques for Definition of Geologic Features Pertinent to Tunnell Location and Design (abs.), in *First Annual Conference on DOT Research and Development in Tunneling,* Williamsburg, Va.

Ewing, G. C. (1950a). Relation Between Band Slicks at the Surface and Internal Waves in the Sea, *Science.* III(2874), 91–94.

Ewing, G. C. (1950b). Slicks, Surface Films, and Internal Waves, *Jour. of Marine Res..* 9(3), 161–187.

Ewing, G. C. (1964). Slithering Isotherms and Thermal Fronts on the Ocean Surface, in *Techniques for Infrared Survey of Sea Temperature,* Bureau of Sport Fisher. and Wildlife Circular No. 202. 92–93. Scripps Inst. of Oceanography, Univ. of Calif., San Diego.

Ewing, G. C. (1969). The Color of the Sea, in *The Oceans from Space* (P. C. Badgley, L. Miloy, and L. Childs, eds.). 46–49. Gulf Pub. Co., Houston, Tex.

Farhat, N. H., and De Cou, A. B. (1969). Relations Between Wave Structure Function Looking Up and Looking Down Through the Atmosphere, *Jour. of Opt. Soc. of Amer.* 59(11), 1489–1490.

Feder, A. M. (1960). Interpreting Natural Terrain from Radar Displays, *Photogram. Engin.* 26(4), 618–630.

Feigelson, E. M., Malkevich, M. S., Kogan, S. Y., Koronatove, T. D., Glazova, K. S., and Kuznetsova, M. A. (1960). *Calculation of the Brightness of Light,* Part 1. Consultants Buréau, New York.

Feynman, R. P. (1963). *The Feynman Lectures on Physics,* Vol. II. Addison-Wesley, Reading, Mass.

Fischer, W. A., et al. (1964). Infrared Survey of Hawaiian Volcanoes, *Science.* 146(3645), 733–742.

Fischer, W. A., Davis, D. A., and Sousa, T. M. (1966). Freshwater Springs of Hawaii from Infrared Images, *U.S.G.S. Hydrol. Invest. Atlas* HA-218.

0-201-04245-2

Flanigan, D. F., and De Long, H. D. (1971). Spectral Absorption Characteristics of the Major Components of Dust Clouds, *Appl. Opt.* 10(1), 51–57.

Flowers, E. C., McCormick, R. A., and Kurfis, K. R. (1969). Atmospheric Turbidity over the U.S., 1961–1966, *Jour. of Appl. Meteor.* 8(6), 955–962.

Foote, R. S. (1969). Radioactive Methods in Mineral Exploration, in *Mining and Groundwater Geophysics.* Geol. Survey of Can. Econ. Geol. Report No. 26, 177–190. Ottawa.

Forestry Remote Sensing Laboratory (1971). Analysis of Remote Sensing Data for Evaluating Vegetation Resources, NASA Contract R-03-038-002. School of Forestry and Conservation, Univ. of Calif., Berkeley.

Fowler, W. B., Reed, E. I., and Blamont, J. E. (1971). Bidirectional Reflectance of the Moonlit Earth, *Appl. Opt.* 10(12), 2657–2660.

Frank, N. H. (1950). *Introduction to Electricity and Optics.* McGraw-Hill, New York.

Fraser, D. S. (1970). DIGHEM: A New Aerial Electromagnetic Prospecting System, unpublished paper. SEG Annual Meeting, November 1970, New Orleans, La.

Fraser, R. S. (1959). *Scattering Properties of Atmospheric Aerosols.* Univ. of Calif., Los Angeles.

Fraser, R. S. (1964a). Apparent Contrast of Objects on the Earth's Surface as Seen from above the Earth's Atmosphere, *Jour. of Opt. Soc. of Amer.* 54(3), 289–300.

Fraser, R. S. (1964b). Computed Intensity and Polarization of Light Scattered Outwards from the Earth and on Overlying Aerosol, *Jour. of Opt. Soc. of Amer.* 54(2), 157–168.

Fraser, R. S., and Walker, W. H. (1968). Effect of Specular Reflection at the Ground on Light Scattered from a Rayleigh Atmosphere, *Jour. of Opt. Soc. of Amer.* 58(5), 636–644.

Fraser, R. S., Bahethi, Om. P., and Al-Abbas, A. M. (1976). Submitted to *Remote Sens. of Environ.*

Frazee, C. J., Myers, V. I., and Westin, F. C. (1971). Remote Sensing for Detection of Soil Limitations in Agricultural Areas, in *Proc. of the Seventh Int'l Symp. on Remote Sens. of Environ.* 327–343. Univ. of Mich., Ann Arbor.

Fried, D. L. (1966a). Limiting Resolution Looking Down Through the Atmosphere, *Jour. of Opt. Soc. of Amer.* 56(10), 1380–1385.

Fried, D. L. (1966b). Optical Resolution Through a Randomly Inhomogenous Medium for Very Long and Very Short Exposures, *Jour. of Opt. Soc. of Amer.* 56(10), 1372–1380.

Fried, D. L. (1967). Optical Heterodyne Detection of an Atmospherically Distorted Signal Wavefront, *Proc. of IEEE.* 55, 57–67.

Friedman, H. P., and Rubon, J. (1967). On Some Invariant Criteria for Grouping Data, *Amer. Stat. Assoc. Jour.* 62(320), 1159–1178.

Friedman, J. D., and Williams, R. S. (1968). Infrared Sensing of Active Geologic Processes, in *Proc. of the Fifth Int'l Symp. on Remote Sens. of Environ.* 787–820. Univ. of Mich., Ann Arbor.

Fritz, N. L. (1967). Optimum Methods for Using Infrared-Sensitive Color Films, *Photogram. Engin.* 33(10), 1128–1138.

Fu, K. S., Landgrebe, D. A., and Phillips, T. L. (1969). Information Processing of Remotely Sensed Agricultural Data, *Proc. of IEEE.* 57(4), 639–653.

Fu, K. S. (1971). On the Applications of Pattern Recognition Techniques to Remote Sensing Problems, Tech. Report EE 77-13. School of Electrical Engineering, Purdue Univ., West Lafayette, Ind.

Fung, A. K., and Chan, H. L. (1969). Backscattering of Waves by Composite Rough Surfaces, *IEEE Trans. Antennas and Propagation.* 17(5), 590–597.

0-201-04245-2

Gallagher, J. L., Reimold, R. J., and Thompson, D. E. (1972). A Comparison of Four Remote Sensing Media, in *Proc. of the Eighth Int'l Symp. on Remote Sens. of Environ.* 1287–1295. Environmental Res. Inst. of Mich., Ann Arbor.

Gates, D. M., and Shawn, C. C. (1960). Infrared Transmission of Clouds, *Jour. of Opt. Soc. of Amer.* 50, 876–882.

Gates, D. M. (1963). The Energy Environment in Which We Live, *Amer. Scientist.* 51(3), 327–348.

Gausman, H. W., Allen, W. A., and Cardenas, R. (1969a). Reflectance of Cotton Leaves and Their Structure, *Remote Sens. of Environ.* 1(1), 19–22.

Gausman, H. W., Allen, W. A., Cardenas, R., and Richardson, A. J. (1969b). Relation of Light Reflectance to Cotton Leaf Maturity (*Gossypium hirsutum* L.), in *Proc. of the Sixth Int'l Symp. on Remote Sens. of Environ.* 1123–1141. Univ. of Mich., Ann Arbor.

Gausman, H. W., Allen, W. A., Myers, V. I., Cardenas, R., and Leamer, R. W. (1969–1970). Reflectance of Single Leaves and Field Plots of Cyocel-Treated Cotton (*Gossypium hirsutum* L.) in Relation to Leaf Structure, *Remote Sens. of Environ.* 1(1), 103–107.

Gausman, H. W., Allen, W. A., Cardenas, R., and Bowen, R. L. (1970). Color Photos, Cotton Leaves and Soil Salinity, *Photogram. Engin.* 36(5), 454–459.

Gausman, H. W., Allen, W. A., Wiegand, C. L., Escobar, D. E., and Rodriquez, R. R. (1971). Leaf Light Reflectance, Transmittance, Absorptance, and Optical and Geometrical Parameters for Eleven Plant Genera with Different Leaf Mesophyll Arrangements, in *Proc. of the Seventh Int'l Symp. on Remote Sens. of Environ.* 1599–1625. Univ. of Mich., Ann Arbor.

Gausman, H. W. (1974). Leaf Reflectance of Near-Infrared, *Photogram. Engin.* 40(2), 183–191.

Gausman, H., Cardenas, R., and Gerbermann, A. H. (1974). Plant Size, etc., and Aerial Films, *Photogram. Engin.* 40(1), 61–67.

Gawarecki, S. J. (1969). Infrared Survey of the Pisgah Crater Area, San Bernardino County, California, U.S.G.S. Interagency Report NASA-99.

Gedney, L., and Van Wormer, D. J. (1973). ERTS-1 Earthquakes and Tectonic Evolution in Alaska, in *Third Earth Resources Tech. Satellite Symp.* NASA SP-351, 1, 745–756. Goddard Space Flight Center, Washington, D.C.

Geiger, R. (1957). *The Climate near the Ground,* Second Printing. Harvard Univ. Press, Cambridge, Mass.

General Electric (1974). *Total Earth Resources System for the Shuttle Era.* Valley Forge, Pa.

Gerryts, E. (1969). Diamond Prospecting by Geophysical Methods—A Review of Current Practice, in *Mining and Groundwater Geophysics.* 1967, 439–446. Geol. Survey of Can. Econ. Geol. Report No. 26. Ottawa.

Geyer, R. A. (1968). Oceanography of the Gulf of Mexico, Progress Report Project 286, Dept. of Oceanography, Texas A & M Univ., College Station.

Gibson, H. L. (1970). *Close-up Photography and Photomacrography.* Eastman Kodak Co., Rochester, N.Y.

Gillerman, E. (1967). Investigation of Cross-polarized Radar on Volcanic Rocks, CRES Report No. 61-25. Univ. of Kansas, Lawrence.

Gillerman, E. (1970). Roselle Lineament of Southeast Missouri, *Geol. Soc. of Amer. Bull.* 81(3), 975–982.

Glaser, A. H., Barnes, J. C., and Beran, D. W. (1968). Apollo Landmark Sighting: An Application of Computer Simulation to a Problem in Applied Meteorology. *Jour. of Appl. Meteor.* 7(5), 768–779.


0-201-04245-2

Goble, A. R., and Baker, D. K. (1962). *Elements of Modern Physics.* Ronald Press, New York.

Goetz, A. F. H., Billingsley, F. C., Elston, D. P., Lucchitta, I., and Shoemaker, E. M. (1973). Geologic Applications of ERTS Images in the Colorado Plateau, Arizona, in *Third Earth Resources Tech. Satellite Symp.* NASA SP-351, 3, 719-744. Goddard Space Flight Center, Washington, D.C.

Goetz, A. F. H., Billingsley, F. C., Gillespie, A. R., Abrams, M. J., and Squires, R. L. (1975). Application of ERTS Images and Image Processing to Regional Geologic Problems and Geologic Mapping in Northern Arizona, NASA Type III Report, NASA-CR-143068.

Gold, D. P., Parizek, R. R., and Alexander, S. A. (1973). Analysis and Application of ERTS-1 Data for Regional Geological Mapping, in *Symp. on Significant Results Obtained from the Earth Resources Tech. Satellite-1.* NASA SP-327, 1, 231–240. Goddard Space Flight Center, New Carrollton, Md.

Goldschmidt, V. M. (1958). Iodine in the Ocean, in *Geochemistry.* 606–608. Oxford Univ. Press, Amen House, London.

Goodman, M. S. (1959). A Technique for the Identification of Farm Crops on Aerial Photographs, *Photogram. Engin.* 25(1), 131–137.

Goodman, M. S. (1964). Criteria for the Identification of Types of Farming on Aerial Photographs, *Photogram. Engin.* 30(6), 984–990.

Goody, R. (1964). *Atmospheric Radiation.* Oxford Univ. Press, New York.

Gramms, L. C., and Boyle, W. C. (1971). Reflectance and Transmittance Characteristics of Several Selected Green and Blue-Green Unialgae, in *Proc. of the Seventh Int'l Symp. on Remote Sens. of Environ.* 1627–1650. Univ. of Mich., Ann Arbor.

Greaves, J. R., Sherr, P. E., and Glaser, A. H. (1970). Cloud Cover Statistics and Their Use in the Planning of Remote Sensing Missions, *Remote Sens. of Environ.* 1(2), 95.

Greaves, J. R., Speigler, D. B., and Willard, J. H. (1971). *Development of a Global Cloud Model for Simulating Earth-Viewing Space Missions* [NASA-CR-61345]. Allied Research Associates, Virginia Road, Concord, Mass.

Greenwood, J. A., Nathan, A., Neumann, G., Pierson, W. J., Jackson, F. C., and Pease, T. E. (1967). Radar Altimetry from a Spacecraft and Its Potential Applications to Geodesy and Oceanography, Lab. Report TR-67-3. Dept. of Geophysical Sciences, New York Univ.

Greenwood, J. G. W. (1969). Role of Satellite Photographs in Photogeology, *Nature.* 224(5218), 506–508.

Griggs, M. (1968). Emissivities of Natural Surfaces in the 8- to 14-Micron Spectral Region, *Jour. of Geophys. Res.* 73(24), 7545–7551.

Grootenboer, J., Eriksson, K., and Truswell, J. (1973). Stratigraphic Subdivision of the Transvaal Dolomite from ERTS Imagery, in *Third Earth Resources Tech. Satellite Symp.* NASA SP-351, 1, 657–664. Goddard Space Flight Center, Washington, D.C.

Guillou, R. B. (1964). The Aerial Radiological Measuring Surveys (ARMS) Program, in *The Natural Radiation Environment* (J. A. S. Adams and W. M. Lowaler, eds.). 705–721. Rice Univ. Semicentennial Series, Univ. of Chicago Press, Ill.

Gunter, G. (1940). Death of Fishes Due to Cold on the Texas Coast, *Ecology.* 22(2), 203–208.

0-201-04245-2

Gupta, J. N., Kettig, D. A., Landgrebe, D. A., and Wintz, P. A. (1973). Machine Boundary Finding and Sample Classification of Remotely Sensed Agricultural Data. Lab. for Applications of Remote Sensing, Information Note 102073. Purdue Univ., West Lafayette, Ind.

Hach, J. P. (1968). A Very Sensitive Airborne Microwave Radiometer Using Two Reference Temperatures, *Trans. IEEE, Microwave Theory and Techniques.* MTT(9), 629–636.

Hackman, R. J. (1967a). Time, Shadows, Terrain, and Photointerpretation, U.S.G.S. Prof. Paper 575B. B155–B160.

Hackman, R. J. (1967b). Geologic Evaluation of Radar Imagery in Southern Utah, U.S.G.S. Prof. Paper 575D. D135–D142.

Haggett, P. (1965). Scale Components in Geographical Problems, in *Frontiers in Geographical Teaching* (R. J. Chorley and P. Haggett, eds.). 164–185. Methuen, London.

Hall, F. F., Jr. (1968). The Effect of Cirrus Clouds on B-13-u Infrared Sky Radiance. *Appl. Opt.* 7(5), 891–898.

Hall, F. G., Bauer, M. E., and Malila, W. A. (1974). First Results from the Crop Identification Technology Assessment for Remote Sensing (Citars), in *Proc. of the Ninth Int'l Symp. on Remote Sens. of Environ.* 1171–1191. Environmental Research Inst. of Mich., Ann Arbor.

Hall, R. C. (1973). Application of ERTS-1 Imagery and Underflight Photography in the Detection and Monitoring of Forest Insect Manifestations in the Sierra Nevada Mountains of California, in *Symp. on Significant Results Obtained from the Earth Resources Tech. Satellite-1.* NASA SP-327, 1, 135–144. Goddard Space Flight Center, New Carrollton, Md.

Halliday, D., and Resnick, R. (1962). *Physics,* Part II. John Wiley, New York.

Hamilton, W. (1971). Recognition on Space Photographs of Structural Elements of Baja, California, U.S.G.S. Prof. Paper 718.

Hannah, J. W. (1967). A Feasibility Study for the Application of Remote Sensors to Selected Urban and Regional Land Use Planning Studies, ONR Contract N00014-67-A-0102-0001. Office of Naval Research, U.S. Navy, Washington, D.C.

Hansen, M. H., and Hurwitz, W. N. (1943). On the Theory of Sampling from Finite Populations, *Ann. Math. Stat.* 14(4), 333–362.

Haralick, R. M. (1967). Pattern Recognition Using Likelihood Functions. Unpublished M.S. Thesis, Univ. of Kansas, Lawrence.

Haralick, R. M., and Kelly, G. L. (1969). Pattern Recognition with Measurement Space and Spatial Clustering for Multiple Images, in *Proc. of IEEE.* 57(4), 654–665.

Haralick, R. M., Caspall, F., and Simonett, D. S. (1970). Using Radar Imagery for Crop Discrimination: A Statistical and Conditional Probability Study, *Remote Sens. of Environ.* 1(2), 131–142.

Haralick, R. M. (1971). On a Texture-Context Feature Extraction Algorithm for Remotely Sensed Imagery, in *Proc. of the IEEE Conference on Decision and Control.* 650–657. Miami Beach, Fla.

Haralick, R. M., and Anderson, D. E. (1971). Texture-Tone Study with Application to Digitized Imagery, Tech. Report 182-2. 1-146. Remote Sensing Lab., Univ. of Kansas, Lawrence.

Haralick, R. M. (1973). Glossary and Index to Remotely Sensed Image Pattern Recognition Concepts, *Pattern Recognition* 5, 391–403.

0-201-04245-2

Haralick, R. M. (1976). Automatic Remote Sensor Image Processing, in *Digital Picture Analysis* (A. Rosenfeld, ed.), 5–63. Springer Verlag, Heidelberg.

Haralick, R. M., and Bosley, R. (1973). Theoretical Analysis of Land Use Classification Using Texture Information on Multi-Images, Technical Memorandum ZZ61-Z, Center for Research in Engineering Science, Univ. of Kansas, Lawrence.

Haralick, R. M., Shanmugan, K., and Dinstein, I. (1973). Textural Features for Image Classification, *IEEE Transactions on Systems, Man, and Cybernetics,* SMC-3(6), 610–621.

Harger, R. O. (1970). *Synthetic Aperture Radar Systems, Theory and Design.* Academic Press, New York.

Harnage, J., and Landgrebe, D. (eds.) (1975). LANDSAT-D Thematic Mapper Technical Working Group, Final Report. JSC-09797. NASA, L.B.J. Space Center. Houston, Tex.

Harnapp, V. R., and Knight, C. G. (1971). Remote Sensing of Tropical Agricultural Systems, in *Proc. of the Seventh Int'l Symp. on Remote Sens. of Environ.* 409–433. Univ. of Mich., Ann Arbor.

Harrison, J. E. (1968). Geologic Interpretation of Infrared Imagery of Pend Orielle Area, Idaho, U.S.G.S. Tech. Letter NASA-72.

Hay, C. M. (1974). Agricultural Inventory Techniques with Orbital and High-Altitude Imagery, *Photogram. Engin.* 40(11), 1283–1293.

Hayt, W. H. (1967). *Engineering Electromagnetics.* McGraw-Hill, New York.

Helbig, M. S. (1972). Investigation of Color Detail, Color Analysis and False-Color Representation in Satellite Photographs, in *Remote Sensing of Earth Resources* (F. Shahrokhi, ed.). 1, 274–292. Univ. of Tenn., Tullahoma.

Heller, R. C., Aldrich, R. C., and Bailey, W. F. (1959). Evaluation of Several Camera Systems for Sampling Forest Insect Damage, *Photogram. Engin.* 25(1), 137-144.

Heller, R. C. Aldrich, R. C., McCambridge, W. F., and Weber, F. P. (1967). The Use of Multispectral Sensing Techniques to Detect Ponderosa Pine Trees under Stress from Insect or Pathogenic Organisms, Annual Progress Report. Sept. 30. Forestry Remote Sensing Lab., Univ. of Calif., Berkeley.

Heller, R. C., Aldrich, R. C., McCambridge, W. F., and Weber, F. P. (1969). The Use of Multispectral Sensing Techniques to Detect Ponderosa Pine Trees under Stress from Insects or Diseases, Annual Progress Report. Forestry Remote Sensing Lab., Univ. of Calif., Berkeley.

Heller, R. C., and Wear, J. F. (1969). Sampling Forest Insect Epidemics with Color Films, in *Proc. of the Sixth Int'l Symp. on Remote Sens. of Environ.* 1157–1167. Univ. of Mich., Ann Arbor.

Heller, R. C. (1975). Evaluation of ERTS-I Data for Forest and Rangeland Surveys. USDA Forest Service Research Paper, PSW-112, 1–67.

Hemphill, W. R., and Danilchik, W. (1968). Geological Interpretation of a Gemini Photography, *Photogram. Engin.* 34(2), 150–154.

Hemphill, W. R., and Stoertz, G. E. (1969). Remote Sensing of Luminescent Materials, in *Proc. of the Sixth Int'l Symp. on Remote Sens. of Environ.* 565–586. Univ. of Mich., Ann Arbor.

Herfindahl, O. C. (1969). *Natural Resources Information for Economic Development* (for Resources for the Future, Inc.). Johns Hopkins Press, Baltimore, Md.

Hickan, G. D., and Hogg, J. E. (1969). Application of an Airborne Pulsed Laser for Near-Shore Bathymetric Measurements, *Remote Sens. of Environ.* 1(1), 47–58.

Higer, A. L., Kolipinski, M. C., Thomson, N. S., and Purkerson, L. (1971). Use of Processed Multispectral Scanner Data with a Digital Simulation Model for Fore-

0-201-04245-2

casting Thermally Induced Changes in Benthic Vegetation in Biscayne Bay, Florida, in *Proc. of the Seventh Int'l Symp. on Remote Sens. of Environ.* 2055–2056. Univ. of Mich., Ann Arbor.

Higgins, J. L., and Deutsch, E. S. (1972). The Effects of Picture Operations in the Fourier Domain and Vice versa, in *Remote Sensing of Earth Resources* (F. Shahrokhi, ed.). 1, 460–480. Univ. of Tenn., Tullahoma.

Hirsch, S. N. (1968). Project Fire Detection, in *Proc. of the Fifth Int'l Symp. on Remote Sens. of Environ.* 447–457. Univ. of Mich., Ann Arbor.

0-201-04245-2

Unusual pagination arises from insertion of some recent key references.

Hirsch, S. N. (1971). Application of Infrared Scanners to Forest Fire Detection, in *Int'l Workshop Earth Resources Survey Systems Proc.* 2, 153–169.

Hodkinson, J. R. (1963). Light Scattering and Extinction by Irregular Particles Larger than the Wavelength, in *Electromagnetic Scattering* (M. Kerker, ed.). 87–100. Pergamon Press, New York.

Hoffer, R. M. (1967). Interpretation of Remote Multispectral Imagery of Agricultural Crops, Res. Bull. No. 831. Lab. for Agricultural Remote Sensing, Purdue Univ. Agricultural Experiment Station, West Lafayette, Ind.

Hoffer, R. M., and Goodrick, F. E. (1971). Variables in Automatic Classification over Extended Remote Sensing Test Sites, in *Proc. of the Seventh Int'l Symp. on Remote Sens. of Environ.* 3, 1967–1981. Univ. of Mich., Ann Arbor.

Hoffer, R. M. (1972). Agricultural and Forest Resource Surveys from Space. Lab. for Applications of Remote Sensing, Information Note 100972. Purdue Univ., West Lafayette, Ind.

Hoffer, R. M., and Fleming, M. D. (1975). Definition of an Optimal Wavelength Band Combination for Forest Cover Type Mapping Using SKYLAB S-192, in SKYLAB Earth Resources Experiment Package, Regional Planning and Development Conference. Paper read, but not mimeographed, Purdue University, West Lafayette, Indiana.

Holland, A. C. (1969). *The Scattering of Polarized Light by Polydisperse Systems of Irregular Particles.* NASA, Washington, D.C.

Holliday, C. T. (1954). The Earth as Seen from Outside the Atmosphere, in *The Earth as a Planet* (G. P. Kuiper, ed.). 713–725. Univ. of Chicago Press, Ill.

Holloway, J. L., Jr. (1958). Smoothing and Filtering of Time Series and Space Fields. U.S. Weather Bureau, Washington, D.C. *Advances in Geophysics.* 4, 351–391. Academic Press, New York.

Holmes, R. A., and MacDonald, R. B. (1969). The Physical Basis of System Design for Remote Sensing in Agriculture, *Proc. of IEEE.* 57(4), 629–639.

Holmes, R. F. (1967). Engineering Materials and Side-Looking Radar, *Photogram. Engin.* 33(7), 767–770.

Holter, M. R., and Legault, R. R. (1965). Orbital Sensors for Oceanography, in *Oceanography from Space* (G. C. Ewing, ed.). 91–109. Woods Hole Oceanographic Inst. Ref. No. 65-10, Massachusetts.

Holter, M. R. (1970). Imaging with Nonphotographic Sensors, in *Remote Sensing with Special Reference to Agriculture and Forestry* (Nat'l Acad. Sciences, ed.). 73–163. Washington, D.C.

Holz, R., Huff, D. L., and Mayfield, R. C. (1969). Urban Spatial Structure Based on Remote Sensing, in *Proc. of the Sixth Int'l Symp. on Remote Sens. of Environ.* 819–830. Univ. of Mich., Ann Arbor.

Hood, P. (1969). Magnetic Surveying Instrumentation–A Review of Recent Advances, in *Mining and Groundwater Geophysics.* 1967, 3–31. Geol. Survey of Can. Econ. Geol. Report No. 26. Ottawa.

Hord, R. M., and Gramenopoulos, N. (1971). Automatic Terrain Classification from Photography, in *Proc. SPIE Symp. on Remote Sens. of Earth Resources and the Environ.* 27, 41–48. Palo Alto, Calif.

Hord, R. M., and Gramenopoulos, N. (1973). Boundary Detection and Regionalized Terrain Classification from Photography. Symp., Electronic Industry Association Committee on Automatic Imagery Pattern Recognition. Washington, D.C.

Hord, R. M., and Brooner, W. G. (1976). Land Use Map Accuracy Criteria, *Photogramm. Eng. and Remote Sensing,* XLII(5), 671–678.

Horton, F., Univ. of Iowa, personal communication.

Hostrop, B. W., and Kawaguchi, T. (1971). Aerial Color and Forestry, *Photogram.*

0-201-04245-2

*Engin.* 38(6), 555–563.
Hotelling, H. (1933). Analysis of a Complex of Statistical Variables in Principal Components, *Jour. of Educ. Psych.* 24, 417–441.
Houston, R. S., Marrs, R. W., Breckenridge, R. M., and Blackstone, D. L., Jr. (1973). Application of the ERTS System to the Study of Wyoming Resources with Emphasis on the Use of Basic Data Products, in *Third Earth Resources Tech. Satellite Symp.* NASA SP-351, 1, 595–620. Goddard Space Flight Center, Washington, D.C.

0-201-04245-2

Unusual pagination arises from insertion of some recent key references.

Hovis, W. A., Jr., Blaine, L. R., and Callaban, W. R. (1968). Infrared Aircraft Spectra over Desert Terrain $8.5\mu$ to $16\mu$, *Appl. Opt.* 7(6), 1137–1140.

Howard, A. D., and Mercado, J. (1970). Low-Sun-Angle Vertical Photography Versus Thermal Infrared Scanning Imagery, *Geol. Soc. of Amer. Bull.* 81(2), 521–523.

Hufnagel, R. E., and Stanley, N. R. (1964), Modulation Transfer Function Associated with Image Transmission Through Turbulent Media. *Jour. of Opt. Soc. of Amer.* 54(1), 52–61.

Hulstrom, R. L. (1973). The Cloud Bright Spot, *Photogram. Engin.* 39(4), 370–377.

International Astronautical Federation (1974). *Ground Systems for Receiving, Analyzing, and Disseminating Earth Resources Satellite Data.* Paris, France.

International Union of Forest Research Organizations, Section 25. (1971). *Application of Remote Sensors in Forestry,* Joint Report by Working Group–IUFRO Congress.

Irvine, W. M., and Peterson, F. W. (1970). Observations of Atmospheric Extinction from 0.315 to 1.06 Microns. *Jour. of Atmos. Sciences.* 27(1), 62–69.

Isachsen, Y. W., Fakundiny, R. H., and Forster, S. W. (1973). Evaluation of ERTS Imagery for Spectral Geological Mapping in Diverse Terrains of New York State, in *Third Earth Resources Tech. Satellite Symp.* NASA SP-351, 1, 691–718. Goddard Space Flight Center, Washington, D.C.

Jackson, H. R., Galway, D., MacDiarmid, S. W., and Wallen, V. R. (1974). Corn Aphid Infestation Computer Analyzed from Color-IR, *Photogramm. Eng.,* XL(8), 943–952.

Jacobowitz, H., and Coulson, K. L. (1973). Effects of Aerosols on the Determination of the Temperature of the Earth's Surface from Radiance Measurements at $11.2\mu m$, NOAA Tech. Report NESS 66, U.S. Dept. Commerce, National Oceanic and Atmospheric Admin., National Envir. Satellite Service, Wash., D.C.

James, W., and Burgess, F. J. (1970). Ocean Outfall Depression, *Photogram. Engin.* 36(12), 1241–2350.

Janza, F. J. (1969). Electromagnetic Sensor Correlation Study, Final Report 29167-5 (ESSA), Contract E-277-68 (N). Ryan Aeronautical Co.

Jefferis, L. H. (1969). A Radar Lineament Analysis of the Grand Canyon Area, northern Arizona (ABS) *Abs. with Programs Pt. 2 South-Cent. Sec.,* 15, Geol. Soc. Am.

Jensen, H. (1945). Geophysical Surveying with the Magnetic Airborne Detector, AN/ASQ-3A. U.S.N.R.L. Report 937-993. U.S. Naval Research Lab., Washington, D.C.

Jerlov, N. G. (1964). Optical Classification of Ocean Water, in *Symp. on the Physical Aspects of Light in the Sea* (J. E. Tyler, ed.). 45–49. Univ. of Hawaii Press, Honolulu.

Johnson, A. W. (1969). Weather Satellites, II, *Scientific Amer.* 220(1), 52–68.

Johnson, C. H. (1970). Recent Advances in Sensitivity and Data Analysis of Airborne Gamma-Ray Spectrometry for Mineral Exploration, *CIM.* Special Vol. 11.

Johnson, C. W., Bowden, L. W., and Pease, R. W. (1969). A System of Regional Agricultural Land Use Mapping Tested Against Small Scale Apollo 9 Infrared Photography of the Imperial Valley (California), U.S.G.S. Contract 14-08-0001-10674. Geographic Applications Program, U.S. Geol. Survey, Washington, D.C.

Johnson, C. W. (1973). Semi-automatic Crop Inventory from Sequential ERTS-1

0-201-04245-2

Imagery, in *Symp. on Significant Results Obtained from the Earth Resources Tech. Satellite-1.* NASA SP-327, 1, 19–26. Goddard Space Flight Center, New Carrollton, Md.

Johnson, P. L., ed. (1969). *Remote Sensing in Ecology.* Univ. of Georgia Press, Athens.

Joseph, R. D., Runge, R. G., and Viglione, S. S. (1969). Design of a Satellite-Born Pattern Classifier, in *Proc. of the Symp. on Information Processing.* 681–700. Purdue Univ., West Lafayette, Ind.

Junge, E. C. (1963). *Air Chemistry and Radioactivity.* Academic Press, New York.

Kaizer, M. (1955). A Quantification of Textures on Aerial Photographs, Tech. Note 121, AD 69484. 1–33. Boston Univ. Research Lab., Mass.

0-201-04245-2

Unusual pagination arises from insertion of some recent key references.

Kanemasu, E. T. (1974). Seasonal Canopy Reflectance Patterns of Wheat, Sorghum, and Soybean, *Remote Sens. of Environ.* 3(1), 43–47.

Katz, A. H. (1969). Let Aircraft Make Earth-Resource Surveys, *Jour. of Astronautics and Aeronautics.* 7(6), 60–68; 7(8), 89–95; 7(10), 78–88.

Keller, G. V. (1969). Application of Resistivity Methods in Mineral and Groundwater Exploration Programs, in *Mining and Groundwater Geophysics.* Geol. Survey of Can. Econ. Geol. Report No. 26. 1967. 51–66. Ottawa.

Kennedy, J. M., Sakamoto, R., and Stogryn, A. P. (1965). Lunar Dust Thickness Determinations Using Microwave Radiometers, in *Trans. Lunar Geol. Field Conferences.* 162–181. Univ. Oregon and N.Y. Acad. of Sciences.

Kennedy, J. M., and Edgerton, A. T. (1967a). Microwave Radiometric Sensing of Soil Moisture Content, Int'l Union Geodesy and Geophysics, 14th General Assembly, Berne.

Kennedy, J. M., and Edgerton, A. T. (1967b). Microwave Sensing of Soils and Sediments. 1–14. Amer. Geophysical Union, 48th Annual Meeting, Washington, D.C.

Kennedy, J. M. (1968). A Microwave Radiometric Study of Buried Karst Topography, *Geol. Soc. of Amer. Bull.* 79(6), 735–742.

Kenney, G. P. (1975). SKYLAB Program Earth Resources Experiment Package, Sensor Performance Evaluation, Final Report 1(5190A), 3(5192), MSC-05546. NASA, L.B.J. Space Center, Houston, Tex.

Kerker, M. (1969). *The Scattering of Light and Other Electromagnetic Radiation.* Academic Press, New York.

Kerr, D. E., ed. (1951). *Propagation of Short Radio Waves,* McGraw-Hill, New York.

Kilinc, I. A., and Lyon, R. J. P. (1970). Geologic Interpretation of Airborne Infrared Thermal Imagery of Goldfield, Nevada, Report 70-3. Remote Sensing Lab., Stanford Univ., Palo Alto, Calif.

Kingston, P., and LaGarde, V. (1972). Spatial Data Massaging Techniques, in *Geographical Data Handling,* Vol. 2 (R. F. Tomlinson, ed.). Int'l. Geographical Union Commission, Ottawa, Canada.

Knauss, J. A. (1957). An Observation of an Oceanographic Front, *Tellus.* 9(2), 234–237.

Knight, C. G., and Harnapp, V. R. (1971). Remote Sensing of Tropical Agricultural Systems, Project No. 1808-821. Dept. of Geography and Center for Research, Inc., Univ. of Kansas, Lawrence.

Knipling, E. B. (1970). Physical and Physiological Basis for the Reflectance of Visible and Near-Infrared Radiation from Vegetation, *Remote Sens. of Environ.* 1(1), 155–159.

Ko, H. C. (1964). *Microwave Scanning Antenna,* Vol. VI, *Aperture.* Academic Press, New York.

Kohout, F. A., and Kolipenski, M. C. (1967). Biological Zonation Related to Groundwater Discharge along the Shore of Biscayne Bay, Miami, Florida, in *Estuaries* (G. H. Lauff, ed.). 488–499. Amer. Assoc. for the Advancement of Science, Washington, D.C.

Kondratyev, K. Ya. (1965). *Radiative Heat Exchange in the Atmosphere.* Pergamon Press, New York.

Kondratyev, K. Ya. (1969). *Radiation in the Atmosphere.* Academic Press, New York.

Kornick, L. J., and McLaren, A. S. (1966). Aeromagnetic Study of the Churchill-Superior Boundary in Northern Manitoba, *Can. Jour. of Earth Science.* 3(4), 547–557.

0-201-04245-2

Kourganoff, V. (1963). *Basic Methods in Transfer Problems.* Dover Publications, New York.

Kover, A. N. (1968). Sidelooking Radar Imagery, in *Proc. of the Fifth Int'l. Symp. on Remote Sens. of Environ.* 781–785. Univ. of Mich., Ann Arbor.

Kozyrev, N. A. (1956). The Luminescence of the Lunar Surface and Intensity of the Solar Corpuscular Radiation, *Izv. Krymskoi Astroizitcheskoi Observatorye.* 16, 148–161.

Kreiss, W. T. (1969). The Influence of Clouds on Microwave Brightness Temperatures Viewing Downward over Open Seas, *Proc. of IEEE.* 57(4), 440–446.

Kriegler, F., Malila, W., Walkpka, R., and Richardson, W. (1969). Preprocessing Transformations and Their Effects on Multispectral Recognition, in *Proc. of the Sixth Int'l Symp. on Remote Sens. of Environ.* 97–106. Univ. of Mich., Ann Arbor.

Kriegler, F. J., Marshall, R. E., Horwitz, H., and Gordon, M. (1972). Adaptive Multispectral Recognition of Agricultural Crops, in *Proc. of the Eighth Int'l. Symp. on Remote Sens. of Environ.* 833–849. Environmental Research Inst. of Mich., Ann Arbor.

Krinsley, D. B. (1973). Preliminary Road Alinement Through the Great Kavir in Iran by Repetitive ERTS-1 Coverage, in *Third Earth Resources Tech. Satellite Symp.* NASA SP-351, 1, 823.Goddard Space Flight Center, Washington, D.C.

Krinsley, D. B. (1974). The Utilization of ERTS-1 Generated Images in the Evaluation of Some Iranian Playas as Sites for Economic and Engineering Development, NASA Order S-70243-AG, Parts 1 and 2. U.S. Geol. Survey, Reston, Va.

Krishen, K. (1968). Reflection from Rough Layers, Technical Report EE-TR-11. Electrical Engineering Dept., Kansas State Univ., Lawrence.

Krishen, K., et al. (1968). Electromagnetic Wave Reflection from Rough Layers, 1968 Wescon Technical Papers. Session 27, Terrain Radar Scatter.

Kronberg, P., Schonfeld, M., Gunther, R., Tsombos, P., and Bannert, D. (1974). ERTS-1 Data in Afar Tectonics, in *Proc. of a Symp. on European Earth Resource Satellite Experiments.* ESRO SP-100, 217–229. Frascati, Italy.

Krumpe, P. F., De Selm, H. R., and Amundsen, C. C. (1971). An Ecological Analysis of Forest Landscape Parameters by Multiband Remote Sensing, in *Proc. of the Seventh Int'l Symp. on Remote Sens. of Environ.* 715–730. Univ. of Mich., Ann Arbor.

Kruse, P. W., McGlauchlin, L. D., and McQuistan, R. B. (1962). *Elements of Infrared Technology.* John Wiley, New York.

Kumar, R., and Silva, L. F. (1974). Statistical Separability of Spectral Classes of Blighted Corn, *Remote Sens. of Environ.* 3(2), 109–115.

Kunde, V. G. (1965). Theoretical Relationship Between Equivalent Blackbody Temperatures and Surface Temperatures Measured by the NIMBUS High Resolution Infrared Radiometer; in *Observations from the NIMBUS I Meteorological Satellite.* Goddard Space Flight Center, Greenbelt, Md.

Kunde, V. G. (1973). Measured Transmittances in the 5.0–25.00 $\mu$m Band through the Entire Atmosphere. Previously unpublished spectra, measured in 1973, and supplied through the courtesy of Mr. V.G. Kunde of Goddard Space Flight Center, Greenbelt, Maryland.

Kunde, V. G., Conrath, B. J., Manel, R. A., McGuire, W. C., Prabhakara, C., and Salomonson, V. (1974). The Nimbus 4 Infrared Spectroscopy Experiment 2. Comparison of Observed Theoretical Radiances from 425–1450 cm$^{-1}$, *Jour. Geophysical Research,* 79, 777–784.

Kylin, H. (1929). Ueber das Vorkommen von Jodiden, Bromiden und Jodidoxydasen bei den Meeresalgen, *Hoppe-Seylers Z. Physiol. Chem.* 186, 50–84.

Kylin, H. (1931). Uber die Jodidspaltende Fahigkeit von Laminaria Digitata, *Hoppe-Seylers Z. Physiol. Chem.* 191, 200–210.

0-201-04245-2

Laboratory for Applications of Remote Sensing (LARS) (1967a). Interpretation of Remote Multispectral Imagery of Agricultural Crops, Res. Bull. 831, Annual Report, Vol. I. Purdue Univ., West Lafayette, Ind.

Laboratory for Applications of Remote Sensing (LARS) (1967b). Remote Multispectral Sensing in Agriculture, Res. Bull. 832, Annual Report, Vol. II. Purdue Univ., West Lafayette, Ind.

Laboratory for Applications of Remote Sensing (1968). Remote Multispectral Sensing in Agriculture, L.A.R.S. 3, Res. Bull. 844. Purdue Univ., West Lafayette, Ind.

La Fond, E. C. (1959). Slicks and Temperature Structure in the Sea, Report 937. U. S. Navy Electronics Lab., San Diego Calif.

Laktionov, A. G. (1964). *Izv. Geophys. Ser.* 6, 953.

Landgrebe, D. A., and staff of LARS (1967). Automatic Identification and Classification of Wheat by Remote Sensing, Agricultural Experiment Station Res. Progress Report No. 279. Lab. for Applications of Remote Sensing, Purdue Univ., West Lafayette, Ind.

Landgrebe, D. A. (1969). Automatic Processing of Earth Resource Data, in *Second Annual Earth Resource Aircraft Program Status Review,* NASA/Manned Spacecraft Center, Houston, Tex.

Landgrebe, D. A., Hoffer, R. H., Goodrick, F. E., et al. (1972). An Early Analysis of ERTS-1 Data, in *Proc. of the Goddard Space Flight Center,* Greenbelt, Md.

Langan, L. (1971). Remote Sensing Measurements of Regional Gaseous Pollution. Joint Conference on Sensing of Environ. Pollutants. AIAA Paper No. 71-1060, Palo Alto, Calif.

Langley, P. G. (1969). New Multi-state Sampling Techniques Using Space and Aircraft Imagery for Forest Inventory, in *Proc. of the Sixth Int'l Symp. on Remote Sens. of Environ.* 1179–1192. Univ. of Mich., Ann Arbor.

Langley, P. G., Van Roessel, J., and Wert, S. (1974). Investigation to Develop a Multistage Forest Sampling Inventory System Using ERTS-1 Imagery, Tech. Report NAS 5-21853. Goddard Space Flight Center, Greenbelt, Md.

Langmuir, I. (1938). Surface Motion of Water Induced by Wind, *Science.* 87(2250), 119–123.

Larson, F. R., Ffolliott, P. F., and Moessner, K. E. (1974). Using Aerial Measurements of Forest Overstory and Topography to Estimate Peak Snowpack, *U.S.D.A. Forest Service Res. Note* RM-267.

Latham, R. P. (1972). Competition Estimator for Forest Trees, *Photogram. Engin.* 38(1), 48–50.

Lathram, E. H. (1973). Analysis of State of Vehicular Scars on Arctic Tundra, Alaska, in *Third Earth Resources Tech. Satellite Symp.* NASA SP-351, 1, 633–642. Goddard Space Flight Center, Washington, D.C.

Lathram, E. H., Tailleur, I. L., and Patton, W. W., Jr. (1973). Preliminary Geologic Application of ERTS-1 Imagery in Alaska, in *Symp. on Significant Results Obtained from the Earth Resources Tech. Satellite-1.* NASA SP-351, 1, 257–264. Goddard Space Flight Center, New Carrollton, Md.

Lattman, L. H. (1963). Geologic Interpretation of Airborne Infrared Imagery, *Photogram. Engin.* 29.

Lattman, L. H. (1963). Geologic Interpretation of Airborne Infrared Imagery. *Proc. 2nd Symp. Remote Sens. of Environ.* 289–294. Univ. of Mich., Ann Arbor.

Laue, E. G., and Drummond, A. J. (1968). Solar Constant: First Direct Measurements, *Science.* 161(3844), 888–891.

Lauer, D. T. (1972). Remote Sensing Can Aid the Forest Land Manager. Forestry Remote Sensing Lab., School of Forestry and Conservation, Univ. of Calif., Berkeley.

0-201-04245-2

La Violette, P. E., and Seim, S. E. (1969). Satellites Capable of Oceanographic Data Acquisition–A Review, U.S.N.O.O. Report TR-215. U.S. Naval Oceanographic Office, Washington, D.C.

Leartek Corporation (1975). *High Altitude Research Vehicle.* Boulder, Colo.

Lee, K. (1969). Infrared Exploration for Shoreline Springs at Mono Lake, California, Test Site, Tech. Report 69-7. Remote Sensing Lab., Stanford Univ., Palo Alto, Calif.

Lee, R. W., and Harp, J. C. (1969). Weak Scattering in Random Media, with Applications to Remote Sensing, *Proc. of IEEE.* 57(4), 375–406.

Lee, W. T. (1922). *The Face of the Earth as Seen from the Air.* The Amer. Geographic Soc., New York.

Leighton, R. B. (1959). *Principles of Modern Physics.* McGraw-Hill, New York.

Leipper, D. F. (1967). Hurricane Hilda, in *Hurricane Symp.,* Pub. No. 1. Amer. Soc. for Oceanography, Houston, Tex.

Lenoir, W. B. (1965). Remote Sounding of the Upper Atmosphere by Microwave Measurements. Ph.D. Dissertation, Mass. Inst. of Tech., Cambridge.

Lent, J. D., and Thorley, G. A. (1969). Some Observations on the Use of Multiband Spectral Reconnaissance for the Inventory of Wildland Resources, *Remote Sens. of Environ.* 1(1), 31–45.

LeSchack, L. A. (1970). ADP of Aerial Imagery for Forest Discrimination. *Am. Soc. Photogramm.* Tech Papers, 36th Annual Meeting, 187–218.

Levin, S. B., and Everett, J. R. (1973). Accelerated Resource Mapping and Map Updating for Latin America by Combined Use of Airborne Radar and Satellite Imagery, in *First Panamer. Symp. on Remote Sens.* 65–76. Panama City, Panama.

Levine, D. (1960). *Radargrammetry.* McGraw-Hill, New York.

Levine, D. (1966). Combinations of Radargrammetric and Photogrammetric Techniques, in *Manual of Photogrammetry* (Amer. Soc. of Photogrammetry, ed.). 1003–1048.

Lewis, G. D. (1971) Land Use Planning Activities in the Forest Service. Reprinted from Great Plains Agricultural Council Publ. 69, Rocky Mountain Forest and Range Experiment Station, Forest Service, U.S. Dept. of Agriculture, Fort Collins, Colo.

Lillestrand, R. L. (1972). Techniques for Change Detection, in *IEEE Trans. on Computers.* C-21(7), 654–660.

Lindner, M. J., and Bailey, J. S. (1968). Distribution of Brown Shrimp (*Penaeus aztecus* Ives) as Related to Turbid Water Photographed from Space, *Fisher. Bull.* 67(2), 289–294.

Lopez, V. M., et al. (1956). Venezuela, in *Handbook of South American Geology, An Explanation of the Geological Map of South America,* Jenks, W. F. (Ed.). *Memoir* 65, *Geol. Soc. Am.*

Lorenz, D. (1968). Temperature Measurements of Natural Surfaces Using Infrared Radiometers, *Appl. Opt.* 7(9), 1705–1710.

Lorenzen, C. J. (1966). A Method for the Continuous Measurement of *in vivo* Chlorophyll Concentration. *Deep Sea Res.* 13(2), 223–227.

Losch, A. (1967). *The Economics of Location.* John Wiley, New York.

Lovell, W. S., and Thiel, L. M. (1968). Interferometric Measurements of the Complex Dielectric Constant of Liquids, Tech. Note 369, Nat'l Bureau of Standards, Washington, D.C.

0-201-04245-2

Lowe, D. S. (1969). Line Scan Devices and Why We Use Them, in *Proc. of the Fifth Int'l Symp. on Remote Sens. of the Environ.* 77–101. Univ. of Mich., Ann Arbor.

Lowman, P. D., Jr. (1965). Space Photography: A Review, in *Oceanography from Space* (G. C. Ewing, ed.). 73–89. Woods Hole Oceanographic Inst. Ref. No. 65-10, Mass.

Lowman, P. D., Jr. (1968). *Space Panorama.* Weltflugbild, R. A. Muller, Feldmeilen. Zurich, Switzerland. 69 plates.

Lowman, P. D., Jr. (1969). Geological Orbital Photography: Experience from the Gemini Program; *Photogrammetria.* 24(3/4), 77–106.

Lowman, P. D., Jr., and Tiedermann, H. A. (1971). Terrain Photography from Gemini Spacecraft, Final Geologic Report X-644-71-15. NASA, Goddard Space Flight Center, Greenbelt, Md.

Lundahl, A. C. (1948). Underwater Depth Determination by Aerial Photography, *Photogram. Engin.* 14(4), 454–462.

Lundberg, H. (1947). Magnetic Surveys with Helicopters, *Trans. Inst. Mining and Metallurgy.* 57(July), 69–75.

Lundien, J. R. (1966). Terrain Analysis by Electromagnetic Means: Radar Responses to Laboratory Prepared Samples, U.S. Army Waterways Experiment Station Tech. Report No. 3-639. Report 2.

Lyon, R. J. P. (1967). Field Infrared Analysis of Terrain, Semiannual Report NGR-05-020-115. Remote Sensing Lab., Dept. of Geophysics, Stanford Univ., Palo Alto, Calif.

Lyon, R. J. P., and Patterson, J. (1969). Airborne Geological Mapping Using Infrared Emission Spectra, in *Proc. of the Sixth Int'l Symp. on Remote Sens. of Environ.* 527–552. Univ. of Mich., Ann Arbor.

Lyon, R. J. P. and Lee, K. (1970). Remote Sensing in Exploration for Mineral Deposits. *Econ. Geol.* 65(7), 785–800.

Lyon, R. J. P. (1972). Infrared Spectral Emittance in Geological Mapping: Airborne Spectrometer Data from Pisgah Crater, California, *Science.* 175(4025), 983–986.

Lyon, R. J. P. (1975). Evaluation of ERTS Multispectral Signatures in Relation to Ground Control Signatures Using a Nested-Sampling Approach, NASA Type III Report, NAS 5-21884. Summary 16, 1–3, 650. Dept. of Geophysics, Stanford Univ., Palo Alto, Calif.

McAlister, E. D. (1964). Infrared-Optical Techniques Applied to Oceanography: 1. Measurement of Total Heat Flow from the Sea Surface, *Appl. Opt.* 3(5), 609–611.

McAlister, E. D., and McLeish, W. L. (1965). Oceanographic Measurements with Airborne Infrared Equipment and Their Limitations, in *Oceanography from Space* (G. C. Ewing, ed.). 189–214. Woods Hole Oceanographic Inst. Ref. No. 65-10, Massachusetts.

McAlister, E. D. (1968). Measurement of the Total Heat Flux from the Sea Surface, Progress Report, in *Proc. of the Ninth Meeting of the Ad Hoc Spacecraft Oceanography Advisory Group.* January. Spacecraft Oceanography Project, NAVOCEANO, Maryland.

McAlister, E. D. (1969). Measurement of the Total Heat Flow from the Sea Surface with an Infrared Two-Wavelength Radiometer, Progress Report, in *The Oceans from Space* (P. C. Badgley, L. Miloy, and L. Childs, eds.). 188–201. Gulf Pub. Co., Houston, Tex.

McAlister, E. D., and McLeish, W. L. (1970). A Radiometric System for Airborne Measurement of the Total Heat Flow from the Sea, Report MPL-U/1670. Scripps Inst. of Oceanography, Univ. of Calif.

0-201-04245-2

McCarthy, H., Hook, J. C., and Knos, D. S. (1956). The Measurement of Association in Industrial Geography, Report 1. 16. Dept. of Geography, Univ. of Iowa, Ames.

McCarthy, J. H., Vaughn, W. W., Learned, R. E., and Meuschke, J. L. (1969). Mercury in Soil Gas and Air–A Potential Tool in Mineral Exploration, U.S.G.S. Circular 609. 1–16.

McClatchey, R. A., Fenn, R. W., Selby, J. E. A., Garing, J. S., and Volz, F. E. (1970). *Optical Properties of the Atmosphere* [AFCRL-71-0279]. Air Force Cambridge Research Lab., Bedford, Mass.

McCluney, W. R. (1974). Ocean Color Spectrum Calculations, *Appl. Opt.* 13(10), 2422–2429.

MacDonald, H. C., Brennan, P. A., and Dellwig, L. F. (1967). Geologic Evaluation by Radar of NASA Sedimentary Test Site, *IEEE Trans. Geoscience Electronics,* GE-5(3), 72-78.

MacDonald, H. C. (1969a). Geologic Evaluation of Radar Imagery from Darien Province, Panama. *Modern Geol.* 1(1), 1–63.

MacDonald, H. C. (1969b). Geologic Interpretation of Radar Imagery from Darien Province, Panama, CRES Tech. Report No. 133-6. Univ. of Kansas, Lawrence.

MacDonald, H. C., Kirk, J. N., Dellwig, L. F., and Lewis, A. J. (1969). The Influence of Radar Look-Direction on the Detection of Selected Geological Features, in *Proc. of Sixth Int'l Symp. on Remote Sens. of Environ.* 637–650. Univ. of Mich., Ann Arbor.

MacDonald, H. C. and Waite, W. P. (1971). Optimum Radar Depression Angles for Geological Analysis, *Modern Geol.* 2, 179–193.

MacDonald, R. B., Bauer, M. E., Allen, R. D., Clifton, J. W., Erickson, J. D., and Landgrebe, D. A. (1972). Results of the 1971 Corn Blight Watch Experiment, in *Proc. of the Eighth Int'l Symp. on Remote Sens. of Environ.* 157–189. Environmental Research Inst. of Mich., Ann Arbor.

McEwen, R. B. (1971). Geometric Calibration of the RBV System for ERTS, in *Proc. of the Seventh Int'l Symp. on Remote Sens. of Environ.* 791–807. Univ. of Mich., Ann Arbor.

Machol, R. E. (1965). *System Engineering Handbook.* McGraw-Hill, New York.

MacKallor, J. A. (1968). A Photomosaic of Western Peru from Gemini Photography, U.S.G.S. Prof. Paper 600C. C169–C173.

McKee, E. D., and Breed, C. S. (1973). An Investigation of Major Sand Seas in Desert Areas Throughout the World, in *Proc. of the Third Earth Resources Tech. Satellite Symp.* NASA SP-351. Goddard Space Flight Center, Washington, D.C.

McLerran, J. H., and Morgan, J. O. (1964). Thermal Mapping of Yellowstone National Park, in *Proc. of the Third Int'l Symp. on Remote Sens. of Environ.* 517–530. Univ. of Mich., Ann Arbor.

McLerran, J. H. (1968). Infrared Sensing of Soils and Rocks, *Materials Res. and Standards.* 8(2), 17–21.

McLerran, J. H. (1969). Remote Sensing and Interpretation of Sea-Ice Features, in *The Oceans from Space* (P. C. Badgley, L. Miloy, and L. Childs, eds.). 159–170. Gulf. Pub. Co., Houston, Tex.

MacNaughton, D. A., and Huckaba, W. A. (1966). Space Photo Points Way to Oil Potential in Amadeus Basin, *Oil and Gas Jour.* 4(24), 132–136.

MacPhail, D. C. (1969). Photomorphic Mapping in Chile. Unpublished manuscript, Univ. of Colorado, Boulder.

MacPhail, D. C. (1970). The Genesis of Photomorphic Analysis in Chile, a paper presented at the National Conference of Latin Americanist Geographers, April 30–May 3. Ball State Univ., Muncie, Ind.

Malila, W. A. (1968). Multispectral Techniques for Image Enhancement and Discrimination. *Photogram. Engin.* 34(6), 566–575.

0-201-04245-2

Malila, W. A., and Nalepka, R. F. (1973). Advanced Processing and Information Extraction Processing and Information Extraction Techniques Applied to ERTS-1 MSS Data, in *Third Earth Resources Tech. Satellite-1 Symp.* I(B), 1943–1772. NASA, Goddard Space Flight Center, Washington, D.C.
Mallon, H. J. (1971). Experimental Applications of Multispectral Data to Natural Resource Inventory and Survey, in *Proc. of the Seventh Int'l Symp. on Remote Sens. of Environ.* 989–1003. Univ. of Mich., Ann Arbor.
Manzer, F. E., and Cooper, G. R. (1967). Aerial Photographic Methods for Potato Disease Detection, Maine Agricultural Experiment Station Bull. 646, Bangor, Me.
Mardon, A. (1965). Application of Microwave Radiometers to Oceanography, in *Oceanography from Space* (G. C. Ewing, ed.). 254–271. Woods Hole Oceanographic Inst. Ref. No. 65-10, Mass.
Marschner, F. J. (1958). Land Use and Its Patterns in the United States, U.S.D.A. Handbook No. 153. U.S. Dept. of Agriculture, Washington, D.C.
Marshall, R. E., Thompson, N., Thompson, F., and Kriegler, F. (1969). Use of Multispectral Recognition Techniques for Conducting Rapid Wide-Area Wheat Surveys, in *Proc. of the Sixth Int'l Symp. on Remote Sens. of Environ.* 3–20. Univ. of Mich., Ann Arbor.
Marshall, R. E., and Kriegler, F. J. (1971). An Operational Multispectral Surveys System, in *Proc. of the Seventh Int'l Symp. on Remote Sens. of Environ.* 3, 2169–2191. Univ. of Mich., Ann Arbor.
Martin, C. D., and Liley, B. (1971). *ERTS Cloud Cover Study* [NASA-CR-121770]. North American Rockwell Corp., Space Division, Downey, Calif.
Martin, J. A., Allen, W. H., Rath, D. L., and Rueff, A. (1973). Summary of an Integrated ERTS-1 Project and Its Results at the Missouri Geological Survey, in *Third Earth Resources Tech. Satellite Symp.* NASA SP-351, 1, 621–632. Goddard Space Flight Center, Washington, D.C.
Maurer, H. (1971). Measurement of Textures of Crop Fields with the Zeiss-Scanning-Microscope-Photometer 05, in *Proc. of the Sixth Int'l Symp. on Remote Sens. of Environ.* 2329–2324. Univ. of Mich., Ann Arbor.
Mazurowskii, M. J., and Sink, D. R. (1965). Attenuation of Photographic Contrast by the Atmosphere, *Jour. of Opt. Soc. of Amer.* 55(1), 26–30.
Meeks, M. L., and Lilley, A. E. (1963). The Microwave Spectrum of Oxygen in the Earth's Atmosphere, *Jour. of Geophys. Res.* 68(6), 1683–1703.
Merifield, P. M., and Rammelkamp, J. (1966). Terrain Seen from TIROS, *Photogram. Engin.* 32(1), 44–54.
Messiah, A. (1966). *Quantum Mechanics,* Vol. 1. John Wiley, New York.
Meyer, M. P., and Maklin, H. A. (1969). P. I. Techniques for Ektachrome IR Transparencies, *Photogram. Engin.* 35(11), 1111–1114.
Meyer, M. P., and Chiang, H. C. (1971). Multiband Reconnaissance of Simulated Insect Defoliation in Corn Fields, in *Proc. of the Seventh Int'l Symp. on Remote Sens. of Environ.* 1231–1234. Univ. of Mich., Ann Arbor.
Meyer, W. D. (1967). Analysis of Radar Calibration Data, Final Report, Goodyear Aerospace Corp. for the U.S. Army Engineering Topographic Lab. 1–84. Litchfield Park, Ariz.
Michael, M., and Lin, W. (1973). Experimental Study of Information Measure and Inter-Intra Class Distance Ratios on Feature Selection and Orderings, *IEEE Trans. on Systems, Man, and Cybernetics,* SMC-3(2), 172–179.
Mikailhov, V. Y. (1960). The Use of Colour Sensitive Films in Aerial Photography in U.S.S.R., *Trans. Can Nat'l Res. Council.*
Miller, L. D. (1966). Location of Anomalously Hot Earth with Infrared Imagery in Yellowstone National Park, in *Proc. of the Fourth Int'l Symp. on Remote Sens. of Environ.* 751–569. Univ. of Mich., Ann Arbor.

0-201-04245-2

Miller, P. R., and Parmeter, J. R. (1963). Ozone Injury to the Foliage of *Pinus Ponderosa, Phytopathology.* 53(9), 1072–1076.

Minnich, R. A., Bowden, L. W., and Pease, R. W. (1969). Mapping Montane Vegetation in Southern California, Tech. Report III, U.S.D.I. Contract 14-08-0001-10674. Dept. of Geography, Univ. of Calif., Riverside.

Mizushima, M., and Hill, R. M. (1954). Microwave Spectrum of $O_2$, *Phys. Rev.* 93(4), 745–748.

Moeckel, E. E. (1969). Let Aircraft Make Earth Resource Surveys: a Criticism, *Astronautics and Aeronautics.* 7(8), 89–91.

Moffat, A. J., and Barringer, A. R. (1969). Recent Progress in the Remote Detection of Vapours and Gaseous Pollutants, in *Proc. of the Sixth Int'l Symp. on Remote Sens. of Environ.* 379–414. Univ. of Mich., Ann Arbor.

Mohr, P. A. (1973). Structural Geology of the African Rift System: Summary of New Data from ERTS-1 Imagery, in *Proc. of the Third Earth Resources Tech. Satellite Symp.* NASA SP-351. Goddard Space Flight Center, Washington, D.C.

Molineux, C. E. (1964). Aerial Reconnaissance of Surface Features with the Multiband Spectral System, in *Proc. of the Third Int'l Symp. on Remote Sens. of Environ.* 339–421. Univ. of Mich., Ann Arbor.

Molineux, C. E. (1965). Multiband Spectral System for Reconnaissance, *Photogram. Engin.* 31, 131–143.

Molnar, P., and Tapponnier, P. (1975). Cenozoic Tectonics of Asia: Effects of a Continental Collision, *Science.* 189(4201), 419–426.

Moore, E. G. (1968). Side-Looking Radar in Urban Research: A Case Study, U.S.G.S. Contract 14-08-0001-10654. Geographic Applications Program, U.S. Geol. Survey, Washington, D.C.

Moore, E. G., and Wellar, B. S. (1969). *Jour. of Amer. Inst. of Planners.* 35, 1.

Moore, R. K., and Pierson, W. J. (1965). Measuring Sea State and Estimating Surface Winds from Polar Orbiting Satellite, in *Proc. of the Int'l Symp. on Electromagnetic Sens. of the Earth from Satellites,* R1–R28.

Moore, R. K. (1966). Radar Scatterometry—An Active Remote Sensing Tool, *Proc. of the Fourth Int'l Symp. on Remote Sens. of Environ.* 339–375. Univ. of Mich., Ann Arbor.

Moore, R. K., and Simonett, D. S. (1967). Potential Research and Earth Resource Studies with Orbiting Radars: Results of Recent Studies, AIAA Tech. Paper No. 67-767.

Moore, R. K., and Biggs, A. W. (1968). Research in Radar Scatterometry and Altimetry, in *Proc. of the Ninth Meeting of the Ad Hoc Spacecraft Oceanography Advisory Group.* January. Spacecraft Oceanography Project, NAVOCEANO, Maryland.

Moore, R. K., Rouse, J. W., and Waite, W. P. (1969). Panchromatic and Polypanchromatic Radar, *Proc. of IEEE,* 57(4), 590–593.

Moore, R. K., and Waite, W. P. (1969). Radar Scatterometry, CRES Tech. Report 118–15. January. Univ. of Kansas, Lawrence.

Moore, R. K., and Pierson, W. J., Jr. (1970). Worldwide Oceanic Wind and Wave Predictions Using a Satellite Radar Radiometer, AIAA Paper No. 70-310.

Moore, R. K., and Thomann, G. C. (1971). Imaging Radars for Geoscience Use, *IEEE Trans. on Geoscience Electronics,* GE-9(3). 155–164.


0-201-04245-2

Morain, S. A., and Simonett, D. S. (1966). Vegetation Analysis with Radar Imagery, in *Proc. of the Fourth Int'l Symp. on Remote Sens. of Environ.* 605–622. Univ. of Mich., Ann Arbor.

Morain, S. A., and Simonett, D. S. (1967). K-band Radar in Vegetation Mapping, *Photogram. Engin.* 33(7), 730–740.

Morain, S. A., and Coiner, J. C. (1970). An Evaluation of Fine Resolution Imagery for Making Agricultural Determinations, Tech. Report 177-7, Center for Research, Inc., Univ. of Kansas, Lawrence.

Morain, S. A. (1971). Recent Advances in Radar Applications to Agriculture. NASA *3rd Annual Earth Resources Program Review,* vol. II, Agriculture, Forestry and Sensor Studies, NASA-MSC doc. 03742.

Morain, S. A., and Williams, D. L. (1973). Estimate of Winter Wheat Yield from ERTS-1, in *Third Earth Resources Tech. Satellite Symp.* NASA SP-351, 1, 21–28. Goddard Space Flight Center, Washington, D.C.

Morain, S. A. (1974). Interpretation and Mapping of Natural Vegetation, in *Remote Sensing Techniques for Environmental Analysis* (J. Estes and L. Senger, eds.). 127–165. Hamilton Press.

Morgan, M. A. (1967). Hardware Models in Geography, in *Models in Geography* (R. J. Chorley and P. Haggett, eds.). 727–774. Methuen, London.

Morley, L. W. (1969). Regional Geophysical Mapping, in *Mining and Groundwater Geophysics.* 1967, 249–258. Geol. Survey of Can. Econ. Geol. Report No. 26. Ottawa.

Morris, W. G. (1970). Photo Inventory of Fine Logging Slash, *Photogram. Engin.* 36(12), 1252–1257.

Morrison, D. F. (1967). *Multivariate Statistical Methods.* McGraw-Hill, New York.

Morrison, R. B., and Cooley, M. E. (1973). Application of ERTS-1 Multispectral Imagery of Monitoring of the Present Episode of Accelerated Erosion in Southern Arizona, in *Symp. on Significant Results Obtained from the Earth Resources Tech. Satellite-1.* NASA SP-327, 1, 283–290. Goddard Space Flight Center, New Carrollton, Md.

Morrison, R. B., and Hallberg, G. R. (1973). Mapping Quaternary Land Forms and Deposits in the Midwest and Great Plains by Means of ERTS-1 Multispectral Imagery, in *Symp. on Significant Results Obtained from the Earth Resources Tech. Satellite-1.* NASA SP-327, 1, 353–362. Goddard Space Flight Center, New Carrollton, Md.

Morton, G. A., and Zworykin, V. K. (1940). *Television–The Electronics of Image Transmission.* John Wiley, New York.

Moura, J. M. (1972). Levantamento dos Recursos Naturais das Regioes Amazonicas e Nordeste do Brazil por Meio de Radar e Otros Sensores, *Revista Brazileira de Cartografia.* 3, 1–4.

Moxham, R. M., and Alcarez, A. (1965). Infrared Survey of Taal Volcano, *Proc. of the Fourth Int'l Symp. on Remote Sens. of Environ.* 827–843. Univ. of Mich., Ann Arbor.

Moxham, R. M., Crendell, D. R., and Marlatt, W. E. (1965a). Thermal Features at Mount Rainier, Washington, as Revealed by Infrared Surveys, U.S.G.S. Prof. Paper 525D. D93–D100.

Moxham, R. M., Foote, R. S., and Bunker, C. M. (1965b). Gamma-Ray Spectrometer Studies of Hydrothermally Altered Rocks, *Econ. Geol.* 60(4), 653–671.

Moxham, R. M. (1968). Aerial Infrared Surveys at the Geysers Geothermal Steam Field, California, U.S.G.S. Interagency Report NASA-123.

Muhm, J. R. (1969). Niger River Geology Derived from Space Photography, *Oil and Gas Jour.* 67(27), 158–164.


0-201-04245-2

Mullens, R. H. (1969). Analysis of Urban Residential Environments Using Color Infrared Aerial Photography: An Examination of Socioeconomic Variables and Physical Characteristics of Selected Areas in Los Angeles Basin, *Studies in Remote Sens. of Southern California and Related Environ.*, Contract 14-09-0001-10674, Tech. Report No. 4, U.S. Dept. of the Interior, Washington, D.C.

0-201-04245-2

Unusual pagination arises from insertion of some recent key references.

Murtha, P. A. (1972). *A Guide to Air Photo Interpretation of Forest Damage in Canada.* Canadian Forestry Service, Publ. 1292. Ottawa, Canada.

Myers, V. I., and Allen, W. A. (1968). Electro-optical Remote Sensing Methods as Nondestructive Testing and Measuring Techniques in Agriculture, *Appl. Opt.* 7(9), 1819–1838.

Myers, V. I., and Heilman, M. D. (1969). Thermal Infrared for Soil Temperature Surveys, *Photogram. Engin.* 35(10), 1024–1032.

Myers, V. K. (1970). Soil, Water, and Plant Relations, in *Remote Sensing: With Special Reference to Agriculture and Forestry.* 253–297. Nat'l. Acad. of Sciences, Washington, D.C.

Nagel, M., and Pivovonsky, N. (1961). *Tables of Blackbody Radiation Functions.* Macmillan, New York.

Nagy, E., Shelton, G., and Tolaba, J. (1971). Procedural Questions in Signature Analysis, in *Proc. of the Seventh Int'l Symp. on Remote Sens. of Environ.* 2, 1387–1401. Univ. of Mich., Ann Arbor.

Nagy, G. (1968). State of the Art in Pattern Recognition, *Proc. of IEEE.* 56(5), 836–862.

Naidu, P. S. (1969). Estimation of Spectrum and Cross-Spectrum of Aeromagnetic Field Using Fast Digital Fourier Transform (FDFT) Techniques, *Geophysical Prospecting,* 8(3), 344–361.

National Aeronautics and Space Administration (NASA) (1971), *Earth Albedo and Emitted Radiation.* Goddard Space Flight Center, Greenbelt, Michigan.

National Aeronautics and Space Administration (NASA) (1972). *ERTS-1 Data User's Handbook.* Goddard Space Center, Greenbelt, Md.

National Aeronautics and Space Administration (NASA) (1973). *Corn Blight Watch Final Report.* I–III. NASA Earth Resources Program, L.B.J. Space Center, Houston, Tex.

National Aeronautics and Space Administration (NASA) (1974). *SKYLAB Earth Resources Data Catalog.* L.B.J. Space Center, Houston, Tex.

National Aeronautics and Space Council (NASC) (1970). *Aeronautics and Space Report to the President.* Government Printing Office, Washington, D.C.

National Council on Marine Resources and Engineering Development (1967). *United States Activities in Spacecraft Oceanography.* Government Printing Office, Washington, D.C.

National Research Council Committee on Remote Sensing for Agricultural Purposes (1970). *Remote Sensing: With Special Reference to Agriculture and Forestry.* Nat'l. Acad. of Sciences, Washington, D.C.

Naumov, A. P. (1968). *Izv. Atmos. Ocean. Physics.* 4, 96.

Nelson, P. H., and Morris, D. B. (1969). Theoretical Response of a Time-Domain, Airborne, Electromagnetic System, *Geophysics.* 34(5), 729–738.

Nelson, R. M. (1969). The Potential Application of Remote Sensing to Selected Ocean Circulation Problems, in *The Oceans from Space* (P. C. Badgley, L. Miloy, and L. Childs, eds.). 38–45. Gulf Pub. Co., Houston, Tex.

New Brunswick Department of Natural Resources (1972). E-Phase, Radiophase, Magnetic, Gamma-Ray Total Count, Gamma-Ray Potassium Count, Caledonia Mt. Area. 21 GIE, 21 H4W, 21 H5 E, W, 21 HG E, W, 21 G 8 E, 21 H10E, W, 21 H12E, 21 H15, E, 21 H16E, W, Fredericton, New Brunswick, Canada.

Newcomb, G. S., and Millan, M. M. (1970). Theory, Applications, and Results of the Long-Line Correlation Spectrometer, *IEEE Trans. on Geoscience Electronics.* GE-8(3), 149–157.

0-201-04245-2

Nichols, J. D., Gialdini, M., and Faakkola, S. (1973). A Timber Inventory Based upon Manual and Automated Analysis of ERTS-1 and Supporting Aircraft Data Using Multistage Probability Sampling, in *Third Earth Resources Tech. Satellite Symp.* NASA SP-351, 1, 145–159. Goddard Space Flight Center, Washington, D.C.

0-201-04245-2

Unusual pagination arises from insertion of some recent key references.

Nicolais, S. M. (1973). Mineral Exploration with ERTS Imagery, in *Third Earth Resources Tech. Satellite Symp.* NASA SP-351, 1, 785–796. Goddard Space Flight Center, Washington, D.C.

Nielsen, U. (1974). Tests of an Airborne Tilt-Indicator, *Photogram. Engin.* 40(8), 953–954.

Nilsson, N. J. (1965). *Learning Machines.* McGraw-Hill, New York.

Noble, V. E., Ketchum, R. D., and Ross, D. B. (1969). Some Aspects of Remote Sensing as Applied to Oceanography, *Proc. of IEEE.* 57(4), 594–604.

Norbeck, W. (1965). *The Law of Allometric Growth.* Mich. Interuniv. Community of Mathematical Geographers. Univ. of Mich., Ann Arbor.

Norwood, V. T. (1969). Optimization of a Multispectral Scanner for ERTS, *Proc. of the Sixth Int'l Symp. on Remote Sens. of Environ.* 227–235, Univ. of Mich., Ann Arbor.

Norwood, V. T. (1974). *Multispectral Scanner (MSS) Improved Resolution Study,* NASA Contract NAS5-21935. Hughes Aircraft Co., Culver City, Calif.

Nunnally, N. R. (1969a). Integrated Landscape Analysis with Radar Imagery. *Remote Sens. of Environ.* 1, 1–6.

Nunnally, N. R. (1969b). Introduction to Remote Sensing–The Physics of Electromagnetic Radiation, U.S.G.S. Contract 14-08-0001-10921. Geographic Applications Program, U.S. Geol. Survey, Washington, D.C.

Nunnally, N. R., and Witmer, R. E. (1970). Remote Sensing for Land-Use Studies, *Photogram. Engin.* 36(5), 449–453.

Oceanographer of the Navy (OCEANAV) (1970). *The Ocean Science Program of the U.S. Navy.* Office of the Oceanographer of the Navy, Alexandria, Va.

Ogilvy, A. A. (1969). Geophysical Studies in Permafrost Regions in the U.S.S.R., in *Mining and Groundwater Geophysics.* 1967, 641–648. Geol. Survey of Can. Econ. Geol. Report No. 26. Ottawa.

Olson, C. E., Jr. (1968). Some Observations on Photographic Interpretations–Multispectral Camera Systems, lecture notes prepared for NSA Summer Inst. on Geographic Applications of Remote Sensing. Univ. of Mich., Ann Arbor.

Olson, C. E., Jr., Ward, J. M., and Rohde, W. G. (1969). Remote Sensing of Changes in Morphology and Physiology of Trees under Stress, Annual Progress Report. Forestry Remote Sensing Lab., Univ. of Calif., Berkeley.

Olson, C. E., Jr. (1970). Remote Sensing of Changes in Morphology and Physiology of Trees under Stress, Tech. Report NASA Contract R-09-038-022. Forestry Remote Sensing Lab., School of Forestry and Conservation. Univ. of Calif., Berkeley.

Olson, C. E., Jr., Rohde, W. G., and Ward, J. M. (1971). Remote Sensing of Changes in Morphology and Physiology of Trees under Stress, Tech. Report, NASA Contract R-09-038-002. Forestry Remote Sensing Lab., School of Forestry and Conservation, Univ. of Calif., Berkeley.

Olson, C. E., Jr. (1972). Remote Sensing of Changes in Morphology and Physiology of Trees under Stress, Final Tech. Report, NASA Contract R-09-033-002. Forestry Remote Sensing Lab., School of Forestry and Conservation, Univ. of Calif., Berkeley.

Ontario Department of Mines and Northern Affairs (1971a). Airborne BCB E-Phase Survey, Uxbridge, Whitchurch, and Markham Townships, Preliminary Map P725. Geophys. Ser.

Ontario Department of Mines and Northern Affairs (1971b). Airborne VLF E-Phase Survey, Uxbridge, Whitchurch, and Markham Townships, Preliminary Map P726. Geophys. Ser.

0-201-04245-2

Ontario Department of Mines and Northern Affairs (1971c). E-Phase, Radiophase, Magnetic, and Gamma-Ray Total Count Survey of Beatty, Warden, Monro, Hislop, Guibord, Coulson, Maps P641 to 664. Geophys. Ser.

Orhaug, T. A. (1965). *Thermal Noise Radiation from the Atmosphere.* Scandinavian Univ. Press, Goteberg, Sweden.

Orme, A. R., Bowden, L. W., and Minnich, R. A. (1971). Remote Sensing of Disturbed Insular Vegetation from Color Infrared Imagery, in *Proc. of the Seventh Int'l. Symp. on Remote Sens. of Environ.* 1235–1243. Univ. of Mich., Ann Arbor.

Orr, D. G. (1968). Multiband-Color Photography, in *Manual of Color Aerial Photography* (J. T. Smith, ed.). 441–450. Amer. Soc. of Photogrammetry, Falls Church, Va.

Ory, T. R. (1964). Line-Scanning Reconnaissance Systems in Land Utilization and Terrain Studies, in *Proc. of the Third Int'l Symp. on Remote Sens. of Environ.* 393–398. Univ. of Mich., Ann Arbor.

Otala, M. (1969). *A New Experimental Airborne Magnetic Gradiometer,* Report S8 1969. Univ. of Oulu, Finland.

Paal, G. (1965). Ore Prospecting Based on VLF-Radio Signals, *Geoexploration.* 3(3), 139–147.

Paarma, H., and Talvitie, J. (1968). High Altitude False Color Photointerpretation in Prospecting, in *Manual of Color Aerial Photography* (J. T. Smith, ed.). 431–440. Amer. Soc. of Photogrammetry, Falls Church, Va.

Packard, M., and Varian, R. (1954). Free Nuclear Induction in the Earth's Magnetic Field (abstract), *Phys. Review.* 93, 941.

Palgen, J. J. O. (1970). Applicability of Pattern Recognition Techniques to the Analysis of Urban Quality from Satellites, *Pattern Recognition.* 2, 255–260.

Paris, J. F. (1968). Preliminary Results from Connvair 240A Mission 50, June 12, 1967, over the Mississippi Delta, Ref. No. 68-6T. Dept. of Oceanography, Texas A & M Univ., College Station.

Paris, J. F. (1969). Microwave Radiometry and its Application to Marine Meteorology and Oceanography, Ref. No. 69-1T. Dept. of Oceanography, Texas A & M Univ., College Station.

Park, Charles F., Jr. (1968). *Affluence in Jeopardy: Minerals and the Political Economy.* Freeman, Cooper and Co., San Francisco, Calif.

Parker, D. C. (1968). Remote Sensing for Engineering Investigations of Terrain-Infrared Systems, in *Proc. of the Fifth Int'l Symp. on Remote Sens. of Environ.* 701–709. Univ. of Mich., Ann Arbor.

Paterson, N. R. (1961). Experimental and Field Data for the Dual-Frequency Phase-shift Method of Airborne Electromagnetic Prospecting, *Geophysics.* 26(5), 601–617.

Paterson, N. R. (1970). Airborne VLF-EM Test, *Can. Mining Jour.* 91(11), 47–50.

Peake, W. H. (1959). Interaction of electromagnetic waves with some natural surfaces, I.R.E. *Trans. on Antennas and Propagation,* S324–S329 (December).

Peake, W. H., and Cost, S. T. (1968). The Bistatic Echo of Terrain at 10 GHz, 1968 Wescon Technical Papers. Session 22, Terrain Radar Scatter.

Pearson, R. L., and Miller, L. D. (1972). Remote Mapping of Standing Crop Biomass for Estimation of the Productivity of the Short-grass Prairie, Pawnee National Grasslands, Colorado, in *Proc. of the Eighth Int'l Symp. on Remote Sens. of Environ.* 1355–1379. Environmental Res. Inst. of Mich., Ann Arbor.

0-201-04245-2

Pease, R. W., and Bowden, L. W. (1969). Making Color Infrared a More Effective High-Altitude Remote Sensor, *Remote Sens. of Environ.* 1(1), 23–30.

Pease, R. W. (1971). Mapping Terrestrial Radiation Emission with a Scanning Radiometer, in *Proc. of the Seventh Int'l Symp. on Remote Sens. of Environ.* 501–521. Univ. of Mich., Ann Arbor.

Peloquin, R. A. (1961). Implementation of an Airborne Oceanographic Platform, in U.S.N.O.O. Informal Oceanographic Manuscript No. IOM. 18–61. U.S. Naval Oceanographic Office, Washington, D.C.

Pemberton, R. H. (1962). Airborne Electromagnetics in Review, *Geophysics.* 27(5), 691–713.

Pemberton, R. H. (1969). Airborne Radiometric Surveying for Mineral Deposits, in *Mining and Groundwater Geophysics.* 1967, 416–424. Geol. Survey of Can. Econ. Geol. Report No. 26. Ottawa.

Penndorf, R. (1957). Tables of the Refractive Index for Standard Air and the Rayleigh Scattering Coefficient for the Spectral Region Between 0.2 and 20.0 Microns and Their Application to Atmospheric Optics. *Jour. of Opt. Soc. of Amer.* 47(2), 176–182.

Peplies, R., and Wilson, D. (1970). Analysis of Spacecraft Photography of Humid-Forested Region: A Case Study of Northern Alabama, unpublished paper. Dept. of Geography, East Tenn. State Univ., Johnson City.

Pestrong, R. (1969). Multiband Photos for a Tidal Marsh, *Photogram. Engin.* 35(5), 453–470.

Peterson, F. (1968). The Utility of Radar and Other Remote Sensing in Thematic Land Use Mapping from Spacecraft, NASA Interagency Report 140. Univ. of Kansas, Lawrence.

Peterson, R. M., Cochrane, G. R., Morain, S. A., and Simonett, D. S. (1969). A Multi-sensor Study of Plant Communities at Horsefly Mountain, Oregon, in *Remote Sensing in Ecology* (P. Johnson, ed.). 63–93. Univ. of Georgia Press, Athens.

Pettinger, L. R. (1969). Analysis of Earth Resources on Sequential High Altitude Multiband Photography, Special Report. Forest Remote Sensing Lab., Univ. of Calif., Berkeley.

Phleger, F. B. (1965). Study of River Effluents from Space Vehicles, in *Oceanography from Space* (G. C. Ewing, ed.). 425–426. Woods Holes Oceanographic Inst. Ref. No. 65–10, Massachusetts.

Pickering, S. M., and Jones, R. C. (1973). Geologic Evaluation and Applications of ERTS-1 Imagery over Georgia, in *Third Earth Resources Tech. Satellite Symp.* NASA SP-351, 1, 857–868. Goddard Space Flight Center, Washington, D.C.

Pickering, S. M., and Jones, R. C. (1974). Geologic Evaluation and Applications of ERTS-2 Imagery over Georgia, in *Third Earth Resources Tech. Satellite Symp.* NASA SP-351, 2, 41–49. Goddard Space Flight Center, Washington, D.C.

Pickett, R. L., and Wilkerson, J. C. (1966). Airborne IR Tracking of Gulf Stream, *Geo-Marine Tech.* 2(5), 8–10.

Pierce, K. L. (1968). Evaluation of Infrared Imagery Applications to Studies of Surficial Geology, Yellowstone Park. U.S.G.S. NASA CR-94789. Denver, Colo.

Pierson, W. J., Jr., Scheps, B. B., and Simonett, D. S. (1965). Some Applications of Radar Return Data to the Study of Terrestrial and Oceanic Phenomena, *Amer. Astronautical Soc., Science and Tech. Ser.* 4, 87–137.

Pierson, W. J., Jr. (1968). Oceanographic Uses of Radar, in *Proc. of the Ninth Meeting of the Ad Hoc Spacecraft Oceanography Advisory Group.* January. Spacecraft Oceanography Project, NAVOCEANO, Maryland.

0-201-04245-2

Pierson, W. J., Jr. (1969). The Sea Surface, in *The Oceans from Space* (P. C. Badgley, L. Miloy, and L. Childs, eds.). 104–130. Gulf Pub. Co., Houston, Tex.
Plass, G. N., and Kattawar, G. W. (1962). Monte Carlo Calculations of Light Scattering from Clouds. *Appl. Opt.* 7(3), 415–419.

0-201-04245-2

Unusual pagination arises from insertion of some recent key references.

Plass, G. N., and Kattawar, G. W. (1968b). Radiance and Polarization of Multiple Scattered Light from Haze and Clouds, *Appl. Opt.* 7(8), 1519–1527.

Plass, G. N., and Kattawar, G. W. (1969). Radiative Transfer in an Atmosphere-Ocean System, *Appl. Opt.* 8(2), 455–466.

Plass, G. N., and Kattawar, G. W. (1972). Effect of Aerosol Variation on Radiance in the Earth's Atmosphere-Ocean System, *Appl. Opt.* 11(7), 1598–1604.

Polcyn, F. C., and Noble, V. E. (1968). Remote Sensing Techniques for the Detection of Doubtful Shoals, in *Proc. of the Ninth Meeting of the Ad Hoc Spacecraft Oceanography Advisory Group.* January. Spacecraft Oceanography Project, NAVOCEANO, Maryland.

Popenoe, P. (1962). Aero-Radioactivity Survey and Aerial Geology of Parts of East Central New York and West Central New England. U.S. AEC Report ARMS-1.

Poulin, A. O., and Harwood, T. A. (1966). Infrared Mapping of Thermal Anomalies in Glaciers, *Can. Jour. of Earth Sciences.* 3(6), 881–885.

Poulton, C. E., Faulkner, D. P., Johnson, J. R., Mouat, D. A., and Schrumpf, B. J. (1971). Inventory and Analysis of Natural Vegetation and Related Resources from Space and High Altitude Photography, Technical Report, NASA Contract R-09-038-002. Forestry Remote Sensing Lab., School of Forestry and Conservation, Univ. of Calif., Berkeley.

Poulton, C. E. (1972). Inventory and Analysis of Natural Vegetation and Related Resources from Space and High Altitude Photography, NASA Contract R-09-038-002. Forestry Remote Sensing Lab., School of Forestry and Conservation, Univ. of Calif., Berkeley.

Powell, W. J., Copeland, C. W., and Drahouzal, J. H. (1970). Delineation of Linear Features and Application to Reservoir Engineering Using Apollo 9 Multispectral Photography. Geol. Survey of Alabama Information Ser. No. 41. Univ. of Alabama, University.

Pozdniak, S. I. (1960). Measurement of Electrical Parameters of a Medium by a Polarization Method. *Radiotekhni a i electronika.* 5(10), 1730–1733. Trans. Brief Commun.

Pratt, W. P. (1968). Infrared Imagery of Lordsburg-Silver City Area, New Mexico, U.S.G.S. Interagency Report, NASA TM-X-61711. NASA Manned Spacecraft Center, Houston, Tex.

Prentice, V., and Olson, C. E., Jr. (1968). Temporal and Spectral Factors in Utilizing Infrared Imagery for Geographic Investigations, unpublished report. Inst. of Science and Tech., Willow Run, Univ. of Mich., Ann Arbor.

Press, N. P. (1974). Remote Sensing to Detect the Toxic Effects of Metals on Vegetation for Mineral Exploration, in *Proc. of the Ninth Int'l Symp. on Remote Sens. of Environ.* 2027–2038. Environmental Research Inst. of Mich., Ann Arbor.

Pressman, A. E. (1968). Geologic Comparison of Ektachrome and Infrared Ektachrome Photography, in *Manual of Color Aerial Photography* (J. T. Smith, ed.). 396–397. Amer. Soc. of Photogrammetry, Falls Church, Va.

Preston, K., Jr. (1972). A Comparison of Analog and Digital Techniques for Pattern Recognition, in *Proc. of IEEE.* 60(10), 1216–1231.

Pruitt, E. L. (1965). Use of Orbiting Research Laboratories for Experiments in Coastal Geography, in *Oceanography from Space* (G. C. Ewing, ed.). 419–421. Woods Hole Oceanographic Inst. Ref. No. 65-10, Massachusetts.

Puranen, M., and Kahma, A. A. (1953). Method for Inductive Prospecting, U.S. Patent 2,642,477.

Puranen, M., and Kahma, A. A. (1956). Method for Inductive Prospecting, U.S. Patent 2,741,736.

0-201-04245-2

Quade, J. G., Chapman, P. E., Brennan, P. A., and Blinn, J. C., III (1970). Multispectral Remote Sensing of an Exposed Volcanic Province, Memo 23-453. Jet Propulsion Lab., Calif. Inst. of Tech., Pasadena.

0-201-04245-2

Unusual pagination arises from insertion of some recent key references.

Raff, A. D., and Mason, R. G. (1961). Magnetic Survey of the West Coast of North America, 40°N. Latitude to 52°N. Latitude, *Bull. Geol. Soc. of Amer.* 72(8), 1267–1270.

Ramapriyan, H. K. (1972). Spatial Frequency Analysis of Multispectral Data, in *Remote Sensing of Earth Resources* (F. Shahrokhi, ed.). 1, 621–644. Univ. of Tenn., Tullahoma.

Ramo, S., Whinnery, J. R., and Van Duzer, T. (1965). *Fields and Waves in Communications and Electronics.* John Wiley, New York.

Ramsey, R. C. (1968). Study of the Remote Measurement of Ocean Color, Final Report. TRW Systems Group for NASA Headquarters, Redondo Beach, California.

Ranz, E., and Schneider, S. (1971). Progress in the Application of Agfacontour Equidensity Film for Geo-scientific Photo Interpretation, in *Proc. of the Seventh Int'l Symp. in Remote Sens. of Environ.* 3, 779–790. Univ. of Mich., Ann Arbor.

Ray, R. G. (1960). Aerial Photographs in Geologic Interpretation and Mapping, U.S.G.S. Prof. Paper 373.

Ready, P. J., Wintz, P. A., Whitsitt, S. J., and Landgrebe, D. A. (1971). Effects of Compression and Random Noise on Multispectral Data, in *Proc. of the Seventh Int'l Symp. on Remote Sens. of Environ.* 1321–1342. Univ. of Mich., Ann Arbor.

Ready, P. J. (1972). *Multispectral Data Compression Through Transform Coding and Block Quantization.* Ph.D. Thesis, Purdue Univ., West Lafayette, Ind.

Reeves, R. G. (1969). Structural Geologic Interpretations from Radar, *Geol. Soc. of Amer. Bull.* 80(11), 2159–2164.

Reeves, R. G. (Ed.) (1975). *Manual of Remote Sensing.* Amer. Soc. Photogramm., Falls Church, Va.

Rice, S. O. (1951). Reflection of Electromagnetic Waves from Slightly Rough Surfaces, *Commun. Pure and Appl. Math.* 4(43), 351–378.

Rich, E. I. (1973). Relation of ERTS-1 Detected Geologic Structure to Known Economic Ore Deposits, in *Symp. on Significant Results Obtained from the Earth Resources Tech. Satellite-1.* NASA SP-327, 1, 395–402. Goddard Space Flight Center, New Carrollton, Md.

Richardson, A. J., Allen, W. A., and Thomas, J. R. (1969). Discrimination of Vegetation by Multispectral Reflectance Measurements, in *Proc. of the Sixth Int'l. Symp. on Remote Sens. of Environ.* 1143–1156. Univ. of Mich., Ann Arbor.

Richardson, A. J., Torline, R. J., and Allen, W. A. (1971). Computer Identification of Ground Pattern from Aerial Photographs, in *Proc. of the Seventh Int'l. Symp. on Remote Sens. of Environ.* 2, 1357–1376. Univ. of Mich., Ann Arbor.

Richardson, W. S. (1957). Airborne Oceanographers, *Oceanus.* 1, 2.

Robinson, N., ed. (1966). *Solar Radiation.* Elsevier, New York.

Rogers, R. H., Peacock, K., and Shah, N. J. (1973). A Technique for Correcting ERTS Data for Solar and Atmospheric Effects, in *Third Earth Resources Tech. Satellite Symp.* I(B), 1787–1804. NASA, Goddard Space Flight Center, Washington, D.C.

Rohde, W. G., and Olson, C. E., Jr. (1970). Detecting Tree Moisture Stress, *Photogram. Engin.* 36(6), 561–566.

Rohde, W. G. (1972). Multispectral Sensing of Forest Tree Species, *Photogram. Engin.* 38(12), 1209–1215.

Rohde, W. G., and Simonett, D. S. (1975). Land Use Studies with SKYLAB Data, *Amer. Astronautical Soc., Science and Tech. Ser.* 38, 127–185.

Roll, H. V. (1965). *Physics of the Marine Atmosphere.* Academic Press, London.

0-201-04245-2

Roscoe, J. H. (1960). Photo Interpretation in Geography, in *Manual of Photographic Interpretation* (Amer. Soc. of Photogrammetry, eds.). 735–792. Washington, D.C.

Rosenfeld, A. (1962). Automatic Recognition of Basic Terrain Types from Aerial Photographs, *Photogram. Engin.* 28(2), 115–132.

Rosenfeld, A. (1969). *Picture Processing by Computer.* Academic Press, New York.

Ross, D. S. (1968). Photographic and Spectral Factors Affecting Depth Analysis by Color Variation, in *Proc. of the Ninth Meeting of the Ad Hoc Spacecraft Oceanography Project.* January. Spacecraft Oceanography Project, NAVOCEANO, Maryland.

Ross, D. S. (1969a). Color Enhancement for Ocean Cartography, in *The Oceans from Space* (P. C. Badgley, L. Miloy, and L. Childs, eds.). 50–63. Gulf Pub. Co., Houston, Tex.

Ross, D. S. (1969b). Experiments in Oceanographic Aerospace Photography, I: Ben Franklin Spectral Filter Tests, Report for Naval Oceanography Project. Philco-Ford Corp., Space and Re-entry Systems Division, Palo Alto, Calif.

Roth, C. B., and Baumgardner, M. F. (1971). Correlation Studies with Ground Truth and Multispectral Data: Effect of Size of Training Field, in *Proc. of the Seventh Int'l Symp. on Remote Sens. of Environ.* 2, 1403–1414. Univ. of Mich., Ann Arbor.

Rouse, J. W., Jr., Haas, R. H., Schell, J. A., Deering, D. W., and Harlan, J. C. (1974). Monitoring the Vernal Advancement and Retrogradation (Green Wave) Effect of Natural Vegetation, Report No. RSC 1974-8. Remote Sensing Center, Texas A & M Univ., College Station.

Rouse, J. W., Jr., Waite, W. P., and Walters, R. L. (1966). Use of Orbital Radars for Geoscience Investigations, CRES Report No. 61-8. Univ. of Kansas, Lawrence.

Rowan, L. C., Offield, T. W., Watson, K., Cannon, P. J., and Watson, R. D. (1970). Thermal Infrared Investigations, Arbuckle Mountains, Oklahoma, *Geol. Soc. of Amer. Bull.* 82(12), 3549–3562.

Rowan, L. C., Wetlaufer, P. H., Billingsley, F. C., and Goetz, A. F. H. (1973). Mapping of Hydrothermal Alternation Zones and Regional Rock Types Using Computer Enhanced ERTS MSS Images, in *Third Earth Resources Tech. Satellite Symp.* NASA SP-351, 1, 807. Goddard Space Flight Center, Washington, D.C.

Rowan, L. C., Wetlaufer, P. H., Goetz, A. F. H., Billingsley, F., and Stewart, J. H. (1974). Discrimination of Rock Types and Detection of Hydrothermally Altered Areas in South-Central Nevada by the Use of Computer-Enhanced ERTS Images. U.S.G.S. Prof. Paper 883, 35.

Rowan, L. C. (1975). Application of Satellites to Geologic Exploration, *Amer. Scientist.* 63(4), 393–403.

Rowan, L. C., and Wetlaufer, P. H. (1975). Iron Absorption Band Analysis for the Discrimination of Iron-Rich Zones, NASA Type III Report. 160. Washington, D.C.

Royer, G. (1967). Two Years' Survey with Cesium Vapour Magnometer, *Geophys. Prospecting.* 15(2), 174–193.

Rudd, R. D., and Highsmith, R. M. (1970). The Use of Air Photo Mosaics as Simulators of Spacecraft Photography in Land Use Mapping, U.S.G.S. Contract 14-08-0001-12009, Geographic Applications Program, U.S. Geol. Survey, Washington, D.C.

Rydstrom, H. O. (1967). Interpreting Local Geology from Radar Imagery, *Geol. Soc. of Amer. Bull.* 78(3), 429–436.

Ryerson, R. A., and Wood, H. A. (1971). Air Photo Analysis of Beef and Dairy Farming, *Photogram. Engin.* 37(2), 157–169.

0-201-04245-2

Sabatini, R. R. (1975). Sea-Surface Wind Speed Estimates from the NIMBUS 5 ESMR, Final Report, Contract N66856-4120-5501. Earth Satellite Corporation, Washington, D.C.

Sabatini, R. R., Hlavka, D. L., and Arcese, R. (1975). Applications of the NIMBUS 5 ESMR to Rainfall Detection over the Oceans and to Sea-Ice Detection, Final Report, Contract No. N66314-73-c1572. Earth Satellite Corporation, Washington, D.C.

Sabins, F. F. (1967). Infrared Imagery and Geologic Aspects, *Photogram. Engin.* 33(7), 743–750.

Sabins, F. F. (1969). Thermal Infrared Imagery and Its Applications to Structural Mapping in Southern California, *Geol. Soc. of Amer. Bull.* 80(3), 397–404.

Sabins, F. F. (1973). Flight Planning and Navigation for Thermal IR Surveys. *Photogram. Engin.* 39(1), 49–58.

Sabins, F. F. (1973). Engineering geology applications of remote sensing in *Geology, Seismicity and Environmental Impact,* Association of Engineering Geologists Special Publication, p. 141–155.

Sabins, F. F. (1976). *Remote Sensing Handbook.* Chevron Oil Field Research Co. Research Report RR76000200, Part 1 of 2, 257 p.

Salas, F., Cabello, O., Alarcon, F., and Ferrer, C. (1973). ERTS-A Multispectral Image Analysis Contribution for the Geomorphological Evaluation of Southern Maracaibo Lake Basin, in *Third Earth Resources Tech. Satellite Symp.* NASA SP-351, 1, 943–954. Goddard Space Flight Center, Washington, D.C.

Sattinger, I. J., ed. (1966). Peaceful Uses of Earth-Observation Spacecraft, I. Univ. of Mich., Ann Arbor.

Saunders, P. M. (1967). Radiance of Sea and Sky in the Infrared Window 800–1200 $cm^{-1}$. *Jour. of Opt. Soc. of Amer.* 58(5), 645–652.

Saur, J. F. T. (1965). Oceanographic Observations from Manned Satellites for Fishery Research and Commercial Fishery Applications, in *Oceanography from Space* (G. C. Ewing, ed.). 313–314. Woods Hole Oceanographic Inst. Ref. No. 65-10, Massachusetts.

Saxton, J. A., and Lane, J. A. (1952). Electrical Properties of Sea Water, *Wireless Engineer.* 29, 269–275.

Sayn-Wittgenstein, L., and Aldred, A. H. (1972). Tree Size from Large-Scale Photos. *Photogram. Engin.* 38(10), 971–972.

Schaber, G. G. (1967). Radar Images—San Francisco Volcanic Field, Arizona—A Preliminary Evaluation, U.S.G.S. Tech. Letter NASA-84. Washington, D.C.

Schell, J. A. (1972). A Comparison of Two Approaches for Category Identification and Classification Analysis from an Agricultural Scene, 374–394, in *Remote Sensing of Earth Resources* (F. Shahrokhi, Ed.), Univ. of Tennessee, Space Institute, Tullahoma.

Schepis, E. L. (1968). Time-Lapse Remote Sensing in Agriculture, *Photogram. Engin.* 34(11), 1166–1179.

Scheps, B. (1963). The History of Radar Geology, *Proc. of the First Int'l Symp. on Remote Sens. of Environ.* 106–111. Univ. of Mich., Ann Arbor.

Schmer, F. A., Ryland, D. W., and Waltz, F. A. (1972). Investigation of Remote Sensing Techniques for Agricultural Feedlot Pollution Detection, in *Proc. of the Eighth Int'l. Symp. on Remote Sens. of Environ.* 603–616. Environmental Res. Inst. of Mich., Ann Arbor.

Schmidt, R. G. (1973). Use of ERTS-1 Images in the Search for Porphyry Copper Deposits in Pakistani Baluchistan, in *Symp. on Significant Results Obtained from the Earth Resources Tech. Satellite-1.* NASA SP-327, 1, 387–394. Goddard Space Flight Center, New Carrollton, Md.

Schmidt, R. G. (1975). Exploration for Porphry Copper Deposits in Pakistan Using Digital Processing of ERTS-1 Data, U.S.G.S. Report No. 75-78, 26–40.

0-201-04245-2

Schule, J. J., Jr., and Wilkerson, J. C. (1967). An Oceanographic Aircraft, U.S.N.O.O. Informal Report IR 66-26. U.S. Naval Oceanographic Office, Washington, D.C.

Schwarz, D. E., and Caspall, F. C. (1968). Use of Radar in the Discrimination and Identification of Agricultural Land Use, in *Proc. of the Fifth Int'l Symp. on Remote Sens. of Environ.* 233–248. Univ. of Mich., Ann Arbor.

Schwarz, D. E., Simonett, D. S., Jenks, G. F., and Ratzlaff, J. R. (1969). The Construction of Thematic Land Use Maps with Spacecraft Photography, unpublished report. Dept. of Geography and Center for Research in Engineering Science, Univ. of Kansas, Lawrence.

Scruton, P. C., and Moore, D. G. (1953). Distribution of Surface Turbidity off the Mississippi Delta, *Bull. of Amer. Soc. of Petroleum Geologists.* 37, 1067–1074.

Sharples, J. A. (1973). The Corn Blight Watch Experiment: Economic Implications for Use of Remote Sens for Collecting Data on Major Crops. Lab. for Applications of Remote Sensing, Information Note 110173. Purdue Univ., West Lafayette, Ind.

Shaw, J. H. (1970). Determination of the Earth's Surface Temperature from Remote Spectral Radiance Observations near 2600 $cm^{-1}$, *Jour. of Atmos. Sciences* 27(6), 950–959.

Sherman, J. W., III (1969). Passive Microwave Sensors for Satellites, in *Proc. of the Sixth Int'l Symp. on Remote Sensing of Environ.* 651–669. Univ. of Mich., Ann Arbor.

Sherr, P. E., Glaser, A. H., Barnes, J. C., and Willard, J. H. (1968). *Worldwide Cloud Cover Distribution for Use in Computer Simulations* [NASA-CR-61226]. Allied Research Associates, Virginia Road, Concord, Mass.

Sherz, J. P. (1971). Monitoring Water Pollution by Remote Sensing, *J. Surv. and Mapp. Div., A.S.C.E.*, 97, 309–320.

Shifrin, K. S., and Shubova, G. L. (1964). *Izv. Geophys. Ser.*, (2) 279.

Shilin, B. V., Gusyev, I. A., Miroshnikov, M. M., and Korzhinski, Ye. Ya. (1971). The Investigation of Natural Resources by Infrared Remote Sensing Methods, in *Proc. of the Seventh Int'l. Symp. on Remote Sens. of Environ.* 133–146. Univ. of Mich., Ann Arbor.

Short, N. M. (1973). Mineral Resources, Geologic Structure, and Landform Surveys, in *Third Earth Resources Tech. Satellite Symp.* NASA SP-351, 3, 33–51. Goddard Space Flight Center, Washington, D.C.

Short, N. M. (1974). Exploration for Fossil and Nuclear Fuels from Orbital Altitudes, Tech. Report X-923-74-322. Goddard Space Flight Center, Greenbelt, Md.

Silver, S. (1949). *Microwave Antenna Theory and Design.* McGraw-Hill, New York.

Simonett, D. S., and Brown, D. A. (1965). Possible Uses of Radar on Spacecraft in Contributing to Antarctic Mapping, Crevasse, Sea Ice, and Mass Budget Studies. CRES Tech. Report 61-4, Center for Research in Engineering Science. 1–18. Univ. of Kansas, Lawrence.

Simonett, D. S., Eagleman, J. E., Erhart, A. B., Rhodes, D. C., and Schwarz, D. E. (1967). The Potential of Radar as a Remote Sensor in Agriculture: A Study with K-Band Imagery in Western Kansas, CRES Report 61-21. Univ. of Kansas, Lawrence.

Simonett, D. S., and Morain, S. A. (1967). Remote Sensing from Spacecraft as a Tool for Investigating Arctic Environment, in *Arctic and Alpine Environments* (H. E. Wright, Jr., and W. H. Osborne, eds.). 295–306. Indiana Univ. Press, Bloomington.

0-201-04245-2

Simonett, D. S. (1968a). Land Evaluation Studies with Remote Sensors in the Infrared and Radar Regions, in *Land Evaluation* (G. A. Stewart, ed.). 349–366. Macmillan, Melbourne, Australia.

Simonett, D. S. (1968b). *Earth Resources Aircraft Program-Status Review 1.* 8-3. Manned Spacecraft Center, Houston, Tex.

Simonett, D. S. (1969a). Thematic Land Use Mapping: Some Potentials and Problems, in *Earth Resources Aircraft Program–Status Review 2.* 1–28. Manned Spacecraft Center, Houston, Tex.

Simonett, D. S. (1969b). Remote Sensing Studies in Geography: A Review, unpublished report. Dept. of Geography and Center for Research in Engineering Science, Univ. of Kansas, Lawrence.

Simonett, D. S., ed. (1969c). The Utility of Radar and Other Remote Sensors in Thematic Land Use Mapping from Spacecraft, Second Annual Report, U.S.G.S. Contract 14-08-0001-10848. Univ. of Kansas, Lawrence.

Simonett, D. S., Cochrane, G. R., Morain, S. A., and Egbert, D. E. (1969a). Environmental Mapping with Spacecraft Photography: A Central Australian Example. U.S. Geological Survey Interagency Report. NASA-172. Washington, D.C.

Simonett, D. S., Henderson, F. M., and Egbert, D. D. (1969b). On the Use of Space Photography for Identifying Transportation Routes: A Summary of Problems, in *Proc. of the Sixth Int'l Symp. on Remote Sens. of Environ.* 2, 855–893. Univ. of Mich., Ann Arbor.

Simonett, D. S., Eagleman, J. R., Marshall, J. R., and Morain, S. A. (1969c). The Complementary Roles of Aerial Photography and Radar Imaging Related to Weather Conditions, U.S. Geol. Survey, Interagency Rept., NASA-172, 1–47.

Simonett, D. S., ed. (1970). The Utility of Space Photography, High Altitude Aerial Photography, Radar Imagery and Other Remote Sensor Imagery in Thematic Land Use Mapping, Final Report, U.S.G.S. Contract 14-08-0001-12077. Univ. of Kansas, Lawrence.

Simonett, D. S., Schuman, R. L., and Williams, D. L. (1970). The Use of Air Photos in a Study of Landslides in New Guinea, Tech. Rept. 5, ONR Contract 538(11), Task 389-133, Dept. of Geography, Univ. of Kansas, Lawrence.

Simonett, D. S., and Coiner, J. C. (1971). Susceptibility of Environments to Low-Resolution Imaging for Land Use Mapping, *Proc. of the Sixth Int'l Symp. on Remote Sens. of Environ.* 373–394. Univ. of Mich., Ann Arbor.

Simonett, D. S., Shotwell, R., and Belknap, N. (1976). Applications of SKYLAB EREP Data for Land Use Management, Final Report, NASA Contract No. NAS9-13314, Earth Satellite Corp., Wash., D.C.

Simons, J. H. (1964). Some Applications of Side-Looking Airborne Radar, in *Proc. of the Third Int'l Symp. on Remote Sens. of Environ.* 563–571. Univ. of Mich., Ann Arbor.

Simpson, R. V. (1969). APQ-97 Imagery of New England, a Geographic Evaluation, in *Proc. of the Sixth Int'l Symp. on Remote Sens. of Environ.* 909–925. Univ. of Mich., Ann Arbor.

Skolnik, M. I. (1962). *Introduction to Radar Systems.* McGraw-Hill, New York.

Skolnik, M. I., ed. (1970). *Radar Handbook.* McGraw-Hill, New York.

Slack, H. A., Lynch, V. M., and Langan, L. (1967). The Geomagnetic Gradiometer, *Geophysics.* 32(5), 877–892.

Smedes, H. W. (1971). Automatic Computer Mapping of Terrain, in *Proc. of the Int'l Workshop on Earth Resources Survey Systems.* 2, 344–406. Univ. of Mich., Ann Arbor.


0-201-04245-2

Smedes, H. W., Linnerud, H. J., Woolaver, L. B., and Hawks, S. J. (1971). Digital Computer Mapping of Terrain by Clustering Techniques Using Color Film as a Three-Band Sensor, in *Proc. of the Seventh Int'l Symp. on Remote Sens. of Environ.* 3, 2073–2094. Univ. of Mich., Ann Arbor.

Smith, J. A., and Oliver, R. E. (1972). Plant Canopy Models for Simulating Composite Scene Spectroradiance in the 0.4 to 1.05 Micrometer Region, in *Proc. of the Eighth Int'l Symp. on Remote Sens. of Environ.* 1333–1339. Environmental Research Inst. of Mich., Ann Arbor.

Smith, J. H. G., and Chiam, Y. C. (1970). Studies of Tree Shocking and Use of Crown-Width Ratios to Refine Estimate of Stand Density, *Photogram. Engin.* 36(10), 1094–1095.

Smith, J. T. (1968). *Manual of Color Aerial Photography,* First Ed. Amer. Soc. of Photogrammetry.

Smith, W. L., Rao, P. K., Koffler, R., and Curtis, W. R. (1970). The Determination of Sea-Surface Temperature from Satellite High Resolution Infrared Window Radiation Measurements, *Monthly Weather Review.* 98(8), 604–611.

Snavely, P. D., and Wagner, H. C. (1966). Geological Evaluation of Radar Imagery, Oregon Coast. U.S.G.S. Tech. Letter NASA-16.

Sobolev, V. V. (1963). *A Treatise on Radiative Transfer.* D. van Nostrand, Princeton, N.J.

Soules, S. D. (1970). Sun Glitter Viewed from Space, *Deep-Sea Res.* 17(1), 191–195.

Squire, J. L., Jr. (1961). Aerial Fish Spotting in the United States Fisheries, *Commercial Fisher. Review.* 23(12), 1–7.

Squire, J. L., Jr. (1965). Airborne Oceanographic Programs of the Tiburon Marine Laboratory and Some Observations on Future Development and Uses of this Technique, in *Oceanography from Space* (G. C. Ewing, ed.). 119–123. Woods Hole Oceanographic Inst. Ref. No. 65-10, Massachusetts.

Staelin, D. H. (1966). Measurement and Interpretation of the Microwave Spectrum of the Terrestrial Atmosphere near 1 cm Wavelength, *Jour. of Geophys. Res.* 71(2), 2875–2881.

Staelin, D. H. (1969). Passive Remote Sensing at Microwave Wavelengths, *Proc. of ICCC.* 57(4), 427–439.

Staelin, D. H. (date unknown). *Microwave Spectral Measurements Applicable to Oceanography.* Mass. Inst. of Tech., Cambridge.

Stamp, L. D. (1935). *The Land Utilization Survey of Britain.* Univ. of London.

Steiner, D. (1969). In *Automatic Interpretation and Classification of Images* (A. Grasselli, ed.). 235–241. Academic Press, New York.

Steiner, D., and Maurer, H. (1969). The Use of Stereo-height as a Discriminating Variable for Crop Classification on Aerial Photographs, *Photogrammetria.* 24(5), 223–241.

Steiner, D. (1970). Time Dimension for Crop Survey from Space, *Photogram. Engin.* 36(2), 187–194.

Steiner, D., ed. (1972). Automatic Processing and Classification of Remote Sensing Data, in *Geographical Data Handling,* Vol. 3 (R. F. Tomlinson, ed.). Second Symp. on Geographical Information Systems, Int'l. Geographical Union, Ottawa, Canada.

Stelzried, C. T. (1968). Microwave Thermal Noise Standards. *IEEE Trans.* on *Microwave Theory and Tech.* MMT-16(9), 646–655.

Sterling, T. D., and Pollock, S. V. (1968). *Introduction to Statistical Data Processing.* Prentice-Hall, Englewood Cliffs, N.J.

Stevenson, R. E., and Nelson, R. M. (1968). An Index of Ocean Features Photographed by Gemini Spacecraft, Contribution No. 253. Bureau of Commercial Fisheries Biological Lab., Galveston, Tex.

0-201-04245-2

Stevenson, R. E. (1969a). New Tool for Studying Ocean Weather, *Ocean Industry* 4(3), 51–55.

Stevenson, R. E. (1969b). The 200-Mile Fishline, in *The Oceans from Space* (P. C. Badgley, L. Miloy, and L. Childs, eds.). 29–37. Gulf Pub. Co., Houston, Tex.

Stevenson, R. E. (1969c). Space Photos Detect Changing Currents along the Texas Coast, *Ocean Industry.* 4(10), 38–39.

Stevenson, R. E., and Terry, R. D. (1969). Coastal Currents Around Biak Island as Seen from Apollo 7, *Australian Fisher.* 28(7).

Stingelin, R. W. (1969). Operational Airborne Thermal Imaging Systems, *Geophysics.* 34(5), 760–771.

Stoiber, R. E., and Rose, W. I., Jr. (1975). An Investigation of the Thermal Anomalies in the Central American Volcanic Chain and Evaluation of the Utility of Thermal Anomaly Monitoring in the Prediction of Volcanic Eruptions, Final Report, NASA Contract NAS-9-13311.

Stommel, H. M., Von Arx, W. S., Richardson, W. S., and Parson, D. (1953). Rapid Aerial Survey of Gulf Stream with Camera and Radiation Thermometer. *Science.* 117(3049), 639–640.

Stone, K. (1968). Multiple-Scale Classification for Rural Settlement Geography, *Acta Geographical.* 20(22), 307–328.

Stoner, E. R., Baumgardner, M. F., and Swain, P. H. (1972a). Determining Density of Maize Canopy, I: Digitized Photography. Lab. for Applications of Remote Sensing, Print 111172. Purdue Univ., West Lafayette, Ind.

Stoner, E. R., Baumgardner, M. F., and Cipra, J. E. (1972b). Determining Density of Maize Canopy, II: Airborne Multispectral Scanner Data. Lab. for Applications of Remote Sensing, Print 111272. Purdue Univ., West Lafayette, Ind.

Stoner, E. R., Baumgardner, M. F., and Cipra, J. E. (1972c). Density of Maize Canopy, III: Temporal Considerations. Lab. for Applications of Remote Sensing, Print 111372. Purdue Univ., West Lafayette, Ind.

Strangway, D. W., and Holmer, R. C. (1964). Infrared Geology, in *Proc. of the Third Int'l Symp. on Remote Sens. of Environ.* 293–319. Univ. of Mich., Ann Arbor.

Stratton, J. (1941). *Electromagnetic Theory.* McGraw-Hill, New York.

Strogryn, A. (1967). The Apparent Temperature of the Sea at Microwave Frequencies, *IEEE Trans. Antennas and Propagation.* AP-15, 278–286.

Strong, A. C., and McLain, E. P. (1969). Sea-State Measurements from Satellites, *Mariners Weather Log.* 13(5), 205.

Studt, F. E. (1968). Geophysical Reconnaissance at Kawerau, New Zealand. *Jour. of Geol. and Geophys.* 1(2), 219–246.

Su, M. Y. (1972). An Unsupervised Classification Technique for Multispectral Remote Sensing Data, in *Int'l Symp. on Remote Sens. of Environ.* 2, 861–879.

Su, M. Y., Jayroe, R. R., and Cummings, R. E. (1972). Unsupervised Classification of Earth Resources Data, in *Remnte Sensing of Earth Resources* (F. Shahrokhi, ed.). 1, 673–694. Univ. of Tenn., Tullahoma.

Suits, G. H. (1972). The Calculation of the Directional Reflectance of a Vegetative Canopy, *Remote Sens. of Environ.* 2(2), 117–125.

Sullivan, A. E., and Brooner, W. G. (1971). Remote Sensing of Chaparral Fire Potential: Case Study in Topanga Canyon, California, Tech. Report T-71-1, U.S.D.I. Contract N00014-69-A-200-5001. Dept. of Geography, Univ. of Calif., Riverside.

0-201-04245-2

Sutherland, D. B. (1969). AFMAG for Electrical Mapping, in *Mining and Groundwater Geophysics.* 1967. 228–237. Geol. Survey of Can Econ. Geol. Report No. 26. Ottawa.
Sverdrup, H. U., Johnson, M. W., and Fleming, R. H. (1942). *The Oceans.* Prentice-Hall, Englewood Cliffs, N.J.
Swain, P. H. (1972). Pattern Recognition: A Basis for Remote Sensing Data Analysis. Lab. for Applications of Remote Sensing, Information Note 111572. Purdue Univ., West Lafayette, Ind.

Tarkington, R. G., and Sorem, A. L. (1963). Color and False-Color Films for Aerial Photography, *Photogram. Engin.* 29(1), 88–95.
Tatarski, V. I. (1961). *Wave Propagation in a Turbulent Medium.* McGraw-Hill, New York.
Taylor, J. I., and Stingelin, R. W. (1969). Infrared Imaging for Water Resource Studies, *Jour. Hydraulics* Div. 95, 175–189.
Technology Application Center, Institute for Social Research and Development (1974). Quarterly Literature Review of the Remote Sensing of Natural Resources, Univ. of New Mexico, Albuquerque.
Texas A & M University, Dept. of Oceanography (1966). Space Oceanography Status Report. February 1966–October 1966. College Station.
Texas A & M University, Dept. of Oceanography (1969). Oceanography Using Remote Sensors, Final Report. January 15, 1969. Report Ref. No. 69-2F. College Station.
Thibault, D., Simonett, D. S., and Garofalo, D. L. (1972). Land Use Indicators of Environmental Quality, Tech. Report, Earth Satellite Corporation, for the Council on Environmental Quality, Washington, D.C.
Thomann, G. C. (1970). Wide Bandwidth Signals in Radar. Ph.D. dissertation, Univ. of Kansas, Lawrence.
Thomson, F. J., Erickson, J. P., Nalepka, R. F., and Webber, J. D. (1974). *Multispectral Scanner Data Application Evaluation,* Tech. Report 10200-40-F. Environmental Research Inst. of Mich., Ann Arbor.
Thompson, W. I., III, and Haroules, G. C. (1969). A Review of Radiometric Measurements of Atmosphere Attenuation at Wave Lengths from 75 Centimeters to 2 Millimeters. NASA, Washington, D.C.
Thompson, W. I., III (1971). *Atmospheric Transmission Handbook* [NASA-CR-117173]. U.S. Department of Transportation, Transportation Systems Center, Cambridge, Mass.
Thrower, N. J. W. and Senger, L. W. (1969). Land Use Mapping of Southwestern United States from Satellite Images, AAS 69-579. American Astronautics Society Nat'l Meeting, New Mexico State University, Las Cruces.
Thrower, N. J. W. (1970). Land Use in the Southwestern United States–From Gemini and Apollo Imagery, *Annals of the Assoc. of Amer. Geographers* 60(1), 208–209, and map supplement.
Tobler, W. (1968). Satellite Confirmation of Settlement Size Coefficient, NASA Interagency Report, NASA-163. Washington, D.C.
Tomlinson, R. D. (ed.) (1972). *Geographical Data Handling,* Symp. Edition. Int'l. Geographical Union Commission, Ottawa, Canada.
Tompson, F. J., et al. (1974). Multispectral Scanner Data Application Evaluation, I. NAS 9-13386-CCA2. NASA, L.B.J. Space Center, Houston, Tex.
Toong, H. D., and Staelin, D. H. (1970). Passive Microwave Spectrum Measurements of Atmospheric Water Vapor and Clouds, *Jour. of Atmos. Sciences.* 781–784.

0-201-04245-2

Townsend, M. R. (1969). Interrogation and Position-Fixing from Spacecraft, in *The Oceans from Space* (P. C. Badgley, L. Miloy, and L. Childs, eds.). 64–81. Gulf Publ. Co., Houston, Tex.

Toy, H. D. (1966). Anthology on NASA 926 and NASA 927 Aircraft as Applied to the Earth Resources Survey Program. NASA, Manned Spacecraft Center, Houston, Tex.

Unusual pagination arises from insertion of some recent key references.

0-201-04245-2

Trembath, C. L., et al. (1968). A Low-Temperature Microwave Noise Standard, *IEEE Trans. Microwave Theory and Tech.* MTT-16(9), 709–714.

Trollinger, W. V. (1968). Surface Evidence of Deep Structure in the Delaware Basin, in *Delaware Basin Exploration,* West Texas Geol. Soc. Guidebook No. 68-55. 87–104.

TRW Systems Group/EARTHSAT (1973). Mission Requirements for a Manned Earth Observatory, Contract NAS8-28013, One Space Park, Redondo Beach, California.

Tryon, R. C., and Bailey, D. E. (1970). *Cluster Analysis.* McGraw-Hill, New York.

Turner, B. J. (1972). Cluster Analysis of Multispectral Scanner Remote Sensor Data, *Remote Sensing of Earth Resources* (F. Shahrokhi, ed.). 1, 538–549, Univ. of Tenn., Tullahoma.

Turner, R. E., and Spencer, M. M. (1972). Atmospheric Model for Correction of Spacecraft Data, in *Int'l. Symp. on Remote Sens. of the Environ.* 2, 895–934.

Twomey, S., Jacobwitz, H., and Howell, H. B. (1966). *Jour. of Atmos. Sciences.* 23, 289.

Tyler, J. E. (1964). *In Situ* Detection and Estimation of Chlorophyll and Other Pigments–Preliminary Results, *Botany,* 51, 671-678.

Ulaby, F. T., and Straiton, A. W. (1969). Atmospheric Attenuation Studies in the 183–325 GHz region, *IEEE Trans. Antennas and Propagation.* AP-17(3), 337–342.

Ulaby, F. T., Moore, R. K., Moe, R., and Holtzman, J. (1972). On Microwave Remote Sensing of Vegetation, in *Proc. of the Eighth Int'l. Symp. on Remote Sens. of Environ.* 1279–1285. Environmental Research Inst. of Mich., Ann Arbor.

Ulliman, J. L., Latham, R. P., and Meyer, M. P. (1970). 70 mm. Quadricamera System, *Photogram. Engin.* 36(1), 49–54.

Ulliman, J. L., and Meyer, M. P. (1971). The Feasibility of Forest Cover Type Interpretation Using Small Scale Aerial Photographs, in *Proc. of the Seventh Int'l Symp. on Remote Sens. of Environ.* 1219–1230. Univ. of Mich., Ann Arbor.

Underwood, R. W. (1968). Color Photography from Space, in *Manual of Color Aerial Photography* (J. T. Smith, ed.). 365–379. Amer. Soc. of Photogrammetry, Falls Church, Va.

U.S. Naval Oceanographic Office (1966). RDT & E at the United States Naval Oceanographic Office–1960–1966. Government Printing Office, Washington, D.C.

U.S. Navy Weather Research Facility (NWRF) 33-0667-125. June 1967. Navy Weather Research Facility, Norfolk, Va.

University of Michigan (1962). *Proc. of the Second Int'l Symp. on Remote Sens. of Environ.* Ann Arbor.

University of Michigan (1964). *Proc. of the Third Int'l Symp. on Remote Sens. of Environ.* Ann Arbor.

University of Michigan (1966). *Proc. of the Fourth Int'l Symp. on Remote Sens. of Environ.* Ann Arbor.

Valley, S. L., ed. (1965). *Handbook of Geophysics and Space Environments.* Air Force Cambridge Research Lab., Bedford, Mass.

Van de Hulst, H. C. (1957). *Light Scattering by Small Particles.* John Wiley, New York.

Van der Meer Mohr, H. E. C. (1969). Geological Interpretation of Hyperaltitude Photographs from Gemini Spacecraft, *Photogrammetria.* 24(3/4), 167–174.


0-201-04245-2

Van Lopik, J. (1968). Infrared Mapping–Basic Technology and Geoscience Applications, *Geoscience News.* January-February, 407, 24–31.

Van Lopik, J., Pressman, A. E., and Ludlum, R. L. (1968). Mapping Pollution with Infrared, *Photogram. Engin.* 34(1), 561-564.

Van Vleck, J. H. (1947). The Absorption of Microwaves by Oxygen, *Phy. Review,* 71(7), 413–424.

Viglione, S. S. (1970). Applications of Pattern Recognition Technology, in *Adaptive, Learning, and Pattern Recognition Systems.* Chapter 4. Academic Press, New York.

Viksne, A., Liston, T. C., and Sapp, C. D. (1969). SLR Reconnaissance of Panama, *Geophysics.* 34(1), 54–64.

Viljoen, R. P. (1973). ERTS-1 Imagery as an Aid to the Understanding of the Regional Setting of Base Metal Deposits in the Northwest Cape Province, South Africa, in *Third Earth Resources Tech. Satellite Symp.* NASA SP-351, 1, 797–806. Goddard Space Flight Center, Washington, D.C.

Vincent, R. K. (1973). Ratio Maps of Iron Ore Deposits, Atlantic City District, Wyoming, in *Symp. on Significant Results Obtained from Earth Resources Tech. Satellite-1.* NASA SP-327, 1, 379–386. Goddard Space Flight Center, New Carrollton, Md.

Von Arx, W. S., and Richardson, W. S. (1953). *The Surface Outcrop of the Gulf Stream Front.* Woods Hole Oceanographic Inst. Ref. 53-24, Massachusetts.

Von Arx, W. S. (1966). Level-Surface Profiles across the Puerto Rican Trench, *Science.* 154(3757), 1651-1653.

Von Steen, D. H., and Wigton, W. H. (1973). Crop Identification and Acreage Measurement Utilizing ERTS Imagery, in *Third Earth Resources Technology Satellite Symp.* NASA SP-351, 1, 87–92. Goddard Space Flight Center, Washington, D.C.

Wacker, A. G., and Landgrebe, D. A. (1970). Boundaries in Multispectral Imagery by Clustering, in *Proc. of the 1970 (Ninth) IEEE Symp. on Adaptive Processes: Decision and Control.* X14.1–X14.8, New York.

Wacker, A. G., and Landgrebe, D. A. (1972). Minimum Distance Classification in Remote Sensing, in *First Canadian Symp. for Remote Sens.*, Ottawa, Canada.

Waelti, H. (1970). Forest Road Planning, *Photogram. Engin.* 36(3), 246–252.

Waite, W. P. (1970). Polypanchromatic Radar. Ph.D. dissertation, Univ. of Kansas, Lawrence.

Walker, T. J. (1965). Detection of Marine Organisms by an Infrared Mapper, in *Oceanography from Space* (G. C. Ewing, ed.). 321–325. Woods Hole Oceanographic Inst. Ref. 65-10, Massachusetts.

Wallace, R. E., and Moxham, R. M. (1967). Use of Infrared Imagery of the San Andreas Fault System, California. U.S.G.S. Prof. Paper 575D. D147–D156.

Walsh, D. (1969a). Characteristic Patterns of River Outflow in the Mississippi Delta, Ref. No. 69-8T. Dept. of Oceanography, Texas A & M Univ., College Station.

Walsh, D. (1969b). The Mississippi River Delta, in *The Oceans from Space* (P. C. Badgley, L. Miloy, and L. Childs, eds.). 171–187. Gulf Pub. Co., Houston, Tex.

Ward, P. L., Endo, E. T., Harlow, D. H., Allen, R., and Eaton, J. P. (1973). A New Method for Monitoring Global Volcanic Activity, in *Third Earth Resources Tech. Satellite Symp.* NASA SP-351, 1, 681–690. Goddard Space Flight Center, Washington, D.C.

Ward, S. H., et al. (1958). Prospecting by Use of Natural Alternating Magnetic Fields of Audio and Sub-Audio Frequencies, *Trans. Can. Inst. Mining and Metallurgy.* 61, 261–268.

Ward, S. H. (1959). AFMAG–Airborne and Ground, *Geophysics.* 24(4), 761–789.

0-201-04245-2

Ward, S. H. (1967). Electromagnetic Theory for Geophysical Applications, in *Mining Geophysics.* 2, 10–196. Soc. of Exploration Geophysics, Tulsa, Okla.
Ward, S. H., O'Brien, D. P., Parry, J. R., and McKnight, B. K. (1968). AFMAG Interpretation, *Geophysics.* 33(4), 621–644.
Ward, S. H. (1969). Airborne Electromagnetic Methods, in *Mining and Groundwater Geophysics.* Geol. Survey of Can. Econ., Geol. Report 26, 81-108. Ottawa.
Warecke, G., McMillin, L. M., and Allison, L. J. (1969). Ocean Current and Sea Surface Temperature Observations from Meteorological Satellites, NASA Tech. Note TN D-5142. NASA, Washington, D.C.
Waters, J. W., and Staelin, D. H. (1968). Statistical Inversion of Radiometric Data, Quart. Progress Report 89. Research Lab. of Electronics, Mass. Inst. of Tech., Cambridge.
Watson, R. D. and Rowan, L. C. (1971). Automated Geologic Mapping Using Rock Reflectances, *Proc. 7th Int'l Symp. Remote Sens. Environ.* 2043–2054. Univ. of Mich., Ann Arbor.
Weber, F. P., and Olson, C. E. (1967). Remote Sensing Implications of Changes in Physiologic Structure and Function of Tree Seedling under Moisture Stress, Annual Progress Report. Forestry Remote Sensing Lab., School of Forestry and Conservation, Univ. of Calif., Berkeley.
Weber, F. P., and Wear, J. F. (1970). The Development of Spectro-Signature Indicators of Root Disease Impacts on Forest Stands, Tech. Report, NASA Contract R-09-038-002. Forestry Remote Sensing Lab., School of Forestry and Conservation, Univ. of Calif., Berkeley.
Weber, F. P., and Polcyn, F. C. (1972). Remote Sensing to Detect Stress in Forest. *Photogram. Engin.* 38(2), 163–175.
Weiner, M. M. (1967). Atmospheric Turbulence in Optical Surveillance Systems, *Appl. Opt.* 6(3), 1984–1991.
Weiss, O. (1967). Method of Aerial Prospecting Which Includes a Step of Analyzing Each Sample for Element Content, Number, and Size of Particles, U.S. Patent 3,309,518.
Weiss, O. (1969). Aerial Prospecting, U.S. Patent 3,462,995.
Weiss, O. (1970). Airborne Geochemical Prospecting, *CIM.* Special Vol. 2.
Welch, R. (1970). Modulation Transfer Functions, *Photogram. Engin.* 37(3), 247–259.
Wellar, B. S. (1968). *Proc. of the Fifth Int'l Symp. on Remote Sens. of Environ.* 913–926. Univ. of Mich., Ann Arbor.
Wellar, B. S. (1969). The Role of Space Photography in Urban and Transportation Data Series, in *Proc. of the Sixth Int'l Symp. on Remote Sens. of Environ.* 831–854. Univ. of Mich., Ann Arbor.
Wert, S. L. (1969). A System for Using Remote Sensing Techniques to Detect and Evaluate Air Pollution Effects on Forest Stands, *Proc. of the Sixth Int'l. Symp. on Remote Sens. of Environ.* 1169–1178. Univ. of Mich., Ann Arbor.
White, L. P. (1974). Natural-Resources Mapping of Ethiopia from ERTS-1 Imagery, in *Proc. of a Symp. on European Earth Resource Satellite Experiment.* ESRO SP-100, 179–184. Frascati, Italy.
White, P. G. (1968). Ocean Color Measurement, in *Proc. of the Ninth Meeting of the Ad Hoc Spacecraft Oceanography Advisory Group.* January. Spacecraft Oceanography Project, NAVOCEANO, Maryland.
Wiegand, C. L., et al. (1969). Spectral Survey of Irrigated Region Crops and Soils, Annual Report. Weslaco, Tex., Agricultural Experiment Station.

0-201-04245-2

Wiegand, C. L., Gausman, H. W., Cuellar, J. A., Gerbermann, A. H., and Richardson, A. J. (1973). Vegetation Density as Deduced from ERTS-1 MSS Response, in *Third Earth Resources Tech. Satellite Symp.* NASA SP-351, 1, 93–116. Goddard Space Flight Center, Washington, D.C.

Wier, C. E., Wobber, F. J., Russell, O. R., Amato, R. V., and Leshendok, T. V. (1973). Relationship of Roof Falls in Underground Coal Mines to Fractures Mapped in ERTS-1 Imagery, in *Third Earth Resources Tech. Satellite Symp.* NASA SP-351, 1, 825–844. Goddard Space Flight Center, Washington, D.C.

Wilcox, S. W. (1944). Sand and Gravel Prospecting by the Earth Resistivity Method, *Geophysics.* 9(11), 36–46.

Wilkerson, J. C. (1966). Airborne Oceanography, *Geo-Marine Tech.* 2(8), 9–15.

Wilkerson, J. C. (1967a). The Gulf Stream from Space. *Oceanus.* 13(2,3), 2–8.

Wilkerson, J. C. (1967b). Meanders in the North Wall of the Gulf Stream, *Mariners Weather Log.* 11, 201–203.

Wilkerson, J. C., and Noble, V. E. (1969). Potential Impact of Satellite Data on Surface Temperature Analysis, U.S.N.O.O. Informal Report IR 69-64. U.S. Naval Oceanographic Office, Washington, D.C.

Williams, D. T., and Kolitz, B. L. (1968). Molecular Correlation Spectrometry, *Appl. Opt.* 7(1), 607–616.

Williams, G. F. (1968). Microwave Radiometry of the Ocean, in *Proc. of the Ad Hoc Spacecraft Oceanography Advisory Group.* January. Spacecraft Oceanography Project, NAVOCEANO, Maryland.

Williams, G. F. (1969). Microwave Radiometry of the Ocean and the Possibility of Marine Wind Velocity Determination from Satellite Observations, *Jour. of Geophys. Res.* 74(18), 4591–4594.

Williams, R. S., and Ory, T. R. (1967). Infrared Imagery Mosaics for Geological Investigations, *Photogram. Engin.* 33(12), 1377–1380.

Williams, R. S., Jr., Bovarsson, A., Fridriksson, S., Palmason, G., Rist, S., Sigtrysson, H., Thorarinsson, S., and Thorssteinsson, I. (1973). Satellite Geological and Geophysical Remote Sensing of Iceland: Preliminary Results from Analysis of MSS Imagery, in *Symp. on Significant Results Obtained from the Earth Resources Tech. Satellite-1.* NASA SP-327, 1, 317–328. Goddard Space Flight Center, New Carrollton, Md.

Williston, S. H. (1964). The Mercury Halo Method of Exploration, *Engin. Mining Jour.* 165, May. 98–101.

Williston, S. H. (1968). Mercury in the Atmosphere, *Jour. of Geophys. Res.* 73, 7051–7055.

Wing, R. S., and Dellwig, L. F. (1970). Radar Expression of Virginia Dale Precambrian Ring-Dike Complex, Wyoming/Colorado, *Geol. Soc. of Amer. Bull.* 81, 293–298.

Wing, R. S. (1971). Structural Analysis from Radar Imagery of Eastern Panama Isthmus. *Modern Geol.* 2(1), 1–21.

Wing, R. S. (in press). Structural Analysis from Radar, Eastern Panama, *Jour. of Modern Geol.*

Wise, D. U. (1967). Radar Geology and Pseudo-Geology on an Appalachian Piedmont Cross-Section, *Photogram. Engin.* 33(7), 752–761.

Wobber, F. J. (1969). Environmental Studies Using Earth Orbital Photography, *Photogrammetria.* 24(3/4), 107–165.

Wobber, F. J. (1970). Orbital Photography Applied to the Environment, *Photogram. Engin.* 36(8), 852–864.

0-201-04245-2

Wobber, F. J., and Martin, K. R. (1973). Exploitation of ERTS-1 Imagery Utilizing Snow Enhancement Techniques, in *Symp. on Significant Results Obtained from the Earth Resources Tech. Satellite-1.* NASA SP-327, 1, 345–352. Goddard Space Flight Center, New Carrollton, Md.

Wolffe, P. M. (1965). Operational Analyses and Forecasting of Ocean Temperature Structure, in *Oceanography from Space* (G. C. Ewing, ed.), 125–147. Woods Hole Oceanographic Inst. Ref. No. 65-10, Massachusetts.

Wolffe, P. M., Laevastu, T., and Tatro, P. (1969). Synoptic Analyses and Prediction of Conditions and Processes in the Surface Layers of the Sea, in *The Oceans from Space* (P. C. Badgley, L. Miloy, and L. Childs, eds.). 131–158. Gulf Pub. Co., Houston, Tex.

Wolfe, E. W. (1971). Thermal IR for Geology, *Photogram. Engin.* 37, 43–52.

Wolfe, W. L., ed. (1965). *Handbook of Military Infrared Technology.* Office of Naval Research, Washington, D.C.

Wolyce, U., and Llunga, S. (1973). Geologic Hypotheses on Lake Tanganyika Region, Zaire, Drawn from ERTS-Imagery, in *Third Earth Resources Tech. Satellite Symp.* NASA SP-351, 1, 955–968. Goddard Space Flight Center, Washington, D.C.

Woods Hole Oceanographic Institution (WHOI) (1965). *Oceanography from Space* (G. C. Ewing, ed.). Ref. No. 65-10. Woods Hole, Massachusetts.

Woolnough, D. F. (1972). Automatic Recognition of Muskeg from Aerial Photographs, *Photogrammetria.* 28(1), 17–25.

Wright, J. W. (1966). Backscattering from Capillary Waves with Application to Sea Clutter, *IEEE Trans. Antennas and Propagation.* AP-14, 749–754.

Wright, J. W. (1968). A New Model for Sea Clutter, *IEEE Trans. Antennas and Propagation.* AP-16, 217–223.

Wukelic, G. E., and Frazier, N. A. (1969). Selected Space Goals and Objectives and Their Relations to National Goals, NASA Contract NASA-1146. Battelle Memorial Inst.

Wulfsberg, K. N. (1964a). Sky Noise Measurements at Millimeter Wavelenths, *Proc. of IEEE.* 52(1), 321–322.

Wulfsberg, K. N. (1964b). Sky Noise Measurements at Millimeter Wavelengths, *Proc. of IEEE.* 52(3), 321–322.

Wyckoff, R. D. (1948). The Gulf Airborne Magnetometer, *Geophysics.* 13, 182–208.

Yamamoto, G., Tanaka, M., and Kamitani, K. (1966). Radiative Transfer in Clouds in the 10-micron Window Regions, *Jour. of Atmos. Sciences.* 23(3), 305–313.

Yamamoto, G., Tanaka, M., and Asano, S. (1970). Radiative Transfer in Water Clouds in the Infrared Region, *Jour. of Atmos. Sciences.* 27(2), 282–292.

Yates, H. W., and Taylor, J. H. (1960). Infrared Transmission of the Atmosphere. U.S.N.R.L. Report 5453, U.S. Naval Research Lab., Washington, D.C.

Yentsch, C. S. (1960). The Influence of Phytoplankton Pigments on the Colour of Sea Water, *Deep Sea Res.* 7, 1–9.

Yost, E., and Wenderoth, S. (1968). Additive Color Aerial Photography, in *Manual of Color Aerial Photography* (J. T. Smith, ed.). 451–471. Amer. Soc. of Photogrammetry, Falls Church, Va.

Yost, E., and Wenderoth, S. (1969). Ecological Applications of Multispectral Color Aerial Photography, in *Remote Sensing in Ecology* (P. L. Johnson, ed.). 46–62. Univ. of Georgia Press, Athens.

0-201-04245-2

Yotko, H. J., and Zaitzeff, J. B. (1967). Oceanographic Ground Truth Requirements, in Earth Resources Program, Ground Truth Session, Test and Operations Office, Science and Applications Directorate, Manned Spacecraft Center, Houston, Tex.

Younes, H. A., Abdel-Aal, R. M., Khodair, M. M., and Abdel-Samie, A. G. (1974). Spectral Reflectance Studies on Mineral Deficiency in Corn Plants, *Proc. of the Ninth Int'l Symp. on Remote Sens. of the Environ.* 1105–1125. Environmental Research Inst. of Mich., Ann Arbor.

0-201-04245-2

# Author Index

Numbers set in *italics* designate pages on which complete literature citations are given.

# Subject Index